Universitext

Springer

New York
Berlin
Heidelberg
Barcelona
Budapest
Hong Kong
London
Milan
Paris
Santa Clara
Singapore
Tokyo

Universitext

Editors (North America): S. Axler, F.W. Gehring, and P.R. Halmos

Aksoy/Khamsi: Nonstandard Methods in Fixed Point Theory
Aupetit: A Primer on Spectral Theory
Booss/Bleecker: Topology and Analysis
Borkar: Probability Theory; An Advanced Course
Carleson/Gamelin: Complex Dynamics
Cecil: Lie Sphere Geometry: With Applications to Submanifolds
Chae: Lebesgue Integration (2nd ed.)
Charlap: Bieberbach Groups and Flat Manifolds
Chern: Complex Manifolds Without Potential Theory
Cohn: A Classical Invitation to Algebraic Numbers and Class Fields
Curtis: Abstract Linear Algebra
Curtis: Matrix Groups
DiBenedetto: Degenerate Parabolic Equations
Dimca: Singularities and Topology of Hypersurfaces
Edwards: A Formal Background to Mathematics I a/b
Edwards: A Formal Background to Mathematics II a/b
Foulds: Graph Theory Applications
Fuhrmann: A Polynomial Approach to Linear Algebra
Gardiner: A First Course in Group Theory
Gårding/Tambour: Algebra for Computer Science
Goldblatt: Orthogonality and Spacetime Geometry
Hahn: Quadratic Algebras, Clifford Algebras, and Arithmetic Witt Groups
Holmgren: A First Course in Discrete Dynamical Systems
Howe/Tan: Non-Abelian Harmonic Analysis: Applications of $SL(2, R)$
Howes: Modern Analysis and Topology
Humi/Miller: Second Course in Ordinary Differential Equations
Hurwitz/Kritikos: Lectures on Number Theory
Jennings: Modern Geometry with Applications
Jones/Morris/Pearson: Abstract Algebra and Famous Impossibilities
Kannan/Krueger: Advanced Real Analysis
Kelly/Matthews: The Non-Euclidean Hyperbolic Plane
Kostrikin: Introduction to Algebra
Luecking/Rubel: Complex Analysis: A Functional Analysis Approach
MacLane/Moerdijk: Sheaves in Geometry and Logic
Marcus: Number Fields
McCarthy: Introduction to Arithmetical Functions
Meyer: Essential Mathematics for Applied Fields
Mines/Richman/Ruitenburg: A Course in Constructive Algebra
Moise: Introductory Problems Course in Analysis and Topology
Morris: Introduction to Game Theory
Porter/Woods: Extensions and Absolutes of Hausdorff Spaces
Ramsay/Richtmyer: Introduction to Hyperbolic Geometry
Reisel: Elementary Theory of Metric Spaces
Rickart: Natural Function Algebras
Rotman: Galois Theory
Rubel/Colliander: Entire and Meromorphic Functions
Sagan: Space-Filling Curves
Samelson: Notes on Lie Algebras

(continued following index)

Michel Simonnet

Measures and Probabilities

Foreword by Charles-Michel Marle

Springer

Michel Simonnet
Department of Mathematics
University of Dakar
Senegal

AMS Subject Classification (1991): 28-02, 28c05, 60Fxx

Library of Congress Cataloging-in-Publication Data
Simonnet, Michel
      Measures and probablilities / Michel Simonnet.
         p.   cm – (Universitext)
      Includes bibliographical references and indexes.
      ISBN 0-387-94644-6 (softcvr : alk. paper)
      1. Measure theory. 2. Probabilities.   I. Title.
   QA312.S54   1996
   515′.783–dc20   95-49240

Printed on acid-free paper.

Production managed by Robert Wexler; manufacturing supervised by Joe Quatela.
Photocomposed copy prepared using the author's LATEX file.
Printed and bound by R.R Donnelley & Sons, Harrisonburg, VA.
Printed in the United States of America.

9 8 7 6 5 4 3 2 1

ISBN 0-387-94644-6 Springer-Verlag New York Berlin Heidelberg SPIN 10523961

Foreword

Integration theory holds a prime position, whether in pure mathematics or in various fields of applied mathematics. It plays a central role in analysis; it is the basis of probability theory and provides an indispensable tool in mathematical physics, in particular in quantum mechanics and statistical mechanics. Therefore, many textbooks devoted to integration theory are already available. The present book by Michel Simonnet differs from the previous texts in many respects, and, for that reason, it is to be particularly recommended.

When dealing with integration theory, some authors choose, as a starting point, the notion of a measure on a family of subsets of a set; this approach is especially well suited to applications in probability theory. Other authors prefer to start with the notion of Radon measure (a continuous linear functional on the space of continuous functions with compact support on a locally compact space) because it plays an important role in analysis and prepares for the study of distribution theory. Starting off with the notion of Daniell measure, Mr. Simonnet provides a unified treatment of these two approaches. Both integration theories (the one with respect to a set-type measure or the other with respect to a Radon measure) are deduced as particular cases of integration with respect to a Daniell measure. Incidentally, theorems on extension of a measure initially defined on a Boolean semiring of subsets of a set are obtained as direct consequences of properties of the integral with respect to a Daniell measure. Thus the author has been able to offer developments of the theory that are of interest to mathematical analysis, and others that are relevant to probability theory. Furthermore, this exposition has the advantage that it leads very quickly to some important results of constant use, such as convergence theorems.

On the other hand, the framework chosen here is fairly more general than the one usually adopted: measures with respect to which one integrates can be real of either sign or even complex; from the outset, the integral is defined for functions taking their values in a Banach space, as well as for real- (or extended real-) valued functions.

Finally, the treatment of the subject given in this book is much more comprehensive than the treatment found in most other texts of the same level. The important aspects of integration theory (L^p spaces, different types of convergence, measure decomposition, Radon-Nikodym derivatives, image measures and product measures) are all treated with great care. The book contains an excellent introduction to probability theory, covering the strong law of large numbers, the central limit theorem, and conditional probabilities. It also provides some applications of integration in analysis, such as the definition of Haar measure on locally compact groups and the convolution of measures.

The book is quite easily accessible thanks to the inclusion by the author of most of the necessary background in topology and analysis. Clarity of the exposition and numerous exercises at the ends of the chapters allow it to be used as a textbook. It will also prove very useful to teachers and will provide a good source of reference for many mathematicians.

CHARLES-MICHEL MARLE
Université Paris VI
Pierre et Marie Curie
Mathématiques

Preface

This book is intended to be an introductory, yet sophisticated, treatment of measure theory. It should provide an in-depth reference for the practicing mathematician. Having said this, I hope that advanced students as well as instructors will find it useful.

The first part of this book (Chapters 1 to 8) should prove useful to both analysts and probabilists. One may treat the second and third parts (Chapters 9 to 14 and 15 to 18) as an introduction to the theory of probability, or use the fourth part as an introduction to analysis.

I generally proceed from first principles and my treatment is, for the most part, self-contained. Other than familiarity with general topology, some functional analysis, and a certain degree of mathematical sophistication, little is required for profitable reading of this text. The topological background needed is covered, for example, in *Modern General Topology* by J. Nagata. *Functional Analysis* by W. Rudin will provide the reader with most of the results on topological vector spaces used throughout this text. All other results may be found in *A Course in Functional Analysis* by J. B. Conway. At the end of each chapter, exercises, which are designed to present some additional material and examples, are provided.

The treatment of integration at the elementary level usually starts with either positive or signed measures on σ-algebras or with complex Radon measures on locally compact Hausdorff spaces. In this book, we first consider complex Daniell measures on a space $\mathcal{H}(\Omega, \mathbf{C})$ of elementary functions. To define the upper integral, $\int^* f \, dV\mu$, of any positive function f with respect to the variation $V\mu$ of a complex Daniell measure μ, a class $\mathcal{J}$ of positive functions, that contains the upper envelopes of increasing sequences of positive elementary functions, is introduced. Given a Banach space F, a function f from Ω into F is μ-integrable if and only if, for each $\varepsilon > 0$, there exists a decomposable function g such that $\int^* |f - g| \, dV\mu$ is less than ε. Then are defined the μ-measurable functions taking their values in a metrizable space. This procedure works, in particular, when $\mathcal{H}(\Omega, \mathbf{C})$ is the set of simple functions on a semiring S. It also works when Ω is a locally compact space—not necessarily σ-compact—and when $\mathcal{H}(\Omega, \mathbf{C})$ is the set of continuous functions on Ω with compact support. Two fairly general and easy to handle integration theories are thus obtained. The first one (Chapters 9 to 14) is especially suited to the needs of probabilists, the second one (Chapters 19 to 24) to those of analysts. Incidentally, note that many major results on Radon measures follow immediately from their counterpart results on abstract measures defined on the semiring of differences of compact sets.

This present volume is strongly influenced by, and owes much to, *Probability and Measures* by P. Billingsley (Wiley, 1985), *Intégration* by N. Bourbaki (Hermann, 1965), *Treatise on Analysis* (volume 2) by J. Dieudonné (Academic Press, 1976), *Measure Theory* by P. Halmos (Springer-Verlag, 1974), *Mesures et probabilités* by C. M. Marle (Hermann, 1974), and *Real and Complex Analysis* by W. Rudin (McGraw–Hill, 1966). The fourth part follows closely N. Bourbaki's treatise.

The courses given by Thi'si Ho Van in measure theory at the University of Liège and the University of Abidjan have provided the motivation and impetus for this exposition of the theory. Note that this list is by no means exhaustive.

I wish to express my gratitude to R. Descombes, J. Dieudonné, C.M. Marle from Paris University, W. Rudin of The University of Wisconsin, and to K. Vo-Khac from Orléans University for their kind remarks and encouragement. I am especially indebted to Professor J. Dieudonné for his longtime support and good counsel and to Professor C. M. Marle for his invaluable suggestions.

Finally I would like to thank P. Toppo and G. Vinel for helping with the corrections, R. Eastaway-Gine and M. Gamar for typesetting the manuscript in LaTeX, and Springer-Verlag for accepting the manuscript. The chapter introductions are due to G. Vinel.

Contents

Part I

Integration Relative to Daniell Measures

1

Riesz Spaces

Let X be the space of continuous functions on $[a, b]$ (a, b reals). The mapping $I : X \ni f \mapsto I(f) = \int_a^b f(t)dt$, where $I(f)$ is the Riemann integral of f over $[a, b]$, is linear. Moreover, if f_n converges uniformly to f on $[a, b]$, then $\int_a^b f_n(t)dt$ converges to $\int_a^b f(t)dt$. As we shall see later, this last condition is a consequence of the fact that the integral of a positive continuous function on $[a, b]$ is a positive real number. In this chapter, we generalize these ideas and define Daniell measures as linear forms on some function spaces. To this end, we first have to define the notion of positive elements in a vector space.

Summary

1.1   This first section introduces the notion of ordered groups and lattices. It might be useful to keep in mind some well known examples, such as $\mathbf{Z}$ or $\mathbf{R}$ or the space $C([a, b])$ of real-valued continuous functions on $[a, b]$ with their natural orderings.

1.2   It is devoted to a short study of Riesz spaces, that is, ordered vector spaces which are lattices. The space X of real-valued continuous functions on $[a, b]$, endowed with its usual ordering, is an example of Riesz space.

1.3   In this section we turn our attention to the duals of Riesz spaces. If E is any ordered vector space, there is a natural notion of positive linear form: a linear form I is said to be positive if the image of any positive element of E under I is a positive real number. Notice that if E is the space of real-valued continuous functions on a closed interval $[a, b]$, equipped with its natural ordering, a positive linear form on E is continuous.

1.4   We finally define Daniell measures. A Daniell measure is a linear form L on a space of functions $\mathcal{H}$ which has a "finite variation" (Definition 1.4.1) and such that $L(f_n) \to 0$ whenever f_n decreases to 0 in $\mathcal{H}^+$. This last condition should be understood as a continuity condition. Among the examples that the reader may want to keep in mind are the Riemann integral considered as a positive linear form on the set of continuous functions on $[a,b]$ and also the positive linear forms ε_c : $X \ni f \mapsto f(c)$ where $c \in [a,b]$.

1.1   Ordered Groups

Definition 1.1.1 Let G be a commutative group written additively. An order structure on G, denoted by $\leq$, is said to be compatible with the group structure of G whenever it satisfies the following axiom: If $x \leq y$, then $x+z \leq y+z$ for all $z \in G$.

The group G, endowed with a compatible order structure, is called an ordered group.

Taking $z = -x$, $z = -y$, and $z = -(x+y)$ successively in Definition 1.1.1, we conclude that, in G, the inequalities $x \leq y$, $0 \leq y - x$, $x - y \leq 0$, and $-y \leq -x$ are equivalent.

For example, $\mathbf{R}$, with its usual addition and its usual order, is an ordered group.

Proposition 1.1.1 Suppose that $(x_i)_{1 \leq i \leq n}$ and $(y_i)_{1 \leq i \leq n}$ are two sequences of n elements in an ordered group G such that $x_i \leq y_i$ for all $1 \leq i \leq n$. Then $x_1 + \ldots + x_n \leq y_1 + \ldots + y_n$. Furthermore, $x_1 + \ldots + x_n < y_1 + \ldots + y_n$ whenever there exists an $1 \leq i \leq n$ for which $x_i < y_i$.

PROOF: First, assume that $n = 2$. The relations $x_1 + x_2 \leq x_1 + y_2$ and $x_1 + y_2 \leq y_1 + y_2$ imply $x_1 + x_2 \leq y_1 + y_2$. On the other hand, if $x_1 + x_2 = y_1 + y_2$, then $x_1 + x_2 = x_1 + y_2 = y_1 + y_2$, hence $x_2 = y_2$ and $x_1 = y_1$. The general case follows by induction on n.   □

For every $z \in G$, the translation $x \to x + z$ preserves the order of G. Thus, if a nonempty family $(x_i)_{i \in I}$ of elements of G has a supremum $\bigvee_{i \in I} x_i$, then the family $(z + x_i)_{i \in I}$ also has a supremum, and $\bigvee_{i \in I}(z + x_i) = z + \bigvee_{i \in I} x_i$. A similar statement holds for the infimum $\bigwedge_{i \in I} x_i$.

Now the mapping $x \to -x$ reverses the order of G so that a nonempty family $(x_i)_{i \in I}$ of elements of G has a supremum if and only if the family $(-x_i)_{i \in I}$ has an infimum, and then $\bigwedge_{i \in I}(-x_i) = -\bigvee_{i \in I} x_i$.

Definition 1.1.2 In an ordered group G, $x \in G$ is said to be positive (respectively, negative) if $x \geq 0$ (respectively, $x \leq 0$); x is said to be strictly positive (respectively, strictly negative) if $x > 0$ (respectively, $x < 0$).

If G is an ordered group, the set of its positive elements will usually be denoted by G^+. So $G^+ \cap (-G^+) = \{0\}$ and $G^+ \supset G^+ + G^+$.

Proposition 1.1.2 *Let a commutative group G be given, and let P be a subset of G such that $P \cap (-P) = \{0\}$ and $P + P \subset P$. The relation $x \leq y$ if and only if $y - x \in P$ defines an order structure compatible with the group structure of G. Moreover, G is totally ordered if and only if $G = P \cup (-P)$.*

PROOF: This is obvious.   □

An ordered set E is said to be directed upward (respectively, downward) whenever, for all $x \in G$, $y \in G$, there exists $z \in G$ satisfying $x \leq z$ and $y \leq z$ (respectively, $x \geq z$ and $y \geq z$). Every ordered group G that is directed upward is also directed downward, and conversely. In this case, G is called a directed group.

Proposition 1.1.3 *An ordered group G is directed if and only if it is generated by G^+, or, equivalently, if every element of G is the difference of two positive elements.*

PROOF: If G is directed, for every $x \in G$, there exists a $z \in G^+$ greater than x; now $x = z - (z - x)$. Conversely, if $x = u - v$ and $y = w - t$ where u, v, w, t are all positive, then $x \leq u + w$ and $y \leq u + w$.   □

Proposition 1.1.4 *Let $(x_i)_{i \in I}$ be a nonempty finite family in a directed group G. There exists a $z \in G^+$ such that $x_i + z$ is positive for all $i \in I$.*

PROOF: If $x_i = u_i - v_i$ for every $i \in I$, where u_i, v_i are positive, take $z = \sum_{i \in I} v_i$.   □

An ordered set E is called a lattice whenever every nonempty finite subset of E has a supremum and an infimum in E.

Proposition 1.1.5 *Let G be an ordered group and x, $y \in G$ be given. A necessary and sufficient condition that $x \wedge y$ exist is that $x \vee y$ exist. Then $x + y = x \wedge y + x \vee y$.*

PROOF: Suppose that $x \wedge y$ exists. Then

$$x + y - x \wedge y = x + y + (-x) \vee (-y) = x \vee y.$$

The proof of the converse is just as easy.   □

Proposition 1.1.6 *Suppose G is an ordered group. Then G is a lattice if and only if $G = G^+ - G^+$ and one of the following conditions holds:*

(a) Every pair of elements of G^+ has a supremum.

(b) Every pair of elements of G^+ has an infimum in G^+.

PROOF: These conditions are obviously necessary. Conversely, under hypothesis (a) (respectively, (b)), each pair of elements $x \in G^+$, $y \in G^+$ has a supremum (respectively, an infimum) in G equal to its supremum a (respectively, its infimum b) in G^+. This is obvious for a; for b, let $z \in G$ be a lower bound for x and y. Then there exists $u \in G^+$ such that $z + u \in G^+$ (because $G = G^+ - G^+$). Now $\inf_{G^+}(x + u, y + u)$ is greater than $b + u$ and can be written $b + c + u$ ($c \geq 0$). Since $b + c \leq x$ and $b + c \leq y$, $c = 0$. Hence $b + u = \inf_{G^+}(x + u, y + u) \geq z + u$, and $z \leq b$.

Next, let $x \in G$, $y \in G$ be arbitrary and $v \in G^+$ be such that $x + v \geq 0$ and $y + v \geq 0$; under hypothesis (a) (respectively, (b)), $x + v$ and $y + v$ have a supremum (respectively, an infimum) in G^+, and therefore in G by what we have just shown. So x, y have a supremum (respectively, an infimum) in G. □

Unless otherwise stated, G will always denote a lattice ordered group.

Theorem 1.1.1 (Decomposition Lemma) *Let* $(x_i)_{1 \leq i \leq p}$, $(y_j)_{1 \leq j \leq q}$ *be two finite sequences in* G^+ *satisfying* $\sum_{1 \leq i \leq p} x_i = \sum_{1 \leq j \leq q} y_j$. *In* G^+, *then, there exists a double sequence* $(z_{i,j})_{1 \leq i \leq p, \, 1 \leq j \leq q}$ *such that* $x_i = \sum_{1 \leq j \leq q} z_{i,j}$ *for all* i *and* $y_j = \sum_{1 \leq i \leq p} z_{i,j}$ *for all* j.

PROOF:

1. First, consider the case $p = q = 2$. Put $z_{1,1} = \sup(0, x_1 - y_2)$ and $z_{2,2} = z_{1,1} - (x_1 - y_2)$. Since $x_1 - y_2 = y_1 - x_2$ is smaller than x_1 and y_1, $z_{1,2} = x_1 - z_{1,1}$ and $z_{2,1} = y_1 - z_{1,1}$ are positive. Moreover, $x_1 = z_{1,1} + z_{1,2}$, $x_2 = z_{2,1} + z_{2,2}$, and $y_1 = z_{1,1} + z_{2,1}$, $y_2 = z_{1,2} + z_{2,2}$.

2. Next, we suppose that the theorem holds for $p < m$ and $q = n$ (with $m > 2$, $n \geq 2$) and prove that it holds for $p = m$ and $q = n$. By hypothesis,

$$\left(\sum_{1 \leq i \leq m-1} x_i \right) + x_m = \sum_{1 \leq j \leq n} y_j.$$

Since the result holds for $p = 2$ and $q = n$, we can find two sequences $(z_j')_{1 \leq j \leq n}, (z_j'')_{1 \leq j \leq n}$ in G^+ satisfying

$$\sum_{1 \leq j \leq n} z_j' = \sum_{1 \leq i \leq m-1} x_i, \quad \sum_{1 \leq j \leq n} z_j'' = x_m,$$

and $y_j = z_j' + z_j''$ for all $1 \leq j \leq n$. Now, since the theorem is true for $p = m - 1$ and $q = n$, there is a double sequence $(u_{i,j})_{1 \leq i \leq m-1, 1 \leq j \leq n}$ in G^+ such that $x_i = \sum_{1 \leq j \leq n} u_{i,j}$ for all $1 \leq i \leq m - 1$ and $z_j' = \sum_{1 \leq i \leq m-1} u_{i,j}$ for all $1 \leq j \leq n$. Put $z_{i,j} = u_{i,j}$ for all $1 \leq i \leq m - 1$ and $z_{m,j} = z_j''$ ($1 \leq j \leq n$). Then the double sequence $(z_{i,j})_{1 \leq i \leq m, 1 \leq j \leq n}$ has the desired properties.

3. Interchanging the roles of the x_i and the y_j, we find similarly that if the theorem holds for $p = m$ and $q < n$ $(m \geq 2, n > 2)$, then it holds for $p = m$ and $q = n$. The theorem now follows by double induction.

$\square$

Corollary *Let $y, x_1, \ldots, x_n$ be elements of G^+ satisfying $y \leq \sum_{1 \leq i \leq n} x_i$. Then we can find $y_i \in G^+ (1 \leq i \leq n)$ such that $y_i \leq x_i$ for all i and $y = \sum_{1 \leq i \leq n} y_i$.*

PROOF: Apply Theorem 1.1.1 to the sequence $(x_i)_{1 \leq i \leq n}$ and the sequence formed from the two elements y and $z = \left(\sum_{1 \leq i \leq n} x_i \right) - y$. $\square$

Definition 1.1.3 For $x \in G$, the element $x \vee 0$ (respectively $(-x) \vee 0$, respectively $(-x) \vee x$) is called the positive part of x (respectively the negative part, respectively the absolute value, or variation, of x) and is written x^+ (respectively x^-, $|x|$).

Note that, according to these definitions, the negative part of x is a positive element. Clearly, $x^- = (-x)^+$ and $|-x| = |x|$. Note also the formulas $x \vee y = x + (y - x)^+$ and $x \wedge y = y - (y - x)^+$. The first follows immediately from the fact that the ordering of G is preserved under translation; the second is a consequence of Proposition 1.1.5.

Theorem 1.1.2

(a) $x = x^+ - x^-$ and $x^+ \wedge x^- = 0$, for every $x \in G$.

(b) *For any expression $x = u - v$ of $x \in G$ as a difference of two positive elements, $u = x^+ + w$ and $v = x^- + w$, where $w = u \wedge v$.*

(c) *$x \leq y$ is equivalent to "$x^+ \leq y^+$ and $x^- \geq y^-$".*

(d) $|x| = x^+ + x^-$.

(e) *$|x + y| \leq |x| + |y|$ for all $x, y \in G$; more generally $|\sum_{1 \leq i \leq n} x_i| \leq \sum_{1 \leq i \leq n} |x_i|$ for any finite sequence $(x_i)_{1 \leq i \leq n}$ in G.*

(f) *$\left| |x| - |y| \right| \leq |x - y|$ for all $x, y \in G$.*

PROOF: We will prove (a) and (b) simultaneously. If $x = u - v$, where u, v are positive, then $u \geq x^+$, and $w = u - x^+$ is positive. Furthermore, $x^+ - x = x \vee 0 - x = (x - x) \vee (-x) = x^-$, hence $x = x^+ - x^-$ and $w = v - x^-$. Let $z \in G$ be smaller than $x^- = x^+ - x$. Then $x \leq x^+ - z$; moreover, if $z \leq x^+$, then $x^+ - z$ is positive, so $x^+ \leq x^+ - z$, and z is negative. Therefore, we find that $x^+ \wedge x^- = 0$ and, by translation, that $u \wedge v = w$.

(c): Let x, $y \in G$ be given. If $x \leq y$, then $x \leq y^+$ and $-y \leq (-x)^+ = x^-$, so $x^+ \leq y^+$ and $x^- \geq y^-$. Conversely, $x = x^+ - x^- \leq y^+ - y^- = y$ whenever $x^+ \leq y^+$ and $x^- \geq y^-$.

(d): $x^+ \vee x^- \geq |x|$, because $x \leq x^+$ and $-x \leq x^-$. Next, if a is an upper bound for x and $-x$, then $a^+ \geq x^+$, $a^+ \geq x^-$, $a^- \leq x^-$, and $a^- \leq x^+$ by (c). Since a^- is positive and since $\inf(x^+, x^-) = 0$, the last two inequalities imply that $a^- = 0$ and $a = a^+$; the first two inequalities then give $a \geq x^+ \vee x^-$. Hence $x^+ \vee x^- = |x|$. Now $x^+ \vee x^- = x^+ + x^-$ by Proposition 1.1.5.

(e): $x + y \leq |x| + |y|$ because $x \leq |x|$ and $y \leq |y|$. Similarly, $-x - y \leq |x| + |y|$. Hence $|x + y| \leq |x| + |y|$.

(f): Replacing x and y in (e) by y and $x - y$, we have $|x| - |y| \leq |x - y|$; similarly, $|y| - |x| \leq |y - x| = |x - y|$. □

Note that $x = x^+ - x^- = 0$ if $|x| = x^+ + x^- = 0$; thus $|x| > 0$ for all $x \in G$, $x \neq 0$.

Proposition 1.1.7 *Let $(x_i)_{i \in I}$ be a nonempty family in G with an infimum y, and let $z \in G$ be arbitrary. Then the family $(z \vee x_i)_{i \in I}$ has an infimum and $\wedge_{i \in I}(z \vee x_i) = z \vee (\wedge_{i \in I} x_i)$.*

PROOF: We may assume that $z = 0$. Now $y^+ \leq x_i^+$ for all i. If a is a lower bound for $(x_i^+)_{i \in I}$, then $a \leq x_i + x_i^-$ for all $i \in I$. Now $x_i^- \leq y^-$ for every $i \in I$; thus $a \leq x_i + y^-$ for every $i \in I$. and $a \leq y + y^- = y^+$. Hence $\wedge_{i \in I} x_i^+ = y^+$, as was to be shown. □

Interchanging $\wedge$ and $\vee$, we obtain a similar proposition. In particular, if $(x_i)_{i \in I}$ is a nonempty family with a supremum x in G, then $x^- = \wedge_{i \in I} x_i^-$, because $(-x)^+ = (\wedge_{i \in I} - x_i)^+ = \wedge_{i \in I}(-x_i)^+$.

Definition 1.1.4 Two elements x, y in G are said to be *disjoint*, or *mutually singular*, whenever $|x| \wedge |y| = 0$. In this case, we shall also say that x is disjoint from y. If A is a subset of G, $x \in G$ is said to be disjoint from A whenever it is disjoint from every element of A. Finally, two subsets A, B of G are said to be disjoint whenever each element of A is disjoint from each element of B. For any subset A of G, the set $\{y \in G : y$ is disjoint from $A\}$ is called the *disjoint complement* of A.

Proposition 1.1.8 *Let $(x_i)_{i \in I}$ and $(y_j)_{j \in J}$ be two nonempty finite families in G^+. Then $(\sum_i x_i) \wedge (\sum_j y_j) \leq \sum_{i,j}(x_i \wedge y_j)$.*

PROOF: By induction on $|I|$ and $|J|$, it suffices to show that $x \wedge (y + z) \leq x \wedge y + x \wedge z$ for all x, y, $z \in G^+$. Now put $t = x \wedge (y + z)$. By Theorem 1.1.1, t can be written $t_1 + t_2$, where $0 \leq t_1 \leq y$ and $0 \leq t_2 \leq z$. Then $t_1 \leq x \wedge y$ and $t_2 \leq x \wedge z$, and the result follows. □

Corollary 1 *If $x \in G$ is disjoint both from y and z, then it is disjoint from $y + z$.*

PROOF: Indeed, $|x| \wedge |y + z| \leq |x| \wedge (|y| + |z|) \leq |x| \wedge |y| + |x| \wedge |z| = 0$.   □

Corollary 2 *If $(x_i)_{1 \leq i \leq m}, (y_j)_{1 \leq j \leq n}$ are two finite families in G such that each x_i is disjoint from each y_j, then $x_1 + \ldots + x_m$ and $y_1 + \ldots + y_n$ are disjoint.*

PROOF: This follows from Corollary 1 by induction on m and n.   □

Corollary 3 *If $x_1, \ldots, x_n$ are mutually disjoint, then $|x_1 + \ldots + x_n| = |x_1| + \ldots + |x_n|$.*

PROOF: If x, y are disjoint, then so are $x^+ + y^+$ and $x^- + y^-$. Thus $x^+ + y^+ = (x + y)^+$ and $x^- + y^- = (x + y)^-$.   □

Proposition 1.1.9 *Let $(x_i)_{i \in I}$ be a nonempty family in G having a supremum x. Then any $y \in G$ that is disjoint from each x_i is also disjoint from x.*

PROOF: $|y| \wedge x_i^+ = 0$ for each $i \in I$, so $|y| \wedge x^+ = \vee_{i \in I}(|y| \wedge x_i^+) = 0$ (Proposition 1.1.7). Moreover,

$$|y| \wedge x^- = |y| \wedge (\wedge_{i \in I} x_i^-) = \wedge_{i \in I}(|y| \wedge x_i^-) = 0,$$

so y is disjoint from both x^+ and x^-, and hence from $|x| = x^+ + x^-$.   □

1.2   Riesz Spaces

Definition 1.2.1 Suppose that E is a real vector space. An order relation $\leq$ on E is said to be compatible with the vector space structure of E if the following two conditions hold:

(a) $x \leq y$ implies $x + z \leq y + z$ for all $z \in E$.

(b) $x \geq 0$ implies $\lambda x \geq 0$ for all real numbers $\lambda > 0$.

A real vector space with a compatible order is called an ordered vector space. By definition, a Riesz space is an ordered vector space which is a lattice.

Let a Riesz space E be given. For any real number $\lambda > 0$, the order of E remains invariant under the homothety $z \mapsto \lambda z$; thus $\sup(\lambda x, \lambda y) = \lambda \cdot \sup(x, y)$ for all $x, y \in E$. In particular, $(\lambda x)^+ = \lambda \cdot x^+$ and $(\lambda x)^- = \lambda \cdot x^-$ for every $x \in E$ and for every $\lambda \geq 0$. On the other hand, $(\lambda x)^+ = (-\lambda x)^- = |\lambda| \cdot x^-$ and $(\lambda x)^- = (-\lambda x)^+ = |\lambda| \cdot x^+$ for all $x \in E$ and all $\lambda \leq 0$. Thus $|\lambda x| = |\lambda| \cdot |x|$ for all $x \in E$ and all $\lambda \in \mathbf{R}$.

If $x \in E$, $y \in E$ are disjoint, then so are x and λy for all $\lambda \in \mathbf{R}$. Indeed, for any given integer $n \geq |\lambda|$, x and ny are disjoint and $|ny| \geq |\lambda y|$. It follows that, if A is a nonempty subset of E, the disjoint complement of A is a vector subspace of E.

Definition 1.2.2 E is said to be Dedekind complete whenever each nonempty subset of E that is bounded above has a supremum. Equivalently, E is Dedekind complete whenever each nonempty subset of E that is bounded below has an infimum.

Proposition 1.2.1 *E is Dedekind complete if and only if one of the following two conditions holds:*

(a) *Every nonempty subset A of E^+ that is directed upward and bounded above has a supremum.*

(b) *Every nonempty subset A of E^+ that is directed downward has an infimum.*

PROOF: Clearly, the conditions are necessary. Conversely, assume that condition (a) holds, and let B be a nonempty subset of E, bounded above. Then the suprema of all nonempty finite parts of B constitute a set C that is directed upward. Let a be one of its elements, and define $C_a = \{x \in C : x \geq a\}$. Then $C_a - a$, that is directed upward and bounded above, has a supremum b. Now $a + b$ is the supremum of C_a, and so of B as well.

In fact, condition (b) implies condition (a). Indeed, if B is a nonempty subset of E^+, directed upward and bounded above, and if c is an upper bound for B, then $c - B$ is directed downward. Denoting by m the infimum of $c - B$, we see that $c - m$ is the supremum of B. $\qquad\square$

Every vector subspace of an ordered vector space F is an ordered vector space with order induced by that on E.

Let $(E_i)_{i \in I}$ be a family of ordered vector spaces. On $E = \prod_{i \in I} E_i$, the relation "$(x_i)_i \leq (y_i)_i$ if $x_i \leq y_i$ for all $i \in I$" is compatible with the vector space structure of E; moreover, if each E_i is also a lattice (respectively, is Dedekind complete), then so is E.

Definition 1.2.3 Let E be an ordered vector space and V, W be two supplementary vector subspaces of E. We say that E is the ordered direct sum of V and W whenever the canonical map $(x, y) \to x + y$ from the ordered vector space $V \times W$ onto the ordered vector space E is an isomorphism. This surely holds if the relations $x \in V$, $y \in W$, and $x + y \geq 0$ imply $x \geq 0$ and $y \geq 0$.

Definition 1.2.4 Let E be a Riesz space. A vector subspace F of E is called a Riesz subspace of E whenever $x \vee y$ (and hence $x \wedge y$) belongs to F for all $x \in F$ and $y \in F$. F is called an ideal whenever $x \in F$, $y \in E$, and $|y| \leq |x|$ imply $y \in F$. F is called a band if it is an ideal and if $z \in F$ whenever z is the supremum in E of a nonempty subset of F.

For example, given a topological space Ω, let $\mathcal{F}(\Omega, \mathbf{R})$ be the vector space of all real-valued functions on Ω. Give $\mathcal{F}(\Omega, \mathbf{R})$ the order "$f \leq g$ if $f(x) \leq g(x)$ for all $x \in \Omega$". Let $\mathcal{B}(\Omega, \mathbf{R})$ (respectively, $\mathcal{C}(\Omega, \mathbf{R})$) be the space of all bounded (respectively, continuous) real-valued functions on Ω. Then $\mathcal{B}(\Omega, \mathbf{R})$ and $\mathcal{C}(\Omega, \mathbf{R})$ are Riesz subspaces of $\mathcal{F}(\Omega, \mathbf{R})$. $\mathcal{B}(\Omega, \mathbf{R})$ is an ideal of $\mathcal{F}(\Omega, \mathbf{R})$, but $\mathcal{C}(\Omega, \mathbf{R})$ is not. Finally, $\mathcal{B}(\Omega, \mathbf{R})$ is not a band in $\mathcal{F}(\Omega, \mathbf{R})$, even though it is Dedekind complete.

For the remainder of this section, E will denote a Dedekind complete Riesz space.

Note that every band B in E is Dedekind complete and that the intersection of a family of bands is a band. Now let A be a subset of E, and define the band generated by A as the smallest band that contains A.

Proposition 1.2.2 *Let D be a nonempty subset of E^+ such that*

(a) $D + D \subset D$ *and*

(b) $x \in D$ *and* $0 \leq y \leq x$ *imply* $y \in D$.

Let M be the set of suprema of those nonempty sets in D that are bounded above. Then every $x \in E^+$ can be written $y + z$, where $y \in M$ is the supremum of $\{v \in D : v \leq x\}$ and $z \in E^+$ is disjoint from M.

PROOF: By Proposition 1.1.9, it suffices to show that $z = x - y$ is disjoint from D or, equivalently, that $u = z \wedge t = 0$ for all $t \in D$. Now, for all $v \in D$ satisfying $v \leq x$, we have $v \leq y$, and $u + v \leq u + y \leq x$. Since $u + v$ belongs to D, by hypothesis, $u + v \leq y$. Now $u + y$ is the supremum of $\{u + v : v \in D,\ v \leq x\}$, and so $u + y \leq y$, whence $u \leq 0$. The proof is complete.   □

Theorem 1.2.1 (F. Riesz Decomposition Theorem) *Let A be a subset of E. Then the disjoint complement A' of A is a band. The disjoint complement A'' of A' is exactly the band generated by A, and E is the ordered direct sum of A', A''.*

PROOF: By Proposition 1.1.9, A' and A'' are bands. Every $x \in E^+$ can be written $y + z$, where $y \in A'$ and $z \in A''$ are positive (Proposition 1.2.2). This means that $E = A' + A''$. Furthermore, $A' \cap A'' = \{0\}$, so that A', A'' are supplementary vector subspaces of E. Finally, since the components in A', A'' of a positive element are positive, we see that E is the ordered direct sum of A', A''.

It remains to be shown that A'' is exactly the band B generated by A. But Proposition 1.2.2 shows that E is the direct sum of B and its disjoint complement B'. Now $A \subset B$, so $B' \subset A'$. On the other hand, $B \subset A''$ and E is the direct sum of A', A''; thus $B = A''$ and $B' = A'$. $\square$

Note, incidentally, that $x \in E$ is disjoint from A if and only if it is disjoint from B (because $A' = B'$).

For any band B in E, the component of $x \in E$ in B relative to the decomposition $E = B \ominus B'$ will simply be called the component of x in B.

Proposition 1.2.3 *Let A be a subset of E and B the band generated by A. Further, let M_1 be the set of those $x \in E^+$ that have an upper bound of the form $\sum_{i \in I} |x_i|$ for a finite family $(x_i)_{i \in I}$ in A. Let M_2 be the class of suprema of those nonempty subsets of M_1 that are bounded above in E. Then $M_2 = B^+$.*

PROOF: Indeed, $M_2 \subset B$. Now, letting B' be the disjoint complement of A (or, equivalently, of B), and taking $D = M_1$ in Proposition 1.2.2, we see that every element of E^+ is the sum of an element of M_2 and an element of B'. Because E is the direct sum of B and B', the proposition follows. $\square$

Theorem 1.2.2 *Let B_a be the band generated by some fixed $a \in E$ and let B'_a be the disjoint complement of B_a. Then, for every $x \in E^+$, the component of x in B_a is equal to $\sup_{n \in \mathbf{Z}^+} \inf(n|a|, x)$.*

PROOF: Assume in Proposition 1.2.3 that $A = \{a\}$, and replace D by M_1 in Proposition 1.2.2; the result follows. $\square$

If $a \in E$, $b \in E$ are disjoint, then the bands A, B generated by a, b, respectively, are disjoint. Indeed, b belongs to the disjoint complement A' of $\{a\}$, so $B \subset A'$.

1.3   Order Dual of a Riesz Space

Definition 1.3.1 A linear form L on an ordered vector space E is said to be *positive* if $L(x) \geq 0$ (in $\mathbf{R}$) for all $x \in E^+$.

If L is positive, the relation $x \leq y$ implies $L(x) \leq L(y)$.

Proposition 1.3.1 *Let E be an ordered vector space and L an additive function from E into $\mathbf{R}$ satisfying $L(x) \geq 0$ for all $x \in E^+$. Then $L(\lambda x) = \lambda \cdot L(x)$ for all $\lambda \in \mathbf{R}$ and all $x \in E^+$.*

PROOF: Additivity of L implies that $L(-z) = -L(z)$ for every $z \in E$, and so we need only treat the case in which $\lambda \geq 0$. Now $L(nz) = n \cdot L(z)$ for all $n \in \mathbf{N}$ and all $z \in E$, hence

$$L\left(\frac{1}{n}z\right) = \frac{1}{n} \cdot L(z).$$

Therefore, $L(rz) = r \cdot L(z)$ for all rational numbers $r \geq 0$ and all $z \in E$. Now, if r, r' are rational numbers satisfying $r \leq \lambda \leq r'$, we have

$$r \cdot L(x) \leq L(\lambda x) \leq r' \cdot L(x).$$

But $r \cdot L(x)$ and $r' \cdot L(x)$ may be taken as close to $\lambda \cdot L(x)$ as desired, and this gives $L(\lambda x) = \lambda \cdot L(x)$.   □

Proposition 1.3.2 *Let E be a real vector space, and let C be a convex cone in E with vertex 0 (i.e., a nonempty subset of E such that $C + C \subset C$ and $\lambda C \subset C$ for all $\lambda > 0$) for which $E = C - C$. Let M be a mapping from C into $\mathbf{R}$ satisfying $M(x + y) = M(x) + M(y)$ for all x, $y \in C$ and $M(\lambda x) = \lambda M(x)$ for every $x \in C$ and every $\lambda > 0$. Then there exists a unique linear extension L of M to E.*

PROOF: Each $z \in E$ can be written $x - y$, where x, y belong to C. But $L(z) = M(x) - M(y)$ does not depend on the decomposition of z; if $z = x' - y'$ is another decomposition of z $(x', y' \in C)$, then $x + y' = x' + y$, so $M(x) + M(y') = M(x') + M(y)$. The mapping $E \ni z \to L(z)$ is obviously a linear form on E.   □

Proposition 1.3.3 *Let E be a directed vector space, and let M be a mapping from E^+ into $\mathbf{R}^+$ such that $M(x + y) = M(x) + M(y)$ for all x, $y \in E^+$. Then M has a unique positive linear extension to the whole of E.*

PROOF: The same argument as that of Proposition 1.3.2 proves that M has a unique additive extension L to E. Proposition 1.3.1 then shows that L is linear.   □

For the remainder of this section, let a Riesz space E be given. The set Q of positive linear forms on E is a subset of the algebraic dual E^* of E (space of all linear forms on E). Clearly, $Q + Q \subset Q$ and $\lambda \cdot Q \subset Q$ for all real numbers $\lambda > 0$; moreover, $Q \cap (-Q) = \{0\}$ because, if L and $-L$ both belong to Q, then $L(x) \geq 0$ and $L(x) \leq 0$ for all $x \geq 0$, so $L = 0$ (Proposition 1.1.3). This means that Q defines an order relation on E^*: $L \leq M$ if and only if $M - L$ is a positive linear form, or $L \leq M$ if and only if $L(x) \leq M(x)$ for all $x \in E^+$. E^* is thus an ordered vector space, and its positive elements are the positive linear forms; this fact accounts for the terminology previously introduced.

Definition 1.3.2 A linear form L on E is said to be order-bounded whenever L is bounded on $\{y \in E : |y| \le x\}$ for every $x \in E^+$. The space of all order-bounded linear forms on E is called the order dual of E.

Theorem 1.3.1

   (a) *A linear form L on E is order-bounded if and only if it is the difference of two positive linear forms.*

   (b) *The order dual Ω of E is a Dedekind complete Riesz space.*

PROOF: If $L = U - V$, where U, V are two positive linear forms on E, then, for all $x \in E^+$, the relation $-x \le y \le x$ implies $-U(x) \le U(y) \le U(x)$ and $-V(x) \le V(y) \le V(x)$. Hence $|L(y)| \le U(x) + V(x)$, and so L is order-bounded.

Conversely, assume that L is order-bounded. We need only find a positive linear form N such that $N(x) \ge L(x)$ for each $x \in E^+$ because, in this case, $L = N - (N - L)$, which is the difference of two positive linear forms. Put $M(x) = \sup_{0 \le y \le x} L(y)$ for all $x \in E^+$. Then $M(x) + M(x') = \sup_{0 \le y \le x, 0 \le y' \le x'} L(y + y') \le M(x + x')$ for all x, $x' \in E^+$. Next, for all z satisfying $0 \le z \le x + x'$, we can find y, $y' \in E^+$ such that $0 \le y \le x$, $0 \le y' \le x'$, and $z = y + y'$ (corollary to Theorem 1.1.1); then $L(z) = L(y) + L(y') \le M(x) + M(x')$, whence $M(x + x') \le M(x) + M(x')$. Since $M(x+x') = M(x)+M(x')$ for all x, $x' \in E^+$, the mapping $x \mapsto \sup_{0 \le y \le x} L(y)$ from E^+ into $\mathbf{R}$ has a positive linear extension M to E. This proves the first part of the theorem and shows, moreover, that M is the supremum of 0 and L in Ω.

It remains to be proven that Ω is Dedekind complete or, equivalently, that every nonempty set of positive linear forms that is directed upward and bounded above (in E^*) has a supremum. All this is, in fact, a consequence of the following lemma. $\qquad\square$

Lemma 1.3.1 *Let H be a nonempty upward-directed subset of E^*. If $\sup_{u \in H} u(x)$ is finite for all $x \in E^+$, then H has a supremum v in E^*, and $v(x) = \sup_{u \in H} u(x)$ for all $x \in E^+$.*

PROOF: For every $x \in E^+$, put $v(x) = \sup_{u \in H} u(x)$. Then $v(\lambda x) = \lambda v(x)$ for all $x \in E^+$ and all $\lambda \ge 0$. Furthermore, $v(x + y) = v(x) + v(y)$ for all x, $y \in E^+$, because H is directed upwards. The lemma now follows from Proposition 1.3.2. $\qquad\square$

Recall that $L^+(x) = \sup_{0 \le y \le x} L(y)$ for all $L \in \Omega$ and all $x \in E^+$. From this we conclude that

$$(L \vee M)(x) = \sup_{y \ge 0, z \ge 0, y+z=x} L(y) + M(z)$$

$$\text{and} \quad (L \wedge M)(x) = \inf_{y \ge 0, z \ge 0, y+z=x} L(y) + M(z) \qquad (1)$$

for all $L, M \in \Omega$ and for every $x \in E^+$. In particular, replacing M by $-L$ in the first of these formulas, we see that

$$|L|(x) = \sup_{y \geq 0, z \geq 0, y + z = x} L(y - z).$$

Observe that $|y - z| \leq x$ in this last formula; also, the relation $|u| \leq x$ implies $L(u) = L(u^+) - L(u^-) \leq |L|(u^+) + |L|(u^-) = |L|(|u|) \leq |L|(x)$, whence $|L|(x) = \sup_{|u| \leq x} L(u)$. In particular, $|L(x)| \leq |L|(|x|)$ for all $x \in E$.

Proposition 1.3.4 *Two positive linear forms L, M on E are disjoint if and only if, for every $x \in E^+$ and every $\varepsilon > 0$, there exist $y, z \in E^+$ such that $x = y + z$ and $L(y) + M(z) \leq \varepsilon$.*

PROOF: In view of the second formula in Eq. (1), this condition means that $L \wedge M = 0$. □

Proposition 1.3.5 *Let L be a positive linear form on E. A positive linear form M on E belongs to the band generated by L in Ω if and only if, for every $x \in E^+$ and every $\varepsilon > 0$, there exists a real number $\delta > 0$ such that the relations $0 \leq y \leq x$ and $L(y) \leq \delta$ imply $M(y) \leq \varepsilon$.*

PROOF: First, we show that the condition is necessary: if $M \geq 0$ belongs to the band generated by L, then $M = \bigvee_{n \geq 0} ((nL) \wedge M)$ by Theorem 1.2.2. Now $U_n = M - (nL) \wedge M$ is positive for all $n \in \mathbf{Z}^+$, and since $\bigwedge_{n \geq 0} U_n = 0$, Lemma 1.3.1 proves that $U_n(x)$ converges to 0 as $n \to +\infty$; thus there exists n such that $U_n(x) \leq \varepsilon/2$. n having been chosen, we see that the relation $0 \leq y \leq x$ implies

$$U_n(y) \leq \varepsilon/2 \quad \text{and} \quad M(y) \leq \varepsilon/2 + ((nL) \wedge M)(y) \leq \varepsilon/2 + nL(y).$$

Thus $M(y) \leq \varepsilon$ whenever $L(y) \leq \varepsilon/(2n)$, as was to be shown.

We now prove that the condition is sufficient. For every positive linear form M on E we can write $M = U + V$, where U belongs to the band generated by L, V is disjoint from L, and U, V are positive (Theorem 1.2.1). Now, if M satisfies the condition of the proposition, then so does $V = M - U$. Given $x \in E^+$ and $\varepsilon > 0$, choose $0 < \delta \leq \varepsilon$ such that the relations $0 \leq u \leq x$ and $L(u) \leq \delta$ imply $V(u) \leq \varepsilon$; for some $y, z \in E^+$, we have $x = y + z$ and $L(y) + V(z) \leq \delta$ (Proposition 1.3.4). Thus $V(x) = V(y) + V(z) \leq \varepsilon + \delta \leq 2\varepsilon$. Since ε is arbitrary, $V(x) = 0$ for all $x \in E^+$. Hence $V = 0$. □

For example, let $\mathcal{F}(\Omega, \mathbf{R})$ be the ordered vector space of all real-valued functions on a set Ω. Given distinct a, b in Ω, we see that the positive linear forms $\varepsilon_a : f \to f(a)$ and $\varepsilon_b : f \to f(b)$ on $\mathcal{F}(\Omega, \mathbf{R})$ are disjoint.

1.4   Daniell Measures

Define $\mathcal{F}(\Omega, \mathbf{C})$ (respectively, $\mathcal{F}(\Omega, \mathbf{R})$) to be the space of complex- (respectively, real-) valued functions on a given nonempty set Ω, and, for $f \in \mathcal{F}(\Omega, \mathbf{C})$, denote by $\mathrm{Re}\, f$ and $\mathrm{Im}\, f$ the real and imaginary parts of f. Let $\mathcal{H}(\Omega, \mathbf{C})$ be a vector subspace of $\mathcal{F}(\Omega, \mathbf{C})$, such that $\mathrm{Re}\, f$, $\mathrm{Im}\, f$, and $|f|$ belong to $\mathcal{H}(\Omega, \mathbf{R}) = \mathcal{H}(\Omega, \mathbf{C}) \cap \mathcal{F}(\Omega, \mathbf{R})$ for any $f \in \mathcal{H}(\Omega, \mathbf{C})$. Since

$$\sup(f_1, f_2) = \frac{f_1 + f_2 + |f_1 - f_2|}{2} \quad \text{and} \quad \inf(f_1, f_2) = \frac{f_1 + f_2 - |f_1 - f_2|}{2}$$

for all f_1, $f_2 \in \mathcal{H}(\Omega, \mathbf{R})$, we see that $\mathcal{H}(\Omega, \mathbf{R})$ is a Riesz subspace of $\mathcal{F}(\Omega, \mathbf{R})$. Moreover, we assume that, for all f_1, f_2 in $\mathcal{H}^+ = \mathcal{H}(\Omega, \mathbf{R})^+$ and all g in $\mathcal{H}(\Omega, \mathbf{C})$ such that $|g| \leq f_1 + f_2$, there exist g_1, g_2 in $\mathcal{H}(\Omega, \mathbf{C})$ satisfying $|g_1| \leq f_1$, $|g_2| \leq f_2$, and $g_1 + g_2 = g$.

Then, fixing a $\mathbf{C}$-linear form μ on $\mathcal{H}(\Omega, \mathbf{C})$ amounts to fixing an $\mathbf{R}$-linear mapping τ from $\mathcal{H}(\Omega, \mathbf{R})$ into $\mathbf{C}$, μ and τ related by $\mu(f) = \tau(\mathrm{Re}\, f) + i\tau(\mathrm{Im}\, f)$ for all $f \in \mathcal{H}(\Omega, \mathbf{C})$. For this reason, we need not distinguish $\mathbf{C}$-linear forms on $\mathcal{H}(\Omega, \mathbf{C})$ from $\mathbf{R}$-linear mappings of $\mathcal{H}(\Omega, \mathbf{R})$ into $\mathbf{C}$.

Define a $\mathbf{C}$-linear form μ on $\mathcal{H}(\Omega, \mathbf{C})$ as real if $\mu(f)$ is real for every f in $\mathcal{H}(\Omega, \mathbf{R})$, and positive if $\mu(f)$ is positive for each f in $\mathcal{H}^+$.

The set of $\mathbf{R}$-linear forms on $\mathcal{H}(\Omega, \mathbf{R})$ will be ordered by the relation $\mu_1 \leq \mu_2$ whenever $\mu_1(f) \leq \mu_2(f)$ for all $f \in \mathcal{H}^+$.

Definition 1.4.1 A $\mathbf{C}$-linear form μ on $\mathcal{H}(\Omega, \mathbf{C})$ is said to be of finite variation whenever $L(f) = \sup_{g \in \mathcal{H}(\Omega, \mathbf{C}), |g| \leq f} |\mu(g)|$ is finite for every $f \in \mathcal{H}^+$.

Theorem 1.4.1 *Let μ be a $\mathbf{C}$-linear form on $\mathcal{H}(\Omega, \mathbf{C})$, of finite variation. Then the mapping $f \mapsto L(f)$ on $\mathcal{H}^+$ has a unique $\mathbf{C}$-linear extension $|\mu|$ (or $V\mu$) to the whole of $\mathcal{H}(\Omega, \mathbf{C})$. Moreover, $|\mu|$ is the smallest of all positive linear forms ν on $\mathcal{H}(\Omega, \mathbf{C})$ satisfying $|\mu(g)| \leq \nu(|g|)$ for all $g \in \mathcal{H}(\Omega, \mathbf{C})$. $|\mu|$ is called the absolute value, or the variation, of μ.*

PROOF: Let $\mathbf{T}$ be the unit circle in the complex plane. Let f_1, $f_2 \in \mathcal{H}^+$. If g_1, $g_2 \in \mathcal{H}(\Omega, \mathbf{C})$ are such that $|g_1| \leq f_1$ and $|g_2| \leq f_2$, then $|g_1 + \zeta g_2| \leq f_1 + f_2$ for all $\zeta \in \mathbf{T}$; but we may choose ζ so that $|\mu(g_1) + \zeta\mu(g_2)| = |\mu(g_1)| + |\mu(g_2)|$. Since $|\mu(g_i)|$ is arbitrarily close to $L(f_i)$, we conclude that $L(f_1) + L(f_2) \leq L(f_1 + f_2)$.

On the other hand, consider a function $g \in \mathcal{H}(\Omega, \mathbf{C})$ such that $|g| \leq f_1 + f_2$. By hypothesis, there exist g_1, $g_2 \in \mathcal{H}(\Omega, \mathbf{C})$ satisfying $|g_1| \leq f_1$, $|g_2| \leq f_2$, and $g = g_1 + g_2$; thus $|\mu(g)| \leq |\mu(g_1)| + |\mu(g_2)| \leq L(f_1) + L(f_2)$. Since $|\mu(g)|$ is arbitrarily close to $L(f_1 + f_2)$, we see that $L(f_1 + f_2) \leq L(f_1) + L(f_2)$, and so $L(f_1 + f_2) = L(f_1) + L(f_2)$.

By Proposition 1.3.3, L has a unique linear extension $|\mu|$ to the whole of $\mathcal{H}(\Omega, \mathbf{C})$ and $|\mu|$ is positive. The rest of the proof now follows easily. $\square$

Theorem 1.4.2 *Let μ be a C-linear form on $\mathcal{H}(\Omega, \mathbf{C})$, and assume that μ is real. Let τ be the R-linear form on $\mathcal{H}(\Omega, \mathbf{R})$ corresponding to μ. Then μ has finite variation if and only if τ is an order-bounded linear form; moreover, $|\mu|(f) = \sup_{g \in \mathcal{H}(\Omega, \mathbf{R}), |g| \leq f} |\mu(g)|$ (so that $|\tau|$ corresponds to $|\mu|$).*

PROOF: The first statement is obvious. Now, for fixed $f \in \mathcal{H}^+$, put

$$a = \sup_{g \in \mathcal{H}(\Omega, \mathbf{R}), |g| \leq f} |\mu(g)|.$$

For every $g \in \mathcal{H}(\Omega, \mathbf{C})$ satisfying $|g| \leq f$, there exists a complex number ζ with modulus 1 such that $|\mu(g)| = \mu(\zeta g)$. Then $|\mu(g)| = \mu(\mathrm{Re}(\zeta g)) \leq a$. So $|\mu|(f) \leq a$. Now the reverse inequality is clearly true, and the result follows. $\square$

The space $Q\mathcal{M}(\Omega, \mathbf{C})$, of those C-linear forms on $\mathcal{H}(\Omega, \mathbf{C})$ that have finite variation, is a vector subspace of the algebraic dual $\mathcal{H}(\Omega, \mathbf{C})^*$ and contains the set of positive linear forms on $\mathcal{H}(\Omega, \mathbf{C})$. Moreover, $|\mu| = \mu$ for any positive linear form (by Theorem 1.4.2), $|\mu_1 + \mu_2| \leq |\mu_1| + |\mu_2|$, and $|\alpha \mu| = |\alpha| \cdot |\mu|$ for all μ_1, μ_2, μ in $Q\mathcal{M}(\Omega, \mathbf{C})$ and all $\alpha \in \mathbf{C}$.

For any linear form μ on $\mathcal{H}(\Omega, \mathbf{C})$, the linear form $\bar{\mu} : f \to \overline{\mu(\bar{f})}$ is called the conjugate of μ; $\mathrm{Re}\,\mu = (\mu + \bar{\mu})/2$ and $\mathrm{Im}\,\mu = (\mu - \bar{\mu})/(2i)$ are called the real and imaginary parts of μ, respectively. Clearly, $(\mathrm{Re}\,\mu)(f) = \mathrm{Re}(\mu(f))$ and $(\mathrm{Im}\,\mu)(f) = \mathrm{Im}(\mu(f))$ for all $f \in \mathcal{H}(\Omega, \mathbf{R})$, but not for all $f \in \mathcal{H}(\Omega, \mathbf{C})$. For every $\mu \in Q\mathcal{M}(\Omega, \mathbf{C})$, $\bar{\mu}$, $\mathrm{Re}\,\mu$ and $\mathrm{Im}\,\mu$ belong to $Q\mathcal{M}(\Omega, \mathbf{C})$. Additionally $|\bar{\mu}| = |\mu|$, $|\mathrm{Re}\,\mu| \leq |\mu|$, $|\mathrm{Im}\,\mu| \leq |\mu|$ and $|\mu| \leq |\mathrm{Re}\,\mu| + |\mathrm{Im}\,\mu|$.

Definition 1.4.2 A μ in $Q\mathcal{M}(\Omega, \mathbf{C})$ is said to be a Daniell measure if $\mu(f_n)$ converges to 0 for every decreasing sequence $(f_n)_{n \geq 1}$ in $\mathcal{H}^+$ admitting 0 as its lower envelope. Equivalently, $\mu \in Q\mathcal{M}(\Omega, \mathbf{C})$ is a Daniell measure if $\mu(f_n)$ tends to $\mu(f)$ for every increasing sequence in $\mathcal{H}^+$ admitting $f \in \mathcal{H}^+$ as its upper envelope.

Call $\mathcal{M}(\Omega, \mathbf{C})$ the vector subspace of $Q\mathcal{M}(\Omega, \mathbf{C})$ consisting of the Daniell measures, and $Q\mathcal{M}(\Omega, \mathbf{R})$ (respectively, $\mathcal{M}(\Omega, \mathbf{R})$) the R-vector space of those μ in $Q\mathcal{M}(\Omega, \mathbf{C})$ (respectively, in $\mathcal{M}(\Omega, \mathbf{C})$) that are real.

Proposition 1.4.1 $\mathcal{M}(\Omega, \mathbf{R})$ *is a band in $Q\mathcal{M}(\Omega, \mathbf{R})$.*

PROOF: Let $\mu \in \mathcal{M}(\Omega, \mathbf{R})$, and let $(f_n)_{n \geq 1}$ be an increasing sequence in $\mathcal{H}^+$ whose upper envelope f belongs to $\mathcal{H}^+$. The sequence $(\mu^+(f_n))_{n \geq 1}$ increases and is bounded, so $a = \sup_{n \geq 1} \mu^+(f_n)$ is finite. For $\varepsilon > 0$, there exists g in $\mathcal{H}^+$, $g \leq f$, such that $\mu^+(f) \leq \mu(g) + \varepsilon/2$. Now the sequence $(\inf(f_n, g))_{n \geq 1}$ increases and has $\inf(f, g) = g$ as its upper envelope. Thus $\mu(\inf(f_n, g))$ tends to $\mu(g)$, and there is an n for which $\mu(g) \leq \mu(\inf(f_n, g)) + \varepsilon/2 \leq \mu^+(f_n) + \varepsilon/2 \leq a + \varepsilon/2$. Finally, $\mu^+(f) \leq a + \varepsilon$, which proves that $\mu^+(f) = a$;

whence we see that μ^+ is a Daniell measure and $\mathcal{M}(\Omega, \mathbf{R})$ is a Riesz subspace of $Q\mathcal{M}(\Omega, \mathbf{R})$.

Next, let A be a nonempty subset of $\mathcal{M}^+ = \mathcal{M}(\Omega, \mathbf{R})^+$, directed upward, bounded above in $Q\mathcal{M}(\Omega, \mathbf{R})$, and put $\mu = \sup A$. Let $(f_n)_{n \geq 1}$ be a decreasing sequence in $\mathcal{H}^+$, admitting 0 as its lower envelope. For $\varepsilon > 0$, there exists $\nu \in A$ such that $(\mu - \nu)(f_1) \leq \varepsilon/2$. Then $(\mu - \nu)(f_n) \leq \varepsilon/2$ for all $n \geq 1$. Now, for n large enough, $\nu(f_n) \leq \varepsilon/2$, so $\mu(f_n) \leq \varepsilon$. Hence μ is a Daniell measure.   □

Observe that $|\mu|$ belongs to $\mathcal{M}(\Omega, \mathbf{R})$ for every $\mu \in \mathcal{M}(\Omega, \mathbf{C})$, because $|\mu| \leq |\operatorname{Re}\mu| + |\operatorname{Im}\mu|$.

In important cases which we shall examine later, if $\mu \in Q\mathcal{M}(\Omega, \mathbf{C})$ and $\nu \in Q\mathcal{M}^+$ are such that $|\mu(h)| \leq \nu(h)$ for all $h \in \mathcal{H}^+$, then $|\mu(g)| \leq \nu(|g|)$ for all $g \in \mathcal{H}(\Omega, \mathbf{C})$. For $\mu \in Q\mathcal{M}(\Omega, \mathbf{C})$, then $|\mu| = \sup_{\alpha \in \mathbf{T}} \operatorname{Re}(\alpha\mu)$.

Exercises for Chapter 1

1 Let E be a Dedekind complete Riesz space. A filter basis $\mathcal{F}$ on E is said to be bounded whenever there exists $X \in \mathcal{F}$ that is bounded above and below. Then we may define the upper limit (respectively, lower limit) of $\mathcal{F}$, written $\limsup \mathcal{F}$ (respectively, $\liminf \mathcal{F}$), by the element $\inf_X(\sup X)$ (respectively, $\sup_X(\inf X)$) of E, where X extends over the class B of bounded subsets $X \in \mathcal{F}$; thus $\liminf \mathcal{F} \leq \limsup \mathcal{F}$. We say that $\mathcal{F}$ has an order limit in E, or is (order-) convergent, if $\liminf \mathcal{F} = \limsup \mathcal{F}$. The common value of these two elements is then denoted by $\lim \mathcal{F}$, and we say that $\mathcal{F}$ converges to this value.

  1. For any bounded filter basis $\mathcal{F}$ on E, prove that
  $$\inf_{X \in B}(\sup X - \inf X) = \limsup \mathcal{F} - \liminf \mathcal{F}.$$

  2. Let f be an ordered isomorphism from E onto an ideal of a Dedekind complete Riesz space F. If a filter basis $\mathcal{F}$ on E converges, prove that the filter basis $f(\mathcal{F})$ on F converges and that $\lim f(\mathcal{F}) = f(\lim \mathcal{F})$.

  3. Let A be a nonempty ordered set, directed upward, and S the filter basis on A consisting of the sets $\{\beta \in A : \beta \geq \alpha\}$ (for $\alpha \in A$). The filter on A generated by S is called the filter of sections of A. Let f be a mapping from A into E. If the filter basis $f(S)$ on E converges to a point $x \in E$, we say that f has limit x along A. Prove that f converges to $\sup_{\alpha \in A} f(\alpha)$ along A, whenever f is increasing and $f(A)$ is bounded above.

  4. Let $(E_i)_{i \in I}$ be a nonempty family of Dedekind complete Riesz spaces, and let E be the ordered vector space $\prod_{i \in I} E_i$. Show that a bounded filter basis $\mathcal{F}$ on E is order-convergent if and only if, for every i, the filter basis $pr_i(\mathcal{F})$ is order-convergent in E_i.

2 Let E be a Dedekind complete Riesz space and let $(x_i)_{i \in I}$ be a family in E. Order by inclusion the class $\mathcal{F}(I)$ of finite subsets of I, and put $s_H = \sum_{i \in H} x_i$ for all $H \in \mathcal{F}(I)$ $(s_\emptyset = 0)$. The family $(x_i)_{i \in I}$ is said to be (order) summable if

$\{s_H : H \in \mathcal{F}(I)\}$ is bounded and if the mapping $H \mapsto s_H$ has a limit along the upward-directed set $\mathcal{F}(I)$. This limit is then called the sum of the family $(x_i)_{i \in I}$ and is written $\sum_{i \in I} x_i$.

1. Prove that a family $(x_i)_{i \in I}$ is summable if and only if the following conditions hold:

    (a) For every $H \in \mathcal{F}(I)$, the set $\{|s_K| : K \in \mathcal{F}(I), K \cap H = \emptyset\}$ has a supremum r_H in E.

    (b) $\inf_{H \in \mathcal{F}(I)} r_H = 0$.

    (Use part 1 of Exercise 1.)

2. Prove that any subfamily of a summable family is summable.

3. Let $(x_i)_{i \in I}$ be a summable family in E and $(I_\lambda)_{\lambda \in L}$ be a partition of I. For each $\lambda \in L$, put $s_\lambda = \sum_{i \in I_\lambda} x_i$. Show that the family $(s_\lambda)_{\lambda \in L}$ is summable and that $\sum_{\lambda \in L} s_\lambda = \sum_{i \in I} x_i$.

4. Let $(x_i)_{i \in I}$ be a family in E and $(I_\lambda)_{\lambda \in L}$ be a finite partition of I. Suppose that the family $(x_i)_{i \in I_\lambda}$ is summable for all $\lambda \in L$. Show that $(x_i)_{i \in I}$ is summable.

5. Let $(x_i)_{i \in I}$ and $(y_i)_{i \in I}$ be two summable families in E with the same index set, I. Prove that $(x_i + y_i)_{i \in I}$ is summable and that $\sum_{i \in I}(x_i + y_i) = \sum_{i \in I} x_i + \sum_{i \in I} y_i$.

6. Show that a family $(x_i)_{i \in I}$ of positive elements is summable if and only if the set $\{s_H : H \in \mathcal{F}(I)\}$ is bounded above and that, in this case, $\sum_{i \in I} x_i = \sup_{H \in \mathcal{F}(I)} s_H$. Let $(y_i)_{i \in I}$ be a family in E such that $0 \le y_i \le x_i$ for all $i \subset I$. Show that the equality $\sum_{i \in I} y_i = \sum_{i \in I} x_i$ holds if and only if $x_i = y_i$ for all $i \in I$.

7. Show that a family $(x_i)_{i \in I}$ is summable in E whenever $(|x_i|)_{i \in I}$ is summable and that, in this case, $|\sum_{i \in I} x_i| \le \sum_{i \in I} |x_i|$.

8. Let $(x_i)_{i \in I}$ be a family in E of mutually disjoint elements, and suppose that $\{|x_i| : i \in I\}$ is bounded above. Show that $(|x_i|)_{i \in I}$ is summable; prove, in fact, that $(\sum_{i \in I} x_i)^+ = \sum_{i \in I} x_i^+$ and $(\sum_{i \in I} x_i)^- = \sum_{i \in I} x_i^-$, so that $|\sum_{i \in I} x_i| = \sum_{i \in I} |x_i|$.

3 Let E be a Dedekind complete Riesz space.

1. Show that there exists a family $(u_i)_{i \in I}$ in E of strictly positive elements with the following properties:

    (a) u_j, u_k are disjoint whenever $j \ne k$.

    (b) For every $x > 0$, there exists $i \in I$ such that $\inf(x, u_i) > 0$.

    (Use Zorn's lemma.)

2. Let $x \in E$ and, for each $i \in I$, let x_i be the component of x in the band B_i generated by u_i. Prove that the family $(x_i)_{i \in I}$ is summable with sum x (first, assume $x \ge 0$).

3. Conversely, let $(x_i)_{i \in I}$ be a family such that x_i belongs to B_i for each $i \in I$, and suppose that $\{|x_i| : i \in I\}$ is bounded above in E. Show that, for every $j \in I$, x_j is the component of $x = \sum_{i \in I} x_i$ in B_j.

4 Let A be a lattice.

1. Prove that $x \wedge (y_1 \vee y_2) = (x \wedge y_1) \vee (x \wedge y_2)$ for all x, y_1, $y_2 \in A$ if and only if $x \vee (y_1 \wedge y_2) = (x \vee y_1) \wedge (x \vee y_2)$ for all x, y_1, $y_2 \in A$ (in this case A is said to be distributive).

2. Suppose the following conditions are satisfied:

    (a) A is distributive.

    (b) A has an infimum, 0, and a supremum, 1.

    (c)   Any $x \in A$ has a complement x' in A, that is, for any $x \in A$ there exists an $x' \in A$ such that $x \wedge x' = 0$ and $x \vee x' = 1$.

    (That is, A is a Boolean algebra.) Prove that every $x \in A$ has a unique complement x' and that $(x \vee y)' = x' \wedge y'$, and $(x \wedge y)' = x' \vee y'$ for all x, $y \in A$.

5 In a topological space, a clopen set is a subset that is both closed and open. A Hausdorff space S is called zerodimensional if the clopen sets form a base for the topology of S; S is then totally disconnected, that is, for every $x \in S$ the connected component of x is $\{x\}$. Conversely, a totally disconnected, compact Hausdorff space is zerodimensional.

1. Let S be a totally disconnected, compact Hausdorff space. Let $b(S)$ be the class of clopen subsets of S, and order $b(S)$ by inclusion. Prove that $b(S)$ is a Boolean algebra.

2. Let S, T be totally disconnected, compact Hausdorff spaces and f a lattice isomorphism from $b(S)$ onto $b(T)$. Prove that there exists a homeomorphism φ of S onto T such that $f(U) = \{\varphi(x) : x \in U\}$ for all $U \in b(S)$ (if $s \in S$, put $B = \{U \in b(S) : s \in U\}$, and show that $\bigcap_{U \in B} f(U)$ is a singleton).

3. Let A be a Boolean algebra. A totally disconnected, compact Hausdorff space T is "a Stone space of A" if A and $b(T)$ are isomorphic lattices. We prove the Stone representation theorem: A has a Stone space (unique up to a homeomorphism, by part 2). We may restrict ourselves to the case where $0 \neq 1$.

    A subset J of A is called a filter if

    (a) $1 \in J$,

    (b) $0 \notin J$, and

    (c)   for all x, $y \in A$, $x \wedge y \in J$ if and only if $x \in J$ and $y \in J$.

    Let S be the class of maximal filters and, for every $x \in A$, put $S_x = \{J \in S : x \in J\}$. Prove that $(S_x)' = S_{x'}$ and that $S_{x \wedge y} = S_x \cap S_y$ and $S_{x \vee y} = S_x \cup S_y$ for all x, y in A. Show that the sets S_x form the base of a Hausdorff topology on S.

4. Let $A_1 \subset A$ be such that $S \neq \bigcup_{x \in A_2} S_x$ for all finite subsets A_2 of A_1. Denote by I the set of those $z \in A$ for which there exists a finite set $A_2 \subset A_1$ such that $z \geq (\sup A_2)'$. Prove that I is a filter, and conclude that S is compact by showing that $S \neq \bigcup_{x \in A_1} S_x$ (apply Zorn's lemma).

5. Show that the map $x \mapsto S_x$ is a lattice isomorphism from A onto $b(S)$ (use the compactness of S).

6 A Boolean algebra A is said to be Dedekind complete if any part of A has a supremum.

1. A Hausdorff space X is said to be extremely disconnected if the following equivalent conditions hold:

   (a) Every open set has open closure.

   (b) Every open set U such that $U = \mathrm{int}(\bar{U})$ is clopen.

   (c) If U, V are disjoint open subsets of X, then $\bar{U} \cap \bar{V} = \emptyset$.

   Prove that if X is extremely disconnected, then X is totally disconnected.

2. Let X be a totally disconnected, compact Hausdorff space. Prove that $b(X)$ is Dedekind complete if and only if X is extremely disconnected.

7 Let X be a topological space. An *LSC* function on X (respectively, a *USC* function on X) is an abbreviation for a lower (respectively, upper) semicontinuous function on X.

If A is a dense subset of X and f a function from A into $\overline{\mathbf{R}}$, then put

$$g(x) = \limsup_{y \mapsto x, y \in A} f(y) = \inf_{V \text{ neighborhood of } x} \sup_{y \in V \cap A} f(y)$$

for any $x \in X$. Then $g : x \mapsto g(x)$ is the smallest of those USC functions φ on X satisfying $\varphi(y) \geq f(y)$ for all $y \in A$, and g is called the USC regularization of f.

Henceforth, assume that X is compact and extremely disconnected.

1. Let f be an LSC function on X and g its USC regularization. For every $a \in \mathbf{R}$, a point $x \in X$ such that $g(x) > a$ does not belong to the closure of $A = \{y \in X : g(y) < a\}$; so g is LSC. Hence g is continuous. For every pair of rational numbers r, s such that $r < s$, show that the closed set $\{x \in X : f(x) \leq r, g(x) \geq s\}$ is nowhere dense, and conclude that $f = g$ outside a meager set.

2. Now, let h be a function with values in $\overline{\mathbf{R}}$, defined and continuous on the complement $X - M$ of a nowhere dense set M. Let f be the LSC function from X into $\overline{\mathbf{R}}$ that agrees with h on $X - \bar{M}$ and is equal to $-\infty$ on $\bar{M}$. Show that the USC regularization g of f agrees with h on $X - \bar{M}$, and so on $X - M$, that is, show that h has a unique continuous extension to X.

3. Denote by $\mathcal{C}^\infty(X, \overline{\mathbf{R}})$ the set of those continuous functions f from X into $\overline{\mathbf{R}}$ for which $f^{-1}(+\infty)$ and $f^{-1}(-\infty)$ are nowhere dense. For all f, $g \in \mathcal{C}^\infty(X, \overline{\mathbf{R}})$, there exists a unique $h \in \mathcal{C}^\infty(X, \overline{\mathbf{R}})$ such that $h(x) = f(x) + g(x)$ whenever this expression has meaning; in this case, we put $h = f + g$. Similarly, we define λf for all $\lambda \in \mathbf{R}$ and all $f \in \mathcal{C}^\infty(X, \overline{\mathbf{R}})$. Then $\mathcal{C}^\infty(X, \overline{\mathbf{R}})$, canonically ordered, is a Riesz space. Show that, if H is a nonempty subset of $\mathcal{C}^\infty(X, \overline{\mathbf{R}})$, bounded above, and if f is its upper envelope ($f(x) = \sup_{h \in H} h(x)$ for all $x \in X$), then the USC regularization

g of f is the supremum of H in $C^\infty(X, \overline{\mathbf{R}})$. Conclude that $C^\infty(X, \overline{\mathbf{R}})$ and its ideal $C(X, \mathbf{R})$ are Dedekind complete.

4. Let A be an open subset of X, and let H consist of the indicator functions 1_U of the clopen subsets contained in A. Prove that $1_{\bar{A}}$ is the supremum of H in $C(X, \mathbf{R})$.

8 Let E be a Dedekind complete Riesz space.

1. If $x \in E$ is such that $\{nx : n \in \mathbf{Z}^+\}$ is bounded above, prove that x is negative.

2. Show that $\inf_{n \in \mathbf{N}} \left((1/n)u \right) = 0$ for every $u \in E^+$, that is, E is Archimedian.

9 Let E be a Dedekind complete Riesz space and B_1, B_2 two bands in E satisfying $B_1 \subset B_2$. Let x be an element of E and x_1, x_2 its components in B_1, B_2. Show that x_1 is the component of x_2 in the band B_1 of B_2.

10 Let $u > 0$ be given in a Dedekind complete Riesz space E. Those $x \in E$ satisfying $|x| \leq nu$, for at least one integer $n \geq 0$, form an ideal C_u of E. For all $x \in C_u$, put $\|x\| = \inf\{\lambda \in \mathbf{R}^{+*} : |x| \leq \lambda u\}$.

1. Show that $|x| \leq \|x\|u$ for all $x \in C_u$, that the function $x \to \|x\|$ is a norm on C_u, and that $\|y + z\| = \sup(\|y\|, \|z\|)$ for any disjoint elements of C_u. Furthermore, show that C_u^+ is closed in the normed vector space C_u.

2. A $c \in E$ is the component of u in a band of C_u if and only if $0 \leq c \leq u$ and $\inf(c, u - c) = 0$. Show that the set A of these elements c is a Dedekind complete Boolean algebra and is closed in the normed space C_u.

3. Fix x in C_u. For every λ in $\mathbf{R}$, let c_λ be the component $\sup_{n \geq 0} \inf \left(n(\lambda u - x)^+, u \right)$ of u in the band of C_u generated by $(\lambda u - x)^+$. Show that $c_\lambda \leq c_\mu$ whenever $\lambda \leq \mu$, that $c_\lambda = 0$ for $\lambda \leq -\|x\|$, and that $c_\lambda = u$ for $\lambda > \|x\|$. For all $\lambda \in \mathbf{R}$, prove that the band B_λ in C_u, generated by c_λ, is exactly the band in C_u generated by $(\lambda u - x)^+$.

4. With the same notation as in part 3, show that the component of x in B_λ is $\lambda c_\lambda - (\lambda u - x)^+$, and so is smaller than λc_λ. If c belongs to A, if $c \leq c_\lambda$, and if B is the band in C_u generated by c, prove that the component of x in B is smaller than λc.

5. Let α, $\beta \in \mathbf{R}$ with $\alpha < \beta$, and let p_α, p_β be the projections of C_u onto the bands generated by c_α and c_β, respectively. Prove that $(p_\beta - p_\alpha)(x) = (p_\beta - p_\alpha) \cdot \left(p_\beta(x) \right)$ is smaller than $\beta(c_\beta - c_\alpha)$. Now denote by B'_α the disjoint complement of $\{c_\alpha\}$ in C_u, and put $B = B'_\alpha \cap B_\beta$. Since the component $(\alpha u - x)^-$ of $x - \alpha u$ in B'_α is positive, the component of $x - \alpha u$ in B is positive as well. Conclude that $\alpha(c_\beta - c_\alpha) \leq p_\beta(x) - p_\alpha(x)$. In short, $\alpha(c_\beta - c_\alpha) \leq p_\beta(x) - p_\alpha(x) \leq \beta(c_\beta - c_\alpha)$.

6. Let $\varepsilon > 0$ and let $\lambda_0, \lambda_1, \ldots, \lambda_n$ be real numbers such that $c_{\lambda_0} = 0$, $c_{\lambda_n} = u$, and $0 < \lambda_i - \lambda_{i-1} \leq \varepsilon$ for all $1 \leq i \leq n$. For every $0 \leq i \leq n$, denote by p_i

the projection from C_u onto the band generated by c_{λ_i}. Then

$$|(p_i - p_{i-1})(x) - \lambda_i(c_{\lambda_i} - c_{\lambda_{i-1}})| \leq \varepsilon(c_{\lambda_i} - c_{\lambda_{i-1}})$$

for all $1 \leq i \leq n$, and so $\|x - \sum_{1 \leq i \leq n} \lambda_i(c_{\lambda_i} - c_{\lambda_{i-1}})\| \leq \varepsilon$. Deduce that the vector subspace V of C_u, generated by A, is dense in C_u.

7. Let S be the Stone space of A, θ_c the indicator function of the clopen subset S_c, for every $c \in A$ (notation as in Exercise 7), and W the vector subspace of $\mathcal{C}(S, \mathbf{R})$ generated by the θ_c. Let $g \in \mathcal{C}(S, \mathbf{R})$ and $\varepsilon > 0$. For the real numbers $\lambda_0, \lambda_1, \ldots, \lambda_n$, such that $\lambda_0 \leq \inf_{t \in S} g(t)$, $\lambda_n > \sup_{t \in S} g(t)$, and $0 < \lambda_i - \lambda_{i-1} \leq \varepsilon$ for every $1 \leq i \leq n$, put $A_i = \{t \in S : g(t) < \lambda_i\}$ for all $1 \leq i \leq n$ and $B_1 = \bar{A}_1$, $B_i = \bar{A}_i \cap C\bar{A}_{i-1}$ for all $2 \leq i \leq n$. Prove that $|g - \sum_{1 \leq i \leq n} \lambda_i 1_{B_i}| \leq \varepsilon$. Thus W is dense in $\mathcal{C}(S, \mathbf{R})$.

8. Prove that a linear combination $\sum_{j \in I} \lambda_j c_j$ of elements $c_j \in A$ is positive if and only if $\sum_{j \in I} \lambda_j \theta_{c_j}$ is positive (assume that I is finite and denote by $\mathcal{F}(I)$ the class of all subsets of I; for every H in $\mathcal{F}(I)$, put

$$z_h = \inf(\{c_i : i \in H\} \cup \{u - c_i : i \notin H\});$$

so the z_H are disjoint and $u = \sum_{H \in \mathcal{F}(I)} z_H$; observe that

$$c_j = \sum_{H \in \mathcal{F}(I), j \in H} z_H \quad \text{and} \quad \theta_{c_j} = \sum_{H \in \mathcal{F}(I), j \in H} \theta_{z_H}$$

for $j \in I$). Thus there is a unique isomorphism θ from the ordered vector space V onto the ordered vector space W such that $\theta(c) = \theta_c$ for all $c \in A$. Show that θ is an isometry from the normed subspace V of C_u onto the normed subspace W of the canonically normed space $\mathcal{C}(S, \mathbf{R})$.

9. Let $(x_n)_{n \geq 1}$ be a Cauchy sequence in C_u, and let $(x_{n_k})_{k \geq 1}$ be a subsequence such that $\|x_{n_{k+1}} - x_{n_k}\| \leq 2^{-k}$ for all $k \geq 1$. Put $y_k = x_{n_{k+1}} - x_{n_k}$ for all $k \geq 1$. Prove that the family $(y_k)_{k \geq 1}$ is order summable in E and show that, if $y = \sum_{k \geq 1} y_k$, then $(x_{n_k})_{k \geq 1}$ converges to $x = y + x_{n_1}$ in the normed space C_u (so C_u is a Banach space). Conclude that θ extends to an isometry from C_u into $\mathcal{C}(S, \mathbf{R})$.

10. Let $x \in C_u$. Prove that $\theta_x = g$ is positive if and only if x is positive (take $\lambda_0 = 0$ in questions 6, 7). Conclude that θ is a Riesz isomorphism from C_u onto $\mathcal{C}(S, \mathbf{R})$ (this result is due to K. Yosida, S. Kakutani, M. and S. Krein, H. Nakano).

11 Let E, $u > 0$, C_u be as in Exercise 10, and let B_u be the band in E generated by u.

1. For every $x \in B_u^+$, put $x_n = \inf(nu, x)$ for all integers $n \geq 0$, and let $\bar{\theta}x$ be the USC regularization of the upper envelope of the θx_n, so that $\bar{\theta}x$ is continuous. If y belongs to C_u^+ and $\theta y \leq \bar{\theta}x$, prove that $\theta y \leq \theta\big(\inf(x, y)\big)$, and therefore that $y \leq x$. Deduce that $\bar{\theta}x$ belongs to $\mathcal{C}^\infty(S, \overline{\mathbf{R}})$ and is the supremum in $\mathcal{C}^\infty(S, \overline{\mathbf{R}})$ of the θx_n. Moreover, $\bar{\theta}x = \sup_{x' \in C_u, 0 \leq x' \leq x} \theta x'$.

2. Show that there is a unique linear map $\bar{\theta}$ from C_u into $\mathcal{C}^{\infty}(S, \overline{\mathbf{R}})$ such that $\bar{\theta}(x) = \bar{\theta}x$ for all $x \in B_u^+$, and prove that $\bar{\theta}$ is an isomorphism from B_u onto an ordered vector subspace of $\mathcal{C}^{\infty}(S, \overline{\mathbf{R}})$ (observe that if $x, y \in B_u^+$ and $\bar{\theta}y \leq \bar{\theta}x$, then $y \leq x$).

3. Let $x \in B_u^+$ and $g \in \mathcal{C}^{\infty}(S, \overline{\mathbf{R}})$ be given, such that $0 \leq g \leq \bar{\theta}x$. For any integer $n \geq 0$, there exists $y_n \in C_u$ such that $\theta_{y_n} = \inf(g, n)$. If $y = \sup y_n$, prove that $\bar{\theta}(y) = g$. So $\bar{\theta}(B_u)$ is an ideal in $\mathcal{C}^{\infty}(S, \overline{\mathbf{R}})$.

4. Prove the Maeda–Ogasawara representation theorem: there exists a Riesz isomorphism $\bar{\theta}$ from B_u onto an ideal of $\mathcal{C}^{\infty}(S, \overline{\mathbf{R}})$, such that $\bar{\theta}(c) = 1_{S_c}$ for all c in A and θ maps C_u onto $\mathcal{C}(S, \mathbf{R})$.

5. Show that V is order dense in C_u (examine the argument in question 6 of Exercise 10). Since $x = \sup_{n \geq 0} \inf(nu, x)$ for every $x \in B_u^+$, $\bar{\theta}$ is the unique isomorphism from the ordered vector space B_u onto an ideal of $\mathcal{C}^{\infty}(S, \overline{\mathbf{R}})$ such that $\bar{\theta}(c) = 1_{S_c}$ for all $c \in A$.

12 Let E be a Dedekind complete Riesz space, $(u_\alpha)_{\alpha \in A}$ a maximal system in E of strictly positive, disjoint elements, and, for each $\alpha \in A$, denote by B_α the band generated by u_α. E is isomorphic to the ideal of $\Pi_{\alpha \in A} B_\alpha$ consisting of those $(x_\alpha)_{\alpha \in A}$ such that $\{|x_\alpha| : \alpha \in A\}$ is bounded above (see Exercise 3). For every $\alpha \in A$, let K_α be the Stone space of $\{x \in E : 0 \leq x \leq \alpha,\ \inf(x, \alpha - x) = 0\}$ and $\bar{\theta}_\alpha$ the isomorphism constructed above from B_α onto an ideal of $\mathcal{C}^{\infty}(K_\alpha, \overline{\mathbf{R}})$ (see Exercise 11).

Let X be the topological sum of the K_α, $\mathcal{C}^{\infty}(X, \overline{\mathbf{R}})$ the space of those continuous functions f from X into $\overline{\mathbf{R}}$ for which $f^{-1}(+\infty)$ and $f^{-1}(-\infty)$ are nowhere dense in X, and $\mathcal{H}(X, \mathbf{R})$ the space of continuous functions from X into $\mathbf{R}$ with compact support. Then we obtain an isomorphism from $\prod_\alpha B_\alpha$ onto an ideal of $\mathcal{C}^{\infty}(X, \overline{\mathbf{R}})$ and, finally, an isomorphism Φ from E onto an ideal of $\mathcal{C}^{\infty}(X, \overline{\mathbf{R}})$ containing $\mathcal{H}(X, \mathbf{R}) = \mathcal{H}$.

1. If B is a nonempty subset of E, bounded above, and if f is the upper envelope of the $\Phi(x)$, for $x \in B$, show that $\Phi(\sup B)$ is the USC regularization of f.

2. Denote by $\mathcal{F}(A)$ the class of finite subsets of A. If $x \in E^+$ and $x = \sum_{\alpha \in A} x_\alpha$ is the decomposition of x along the B_α, then, for all $\alpha \in A$, we have $x_\alpha = \sup_{n \geq 0} \inf(nu_\alpha, x_\alpha)$, where $\inf(nu_\alpha, x_\alpha)$ belongs to $\Phi^{-1}(\mathcal{H}^+)$. Hence $x = \sup_{\alpha \in A} x_\alpha$ is the supremum of the upward-directed set B consisting of the $\sup_{\alpha \in J} \inf(nu_\alpha, x_\alpha)$ ($J \in \mathcal{F}(A)$, n positive integer). Deduce that the filter $\mathcal{G}$ of sections of B order converges to x and contains $\Phi^{-1}(\mathcal{H}^+)$.

13 Let E be a Riesz space, and $(U_i)_{i \in I}$ a nonempty family of positive linear forms on E. Assume that the topology T, defined by the seminorms $U_i(|x|)$, is a Hausdorff topology.

1. Show that the mapping $x \mapsto |x|$ is uniformly continuous from E into E and that E^+ is closed in the topology T.

2. Suppose that E is Dedekind complete. Let B be a band in E and B' its disjoint complement. Show that E is the topological direct sum of B and B'.

3. Henceforth, we suppose that the topological vector space E is complete. Now let A be a nonempty upward-directed subset of E, such that each $U_i(A)$ is bounded above in $\mathbf{R}$, and denote by $\mathcal{F}$ the filter of sections of A. Show that $\mathcal{F}$ converges (to the supremum of A) in the topological space E. For any continuous and increasing function f from A into $\mathbf{R}$, show that $\sup_{x \in A} f(x) = f(\sup A)$. Conclude that E is Dedekind complete.

4. Let $\mathcal{F}$ be an order-convergent filter on E with limit x. Let A be the set of the $\sup X - \inf X$, where X entends over the class of bounded sets in $\mathcal{F}$. Note that A is directed downward, and conclude from part 3 that $\mathcal{F}$ converges to x in the topological space E.

2

Measures on Semirings

The theory of abstract measures, which is introduced in this chapter, is essential, for example in the theory of probability.

Summary

2.1   Given a nonempty set Ω, the power set of Ω, equipped with symmetric difference and intersection, is a ring. A nonempty subring is called a ring. A σ-ring is a ring which is stable under countable unions. The subset S of semiclosed subintervals $]a, b]$ is not a ring but merely a semiring (Definition 2.1.2). The σ-ring of Halmos sets and the σ-algebra of Borel sets (Definition 2.1.6), in a topological space Ω, are among the most important examples of σ-rings and algebras and will be widely used throughout this book.

2.2   In this section we first define quasi-measures, then measures on semirings. A function μ on a semiring S is a measure if and only if, for any sequence of disjoints sets $A_i \in S$ such that $\bigcup A_i \subset A \in S$, the series $\sum \mu(A_i)$ converges, and if $\mu(A) = \sum \mu(A_i)$ whenever $A = \bigcup A_i$ (Theorem 2.2.3). Any complex measure on a σ-ring is bounded (Proposition 2.2.3). Theorem 2.2.5 gives a basic relationship between abstract measures defined on a semiring S and Daniell measures defined on S-simple functions ("step functions").

2.3   We introduce Lebesgue measure as a measure defined on the semiring of all semiclosed intervals $]\alpha, \beta]$ included in some interval I.

2.1   Semirings, Rings, and σ-Rings

Let $\mathcal{P}(\Omega)$ be the power set of a nonempty set Ω. For all $A, B \in \mathcal{P}(\Omega)$, the difference of A and B, $A - B$, is defined as $A \cap B^c$; the symmetric difference

$A \triangle B = (A \cap B^c) \cup (A^c \cap B)$ consists of the elements that lie in A or B but not in both.

For each $A \in \mathcal{P}(\Omega)$, denote by φ_A the mapping from Ω into the ring $\mathbf{Z}/2\mathbf{Z}$ equal to 1 on A, to 0 on A^c. Then $\varphi_{A \triangle B} = \varphi_A + \varphi_B$ and $\varphi_{A \cap B} = \varphi_A \cdot \varphi_B$ for all A, $B \in \mathcal{P}(\Omega)$. So the operations $\triangle$ and $\cap$ give $\mathcal{P}(\Omega)$ the structure of a unitary ring.

Definition 2.1.1 A class $\mathcal{R} \subset \mathcal{P}(\Omega)$ is called a ring of subsets of Ω (or simply a ring in Ω) if it is a subring of $\mathcal{P}(\Omega)$ in the usual sense. Equivalently, $\mathcal{R}$ is a ring when it is nonempty and when $A \triangle B$, $A \cap B$ lie in $\mathcal{R}$ for all A, $B \in \mathcal{R}$.

Proposition 2.1.1 *A nonempty class $\mathcal{R} \subset \mathcal{P}(\Omega)$ is a ring if and only if $A \cap B^c$ and $A \cup B$ lie in $\mathcal{R}$ for all A, $B \in \mathcal{R}$.*

PROOF: The condition is necessary because $A \cup B = (A \triangle B) \triangle (A \cap B)$ and $A \cap B^c = A \triangle (A \cap B)$; it is sufficient because $A \triangle B = (A \cap B^c) \cup (A^c \cap B)$ and $A \cap B = (A \cup B) \triangle (A \triangle B)$. □

Definition 2.1.2 A nonempty class $S \subset \mathcal{P}(\Omega)$ is called a semiring if, for all A, $B \in S$, there is a finite collection of mutually disjoint S-sets whose union is $A \cap B$, and if the same property holds for $A \cap B^c$. This means, of course, that the empty set necessarily lies in S.

Now the intersection of a collection of rings is a ring. So we may define the ring generated by $\mathcal{C}$ to be the smallest ring containing a given subclass $\mathcal{C}$ of $\mathcal{P}(\Omega)$.

Proposition 2.1.2 *Let S be a semiring in Ω. Then $E \in \mathcal{P}(\Omega)$ lies in the ring $\mathcal{R}$ generated by S if and only if it is a finite union of S-sets, which may be taken disjoint.*

PROOF: Let $\mathcal{R}'$ be the class of those subsets of Ω which can be finitely partitioned by S-sets. We first show that every union of a finite family of S-sets (not necessarily disjoint) lies in $\mathcal{R}'$. The proof is by induction on the number of elements of the family; suppose it is true if this number is $n - 1$ $(n \geq 2)$, and let $(X_i)_{1 \leq i \leq n}$ be a family in S with n elements. We can write

$$Z = \bigcup_{1 \leq i \leq n} X_i = [(\bigcup_{1 \leq i \leq n-1} X_i) \cap X_n^c] \cup X_n.$$

Now, by the induction hypothesis, $\bigcup_{1 \leq i \leq n-1} X_i = \bigcup_{1 \leq j \leq p} Y_j$, where $Y_j \in S$ and $Y_j \cap Y_k = \emptyset$ if $j \neq k$. Moreover, $Y_j \cap X_n^c = \bigcup_{1 \leq \alpha \leq p_j} Z_{j,\alpha}$, where $Z_{j,\alpha} \in S$ and $Z_{j,\alpha} \cap Z_{j,\beta} = \emptyset$ if $\alpha \neq \beta$. Therefore, $Z = [\bigcup_{1 \leq j \leq p} \bigcup_{1 \leq \alpha \leq p_j} Z_{j,\alpha}] \cup X_n$, and we see that $Z_{j,\alpha} \cap Z_{k,\beta} = \emptyset$ if $j \neq k$ or $\alpha \neq \beta$, and that $Z_{j,\alpha} \cap X_n = \emptyset$ for all j, α. This means that Z lies in $\mathcal{R}'$, and the proof is complete.

Now $A \cup B$ and $A \cap B$ lie in $\mathcal{R}'$ for all A, $B \in \mathcal{R}'$. It remains to be proven that $A \cap B^c$ belongs to $\mathcal{R}'$. By the preceding argument, it suffices to consider the case in which A lies in S. Then let A be an S-set and $B = \bigcup_{1 \leq i \leq n} B_i$ be a finite union of S-sets. $A \cap B^c = \bigcap_{1 \leq i \leq n} (A \cap B_i^c)$, where $A \cap B_i^c$ lies in $\mathcal{R}'$; hence $A \cap B^c$ is an $\mathcal{R}'$-set.   □

Proposition 2.1.3 *Let S be a semiring in Ω. For any nonempty finite family $(A_i)_{i \in I}$ of S-sets, there exist disjoint S-sets B_k $(k \in K$, K finite) such that each A_i is a union of some of the sets B_k.*

PROOF: $E_J = (\bigcap_{i \in J} A_i) \cap (\bigcap_{i \notin J} A_i^c)$ lies in $\mathcal{R}$ for any nonempty subset J of I, so each E_J can be finitely partitioned by S-sets. Moreover, each A_i is the union of those E_J for which $i \in J$.   □

Proposition 2.1.4 *Let C be a subclass of $\mathcal{P}(\Omega)$ and denote by Ψ the class of all subsets of Ω which can be written $\bigcap_{i \in I} A_i$, where I is finite and nonempty and either A_i or A_i^c lies in C for each $i \in I$, with at least one A_i in C. Then Ψ is a semiring, and the ring generated by Ψ is exactly the ring generated by C.*

PROOF: For all $E = \bigcap_{1 \leq j \leq n_1} A_{1,j}$ and $F = \bigcap_{1 \leq j \leq n_2} A_{2,j}$ in Ψ, the set $E \cap F$ clearly lies in Ψ. On the other hand,

$$\bigcup_{1 \leq j \leq n_2} A_{2,j}^c = \bigcup_{1 \leq j \leq n_2} [A_{2,j}^c \cap (\bigcap_{1 \leq k \leq j-1} A_{2,k})],$$

where the $A_{2,j}^c \cap (\bigcap_{1 \leq k \leq j-1} A_{2,k})$ are disjoint; so

$$E \cap F^c = \bigcup_{1 \leq j \leq n_2} \bigcap_{1 \leq j \leq n_1} [(\bigcap A_{1,j}) \cap A_{2,j}^c \cap (\bigcap_{1 \leq k \leq j-1} A_{2,k})]$$

is the finite union of disjoint Ψ-sets.   □

Let C and Ψ be as in Proposition 2.1.4. Denote by $\mathcal{D}$ the class of finite intersections of C-sets. Let $(A_i)_{i \in I}$ be a finite nonempty family of subsets of Ω, such that at least one A_i belongs to C, and such that either A_i or A_i^c lies in C for all $i \in I$. Put $J = \{i \in I : A_i \in C\}$. Then

$$1_{\bigcap_{i \in I} A_i} = \left(\prod_{i \in J} 1_{A_i} \right) \cdot \prod_{i \in I-J} (1 - 1_{A_i^c})$$

$$= \sum_{K \subset I-J} (-1)^{|K|} \cdot 1_{(\bigcap_{i \in J} A_i) \cap (\bigcap_{i \in K} A_i^c)}$$

is a linear combination of the 1_B $(B \in \mathcal{D})$. If F is an **R**-vector space and f a mapping from Ω into F of the form $\sum_{i \in I} c_i \cdot 1_{A_i}$ (I finite, $c_i \in F$, $A_i \in \Psi$), then, by the preceding remark, f is also of the form $\sum_{j \in J} d_j \cdot 1_{B_j}$ (J finite, $d_j \in F$, $B_j \in \mathcal{D}$).

Definition 2.1.3 A subclass $\mathcal{R}$ of $\mathcal{P}(\Omega)$ is said to be a σ-ring whenever it is a ring and a countable union of $\mathcal{R}$-sets is an $\mathcal{R}$-set.

Hence, $\mathcal{R} \subset \mathcal{P}(\Omega)$ is a σ-ring if it is nonempty, if $A \cap B^c$ lies in $\mathcal{R}$ for all $A, B \in \mathcal{R}$, and if a countable union of $\mathcal{R}$-sets is an $\mathcal{R}$-set. In this case, a countable intersection of $\mathcal{R}$-sets is also an $\mathcal{R}$-set, because

$$\bigcap_{i \geq 0} A_i = A \cap \left(\bigcup_{i \geq 0} (A \cap A_i^c) \right)^c ,$$

where $A = \bigcup_{i \geq 0} A_i$, for any sequence $(A_i)_{i \geq 0}$ of $\mathcal{R}$-sets.

As before, there is a smallest σ-ring containing a given subclass C of $\mathcal{P}(\Omega)$, called the σ-ring generated by C and often written $\sigma(C)$ or $\tilde{C}$.

Proposition 2.1.5 *Let C be a subclass of $\mathcal{P}(\Omega)$. If E is any set in $\sigma(C)$, then there exists a countable subclass D of C such that $E \in \sigma(D)$.*

PROOF: The union of those σ-subrings of $\sigma(C)$ that are generated by some countable subclass of C is a σ-ring containing C and contained in $\sigma(C)$; it is therefore identical to $\sigma(C)$. □

With notation as in Proposition 2.1.5, the class of those sets in Ω covered by a countable union of C-sets is a σ-ring. So every element in $\sigma(C)$ is contained in a countable union of C-sets.

Definition 2.1.4 A π-system in Ω is a class P of subsets of Ω such that $A \cap B$ is a finite or countable union of disjoint P-sets, for all P-sets A, B.

Definition 2.1.5 A λ-system in Ω is a class $\mathcal{L}$ of subsets of Ω with the following properties:

(a) $F - E$ lies in $\mathcal{L}$, for all $\mathcal{L}$-sets E, F such that $E \subset F$.

(b) A countable union of disjoint $\mathcal{L}$-sets is an $\mathcal{L}$-set.

Observe that if a λ-system is closed with respect to finite intersections, then it is a σ-ring.

Theorem 2.1.1 ($\pi - \lambda$ Theorem) *Let P be a π-system and $\mathcal{L}$ the λ-system generated by P (intersection of all λ-systems containing P). Then $\mathcal{L}$ is the σ-ring generated by P.*

PROOF: For every $A \in \mathcal{L}$, $\mathcal{L}_A = \{B \in \mathcal{L} : A \cap B \in \mathcal{L}\}$ is a λ-system. If B is a P-set, then $P \subset \mathcal{L}_B$, so $\mathcal{L}_B = \mathcal{L}$. Hence $A \cap B$ is an $\mathcal{L}$-set for all $A \in \mathcal{L}$ and all $B \in P$. Now $\mathcal{L}_A$ contains P, for each $A \in \mathcal{L}$; thus $\mathcal{L}_A = \mathcal{L}$, which proves that $\mathcal{L}$ is closed with respect to finite intersections. □

Given a semiring S in Ω, the elements of $\sigma(S)$ will henceforth be called the S-Borel sets, or simply the Borel sets when there is no possibility of confusion.

A semiring (respectively, a ring, a σ-ring) $\mathcal{R}$ in Ω is said to be a semialgebra (respectively, an algebra, a σ-algebra) whenever Ω belongs to $\mathcal{R}$. The smallest algebra (respectively, σ-algebra) containing a given subclass $\mathcal{C}$ of $\mathcal{P}(\Omega)$ is called the algebra (respectively, σ-algebra) generated by $\mathcal{C}$: its elements are the $X \in \mathcal{R}$ and the X^c, for $X \in \mathcal{R}$, where $\mathcal{R}$ is the ring (respectively, σ-ring) generated by $\mathcal{C}$.

Definition 2.1.6 In a topological space Ω, the σ-ring $\mathcal{B}$ (respectively, $\mathcal{B}_K$), generated by the class of all open sets (respectively, of all compact sets), is called the Borel σ-algebra of Ω (respectively, the Halmos σ-ring of Ω), and its elements are the Borel sets (respectively, the Halmos sets) in Ω.

Proposition 2.1.6 *Let Ω be a Hausdorff space, and let $\mathcal{B}$, $\mathcal{B}_K$ be as in Definition 2.1.6. Then a subset A of Ω lies in $\mathcal{B}_K$ if and only if A lies in $\mathcal{B}$ and is contained in a countable union of compact sets.*

PROOF: $\mathcal{L} = \{A \subset \Omega : A \cap X \in \mathcal{B}_K \text{ for all } X \in \mathcal{B}_K\}$ is a σ-algebra and contains the closed subsets of Ω, so $\mathcal{L} \supset \mathcal{B}$. If A belongs to $\mathcal{B}$ and is contained in the countable union of compact sets K_i $(i \in I)$, then $A = \bigcup_{i \in I} A \cap K_i$ lies in $\mathcal{B}_K$, because each $A \cap K_i$ belongs to $\mathcal{B}_K$.

Conversely, the class of those Borel sets contained in countable unions of compact sets is a σ-ring and contains $\mathcal{B}_K$.   □

2.2   Measures on Semirings

Let S be a semiring in a nonempty set Ω and $\mathcal{R}$ the ring generated by S.

Definition 2.2.1 A mapping μ from S into $\mathbf{C}$ is said to be additive whenever $\mu(\bigcup_{i \in I} A_i) = \sum_{i \in I} \mu(A_i)$ for every finite family $(A_i)_{i \in I}$ of mutually disjoint S-sets whose union lies in S. We define $\mu(\emptyset)$ to be 0.

Theorem 2.2.1 *Given the additive function $\mu : S \to \mathbf{C}$, there exists a unique additive function μ_1 from $\mathcal{R}$ into $\mathbf{C}$ whose restriction to S is μ.*

PROOF: Each $Z \in \mathcal{R}$ is of the form $Z = \bigcup_{1 \leq i \leq n} X_i$, with $X_i \in S$ and $X_i \cap X_k = \emptyset$ whenever $i \neq k$. Hence we may put $\mu_1(Z) = \sum_{1 \leq i \leq n} \mu(X_i)$. To prove that μ_1 is well-defined, we must show that, if Z is also of the form $Z = \bigcup_{1 \leq j \leq p} Y_j$, with $Y_j \in S$ and $Y_j \cap Y_l = \emptyset$ whenever $j \neq l$, then $\sum_{1 \leq i \leq n} \mu(X_i) = \sum_{1 \leq j \leq p} \mu(Y_j)$. Now, for every $1 \leq i \leq n$ and every $1 \leq j \leq p$, $X_i \cap Y_j$ is the disjoint union of S-sets $Z_{i,j,\alpha}$ (with $1 \leq \alpha \leq m(i,j)$), so

$$\mu(X_i) = \sum_{1 \leq j \leq p} \sum_{1 \leq \alpha \leq m(i,j)} \mu(Z_{i,j,\alpha}) \quad \text{for all } 1 \leq i \leq n$$

and

$$\mu(Y_j) = \sum_{1\le i\le n}\sum_{1\le\alpha\le m(i,j)} \mu(Z_{i,j,\alpha}) \quad \text{for all } 1 \le j \le p,$$

whence we see that $\sum_{1\le i\le n}\mu(X_i) = \sum_{1\le j\le p}\mu(Y_j)$.

We prove now that μ_1 is additive on $\mathcal{R}$: Let $(X_i)_{1\le i\le n}$ be a finite family of mutually disjoint $\mathcal{R}$-sets, and put $Z = \bigcup_{1\le i\le n} X_i$. We have $Z = \bigcup_{1\le j\le p} Y_j$, with $Y_j \in S$ and $Y_j \cap Y_l = \emptyset$ whenever $j \ne l$. Each $X_i \cap Y_j$ is the disjoint union of S-sets $Z_{i,j,\alpha}(1 \le \alpha \le m(i,j))$, so $Y_j = \bigcup_{1\le i\le n}\bigcup_{1\le\alpha\le m(i,j)} Z_{i,j,\alpha}$ for all $1 \le j \le p$, and

$$\mu(Y_j) = \sum_{1\le i\le n}\sum_{1\le\alpha\le m(i,j)} \mu(Z_{i,j,\alpha}) = \sum_{1\le i\le n} \mu_1(X_i \cap Y_j).$$

Finally,

$$\begin{aligned}
\mu_1(Z) &= \sum_{1\le j\le p}\sum_{1\le i\le n} \mu_1(X_i \cap Y_j)\\
&= \sum_{1\le i\le n}\sum_{1\le j\le p} \mu_1(X_i \cap Y_j)\\
&= \sum_{1\le i\le n} \mu_1(X_i),
\end{aligned}$$

because $\mu_1(X_i) = \sum_{1\le j\le p}\sum_{1\le\alpha\le m(i,j)} \mu(Z_{i,j,\alpha}) = \sum_{1\le j\le p}\mu_1(X_i \cap Y_j)$ for all $1 \le i \le n$. $\qquad\square$

Proposition 2.2.1 *Suppose that S is a ring and let $\mu : S \to \mathbf{C}$ be an additive function. Then*

$$\mu\left(\bigcup_{i\in I} A_i\right) = \sum_{J\in\mathcal{F}(I),J\ne\emptyset} (-1)^{|J|-1}\cdot\mu\left(\bigcap_{i\in J} A_i\right)$$

for any finite family $(A_i)_{i\in I}$ of S-sets, where $\mathcal{F}(I)$ is the power set of I (the so-called inclusion-exclusion formula).

PROOF: The result is obvious for $|I| = 1$. Let $n \ge 1$, and assume the formula is valid for $|I| \le n$. Now suppose that $|I| = n+1$, and choose an element k of I. Then

$$\mu\left(\bigcup_{i\in I} A_i\right) = \mu(A_k) + \mu\left(\bigcup_{i\in I-\{k\}} A_i\right) - \mu\left(\bigcup_{i\in I-\{k\}}(A_k \cap A_i)\right),$$

because $\mu(A \cup B) = \mu(A) + \mu(B) - \mu(A \cap B)$ for all A, B in S. Next, use the inclusion-exclusion formula to write $\mu\left(\bigcup_{i\in I-\{k\}} A_i\right)$ and $\mu\left(\bigcup_{i\in I-\{k\}} A_k \cap A_i\right)$ in the desired form. $\qquad\square$

Definition 2.2.2 A quasi-measure on the semiring S is an additive function from S into $\mathbf{C}$ such that $|\mu|(A) = \sup \sum_{i \in I} |\mu(A_i)|$ is finite for every $A \in S$, the supremum being taken over the class of finite partitions $(A_i)_{i \in I}$ of A into S-sets. The mapping $|\mu| : A \to |\mu|(A)$ from S into $\mathbf{R}^+$ is then called the absolute value (or the variation) of μ; it is also written $V\mu$.

Theorem 2.2.2 $V\mu$ *is additive for any quasi-measure μ on S.*

PROOF: Let $A \in S$ and $(A_i)_{i \in I}$ be a finite partition of A into S-sets. Clearly, $\sum_{i \in I} V\mu(A_i) \leq V\mu(A)$. Conversely, let $(B_j)_{j \in J}$ be an arbitrary finite partition of A into S-sets. For every $(i,j) \in I \times J$, $A_i \cap B_j$ is the union of a finite family $Z_{i,j,k}$ (where $k \in K_{i,j}$) of disjoint S-sets. Hence $|\mu(B_j)| \leq \sum_{i \in I} \sum_{k \in K_{i,j}} |\mu(Z_{i,j,k})|$ for all $j \in J$, and

$$\sum_{j \in J} |\mu(B_j)| \leq \sum_{i \in I} \left(\sum_{j \in J} \sum_{k \in K_{i,j}} |\mu(Z_{i,j,k})| \right) \leq \sum_{i \in I} V\mu(A_i),$$

whence we see that $V\mu(A) \leq \sum_{i \in I} V\mu(A_i)$. $\square$

$V\mu$ is the smallest of those additive functions ν from S into $\mathbf{R}^+$ satisfying $|\mu(A)| \leq \nu(A)$ for all $A \in S$.

If μ is a quasi-measure on S, then $\sum_{i \geq 1} V\mu(A_i) \leq V\mu(A)$ for every sequence $(A_i)_{i \geq 1}$ of disjoint S-sets whose union is contained in an S-set A. Indeed, for each integer $n \geq 1$, $A \cap \left(\bigcup_{1 \leq i \leq n} A_i \right)^c$ can be partitioned into finitely many S-sets $A'_k (k \in K)$; for any finite partition of A_i $(1 \leq i \leq n)$ into S-sets $A_{i,j}$ $(j \in J_i)$, the $A_{i,j}$ $(1 \leq i \leq n, j \in J_i)$ and the A'_k $(k \in K)$ form a partition of A, so

$$\sum_{1 \leq i \leq n} \sum_{j \in J_i} |\mu(A_{i,j})| \leq \sum_{1 \leq i \leq n} \sum_{j \in J_i} |\mu(A_{i,j})| + \sum_{k \in K} |\mu(A'_k)| \leq V\mu(A),$$

and finally $\sum_{1 \leq i \leq n} V\mu(A_i) \leq V\mu(A)$.

Definition 2.2.3 A function μ from S into $\mathbf{C}$ is a measure whenever

(a) it is a quasi-measure;

(b) it is σ-additive (i.e., $\mu(A) = \sum_{i \geq 1} \mu(A_i)$ for every sequence $(A_i)_{i \geq 1}$ of disjoint S-sets whose union A is an S-set).

Theorem 2.2.3 *A function μ from S into $\mathbf{C}$ is a measure if (and only if), for any sequence $(A_i)_{i \geq 1}$ of disjoint S-sets contained in an $A \in S$, the series $\sum_{i \geq 1} \mu(A_i)$ converges, and if $\mu(A) = \sum_{i \geq 1} \mu(A_i)$ whenever $A = \bigcup_{i \geq 1} A_i$.*

PROOF: For any sequence $(A_i)_{i \geq 1}$ of disjoint S-sets contained in an $A \in S$, the series $\sum_{i \geq 1} \mu(A_i)$ is commutatively convergent, so $\sum_{i \geq 1} |\mu(A_i)|$ must be finite.

For every S-set A, put $V\mu(A) = \sup \sum_{i \in I} |\mu(A_i)|$, where the supremum is taken over the class of finite partitions $(A_i)_{i \in I}$ of A by S-sets. We have to show that $V\mu(A)$ is finite.

First, observe that, if this property is true for all elements A_i of one finite partition of some A by S-sets, then it holds for A, and $V\mu(A) \leq \sum_{i \in I} V\mu(A_i)$ (by the argument of Theorem 2.2.2). Now assume that the property is false for one $A \in S$. Then there exists a finite partition $A_i^1 (i \in I_1)$ of A into S-sets such that $\sum_{i \in I_1} |\mu(A_i^1)| \geq 2 + |\mu(A)|$, and, by the previous observation, the property is false for at least one $A_{i_1}^1$ $(i_1 \in I_1)$. Moreover,

$$\sum_{i \in I_1 - \{i_1\}} |\mu(A_i^1)| = \sum_{i \in I_1} |\mu(A_i^1)| - |\mu(A_{i_1}^1)|$$

$$\geq 2 + |\mu(A)| - |\mu(A) - \sum_{i \in I_1 - \{i_1\}} \mu(A_i^1)|$$

$$\geq 2 + |\mu(A)| - |\mu(A)| - \sum_{i \in I_1 - \{i_1\}} |\mu(A_i^1)|,$$

so $\sum_{i \in I_1 - \{i_1\}} |\mu(A_i^1)| \geq 1$. For the same reason, $A_{i_1}^1$ can be partitioned by S-sets A_i^2 $(i \in I_2)$ so that the property is false for some $A_{i_2}^2$ satisfying $\sum_{i \in I_2 - \{i_2\}} |\mu(A_i^2)| \geq 1$. Repeating this process, we find disjoint S-sets A_i^k $(k \geq 1, i \in I_k - \{i_k\})$ such that the series $\sum_{k \geq 1} \sum_{i \in I_k - \{i_k\}} |\mu(A_i^k)|$ diverges, which is a contradiction. $\qquad\square$

Theorem 2.2.4 $V\mu$ *is a measure for every measure μ on S.*

PROOF: Let $(A_i)_{i \geq 1}$ be a sequence of disjoint S-sets whose union A lies in S. We need only show that $V\mu(A) \leq \sum_{i \geq 1} V\mu(A_i)$. But, if $(B_j)_{j \in J}$ is an arbitrary finite partition of A into S-sets, each $A_i \cap B_j$ $(i \geq 1, j \in J)$ is the disjoint union of S-sets $Z_{i,j,k}$ $(k \in K_{i,j}, K_{i,j}$ finite$)$, so

$$\sum_{j \in J} |\mu(B_j)| = \sum_{j \in J} \left| \sum_{i \geq 1} \sum_{k \in K_{i,j}} \mu(Z_{i,j,k}) \right|$$

$$\leq \sum_{i \geq 1} \sum_{j \in J} \sum_{k \in K_{i,j}} |\mu(Z_{i,j,k})|$$

$$\leq \sum_{i \geq 1} V\mu(A_i).$$

$\qquad\square$

Proposition 2.2.2 *Let μ be a quasi-measure (respectively, a measure) on S and μ_1 the unique additive extension of μ to $\mathcal{R}$. Then μ_1 is a quasi-measure (respectively, a measure) and $V\mu_1$ extends $V\mu$.*

PROOF: Denote by ν_1 the additive extension of $V\mu$ to $\mathcal{R}$. Then $|\mu_1(A)| \leq \nu_1(A)$ for all $A \in \mathcal{R}$, so μ_1 is a quasi-measure and $V\mu_1 \leq \nu_1$. Moreover, $V\mu_1(A) \geq V\mu(A) = \nu_1(A)$ for all $A \in S$, so $V\mu_1(A) = \nu_1(A)$ for all $A \in S$. Hence $V\mu_1$ and ν_1, which agree on S, are identical.

Next, we assume that μ is a measure and prove that μ_1 is also a measure. Let $(A_i)_{i \geq 1}$ be a sequence of disjoint $\mathcal{R}$-sets whose union $Z = \bigcup_{i \geq 1} A_i$ belongs to $\mathcal{R}$, and let $(B_j)_{j \in J}$ be an arbitrary finite partition of Z into S-sets. Each $A_i \cap B_j$ ($i \geq 1$, $j \in J$) is the union of a family $(Z_{i,j,k})_{k \in K_{i,j}}$ ($K_{i,j}$ finite) of disjoint S-sets. Then $B_j = \bigcup_{i \geq 1} \bigcup_{k \in K_{i,j}} Z_{i,j,k}$, whence $\mu(B_j) = \sum_{i \geq 1} \sum_{k \in K_{i,j}} \mu(Z_{i,j,k}) = \sum_{i \geq 1} \mu_1(A_i \cap B_j)$, for every $j \in J$. Therefore, $\mu_1(Z) = \sum_{j \in J} \sum_{i \geq 1} \mu_1(A_i \cap B_j)$. Since each family $(\mu_1(A_i \cap B_j))_{i \geq 1}$ is summable, so is $(\mu_1(A_i \cap B_j))_{i \geq 1, j \in J}$. Thus $\mu_1(Z) = \sum_{i \geq 1} \sum_{j \in J} \mu_1(A_i \cap B_j) = \sum_{i \geq 1} \mu_1(A_i)$, and μ_1 is σ-additive. $\square$

Let μ be a measure on S and $(A_n)_{n \geq 1}$ a sequence of S-sets decreasing to A in S. Then $\mu(A_n) \to \mu(A)$ as $n \to +\infty$; indeed, $\mu_1(A_1 - A_n) = \sum_{1 \leq i \leq n-1} \mu_1(A_i - A_{i+1})$ converges to $\mu_1(A_1 - A) = \sum_{i \geq 1} \mu_1(A_i - A_{i+1})$ as $n \to +\infty$. Similarly, if $(A_n)_{n \geq 1}$ is an increasing sequence of S-sets whose union A belongs to S, then $\mu(A_n) \to \mu(A)$ as $n \to +\infty$.

Conversely, assume that S is a ring, and let μ be a quasi-measure on S. If $\mu(A_n) \to 0$ for every sequence $(A_n)_{n \geq 1}$ of S-sets which decreases to $\emptyset$, then μ is a measure.

Proposition 2.2.3 *Suppose that S is a σ-ring. Then all measures μ on S are bounded (i.e., $\sup_{A \in S} V\mu(A) < +\infty$).*

PROOF: Suppose there is a measure μ on S which is not bounded. Then, for each integer $n \geq 1$, there is a B_n in S such that $V\mu(B_n) \geq n$; now, putting $B = \bigcup_{n \geq 1} B_n$, we have $V\mu(B) \geq n$ for every n, and so $V\mu(B) = +\infty$. $\square$

A positive function μ on a semiring S is a measure if and only if it is σ-additive. For any sequence $(A_i)_{i \geq 1}$ of S-sets (not necessarily disjoint) and for any S-set B included in $\bigcup_{i \geq 1} A_i$, in this case, $\mu(B) \leq \sum_{i \geq 1} \mu(A_i)$. Indeed, let μ_1 be the unique measure extending μ to $\mathcal{R}$, and, for every integer $n \geq 1$, put $B_n = B \cap A_n \cap \left(\bigcup_{1 \leq i \leq n-1} A_i\right)^c$. Then, since B is the disjoint union of the B_n, we have $\mu(B) = \sum_{n \geq 1} \mu_1(B_n) \leq \sum_{n \geq 1} \mu_1(A_n)$.

Definition 2.2.4 Let F be a real vector space. A mapping f from Ω into F is S-simple if it is of the form $f = \sum_{j \in J} c_j \cdot 1_{A_j}$, where J is finite, A_j lies in S for every $j \in J$, and c_j belongs to F. The set of all S-simple mappings from Ω into F will be denoted by $St(S, F)$.

Proposition 2.2.4 *For any (nonempty) finite family $(f_i)_{i \in I}$ in $St(S, F)$, there exists a finite family $(B_k)_{k \in K}$ of disjoint S-sets such that each f_i can be written $f_i = \sum_{k \in K} \alpha_{i,k} \cdot 1_{B_k}$ for suitable members $\alpha_{i,k}$ of F.*

PROOF: Each f_i can be written $f_i = \sum_{j \in J_i} c_{i,j} \cdot 1_{A_{i,j}}$. By Proposition 2.1.3, there exist disjoint S-sets B_k ($k \in K$, K finite) such that each $A_{i,j}$ (i in I, j in J_i) is a union of some of the B_k: $A_{i,j} = \bigcup_{k \in K_{i,j}} B_k$. Then

$$f_i = \sum_{j \in J_i} c_{i,j} \cdot \left(\sum_{k \in K_{i,j}} 1_{B_k} \right)$$

$$= \sum_{k \in K} \left(\sum_{\{j \in J_i : k \in K_{i,j}\}} c_{i,j} \right) \cdot 1_{B_k}.$$

$\square$

Now let $\mu : S \to \mathbf{C}$ be additive, and let F be a Banach space (real or complex, as μ takes on real or complex values). If $f = \sum_{i \in I} c_i 1_{A_i}$ is an S-simple mapping from Ω into F, we define the μ-integral of f to be the vector $\mu(f) = \int f d\mu = \sum_{i \in I} c_i \mu(A_i)$. This definition makes sense because the vector $\sum_{i \in I} c_i \mu(A_i)$ is independent of the representation $\sum_{i \in I} c_i 1_{A_i}$ of f. Indeed, if $f = \sum_{i \in I} c_i 1_{A_i} = \sum_{j \in J} c'_j 1_{A'_j}$, there exist disjoint nonempty S-sets B_k ($k \in K$, K finite) such that each A_i and each A'_j is a union of some of the B_k. Then

$$\sum_{i \in I} c_i 1_{A_i} = \sum_{i \in I} c_i \cdot \left(\sum_{B_k \subset A_i} 1_{B_k} \right) = \sum_{k \in K} \left(\sum_{A_i \supset B_k} c_i \right) \cdot 1_{B_k}$$

and

$$\sum_{j \in J} c'_j \cdot 1_{A'_j} = \sum_{j \in J} c'_j \cdot \left(\sum_{B_k \subset A'_j} 1_{B_k} \right) = \sum_{k \in K} \left(\sum_{A'_j \supset B_k} c'_j \right) \cdot 1_{B_k},$$

so $\sum_{A_i \supset B_k} c_i = \sum_{A'_j \supset B_k} c'_j$ for every k, and

$$\sum_{i \in I} c_i \mu(A_i) = \sum_{i \in I} c_i \cdot \sum_{B_k \subset A_i} \mu(B_k)$$

$$= \sum_{k \in K} \left(\sum_{A_i \supset B_k} c_i \right) \cdot \mu(B_k)$$

$$= \sum_{k \in K} \left(\sum_{A'_j \supset B_k} c'_j \right) \cdot \mu(B_k)$$

$$= \sum_{j \in J} c'_j \cdot \sum_{B_k \subset A'_j} \mu(B_k)$$

$$= \sum_{j \in J} c'_j \cdot \mu(A'_j).$$

The function $\tilde{\mu} : f \mapsto \int f d\mu$ from $St(S, \mathbf{C})$ into $\mathbf{C}$ is the unique linear form u on $St(S, \mathbf{C})$ such that $u(1_A) = \mu(A)$ for all $A \in S$. This means that, letting E be the vector space of additive functions from S into $\mathbf{C}$, we can identify E and the algebraic dual $St(S, \mathbf{C})^*$ of $St(S, \mathbf{C})$. Henceforth, we consider only $St(S, \mathbf{C})$ instead of the more general $\mathcal{H}(\Omega, \mathbf{C})$ of Section 1.4.

Theorem 2.2.5 *An additive function μ from S into $\mathbf{C}$ is a quasi-measure (respectively, a measure) if and only if $\tilde{\mu}$ has finite variation (respectively, is a Daniell measure). Moreover, $|\tilde{\mu}| = \widetilde{V\mu}$.*

PROOF: First, suppose that $\tilde{\mu}$ has finite variation. Let $(A_i)_{i \in I}$ be a finite partition of $A \in S$ into S-sets. For every $i \in I$, there exists a complex number c_i with modulus 1 satisfying $|\mu(A_i)| = c_i \cdot \mu(A_i)$. Then $f = \sum_{i \in I} c_i 1_{A_i}$ satisfies $|f| \leq 1_A$, so $\sum_{i \in I} |\mu(A_i)| = \tilde{\mu}(f) = |\tilde{\mu}(f)| \leq |\tilde{\mu}|(1_A)$. Thus μ is a quasi-measure and $V\mu(A) \leq |\tilde{\mu}|(1_A)$ for all $A \in S$.

Conversely, suppose that μ is a quasi-measure. Clearly, $|\int g d\mu| \leq \int |g| dV\mu$ for all $g \in St(S, \mathbf{C})$, and $\int f dV\mu \geq 0$ for all $f \in St^+(S) = St(S, \mathbf{R})^+$ (Proposition 2.2.4). Then $|\tilde{\mu}|(f) = \sup_{g \in St(S,C), |g| \leq f} |\int g d\mu| \leq \int f dV\mu$ for every $f \in St^+(S)$. So $\tilde{\mu}$ has finite variation, and $|\tilde{\mu}|(1_A) = V\mu(A)$ for all $A \in S$. This proves that $|\tilde{\mu}| = \widetilde{V\mu}$.

Next, assume that $\tilde{\mu}$ is a Daniell measure, and let $(A_i)_{i \geq 1}$ be a sequence of disjoint S-sets whose union A lies in S. For every integer $n \geq 1$, put $f_n = 1_A - \sum_{1 \leq i \leq n} 1_{A_i}$. Since $(f_n)_{n \geq 1}$ decreases to 0, $\tilde{\mu}(f_n) = \mu(A) - \sum_{1 \leq i \leq n} \mu(A_i)$ tends to 0 as $n \to +\infty$, and hence μ is a measure.

Finally, assume that μ is a measure, and let μ_1 be its unique additive extension to $\mathcal{R}$, so that $\int f dV\mu = \int f dV\mu_1$ for $f \in St(S, \mathbf{C})$. Let $(f_n)_{n \geq 1}$ be a sequence in $St^+(S)$ decreasing to 0, and fix $\varepsilon > 0$. Then

$$B_n = \{x \in \Omega : f_n(x) > \varepsilon\}$$

lies in $\mathcal{R}$ for all $n \geq 1$, and the sequence $(B_n)_{n \geq 1}$ decreases to the empty set, so $V\mu_1(B_n) \to 0$ as $n \to +\infty$. Now

$$\int f_n dV\mu \leq M \cdot V\mu_1(B_n) + \varepsilon \cdot V\mu_1(A),$$

where $M = \sup_{x \in \Omega} f_1(x)$ and $A = \{x \in \Omega : f_1(x) > 0\}$; thus

$$\lim_{n \to +\infty} \widetilde{V\mu}(f_n) \leq \varepsilon \cdot V\mu_1(A),$$

whence we have

$$\lim_{n \to +\infty} \widetilde{V\mu}(f_n) = 0.$$

$\square$

Call $QM(S, \mathbf{C})$ (respectively, $M(S, \mathbf{C})$) the complex vector space of quasi-measures (respectively, measures) on S. By Theorem 1.3.1 and Proposition 1.4.1, the ordered vector space $QM(S, \mathbf{R})$ of real quasi-measures on S is a Dedekind complete Riesz space, and $M(S, \mathbf{R}) = QM(S, \mathbf{R}) \cap M(S, \mathbf{C})$ is a band in $QM(S, \mathbf{R})$. Moreover, for every $\mu \in QM(S, \mathbf{R})$ and for every $A \in S$,

$$
\begin{aligned}
\mu^+(A) &= \frac{1}{2}\Big(\mu(A) + V\mu(A)\Big) \\
&= \frac{1}{2} \sup_{P(A)} \sum_{B \in P(A)} \Big(\mu(B) + |\mu(B)|\Big) \\
&= \sup_{P(A)} \sum_{B \in P(A)} \mu(B)^+,
\end{aligned}
$$

where $P(A)$ extends over the class of all finite partitions of A into S-sets; similarly, $\mu^-(A) = \sup_{P(A)} \sum_{B \in P(A)} \mu(B)^-$. A nonempty upward-directed subset H of $QM(S, \mathbf{R})$ has a supremum ν in $QM(S, \mathbf{R})$ if and only if $\sup_{\mu \in H} \mu(A)$ is finite for each $A \in S$, and then $\nu(A) = \sup_{\mu \in H} \mu(A)$ (see Lemma 1.3.1).

If $\mu \in QM(S, \mathbf{C})$, if $\nu \in QM(S, \mathbf{R})^+$, and if $|\int f d\mu| \le \int f d\nu$ for every $f \in St^+(S)$, then $|\mu(A)| \le \nu(A)$ for every $A \in S$, so $V\mu \le \nu$ and $|\int g d\mu| \le \int |g| dV\mu \le \int |g| d\nu$ for all $g \in St(S, \mathbf{C})$. Hence even the last condition of Section 1.4, is satisfied when we consider $St(S, \mathbf{C})$ (instead of $\mathcal{H}(\Omega, \mathbf{C})$).

The cone of positive measures on S is written $M^+(S)$.

For the remainder of this section, let μ be a positive measure on S. Denote by $\mathcal{J}^+$ the set of upper envelopes of increasing sequences in $\mathcal{H}^+ = St^+(S)$, and put $\mu^*(f) = \sup_{g \in \mathcal{H}^+, g \le f} \mu(g)$ for all $f \in \mathcal{J}^+$.

Proposition 2.2.5 $f - g$ belongs to $\mathcal{J}^+$ for all $f \in \mathcal{J}^+$ and $g \in \mathcal{H}^+$ satisfying $g \le f$. $\inf(f, g)$, $\sup(f, g)$, $f + g$, and λh belong to $\mathcal{J}^+$, for all $f, g, h \in \mathcal{J}^+$ and for all $\lambda > 0$; moreover, $\mu^*(f + g) = \mu^*(f) + \mu^*(g)$ and $\mu^*(\lambda h) = \lambda \cdot \mu^*(h)$. Finally, every increasing sequence $(f_n)_{n \ge 1}$ in $\mathcal{J}^+$ has its upper envelope f in $\mathcal{J}^+$, and $\mu^*(f) = \sup_{n \ge 1} \mu^*(f_n)$.

PROOF: Every $f \in \mathcal{J}^+$ is the upper envelope of an increasing sequence $(f_n)_{n \ge 1}$ in $\mathcal{H}^+$. For every $g \in \mathcal{H}^+$ satisfying $g \le f$, the sequence $(f_n - \inf(f_n, g))_{n \ge 1}$ increases to $f - g$, so $f - g$ belongs to $\mathcal{J}^+$; next the sequence $(\inf(f_n, g))_{n \ge 1}$ increases to g, so

$$
\mu(g) = \lim_{n \to +\infty} \mu(\inf(f_n, g)) \le \lim_{n \to +\infty} \mu(f_n);
$$

therefore,

$$
\sup_{g \in \mathcal{H}^+, \ g \le f} \mu(g) = \lim_{n \to +\infty} \mu(f_n).
$$

Now let $(f_n)_{n \ge 1}$ be an increasing sequence in $\mathcal{J}^+$ and f its upper envelope. For each $n \ge 1$, let $(g_{n,m})_{m \ge 1}$ be a sequence in $\mathcal{H}^+$ increasing to f_n, and put $h_m = \sup(g_{1,m}, g_{2,m}, \dots, g_{m,m})$ for all integers $m \ge 1$. Then $g_{i,m} \le h_m \le f_m$ for all $1 \le i \le m$, so $f_i = \sup_{m \ge i} g_{i,m} \le \sup_{m \ge 1} h_m$ for every $i \ge 1$, and $f \le$

$\sup_{m\geq 1} h_m$. This proves that the sequence $(h_m)_{m\geq 1}$ increases to f. Moreover, $\sup_{m\geq 1} \mu^*(f_m) \geq \sup_{m\geq 1} \mu(h_m) = \mu^*(f)$. The reverse inequality is obvious.

□

2.3   Lebesgue Measure on an Interval

Let I be a nonempty interval of $\mathbf{R}$. Put $a = \inf(I)$ and $b = \sup(I)$, so that $I = \langle a, b \rangle$. Consider the semiring S in I consisting of all intervals $]\alpha, \beta]$ such that $\alpha \leq \beta$ in I, and, when a belongs to I, of all intervals $[a, \beta]$ such that β belongs to I. S is called the natural semiring in I. The σ-ring generated by S is easily seen to be the Borel σ-algebra in I.

Definition 2.3.1 Let μ be a complex measure on S. A function F from I into $\mathbf{C}$ is called an indefinite integral of μ whenever $F(\beta) - F(\alpha) = \mu(]\alpha, \beta])$ for all $\alpha, \beta \in I$ satisfying $\alpha \leq \beta$.

If μ is a measure on S, for all $\alpha, \beta \in I$ we put $\int_\alpha^\beta d\mu = \mu(]\alpha, \beta])$ or $\int_\alpha^\beta d\mu = -\mu(]\beta, \alpha])$, as $\alpha \leq \beta$ or $\alpha \geq \beta$; then

$$\int_\alpha^\beta d\mu + \int_\beta^\gamma d\mu + \int_\gamma^\alpha d\mu = 0$$

for all $\alpha, \beta, \gamma \in I$. For each $x_0 \in I$, the function $x \to \int_{x_0}^x d\mu$ from I into $\mathbf{C}$ is evidently an indefinite integral of μ. Now $F(x) = F(x_0) + \int_{x_0}^x d\mu$ for all $x \in I$ and for any other indefinite integral of μ; so we see that indefinite integrals of μ are equal, up to a constant.

Proposition 2.3.1 Let $F : I \to \mathbf{R}$ be increasing, and right-continuous on $\langle a, b[$. Then there is a unique measure μ on S admitting F as an indefinite integral and satisfying $\mu(\{a\}) = 0$ whenever a belongs to I.

PROOF: The uniqueness of such a measure is obvious; to prove its existence, we define a function μ on S by $\mu(\langle \alpha, \beta]) = F(\beta) - F(\alpha)$ for every $\langle \alpha, \beta]$ of S. Clearly, μ is additive. Let $\langle \alpha, \beta] \in S$ ($\alpha < \beta$ in I) and $(\langle x_n, \beta_n])_{n\geq 1}$ ($\alpha_n < \beta_n$ in I) be a sequence of disjoint S-sets whose union is $\langle \alpha, \beta]$. Then

$$\sum_{n\geq 1} \mu(\langle \alpha_n, \beta_n]) \leq \mu(\langle \alpha, \beta]),$$

because μ is a quasi-measure. Let $\varepsilon > 0$ be given. As F is right-continuous at α, there exists $0 < \delta < \beta - \alpha$ such that $\mu(J) \geq \mu(\langle \alpha, \beta]) - \varepsilon/2$, if we put $J =]\alpha + \delta, \beta]$. Additionally, for every $n \geq 1$, there exists $\delta_n > 0$ such that $\mu(J_n) \leq \mu(\langle \alpha_n, \beta_n]) + \varepsilon/2^{n+1}$, if we put $J_n = \langle \alpha_n, \beta_n + \delta_n] \cap I$. Now the interior J_n° of J_n relative to I contains $\langle \alpha_n, \beta_n]$, and, since the closure of J is included in $\langle \alpha, \beta]$ and so in $\bigcup_{n \geq 1} J_n^\circ$, there exist $n_1 \geq 1, \ldots, n_k \geq 1$ such that $J \subset J_{n_1} \cup \cdots \cup J_{n_k}$. From $\mu(J) \leq \mu(J_{n_1}) + \ldots + \mu(J_{n_k})$ follows now $\mu(\langle \alpha, \beta]) \leq \sum_{n \geq 1} \mu(\langle \alpha_n, \beta_n]) + \varepsilon$. As ε is arbitrary, $\mu(\langle \alpha, \beta]) = \sum_{n \geq 1} \mu(\langle \alpha_n, \beta_n])$, and so μ is a measure. $\square$

Definition 2.3.2 The unique measure μ on S such that $\mu(]\alpha, \beta]) = \beta - \alpha$ for all $\alpha, \beta \in I$ satisfying $\alpha \leq \beta$, and such that $\mu(\{a\}) = 0$ whenever a belongs to I, is called Lebesgue measure on I.

To see that μ exists, take $F(x) = x$ in Proposition 2.3.1.

Exercises for Chapter 2

1   In $\mathbf{R}^k$, a rectangle $\prod_{1 \leq i \leq k}]a_i, b_i]$ $(a_i < b_i$ for $1 \leq i \leq k)$ is called a square whenever $b_i - a_i$ is independent of i. Let S be the class consisting of the empty set and of those squares whose vertices have rational coordinates. Show that S is a semiring, even though $A \cap B$ may not lie in S for arbitrary $A \in S$, $B \in S$.

2   Call S a strong semiring if it contains the empty set and is closed under the formation of finite intersections, and if $A, B \in S$ and $A \subset B$ imply that there exist S-sets $A_0, \ldots A_n$ such that $A = A_0 \subset A_1 \subset \ldots \subset A_n = B$ and $A_k - A_{k-1} \in S$ for all $1 \leq k \leq n$. Let $\mu : S \to \mathbf{C}$ be a function such that $\mu(A \cup B) = \mu(A) + \mu(B)$ whenever $A, B \in S$, $A \cup B \in S$, and $A \cap B = \emptyset$. A finite partition $(E_i)_{i \in I}$ of $E \in S$ into S-sets is called a μ-partition if $\mu(A \cap E) = \sum_{i \in I} \mu(A \cap E_i)$ for all $A \in S$. If $(E_i)_{i \in I}$ and $(F_j)_{j \in J}$ are finite partitions of $E \in S$ into S-sets, then $(E_i)_{i \in I}$ is called a subpartition of $(F_j)_{j \in J}$ whenever each set E_i is contained in one of the sets F_j.

1.   Show that, if a subpartition of a partition $(E_i)_{i \in I}$ is a μ-partition, then $(E_i)_{i \in I}$ is a μ-partition.

2.   If $(E_i)_{i \in I}$ and $(F_j)_{j \in J}$ are two μ-partitions of E, show that $(E_i \cap F_j)_{(i,j) \in I \times J}$ is a μ-partition.

3.   If $E = C_0 \subset C_1 \subset \ldots \subset C_n = F$, where $C_i \in S$ for all $0 \leq i \leq n$, and if $C_i - C_{i-1} = D_i$ lies in S for all $1 \leq i \leq n$, show that $\{E, D_1, \ldots, D_n\}$ is a μ-partition of F.

4.   Show that every finite partition of an S-set E by S-sets is a μ-partition, and conclude that μ is additive on S.

3   Let $\mathcal{R}$ be the ring consisting of the finite and cofinite sets in an uncountable Ω. Define φ on $\mathcal{R}$ by taking $\varphi(A)$ to be the number of points in A if A is finite, and $-n$ where n is the number of points in A^c if A is cofinite. Show that φ is σ-additive on $\mathcal{R}$, but is not a measure.

3

Integrable and Measurable Functions

This chapter introduces the fundamental notions of negligible, integrable, and measurable functions with respect to a given measure. The class of Riemann integrable functions is relatively small (cf. Section 7.6) as was discovered in the nineteenth century through the study of Fourier series, for instance. In contrast, the class of measurable functions for Lebesgue measure is quite large. Indeed, the existence of nonmeasurable functions depends on the axiom of choice.

Here we obtain the first substantial results of the theory. It could be said that this is the heart of the matter and it is essential to understand this chapter fully for a profitable reading of the following chapters.

Summary

3.1 Given a Daniell measure μ, the upper integral of a positive function with respect to $V\mu$ is defined. A positive function is said to be μ-negligible if its upper integral is null. This allows us to define negligible sets and the notion of property true "almost everywhere". We then prove a few important results such as Beppo Levi's theorem (Theorem 3.1.1), Fatou's lemma (Proposition 3.1.2), and the Riesz–Fischer theorem (Theorem 3.1.3) on the completeness of $\mathcal{L}^1(\mu)$.

3.2 One of the main limitations of Riemann's integral lies in the fact that it does not yield significant results when it comes to the integral of sequences or series of functions. For example, if a uniformly bounded sequence of Riemann integrable functions converges pointwise on $[a, b]$ to a (necessarily bounded) function f, then f may not be Riemann integrable. The monotone convergence theorem (Theorem 3.2.1), Fatou's theorem (Proposition 3.2.1), and the Lebesgue dominated convergence theorem (Theorem 3.2.2) will answer some of our questions. For a first lecture, the reader may, without loosing too much, translate "filters" into "sequences". The next two results, continuity and differentiability of an integral with respect to a parameter (Proposition 3.2.3 and Theorem 3.2.3), are of constant use in analysis.

3.3   In this short section we focus our attention to sets rather than functions and define μ-integrable sets and μ-moderate sets.

3.4   This section is devoted to the definition of σ-measurable spaces, that is, sets endowed with a σ-ring, and the notion of measurable mapping between two such spaces. It is important to notice that this notion of measurability does not depend on a measure.

3.5   The main result of this section is the following form of Egorov's theorem (Theorem 3.5.1): a sequence of μ-measurable mappings from Ω into a metrizable space which converges locally almost everywhere to f converges uniformly to f on the complement of an arbitrarily small integrable set. Finally, we prove that a μ-measurable function from Ω into a Banach space is integrable if and only if its upper integral with respect to $V\mu$ is finite (Theorem 3.5.3).

3.6   The essential integral of functions is defined, as well as bounded measures.

3.7   First, we define the upper and lower integral of a positive function: f is integrable if and only if its upper and lower integrals are finite and equal. Notice that if f is integrable so is $|f|$: in this theory, there is no "improper integral". We then prove Jensen's inequality (Theorem 3.7.3) which is an important tool both in the theory of probability and in analysis.

3.8   Intuitively, an atom for a measure μ is a μ-integrable set which has no smaller proper subset (smaller and proper both in the sense of measure theory). A measure without atom is said to be diffuse; Lebesgue measure is an example of such a measure. On the other hand, a measure is said to be atomic if each nonnegligible integrable set contains an atom. The counting measure on the semiring of finite subsets of $\mathbf{N}$ is an example (cf. Section 6.3). If μ is atomic and f is a function from Ω into a metrizable space, f is measurable if and only if it is constant a.e. on each atom (Theorem 3.8.1).

3.9   A Daniell measure μ defines a measure $\hat{\mu}$ on the ring $\hat{\mathcal{R}}$ of integrable sets, called the main prolongation of μ. Similarly, μ defines a measure $\bar{\mu}$, called the essential prolongation of μ, on the ring $\bar{\mathcal{R}}$ of essentially integrable sets. We then study various relationships between these measures.

3.1   Upper Integral of a Positive Function

Let Ω be a nonempty set, $\mathcal{F}(\Omega, \mathbf{C})$ the space of complex-valued functions on Ω, and $\mathcal{H}(\Omega, \mathbf{C})$ a vector subspace of $\mathcal{F}(\Omega, \mathbf{C})$ such that $\mathrm{Re}\, f$, $\mathrm{Im}\, f$, and $|f|$ belong to $\mathcal{H}(\Omega, \mathbf{C})$ for all $f \in \mathcal{H}(\Omega, \mathbf{C})$. Assume that, for all $f_1, f_2 \in \mathcal{H}^+$ and for all $g \in \mathcal{H}(\Omega, \mathbf{C})$ satisfying $|g| \leq f_1 + f_2$, there exist $g_1, g_2 \in \mathcal{H}(\Omega, \mathbf{C})$ such that $|g_1| \leq f_1$, $|g_2| \leq f_2$, and $g = g_1 + g_2$. Moreover, suppose that $\inf(f, 1)$ lies in $\mathcal{H}^+$ for all $f \in \mathcal{H}^+$ (Stone's condition).

Now let $\mathcal{J}^+$ be a set of functions from Ω into $[0, +\infty]$ with the following properties:

(a) $\mathcal{J}^+ \supset \mathcal{H}^+$ and $f - g$ belongs to $\mathcal{J}^+$ for all $f \in \mathcal{J}^+$ and $g \in \mathcal{H}^+$ such that $g \leq f$.

(b) $\inf(f, g)$, $\sup(f, g)$, and $f + g$ belong to $\mathcal{J}^+$ for all $f, g \in \mathcal{J}^+$.

(c) $\lambda f \in \mathcal{J}^+$ for every $\lambda > 0$ and every $f \in \mathcal{J}^+$.

(d) Every increasing sequence in $\mathcal{J}^+$ has its upper envelope in $\mathcal{J}^+$.

(e) For every Daniell measure μ on $\mathcal{H}(\Omega, \mathbf{C})$, if we put

$$V\mu^*(f) = \sup_{g \in \mathcal{H}^+, g \leq f} V\mu(g)$$

for all $f \in \mathcal{J}^+$, then

(i) $V\mu^*(f + g) = V\mu^*(f) + V\mu^*(g)$ for all $f, g \in \mathcal{J}^+$;

(ii) $V\mu^*(\sup_{n \geq 1} f_n) = \sup_{n \geq 1} V\mu^*(f_n)$ for every increasing sequence in $\mathcal{J}^+$.

We encountered such a class $\mathcal{J}^+$ in Section 2.2.

Now let μ be a Daniell measure on $\mathcal{H}(\Omega, \mathbf{C})$.

Definition 3.1.1 For every function f from Ω into $[0, +\infty]$, the number $\inf_{h \in \mathcal{J}^+, h \geq f} V\mu^*(h)$ ($+\infty$ if there is no $h \in \mathcal{J}^+, h \geq f$) is called the upper integral of f with respect to $V\mu$ and is written $V\mu^*(f)$, $\int^* f \cdot dV\mu$, $\int^* f(x)dV\mu(x)$, or $N_1(f)$.

If f, g are two functions from Ω into $[0, +\infty]$ such that $f \leq g$, then $V\mu^*(f) \leq V\mu^*(g)$. If f_1, f_2, f are three functions from Ω into $[0, +\infty]$ and $\lambda > 0$ is a real number, then $V\mu^*(f_1 + f_2) \leq V\mu^*(f_1) + V\mu^*(f_2)$ and $V\mu^*(\lambda f) = \lambda \cdot V\mu^*(f)$.

Theorem 3.1.1 (Beppo Levi's Theorem) *For every increasing sequence* $(f_n)_{n \geq 1}$ *of functions from* Ω *into* $[0, +\infty]$,

$$V\mu^*(\sup f_n) = \sup V\mu^*(f_n).$$

PROOF: It suffices to show that $V\mu^*(\sup f_n) \leq \sup V\mu^*(f_n)$. The inequality clearly holds if $\sup V\mu^*(f_n) = +\infty$. Therefore, we may assume that $\sup V\mu^*(f_n) < +\infty$.

We show that, for each $\varepsilon > 0$, there exists an increasing sequence $(g_n)_{n \geq 1}$ in $\mathcal{J}^+$ such that $f_n \leq g_n$ and $V\mu^*(g_n) \leq V\mu^*(f_n) + \varepsilon$. If g is the upper envelope of the sequence $(g_n)_{n \geq 1}$, we have $V\mu^*(g) = \sup V\mu^*(g_n)$, so $V\mu^*(g) \leq \sup V\mu^*(f_n) + \varepsilon$. Since $\sup f_n \leq g$ and ε is arbitrary, the theorem will follow.

For this, observe that there exists $h_n \in \mathcal{J}^+$ such that $f_n \leq h_n$ and $V\mu^*(h_n) \leq V\mu^*(f_n) + \varepsilon/2^n$. The functions $g_n = \sup(h_1, h_2, \ldots, h_n)$ lie in $\mathcal{J}^+$, form an increasing sequence, and satisfy $f_n \leq g_n$ for all $n \geq 1$.

By induction on n, we prove that $V\mu^*(g_n) \leq V\mu^*(f_n) + \varepsilon(1 - 1/2^n)$. If $n = 1$, the result is clearly true. Otherwise, $g_n = \sup(g_{n-1}, h_n)$, $g_{n-1} \geq$

f_{n-1}, and $h_n \geq f_n \geq f_{n-1}$, so $\inf(g_{n-1}, h_n) \geq f_{n-1}$; since $\inf(g_{n-1}, h_n) + \sup(g_{n-1}, h_n) = g_{n-1} + h_n$, it follows that

$$
\begin{aligned}
V\mu^*(g_n) &= V\mu^*(g_{n-1}) + V\mu^*(h_n) - V\mu^*\big(\inf(g_{n-1}, h_n)\big) \\
&\leq V\mu^*(g_{n-1}) + V\mu^*(h_n) - V\mu^*(f_{n-1}) \\
&\leq V\mu^*(f_n) + \frac{\varepsilon}{2^n} + \varepsilon\Big(1 - \frac{1}{2^{n-1}}\Big) \\
&\leq V\mu^*(f_n) + \varepsilon\Big(1 - \frac{1}{2^n}\Big).
\end{aligned}
$$

Hence $V\mu^*(g_n) \leq V\mu^*(f_n) + \varepsilon$ for all $n \geq 1$, as desired.   □

Proposition 3.1.1 *For every sequence $(f_n)_{n \geq 1}$ of functions from Ω into $[0, +\infty]$, we have $V\mu^*\big(\sum_{k \geq 1} f_k\big) \leq \sum_{k \geq 1} V\mu^*(f_k)$.*

PROOF: If we put $g_n = \sum_{1 \leq k \leq n} f_k$ for every $n \geq 1$, then

$$
V\mu^*(g_n) \leq \sum_{1 \leq k \leq n} V\mu^*(f_k),
$$

so

$$
V\mu^*\Big(\sum_{k \geq 1} f_k\Big) = \sup_{n \geq 1} V\mu^*(g_n) \leq \sum_{k \geq 1} V\mu^*(f_k).
$$

□

Proposition 3.1.2 (Fatou's Lemma) *Let $(f_n)_{n \geq 1}$ be a sequence of functions from Ω into $[0, +\infty]$. Then $V\mu^*(\liminf_{n \to +\infty} f_n) \leq \liminf V\mu^*(f_n)$.*

PROOF: For all integers $n \geq 1$, put $g_n = \inf_{p \geq 0} f_{n+p}$. The sequence $(g_n)_{n \geq 1}$ increases, and $\sup_{n \geq 1} g_n = \liminf f_n$. For each $n \geq 1$, since $g_n \leq f_{n+p}$ for $p \geq 0$, we have $V\mu^*(g_n) \leq \inf_{p \geq 0} V\mu^*(f_{n+p})$. Thus

$$
V\mu^*(\liminf f_n) = \sup V\mu^*(g_n) \leq \liminf V\mu^*(f_n).
$$

□

Definition 3.1.2 A function f from Ω into $[0, +\infty]$ is said to be μ-negligible whenever $V\mu^*(f) = 0$. A subset E of Ω is said to be μ-negligible, or of $V\mu$-measure 0, if 1_E is μ-negligible.

$\sum_{n \geq 1} f_n$, and so $\sup f_n$, is μ-negligible for every sequence $(f_n)_{n \geq 1}$ of μ-negligible functions from Ω into $[0, +\infty]$. Every subset of a μ-negligible set is μ-negligible, and a union of countably many μ-negligible sets is μ-negligible.

Definition 3.1.3 A property is said to be true almost everywhere for μ (abbreviated [a.e]-μ, or a.e.) if and only if it is true outside a μ-negligible set.

Proposition 3.1.3 *A necessary and sufficient condition that a function f from Ω into $[0, +\infty]$ be μ-negligible is that it vanish almost everywhere (with respect to μ).*

PROOF: The condition is necessary. Indeed, suppose that f is negligible, and let E be the set of points $x \in \Omega$ such that $f(x) \neq 0$; then $1_E \leq \sup_n(nf)$, so, 1_E is negligible.

The condition is sufficient. Indeed, suppose that the set E of points x where f does not vanish is negligible; then $f \leq \sup_{n \geq 1} n 1_E$, so f is negligible.   □

Proposition 3.1.4 *Let f and g be two functions from Ω into $[0, +\infty]$. If $f = g$ a.e., then $V\mu^*(f) = V\mu^*(g)$.*

PROOF: Let $E = \{x \in \Omega : f(x) \neq g(x)\}$. Since $\inf(f, g) = \sup(f, g)$ on E^c, it suffices to prove the proposition when $f \leq g$. Then let h be the function equal to $+\infty$ on E and to 0 on E^c; we have $f \leq g \leq f + h$, and so

$$(V\mu^*)(f) \leq (V\mu^*)(g) \leq (V\mu^*)(f + h) \leq (V\mu^*)(f) + (V\mu^*)(h)$$
$$= (V\mu^*)(f).$$

□

Proposition 3.1.5 *If $f : \Omega \to [0, +\infty]$ is such that $(V\mu^*)(f) < +\infty$, then f is finite a.e.*

PROOF: Put $E = f^{-1}(+\infty)$. For every integer $n \geq 1$, we have $n 1_E \leq f$, so $n V\mu^*(1_E) \leq V\mu^*(f)$. Since this is true for arbitrarily large n, $V\mu^*(1_E) = 0$.

□

Henceforth, F will denote a real Banach space. A mapping f from Ω into F is said to be decomposable if it can be written $f = \sum_{i \in I} f_i a_i$, for a finite set I, functions $f_i \in \mathcal{H}(\Omega, \mathbf{R})$, and elements a_i of F. Such a representation is called a decomposition of f. Let $\mathcal{H}(\Omega, \mathbf{R}) \otimes F$ be the vector space of decomposable mappings from Ω into F. Now suppose that F is a real or complex Banach space as μ is real or complex, and let F' be its dual (the space of continuous linear forms on F). If $\sum_{i \in I} f_i a_i$ is a decomposition of $f \in \mathcal{H}(\Omega, \mathbf{R}) \otimes F$, then $\sum_{i \in I} \mu(f_i) a_i$ does not depend on the particular decomposition of f, for $z'\left(\sum_{i \in I} \mu(f_i) a_i\right) = \mu(z' \circ f)$ for all $z' \in F'$ (W. Rudin, *Functional Analysis*, Chapter 3, Corollary to Theorem 3.4). Hence we may denote the vector $\sum_{i \in I} \mu(f_i)a_i$ by $\int f d\mu$, called the μ-integral of f. Furthermore, $|\int f d\mu| = \sup_{z' \in F', \|z'\| \leq 1} |z'(\int f d\mu)|$ (ibid., Corollary to Theorem 3.3), hence

$$\left| \int f d\mu \right| = \sup_{z' \in F', \|z'\| \leq 1} \left| \int (z' \circ f) d\mu \right| \leq \sup_{z'} \int |z' \circ f| dV\mu \leq \int^* |f| dV\mu,$$

where $\int^* |f| dV\mu \leq \sum_{i \in I} |a_i| \cdot V\mu(|f_i|)$ is finite.

In what follows, as soon as we consider the expression $\int f \, d\mu$ (or the similar expression $\bar{\int} f \, d\mu$ of Section 3.6), it shall be understood that F is a real or complex Banach space, as μ is real or complex.

The mappings f from Ω into F such that $\int^* |f| dV\mu < +\infty$ form a vector space $\mathcal{F}_F^1(\mu)$, and the function $f \mapsto \int^* |f| dV\mu = N_1(f)$ is a seminorm on $\mathcal{F}_F^1(\mu)$.

Definition 3.1.4 Denote by $\mathcal{L}_F^1(\mu)$ the closure of $\mathcal{H}(\Omega, \mathbf{R}) \otimes F$ in the seminormed space $\mathcal{F}_F^1(\mu)$. Its elements are the μ-integrable mappings from Ω into F. Now consider $\mathcal{L}_F^1(\mu)$ as a vector subspace of the seminormed space $\mathcal{F}_F^1(\mu)$. Then the continuous linear map $f \mapsto \int f d\mu$ from $\mathcal{H}(\Omega, \mathbf{R}) \otimes F$ into F can be uniquely extended to a continuous linear map from $\mathcal{L}_F^1(\mu)$ into F; we still write $\int f d\mu$ or $\int f(x) d\mu(x)$ for this extension, and call $\int f d\mu$ the μ-integral of f.

Observe that $|\int f d\mu| \leq N_1(f)$ for all $f \in \mathcal{L}_F^1(\mu)$.

A filter that converges in $\mathcal{F}_F^1(\mu)$ is said to converge in the mean.

Proposition 3.1.6 *If U is a continuous linear map from F into a Banach space G and if $f \in \mathcal{L}_F^1(\mu)$, then $U \circ f$ lies in $\mathcal{L}_G^1(\mu)$ and $\int (U \circ f) d\mu = U(\int f d\mu)$.*

Proposition 3.1.7 *Let $(f_n)_{n \geq 1}$ be a sequence in $\mathcal{F}_F^1(\mu)$ such that*

$$\sum_{n \geq 1} N_1(f_n) < +\infty.$$

The series $\sum_{n \geq 1} f_n(x)$ converges absolutely a.e. If we put $f(x) = \sum_{n \geq 1} f_n(x)$ at almost all points where this series converges, and take $f(x) \in F$ arbitrarily elsewhere, then the mapping f lies in $\mathcal{F}_F^1(\mu)$, and

$$N_1\left(f - \sum_{1 \leq k \leq n} f_k \right) \leq \sum_{k \geq n+1} N_1(f_k)$$

for all $n \geq 0$. Therefore, the series $\sum_{n \geq 1} f_n$ converges to f in the mean.

PROOF: Consider the function $g : x \mapsto \sum_{n \geq 1} |f_n(x)|$ of Ω into $[0, +\infty]$. Since $N_1(g) \leq \sum_{n \geq 1} N_1(f_n) < +\infty$, g is finite a.e., and the series $\sum_{n \geq 1} f_n(x)$ converges absolutely a.e. Since F is complete, this series converges a.e., and

$|f(x)| \leq g(x)$ a.e. Hence $N_1(f) \leq N_1(g) < +\infty$, and f lies in $\mathcal{F}_F^1(\mu)$. On the other hand, $\left| f(x) - \sum_{1 \leq k \leq n} f_k(x) \right| \leq \sum_{k \geq n+1} |f_k(x)|$ almost everywhere, for every integer $n \geq 0$. Thus $N_1\left(f - \sum_{1 \leq k \leq n} f_k \right) \leq \sum_{k \geq n+1} N_1(f_k)$. $\square$

Theorem 3.1.2 (Transition Lemma) *Let* $(f_i)_{i \geq 1}$ *be a Cauchy sequence in* $\mathcal{F}_F^1(\mu)$. *There exists a subsequence* $(f_{i_n})_{n \geq 1}$ *of* $(f_i)_{i \geq 1}$ *with the following properties:*

(a) $\sum_{n \geq 1} N_1(f_{i_{n+1}} - f_{i_n})$ *is finite.*

(b) *The sequence* $(f_{i_n}(x))_{n \geq 1}$ *converges almost everywhere.*

(c) *If we put* $f(x) = \lim_{n \to +\infty} f_{i_n}(x)$ *at almost all points* x *where this limit exists, and take* $f(x)$ *arbitrarily otherwise, then* f *belongs to* $\mathcal{F}_F^1(\mu)$ *and the sequence* $(f_i)_{i \geq 1}$ *converges to* f *in the mean.*

(d) *For each* $\varepsilon > 0$, *there exists* $Z \subset \Omega$ *such that* $V\mu^*(Z) \leq \varepsilon$ *and the sequence* $(f_{i_n})_{n \geq 1}$ *converges uniformly to* f *on* $\Omega - Z$.

(e) *There exists a function* g *from* Ω *into* $[0, +\infty]$ *such that* $N_1(g) < +\infty$ *and* $|f_{i_n}(x)| \leq g(x)$ *for all* $x \in \Omega$.

PROOF: By induction on n, define a strictly increasing sequence $(i_n)_{n \geq 1}$ of integers ≥ 1, such that $N_1(f_i - f_{i_n}) \leq 2^{-2n}$ for each $n \geq 1$ and for each $i \geq i_n$. By Proposition 3.1.7, the series with general term $f_{i_{n+1}} - f_{i_n}$ converges almost everywhere, and in the mean, to a mapping $h \in \mathcal{F}_F^1(\mu)$. Thus the sequence $(f_{i_n})_{n \geq 1}$ converges almost everywhere and in the mean to $f_{i_1} + h$. Since the subsequence $(f_{i_n})_{n \geq 1}$ of the Cauchy sequence $(f_i)_{i \geq 1}$ converges, the sequence $(f_i)_{i \geq 1}$ itself converges. This proves (a), (b), (c). Now, for every n in $\mathbf{N}$, put $Y_n = \{x \in \Omega : |f_{i_{n+1}} - f_{i_n}|(x) \geq 1/2^n\}$. From $(1/2^n) \cdot 1_{Y_n} \leq |f_{i_{n+1}} - f_{i_n}|$, it follows that $(1/2^n) \cdot V\mu^*(Y_n) \leq N_1(f_{i_{n+1}} - f_{i_n})$, whence $V\mu^*(Y_n) \leq 2^{-n}$. Put $Z_n = \bigcup_{m \geq n} Y_m$. By construction, $\sup_{x \in \Omega - Z_n} |f_{i_{k+1}} - f_{i_k}| \leq 1/2^k$ for all $k \geq n$, so the series with general term $f_{i_{k+1}} - f_{i_k}$ $(k \geq 1)$ converges uniformly on $\Omega - Z_n$. Since $V\mu^*(Z_n) \leq 1/2^{n-1}$, for every $\varepsilon > 0$ we can choose n such that $V\mu^*(Z_n) \leq \varepsilon$. Then, if N is a μ-negligible set such that $(f_{i_k})_{k \geq 1}$ converges to f on $\Omega - N$, we may take $Z = Z_n \cup N$ and $g = |f_{i_1}| + \sum_{k \geq 1} |f_{i_{k+1}} - f_{i_k}|$, which proves (d). $\square$

Theorem 3.1.3 (Riesz–Fischer Theorem) $\mathcal{F}_F^1(\mu)$ *and* $\mathcal{L}_F^1(\mu)$ *are complete.*

PROOF: $\mathcal{F}_F^1(\mu)$ is complete by Theorem 3.1.2, and so is $\mathcal{L}_F^1(\mu)$ which is closed in $\mathcal{F}_F^1(\mu)$. $\square$

Proposition 3.1.8 *If a Cauchy sequence $(f_i)_{i \geq 1}$ in $\mathcal{L}_F^1(\mu)$ converges almost everywhere to f, then $f \in \mathcal{L}_F^1(\mu)$ and $(f_i)_{i \geq 1}$ converges to f in the mean.*

PROOF: By Theorem 3.1.2, there exists a subsequence $(f_{i_n})_{n \geq 1}$ of $(f_i)_{i \geq 1}$ which converges a.e. to a mapping $g \in \mathcal{L}_F^1(\mu)$, such that $(f_i)_{i \geq 1}$ converges to g in the mean. Hence $f = g$ a.e. □

Proposition 3.1.9 *Let H be a dense subset of $\mathcal{L}_F^1(\mu)$. For every $f \in \mathcal{L}_F^1(\mu)$, there is a sequence $(g_n)_{n \geq 1}$ in H with the following properties:*

(a) $(g_n)_{n \geq 1}$ converges to f in the mean.

(b) $(g_n)_{n \geq 1}$ converges to f almost everywhere.

PROOF: This follows from Theorem 3.1.2. □

Lemma 3.1.1 *Let E be a normed real vector space of finite dimension. For every $\varepsilon > 0$, there exist linear forms $u_1, \ldots, u_p$ on E such that $(1 - \varepsilon) \cdot |z| \leq \sup_{1 \leq i \leq p} |u_i(z)| \leq |z|$ for all $z \in E$.*

PROOF: Put $B = \{z \in E : |z| < 1\}$ and $S = \{z \in E : |z| = 1\}$. B is open convex, and λz lies in B for each $z \in B$ and each $\lambda \in \mathbf{R}$ satisfying $|\lambda| \leq 1$. By the Hahn–Banach theorem, for every $y \in S$, there exists a linear form u_y on E such that $|u_y(z)| < 1$ for all $z \in B$ and $u_y(y) = 1$ (W. Rudin, *Functional Analysis*, Chapter 3, Theorem 3.4). Then $1 - \varepsilon \leq |u_y(z)| \leq 1$ for all $z \in S$ in some suitably chosen open neighborhood V_y of y. Since S is compact, we can find $y_1, \ldots, y_p$ in S so that $\bigcup_{1 \leq i \leq p} V_{y_i}$ contains S, and put $u_i = u_{y_i}$ for each $1 \leq i \leq p$. Then $1 - \varepsilon \leq \sup_{1 \leq i \leq p} |u_i(z)| \leq 1$ for all $z \in S$, because every point of S lies in one of the $V_i = V_{y_i}$. □

Lemma 3.1.2 *If f is a positive μ-integrable function, then $\int f \cdot dV\mu = V\mu^*(f)$.*

PROOF: There exists a sequence $(f_n)_{n \geq 1}$ in $\mathcal{H}(\Omega, \mathbf{R})$ that converges to f in the mean. Replacing f_n by f_n^+, if necessary, we may suppose that each f_n lies in $\mathcal{H}^+$. Then

$$\int^* f \cdot dV\mu = \lim_{n \to +\infty} \int^* f_n \cdot dV\mu = \lim \int f_n \cdot dV\mu = \int f \cdot dV\mu.$$

  □

Proposition 3.1.10 *For $f \in \mathcal{L}_F^1(\mu)$, $|f|$ belongs to $\mathcal{L}_{\mathbf{R}}^1(\mu)$, and $|\int f d\mu| \leq \int |f| \cdot dV\mu$.*

PROOF: To begin, suppose that f lies in $\mathcal{H}(\Omega, \mathbf{R}) \otimes F$, and so has the form $\sum_{1 \leq j \leq n} f_j \cdot a_j$, where $f_j \in \mathcal{H}(\Omega, \mathbf{R})$ and $a_j \in F$. Let $\varepsilon > 0$. By Lemma 3.1.1, there exist linear forms $u_1, \ldots, u_p$ on the vector subspace E of F generated by $a_1, \ldots, a_n$, such that $(1 - \varepsilon) \cdot |z| \leq \sup_{1 \leq i \leq p} |u_i(z)| \leq |z|$ for all $z \in E$. Thus

$$(1 - \varepsilon) \cdot |f| \leq \sup_{1 \leq i \leq p} |u_i \circ f| \leq |f|.$$

But the $u_i \circ f$ belong to $\mathcal{H}(\Omega, \mathbf{R})$, so $\varphi_\varepsilon = \sup_{1 \leq i \leq p} |u_i \circ f|$ lies in $\mathcal{H}^+$. Moreover, $||f| - \varphi_\varepsilon| \leq \varepsilon |f|$ gives $\int^* ||f| - \varphi_\varepsilon| \cdot dV\mu \leq \varepsilon \cdot V\mu^*(|f|)$. Therefore, $|f|$ is μ-integrable.

It remains to treat the case where f is not necessarily decomposable. For this, we let $(f_n)_{n \geq 1}$ be a family in $\mathcal{H}(\Omega, \mathbf{R}) \otimes F$ converging to f in the mean. For every $n \geq 1$, we have $||f| - |f_n|| \leq |f - f_n|$, so $\int^* ||f| - |f_n|| \cdot dV\mu \leq \int^* |f - f_n| \cdot dV\mu$. Hence the sequence $(|f_n|)_{n \geq 1}$ converges to $|f|$ in the mean, and $|f|$ is μ-integrable because $\mathcal{L}^1_{\mathbf{R}}(\mu)$ is closed in $\mathcal{F}^1_{\mathbf{R}}(\mu)$.

Finally, $\left| \int f d\mu \right| \leq N_1(f)$ for every $f \in \mathcal{L}^1_F(\mu)$, and $N_1(f) = \int |f| dV\mu$ as shown in Lemma 3.1.2. □

We can define on $\mathcal{L}^1_F(\mu)$ the equivalence relation "$f \sim g$ if $f = g$ almost everywhere", and give the quotient space $L^1_F(\mu)$ of $\mathcal{L}^1_F(\mu)$ the norm $\dot{f} \mapsto \int |f| dV\mu = N_1(f)$. Then $L^1_F(\mu)$ is a Banach space. For all practical purposes, we need not distinguish $\mathcal{L}^1_F(\mu)$ and $L^1_F(\mu)$.

A function f from Ω into $\overline{\mathbf{R}}$ is called μ-integrable if it is equal a.e. to a μ-integrable function g from Ω into $\mathbf{R}$. Clearly, if g and g' are two such functions, then $\int g d\mu = \int g' d\mu$, so it makes sense to define the complex number $\int g d\mu$ as the integral of f, written $\int f d\mu$. Observe that any function g from Ω into $\mathbf{R}$ equal to f a.e. on $\Omega - (f^{-1}(-\infty) \cup f^{-1}(+\infty))$ is μ-integrable. Letting $\mathcal{L}^1(\mu; \overline{\mathbf{R}})$ be the set of μ-integrable functions from Ω into $\overline{\mathbf{R}}$, a filter $\mathcal{F}$ on $\mathcal{L}^1(\mu; \overline{\mathbf{R}})$ is said to converge to $f \in \mathcal{L}^1(\mu; \overline{\mathbf{R}})$ in the mean if, for g as above, $\int^* |h - g| \cdot dV\mu$ tends to 0 as h extends over the filter $\mathcal{F}$; clearly, this definition does not depend on the choice of g.

Theorem 3.1.4 $f \in \mathcal{J}^+$ is μ-integrable if and only if $\int^* f \cdot dV\mu < +\infty$.

PROOF: Suppose $\int^* f \cdot dV\mu$ is finite. Since $V\mu^*(f) = \sup_{g \in \mathcal{H}^+, g \leq f} V\mu(g)$, for every $\varepsilon > 0$ there exists $g \in \mathcal{H}^+$, $g \leq f$, such that $V\mu^*(f) \leq V\mu(g) + \varepsilon$. Then $V\mu^*(f) = V\mu^*(f - g) + V\mu(g)$, so $V\mu^*(f - g) \leq \varepsilon$. □

Proposition 3.1.11 $\sup(f_1, f_2)$ and $\inf(f_1, f_2)$ lie in $\mathcal{L}^1(\mu; \overline{\mathbf{R}})$, for all f_1, f_2 in $\mathcal{L}^1(\mu; \overline{\mathbf{R}})$.

PROOF: If f_1, f_2, g_1, g_2 are real-valued functions on Ω, then

$$\left| \sup(f_1, f_2) - \sup(g_1, g_2) \right|(x) \leq \sup \left(|f_1 - g_1|(x), |f_2 - g_2|(x) \right)$$

for every $x \in \Omega$. Indeed, this inequality evidently holds if there exists i in $\{1, 2\}$ such that $f_i(x) = \sup(f_1, f_2)(x)$ and $g_i(x) = \sup(g_1, g_2)(x)$. Therefore, suppose that $\sup(f_1, f_2)(x) = f_i(x)$ and $\sup(g_1, g_2)(x) = g_j(x)$ for distinct i, j; if $f_i(x) \geq g_j(x)$, we have $g_i(x) \leq g_j(x) \leq f_i(x)$, so $|f_i - g_j|(x) \leq |f_i - g_i|(x)$; on the other hand, if $f_i(x) \leq g_j(x)$, we have $f_j(x) \leq f_i(x) \leq g_j(x)$, and $|f_i - g_j|(x) \leq |f_j - g_j|(x)$, whence the desired inequality. Now let f_1, f_2 be two μ-integrable functions from Ω into $\mathbf{R}$. For every $\varepsilon > 0$, there exist g_1 in $\mathcal{H}(\Omega, \mathbf{R})$ and g_2 in $\mathcal{H}(\Omega, \mathbf{R})$ such that $\int^* |f_1 - g_1| \cdot dV\mu \leq \varepsilon/2$ and $\int^* |f_2 - g_2| \cdot dV\mu \leq \varepsilon/2$. Then

$$\int^* \left| \sup(f_1, f_2) - \sup(g_1, g_2) \right| \cdot dV\mu \leq \int^* \left(|f_1 - g_1| + |f_2 - g_2| \right) \cdot dV\mu \leq \varepsilon,$$

and so $\sup(f_1, f_2)$ is μ-integrable. $\qquad\square$

3.2   Convergence Theorems

Theorem 3.2.1 (Monotone Convergence Theorem)   *Let $(f_n)_{n\geq 1}$ be an increasing (respectively, a decreasing) sequence in $\mathcal{L}^1(\mu; \overline{\mathbf{R}})$. Then $f = \sup_{n\geq 1} f_n$ (respectively, $f = \inf_{n\geq 1} f_n$) is integrable if and only if $\sup \int f_n dV\mu < +\infty$ (respectively, $\inf \int f_n dV\mu > -\infty$). In this case, $(f_n)_{n\geq 1}$ converges to f in the mean.*

PROOF: We consider the case of an increasing sequence $(f_n)_{n\geq 1}$ in $\mathcal{L}^1(\mu; \mathbf{R})$, and suppose that $\sup_{n\geq 1} \int f_n dV\mu$ is finite. The sequence $(\int f_n \cdot dV\mu)_{n\geq 1}$ increases, so it converges in $\mathbf{R}$. Therefore, $\int |f_p - f_q| dV\mu = \pm \int (f_p - f_q) dV\mu$ converges to 0 as $p, q \to +\infty$. Since $\mathcal{L}^1_{\mathbf{R}}(\mu)$ is complete, there exists a μ-integrable function g from Ω into $\mathbf{R}$ such that $\int |f_n - g| \cdot dV\mu \to 0$ as $n \to +\infty$. But g is the limit a.e. of a subsequence of $(f_n)_{n\geq 1}$, so $g = f$ a.e., and f is μ-integrable. $\qquad\square$

Proposition 3.2.1 (Fatou's Theorem)   *Let $(f_n)_{n\geq 1}$ be a sequence in $\mathcal{L}^1(\mu; \overline{\mathbf{R}})$ such that $\liminf_{n\to+\infty} \int f_n \cdot dV\mu < +\infty$ (respectively, $\limsup_{n\to+\infty} \int f_n \cdot dV\mu > -\infty$). Assume there exists g in $\mathcal{L}^1(\mu; \overline{\mathbf{R}})$ such that $g \leq f_n$ a.e. (respectively, $g \geq f_n$ a.e.) for all $n \geq 1$. Then $f = \liminf f_n$ (respectively, $f = \limsup f_n$) is μ-integrable, and*

$$\int g \cdot dV\mu \leq \int f \cdot dV\mu \leq \liminf_{n\to+\infty} \int f_n \cdot dV\mu$$

(respectively, $\int g \cdot dV\mu \geq \int f \cdot dV\mu \geq \limsup_{n\to+\infty} \int f_n \cdot dV\mu$).

PROOF: We will prove only the first statement. For each fixed n, the sequence $(g_{n,k})_{k \geq n}$, where $g_{n,k} = \inf_{n \leq j \leq k} f_j$, decreases, and we have $\int g_{n,k} \cdot dV\mu \geq \int g \cdot dV\mu$ for all $k \geq n$. By the monotone convergence theorem, $g_n = \inf_{k \geq n} g_{n,k}$ is μ-integrable. Likewise, the sequence $(g_n)_{n \geq 1}$ increases, and we have $\int g_n \cdot dV\mu \leq \inf_{k \geq n} \int f_k \cdot dV\mu \leq c$, where $c = \liminf_{n \to +\infty} \int f_n \cdot dV\mu$. Thus $\sup_{n \geq 1} g_n = \liminf_{n \to +\infty} f_n$ is μ-integrable, and $\int g \cdot dV\mu \leq \int f \cdot dV\mu \leq c$. □

Theorem 3.2.2 (Lebesgue's Dominated Convergence Theorem) *If a sequence $(f_n)_{n \geq 1}$ in $\mathcal{L}_F^1(\mu)$ converges a.e. to a mapping f from Ω into F (where F is a Banach space), and if there exists a function g from Ω into $[0, +\infty]$ such that $|f_n| \leq g$ a.e. for all $n \geq 1$ and $\int^* g dV\mu < +\infty$, then f is μ-integrable and $(f_n)_{n \geq 1}$ converges to f in the mean; consequently, $\int f_n d\mu$ converges to $\int f d\mu$.*

PROOF: It suffices to prove that $\int |f_p - f_q| \cdot dV\mu$ converges to 0 as p and q tend to $+\infty$. Fix $n \geq 1$, and consider the sequence $(g_{n,k})_{k \geq n}$ where $g_{n,k} = \sup_{n \leq p,q \leq k} |f_p - f_q|$. Now $(g_{n,k})_{k \geq n}$ is an increasing sequence of μ-integrable functions, such that $\int g_{n,k} dV\mu \leq 2 \int^* g dV\mu$ for all $k \geq n$. By the monotone convergence theorem, $(g_{n,k})_{k \geq n}$ converges almost everywhere to the μ-integrable function $g_n = \sup_{k \geq n} g_{n,k} = \sup_{p,q \geq n} |f_p - f_q|$.

Next, the sequence $(g_n)_{n \geq 1}$ decreases and converges to 0 a.e. Hence $\int g_n dV\mu$ tends to 0 as $n \to +\infty$. □

Corollary *Let $\mathcal{F}$ be a filter with a countable basis on a set of indices A, $(f_\alpha)_{\alpha \in A}$ be a family in $\mathcal{L}_F^1(\mu)$, and $f : \Omega \to F$ be a mapping. Assume that, for almost all $x \in \Omega$, $\big(f_\alpha(x)\big)_\alpha$ converges to $f(x)$ along $\mathcal{F}$. Now suppose there is a function g from Ω into $[0, +\infty]$ such that $\int^* g.dV\mu < +\infty$ and $|f_\alpha| \leq g$ outside a negligible set (depending on α) for each $\alpha \in A$. Then $f \in \mathcal{L}_F^1(\mu)$ and $(f_\alpha)_\alpha$ converges to f in the mean.*

PROOF: Let $(A_n)_{n \geq 1}$ be a countable decreasing basis of $\mathcal{F}$ and, for every $n \geq 1$, let α_n be an arbitrary element of A_n. By Theorem 3.2.2, f is μ-integrable and $(f_{\alpha_n})_{n \geq 1}$ converges to f in the mean. By contradiction, it follows easily that, along $\mathcal{F}$, $(f_\alpha)_\alpha$ converges to f in the mean. □

Proposition 3.2.2 *In order that $f : \Omega \to F$ be μ-integrable, it is necessary and sufficient that it be a.e. the limit of a sequence in $\mathcal{H}(\Omega, \mathbf{R}) \otimes F$ and that $\int^* |f| dV\mu < +\infty$.*

PROOF: Suppose the conditions are satisfied. Let $(f_n)_{n \geq 1}$ be a sequence in $\mathcal{H}(\Omega, \mathbf{R}) \otimes F$, a.e. converging to f, and let $g \geq 0$, $g \in \mathcal{L}^1(\mu; \mathbf{R})$, be such that $|f| \leq g$ everywhere (Theorem 3.1.4).

Now each f_n can be written $f_n = \sum_{1 \le i \le k_n} \varphi_{i,n} \cdot y_{i,n}$, where $\varphi_{i,n} \in \mathcal{H}^+$ and $y_{i,n} \in F$. Putting $h_{i,n,m} = \inf \left(\varphi_{i,n}, m\left(2g - \inf(2g, |f_n|)\right) \right)$, for each integer $m \ge 1$, the $h_{i,n,m}$ are positive, μ-integrable, and bounded above by $\varphi_{i,n}$. The sequence $(h_{i,n,m})_{m \ge 1}$ (the indices i and n being fixed) converges to the function $\psi_{i,n}$ given by

$$\psi_{i,n}(x) = \begin{cases} \varphi_{i,n}(x) & \text{if } |f_n(x)| < 2g(x) \\ 0 & \text{if } |f_n(x)| \ge 2g(x). \end{cases}$$

By the dominated convergence theorem, $\psi_{i,n}$ is μ-integrable, so $g_n = \sum_{1 \le i \le k_n} \psi_{i,n} \cdot y_{i,n}$ lies in $\mathcal{L}_F^1(\mu)$ (Proposition 3.1.6). Further,

$$g_n(x) = \begin{cases} f_n(x) & \text{if } |f_n(x)| < 2g(x) \\ 0 & \text{if } |f_n(x)| \ge 2g(x). \end{cases}$$

Next, the sequence $(g_n)_{n \ge 1}$ converges a.e. to f, and $|g_n| \le 2g$ for every $n \ge 1$. The dominated convergence theorem then shows that f is integrable. □

Proposition 3.2.3 (Continuity with Respect to a Parameter) *Let A be a topological space, α_0 be a point in A, and suppose that the filter of neighborhoods of α_0 has a countable basis. Let f be a mapping from $\Omega \times A$ into F satisfying the following conditions:*

(a) *For each $\alpha \in A$, the mapping $x \mapsto f(x, \alpha)$ is μ-integrable.*

(b) *For almost all $x \in \Omega$, the mapping $\alpha \mapsto f(x, \alpha)$ is continuous at α_0.*

(c) *There exist a neighborhood U of α_0 and a function g from Ω into $[0, +\infty]$ such that $\int^* g \cdot dV\mu < +\infty$ and that, for every $\alpha \in U$, $|f(x, \alpha)| \le g(x)$ outside a negligible set (depending on α).*

Then the mapping $\alpha \rightarrow \int f(x, \alpha) d\mu(x)$ from A into F is continuous at α_0.

PROOF: This follows immediately from the corollary to Theorem 3.2.2. □

At this point, we introduce some notation and make some definitions. If E, F are real (or complex) normed spaces, for every integer $n \ge 1$, we call $L^n(E, F)$ the normed space of n linear continuous mappings from E^n into F, and we put $L^0(E, F) = F$. On the other hand, we inductively define a normed space $L^{(n)}(E, F)$ as follows: $L^{(0)}(E, F) = F$, and $L^{(n)}(E, F) = L(E, L^{(n-1)}(E, F))$ for $n \ge 1$. Then there is an isometry j_n of $L^{(n)}(E, F)$ onto $L^n(E, F)$, defined by

$$(j_n(\gamma))(u_1, \ldots, u_n) = \gamma(u_n) \cdot u_{n-1} \cdot \ldots \cdot u_1 = \left((\gamma(u_n) \cdot u_{n-1}) \ldots \right) \cdot u_1.$$

Now let X be an open subset of E, f a mapping from X into F, a a point of X, and $n \ge 1$ an integer. f is said to be n times differentiable at a and to have $D^{(n)}f(a)$ as derivative of order n at a, if and only if

(a) f is $(n-1)$ times differentiable at each point of an open neighborhood V of a (with $V \subset X$);

(b) the mapping $x \mapsto D^{(n-1)}f(x)$ from V into $L^{(n-1)}(E, F)$ is differentiable at a, and $D^{(n)}f(a)$ is its derivative at this point.

$D^n f(a) = j_n(D^{(n)}f(a))$ is also called the nth derivative of f at a.

f is said to be of class C^n if it is n times differentiable at each point of X and if the mappings $D^k f$ $(0 \leq k \leq n)$, $x \mapsto D^k f(x)$, are continuous on X. When $E = K$ (with $K = \mathbf{R}$ or $\mathbf{C}$), f is n times differentiable at a if and only if it has a derivative, $f^{(n)}(a)$, of order n at a (in the usual sense), and then $f^{(n)}(a) = D^n f(a)(1, \ldots, 1)$.

We now return to the situation where F is a fixed Banach space.

Theorem 3.2.3 (Differentiation under the Integral Sign) *Let A be an open subset of a normed real (or complex) vector space E, and let f be a mapping from $(\Omega - N) \times A$ into F, where N is a μ-negligible subset of Ω. For a given integer $n \geq 1$, suppose that the following conditions are satisfied:*

(a) *For every $x \in \Omega - N$, the mapping $f(x, \cdot) : z \mapsto f(x, z)$ is of class C^n on A.*

(b) *For each $z \in A$, the mappings $x \mapsto D^k(f(x, \cdot))(z)$ from $\Omega - N$ into $L^k(E, F)$ are all μ-integrable $(0 \leq k \leq n)$.*

(c) *For each $a \in A$, there exists an integrable function g_n from $\Omega - N$ into $[0, +\infty[$ such that, for z sufficiently close to a,*

$$\left\| D^n(f(x, \cdot))(z) \right\| \leq g_n(x)$$

almost everywhere in $\Omega - N$.

Then the mapping $h : z \mapsto \int f(x, z) d\mu(x)$ from A into F is of class C^n, and $D^k h(z) = \int D^k(f(x, \cdot))(z) d\mu(x)$ for all $0 \leq k \leq n$ and $z \in A$.

PROOF: Let $a \in A$ and g_n be as above, and let $r > 0$ be a real number such that A contains the closed ball B with center a and radius r, and such that, for all $z \in B$, $\|D^n(f(x, \cdot))(z)\| \leq g_n(x)$ a.e. in $\Omega - N$.

Let $0 \leq k \leq n - 1$ be fixed for the moment, and let g_k be the function

$$x \mapsto \sum_{0 \leq p \leq n-k-1} \left\| D^{k+p}(f(x, \cdot))(a) \right\| \cdot \frac{r^p}{p!} + g_n(x) \cdot \frac{r^{n-k}}{(n-k)!}$$

from $\Omega - N$ into $[0, +\infty[$. Evidently, g_k is integrable. For any given $z \in B$, Taylor's formula shows that

$$D^k(f(x, \cdot))(z) = \sum_{0 \leq p \leq n-k-1} \frac{1}{p!} \cdot D^{k+p}(f(x, \cdot))(a)(\cdot, \ldots, (z-a)^p)$$

$$+ \left(\int_0^1 \frac{(1-\theta)^{n-k-1}}{(n-k-1)!} \cdot D^n(f(x, \cdot))(a + \theta(z-a)) \cdot d\theta \right) \cdot (\cdot, \ldots, (z-a)^{n-k})$$

for all $x \in \Omega - N$. But

$$\int_0^1 \frac{(1-\theta)^{n-k-1}}{(n-k-1)!} \cdot D^n\big(f(x,\cdot)\big)\big(a + \theta(z-a)\big) \cdot d\theta$$

$$= \lim_{u \to +\infty} \frac{1}{u} \cdot \sum_{1 \le v \le u} \frac{(1-v/u)^{n-k-1}}{(n-k-1)!} \cdot D^n\big(f(x,\cdot)\big)\Big(a + \frac{v}{u}(z-a)\Big),$$

and there exists a negligible subset N_z of Ω containing N such that, for every $u \ge 1$ and for every $1 \le v \le u$, $\big\|D^n\big(f(x,\cdot)\big)\big(a + (v/u)(z-a)\big)\big\| \le g_n(x)$ for all x of $\Omega - N_z$. Thus

$$\left\| \int_0^1 \frac{(1-\theta)^{n-k-1}}{(n-k-1)!} \cdot D^n\big(f(x,\cdot)\big)\big(a + \theta(z-a)\big) \cdot d\theta \right\| \le \frac{1}{(n-k)!} g_n(x)$$

for all x of $\Omega - N_z$. In short, for every $z \in B$, we have $\big\|D^k\big(f(x,\cdot)\big)(z)\big\| \le g_k(x)$ a.e. in $\Omega - N$. Now, for each $0 \le k \le n$, denote by $D^{(k)}h$ the mapping $z \mapsto \int D^{(k)}\big(f(x,\cdot)\big)(z) \cdot d\mu(x)$ from A into $L^{(k)}(E,F)$. By the preceding argument and Proposition 3.2.3, $D^{(k)}h$ is continuous at a.

Next, let $1 \le k \le n$ and $z \in B$ be fixed, and $t \ne 0$ be a scalar such that $|t| \le 1$. Then

$$\frac{D^{(k-1)}\big(f(x,\cdot)\big)\big(a + t(z-a)\big) - D^{(k-1)}\big(f(x,\cdot)\big)(a)}{t}$$

$$= \left(\int_0^1 D^{(k)}\big(f(x,\cdot)\big)\big(a + \theta t(z-a)\big) \cdot d\theta\right) \cdot (z-a),$$

and the same argument as above shows that

$$\left\| \frac{D^{(k-1)}\big(f(x,\cdot)\big)\big(a + t(z-a)\big) - D^{(k-1)}\big(f(x,\cdot)\big)(a)}{t} \right\| \le g_k(x) \cdot \|z - a\|$$

a.e. in $\Omega - N$. The dominated convergence theorem then proves that $\left(D^{(k-1)}h\big(a + t(z-a)\big) - D^{(k-1)}h(a)\right)\big/t$ converges to $D^{(k)}h(a) \cdot (z-a)$ as $t \to 0$, so $D^{(k-1)}h$ is differentiable at a and its derivative is $D^{(k)}h(a)$. □

We note that other versions of this theorem exist, but we shall require only Theorem 3.2.3.

We consider the following examples, which are intended to illustrate the ideas we have just presented.

If ν is a measure on a semiring S in Ω, and if μ is the linear form $f \mapsto \int f d\nu$ on $St(S, \mathbf{C})$, then we put $\int^* f dV\mu = \int^* f dV\nu$ for every $f : \Omega \to [0, +\infty]$, and so on ($\mathcal{J}^+$ taken as in Chapter 2).

Given a nonempty interval $I = \langle a, b \rangle$ of $\mathbf{R}$ with left endpoint a and right endpoint b (a, b in $\overline{\mathbf{R}}$), let λ be Lebesgue measure on I, and let F be a real Banach space.

If α and β are two points of $\overline{\mathbf{R}}$ such that at least one of the intervals with endpoints α, β is included in I and nonempty, and if $f : I \to F$ is such that $f1_H$ is λ-integrable for one (and so every) interval $H \subset I$ with endpoints α, β, we put

$$\int_\alpha^\beta f(t)dt = +\int f1_H d\lambda$$

or

$$\int_\alpha^\beta f(t)dt = -\int f1_H d\lambda,$$

as $\alpha \leq \beta$ or $\alpha \geq \beta$.

$f : I \to F$ is said to be a step function if there exists a partition of I into finitely many intervals J_k such that f is constant on each J_k. Thus the step functions from I into F form a real vector space.

$f : I \to F$ is said to be regulated whenever, on each compact subinterval of I, it is the uniform limit of step functions. It is easily seen that $f : I \to F$ is regulated if and only if it has a right-hand limit at each point of $\langle a, b[$ and a left-hand limit at each point of $]a, b\rangle$. In particular, any continuous function from I into F is regulated.

If $f : I \to F$ is regulated, then $\int_\alpha^\beta f(t)dt$ makes sense for all α, β in I, because $f \cdot 1_{[\alpha,\beta]}$ is the uniform limit of step functions vanishing outside $[\alpha, \beta]$, and hence is λ-integrable.

Let $f : I \to F$ be regulated. Given $x_0 \in I$, the function $x \mapsto \int_{x_0}^x f(t)dt$ is continuous on I by the corollary to the dominated convergence theorem, and it has a right-hand derivative (respectively, a left-hand derivative) equal to $f(x+) = \lim_{y>x, \, y\to x} f(y)$ (respectively, to $f(x-) = \lim_{y<x, \, y\to x} f(y)$) at each point x of $\langle a, b[$ (respectively, of $]a, b\rangle$). In particular, it is continuous and it has a derivative equal to $f(x)$ at all points x that do not belong to some countable subset of I. Therefore, $x \mapsto \int_{x_0}^x f(t)dt$ is a primitive of f, and we conclude that $\int_\alpha^\beta f(t)dt = G(\beta) - G(\alpha)$ for all α, β in I and all primitives G of f. This makes it possible to compute $\int_\alpha^\beta f(t)dt$ for many regulated functions f.

Finally, a regulated function $f : I \longrightarrow F$ is integrable if and only if $\int_\alpha^\beta |f(t)|dt = \int^* |f| \cdot 1_{[\alpha,\beta]}d\lambda$ has a finite limit as $\alpha \to a$ and $\beta \to b$ in I, in which case

$$\int f d\lambda = \lim_{\substack{\alpha\to a, \, \beta\to b \\ \alpha\in I, \, \beta\in I}} \int_\alpha^\beta f(t)dt$$

(by the corollary to the dominated convergence theorem). Taking $I = [0, +\infty[$, therefore, $f : x \mapsto \sin x/x$ is not λ-integrable, even though $\int_0^x \sin t/t\, dt$ has a limit as $x \to +\infty$.

Now we give an example where the dominated convergence theorem does not apply. Let λ be Lebesgue measure on $]0, 1]$. Given $\alpha > 0$, put $f_n = n^\alpha \cdot 1_{]0,1/n]}$ for each $n \in \mathbf{N}$. Then $f :]0, 1] \ni x \mapsto [1/x]^\alpha$ is the supremum of the f_n (where $[1/x]$ designates the integral part of $1/x$). If $\alpha \geq 1$, there is no λ-integrable function which dominates all the f_n. If $\alpha = 1$, $\int f_n \, d\lambda$ does not converge to 0 as $n \to +\infty$.

Next, let A be the set of all finite subsets of $]0,1]$. Order A by inclusion, and let $\mathcal{F}$ be the filter of sections of A. For each finite subset J of $]0,1]$, denote by 1_J the indicator function of J on $]0,1]$. Then 1_J converges pointwise to 1 along $\mathcal{F}$, but $\int 1_J d\lambda = 0$ does not tend to $\int 1 d\lambda = 1$. Compare this result with the corollary to the dominated convergence theorem.

In our last example, we apply Theorem 3.2.3. Let P_1 (respectively, P_2) be the function on $\mathbf{R}$, periodic with period 1, defined by $P_1(x) = x - 1/2$ (respectively, $P_2(x) = x^2 - x + 1/6$) for $x \in [0,1[$. Hence P_1 is regulated and P_2 is continuous. Now let $\Omega_0 = \{z \in \mathbf{C} : \mathrm{Re}\, z > 0\}$ and $\Omega_1 = \{z \in \mathbf{C} : \mathrm{Re}\, z > 1\}$. Given $s \in \Omega_1$, consider the function $f : x \mapsto (x+1)^{-s}$ on $[0, +\infty[$. Write $[x]$ for the integral part of x. Then, for every $n \in \mathbf{N}$,

$$\int_0^n \left(x - [x] - \frac{1}{2} \right) \cdot f'(x) dx$$

$$= \sum_{0 \le m < n} \int_0^1 \left(t - \frac{1}{2} \right) \cdot f'(m+t) dt$$

$$= \sum_{0 \le m < n} \left(t - \frac{1}{2} \right) \cdot f(m+t) \Big|_0^1 - \sum_{0 \le m < n} \int_0^1 f(m+t) dt$$

$$= \sum_{0 \le m < n} \left(\frac{1}{2} f(m+1) + \frac{1}{2} f(m) \right) - \int_0^n f(x) dx$$

$$= \frac{1}{2} f(0) + f(1) + \ldots + f(n-1) + \frac{1}{2} f(n) - \int_0^n f(x) dx,$$

hence

$$\sum_{0 \le m \le n} f(m) = \int_0^n f(x) dx + \frac{1}{2}[f(0) + f(n)] + \int_0^n P_1(x) \cdot f'(x) dx.$$

Letting n tend to $+\infty$,

$$\zeta(s) = \sum_{m \ge 0} (m+1)^{-s} = \frac{1}{s-1} + \frac{1}{2} - s \cdot \int_1^{+\infty} P_1(t) \cdot t^{-s-1} dt,$$

or

$$\zeta(s) - \frac{1}{s-1} = \frac{1}{2} - s \cdot \int_1^{+\infty} P_1(t) \cdot t^{-s-1} dt.$$

Next, the function $s \mapsto \int_1^{+\infty} P_1(t) \cdot t^{-s-1} dt$ is defined and holomorphic in Ω_0. Therefore, there is a holomorphic extension of ζ to $\Omega_0 - \{1\}$, still written ζ, and

$$\zeta'(s) = -(s-1)^{-2} + \int_1^{+\infty} P_1(t) \cdot t^{-s-1}(s \cdot \log t - 1) dt$$

for all $s \in \Omega_0 - \{1\}$. Given $s \in]0, 1[$, let g be the function

$$t \longmapsto t^{-s-1}(s \cdot \log t - 1)$$

from $]0, +\infty[$ into $\mathbf{R}$. Since $b_2 - P_2$ is positive (where $b_2 = 1/6$) and $g'(t) \le 3 \cdot n^{-2}$ for all $t \in [n, n+1]$,

$$\int_n^{n+1} P_1(t)g(t)dt = -\frac{1}{2} \cdot \int_n^{n+1} (b_2 - P_2)'(t) \cdot g(t)dt$$

$$= \frac{1}{2} \int_n^{n+1} (b_2 - P_2)(t) \cdot g'(t)dt$$

is less than $n^{-2}/4$ for each $n \in \mathbf{N}$. Hence $\int_1^{+\infty} P_1(t)g(t)dt$ is less than $\pi^2/24$, and $\zeta'(s)$ is less than $-1 + \pi^2/24$. This shows that ζ is strictly decreasing on $]0, 1[$.

3.3   Integrable Sets

Let μ be a Daniell measure on a space $\mathcal{H}(\Omega, \mathbf{C})$.

Definition 3.3.1 A subset E of Ω is said to be μ-integrable if its indicator function 1_E is μ-integrable. Then we put $\mu(E) = \int 1_E d\mu$.

Call $\widehat{\mathcal{R}}$ the class of integrable sets. For all $A, B \in \widehat{\mathcal{R}}$

$$1_{A \cap B^c} = 1_A - \inf(1_A, 1_B) \quad \text{and} \quad 1_{A \cup B} = \sup(1_A, 1_B),$$

so $A \cap B^c$ and $A \cup B$ belong to $\widehat{\mathcal{R}}$. Hence $\widehat{\mathcal{R}}$ is a ring. Moreover, for every sequence $(A_n)_{n \ge 1}$ in $\widehat{\mathcal{R}}$, $\bigcap_{n \ge 1} A_n$ belongs to $\widehat{\mathcal{R}}$, because

$$1_{\bigcap_{n \ge 1} A_n} = \inf_{n \ge 1} 1_{A_1 \cap \dots \cap A_n}$$

(Theorem 3.2.1). If $(A_n)_{n \ge 1}$ is a sequence in $\widehat{\mathcal{R}}$ such that $\sum_{n \ge 1} V\mu(A_n) < +\infty$, then $\bigcup_{n \ge 1} A_n$ lies in $\widehat{\mathcal{R}}$ by the monotone convergence theorem, because

$$\int 1_{\bigcup_{1 \le k \le n} A_k} dV\mu \le \sum_{1 \le k \le n} \int 1_{A_k} \cdot dV\mu \le \sum_{k \ge 1} V\mu(A_k)$$

for all integers $n \ge 1$. For any sequence $(A_n)_{n \ge 1}$ of integrable subsets of a given integrable set, $A = \bigcup_{n \ge 1} A_n$ is μ-integrable; moreover, $\mu(A) = \sum_{n \ge 1} \mu(A_n)$ if the A_n are disjoint.

Proposition 3.3.1 *Suppose μ is real. For every integrable set E, there exist integrable subsets E^-, E^+ of E such that $\mu(E^-) = \inf_{G \subset E} \mu(G)$ and $\mu(E^+) = \sup_{G \subset E} \mu(G)$, where G extends over the class of all integrable subsets of E. Furthermore, we may choose E^-, E^+ so that $E^- \cap E^+ = \emptyset$ and $E = E^+ \cup E^-$.*

PROOF: Let $(F_m)_{m\geq 1}$ be a sequence of integrable subsets of E, such that $\mu(F_m)$ converges to $\sup_{G\subseteq E}\mu(G)$ as $m \to +\infty$, and put $F = \bigcup_{m\geq 1} F_m$. For each fixed integer $m \geq 1$, the sets F_I^m $(I \subset I_m = \{1, 2, \ldots, m\})$, where

$$F_I^m = F \cap \left(\bigcap_{i\in I} F_i\right) \cap \left(\bigcap_{i\in I_m - I} F_i^c\right),$$

form a finite partition $P_m(F)$ of F into integrable sets. Observe that $F_I^m = (F_I^m \cap F_{m+1}^c) \cup (F_I^m \cap F_{m+1}) = F_I^{m+1} \cup F_{I\cup\{m+1\}}^{m+1}$, so that each set in $P_m(F)$ can be partitioned into sets in $P_{m+1}(F)$.

Denote by E_m the union of those F_I^m $(I \subset I_m)$ such that $\mu(F_I^m) \geq 0$. Then, for every $1 \leq i \leq m$,

$$\mu(F_i) = \sum_{F_I^m \subset F_i} \mu(F_I^m) \leq \sum_{F_I^m \subset F_i,\ \mu(F_I^m)\geq 0} \mu(F_I^m) \leq \mu(E_m).$$

Moreover,

$$\mu(E_m) \leq \mu(E_m \cup E_{m+1}) \leq \ldots \leq \mu\left(\bigcup_{j\geq m} E_j\right),$$

because each $E_m \cup E_{m+1} \cup \ldots \cup E_{m+k}$ is partitioned by $E_m \cup \ldots \cup E_{m+k-1}$ and by those F_I^{m+k} of positive measure disjoint from $E_m \cup \ldots \cup E_{m+k-1}$. This implies that

$$\mu(F_i) \leq \lim_{m\to+\infty} \mu\left(\bigcup_{j\geq m} E_j\right) = \mu\left(\bigcap_{m\geq 1}\bigcup_{j\geq m} E_j\right)$$

for every $i \geq 1$. Put

$$E^+ = \limsup_{m\to+\infty} E_j$$
$$= \bigcap_{m\geq 1}\bigcup_{j\geq m} E_j.$$

Then

$$\sup_{G\subseteq E} \mu(G) = \lim_{i\to+\infty} \mu(F_i) \leq \mu(E^+),$$

and since $E^+ \subset E$ this is an equality.

Finally, $\inf_{G\subseteq E}\mu(G) = \inf_{G\subseteq E}\mu(E - G) = \mu(E) - \mu(E^+) = \mu(E - E^+)$, and so we may take $E^- = E - E^+$.   □

Observe that $\mu(G) \geq 0$ for $G \subset E^+$ and $\mu(G) \leq 0$ for $G \subset E^-$. Indeed, if $G \subset E^+$ for example, then

$$\mu(G) = \mu(E^+) - \mu(E^+ - G) \geq \mu(E^+) - \mu(E^+) = 0.$$

Note that Stone's condition has not yet been used.

Proposition 3.3.2 *Let f be an integrable function from Ω into $\mathbf{R}^+ = [0, +\infty[$. Then $\inf(f, 1)$ is integrable, and $f^{-1}(]r, +\infty[)$ is an integrable set for every real number $r > 0$.*

PROOF: There is a sequence $(f_n)_{n \geq 1}$ in $\mathcal{H}^+$ converging to f in the mean. Therefore, $\int^* |\inf(f_n, 1) - \inf(f, 1)| \cdot dV\mu \leq \int^* |f_n - f| \cdot dV\mu$ goes to 0 as $n \to +\infty$, and $\inf(f, 1)$ is integrable. Now, letting $r > 0$ be a real number and putting $A = f^{-1}(]r, +\infty[)$, the functions

$$h_n = \inf\left(1, n \cdot \left[f - r \cdot \inf\left(\frac{f}{r}, 1 \right) \right]\right)$$

are integrable, dominated by f/r, and converge pointwise to 1_A. Thus 1_A is integrable.   □

Proposition 3.3.3   *Let $f : \Omega \longrightarrow [0, +\infty]$ be a function. For every integer $n \geq 0$, put $f_n = \sum_{1 \leq k \leq n \cdot 2^n + 1} ((k-1)/2^n) \cdot 1_{A_{k,n}}$, where $A_{k,n} = f^{-1}([(k-1)/2^n, k/2^n[)$ for $1 \leq k \leq n \cdot 2^n$ and $A_{n \cdot 2^n + 1, n} = f^{-1}([n, +\infty])$. Then the sequence $(f_n)_{n \geq 1}$ increases to f.*

PROOF: Obvious.   □

Theorem 3.3.1 *Let F be a real Banach space. Then $St(\widehat{\mathcal{R}}, F)$ is dense in $\mathcal{L}^1_F(\mu)$.*

PROOF: First, let $f : \Omega \to \mathbf{R}^+$ be integrable. In the notation of Proposition 3.3.3, the functions f_n lie in $St^+(\widehat{\mathcal{R}})$. Since the sequence $(f_n)_{n \geq 1}$ increases, it converges in the mean to f.

Now, since the set of mappings $\sum_{i \in I} a_i g_i$ (I finite, $a_i \in F$, $g_i \in \mathcal{H}^+$) is dense in $\mathcal{L}^1_F(\mu)$, $St(\widehat{\mathcal{R}}, F)$ itself is dense in $\mathcal{L}^1_F(\mu)$.   □

Now let F be a real Banach space, $f \in \mathcal{L}^1_F(\mu)$, and let A be a μ-integrable set. For every $\varepsilon > 0$, there exists $g \in St(\widehat{\mathcal{R}}, F)$ such that $\int |f - g| dV\mu \leq \varepsilon$. Then $\int^* |f 1_A - g 1_A| dV\mu \leq \varepsilon$ and $g 1_A$ lies in $St(\widehat{\mathcal{R}}, F)$. We may conclude that $f 1_A$ is μ-integrable.

Proposition 3.3.4 $\int 1_A dV\mu = \sup_{g \in \mathcal{H}(\Omega, \mathbf{C}), |g| \leq 1} |\int g 1_A d\mu|$ *for every integrable set A.*

PROOF: Given $\varepsilon > 0$, let $f \in \mathcal{H}^+$ be such that $\int^* |1_A - f| \cdot dV\mu \leq \varepsilon/2$. Replacing f by $\inf(f, 1)$, if necessary, we may suppose that $f \leq 1$. Then let $g \in \mathcal{H}(\Omega, \mathbf{C})$ be such that $|g| \leq f$ and $|\int g d\mu| \geq \int f dV\mu - \varepsilon/2$. In this case,

$$\left| \int g 1_A \cdot d\mu \right| \geq \left| \int g d\mu \right| - \left| \int g 1_{\Omega - A} d\mu \right|$$

$$\geq \left(\int f dV\mu - \frac{\varepsilon}{2} \right) - \int f 1_{\Omega - A} dV\mu = \int f 1_A dV\mu - \frac{\varepsilon}{2}.$$

But

$$\int f 1_A dV\mu = \int (f - 1_A)1_A dV\mu + \int 1_A dV\mu \geq \int 1_A dV\mu - \frac{\varepsilon}{2},$$

so

$$\left| \int g 1_A d\mu \right| \geq \int 1_A dV\mu - \varepsilon.$$

$\square$

Proposition 3.3.5 $\int 1_A dV\mu = \sup_{\substack{h \in St(\widehat{\mathcal{R}}, \mathbf{C}), \\ |h| \leq 1}} \left| \int h 1_A d\mu \right|$ *for every integrable set A.*

PROOF: Given $\varepsilon > 0$, let $g \in \mathcal{H}(\Omega, \mathbf{C})$ be such that $|g| \leq 1$ and $|\int g 1_A d\mu| \geq \int 1_A dV\mu - \varepsilon/2$ (Proposition 3.3.4). Since $St(\widehat{\mathcal{R}}, \mathbf{C})$ is dense in $\mathcal{L}^1_{\mathbf{C}}(\mu)$, there exists $h \in St(\widehat{\mathcal{R}}, \mathbf{C})$ such that $\int |g - h| \cdot dV\mu \leq \varepsilon/2$. Replacing h by $p \circ h$, if necessary, where p is the projection from $\mathbf{C}$ onto the closed convex set $\bar{D}(0,1) = \{z \in \mathbf{C} : |z| \leq 1\}$, we may suppose $|h| \leq 1$. Then

$$\left| \int h 1_A d\mu \right| \geq \left| \int g 1_A d\mu \right| - \left| \int (g - h)1_A d\mu \right| \geq \int 1_A \cdot dV\mu - \varepsilon;$$

whence the result.

$\square$

Theorem 3.3.2 *For every integrable set A, $V\mu(A) = \sup_{(A_i)_i} \sum_{i \in I} |\mu(A_i)|$, where $(A_i)_{i \in I}$ extends over the class of all finite partitions of A into integrable sets.*

PROOF: Given $\varepsilon > 0$, let $h \in St(\widehat{\mathcal{R}}, \mathbf{C})$ be such that $|h| \leq 1$ and $|\int h 1_A d\mu| \geq \int 1_A dV\mu - \varepsilon$. Now h can be written $\sum_{1 \leq i \leq n} 1_{B_i} \cdot y_i$, where the B_i are disjoint integrable sets and the y_i belong to $\bar{D}(0, 1)$. Then

$$\sum_{1 \leq i \leq n} |\mu(A \cap B_i)| \geq \left| \sum_{1 \leq i \leq n} \mu(A \cap B_i) y_i \right| = \left| \int h 1_A d\mu \right| \geq \int 1_A dV\mu - \varepsilon.$$

Hence $\sup_{(A_i)_i} \sum_{i \in I} |\mu(A_i)| \geq \int 1_A \cdot dV\mu$, and the reverse inequality is obvious.

$\square$

Theorem 3.3.2 will be used when we study atomic measures.

Definition 3.3.2 A subset A of Ω is called μ-moderate if it is contained in a countable union of μ-integrable sets. A mapping from Ω into $\overline{\mathbf{R}}$ or a vector space is μ-moderate if it vanishes outside a μ-moderate set.

Proposition 3.3.6 *If $f : \Omega \to [0, +\infty]$ is such that $\int^* f \cdot dV\mu < +\infty$, then f is moderate.*

PROOF: Let h be an integrable function from Ω into $[0, +\infty]$ such that $f \leq h$ (Theorem 3.1.4). Then $h^{-1}(]1/n, +\infty])$ is integrable for every $n \geq 1$, and f vanishes outside $\bigcup_{n \geq 1} h^{-1}(]1/n, +\infty])$. □

3.4 σ-Measurable Spaces

The pair $(\Omega, \mathcal{F})$ of a nonempty set Ω and a σ-ring $\mathcal{F}$ in Ω is called a σ-measurable space.

Definition 3.4.1 Let $(\Omega, \mathcal{F})$ and $(\Omega', \mathcal{F}')$ be two σ-measurable spaces. A mapping $f : \Omega \to \Omega'$ is said to be measurable $\mathcal{F}/\mathcal{F}'$ whenever $f^{-1}(A') \in \mathcal{F}$ for each $A' \in \mathcal{F}'$.

If $(\Omega, \mathcal{F})$, $(\Omega', \mathcal{F}')$, and $(\Omega'', \mathcal{F}'')$ are σ-measurable spaces, f a mapping measurable $\mathcal{F}/\mathcal{F}'$, and g a mapping measurable $\mathcal{F}'/\mathcal{F}''$, then, clearly, $g \circ f$ is measurable $\mathcal{F}/\mathcal{F}''$.

Proposition 3.4.1 *Let Ω, Ω' be two nonempty sets and $f : \Omega \to \Omega'$ a mapping.*

1. *Let $\mathcal{F}$ be a σ-ring in Ω. The class $\mathcal{F}_f$ of all subsets A' of Ω' for which $f^{-1}(A')$ lies in $\mathcal{F}$ is a σ-ring, called the image σ-ring of $\mathcal{F}$ under f. $\mathcal{F}_f$ is the largest σ-ring $\mathcal{F}'$ (the ordering is inclusion) in Ω' such that f is measurable $\mathcal{F}/\mathcal{F}'$.*

2. *Let $\mathcal{F}'$ be a σ-ring in Ω'. The class $f^{-1}(\mathcal{F}') = \{f^{-1}(A') : A' \in \mathcal{F}'\}$ is a σ-ring in Ω, called the σ-ring generated by f. $f^{-1}(\mathcal{F}')$ is the smallest σ-ring $\mathcal{F}$ in Ω such that f is measurable $\mathcal{F}/\mathcal{F}'$.*

PROOF: Obvious. □

Proposition 3.4.2 *Let $(\Omega, \mathcal{F})$, $(\Omega', \mathcal{F}')$ be two σ-measurable spaces, and let C' be a collection of subsets of Ω' such that $\mathcal{F}'$ is the σ-ring generated by C'. Then f is measurable $\mathcal{F}/\mathcal{F}'$ whenever $f^{-1}(A')$ lies in $\mathcal{F}$ for all $A' \in C'$.*

PROOF: In the notation of Proposition 3.4.1, $\mathcal{F}_f$ contains C' and, thus, $\mathcal{F}'$. □

Definition 3.4.2 Let Ω be a nonempty set, $((\Omega_i, \mathcal{F}_i))_{i \in I}$ be a nonempty family of σ-measurable spaces, and, for each $i \in I$, let f_i be a mapping from Ω into Ω_i. The σ-ring $\mathcal{F}$ in Ω generated by $\{f_i^{-1}(A_i) : i \in I, A_i \in \mathcal{F}_i\}$ is called the σ-ring generated by the family $(f_i)_{i \in I}$.

Proposition 3.4.3 *In the notation of Definition 3.4.2, $\mathcal{F}$ is the smallest of those σ-rings $\mathcal{G}$ in Ω such that each f_i is measurable $\mathcal{G}/\mathcal{F}_i$. Let $(\Omega', \mathcal{F}')$ be a σ-measurable space and h be a mapping from Ω' into Ω. Then h is measurable $\mathcal{F}'/\mathcal{F}$ if and only if $f_i \circ h$ is measurable $\mathcal{F}'/\mathcal{F}_i$ for each i.*

PROOF: By Proposition 3.4.2, h is measurable $\mathcal{F}'/\mathcal{F}$ if and only if $(f_i \circ h)^{-1}(A_i)$ lies in $\mathcal{F}'$ for every $i \in I$ and every $A_i \in \mathcal{F}_i$.   □

If $\mathcal{F}_i$ is the σ-ring generated by a class $\mathcal{C}_i$ of subsets of Ω_i, then $\mathcal{F}$ is the σ-ring generated by $\left\{ f_i^{-1}(A_i) : i \in I, \ A_i \in \mathcal{C}_i \right\}$.

Definition 3.4.3 Let $(\Omega, \mathcal{F})$ be a σ-measurable space and E a nonempty subset of Ω. The σ-ring $\mathcal{F}/E$ generated by the canonical injection from E into Ω is called the σ-ring induced by $\mathcal{F}$ on E.

$\mathcal{F}/E$ is simply the class consisting of the traces of the $\mathcal{F}$-sets on E.

When Ω is a topological space and $\mathcal{F}$ its Borel σ-algebra, $\mathcal{F}/E$ is the Borel σ-algebra of E.

Definition 3.4.4 Let $\left((\Omega_i, \mathcal{F}_i) \right)_{i \in I}$ be a nonempty family of σ-measurable spaces. Consider the product set $\Omega = \prod_{i \in I} \Omega_i$ and the canonical projection p_i from Ω onto the factor Ω_i. The σ-ring $\mathcal{F}$ in Ω generated by the family $(p_i)_{i \in I}$ of mappings is called the product σ-ring of the $\mathcal{F}_i$ and is written $\bigotimes_{i \in I} \mathcal{F}_i$.

Theorem 3.4.1 *Let $(\Omega_i)_{i \in I}$ be a family of topological spaces. For each i, let $\mathcal{B}_i$ be the Borel σ-algebra of Ω_i. Also, let Ω be the product topological space $\prod_{i \in I} \Omega_i$ and $\mathcal{B}$ its Borel σ-algebra. Then $\bigotimes_{i \in I} \mathcal{B}_i \subset \mathcal{B}$. In particular, if I is (at most) countable and each topological space Ω_i has a countable basis, then $\mathcal{B} = \bigotimes_{i \in I} \mathcal{B}_i$.*

PROOF: Since p_i is measurable $\mathcal{B}/\mathcal{B}_i$ for each i, the identity of Ω is measurable $\mathcal{B}/\bigotimes_{i \in I} \mathcal{B}_i$, and so $\bigotimes_{i \in I} \mathcal{B}_i \subset \mathcal{B}$.

Now suppose that I is countable and that the topology of each Ω_i has a countable basis $\mathcal{U}_i$. The subsets of Ω of the form $p_{i_1}^{-1}(V_{i_1}) \cap \ldots \cap p_{i_n}^{-1}(V_{i_n})$ ($i_1, \ldots, i_n \in I$, $V_{i_k} \in \mathcal{U}_{i_k}$ for every $1 \le k \le n$) constitute a basis for the product topology; this (countable) basis is contained in $\bigotimes_{i \in I} \mathcal{B}_i$. Now every open subset of Ω is a union (necessarily countable) of distinct elements of this basis, so it lies in $\bigotimes_{i \in I} \mathcal{B}_i$.   □

The pair of a nonempty set Ω and a σ-algebra $\mathcal{F}$ in Ω is called a measurable space.

Definition 3.4.5 Let $(\Omega, \mathcal{F})$ be a measurable space, Ω' a topological space, and $\mathcal{B}'$ its Borel σ-algebra. $f : \Omega \to \Omega'$ is said to be measurable $\mathcal{F}$ if it is measurable $\mathcal{F}/\mathcal{B}'$.

Theorem 3.4.2 Let $(\Omega, \mathcal{F})$ be a measurable space, Ω' a metrizable space, and $(f_n)_{n \geq 1}$ a sequence of mappings from Ω into Ω' which converges pointwise to a mapping f. Suppose that each f_n is measurable $\mathcal{F}$. Then f is measurable $\mathcal{F}$.

PROOF: Consider on Ω' a distance d' compatible with the topology of Ω'. First, we prove that, for every open subset V of Ω',

$$f^{-1}(V) \subset \liminf f_n^{-1}(V) \subset f^{-1}(\bar{V}), \tag{1}$$

where $\liminf f_n^{-1}(V) = \bigcup_{n \geq 1} \bigcap_{p \geq n} f_p^{-1}(V)$.

Indeed, if $x \in f^{-1}(V)$, since $f(x) = \lim_{p \to +\infty} f_p(x)$ and V is open, $f_p(x)$ belongs to V for p large enough, and the first inclusion obtains. Similarly, if $x \in \liminf f_n^{-1}(V)$, there exists an integer n such that $f_p(x)$ belongs to V for all $p \geq n$, so $f(x) = \lim_{p \to +\infty} f_p(x)$ lies in $\bar{V}$. This gives the second inclusion.

Next, let F be a closed subset of Ω', and let B_k be the union of those open balls of radius $1/k$ (where $k \in \mathbf{N}$) centered at points of F. B_k is open, and $F = \bigcap_{k \geq 1} B_k = \bigcap_{k \geq 1} \bar{B}_k$. For, if $x \in \bigcap_{k \geq 1} \bar{B}_k$, then $d(x, F) \leq 1/k$ for all $k \in \mathbf{N}$, and hence $x \in \bar{F} = F$. Now

$$\bigcap_{k \geq 1} f^{-1}(B_k) = f^{-1}\left(\bigcap_{k \geq 1} B_k\right) = f^{-1}(F) = f^{-1}\left(\bigcap_{k \geq 1} \bar{B}_k\right) = \bigcap_{k \geq 1} f^{-1}(\bar{B}_k).$$

Applying relation (1) to B_k, we obtain the inclusion

$$f^{-1}(B_k) \subset \liminf f_n^{-1}(B_k) \subset f^{-1}(\overline{B_k}).$$

So

$$f^{-1}(F) = \bigcap_{k \geq 1} f^{-1}(B_k) \subset \bigcap_{k \geq 1} \left(\liminf f_n^{-1}(B_k)\right) \subset \bigcap_{k \geq 1} f^{-1}(\overline{B_k}) = f^{-1}(F),$$

which says $f^{-1}(F) = \bigcap_{k \geq 1} \left(\liminf f_n^{-1}(B_k)\right)$, and this lies in $\mathcal{F}$.   □

3.5 Measurable Mappings

Let μ and $\mathcal{J}^+$ be as in Section 3.1.

Definition 3.5.1 A subset A of Ω is said to be μ-measurable if $A \cap B$ is μ-integrable for all μ-integrable sets B.

The class $\mathcal{M}$ of μ-measurable sets is obviously a σ-algebra.

Definition 3.5.2 A subset N of Ω is said to be locally μ-negligible if $N \cap E$ is negligible for every integrable set E. A property is true locally almost everywhere (l.a.e.) whenever it is true outside a locally negligible set.

Definition 3.5.3 Let F be a metrizable space. A mapping f from Ω into F is said to be μ-measurable if

(a) f is measurable $\mathcal{M}$;

(b) for every integrable set E, there exists a negligible set $N \subset E$ such that the topological space $f(E - N)$ is separable.

Observe that every mapping from Ω into F which agrees with f l.a.e. is μ-measurable.

A necessary and sufficient condition that a function f from Ω into $\overline{\mathbf{R}}$ be μ-measurable is that $f^{-1}(]r, +\infty])$ lie in $\mathcal{M}$ for all real numbers r. Indeed, the Borel σ-algebra of $\overline{\mathbf{R}}$ is the σ-ring generated by the class consisting of $\overline{\mathbf{R}}$ and the sets $]r, +\infty]$.

Proposition 3.5.1 *Each $f \in \mathcal{J}^+$ is μ-measurable.*

PROOF: Let $r > 0$ be a real number and let $E \in \widehat{\mathcal{R}}$. There exists $g \in \mathcal{J}^+$ such that $g \geq 1_E$ and $\int^* g \cdot dV\mu < +\infty$. If $n > r$ is an integer, $\inf(f, ng)$ belongs to $\mathcal{J}^+$ and is integrable (Theorem 3.1.4). Therefore, $f^{-1}(]r, +\infty]) \cap E = E \cap \inf(f, ng)^{-1}(]r, +\infty])$ is μ-integrable. $\square$

Definition 3.5.4 Let E be a nonempty set and S a semialgebra in E (i.e., a semiring containing E). A mapping f from E into a set Ω' is said to be S-simple whenever there is a finite partition $(E_i)_{i \in I}$ of E into S-sets such that f is constant on each E_i.

Proposition 3.5.2 *A mapping f from Ω into a metrizable space F is μ-measurable if and only if, for each integrable subset E of Ω, f is the limit a.e. in E of a sequence of $\widehat{\mathcal{R}}/E$-simple mappings.*

PROOF: Let d be a distance on F (compatible with the topology of F). If f is μ-measurable and E is an integrable set, there exists a negligible set $N \subset E$ such that $f(E-N)$ is separable. Then, letting $\{y_i : i \geq 1\}$ be a countable dense subset of $f(E-N)$, we have, for every integer $n \geq 1$, $f(E - N) \subset \bigcup_{i \geq 1} B(y_i, 1/n)$, or $E - N \subset \bigcup_{i \geq 1} f^{-1}(B(y_i, 1/n))$, where $B(y_i, 1/n)$ designates the open ball with center y_i and radius $1/n$. Hence there exists an integer k_n such that the $V\mu$-measure of $Y_n = E - \bigcup_{1 \leq i \leq k_n} f^{-1}(B(y_i, 1/n))$ is less than $1/2^n$. Now put $Z_n = \bigcup_{m \geq n} Y_m$, and define an $\widehat{\mathcal{R}}/E$-simple mapping ψ_n:

$$\psi_n(x) = \begin{cases} y_0 \in F & \text{if } x \in A_0 = Z_n, \quad \text{where } y_0 \text{ is fixed but arbitrary} \\ y_j & \text{if } x \in A_j = E \cap \left(\bigcup_{0 \leq i \leq j-1} A_i\right)^c \cap f^{-1}(B(y_j, 1/n)) \\ & \text{for } 1 \leq j \leq k_n. \end{cases}$$

Then the sequence $(\psi_n)_{n \geq 1}$ converges to f at each point of $E - \bigcap_{n \geq 1} Z_n$, and $V\mu(\bigcap_{n \geq 1} Z_n) = 0$.

Conversely, suppose that the condition is satisfied, and let E be an integrable set. We can find a sequence $(\varphi_i)_{i \geq 1}$ of $\widehat{\mathcal{R}}/E$-simple mappings from E into F, which converges to f/E outside a negligible set $N \subset E$. Fix $y \in F$ and, for each $i \geq 1$, let φ_i' be the mapping from E into F which agrees with φ_i on $E - N$ and which is equal to y on N. Then $(\varphi_i')_{i \geq 1}$ converges to the mapping g from E into F which agrees with f on $E - N$ and is equal to y on N, so g, and hence f/E, is measurable $\widehat{\mathcal{R}}/E$. Moreover, $f(E-N)$ is separable, because it is contained in the closure of the countable set $\bigcup_{i \geq 1} \varphi_i(E - N)$. Finally, for every Borel set B in F, $f^{-1}(B) \cap E$ is integrable for all $E \in \widehat{\mathcal{R}}$; so $f^{-1}(B)$ lies in $\mathcal{M}$.   □

Proposition 3.5.3 *Let f_n be a μ-measurable mapping from Ω into a metrizable space F_n, for each integer $n \geq 1$, and put $f = [f_n]_{n \geq 1}$. If u is a continuous mapping from $f(\Omega)$ into a metrizable space G, then $u \circ f$ is μ-measurable.*

PROOF: Let E be an integrable set, and let N be a negligible subset of E such that $f_n(E - N)$ is separable for every $n \geq 1$. For each $n \geq 1$, let $\mathcal{B}_n$ be the Borel σ-algebra of $f_n(E-N)$. If W is an open subset of G, there exists an open subset V of $\Pi_{n \geq 1} f_n(E - N)$ such that $f(E - N) \cap u^{-1}(W) = f(E - N) \cap V$. But V belongs to $\bigotimes_{n \geq 1} \mathcal{B}_n$ (Theorem 3.4.1) and each $f_n/E - N$ is measurable $(\widehat{\mathcal{R}}/E - N)/\mathcal{B}_n$; so $(E - N) \cap f^{-1}(V) = (E - N) \cap (u \circ f)^{-1}(W)$ lies in $\widehat{\mathcal{R}}$. Hence $(u \circ f)^{-1}(W)$ lies in $\mathcal{M}$, and $u \circ f$ is measurable $\mathcal{M}$.

Moreover, $(u \circ f)(E - N)$, like $f(E - N)$, is separable.   □

Theorem 3.5.1 (Egorov's Theorem)   *Let $(f_n)_{n \geq 1}$ be a sequence of μ-measurable mappings from Ω into a metrizable uniform space F, converging l.a.e. to a mapping f. Then f is μ-measurable. Moreover, for every integrable set E and every $\varepsilon > 0$, there exists an integrable set $B \subset E$ such that $V\mu(B) \leq \varepsilon$ and such that the sequence $(f_n)_{n \geq 1}$ converges uniformly to f on $E - B$.*

PROOF: Let N be a locally negligible set such that $(f_n)_{n \geq 1}$ converges to f outside N, and let d be a distance on F. For every pair of integers $n \geq 1$, $r \geq 1$, let $B_{n,r}$ be the set of points $x \in E - N$ such that $d(f_\alpha(x), f_\beta(x)) \geq 1/r$ for at least one pair of integers $\alpha \geq n$, $\beta \geq n$. For fixed $\alpha \geq n$, $\beta \geq n$, the set of those points $x \in E - N$ such that $d(f_\alpha(x), f_\beta(x)) \geq 1/r$ is μ-measurable (by Proposition 3.5.3), and hence is integrable. Consequently, $B_{n,r}$ is the countable union of integrable sets contained in $E - N$, and $B_{n,r}$ is integrable. If we fix r, the intersection of the decreasing sequence $(B_{n,r})_{n \geq 1}$ is empty, because $f_\alpha(x)$ converges to $f(x)$ as $\alpha \to +\infty$ for every $x \in E - N$. Therefore, $\lim_{n \to +\infty} V\mu(B_{n,r}) = 0$, and we can find an integer n_r such that $V\mu(B_{n_r,r}) \leq \varepsilon/2^r$. The union, B, of $E \cap N$ and the sets $B_{n_r,r}$ $(r \geq 1)$

is integrable, and $V\mu(B) \leq \sum_{r\geq 1} V\mu(B_{n_r,r}) \leq \varepsilon$. By construction, $(f_\alpha)_\alpha$ converges uniformly to f on $E-B$ as $\alpha \to +\infty$. Finally, $f/E-N$ is measurable $\widehat{\mathcal{R}}/E - N$, hence f is measurable $\mathcal{M}$. Now we can choose N so that each $f_n(E-N)$ is separable, and hence $f(E-N)$ is separable. $\square$

Proposition 3.5.4 *Let $(f_m)_{m\geq 1}$ be a sequence of μ-measurable mappings from Ω into a metrizable space F, converging l.a.e. to a mapping f. For each $m \geq 1$, let $(f_{m,n})_{n\geq 1}$ be a sequence of μ-measurable mappings from Ω into F, converging l.a.e. to f_m. Then, for each moderate subset A, there exists a strictly increasing sequence $(n(p))_{p\geq 1}$ of integers ≥ 1, such that $\left(f_{p,n(p)}\right)_{p\geq 1}$ converges to f a.e. on A.*

PROOF: First, suppose that A is integrable. Let $\varepsilon > 0$ be given. For every $m \geq 1$, there exists an integrable subset Z_m of A such that $V\mu(Z_m) \leq \varepsilon/2^m$ and such that the sequence $(f_{m,n})_{n\geq 1}$ converges uniformly to f_m on $A - Z_m$; then, putting $Z = \bigcup_{m\geq 1} Z_m$, we have $V\mu(Z) \leq \varepsilon$, and each sequence $(f_{m,n})_{n\geq 1}$ converges uniformly to f_m on $A - Z$.

Now we pass to the case where A is the union of a sequence $(A_i)_{i\geq 1}$ of integrable sets, which we may take disjoint. By what we have just shown, there exists a decreasing sequence $(Y_{i,p})_{p\geq 1}$ of integrable subsets of A_i, such that

(a) $\bigcap_{p\geq 1} Y_{i,p}$ is negligible;

(b) for every $m \geq 1$, the sequence $(f_{m,n})_{n\geq 1}$ converges uniformly to f_m on each $A_i - Y_{i,p}$ (with $p \geq 1$).

Put $Y_p = \bigcup_{i\geq 1} Y_{i,p}$, for all $p \geq 1$, so that $\bigcap_{p\geq 1} Y_p$ is negligible. By induction on p, we construct a strictly increasing sequence $(n(p))_{p\geq 1}$ of integers ≥ 1 such that $d\left(f_{p,n(p)}, f_p\right) \leq 1/p$ on $\bigcup_{1\leq i\leq p}(A_i - Y_{i,p})$. Next, let N be a negligible subset of A such that $(f_m)_{m\geq 1}$ converges to f on $A - N$. Given x in $A - [N \cup (\bigcap_{p\geq 1} Y_p)]$, x belongs to A_i for a suitable $i \geq 1$, and there exists $p_0 \geq i$ such that x does not lie in Y_{p_0}. Then, for all $p \geq p_0$, we have $d\left(f_{p,n(p)}(x), f_p(x)\right) \leq 1/p$. Hence $f(x)$ is the limit of the $f_{p,n(p)}(x)$ as $p \to +\infty$, and the proof is complete. $\square$

Proposition 3.5.5 *Let F be a real Banach space. A mapping f from Ω into F is μ-measurable and μ-moderate if and only if it is the limit a.e. of a sequence in $St(\widehat{\mathcal{R}}, F)$.*

PROOF: By Proposition 3.5.2, the condition is sufficient. Conversely, suppose that f is measurable and moderate, and let $(A_i)_{i\geq 1}$ be a sequence of disjoint integrable sets, such that f vanishes outside $\bigcup_{i\geq 1} A_i$. For every $i \geq 1$, f/A_i is a.e. in A_i the limit of a sequence $(f_{i,n})_{n\geq 1}$ of $\widehat{\mathcal{R}}/A_i$-simple mappings (Proposition 3.5.2). For every $n \geq 1$, define g_n to be the mapping from Ω into F

that agrees with $f_{i,n}$ on each A_i for which $i \leq n$, and that vanishes outside $\bigcup_{1 \leq i \leq n} A_i$. Then $(g_n)_{n \geq 1}$ converges to f a.e. $\square$

Theorem 3.5.2 *Let V be a dense vector subspace of $\mathcal{L}_F^1(\mu)$, where F is a real Banach space. A mapping f from Ω into F is μ-measurable and μ-moderate if and only if it is the limit a.e. of a sequence in V.*

PROOF: By Proposition 3.5.5, Theorem 3.3.1, and Proposition 3.1.9, each f in $\mathcal{L}_F^1(\mu)$ is μ-measurable; thus the condition is sufficient.

Conversely, let $f : \Omega \to F$ be measurable and moderate. By Proposition 3.5.5, f is the limit a.e. of a sequence $(f_m)_{m \geq 1}$ in $St(\widehat{\mathcal{R}}, F)$. But each f_m is the limit a.e. of a sequence in V (Proposition 3.1.9), so f is the limit a.e. of a sequence in V (Proposition 3.5.4). $\square$

Theorem 3.5.3 *A necessary and sufficient condition that a mapping f from Ω into a real Banach space F be μ-integrable is that f be μ-measurable and $\int^* |f| \cdot dV\mu$ be finite.*

PROOF: We know (Proposition 3.2.2) that f is μ-integrable if and only if $\int^* |f| \cdot dV\mu < +\infty$ and f is the limit a.e. of a sequence in $\mathcal{H}(\Omega, \mathbf{R}) \otimes F$. So Theorem 3.5.3 follows from Theorem 3.5.2. $\square$

Theorem 3.5.4 *Let $(f_n)_{n \geq 1}$ be a sequence of μ-measurable functions from Ω into $[0, +\infty]$. Then $\int^* (\sum_{n \geq 1} f_n) \cdot dV\mu = \sum_{n \geq 1} \int^* f_n \cdot dV\mu$.*

PROOF: For all μ-measurable functions f, g from Ω into $[0, +\infty]$, we have $\int^* (f+g) dV\mu = \int^* f \cdot dV\mu + \int^* g \cdot dV\mu$; indeed, this is clear if $\int^* f \cdot dV\mu = +\infty$ or $\int^* g \cdot dV\mu = +\infty$, and also if f and g are μ-integrable. Then Theorem 3.5.4 follows from Beppo Levi's theorem. $\square$

3.6   Essentially Integrable Mappings

By convention, we set $0 \times (+\infty) = 0 \times (-\infty) = 0$.

Definition 3.6.1 For all functions f from Ω into $[0, +\infty]$,

$$\int^\bullet f dV\mu = \sup_{A \in \widehat{\mathcal{R}}} \int^* f 1_A dV\mu$$

is called the upper essential integral of f (with respect to $V\mu$). Consequently, $\int^\bullet f \cdot dV\mu \leq \int^* f \cdot dV\mu$ and $\int^\bullet f \cdot dV\mu = \sup_A \int^* f 1_A dV\mu$, where A extends over the class of μ-measurable and μ-moderate sets.

If E is a subset of Ω, $\int^\bullet 1_E dV\mu = 0$ means that E is locally μ-negligible. If f and g are two functions from Ω into $[0, +\infty]$ which agree l.a.e., then $\int^\bullet f \cdot dV\mu = \int^\bullet g \cdot dV\mu$. If f, g, and h are three functions from Ω into $[0, +\infty]$ and $\lambda > 0$ is a real number, then

$$\int^\bullet (f + g) \cdot dV\mu \le \int^\bullet f \cdot dV\mu + \int^\bullet g \cdot dV\mu$$

and

$$\int^\bullet \lambda h \cdot dV\mu = \lambda \cdot \int^\bullet h \cdot dV\mu.$$

If $(f_n)_{n \ge 1}$ is an increasing sequence of functions from Ω into $[0, +\infty]$ and if $f = \sup_{n \ge 1} f_n$, then

$$\int^\bullet f \cdot dV\mu = \sup_{n \ge 1} \int^\bullet f_n \cdot dV\mu.$$

Proposition 3.6.1 *Let f, g, and h be three functions from Ω into $[0, +\infty]$, with g and h μ-measurable. Then $\int^\bullet f(g+h) \cdot dV\mu = \int^\bullet fg \cdot dV\mu + \int^\bullet fh \cdot dV\mu$.*

PROOF: It suffices to prove that

$$\int^* f(g + h) \cdot dV\mu = \int^* fg \cdot dV\mu + \int^* fh \cdot dV\mu.$$

Since $f(g + h) = fg + fh$, we have

$$\int^* f(g + h) \cdot dV\mu \le \int^* fg \cdot dV\mu + \int^* fh \cdot dV\mu,$$

and it remains to be shown that

$$\int^* fg \cdot dV\mu + \int^* fh \cdot dV\mu \le \int^* f(g + h) \cdot dV\mu.$$

We need only consider the case in which $\int^* f(g+h) \cdot dV\mu < +\infty$. Then there exists $u \in \mathcal{J}^+$ such that $u \ge f(g + h)$.

Put $v = u/(g+h)$ whenever $g + h > 0$ and $u < +\infty$, and put $v = +\infty$ where $g + h = 0$ or $u = +\infty$; then $u \ge v(g + h)$, so $\int^* v(g + h) \cdot dV\mu \le \int^* u \cdot dV\mu$. Moreover, $E = \{x \in \Omega : (g+h)(x) > 0 \text{ and } u(x) < +\infty\}$ is μ-measurable and the function $(a, b) \mapsto a/b$ from $[0, +\infty[\times]0, +\infty]$ into $\bar{R}$ is continuous, and hence v/E is measurable $\mathcal{M}/E$, which shows that v is μ-measurable. Also, vg and vh are μ-measurable. Since $v \ge f$, we have

$$\int^* fg \cdot dV\mu + \int^* fh \cdot dV\mu \le \int^* vg \cdot dV\mu + \int^* vh \cdot dV\mu$$

$$= \int^* v(g + h) \cdot dV\mu \le \int^* u \cdot dV\mu,$$

and the proposition is proved.   □

Theorem 3.6.1 *Let f be a function from Ω into $[0, +\infty]$.*

(a) *If f is not moderate, $\int^* f dV \mu = +\infty$.*

(b) *If f is moderate, $\int^\bullet f dV \mu = \int^* f dV \mu$.*

(c) *If $\int^\bullet f dV \mu < +\infty$, there exists a set A, that is a countable union of integrable sets, such that f vanishes l.a.e. outside A.*

PROOF: (a) and (b) have been treated previously. Now suppose that $\int^\bullet f dV \mu$ is finite. There exists an increasing sequence $(A_n)_{n \geq 1}$ of integrable sets such that $\int^\bullet f dV \mu = \sup_{n \geq 1} \int^* f 1_{A_n} dV \mu$. Put $A = \bigcup_{n \geq 1} A_n$. Then

$$\sup_{n \geq 1} \int^* f 1_{A_n} dV \mu = \int^* f 1_A dV \mu = \int^\bullet f 1_A dV \mu.$$

But

$$\int^\bullet f dV \mu = \int^\bullet f 1_A dV \mu + \int^\bullet f 1_{A^c} dV \mu$$

(Proposition 3.6.1), so $\int^\bullet f 1_{A^c} dV \mu = 0$.   □

Proposition 3.6.2 $\int^\bullet f dV \mu = \int^* f dV \mu$ *for every $f \in \mathcal{J}^+$.*

PROOF: If $g \in \mathcal{H}^+$ and $g \leq f$, then the set $g^{-1}(]0, +\infty[) = A$ is moderate, so $\int g dV \mu \leq \int^* f 1_A dV \mu \leq \int^\bullet f dV \mu$; whence we see that $\int^* f dV \mu \leq \int^\bullet f dV \mu$.   □

Now let F be a Banach space, and let $\overline{\mathcal{F}_F^1}(\mu)$ be the set of mappings f from Ω into F such that $\overline{N_1}(f) = \int^\bullet |f| \cdot dV \mu < +\infty$. Endow $\overline{\mathcal{F}_F^1}(\mu)$ with the seminorm $\overline{N_1} : f \mapsto \int^\bullet |f| \cdot dV \mu$. The closure of $\mathcal{H}(\Omega, \mathbf{R}) \otimes F$ into $\overline{\mathcal{F}_F^1}(\mu)$ is written $\overline{\mathcal{L}_F^1}(\mu)$, and its elements are the essentially μ-integrable mappings from Ω into F. The continuous linear map $f \mapsto \int f d\mu$ from $\mathcal{H}(\Omega, \mathbf{R}) \otimes F$ into F has a unique continuous linear extension to $\overline{\mathcal{L}_F^1}(\mu)$, written $f \mapsto \bar{\int} f d\mu$.

Theorem 3.6.2 f *belongs to $\overline{\mathcal{F}_F^1}(\mu)$ (respectively, $\overline{\mathcal{L}_F^1}(\mu)$) if and only if it is equal l.a.e. to a $g \in \mathcal{F}_F^1(\mu)$ (respectively, $g \in \mathcal{L}_F^1(\mu)$; in this case, $\bar{\int} f d\mu = \int g d\mu$). $f : \Omega \to F$ is essentially μ-integrable if and only if it is μ-measurable and $\int^\bullet |f| dV \mu < +\infty$. f is μ-integrable if and only if it is μ-moderate and essentially μ-integrable.*

PROOF: All statements follow from Theorem 3.6.1.   □

Since $\bar{\int} f d\mu = \int f d\mu$ for all $f \in \mathcal{L}_F^1(\mu)$, the essential integral $\bar{\int} f d\mu$ of any $f \in \overline{\mathcal{L}_F^1}(\mu)$ is usually written $\int f d\mu$. For every μ-measurable set A, $\int f 1_A d\mu$ is then written $\int_A f d\mu$.

Observe that, if a set A is essentially integrable (i.e., if 1_A is), Theorem 3.6.1 shows that A is a union of an integrable set and a locally negligible set.

Now, put $\|\mu\| = \sup_{g \in \mathcal{H}(\Omega, \mathbf{C}), |g| \le 1} |\mu(g)| = \sup_{f \in \mathcal{H}^+, f \le 1} V\mu(f)$, so that $\|\mu\| = \|V\mu\|$.

Definition 3.6.2 μ is said to be bounded whenever $\|\mu\|$ is finite.

Proposition 3.6.3 $\|\mu\| = \int^{\bullet} 1 \cdot dV\mu$. So μ is bounded if and only if Ω is essentially μ-integrable.

PROOF:

$$\int 1_A \cdot dV\mu = \sup_{\substack{g \in \mathcal{H}(\Omega, \mathbf{C}) \\ |g| \le 1}} \left| \int g 1_A d\mu \right| \le \sup_g \int |g| dV\mu \le \sup_{\substack{f \in \mathcal{H}^+ \\ f \le 1}} V\mu(f) = \|\mu\|$$

for every integrable set A, so $\int^{\bullet} 1 \cdot dV\mu \le \|\mu\|$ (Proposition 3.3.4).

Conversely, $V\mu(f) \le \int^{\bullet} 1 \cdot dV\mu$ for all $f \in \mathcal{H}^+$ satisfying $f \le 1$. Hence $\|\mu\| \le \int^{\bullet} 1 \cdot dV\mu$. $\qquad\square$

Proposition 3.6.4 *Suppose μ is positive. If D is a closed convex subset of a real Banach space F, then $\int f g d\mu / \int g d\mu$ lies in D for every mapping f from Ω into D and every essentially integrable function g from Ω into $[0, +\infty[$ such that $\int g d\mu > 0$ and fg is essentially integrable. If $\int^{\bullet} 1 d\mu = 1$, then $\int f d\mu$ lies in the closed convex envelope of $f(\Omega)$, for every essentially integrable mapping f from Ω into a real Banach space.*

PROOF: We need only prove the first claim. Denote by F' the dual space of F. Let $\langle z, a' \rangle = a'(z) \le \alpha$ $(a' \in F', \alpha \in \mathbf{R})$, a relation defining in F a closed half-space E which contains D. Then $\langle f(x)g(x), a' \rangle \le \alpha g(x)$ for all $x \in \Omega$, so $\langle \int f g d\mu, a' \rangle = \int \langle fg, a' \rangle \cdot d\mu \le \alpha \int g d\mu$, and $\int f g d\mu / \int g d\mu$ belongs to E. By the Hahn–Banach theorem, D is the intersection of the closed half-spaces which contain it, and the proof is complete. $\qquad\square$

Proposition 3.6.5 *Suppose μ is positive, and let $f : \Omega \to [0, +\infty]$ be μ-measurable. Put $g(t) = \mu^{\bullet}\left(f^{-1}(]t, +\infty])\right)$ for every real number $t > 0$, and consider λ, Lebesgue measure on $]0, +\infty[$. Then $\int^{\bullet} f d\mu = \int^{*} g(t) d\lambda(t)$.*

PROOF: First, suppose that f has a finite, positive range, namely, $\{t_1, \ldots, t_k\}$ or $\{0, t_1, \ldots, t_k\}$ (with $0 < t_1 < t_2 < \ldots < t_k$). Then

$$\int^{\bullet} f \cdot d\mu = \int^{\bullet} \sum_{1 \le i \le k} t_i \cdot 1_{(f = t_i)} d\mu = \sum_{1 \le i \le k} t_i \cdot \mu^{\bullet}(f = t_i).$$

On the other hand,

$$\int^* g(t)d\lambda(t) = t_1 \cdot \mu^\bullet(f \geq t_1) + \sum_{2 \leq i \leq k} (t_i - t_{i-1}) \cdot \mu^\bullet(f \geq t_i),$$

because g takes the value $\mu^\bullet(f \geq t_1)$ on $]0, t_1[$, the value $\mu^\bullet(f \geq t_2)$ on $[t_1, t_2[$, ... If all the $\mu^\bullet(f = t_i)$ are finite, the sets $(f = t_i)$ are essentially integrable, and

$$\sum_{1 \leq i \leq k} t_i \cdot \mu(f = t_i) = \sum_{1 \leq i \leq k-1} t_i[\mu(f \geq t_i) - \mu(f \geq t_{i+1})] + t_k \cdot \mu(f \geq t_k)$$

$$= t_1 \cdot \mu(f \geq t_1) + \sum_{2 \leq i \leq k} (t_i - t_{i-1}) \cdot \mu(f \geq t_i);$$

hence $\int^\bullet f d\mu = \int^* g(t) \cdot d\lambda(t)$. Evidently, this equality remains true if one of the $\mu^\bullet(f = t_i)$ is infinite.

We now pass to the general case: suppose f is measurable, and consider the same function f_n as in Proposition 3.3.3. For every $t > 0$, the sequence of sets $((f_n > t))_{n \geq 1}$ increases to $(f > t)$, so $\mu^\bullet((f_n > t))$ converges to $\mu^\bullet((f > t))$ as $n \to +\infty$. Therefore,

$$\int^\bullet f \cdot d\mu = \lim_{n \to +\infty} \int^\bullet f_n \cdot d\mu$$

$$= \lim \int^* \mu^\bullet((f_n > t)) \cdot d\lambda(t)$$

$$= \int^* g(t) \cdot d\lambda(t).$$

$\square$

3.7   Upper and Lower Integrals

Let $\mathcal{H}(\Omega, \mathbf{C})$ and $\mathcal{J}^+$ be as in Section 3.1. Denote by $\mathcal{J}$ the set of those functions h from Ω into $\overline{\mathbf{R}}$ with the property that there exists $g \in \mathcal{H}(\Omega, \mathbf{R})$ such that $h - g$ belongs to $\mathcal{J}^+$. Let h be a positive element of $\mathcal{J}$, and let $g \in \mathcal{H}(\Omega, \mathbf{R})$ be such that $h - g$ belongs to $\mathcal{J}^+$; then $g^+ \leq h$, and $g^+ - g \leq h - g$; thus $h - g^+ = (h - g) - (g^+ - g)$ lies in $\mathcal{J}^+$, and also $h = (h - g^+) + g^+$. This means that $\mathcal{J}^+$ is the set of positive elements of $\mathcal{J}$.

Now, for every $h \in \mathcal{J}$, if $g \in \mathcal{H}(\Omega, \mathbf{R})$ is such that $g \leq h$, then $h - g$ lies in $\mathcal{J}$, and hence lies in $\mathcal{J}^+$. Also, $h_1 + h_2$ belongs to $\mathcal{J}$ for all $h_1, h_2 \in \mathcal{J}$, and $\lambda h \in \mathcal{J}$ for all $h \in \mathcal{J}$ and $\lambda \in \mathbf{R}^+$. Furthermore, if h_1 and h_2 belong to $\mathcal{J}$ and g in $\mathcal{H}(\Omega, \mathbf{R})$ is dominated by $\inf(h_1, h_2)$, then $\sup(h_1, h_2) - g = \sup(h_1 - g, h_2 - g)$ and $\inf(h_1, h_2) - g$ lie in $\mathcal{J}^+$. This proves that $\mathcal{J}$ is a lattice.

Henceforth, μ will always denote a positive Daniell measure on $\mathcal{H}(\Omega, \mathbf{C})$.

Definition 3.7.1 For every $h \in \mathcal{J}$, $\mu^*(h) = \sup_{g \in \mathcal{H}(\Omega, \mathbf{R}), g \leq h} \mu(g)$ is called the upper integral of h.

Observe that, when h lies in $\mathcal{J}^+$, $\mu^*(h)$ is just the quantity $\int^* h d\mu$ which we encountered in Section 3.1.

If $h \in \mathcal{J}$ and $g \in \mathcal{H}(\Omega, \mathbf{R})$ are such that $g \leq h$, then $\mu^*(h) = \mu^*(h - g) + \mu(g)$. Hence $\mu^*(h_1 + h_2) = \mu^*(h_1) + \mu^*(h_2)$ for all h_1 and h_2 in $\mathcal{J}$, and $\mu^*(\lambda h) = \lambda \cdot \mu^*(h)$ for every $\lambda \in \mathbf{R}^+$ and every $h \in \mathcal{J}$. Similarly, $\mu^*(\sup h_n) = \sup_{n \geq 1} \mu^*(h_n)$ for every increasing sequence $(h_n)_{n \geq 1}$ in $\mathcal{J}$.

Proposition 3.7.1 *A necessary and sufficient condition that $h \in \mathcal{J}$ be μ-integrable is that $\mu^*(h) < +\infty$. In this case, $\mu^*(h) = \int h \cdot d\mu$.*

PROOF: Suppose that $\mu^*(h) < +\infty$. For every $\varepsilon > 0$, there exists g in $\mathcal{H}(\Omega, \mathbf{R})$, $g \leq h$, such that $\mu(g) \geq \mu^*(h) - \varepsilon$. Then $\mu^*(h - g) = \mu^*(h) - \mu(g) \leq \varepsilon$, which proves that h is μ-integrable. Moreover, $\int h d\mu \leq \mu(g) + \varepsilon \leq \mu^*(h) + \varepsilon$, so $\int h d\mu \leq \mu^*(h)$ and, in fact, $\int h d\mu = \mu^*(h)$. $\square$

Put $-\mathcal{J} = \{-h : h \in \mathcal{J}\}$. Any positive function of $-\mathcal{J}$ takes only finite values and is μ-integrable.

Proposition 3.7.2 *A function f from Ω into $\overline{\mathbf{R}}$ is μ-integrable if and only if, for every $\varepsilon > 0$, there exist μ-integrable functions $g \in -\mathcal{J}$ and $h \in \mathcal{J}$ such that $g \leq f \leq h$ and $\int (h - g) \cdot d\mu \leq \varepsilon$. We may take g positive when f is positive.*

PROOF: Obviously, the condition is sufficient.

Now suppose that f is μ-integrable and observe that we may assume f is positive. Given $\varepsilon > 0$, there exist $u \in \mathcal{H}^+$ such that $\int^* |f - u| \cdot d\mu \leq \varepsilon/4$, and $v \in \mathcal{J}^+$ such that $|f - u| \leq v$ and $\int^* v \cdot d\mu \leq \varepsilon/2$. Now $-v \leq f - u \leq v$, whence $u - v \leq f \leq u + v$ and $(u - v)^+ \leq f \leq u + v$. Then it suffices to take $g = (u - v)^+$ and $h = u + v$ (because $u + v - (u - v)^+ \leq 2v$). $\square$

Proposition 3.7.3 *For every integrable function f from Ω into $\overline{\mathbf{R}}$ (respectively, into $[0, +\infty]$), there exist an increasing sequence $(g_n)_{n \geq 1}$ of integrable functions of $-\mathcal{J}$ (respectively, of positive functions of $-\mathcal{J}$) and a decreasing sequence $(h_n)_{n \geq 1}$ of integrable functions of $\mathcal{J}$ such that $g_n \leq f \leq h_n$ for all $n \geq 1$ and $f = \sup_n g_n = \inf_n h_n$ almost everywhere.*

PROOF: First, suppose that f is positive. For every integer $n \geq 1$, there exist an integrable $v_n \in \mathcal{J}$ and a positive u_n of $-\mathcal{J}$ such that $u_n \leq f \leq v_n$ and $\int (v_n - u_n) \cdot d\mu \leq 1/n$. If we put $g_n = \sup(u_1, \ldots, u_n)$ and $h_n = \inf(v_1, \ldots, v_n)$, we see that the sequences $(g_n)_{n \geq 1}$ and $(h_n)_{n \geq 1}$ have the desired properties. Indeed, since $g = \sup_{n \geq 1} g_n \leq f$, g is μ-integrable (by the monotone convergence theorem); and since $\int (f - g_n) \cdot d\mu \leq \int (v_n - u_n) \cdot d\mu \leq$

$1/n$, we have $\int f d\mu - \int g d\mu = \lim_{n\to+\infty} \int (f - g_n) d\mu = 0$, which proves that f and g are equal almost everywhere. The argument is the same for $(h_n)_{n\geq 1}$.

In general, we may apply the preceding argument to f^+ and f^-: so there are two increasing sequences $(g'_n)_{n\geq 1}$, $(g''_n)_{n\geq 1}$ of integrable functions of $-\mathcal{J}$ and two decreasing sequences $(h'_n)_{n\geq 1}$, $(h''_n)_{n\geq 1}$ of integrable functions of $\mathcal{J}$ such that

(a) $0 \leq g'_n \leq f^+ \leq h'_n$ and $g''_n \leq -f^- \leq h''_n \leq 0$;

(b) $f^+ = \sup g'_n = \inf h'_n$ almost everywhere;

(c) $-f^- = \sup g''_n = \inf h''_n$ almost everywhere.

Then put $g_n = g'_n + g''_n$ and $h_n = h'_n + h''_n$. Clearly, the sequences $(g_n)_{n\geq 1}$ and $(h_n)_{n\geq 1}$ have the stated properties.   □

Definition 3.7.2 For every function $f : \Omega \to \overline{\mathbf{R}}$, $\mu^*(f) = \inf_{h\in\mathcal{J}, h\geq f} \mu^*(h)$ and $\mu_*(f) = -\mu^*(-f)$ are called the upper integral and the lower integral of f.

Observe that $\mu^*(f)$ is the same quantity as that in Definition 3.1.1, when f is positive.

Proposition 3.7.4 $\mu^*(f) = \inf_{h\geq f,\, h\ integrable} \int h d\mu$ for every function f from Ω into $\overline{\mathbf{R}}$

PROOF: By definition, $\mu^*(f) = \inf \{\int h d\mu : h \geq f, h \in \mathcal{J}, h\ \text{integrable}\}$, so $\inf_{h\geq f, h\ integrable} \int h d\mu \leq \mu^*(f)$. Thus we may confine our attention to the case where $\inf_{h\geq f, h\ integrable} \int h d\mu$ is not $+\infty$, and let $r \in \mathbf{R}$ be strictly greater than this number. There exists $h \geq f$, h integrable, such that $\int h d\mu < r$, and, by Proposition 3.7.3, there also exists $h' \in \mathcal{J}$, h' integrable, $h' \geq h$, such that $\int h' d\mu \leq r$. Therefore, $\mu^*(f) \leq r$, and the proposition follows.   □

Similarly, $\mu_*(f) = \sup \{\int g d\mu : g \leq f, g\ \text{integrable}\}$; hence $\mu_*(f) \leq \mu^*(f)$.

Proposition 3.7.5 If f_1 and f_2 are two functions from Ω into $\overline{\mathbf{R}}$ such that $f_1 + f_2$ is defined almost everywhere and if $\mu^*(f_1) + \mu^*(f_2)$ makes sense, then $\mu^*(f_1 + f_2) \leq \mu^*(f_1) + \mu^*(f_2)$. If $(f_n)_{n\geq 1}$ is an increasing sequence of functions from Ω into $\overline{\mathbf{R}}$ such that $\mu^*(f_n) > -\infty$ for n large enough, then $\mu^*(\sup_{n\geq 1} f_n) = \sup_{n\geq 1} \mu^*(f_n)$.

PROOF: To prove the first assertion, it suffices to consider the case where $\mu^*(f_1) < +\infty$ and $\mu^*(f_2) < +\infty$. If h_1, h_2 are two integrable functions such that $h_1 \geq f_1$, $h_2 \geq f_2$, let $\hat{h}_1$, $\hat{h}_2$ be two real-valued functions on Ω equal to h_1, h_2 almost everywhere. Then $\hat{h}_1 + \hat{h}_2 \geq f_1 + f_2$ almost everywhere, so $\mu^*(f_1 + f_2) \leq \mu(\hat{h}_1 + \hat{h}_2) = \mu(h_1) + \mu(h_2)$, and $\mu^*(f_1 + f_2) \leq \mu^*(f_1) + \mu^*(f_2)$.

For the second assertion, we argue as in Theorem 3.1.1, supposing that $h_n \in \mathcal{J}$ is integrable and that $h_n \geq f_n$. $\qquad\qquad\qquad\qquad$ $\square$

Theorem 3.7.1 *When $\mu^*(f)$ is finite, there exists an integrable function f_1 such that $f_1 \geq f$ and $\mu(f_1) = \mu^*(f)$; if f_2 is a second integrable function such that $f_2 \geq f$ and $\mu(f_2) = \mu^*(f)$, then $f_1 = f_2$ almost everywhere. A necessary and sufficient condition that f be integrable is that $\mu_*(f)$ and $\mu^*(f)$ be finite and equal.*

PROOF: Suppose $\mu^*(f)$ is finite. For every $n \geq 1$, there exists an integrable function h_n such that $h_n \geq f$ and $\mu^*(f) \leq \mu(h_n) \leq \mu^*(f) + 1/n$. Clearly, $h = \inf_{n \geq 1} h_n = \inf_{n \geq 1} \inf(h_1, \ldots, h_n)$ is integrable and $\mu^*(f) = \mu(h)$. Now let f_1, f_2 be as in the first assertion. We prove that $f_1 = f_2$ almost everywhere: there is no restriction in assuming that $f_1 \leq f_2$. For every $i \in \{1, 2\}$, let $\hat{f}_i$ be a real-valued function on Ω, equal to f_i at points where f_1 and f_2 are finite, and suppose that $\hat{f}_1 \leq \hat{f}_2$. We have $\mu(\hat{f}_1) = \mu(\hat{f}_2)$, or $\mu(\hat{f}_2 - \hat{f}_1) = 0$, so $\hat{f}_1 = \hat{f}_2$ a.e., and $f_1 = f_2$ a.e.

Finally, let f be a function from Ω into $\overline{\mathbf{R}}$ such that $\mu_*(f)$, $\mu^*(f)$ are finite and equal. There exist integrable functions g, h such that $g \leq f \leq h$ and $\mu(g) = \mu(h)$. Then $g = h = f$ almost everywhere, hence f is integrable. $\qquad$ $\square$

Proposition 3.7.6 *If f_1 and f_2 are two functions from Ω into $\overline{\mathbf{R}}$ such that $f_1 + f_2$ is defined a.e. and if $\mu_*(f_1) + \mu^*(f_2)$ makes sense, then $\mu_*(f_1 + f_2) \leq \mu_*(f_1) + \mu^*(f_2) \leq \mu^*(f_1 + f_2)$. If $f_1 : \Omega \to \overline{\mathbf{R}}$ is integrable and $f_2 : \Omega \to \overline{\mathbf{R}}$ is arbitrary, then $\mu^*(f_1 + f_2) = \mu(f_1) + \mu^*(f_2)$ and $\mu_*(f_1 + f_2) = \mu(f_1) + \mu_*(f_2)$.*

PROOF: To prove that $\mu_*(f_1 + f_2) \leq \mu_*(f_1) + \mu^*(f_2)$, we may suppose $\mu^*(f_2) < +\infty$. Given an integrable function $h_2 \geq f_2$, let $\hat{h}_2$ be a real-valued function equal to h_2 a.e. If g is an integrable function such that $g \leq f_1 + f_2$, we have $g \leq f_1 + \hat{h}_2$ a.e., or $g - \hat{h}_2 \leq f_1$ a.e. It follows that $\mu(g) - \mu(h_2) \leq \mu_*(f_1)$. Therefore, $\mu_*(f_1 + f_2) - \mu(h_2) \leq \mu_*(f_1)$, or $\mu_*(f_1 + f_2) \leq \mu_*(f_1) + \mu(h_2)$, and finally $\mu_*(f_1 + f_2) \leq \mu_*(f_1) + \mu^*(f_2)$.

To establish the inequality $\mu_*(f_1) + \mu^*(f_2) \leq \mu^*(f_1 + f_2)$, consider $-f_1$ and $-f_2$. The last assertion now follows easily. $\qquad\qquad\qquad$ $\square$

Proposition 3.7.7 *Let $f : \Omega \to \overline{\mathbf{R}}$ be such that $\mu_*(f)$ (respectively, $\mu^*(f)$) is finite. If g (respectively, h) is an integrable function from Ω into $\overline{\mathbf{R}}$ such that $g \leq f$ and $\mu(g) = \mu_*(f)$ (respectively, $h \geq f$ and $\mu(h) = \mu^*(f)$), then $\mu_*(f - g) = 0$ and $\mu^*(f - g) = \mu^*(f) - \mu_*(f)$ (respectively, $\mu_*(h - f) = 0$ and $\mu^*(h - f) = \mu^*(f) - \mu_*(f))$.*

PROOF: By Proposition 3.7.6, $\mu^*(f - g) = \mu^*(f) - \mu(g)$ and $\mu_*(f - g) = \mu_*(f) - \mu(g) = 0$. Similarly, $\mu^*(h - f) = \mu(h) - \mu_*(f)$ and $\mu_*(h - f) = \mu(h) - \mu^*(f) = 0$. $\qquad\qquad\qquad\qquad\qquad$ $\square$

Proposition 3.7.8 *Let f be an integrable function from Ω into $\overline{\mathbf{R}}$, and let g be a function from Ω into $\overline{\mathbf{R}}$ such that $\mu^*(g)$ is finite. Then a necessary and sufficient condition that g be integrable is that $\mu(f) = \mu^*(g) + \mu^*(f - g)$.*

PROOF: Suppose that $\mu(f) = \mu^*(g) + \mu^*(f - g)$. Since f is integrable,

$$\mu^*(f - g) = \mu(f) + \mu^*(-g) = \mu(f) - \mu_*(g).$$

But $\mu^*(f - g) = \mu(f) - \mu^*(g)$ by hypothesis, so $\mu^*(g) = \mu_*(g)$, and g is integrable.   □

Definition 3.7.3 For all subsets A of Ω, $\mu^*(A) = \mu^*(1_A)$ and $\mu_*(A) = \mu_*(1_A)$ are called the outer measure and the inner measure of A.

Note that we have already dealt with the quantity $\mu^*(A) \equiv \int^* 1_A d\mu$ in previous work.

Proposition 3.7.9 *For $A \subset \Omega$, $\mu_*(A) = \sup\{\mu(B) : B \subset A, B \text{ integrable}\}$ and $\mu^*(A) = \inf\{\mu(B) : B \supset A, B \text{ integrable}\}$.*

PROOF: Let $r < \mu_*(A)$ be a real number. There exists an integrable function f such that $0 \leq f \leq 1_A$ and $r < \int f d\mu$. For all integers $n \geq 1$, $B_n = \{x \in \Omega : f(x) \geq 1/n\}$ is integrable. By the dominated convergence theorem, $\int f 1_{B_n} d\mu$ converges to $\int f d\mu$ as $n \to +\infty$, so there exists $n \geq 1$ such that $\int f 1_{B_n} d\mu \geq r$. For this n, put $B_n = B$. Since $f \leq 1_A$, we have $\mu(B) = \int 1_B d\mu \geq \int f 1_B d\mu \geq r$, which proves the first assertion.

To prove the second assertion, suppose that $\mu^*(A)$ is finite and let $r > \mu^*(A)$ be a real number. There exists $f \in \mathcal{J}^+$ such that $f \geq 1_A$ and $\int f d\mu \leq r$. Then $B = \{x \in \Omega : f(x) \geq 1\}$ is integrable, contains A, and $\mu(B) \leq \int f d\mu \leq r$. Therefore, $\mu^*(A) = \inf\{\mu(B) : B \supset A, B \text{ integrable}\}$.   □

Proposition 3.7.10 *For every subset A of Ω such that $\mu_*(A) < +\infty$ (respectively, $\mu^*(A) < +\infty$), there exists an integrable set $A_1 \subset A$ (respectively, $A_2 \supset A$) such that $\mu(A_1) = \mu_*(A)$ (respectively, $\mu(A_2) = \mu^*(A)$). For every A_1 (respectively, A_2) of this type, $\mu_*(A - A_1) = 0$ and $\mu^*(A - A_1) = \mu^*(A) - \mu_*(A)$ (respectively, $\mu_*(A_2 - A) = 0$ and $\mu^*(A_2 - A) = \mu^*(A) - \mu_*(A)$).*

PROOF: The existence of A_1, A_2 follows from Proposition 3.7.9, and the other assertions are consequences of Proposition 3.7.7.   □

Proposition 3.7.11 *Let A, B be two disjoint subsets of Ω and C their union. Then*

$$\mu_*(A) + \mu_*(B) \leq \mu_*(C) \leq \mu_*(A) + \mu^*(B) \leq \mu^*(C) \leq \mu^*(A) + \mu^*(B).$$

PROOF: The middle inequalities follow from Proposition 3.7.6; the first and last from Proposition 3.7.5. □

Proposition 3.7.12 *Let A, B be two disjoint subsets of Ω such that $\mu^*(A) < +\infty$, $\mu^*(B) < +\infty$, and let C be their union. Then*

$$\mu_*(C) - \mu_*(A) - \mu_*(B) \leq \mu^*(A) + \mu^*(B) - \mu^*(C).$$

PROOF: First, suppose that $\mu_*(A) = \mu_*(B) = 0$, and let A_2, B_2 be integrable sets containing A, B respectively, such that $\mu(A_2) = \mu^*(A)$, $\mu(B_2) = \mu^*(B)$. If f is an integrable function such that $0 \leq f \leq 1_C$, we have $f 1_{A_2 - A_2 \cap B_2} \leq 1_A$. Thus $\int f 1_{A_2} d\mu = \int f 1_{A_2 \cap B_2} d\mu$. Similarly, $\int f 1_{B_2} d\mu = \int f 1_{A_2 \cap B_2} d\mu$. We see now that

$$
\begin{aligned}
\int f \cdot d\mu &= \int f 1_C \cdot d\mu \\
&= \int f 1_{A_2} d\mu + \int f 1_{B_2} d\mu - \int f 1_{A_2 \cap B_2} d\mu \\
&= \int f 1_{A_2 \cap B_2} d\mu \leq \mu(A_2 \cap B_2).
\end{aligned}
$$

Therefore,

$$
\begin{aligned}
\mu_*(C) \leq \mu(A_2 \cap B_2) &= \mu(A_2) + \mu(B_2) - \mu(A_2 \cup B_2) \\
&\leq \mu^*(A) + \mu^*(B) - \mu^*(C).
\end{aligned}
$$

We now pass to the general case, and let A_1, B_1 be two integrable subsets of A, B respectively, such that $\mu(A_1) = \mu_*(A)$ and $\mu(B_1) = \mu_*(B)$. We know that $\mu_*(A - A_1) = \mu_*(B - B_1) = 0$. By what has just been proved,

$$\mu_*[(A - A_1) \cup (B - B_1)] \leq \mu^*(A - A_1) + \mu^*(B - B_1) - \mu^*[(A - A_1) \cup (B - B_1)].$$

But

$$\mu_*(A \cup B) = \mu_*[(A - A_1) \cup (B - B_1)] + \mu(A_1) + \mu(B_1)$$

by Proposition 3.7.6, and

$$\mu^*[(A - A_1) \cup (B - B_1)] + \mu(A_1) + \mu(B_1) = \mu^*(A \cup B).$$

Therefore,

$$
\begin{aligned}
\mu_*(C) - \mu_*(A) - \mu_*(B) \leq [\mu^*(A) - \mu_*(A)] &+ [\mu^*(B) - \mu_*(B)] \\
&+ \mu_*(A) + \mu_*(B) - \mu^*(C),
\end{aligned}
$$

or

$$\mu_*(C) - \mu_*(A) - \mu_*(B) \leq \mu^*(A) + \mu^*(B) - \mu^*(C).$$

□

Theorem 3.7.2 *A necessary and sufficient condition that a subset A of Ω be μ-measurable is that it split μ^* (i.e., $\mu^*(B) = \mu^*(A \cap B) + \mu^*(A^c \cap B)$ for all subsets B of Ω).*

PROOF: Suppose A is μ-measurable, and let B be an arbitrary subset of Ω. We already know that $\mu^*(B) \leq \mu^*(A \cap B) + \mu^*(A^c \cap B)$. If $\mu^*(B)$ is finite, we let B_2 be an integrable set containing B such that $\mu^*(B) = \mu(B_2)$. Then

$$\mu^*(B) = \mu(B_2) = \mu(A \cap B_2) + \mu(A^c \cap B_2) \geq \mu^*(A \cap B) + \mu^*(A^c \cap B),$$

so $\mu^*(B) = \mu^*(A \cap B) + \mu^*(A^c \cap B)$.

Conversely, suppose that $\mu^*(B) = \mu^*(A \cap B) + \mu^*(A^c \cap B)$ for all integrable sets B. Put $f = 1_B$ and $g = 1_{A \cap B}$. Since $\mu(f) = \mu^*(g) + \mu^*(f - g)$, we see that g is integrable (Proposition 3.7.8). Hence A is μ-measurable. $\square$

Definition 3.7.4 A function f from Ω into $\overline{\mathbf{R}}$ is quasi-integrable (with respect to μ) if and only if it is μ-measurable and $\mu^*(f) = \mu_*(f)$. In this case, we put $\mu(f) = \mu^*(f) = \mu_*(f)$.

This implies that f is integrable if and only if it is quasi-integrable and $\mu(f)$ is finite.

Proposition 3.7.13 *Let f, g be two quasi-integrable functions from Ω into $\overline{\mathbf{R}}$, and suppose that $\mu(f) + \mu(g)$ makes sense. Then $f + g$ is defined almost everywhere, is quasi-integrable, and $\mu(f + g) = \mu(f) + \mu(g)$.*

PROOF: First, assume that $\mu(f)$ is finite. Then f is integrable, and we may suppose that f takes on finite values. Further,

$$\begin{aligned}
\mu_*(f + g) &= \sup\{\mu(k) : k \leq f + g, \quad k \text{ integrable}\} \\
&= \sup\{\mu(f + h) : h \leq g, \quad h \text{ integrable}\} \\
&= \mu(f) + \mu_*(g),
\end{aligned}$$

and similarly $\mu^*(f + g) = \mu(f) + \mu^*(g)$, so $f + g$ is quasi-integrable in this case.

Next, assume that $\mu_*(f) = +\infty = \mu_*(g)$. There exist integrable functions f_1, g_1 such that $f_1 \leq f$, $g_1 \leq g$, and that $\mu(f_1)$ and $\mu(g_1)$ are arbitrarily large. We have $f > -\infty$ a.e. and $g > -\infty$ a.e. Hence $f + g$ is defined a.e., and $f_1 + g_1 \leq f + g$ a.e. Since $\mu(f_1) + \mu(g_1) = \mu(f_1 + g_1) \leq \mu_*(f + g)$, it follows that $\mu_*(f + g) = +\infty = \mu^*(f + g)$.

The case $\mu^*(f) = -\infty = \mu^*(g)$ is treated just as easily. $\square$

Proposition 3.7.14 *For a quasi-integrable function f, at least one of the numbers $\mu^*(f^+)$ and $\mu^*(f^-)$ is finite, and $\mu(f) = \mu^*(f^+) - \mu^*(f^-)$. Conversely, let f be a function from Ω into $\overline{\mathbf{R}}$, μ-measurable and μ-moderate, such that at least one of the numbers $\mu^*(f^+)$ and $\mu^*(f^-)$ is finite; then f is quasi-integrable.*

PROOF: First, suppose that f is quasi-integrable. If $\mu(f)$ is finite, then f is integrable, $\mu^*(f^+)$ and $\mu^*(f^-)$ are finite, and $\mu(f) = \mu(f^+) - \mu(f^-)$. If $\mu_*(f) = +\infty$, there exists an integrable function $g \leq f$; since $f^- \leq g^-$, we have $\mu^*(f^-) < +\infty$, whereas $\mu_*(f^+) \geq \mu_*(f) = +\infty$. If $\mu^*(f) = -\infty$, then $\mu^*(f^+)$ is finite, but $\mu_*(f^-) \geq \mu_*(-f) = -\mu^*(f) = +\infty$.

Now let f be as in the second assertion. We prove that f is quasi-integrable.

First, suppose that f is positive and bounded. If $(A_n)_{n \geq 1}$ is an increasing sequence of integrable sets such that f vanishes outside $\bigcup_{n \geq 1} A_n$, then $\mu^*(f) = \sup_{n \geq 1} \mu^*(f 1_{A_n})$. But $f 1_{A_n}$ is μ-integrable, so $\mu(f 1_{A_n}) \leq \mu_*(f)$ for every $n \geq 1$ and $\mu^*(f) \leq \mu_*(f)$.

Next, suppose that f is positive. Then $\mu^*(\inf(f, n)) = \mu_*(\inf(f, n))$ for all integers $n \geq 1$, by what we have just shown. Hence $\mu^*(f) = \sup \mu^*(\inf(f, n)) \leq \mu_*(f)$.

We now pass to the general case. We may suppose that $\mu^*(f^-)$ is finite, so that f^- is integrable. By the above argument, f^+ is quasi-integrable, and Proposition 7.13 shows that $f = f^+ - f^-$ is quasi-integrable and that $\mu(f) = \mu^*(f^+) - \mu^*(f^-)$. □

Corollary Let $A \subset \Omega$ be μ-measurable and μ-moderate. Then $\mu_*(A) = \mu^*(A)$, and $\mu^*(A) = \mu^*(B) + \mu_*(A - B)$ for all subsets B of A.

PROOF: $\mu_*(A) = \mu^*(A)$ by Proposition 3.7.14. Then

$$\mu^*(A) = \mu^*(B) + \mu_*(A - B)$$

for all subsets B of A, by Proposition 3.7.11. □

If A is a μ-measurable set which is not μ-moderate, it is possible that $\mu_*(A) \neq \mu^*(A)$ (see Chapter 7, Exercise 4).

Lemma 3.7.1 Let D be a convex set in a real locally convex space F, and let φ be a convex function on D (such that $\varphi(\theta y + (1-\theta)z) \leq \theta \cdot \varphi(y) + (1-\theta) \cdot \varphi(z)$ for $y, z \in D$ and $0 \leq \theta \leq 1$). Let L be the class of continuous affine functions ψ on F such that $\psi/D \leq \varphi$. If D is closed and φ lower semicontinuous, then φ is the upper envelope of $\{\psi/D : \psi \in L\}$. The same result holds if D is open and φ continuous.

PROOF: Let y be arbitrary in D, and let a be a real number such that $a < \varphi(y)$. Put $A = \{(z, t) \in D \times \mathbf{R} : t \geq \varphi(z)\}$ if D is closed and φ lower semicontinuous, and put $A = \{(z, t) \in D \times \mathbf{R} : t > \varphi(z) - \varphi(y) + a\}$ if D is open and φ continuous. In the first case A is a closed convex set in $F \times \mathbf{R}$, and in the second case it is an open convex subset of $F \times \mathbf{R}$. By the Hahn–Banach theorem, there exists in $F \times \mathbf{R}$ a closed hyperplane containing (y, a) which does not meet A, namely $H = \{(z, t) : u(z) + \lambda t = \alpha\}$, where u is a continuous linear form on F and λ, α are two real numbers. In fact, $H = \{(z, t) : u(z - y) + \lambda(t - a) = 0\}$, because it contains (y, a). Now $\lambda \neq 0$, because $(y, \varphi(y))$ belongs to A, and so does not lie in H. Therefore, replacing u by $u/(-\lambda)$, we may suppose that $H = \{(z, t) \in F \times \mathbf{R} : u(z - y) = t - a\}$. Since $u(y) - \varphi(y) < u(y) - a$, we have $u(z) - \varphi(z) < u(y) - a$ for every $z \in D$, or $u(z - y) + a < \varphi(z)$, which proves the lemma. $\qquad \square$

Theorem 3.7.3 (Jensen's Inequality) *Assume that $\mu^*(\Omega) = 1$. Let F be a real Banach space, D a convex subset of F, φ a convex function on D, and f an essentially μ-integrable mapping from Ω into F with values in D. If D is closed and φ lower semicontinuous, then $\int f d\mu$ lies in D, $\varphi \circ f$ is quasi-μ-integrable, and $\varphi(\int f d\mu) \leq \int (\varphi \circ f) d\mu$. The same result is true when D is open and φ continuous.*

PROOF: In the first case, we know that $\int f d\mu$ lies in D (Proposition 3.6.4). In the second case, let $p \in D^c$. There exists a continuous linear form u on F and a real number α such that $u(p) = \alpha$ and $u(y) > \alpha$ for all $y \in D$. Then $u(\int f d\mu) - \alpha = \int (u \circ f - \alpha) d\mu > 0$, so $\int f d\mu \neq p$. This proves that $\int f d\mu$ lies in D.

Now, in one case or the other, with the same notation as in Lemma 3.7.1, $\psi(\int f d\mu) = \int (\psi \circ f) d\mu \leq \int_* (\varphi \circ f) d\mu$ for all $\psi \in L$. Therefore, $\varphi(\int f d\mu) \leq \int (\varphi \circ f) d\mu$. $\qquad \square$

3.8   Atoms

Let μ be a Daniell measure on a space $\mathcal{H}(\Omega, \mathbf{C})$.

Definition 3.8.1 A μ-integrable set A is a μ-atom if $V\mu(A) > 0$ and if $V\mu(B)$ is equal either to $V\mu(A)$ or to 0 for every μ-integrable subset B of A.

$V\mu(A) = |\mu(A)|$ for every atom A, by Theorem 3.3.2.

Define an equivalence relation on the class $\mathcal{C}$ of atoms as follows: $A_1 \sim A_2$ if and only if $1_{A_1} = 1_{A_2}$ almost everywhere. Observe that, if A_1, A_2 lie in $\mathcal{C}$, then either $A_1 \cap A_2$ is negligible or $A_1 \sim A_2$. Let ρ be the canonical mapping from $\mathcal{C}$ onto the quotient space $\mathcal{C}/\sim$.

Proposition 3.8.1 *If E is an integrable set and $\mathcal{C}_E$ the class of those atoms which are contained in E, then $\rho(\mathcal{C}_E)$ is at most countable.*

PROOF: For every $n \geq 1$, let $\mathcal{C}_n$ be the set of those $A \in \mathcal{C}_E$ such that $V\mu(A) > (1/(n+1)) \cdot V\mu(E)$. Assume that $\rho(\mathcal{C}_n)$ has $n + 1$ distinct elements $\rho(A_1), \ldots, \rho(A_{n+1})$ for some integer $n \geq 1$. Then $A_i \cap A_j$ is negligible for distinct i, j, so $V\mu(A_1 \cup \ldots \cup A_{n+1}) = \sum_{1 \leq i \leq n+1} V\mu(A_i)$ by the inclusion-exclusion formula, whence $V\mu(A_1 \cup \ldots \cup A_{n+1}) > V\mu(E)$, a contradiction.

Hence $\rho(\mathcal{C}_n)$ has at most n elements, for every $n \geq 1$, and $\rho(\mathcal{C}_E)$ is countable.
□

Proposition 3.8.2 *Suppose μ is real. Let E be an integrable set containing no atom. Then $\{\mu(F) : F$ integrable subset of $E\}$ is a compact interval in $\mathbb{R}$.*

PROOF: Let E^+, E^- be as in Proposition 3.3.1. Observe that $V\mu(F) = \mu(F)$ (respectively, $V\mu(F) = -\mu(F)$) for all integrable subsets F of E^+ (respectively, of E^-), by Theorem 3.3.2.

If we can prove that the Proposition holds for E^+ and E^-, then we conclude that $\{\mu(F) : F$ integrable, $F \subset E\} = [\mu(E^-), \mu(E^+)]$. Therefore, it suffices to consider the case in which $\mu(E) > 0$ and $\mu(F) \geq 0$ for all integrable subsets F of E.

Let $0 < \beta < \mu(E)$ be a real number. We prove that there exists an integrable subset F of E such that $\mu(F) = \beta$.

By hypothesis, there exists an integrable subset A of E such that $0 < \mu(A) < \mu(E)$, and, replacing A by $E - A$, if necessary, we may suppose $0 < \mu(A) \leq \mu(E)/2$. Now, by induction, we produce decreasing integrable subsets A_n of E ($n \geq 1$) such that $0 < \mu(A_n) \leq \mu(E)/2^n$. Hence, for every $\varepsilon > 0$, we can find an integrable subset A of E such that $0 < \mu(A) \leq \varepsilon$.

Let γ be the supremum of the $\mu(A)$, where A extends over the class of integrable subsets of E such that $\mu(A) \leq \beta$.

Suppose $\gamma \leq \beta/2$. If A_1, A_2 are two integrable subsets of E such that $\mu(A_1) \leq \gamma$ and $\mu(A_2) \leq \gamma$, then $\mu(A_1 \cup A_2) \leq \beta$, and so $\mu(A_1 \cup A_2) \leq \gamma$. Consequently, there is an increasing sequence $(A_n)_{n \geq 1}$ of integrable subsets of E such that $\mu(A_n)$ converges to γ as $n \to +\infty$. If we put $A = \bigcup_{n > 1} A_n$, then $\mu(A) = \gamma$. Now there exists an integrable subset A' of $E - A$ such that $0 < \mu(A') \leq \beta - \gamma$; we have $\mu(A \cup A') \leq \beta$ and $\mu(A \cup A') > \gamma$, which contradicts the definition of γ. Hence $\gamma > \beta/2$.

Now, if C is an integrable subset of E such that $\mu(C) \leq \beta$, then $\mu(C) \leq \gamma$. So $\mu(E - C) > \beta - \gamma$, and (by the preceding argument) there is an integrable subset D of $E - C$ for which $(\beta - \gamma)/2 \leq \mu(D) \leq \beta - \gamma$. From the fact that $\mu(C \cup D) \leq \beta$, we have $\mu(C \cup D) \leq \gamma$, and $\mu(C) + (\beta - \gamma)/2 \leq \gamma$. Hence $\gamma + (\beta - \gamma)/2 \leq \gamma$, which proves that $\gamma = \beta$.

By induction on n, we construct an increasing sequence $(F_n)_{n \geq 1}$ of integrable subsets of E such that $\beta(1 - 1/2^n) \leq \mu(F_n) \leq \beta$. Indeed, suppose that $F_1, \ldots, F_n$ have been obtained for some $n \geq 1$. If $\mu(F_n) \geq \beta(1 - 1/2^{n+1})$, we take $F_{n+1} = F_n$. On the other hand, if $\mu(F_n) < \beta(1 - 1/2^{n+1})$, there exists an integrable subset F_n' of $E - F_n$ such that $\beta(1 - 1/2^{n+1}) - \mu(F_n) \leq \mu(F_n') \leq \beta - \mu(F_n)$, and we take $F_{n+1} = F_n \cup F_n'$. Then $\mu(\bigcup_{n \geq 1} F_n) = \beta$.
□

Proposition 3.8.3 *Let f be a μ-measurable mapping from Ω into a metrizable space. Then f is constant a.e. in each atom A.*

PROOF: f is the limit, a.e. in A, of a sequence $(f_n)_{n\geq1}$ of $\widehat{\mathcal{R}}/A$-simple mappings. Since A is an atom, each f_n is constant a.e. in A, and there exists a negligible subset N_n of A such that f_n takes the constant value y_n on $A - N_n$. Let N be a negligible subset of A containing all the N_n, such that $(f_n)_{n\geq1}$ converges to f on $A-N$. For all $x \in A-N$, the sequence $(f_n(x))_{n\geq1} = (y_n)_{n\geq1}$ converges to $f(x)$, and so f is constant on $A - N$. $\square$

Definition 3.8.2 μ is said to be diffuse if it has no atoms. μ is said to be atomic if each integrable set which is not negligible contains an atom.

Theorem 3.8.1 *Suppose μ is atomic. In order that a mapping from Ω into a metrizable space be μ-measurable, it is necessary and sufficient that it assume a constant value a.e. in each atom. $\int^{\bullet} f \cdot dV\mu = \sum_{\rho(A)\in\mathcal{F}} \int^{*} f1_A dV\mu$ for all functions f from Ω into $[0, +\infty]$, where $\rho(A)$ extends over the set $\mathcal{F}$ of classes of atoms. $\int f d\mu = \sum_{\rho(A)} \int f1_A d\mu$ for every essentially μ-integrable mapping from Ω into a Banach space.*

PROOF: Let f be a mapping from Ω into a metrizable space, and suppose that f is constant a.e. in each atom. Fix an integrable set E, and let $(A_i)_{i\in I}$ be a family of atoms contained in E, such that each atom contained in E is equivalent to one and only one A_i. The set $E - \bigcup_{i\in I} A_i$ contains no atom and hence is negligible. For every $i \in I$, there is a negligible subset N_i of A_i such that f is constant on $A_i - N_i$. Now let N be a negligible subset of E, containing $E-\bigcup_{i\in I} A_i$, $\bigcup_{i\in I} N_i$, and $\bigcup_{(i,j)\in I\times I, i\neq j} A_i \cap A_j$. Clearly, f is the limit on $E - N$ of a sequence of $\widehat{\mathcal{R}}/E$-simple mappings, which proves that f is μ-measurable.

Now let f be a function from Ω into $[0, +\infty]$. Then

$$\int^{\bullet} f \cdot dV\mu \leq \sum_{\rho(A)\in\mathcal{F}} \int^{*} f1_A \cdot dV\mu,$$

because $f1_E \leq \sum_{\rho(A)\in\mathcal{F}_E} f1_A$ almost everywhere for each integrable set E ($\mathcal{F}_E$ is the set of classes of those atoms that are contained in E).

Suppose that f is μ-measurable and vanishes outside a μ-measurable and μ-moderate set E. Then $\int^{*} f \cdot dV\mu = \sum_{\rho(A)\in\mathcal{F}_E} \int^{*} f1_A dV\mu$ by Beppo Levi's theorem. Hence $\int^{*} f \cdot dV\mu = \sum_{\rho(A)\in\mathcal{F}} \int^{*} f1_A \cdot dV\mu$.

Next, assume that $\int^{*} f \cdot dV\mu < +\infty$. Then, for every $\varepsilon > 0$, there exists $g \in \mathcal{J}^{+}$ such that $g \geq f$ and $\int^{*} g \cdot dV\mu \leq \int^{*} f \cdot dV\mu + \varepsilon$. But g is μ-measurable, so

$$\sum_{\rho(A)\in\mathcal{F}} \int^{*} f1_A \cdot dV\mu \leq \sum_{\rho(A)\in\mathcal{F}} \int^{*} g1_A \cdot dV\mu = \int^{*} g \cdot dV\mu \leq \int^{*} f \cdot dV\mu + \varepsilon,$$

which proves that $\int^{*} f \cdot dV\mu \geq \sum_{\rho(A)\in\mathcal{F}} \int^{*} f1_A \cdot dV\mu$.

Now suppose that $\int^\bullet f \cdot dV\mu < +\infty$. There exists a μ-measurable and μ-moderate set E such that $f = f1_E$ locally almost everywhere. So

$$\int^\bullet f \cdot dV\mu = \int^* f1_E \cdot dV\mu \geq \sum_{\rho(A)\in\mathcal{F}} \int^* f1_{A\cap E} \cdot dV\mu$$

$$= \sum_{\rho(A)\in\mathcal{F}} \int^* f1_A \cdot dV\mu.$$

We have therefore proved that $\int^\bullet f \cdot dV\mu = \sum_{\rho(A)\in\mathcal{F}} \int^* f1_A \cdot dV\mu$ for every $f : \Omega \to [0,+\infty]$.

Finally, let f be an essentially μ-integrable mapping from Ω into a Banach space, and let E be a μ-measurable and μ-moderate set such that $f = f1_E$ locally almost everywhere. Then $\int f d\mu = \int f1_E d\mu = \sum_{\rho(A)\in\mathcal{F}_E} \int f1_A d\mu = \sum_{\rho(A)\in\mathcal{F}} \int f1_A d\mu$ by the dominated convergence theorem. $\square$

3.9 Prolongations of μ

If λ is a measure on a semiring S in Ω, and μ is the linear form $f \mapsto \int f d\lambda$ on $St(S,\mathbf{C})$, then we put $\int^* f \cdot dV\mu = \int^* f \cdot dV\lambda$ and $\int^\bullet f \cdot dV\mu = \int^\bullet f dV\lambda$ for every $f : \Omega \to [0,+\infty]$, and so on ($\mathcal{J}^+$ defined as in Chapter 2).

Now consider a general Daniell measure μ and an arbitrary $\mathcal{J}^+$. Define $\widehat{\mathcal{R}}$ (respectively, $\overline{\mathcal{R}}$) as the ring of μ-integrable sets (respectively, of essentially μ-integrable sets).

The function $\hat{\mu} : A \mapsto \int 1_A d\mu$ from $\widehat{\mathcal{R}}$ into $\mathbf{C}$ is a quasi-measure, and hence is a measure, by Theorem 3.3.2. Moreover, $V\hat{\mu}(A) = \int 1_A dV\mu$ for all $A \in \widehat{\mathcal{R}}$.

Definition 3.9.1 $\hat{\mu}$ is called the main prolongation of μ.

Theorem 3.9.1 $\int^* f \cdot dV\mu = \int^* f \cdot dV\hat{\mu}$ for every function f from Ω into $[0,+\infty]$. $\mathcal{L}^1_F(\mu) = \mathcal{L}^1_F(\hat{\mu})$ for every Banach space F, and $\int f d\mu = \int f d\hat{\mu}$ for all $f \in \mathcal{L}^1_F(\mu)$. A mapping from Ω into a metrizable space is μ-measurable if and only if it is $\hat{\mu}$-measurable.

PROOF: If $(g_n)_{n\geq 1}$ is an increasing sequence in $St^+(\widehat{\mathcal{R}})$ and $g = \sup_{n\geq 1} g_n$, then $\int^* g \cdot dV\hat{\mu} = \sup_n \int g_n \cdot dV\hat{\mu} = \sup_n \int g_n \cdot dV\mu = \int^* g \cdot dV\mu$. This shows that $\int^* f \cdot dV\mu \leq \int^* f \cdot dV\hat{\mu}$ for every function f from Ω into $[0,+\infty]$. For the reverse inequality, observe that every $g \in \mathcal{J}^+$ such that $\int^* g \cdot dV\mu < +\infty$ is μ-integrable, and hence is the upper envelope of an increasing sequence in $St^+(\widehat{\mathcal{R}})$ (Proposition 3.3.3). Thus $\int^* g \cdot dV\mu = \int^* g \cdot dV\hat{\mu}$. It follows easily that $\int^* f \cdot dV\hat{\mu} \leq \int^* f \cdot dV\mu$ for every $f : \Omega \to [0,+\infty]$.

Next, let F be a Banach space. On $\mathcal{F}^1_F(\hat{\mu}) = \mathcal{F}^1_F(\mu)$ define the seminorm $f \mapsto \int^* |f| dV\mu = \int^* |f| dV\hat{\mu}$. Since $St(\mathcal{R}, F)$ is dense in the closed subspace

$\mathcal{L}_F^1(\mu)$ of $\mathcal{F}_F^1(\mu)$, $\mathcal{L}_F^1(\mu)$ is the closure of $St(\widehat{\mathcal{R}}, F)$ in $\mathcal{F}_F^1(\hat{\mu})$; thus $\mathcal{L}_F^1(\mu) = \mathcal{L}_F^1(\hat{\mu})$. The linear mappings $f \mapsto \int f\, d\mu$ and $f \mapsto \int f\, d\hat{\mu}$ from $\mathcal{L}_F^1(\mu)$ into F, which are continuous and agree on $St(\widehat{\mathcal{R}}, F)$, are in fact identical. $\square$

Now, for $A \in \overline{\mathcal{R}}$, $V\mu(A) = \sup_{(A_i)} \sum_{i \in I} |\mu(A_i)|$, $(A_i)_{i \in I}$ ranging over the class of finite partitions of A into $\overline{\mathcal{R}}$-sets. Indeed, every $E \in \overline{\mathcal{R}}$ is a union of an integrable set and a locally negligible set. Therefore, $\bar{\mu} : A \mapsto \int 1_A\, d\mu$ is a measure on $\overline{\mathcal{R}}$, and $V\bar{\mu}(A) = \int 1_A\, dV\mu$ for all $A \in \overline{\mathcal{R}}$.

Definition 3.9.2 $\bar{\mu}$ is called the essential prolongation of μ.

Theorem 3.9.2 $\int^* f \cdot dV\bar{\mu} = \int^{\bullet} f \cdot dV\mu$ for every function f from Ω into $[0, +\infty]$. $\mathcal{L}_F^1(\bar{\mu}) = \overline{\mathcal{L}_F^1}(\mu)$ for every Banach space F, and $\int f\, d\bar{\mu} = \bar{\int} f\, d\mu$ for all $f \in \mathcal{L}_F^1(\mu)$. A mapping from Ω into a metrizable space is $\bar{\mu}$-measurable if and only if it is μ-measurable.

PROOF: Let f be a function from Ω into $[0, +\infty]$. By the definition of $\int^* f \cdot dV\bar{\mu}$ and $\int^* f \cdot dV\hat{\mu}$, we see that $\int^* f \cdot dV\bar{\mu} \leq \int^* f \cdot dV\hat{\mu}$, and that $\int^* f \cdot dV\hat{\mu} \leq \int^* f \cdot dV\bar{\mu}$ if f is moderate.

Now $\int^* f1_A \cdot dV\mu = \int^* f1_A \cdot dV\hat{\mu} = \int^* f1_A \cdot dV\bar{\mu} \leq \int^* f \cdot dV\bar{\mu}$ for every moderate set A, so $\int^{\bullet} f \cdot dV\mu \leq \int^* f \cdot dV\bar{\mu}$.

Next, suppose $\int^{\bullet} f \cdot dV\mu < +\infty$. There exists a subset A of Ω, measurable and moderate, such that $f = f1_A$ l.a.e. But any locally negligible set is $\bar{\mu}$-negligible; hence

$$\int^* f \cdot dV\bar{\mu} = \int^* f1_A \cdot dV\bar{\mu} = \int^* f1_A \cdot dV\hat{\mu} = \int^* f1_A \cdot dV\mu$$
$$\leq \int^{\bullet} f \cdot dV\mu.$$

This shows that $\int^* f \cdot dV\bar{\mu} = \int^{\bullet} f \cdot dV\mu$.

If F is a Banach space, then $\mathcal{L}_F^1(\bar{\mu}) = \bar{\mathcal{L}}_F^1(\mu)$ by the same argument as that of Theorem 3.9.1. The remaining assertions are obvious. $\square$

Proposition 3.9.1 $\hat{\hat{\mu}} = \hat{\mu}$, $\hat{\bar{\mu}} = \bar{\mu}$, $\bar{\hat{\mu}} = \bar{\mu}$, and $\bar{\bar{\mu}} = \bar{\mu}$.

PROOF: These facts are immediate from Theorems 3.9.1 and 3.9.2. $\square$

Of particular interest is the following problem. Given a measure μ_Φ on a semiring Φ in Ω, when is it true that $\hat{\mu} = \hat{\mu}_\Phi$ (respectively, $\bar{\mu} = \bar{\mu}_\Phi$)? A partial answer is given in Propositions 3.9.2 and 3.9.3.

Proposition 3.9.2 Let Φ be a semiring of essentially μ-integrable sets, and denote by μ_Φ the measure $E \to \int 1_E d\mu$ on Φ. Suppose that *(a)* $St(\Phi, \mathbf{C})$ *is dense in* $\bar{\mathcal{L}}^1_{\mathbf{C}}(\mu)$;

(b) *any set* $A \subset \Omega$ *whose trace* $A \cap E$ *on* E *is locally* μ-*negligible for every* $E \in \Phi$ *is itself locally* μ-*negligible.*

Then $\bar{\mathcal{L}}^1_{\mathbf{C}}(\mu_\Phi) \subset \bar{\mathcal{L}}^1_{\mathbf{C}}(\mu)$; besides, $\int f \cdot dV\mu_\Phi = \int f \cdot dV\mu$ and $\int f d\mu_\Phi = \int f \cdot d\mu$ for all $f \in \bar{\mathcal{L}}^1_{\mathbf{C}}(\mu_\Phi)$. Finally, $\bar{\mu}_\Phi = \bar{\mu}$ if (and only if) each locally μ-negligible set is locally μ_Φ-negligible.

PROOF: Since $St(\Phi, \mathbf{C})$ is dense in $\bar{\mathcal{L}}^1_{\mathbf{C}}(\mu)$, the same argument as that of Proposition 3.3.5 and Theorem 3.3.2 gives $V\mu_\Phi(E) = \int 1_E \cdot dV\mu$ for all $E \in \Phi$. Then $\int^\bullet h \cdot dV\mu = \int^* h \cdot dV\mu_\Phi$ for all upper envelopes h of increasing sequences in $St^+(\Phi)$, and so $\int^\bullet f \cdot dV\mu \le \int^* f \cdot dV\mu_\Phi$ for all functions f from Ω into $[0, +\infty]$. We conclude that $\bar{\mathcal{L}}^1_{\mathbf{C}}(\mu_\Phi) \subset \bar{\mathcal{L}}^1_{\mathbf{C}}(\mu)$, and that $\int f \cdot dV\mu_\Phi = \int f \cdot dV\mu$ and $\int f d\mu_\Phi = \int f d\mu$ for all $f \in \bar{\mathcal{L}}^1_{\mathbf{C}}(\mu_\Phi)$.

If a set A is locally μ_Φ-negligible, its trace $A \cap E$ on each $E \in \Phi$ is locally μ-negligible, so A is locally μ-negligible by hypothesis (b). Hence $\bar{\mathcal{L}}^1_{\mathbf{C}}(\mu_\Phi) \subset \bar{\mathcal{L}}^1_{\mathbf{C}}(\mu)$.

To finish the proof, suppose that each locally μ-negligible set is locally μ_Φ-negligible. If f belongs to $\bar{\mathcal{L}}^1_{\mathbf{C}}(\mu)$, it is locally μ-almost everywhere the limit of a Cauchy sequence $(f_n)_{n \ge 1}$ in $St(\Phi, \mathbf{C})$ (which may be regarded either as a subspace of $\bar{\mathcal{L}}^1_{\mathbf{C}}(\mu_\Phi)$ or of $\bar{\mathcal{L}}^1_{\mathbf{C}}(\mu)$). Then, since f is locally μ_Φ-almost everywhere the limit of $(f_n)_{n \ge 1}$, it belongs to $\bar{\mathcal{L}}^1_{\mathbf{C}}(\mu_\Phi)$. □

Proposition 3.9.3 *Let* Φ *be a semiring of* μ-*integrable sets and let* μ_Φ *be the measure* $E \rightarrow \int 1_E \cdot d\mu$ *on* Φ. *Suppose that* $St(\Phi, \mathbf{C})$ *is dense in* $\mathcal{L}^1_{\mathbf{C}}(\mu)$. *Then* $\mathcal{L}^1_{\mathbf{C}}(\mu_\Phi) \subset \mathcal{L}^1_{\mathbf{C}}(\mu)$, *besides,* $\int f dV\mu_\Phi = \int f \cdot dV\mu$ *and* $\int f d\mu_\Phi = \int f d\mu$ *for all* $f \in \mathcal{L}^1_{\mathbf{C}}(\mu_\Phi)$. *Finally,* $\hat{\mu}_\Phi = \hat{\mu}$ *if and only if each* μ-*negligible set is* μ_Φ-*negligible.*

PROOF: Argue as in Proposition 3.9.2. □

The next result, motivated by Proposition 3.9.3, deals with semirings Φ for which $St(\Phi, \mathbf{C})$ is dense in $\bar{\mathcal{L}}^1_{\mathbf{C}}(\mu)$.

Proposition 3.9.4 *Let* S, Φ *be two semirings of* μ-*integrable sets, and let* $\tilde{\Phi}$ *be the* σ-*ring generated by* Φ. *Suppose that, for each* $E \in S$, *there exists* $B \in \tilde{\Phi}$ *for which* $B \triangle E$ *is* μ-*negligible. If* $St(S, \mathbf{C})$ *is dense in* $\mathcal{L}^1_{\mathbf{C}}(\mu)$, *then so is* $St(\Phi, \mathbf{C})$.

PROOF: Let V be the closure of $St(\Phi, \mathbf{C})$ in $\mathcal{L}^1_{\mathbf{C}}(\mu)$. Given $A \in \Phi$, define T_A to be the class of the $B \subset A$ whose indicator function lies in V. For every $B \in T_A$ and for every $\varepsilon > 0$, there exists $f \in St^+(\Phi)$ such that $\int^* |1_B - f| \cdot dV\mu \le \varepsilon$, and therefore such that $\int^* |1_{A-B} - (1_A - f)| \cdot dV\mu \le \varepsilon$; this proves that $A - B$

lies in T_A. Similarly, if B_1, B_2 are elements of T_A, there exist f_1, $f_2 \in St^+(\Phi)$ such that $\int^* |1_{B_i} - f_i| \cdot dV\mu \leq \varepsilon/2$ for every $1 \leq i \leq 2$, and then

$$\int^* \left| \sup(1_{B_1}, 1_{B_2}) - \sup(f_1, f_2) \right| \cdot dV\mu \leq \varepsilon,$$

which shows that $B_1 \cup B_2$ lies in T_A. Finally, let $(B_n)_{n \geq 1}$ be an increasing sequence in T_A, and put $B = \bigcup_{n \geq 1} B_n$; the sequence $(1_{B_n})_{n \geq 1}$ converges to 1_B in $\mathcal{L}_{\mathbf{C}}^1(\mu)$, so 1_B belongs to V. Now T_A is a σ-algebra in A. This means that the class of those sets $E \subset \Omega$ for which $1_{E \cap A}$ lies in V for every $A \in \Phi$ is a σ-ring containing $\tilde{\Phi}$.

Now let $E \in S$ and $B \in \tilde{\Phi}$ as in the statement of the proposition. There is a sequence $(A_n)_{n \geq 1}$ of disjoint Φ-sets such that $B \subset \bigcup_{n \geq 1} A_n$. The functions $\sum_{1 \leq k \leq n} 1_{A_k \cap B}$ lie in V and they converge pointwise to 1_B as $n \to +\infty$. The monotone convergence theorem then shows that 1_B, and hence 1_E, lies in V. Finally, observe that $St(S, \mathbf{C}) \subset V$ and $V = \mathcal{L}_{\mathbf{C}}^1(\mu)$. $\square$

Exercises for Chapter 3

1 Let μ be a positive Daniell measure on $\mathcal{H}(\Omega, \mathbf{C})$. Let $(f_n)_{n \geq 1}$ be a sequence of integrable functions from Ω into $[0, +\infty]$, and let f be an integrable function from Ω into $[0, +\infty]$ such that $f \leq \liminf_{n \to +\infty} f_n$ almost everywhere.

   If $\int f_n d\mu$ converges to $\int f d\mu$ as $n \to +\infty$, prove that $(f_n)_{n \geq 1}$ converges to f in the mean.

2 Let μ be a positive Daniell measure on $\mathcal{H}(\Omega, \mathbf{C})$, F a real Banach space, and $(f_n)_{n \geq 1}$ a sequence in $\mathcal{L}_F^1(\mu)$ converging almost everywhere to f (f in $\mathcal{L}_F^1(\mu)$). Suppose that $N_1(f_n)$ converges to $N_1(f)$ as $n \to +\infty$. Then prove that $(f_n)_{n \geq 1}$ converges to f in the mean (observe that $\big| \|f_n\| - |f_n - f| \big| \leq |f|$).

3 Let μ be a positive Daniell measure on $\mathcal{H}(\Omega, \mathbf{C})$. Let $(a_n)_{n \geq 1}$, $(b_n)_{n \geq 1}$, $(f_n)_{n \geq 1}$ be three sequences in $\mathcal{L}^1(\mu; \overline{\mathbf{R}})$, converging almost everywhere to functions a, b, and f respectively, and such that $a_n \leq f_n \leq b_n$ almost everywhere. Suppose that a, b are integrable and that $\int a_n d\mu$, $\int b_n d\mu$ go to $\int a d\mu$ and $\int b d\mu$, respectively, as $n \to +\infty$.

1. Show that $\int_* f d\mu = \int^* f d\mu$ (use Fatou's lemma). Deduce that f is integrable and that $\int f_n d\mu$ tends to $\int f d\mu$ as $n \to +\infty$.

2. Prove that, if $(a_n)_{n \geq 1}$ converges to a in the mean, then $(f_n)_{n \geq 1}$ converges to f in the mean (use Exercise 2).

4 Let μ be a positive Daniell measure on $\mathcal{H}(\Omega, \mathbf{C})$. Let $(A_n)_{n \geq 1}$ be a sequence of subsets of Ω. Suppose there is a sequence $(B_n)_{n \geq 1}$ of disjoint μ-measurable sets such that $A_n \subset B_n$ for every $n \geq 1$. Prove that $\mu^*\left(\bigcup_{n \geq 1} A_n\right) = \sum_{n \geq 1} \mu^*(A_n)$ and that $\mu_*\left(\bigcup_{n \geq 1} A_n\right) = \sum_{n \geq 1} \mu_*(A_n)$.

5 Let μ be a real Daniell measure on $\mathcal{H}(\Omega, \mathbf{C})$ and let E be an integrable set. Show that $\{\mu(B) : B \text{ integrable}, B \subset E\}$ is compact (use Proposition 3.8.2).

4

Lebesgue Measure on **R**

At this point, it seems useful to look at what we have done in the light of some examples. Doing so, will also provide us with many counterexamples for the next chapters.

Summary

4.1 This is a short review of base-b expansions of a real number. These expansions will be often used throughout the text and the exercises.

4.2 We now define the famous Cantor singular function: continuous, increasing, but its derivative is 0 almost everywhere (for Lebesgue measure).

4.3 We prove the existence of "pathological" sets. In particular, using Cantor singular function we prove the existence of a non-Borelian negligible set.

4.1   Base-b Expansions of a Real Number

Let $b > 1$ be an integer and x a real number. For each integer $n \geq 0$, let p_n be the greatest integer such that $r_n = p_n \cdot b^{-n}$ is less than x; then $x - b^{-n} < r_n \leq x$. Since $bp_{n-1} \cdot b^{-n} \leq x$, and so $r_{n-1} \leq r_n$, for every $n \geq 1$, the sequence $(r_n)_{n \geq 0}$ increases to x. For each integer $n \geq 1$, put $u_n = p_n - bp_{n-1}$. Since $bp_{n-1} \cdot b^{-n} \leq x < b(p_{n-1} + 1) \cdot b^{-n}$, we have $0 \leq u_n \leq b - 1$. Now $r_n = p_0 + \sum_{1 \leq k \leq n} u_k/b^k$, hence $x = p_0 + \sum_{k \geq 1} u_k/b^k$.

Definition 4.1.1 With notation as above, $(p_0, (u_n)_{n \geq 1})$ is called the proper expansion of x (with respect to the base b).

Put $A = \{0, 1, 2, \ldots, b-1\}$. By what we have just shown, the function $\varphi : (q_0, (v_n)_{n \geq 1}) \mapsto q_0 + \sum_{n \geq 1} v_n/b^n$ sends $\mathbf{Z} \times A^{\mathbf{N}}$ onto $\mathbf{R}$. We will prove that, if $x \in \mathbf{R}$ is not of the form k/b^n (with $k \in \mathbf{Z}$, $n \geq 0$), it is the image under φ of its sole proper expansion. On the other hand, if x has the form k/b^n, we will show that it is the image under φ of two elements, namely, its proper expansion and another expansion which we call improper.

So let $(q_0, (v_n)_{n \geq 1})$ be such that $x = q_0 + \sum_{n \geq 1} v_n/b^n$ and, for every integer $m \geq 0$, put $s_m = q_0 + \sum_{1 \leq n \leq m} v_n/b^n$. Clearly, $s_m \leq x \leq s_m + 1/b^m$, and $x < s_m + 1/b^m$ unless $v_n = b-1$ for any $n > m$. We conclude that $r_m = s_m$ or $r_m = s_m + 1/b^m$, the latter equality holding only when $v_n = b-1$ for all integers $n > m$. If the number of integers $n \geq 1$ such that $v_n < b-1$ is finite, and if m is the smallest positive integer such that $v_n = b-1$ for $n > m$, then $x = q_0 + \sum_{1 \leq n \leq m} v_n/b^n + 1/b^m$ has the form k/b^m ($k \in \mathbf{Z}$). Thus, if x is not of the form k/b^n, $(q_0, (v_n)_{n \geq 1})$ is necessarily the proper expansion of x.

Now suppose that x has the form k/b^n. There are two possibilities:

1. $v_n < b-1$ for infinitely many indices n. Then $(q_0, (v_n)_{n \geq 1})$ is the proper expansion of x.

2. $v_n < b-1$ for only finitely many indices n. Then let m be the smallest positive integer such that $v_n = b-1$ for $n > m$, so that x has the form k/b^m (with $k \in \mathbf{Z}$). When $m \geq 1$, k is not divisible by b. Thus m is the smallest of those integers $m' \geq 0$ such that x has the form $k'/b^{m'}$ ($k' \in \mathbf{Z}$). Let $(p_0, (u_n)_{n \geq 1})$ be the proper expansion of x. If $m = 0$, then $p_0 = q_0 + 1$ and $u_n = 0$ for $n \geq 1$. On the other hand, if $m \geq 1$, then $p_0 = q_0$, $u_n = v_n$ for $1 \leq n < m$, $u_m = v_m + 1$, and $u_n = 0$ for $n > m$.

In short, when x has the form k/b^n, it is the image under φ of at most two elements, and in fact of exactly two elements, and our claim is true.

A real number x has a "terminating expansion" (such that $v_n = 0$ for n large enough) if and only if it has the form k/b^n ($n \geq 0$, $k \in \mathbf{Z}$), and in that case its terminating expansion is its proper expansion.

Next, give $A^{\mathbf{N}}$ the lexicographic ordering ($(u_n)_{n \geq 1} < (v_n)_{n \geq 1}$ if $u_m < v_m$ for the smallest, m, of those integers n such that $u_n \neq v_n$). Then the function $G : (v_n)_{n \geq 1} \mapsto \sum_{n \geq 1} v_n/b^n$ from $A^{\mathbf{N}}$ onto $[0, 1]$ increases. Furthermore, for u, $v \in A^{\mathbf{N}}$ satisfying $v < u$, the equality $G(u) = G(v)$ holds if and only if u, v are, respectively, the proper expansion and the improper expansion of a same $x \in {]0, 1[} \cap B$, where $B = \{k/b^n : n \geq 1, \ k \in \mathbf{Z}\}$.

Definition 4.1.2 When $b = 2$ (respectively, $b = 3$, $b = 10$), the proper base-b expansion of x is called its proper dyadic (respectively, triadic, decimal) expansion. The improper base-b expansion of x, when x has the form k/b^n, is called its improper dyadic (respectively, triadic, decimal) expansion.

4.2   The Cantor Singular Function

For any nonempty subinterval I of **R**, we shall denote by $\alpha(I)$ its left endpoint and by $\beta(I)$ its right endpoint.

Define, by induction on n, a family $(I_{n,p})_{n,p}$ of mutually disjoint, open intervals, as follows. The integer n takes all positive values; for each $n \geq 0$, p takes the values $1, 2, 3, \ldots, 2^n$. All intervals $I_{n,p}$ are contained in $]0, 1[$, and we take $I_{0,1} =]1/3, 2/3[$. Next, suppose the $2^{n+1} - 1$ intervals $I_{m,p}$ have been defined for $0 \leq m \leq n$, in such a way that, if J_n is their union, then $[0, 1] \cap J_n^c$ is the union of 2^{n+1} closed intervals $K_{n,p}$ (with $1 \leq p \leq 2^{n+1}$), mutually disjoint, all of length $1/3^{n+1}$, and such that $\alpha(K_{n,1}) < \alpha(K_{n,2}) < \ldots < \alpha(K_{n,2^{n+1}})$. If $K_{n,p} = [a, b]$, then we take for $I_{n+1,p}$ the open interval $]a + (b-a)/3, b - (b-a)/3[$.

Definition 4.2.1 The complement K in $I = [0, 1]$ of the union of the $I_{n,p}$ is called the Cantor set. The $I_{n,p}$ are called the intervals contiguous to K.

Let μ be Lebesgue measure on I.

Since $\mu(I - J_n) = (2/3)^{n+1}$ for every $n \geq 1$, the Cantor set $K = \bigcap_{n \geq 0}(I - J_n)$ is μ-negligible, hence totally disconnected. If n, p are two integers such that $n \geq 0$ and $1 \leq p \leq 2^{n+1}$, then $2^{n+1} - p$ can be written $\lambda_1 2^n + \ldots + \lambda_{n+1} 2^0$, where $\lambda_1, \ldots, \lambda_{n+1}$ are uniquely determined elements of $\{0, 1\}$. By induction on n, it is easily shown that

(a) $\alpha(K_{n,p}) = \sum_{1 \leq k \leq n+1} 2(1 - \lambda_k)/3^k$;

(b) If we set $p_j = 2^{j+1} - (\lambda_1 2^j + \ldots + \lambda_{j+1} 2^0)$ for all $0 \leq j \leq n - 1$, then $K_{n,p} \subset K_{n-1, p_{n-1}} \subset \ldots \subset K_{0, p_0}$.

Denote by ψ the function $(x_k)_{k \geq 1} \mapsto \sum_{k \geq 1} x_k/3^k$ from $B = \{0, 1, 2\}^{\mathbf{N}}$ into I.

Given integers $n \geq 0$ and $1 \leq p \leq 2^{n+1}$, let $(\nu_k)_{k \geq 1}$ be the proper triadic expansion of $\alpha(K_{n,p})$. Now let $(x_k)_{k \geq 1}$ be a point of B and x its image under ψ. In order that x lie in $K_{n,p}$ it is necessary and sufficient that one of the following three conditions hold:

(a) $x_k = \nu_k$ for every $1 \leq k \leq n + 1$.

(b) $\alpha(K_{n,p}) \neq 0$ and $(x_k)_{k \geq 1}$ is its improper triadic expansion.

(c) $\beta(K_{n,p}) \neq 1$ and $(x_k)_{k \geq 1}$ is its proper triadic expansion.

In the particular case where $(x_k)_{k \geq 1}$ belongs to $C = \{0, 2\}^{\mathbf{N}}$, it follows that x lies in $K_{n,p}$ if and only if $x_k = \nu_k$ for all $1 \leq k \leq n + 1$.

Next, give C the lexicographic ordering. Clearly, the restriction of ψ to C is strictly increasing and $\psi(C)$ is contained in K. Conversely, let $x \in K$. For each $n \geq 0$, write α_n for the origin of the $K_{n,p}$ containing x. Then $\alpha_{n+1} - \alpha_n$

has the form $x_{n+2}/3^{n+2}$ for a suitable x_{n+2} in $\{0,2\}$, whence we deduce that x lies in $\psi(C)$. Therefore, $K = \psi(C)$ has the cardinality of the continuum.

Now denote by $\hat{K}$ the set of the $\beta(K_{n,p})$. If x lies in $\hat{K}$, then its improper triadic expansion belongs to C. On the other hand, if $(x_k)_{k\geq 1}$ is an element of C and if there exists an integer $n \geq 0$ such that $x_k = 2$ for all $k \geq n+2$, then $\psi((x_k)_k) = \beta(K_{n,p})$ for a suitable $1 \leq p \leq 2^{n+1}$; thus, if x is a point of $K - \hat{K}$, then its proper triadic expansion belongs to C. Finally, a necessary and sufficient condition that $x \in K$ be the image under ψ of two distinct elements of B is that it be an endpoint of some interval contiguous to K.

Proposition 4.2.1   *Let f be the function from K into $[0,1]$ such that $f\left(\psi((x_k)_k)\right) = \sum_{k\geq 1} x_k/2^{k+1}$ for each $(x_k)_k$ of C. There exists a function g from I into I which extends f and is constant over each interval $\bar{I}_{n,p}$ (with $n \geq 0, 1 \leq p \leq 2^n$).*

PROOF: Clearly, f is surjective and increasing. The unique pairs (x,y) in $K \times K$ such that $x < y$ and $f(x) = f(y)$ are the pairs $(\alpha(I_{n,p}), \beta(I_{n,p}))$, whence we deduce the existence of g. □

Definition 4.2.2 g is called the Cantor singular function.

Now let $(x_k)_k$ be an element of B and $x = \psi((x_k)_k)$. Define m as the supremum of those integers $k \geq 0$ such that $x_1, \ldots, x_k$ are all even (m may be equal to $+\infty$). If x lies in $I - K$, and if we put $q = \sum_{1\leq k\leq m}(x_k/2)\cdot 2^{m-k}+1$, then x belongs to $I_{m,q}$ and $g(x) = f(\beta(I_{m,q}))$, whence we see that

$$g(x) = \sum_{1\leq k\leq m} \frac{x_k}{2^{k+1}} + \frac{1}{2^{m+1}}.$$

This last equality is true even when x lies in K. Hence we have obtained the explicit expression of $g(x)$ for all $x \in I$. Observe that g is equal to $(2p-1)/2^{n+1}$ on each $I_{n,p}$.

Proposition 4.2.2 *For every $n \geq 0$, put $K_n = \bigcup_{1\leq p\leq 2^{n+1}} K_{n,p}$, and write 1_{K_n} for the indicator function of K_n on I. Also, set $h_n = (3/2)^{n+1}\cdot 1_{K_n}$, and let g_n be the function $x \mapsto \int_0^x h_n(t)dt = \int h_n 1_{[0,x]}d\mu$ on I. Then the functions g_n converge uniformly to the Cantor singular function g.*

PROOF: If x is a point of some $I_{m,p}$ ($m \geq 0, 1 \leq p \leq 2^m$), then $g_{m+k}(x) = (2p-1)/2^{m+1}$ for all integers $k \geq 0$, so $(g_{n+1} - g_n)(x) = 0$ for each integer $n \geq m$. On the other hand, if x is a point of some $K_{n,p}$ ($n \geq 0, 1 \leq p \leq 2^{n+1}$), then $g_n(\alpha(K_{n,p})) \leq g_n(x) \leq g_n(\beta(K_{n,p}))$, whence we deduce that $g_n(x)$ and, similarly, $g_{n+1}(x)$ lie in $[(p-1)/2^{n+1}, p/2^{n+1}]$. So $|g_{n+1} - g_n| \leq 1/2^{n+1}$ for all integers $n \geq 0$.

Clearly, the limit of the g_n is constant on each connected component of $I - K$.

Now let $(x_k)_k$ be an element of C and $x = \psi((x_k)_k)$. For any integer $n \geq 0$, x lies in $K_{n,p}$ for some p (with $1 \leq p \leq 2^{n+1}$), and we can compute p (we showed this after defining the function ψ). It follows that

$$\sum_{1 \leq k \leq n+1} \frac{x_k}{2^{k+1}} \leq g_n(x) \leq \sum_{1 \leq k \leq n+1} \frac{x_k}{2^{k+1}} + \frac{1}{2^{n+1}}.$$

We conclude that $(g_n)_{n \geq 0}$ converges (uniformly) to g. □

In short, g is continuous, increasing, and has derivative zero μ-almost everywhere, and yet is the uniform limit of the absolutely continuous functions g_n. By Proposition 12.1.4, g has derivative $+\infty$ at uncountably many points of I.

4.3   Example of a Nonmeasurable Set

For any subset E of **R** and any real number a, we define $E + a$ as the set $\{x + a : x \in E\}$. More generally, for any subsets E, F of **R**, we let $E + F$ be the set $\{x + y : x \in E, y \in F\}$. Finally, we write λ for Lebesgue measure on **R**.

Proposition 4.3.1 *If E is a λ-integrable set with measure $\lambda(E) > 0$, then $D(E) = \{x - y : x, y \in E\}$ is a neighborhood of 0.*

PROOF: For any real number $0 < \alpha < 1$, there is an open set U containing E, such that $\alpha \cdot \lambda^*(U) \leq \lambda(E)$ (see Theorem 6.1.1). Then $\lambda(E \cap J) \geq \alpha \cdot \lambda(J)$ for at least one connected component of U.

Now take $\alpha = 3/4$, and let J be a bounded open subinterval of **R** such that $\lambda(E \cap J) \geq 3\lambda(J)/4$. If x is a point of $]-\lambda(J)/2, \lambda(J)/2[$, then $\lambda(J \cup (J+x)) < 3\lambda(J)/2$, so $E \cap J$ and $(E \cap J) + x$ cannot be disjoint. Hence $D(E)$ contains $]-\lambda(J)/2, \lambda(J)/2[$. □

Proposition 4.3.2 *There is a (nonmeasurable) subset M of **R** such that $\lambda_*(M \cap E) = 0$ and $\lambda^*(M \cap E) = \lambda^*(E)$ for every λ-measurable subset E of **R**.*

PROOF: Let θ be an irrational number, and let B (respectively, C) be the set of numbers of the form $n + m\theta$, where $m \in \mathbf{Z}$ and n is an even (respectively, odd) integer. For each integer $i \geq 0$, there is a unique even integer $n_i \in \mathbf{Z}$ such that $0 \leq n_i + i\theta < 2$; put $x_i = n_i + i\theta$. Now, if J is a bounded open interval, let $k \in \mathbf{N}$ be such that $\lambda(J) \geq 2/k$; among the numbers $x_0, \ldots, x_k$, there are at least two, say x_i and x_j, such that $|x_i - x_j|$ is strictly less than $2/k$, and then there exists $r \in \mathbf{Z}$ for which $r(x_i - x_j)$ belongs to J. This proves that B is dense in $\mathbf{R}$. Clearly, $C = B + 1$ is also dense in $\mathbf{R}$.

Now define an equivalence relation on $\mathbf{R}$ as follows: $x \sim y$ whenever $x - y$ belongs to $A = \mathbf{Z} + \theta\mathbf{Z}$. Let L be a subset of $\mathbf{R}$ containing exactly one point of each equivalence class, and put $M = L + B$. If E is any λ-integrable subset of M, then $D(E)$ contains no point of the dense set C; therefore, $\lambda(E) = 0$ by Proposition 4.3.1. We conclude that $\lambda_*(M) = 0$. Likewise, $\lambda_*(M^c) = \lambda_*(L + C) = 0$. Now $\lambda^*(M \cap E) = \lambda^*(E)$ for every λ-measurable subset E of $\mathbf{R}$, by Proposition 3.7.11. Finally, M cannot be λ-measurable, because $M \cup (M + 1) = \mathbf{R}$ is not λ-negligible.   □

Proposition 4.3.3 *There exists a λ-negligible set which is not a Borel set.*

PROOF: Let $\bar{g}$ be the function that agrees with the Cantor singular function g on $I = [0, 1]$ and that is equal to 0 on $] - \infty, 0]$ and to 1 on $[1, +\infty[$. The continuous function $h : x \mapsto (x + \bar{g}(x))/2$ is strictly increasing on $\mathbf{R}$; hence it is a homeomorphism, with inverse h^{-1}. Since $\lambda\big(h(I_{n,p})\big) = \lambda(I_{n,p})/2$ for all integers $n \geq 0$ and $1 \leq p \leq 2^n$, we see that $\lambda\big(h(I - K)\big) = 1/2$. Therefore, $\lambda\big(h(K)\big) = 1/2$, even though the Cantor set K is λ-negligible.

With notation as in Proposition 4.3.2, $E = h^{-1}\big(M \cap h(K)\big)$ is λ-negligible, because it is included in K, but its image $M \cap h(K)$ under h is not λ-measurable. Finally, E is not a Borel set, because $h(E)$ is not a Borel set.   □

Proposition 4.3.4 *Let M be as in Proposition 4.3.2 and let $\mathcal{M}$ be the class of λ-measurable sets. Then the class $\mathcal{F}$ of the sets $(M \cap E_1) \cup (M^c \cap E_2)$, for E_1, E_2 in $\mathcal{M}$, is a σ-algebra containing $\mathcal{M}$, and*

$$\rho\big((M \cap E_1) \cup (M^c \cap E_2)\big) = \frac{1}{2}\lambda^*(E_1) + \frac{1}{2}\lambda^*(E_2)$$

consistently defines a σ-additive function ρ on $\mathcal{F}$ such that $\rho(E) = \lambda^*(E)$ for all $E \in \mathcal{M}$.

PROOF: Since $\lambda^*(M \cap E_1) = \lambda^*(E_1)$ and $\lambda^*(M^c \cap E_2) = \lambda^*(E_2)$ for all E_1, E_2 in $\mathcal{M}$, the proposition is obvious.   □

Thus $\hat{\lambda}$ is not a maximal extension of λ.

Exercises for Chapter 4

1 Let g be the function $x \mapsto x - [x + 1/2]$ on $\mathbf{R}$, where $[x + 1/2]$ designates the integral part of $x + 1/2$. Then, for every $n \in \mathbf{N}$, the function $x \mapsto g(nx)$ has a discontinuity at each $(2k + 1)/(2n)$ (with $k \in \mathbf{Z}$).

1. Show that $f : x \mapsto \sum_{n \geq 1} g(nx)/n^2$ is regulated and that it is continuous at every point which is not of the form $(2k+1)/(2n)$.

2. Let a be a point of the form $(2k'+1)/(2n')$, and express a in lowest terms: $a = (2k+1)/(2n)$ (with $n \geq 1$, $k \in \mathbf{Z}$). Show that $f(a+) - f(a) = 0$ and that $f(a-) - f(a) = \left(\sum_{i \geq 0} 1/(2i+1)^2\right)/n^2 = \pi^2/(8n^2)$. Conclude that f has a dense set of points of discontinuity.

2 Let $b > 1$ be an integer.

1. Let x be a point of $[0, 1[$ and $\left(0, (u_k)_{k \geq 1}\right)$ its proper base-b expansion. Show that x is rational if and only if there exist integers $n_0 \geq 1$ and $r \geq 1$ such that $u_{n+r} = u_n$ for all $n \geq n_0$.

2. Deduce from part 1 that $x \in \mathbf{R}$ is rational if and only if its proper base-b expansion is periodic.

3 Prove that the Cantor singular function g induces a bijection from $K \cap (\mathbf{R} - \mathbf{Q})$ onto $[0,1] \cap (\mathbf{R} - \mathbf{Q})$ (use Exercise 2).

4 Let K be the Cantor set.

1. Show that $D(K) = \{y - z : y \in K, z \in K\}$ contains $[0, 1]$, and hence that $D(K) = [-1, 1]$.

2. Compare the conclusion in part 1 with Proposition 4.3.1.

5 In the notation of Proposition 4.3.2, if $a_1, \ldots, a_p$ are points of A, then $\lambda_* \left(\bigcup_{1 \leq i \leq p} (L + a_i)\right) = 0$. Indeed, $A \cap D\left(\bigcup_{1 \leq i \leq p} (L + a_i)\right)$ contains no other points than the $a_i - a_j$ $(1 \leq i, j \leq p)$.

1. Let $(a_n)_{n \geq 1}$ be an enumeration of the points in A. For each integer $p \geq 0$, put $F_p = \bigcup_{n > p} (L + a_n)$. Show that $(F_p)_{p \geq 1}$ decreases to the empty set but that $\lambda^*(E \cap F_p) = \lambda^*(E)$ for every λ-measurable set E and every $p \geq 1$.

2. Compare the result in part 1 with Beppo Levi's theorem.

6 Denote by S the natural semiring of $\mathbf{R}$.

1. Let Ω be a nonempty set and μ a measure on a semiring in Ω. Also, let E be a μ-integrable set and f a function from Ω into $\mathbf{R}$ such that $f1_E$ is μ-integrable. For every $\varepsilon > 0$, prove the existence of $\delta > 0$ such that, for all countable partitions $(J_k)_{k \in L}$ of $\mathbf{R}$ into S-sets of length less than δ, we have $\left| \int f1_E d\mu - \sum_{k \in L} c_k \cdot \mu\left(E \cap f^{-1}(J_k)\right) \right| \leq \varepsilon$, independent of the choice of c_k in J_k.

2) When μ is Lebesgue measure on an interval, give a geometric interpretation of this result.

5

L^p Spaces

A measurable function is said to be in $\mathcal{L}^p(\mu)$ if the pth power of $|f|$ is integrable. $\mathcal{L}^p$ has the structure of a vector space and $f \mapsto N_p(f) = (\int |f|^p d\mu)^{1/p}$ is a seminorm when $p \geq 1$. L^p, which is the quotient of $\mathcal{L}^p$ with respect to the relation $f = g$ μ-a.e., is a Banach space ($p \geq 1$). In this chapter we will extend some of the results of Section 3.2 (convergence theorems) to these spaces. The Fischer–Riesz theorem (completeness of L^p for $p \geq 1$) may be considered, after the dominated convergence theorem, as the second fundamental theorem of measure theory in that it allows us to use all the tools of Banach spaces theory.

Summary

5.1   In this section we prove several fundamental inequalities. For example, if p, q, and r with $1/p + 1/q = 1/r$ belong to $[0, +\infty]$, and if f, g are two functions on Ω such that $N_p(f)$ and $N_q(g)$ are finite, then $N_r(fg) \leq N_p(f)N_q(g)$ (Proposition 5.1.2). Theorem 1.1 is a generalization of Minkowski's inequality: if f and g are two functions on Ω and $p \geq 1$, then $N_p(f + g) \leq N_p(f) + N_p(g)$.

5.2   We now generalize to $L^p(\mu)$ some of the convergence theorems established in Section 3.2, for example the dominated convergence theorem, and prove the Fischer–Riesz theorem (Proposition 5.2.3). Next, we analyze the duality of L^q and L^q when p and q are conjugate, that is, when $1/p + 1/q = 1$. If F is a Banach space, and p and q are conjugate, then, for every $g \in L^q_{F'}(\mu)$, $f \mapsto \int fg d\mu$ is a continuous linear functional on $L^p_F(\mu)$ with norm $N_q(g)$ (Theorem 5.2.5). The converse will be dealt with in Chapter 10.

5.3   The notion of convergence in measure is introduced. This section requires some knowledge of uniform spaces.

5.4   This section, which may be omitted, deals with uniformly integrable sets.

5.1   Definition of L^p Spaces

Lemma 5.1.1 *Let α, $\beta \in\,]0,1[$ total 1. Then $a^\alpha b^\beta \leq \alpha a + \beta b$ for all positive real numbers a, b, equality holding if and only if $a = b$.*

PROOF: The inequality is obvious for $a = 0$ or $b = 0$. Therefore, suppose that $0 < a < b$, and consider the function $x \mapsto x^\beta$ from $]0, +\infty[$ into **R**. There exists $c \in\,]a, b[$ such that $b^\beta - a^\beta = (b - a)\beta \cdot c^{\beta-1}$. Thus $b^\beta - a^\beta < (b - a) \cdot \beta a^{\beta-1}$, and $a^\alpha b^\beta < a^\alpha a^\beta + (b - a)\beta a^{\alpha+\beta-1} = a + (b - a)\beta = \alpha a + \beta b$.   □

In what follows, let Ω be a nonempty set and μ a Daniell measure on a space $\mathcal{H}(\Omega, \mathbf{C})$, which will initially be positive.

Proposition 5.1.1 *Let α, $\beta \in\,]0,1[$ satisfy $\alpha + \beta = 1$, and let f, g be two functions from Ω into $[0, +\infty[$ such that $\int^* f d\mu < +\infty$ and $\int^* g d\mu < +\infty$. Then $\int^* f^\alpha g^\beta d\mu \leq (\int^* f d\mu)^\alpha \cdot (\int^* g d\mu)^\beta$. A necessary and sufficient condition that the equality hold, when f, g are μ-measurable but not μ-negligible, is that there exist $a > 0$ and $b > 0$ such that $af = bg$ μ-almost everywhere.*

PROOF: We may suppose that $A = \int^* f d\mu > 0$ and $B = \int^* g d\mu > 0$. From the inequality $(f/A)^\alpha \cdot (g/B)^\beta \leq (\alpha f)/A + (\beta g)/B$, it follows that

$$\frac{1}{A^\alpha \cdot B^\beta} \int^* f^\alpha g^\beta \, d\mu \leq \frac{\alpha}{A} \int^* f d\mu + \frac{\beta}{B} \int^* g \, d\mu = \alpha + \beta = 1,$$

whence the desired inequality. When, additionally, f and g are μ-measurable, the equality holds if and only if

$$\int^* \left[\frac{\alpha f}{A} + \frac{\beta g}{B} - \left(\frac{f}{A}\right)^\alpha \left(\frac{g}{B}\right)^\beta \right] \cdot d\mu = 0,$$

that is, if and only if

$$\left(\frac{f}{A}\right)^\alpha \cdot \left(\frac{g}{B}\right)^\beta = \frac{\alpha f}{A} + \frac{\beta g}{B} \quad \text{almost everywhere,}$$

or, if and only if $f/A = g/B$ almost everywhere.   □

For all real numbers $p > 0$ and all functions f from Ω into $[0, +\infty]$, we put $N_p(f) = (\int^* f^p d\mu)^{1/p}$. For every function f from Ω into $\overline{\mathbf{R}}$, we denote by $M_\infty(f)$ the smallest of those $c \in \overline{\mathbf{R}}$ such that $f(x) \leq c$ locally μ-almost everywhere, and we put $m_\infty(f) = -M_\infty(-f)$. Finally, we write $N_\infty(f)$ for the smallest of those $c \in [0, +\infty]$ such that $|f(x)| \leq c$ locally μ-almost everywhere. Thus $m_\infty(f) \leq M_\infty(f)$ and $N_\infty(f) = M_\infty(|f|)$ when $\mu \neq 0$.

Proposition 5.1.2 *If p, q, r are in $]0, +\infty]$, such that $1/r = 1/p + 1/q$, and if f, g are two functions from Ω into $[0, +\infty[$ such that $N_p(f)$ and $N_q(g)$ are* *finite, then* $N_r(fg) \leq N_p(f)N_q(g)$. *A necessary and sufficient condition that the equality hold, when p and q are finite and f and g μ-measurable but not μ-negligible, is that there exist $a > 0$, $b > 0$ such that $a \cdot f^p = b \cdot g^q$ μ-almost everywhere.*

PROOF: First, suppose that p and q are finite. Then

$$\int^* f^r \cdot g^r d\mu \leq \left(\int^* f^p d\mu \right)^{r/p} \cdot \left(\int^* g^q d\mu \right)^{r/q}$$

by Proposition 5.1.1. Thus $N_r(fg) \leq N_p(f)N_q(g)$. Clearly, the latter inequality remains true when $p = +\infty$ or $q = +\infty$. □

In particular, $N_1(fg) \leq N_p(f)N_q(g)$ when p, $q \in [1, +\infty]$ are conjugate (i.e., such that $1/p + 1/q = 1$). This is Hölder's inequality.

Proposition 5.1.3 *Let $g : \Omega \to [0, +\infty[$ be μ-integrable and let $f : \Omega \to \mathbf{R}$ be μ-measurable. If $N_\infty(f)$ is finite, then fg is μ-integrable, and*

$$m_\infty(f) \cdot \int g d\mu \leq \int fg d\mu \leq M_\infty(f) \cdot \int g d\mu$$

whenever $\mu \neq 0$.

PROOF: Since fg is μ-measurable and since

$$m_\infty(f)g(x) \leq f(x)g(x) \leq M_\infty(f)g(x)$$

almost everywhere (when $\mu \neq 0$), the result is obvious. □

Proposition 5.1.4 *Let f be a function from Ω into $[0, +\infty[$ which is not μ-negligible. The set I of those $p \in]0, +\infty]$ for which $N_p(f)$ is finite is either empty or an interval of $\overline{\mathbf{R}}$, and, when $J = I \cap \mathbf{R}$ is nonempty, $1/p \mapsto \log N_p(f)$ is convex on $\{1/p : p \in J\}$. Now assume that f is μ-measurable; then the function $p \mapsto N_p(f)$ is continuous on the closure $\bar{I}$ of I with respect to $]0, +\infty]$, and it is infinitely differentiable in the interior J° of J when J° is nonempty.*

PROOF: Let r, s be two distinct elements of J and let $0 < t < 1$ be a real number. Put $1/p = t/r + (1-t)/s$. Then $\alpha = tp/r$ and $\beta = (1-t)p/s$ total 1. By Proposition 5.1.1, $\int^* f^{r\alpha} \cdot f^{s\beta} d\mu \leq \left(\int^* f^r d\mu \right)^\alpha \cdot \left(\int^* f^s d\mu \right)^\beta$, that is, $N_p(f) \leq \left(N_r(f) \right)^t \cdot \left(N_s(f) \right)^{1-t}$, and $\log N_p(f) \leq t \cdot \log N_r(f) + (1-t) \cdot \log N_s(f)$. Therefore, J is either empty or an interval of $\mathbf{R}$, and the function $1/p \mapsto \log \left(N_p(f) \right)$ is convex on $\{1/p : p \in J\}$.

Next, suppose that J is nonempty and fix $s \in J$. From $f^p = f^s f^{p-s}$ follows $N_p(f) \leq \left(N_s(f) \right)^{s/p} \cdot N_\infty(f)^{(p-s)/p}$ for all real numbers $p > s$, and

$\limsup_{p \to +\infty} N_p(f) \leq N_\infty(f)$. Hence, if $+\infty$ belongs to I. J contains arbitrarily large numbers, which proves that I is an interval of $\overline{\mathbf{R}}$. Now suppose that J is not a point, and let r be its origin and s its endpoint ($r < s \leq +\infty$). As $1/p \mapsto \log N_p(f)$ is convex on $\{1/p : p \in J\}$, $p \mapsto N_p(f)$ is continuous on $]r, s[$. Put $A = \{x \in \Omega : f(x) \geq 1\}$. When $p \in J$ tends to r, $f^p 1_A$ decreases to $f^r 1_A$ and $f^p 1_{A^c}$ increases to $f^r 1_{A^c}$, so $\int f^p 1_A d\mu$ decreases to $\int f^r 1_A d\mu$ by the monotone convergence theorem and $\int f^p 1_{A^c} d\mu$ increases to $\int^* f^r 1_{A^c} d\mu$. Finally, $\int f^p d\mu$ tends to $\int f^r \cdot 1_A d\mu + \int^* f^r \cdot 1_{A^c} d\mu = \int^* f^r d\mu$, which proves that the function $p \mapsto N_p(f)$ is continuous at r (whenever $r > 0$). The same argument works for s, if $s < +\infty$. Now assume that $s = +\infty$ and let a be a real number such that $0 < a < N_\infty(f)$. Since $N_p(f) < +\infty$ for some $p < +\infty$, the set $A = \{x \in \Omega : f(x) \geq a\}$ is integrable, because now $1_A \leq (f/a)^p$. Moreover, $a \cdot \mu(A)^{1/p} \leq N_p(f)$. Letting p tend to $+\infty$, we have $\liminf_{p \to +\infty} N_p(f) \geq a$. As we have already seen that $N_\infty(f) \geq \limsup_{p \to +\infty} N_p(f)$, we conclude that $p \mapsto N_p(f)$ is continuous at s.

Finally, suppose that J° is nonempty. We prove that $p \to \int f^p d\mu$ is infinitely differentiable in J° and that its nth derivative is $p \to \int f^p \cdot \log^n f \cdot d\mu$.

Put $A = \{x \in \Omega : 0 < f(x) < 1\}$ and $B = \{x \in \Omega : f(x) \geq 1\}$. For a fixed $p \in J^\circ$, let r, s be elements of J such that $r < p < s$, and choose a real number $\delta > 0$ so that $r < p - \delta < p + \delta < s$. The functions $t \mapsto |t^\delta \cdot \log^n t|$ from $]0, 1[$ into $\mathbf{R}$ and $t \mapsto t^{-\delta} \cdot \log^n t$ from $[1, +\infty[$ into $\mathbf{R}$ are bounded by some $c > 0$. Then

$$|1_A \cdot f^q \cdot \log^n f| = |1_A \cdot f^{q-\delta} \cdot f^\delta \cdot \log^n f| \leq c \cdot f^{q-\delta} \cdot 1_A \leq c f^r 1_A$$

and

$$1_B \cdot f^q \cdot \log^n f = 1_B \cdot f^{q+\delta} \cdot f^{-\delta} \cdot \log^n f \leq c \cdot f^{q+\delta} \cdot 1_B \leq c f^s 1_B,$$

so $|f^q \cdot \log^n f| \leq c f^r 1_A + c f^s 1_B$ for all q satisfying $r \leq q - \delta < q + \delta \leq s$. The desired conclusion follows by the standard results on differentiation under the integral sign (Theorem 3.2.3).   □

Proposition 5.1.5 *Suppose that $\int^* 1_\Omega d\mu = 1$. Let $f : \Omega \to [0, +\infty[$ be μ-measurable and not negligible. If $J = \{p \in]0, +\infty[: N_p(f) < +\infty\}$ is nonempty, its infimum is 0. Moreover, the function $p \mapsto N_p(f)$ is increasing on J, $\log(f)$ is quasi-integrable, and $N_p(f)$ tends to $\exp\left(\int^* \log(f) d\mu\right)$ as $p \to 0$.*

PROOF: Let $s \in J$ and $p \in]0, s[$. Then s/p and $s/(s-p)$ are conjugate exponents, so $N_1(f^p) \leq N_{s/p}(f^p) N_{s/(s-p)}(1)$ and $N_p(f) \leq N_s(f)$. Thus, if J is nonempty, its infimum is 0 and the function $p \mapsto N_p(f)$ is increasing on J.

Put $A = \{x \in \Omega : 0 \leq f(x) < 1\}$ and $B = \{x \in \Omega : f(x) \geq 1\}$. For every u of $[0, +\infty[$, $(u^p - 1)/p$ decreases to $\log u$ as p goes to 0 in $]0, +\infty[$. Indeed, for r, $s \in]0, +\infty[$ such that $r \leq s$,

$$\frac{u^s - 1}{s} = \frac{1}{s} \int_0^{s \cdot \log u} e^x dx$$

$$= \frac{1}{s} \int_0^{r \cdot \log u} e^{(s/r)y} \cdot \frac{s}{r} dy$$

$$= \frac{1}{r} \int_0^{r \cdot \log u} e^{(s/r)y} dy$$

exceeds $(1/r) \cdot \int_0^{r \cdot \log u} e^y dy = (u^r - 1)/r$. Now let s be in J. By the monotone convergence theorem, $\int_B (f^p - 1)/p \, d\mu$ converges to $\int_B \log(f) d\mu$ as $p \longrightarrow 0$ in $]0, s]$. On the other hand, $\int_A (f^p - 1)/p \, d\mu$ converges to $\int_* 1_A \cdot \log(f) d\mu$ (Proposition 3.7.5). Thus $\int (f^p - 1)/p \, d\mu$ converges to $\int \log^+(f) d\mu - \int^* \log^-(f) d\mu = \int^* \log(f) d\mu$ (Proposition 3.7.14). Since $\log u = \int_1^u (1/x) dx \le u - 1$ for all $u \in]0, +\infty[$,

$$\log N_p(f) = \frac{1}{p} \cdot \log\left(\int f^p d\mu \right) \le \frac{1}{p} \left(\int f^p d\mu - 1 \right) = \int \frac{f^p - 1}{p} d\mu$$

for all $p \in]0, s]$, which leads to

$$\log(\lim N_p(f)) = \lim_{p \to 0} \log(N_p(f)) \le \int^* \log(f) d\mu$$

and $\lim_{p \to 0} N_p(f) \le \exp(\int^* \log(f) d\mu)$.

If $\int^* \log(f) d\mu = -\infty$, the inequality $\exp(\int^* \log(f) d\mu) \le \lim_{p \to 0} N_p(f)$ is obvious. Otherwise, $\log(f)$ is integrable by the preceding argument, and we may further suppose that f is strictly positive everywhere. Since, for every p in $]0, s]$, the function $x \mapsto e^{px}$ from $\mathbf{R}$ into $\mathbf{R}$ is convex, we have $\exp(p \cdot \int \log f \cdot d\mu) \le \int \exp(p \cdot \log f) d\mu$ by Jensen's inequality, and $\exp(\int \log f \cdot d\mu) \le N_p(f)$. A moment's reflection now shows that $\exp\left(\int^* \log(f) d\mu \right) = \lim_{p \to 0} N_p(f)$. $\square$

Lemma 5.1.2 *Let a, b be two positive real numbers and p be a strictly positive real number. Then $(a + b)^p \le a^p + b^p$ if $0 < p \le 1$, and $a^p + b^p \le (a + b)^p$ if $p \ge 1$.*

PROOF: Suppose that $p \in]0, 1]$. The inequality $(a+b)^p \le a^p + b^p$ is obvious for $a = b = 0$; for $a + b > 0$, the above may be written $(a/(a+b))^p + (b/(a+b))^p \ge 1$, which follows from the fact that $(a/(a + b))^p \ge a/(a+b)$, $(b/(a+b))^p \ge b/(a + b)$, and $a/(a + b) + b/(a + b) = 1$. The case where $p \ge 1$ is treated similarly. $\square$

Proposition 5.1.6 *Let $0 < p < 1$ be a real number and f, g two μ-measurable functions from Ω into $[0, +\infty[$. Then $N_p(f) + N_p(g) \le N_p(f + g)$.*

PROOF: Assume that $0 < N_p(f + g) < +\infty$. Let h be the function from Ω into $[0, +\infty]$ which is equal to $1/(f + g)$ on $\{x \in \Omega : (f + g)(x) > 0\}$ and to

1 outside this set. Since $\int f^p d\mu = \int (f+g)^{p(1-p)} \cdot h^{p(1-p)} \cdot f^p d\mu$ and since $N_{1/(1-p)}\big((f+g)^{p(1-p)}\big) < +\infty$ and $N_{1/p}\big((fh \cdot h^{-p})^p\big) < +\infty$, we have

$$\int f^p d\mu \le \left(\int (f+g)^p d\mu\right)^{1-p} \cdot \left(\int h^{1-p} f d\mu\right)^p$$

by Hölder's inequality, or $\int h^{1-p} \cdot f d\mu \ge N_p(f) \cdot N_p(f+g)^{p-1}$. Similarly, $\int h^{1-p} \cdot g d\mu \ge N_p(g) \cdot N_p(f+g)^{p-1}$. Adding the last two inequalities, we get $\int (f+g)^p d\mu \ge \big(N_p(f) + N_p(g)\big) \cdot N_p(f+g)^{p-1}$, and therefore $N_p(f) + N_p(g) \le N_p(f+g)$.   □

Proposition 5.1.7 *Let $p \ge 1$ be a real number and f, g two functions from Ω into $[0, +\infty[$. Then $N_p(f+g) \le N_p(f) + N_p(g)$ (Minkowski's inequality).*

PROOF: Assume that $p > 1$, and let q be its conjugate exponent. To prove Minkowski's inequality, we may suppose that $N_p(f+g) > 0$, $N_p(f) < +\infty$ and $N_p(g) < +\infty$. As the function $x \mapsto x^p$ from $[0, +\infty[$ into $\mathbf{R}$ is convex, we have $(f+g)^p = 2^p (f/2 + g/2)^p \le 2^{p-1} \cdot (f^p + g^p)$, and so $N_p(f+g)$ is finite. Since $N_q\big((f+g)^{p-1}\big) < +\infty$,

$$N_1\big(f \cdot (f+g)^{p-1}\big) \le N_p(f) \cdot \left(\int^* (f+g)^p d\mu\right)^{1/q}.$$

Similarly,

$$N_1\big(g \cdot (f+g)^{p-1}\big) \le N_p(g) \cdot \left(\int^* (f+g)^p d\mu\right)^{1/q}.$$

From the fact that $N_1\big((f+g)^p\big) \le \big(N_p(f) + N_p(g)\big) \cdot \big(\int^* (f+g)^p d\mu\big)^{1/q}$, it now follows that $N_p(f+g) \le N_p(f) + N_p(g)$.   □

Theorem 5.1.1 (Inequality of Countable Convexity) *Let $(f_n)_{n \ge 1}$ be a sequence of functions from Ω into $[0, +\infty]$, and let $p \ge 1$ be a real number. Then $N_p(\sum_{n \ge 1} f_n) \le \sum_{n \ge 1} N_p(f_n)$.*

PROOF: Put $g_n = \sum_{1 \le k \le n} f_k$ for every $n \ge 1$.

Then $N_p(g_n) \le \sum_{1 \le k \le n} N_p(f_k)$ by Minkowski's inequality. Therefore, $N_p(\sum_{k \ge 1} f_k) = \lim_{n \to +\infty} N_p(g_n)$ is less than $\sum_{k \ge 1} N_p(f_k)$.   □

Definition 5.1.1 Let F be a real Banach space and $p \ge 1$ a real number. We define by $\mathcal{F}_F^p(\mu)$ (respectively, $\mathcal{L}_F^p(\mu)$) the space of those mappings f from Ω into F such that $N_p(f) = \big(\int^* |f|^p d\mu\big)^{1/p}$ is finite (respectively, such that f is μ-measurable and $N_p(f)$ is finite). In particular, when $p = 2$, any element of $\mathcal{L}_F^2(\mu)$ is called a square-integrable mapping.

Clearly, $\mathcal{F}_F^p(\mu)$ and $\mathcal{L}_F^p(\mu)$ are vector spaces, and the function $f \mapsto N_p(f)$ defines a seminorm on $\mathcal{F}_F^p(\mu)$. The relation $f \sim g$ if and only if $f = g$ μ-almost everywhere is an equivalence relation on $\mathcal{F}_F^p(\mu)$, and we shall often fail to distinguish $\mathcal{F}_F^p(\mu)$ from its quotient space and $\mathcal{L}_F^p(\mu)$ from $L_F^p(\mu)$.

For every function $f : \Omega \to F$, set $\overline{N_p}(f) = \left(\int^{\bullet} |f|^p \, d\mu \right)^{1/p}$. We denote by $\overline{\mathcal{F}_F^p}(\mu)$ (respectively, $\overline{\mathcal{L}_F^p}(\mu)$) the space of functions from Ω into F (respectively, of μ-measurable functions f) for which $\overline{N_p}(f)$ is finite.

For every real number $\alpha > 0$, $\varphi : z \mapsto |z|^{\alpha-1} \cdot z$ is defined and continuous on $F - \{0\}$. Since $|\varphi(z)| = |z|^\alpha$, $\varphi(z)$ converges to 0 as $z \to 0$ in $F - \{0\}$, and φ has a unique continuous extension to the whole of F, which is still written $z \mapsto |z|^{\alpha-1} \cdot z$. Now let ψ be the mapping $z \mapsto |z|^{1/\alpha-1} \cdot z$ from F into F. Then $\psi \circ \varphi = id_F$, so φ is a homeomorphism from F into F. For all real numbers p, q in $[1, +\infty[$, a mapping f belongs to $\mathcal{F}_F^p(\mu)$ (respectively, $\mathcal{L}_F^p(\mu)$) if and only if $|f|^{p/q-1} f$ belongs to $\mathcal{F}_F^q(\mu)$ (respectively, $\mathcal{L}_F^q(\mu)$).

5.2 Convergence Theorems

Let $p \geq 1$ be a real number.

Proposition 5.2.1 *Let F be a real Banach space, and let $(f_n)_{n \geq 1}$ be a sequence in $\mathcal{F}_F^p(\mu)$ such that $\sum_{n \geq 1} N_p(f_n) < +\infty$. The series $\sum_{n \geq 1} f_n(x)$ converges absolutely a.e. If we put $f(x) = \sum_{n \geq 1} f_n(x)$ at almost all points x where this series converges, and take $f(x) \in F$ arbitrarily elsewhere, then f belongs to $\mathcal{F}_F^p(\mu)$ and $N_p\left(f - \sum_{1 \leq k \leq n} f_k \right) \leq \sum_{k \geq n+1} N_p(f_k)$ for all $n \geq 0$. Therefore, the series $\sum_{n \geq 1} f_n$ converges to f in $\mathcal{F}_F^p(\mu)$.*

PROOF: Argue as in Proposition 3.1.7.   □

Theorem 5.2.1 *Let $(f_i)_{i \geq 1}$ be a Cauchy sequence in $\mathcal{F}_F^p(\mu)$. There exists a subsequence $(f_{i_n})_{n \geq 1}$ of $(f_i)_{i \geq 1}$ with the following properties:*

(a) *$\sum_{n \geq 1} N_p(f_{i_{n+1}} - f_{i_n})$ is finite.*

(b) *The sequence $\left(f_{i_n}(x) \right)_{n \geq 1}$ converges a.e.*

(c) *If we set $f(x) = \lim_{n \to +\infty} f_{i_n}(x)$ at almost all points x where this limit exists and take $f(x)$ arbitrarily elsewhere, then f belongs to $\mathcal{F}_F^p(\mu)$ and the sequence $(f_i)_{i \geq 1}$ converges to f in $\mathcal{F}_F^p(\mu)$.*

(d) *For every $\varepsilon > 0$, there exists $Z \subset \Omega$ such that $\mu^*(Z) \leq \varepsilon$ and the sequence $(f_{i_n})_{n \geq 1}$ converges uniformly to f on $\Omega - Z$.*

(e) *There exists a function g from Ω into $[0, +\infty]$ such that $N_p(g) < +\infty$ and $|f_{i_n}(x)| \leq g(x)$ for all $x \in \Omega$.*

PROOF: By induction on n, define a strictly increasing sequence $(i_n)_{n \geq 1}$ in $\mathbf{N}$ such that $N_p(f_i - f_{i_n}) \leq 2^{-2n}$ for all $n \geq 1$ and all $i \geq i_n$. For each integer $n \geq 1$, put $Y_n = \{x \in \Omega : |f_{i_{n+1}} - f_{i_n}|(x) \geq 1/2^n\}$. Now $(1/2^n) \cdot 1_{Y_n} \leq |f_{i_{n+1}} - f_{i_n}|$, and $(1/2^{np}) \cdot 1_{Y_n} \leq |f_{i_{n+1}} - f_{i_n}|^p$. It follows that $(1/2^n) \cdot \mu^*(Y_n)^{1/p} \leq N_p(f_{i_{n+1}} - f_{i_n}) \leq 2^{-2n}$, and $\mu^*(Y_n)^{1/p} \leq 2^{-n}$. Hence $\mu^*(Y_n) \leq 2^{-n}$. Now, arguing as in Theorem 3.1.2, we obtain the result.   □

By Theorem 5.2.1, $\mathcal{F}_F^p(\mu)$ and $\mathcal{L}_F^p(\mu)$ are complete. Also, Propositions 3.1.8 and 3.1.9 remain true if we replace $\mathcal{L}_F^1(\mu)$ by $\mathcal{L}_F^p(\mu)$.

Theorem 5.2.2 (Dominated Convergence Theorem in $\mathcal{L}_F^p(\mu)$)   Let $(f_n)_{n \geq 1}$ be a sequence in $\mathcal{L}_F^p(\mu)$, converging almost everywhere to some function f. Suppose there exists a function g from Ω into $[0, +\infty]$ such that $N_p(g) < +\infty$ and $|f_n| \leq g$ a.e. for every $n \geq 1$. Then f belongs to $\mathcal{L}_F^p(\mu)$ and $(f_n)_{n \geq 1}$ converges to f in $\mathcal{L}_F^p(\mu)$.

PROOF: f is μ-measurable. Since $|f| \leq g$ a.e., $N_p(f)$ is finite, and so f belongs to $\mathcal{L}_F^p(\mu)$. The functions $|f_n - f|^p$ are μ-integrable and dominated a.e. by $(2g)^p$. Theorem 3.2.2 now shows that $\int |f_n - f|^p d\mu$ converges to 0 as $n \to +\infty$.   □

Theorem 5.2.3 $\mathcal{L}_F^p(\mu)$ is the closure of $St(\hat{\mathcal{R}}, F)$ in $\mathcal{F}_F^p(\mu)$, where $\hat{\mathcal{R}}$ is the ring of μ-integrable sets.

PROOF: Let $f \in \mathcal{L}_F^p(\mu)$. Since f is μ-measurable and μ-moderate, it is the limit a.e. of a sequence $(f_n)_{n \geq 1}$ of elements of $St(\hat{\mathcal{R}}, F)$. For every integer $n \geq 1$, define $g_n : \Omega \to F$ by $g_n(x) = f_n(x)$ if $|f_n(x)| \leq 2|f(x)|$ and $g_n(x) = 0$ if $|f_n(x)| > 2|f(x)|$. Then g_n belongs to $St(\hat{\mathcal{R}}, F)$ and $|g_n| \leq 2|f|$. Moreover, $(g_n)_{n \geq 1}$ converges a.e. to f. Therefore, Theorem 5.2.2 shows that $(g_n)_{n \geq 1}$ converges to f in $\mathcal{L}_F^p(\mu)$ □

Denote by $\mathcal{L}^p(\mu, \overline{\mathbf{R}})$ the set of functions f from Ω into $\overline{\mathbf{R}}$ with the following property: there exists $g \in \mathcal{L}_{\mathbf{R}}^p(\mu)$ such that $g = f$ a.e. A function f from Ω into $\overline{\mathbf{R}}$ thus lies in $\mathcal{L}^p(\mu, \overline{\mathbf{R}})$ if and only if it is μ-measurable and $N_p(f) = N_p(|f|) < +\infty$.

Theorem 5.2.4 Let $(f_n)_{n \geq 1}$ be an increasing sequence in $\mathcal{L}_{\mathbf{R}}^p(\mu)$. Then $f = \sup_{n \geq 1} f_n$ belongs to $\mathcal{L}^p(\mu, \overline{\mathbf{R}})$ if and only if $\sup_{n \geq 1} N_p(f_n) < +\infty$. In this case, $N_p(f - f_n)$ converges to 0 as $n \to +\infty$.

PROOF: Since $f - f_1 = \sup_{n \geq 1}(f_n - f_1)$, we may suppose that the f_n are positive. For all integers m, n such that $1 \leq m \leq n$, we have

$$f_m^p + (f_n - f_m)^p \leq f_n^p$$

(Lemma 5.1.2) and

$$N_p(f_n - f_m)^p \leq N_p(f_n)^p - N_p(f_m)^p.$$

Hence, if $\sup_{n\geq 1} N_p(f_n) < +\infty$, then $(f_n)_{n\geq 1}$ is a Cauchy sequence in $\mathcal{L}_{\mathbf{R}}^p(\mu)$. Our theorem now follows immediately from Theorem 5.2.1. $\qquad\square$

Proposition 5.2.2 *Define an order on $L_{\mathbf{R}}^p(\mu)$ as follows: $\dot{f} \leq \dot{g}$ if and only if $f \leq g$ a.e., and let H be an upward-directed subset of $L_{\mathbf{R}}^p(\mu)$, consisting of positive elements. H has a supremum b in $L_{\mathbf{R}}^p(\mu)$ whenever $M = \sup_{u\in H} N_p(u)$ is finite. In this case, the filter of sections of H converges to b in $L_{\mathbf{R}}^p(\mu)$; moreover, there exists an increasing sequence $(u_n)_{n\geq 1}$ in H such that, if g_n is a representative of u_n, then the upper envelope of the g_n is equal a.e. to any representative of b. $L_{\mathbf{R}}^p(\mu)$ is a Dedekind complete Riesz space.*

PROOF: Assume that M is finite. Since H is directed upward, there exists an increasing sequence $(u_n)_{n\geq 1}$ in H such that $M^p - 1/2^{np} \leq N_p(u_n)^p$ for all $n \geq 1$. Let g_n be a representative of u_n. We can choose the functions g_n so that $g_n \geq 0$, the sequence $(g_n)_{n\geq 1}$ increases, and $g = \sup g_n$ is finite (Theorem 5.2.4).

Given $u \in H$, let $h \geq 0$ be a representative of u. From $(h - g_n)^+ = \sup(h, g_n) - g_n$, it follows that $N_p((h-g_n)^+)^p \leq N_p(\sup(h,g_n))^p - N_p(g_n)^p \leq M^p - N_p(g_n)^p \leq 1/2^{np}$ for all integers $n \geq 1$. Since $N_p(\sum_{n\geq 1}(h - g_n)^+) < +\infty$, the sum $\sum_{n\geq 1}(h - g_n)^+$ is finite a.e., and $(h - g_n)^+$ converges to 0 a.e. as $n \to +\infty$. Therefore, $(h - g)^+ = 0$ a.e., and $h \leq g$ a.e. This proves that $b = \dot{g}$ is an upper bound for H. If a is another upper bound for H and f is a representative of a, then $g_n \leq f$ a.e. for every $n \geq 1$, so $g \leq f$ a.e. and $b \leq a$. Hence $b = \sup H$. For each $\varepsilon > 0$, there exists $n \geq 1$ such that $N_p(g - g_n) \leq \varepsilon$, and, if $v \in H$ is greater than u_n, then $N_p(b - v) \leq \varepsilon$. Therefore, the filter of sections of H converges to b.

Finally, $L_{\mathbf{R}}^p(\mu)$ is a Dedekind complete Riesz space by Proposition 1.2.1. $\qquad\square$

Observe that, in general, $\mathcal{L}_{\mathbf{R}}^p(\mu)$ is not Dedekind complete. For instance, if we take for μ Lebesgue measure on $[0, 1]$ and let E be a subset of $[0, 1]$ which is not μ-measurable, then the set H of indicator functions of finite subsets of E has no supremum in $\mathcal{L}_{\mathbf{R}}^p(\mu)$.

For every real Banach space F, denote by $\mathcal{L}_F^\infty(\mu)$ the space of μ-measurable mappings from Ω into F such that $N_\infty(f) = N_\infty(|f|)$ is finite, and equip $\mathcal{L}_F^\infty(\mu)$ with the seminorm $f \mapsto N_\infty(f)$. Write $L_F^\infty(\mu)$ for the quotient space of $\mathcal{L}_F^\infty(\mu)$ by the equivalence relation: $f \sim g$ if and only if $f = g$ locally μ-almost everywhere.

Proposition 5.2.3 *$L_F^\infty(\mu)$ is a Banach space. A necessary and sufficient condition that a sequence $(f_n)_{n\geq 1}$ converge in $\mathcal{L}_F^\infty(\mu)$ is that there exists a locally μ-negligible set N such that $(f_n)_{n\geq 1}$ converges uniformly on $\Omega - N$.*

PROOF: Let $(f_n)_{n\geq 1}$ be a Cauchy sequence in $\mathcal{L}_F^\infty(\mu)$, and put

$$B = \{x \in \Omega : |f_n(x)| \leq N_\infty(f_n) \text{ for every } n \geq 1\}.$$

For every $n \in \mathbf{N}$, there exists an integer k_n such that $N_\infty(f_r - f_s) \leq 1/n$ for all $r \geq k_n$ and $s \geq k_n$. We can find a locally μ-negligible set A_n, $A_n \supset B^c$, such that $|f_r(x) - f_s(x)| \leq 1/n$ for all $x \in A_n^c$ and all integers $r \geq k_n$, $s \geq k_n$. Now put $A = \bigcup_{n \geq 1} A_n$ and $g_n = f_n 1_{A^c}$. Then the sequence $(g_n)_{n \geq 1}$ converges uniformly to a function g, which is therefore μ-measurable. Clearly, g is bounded. Finally, $N_\infty(f_n - g) = N_\infty(g_n - g)$ converges to 0 as $n \to +\infty$, which proves that $\mathcal{L}_F^\infty(\mu)$ is complete. The second assertion is obvious. $\square$

Proposition 5.2.4 *Let γ be a continuous bilinear mapping from $F \times G$ into H, where F, G, H are three Banach spaces, and assume that $\|\gamma\| \leq 1$. Let p, q be two conjugate exponents in $[0, +\infty]$, and f, g two elements of $\mathcal{L}_F^p(\mu)$ and $\mathcal{L}_G^q(\mu)$, respectively. Then $\gamma \circ [f, g]$ belongs to $\mathcal{L}_H^1(\mu)$ and $N_1(\gamma \circ [f, g]) \leq N_p(f) N_p(g)$.*

PROOF: Obvious. $\square$

In particular, we can define the inner product $(\dot{f}, \dot{g}) \mapsto \int f \bar{g} d\mu$ $(\mu \geq 0)$ on $L_\mathbf{C}^2(\mu)$. Then $L_\mathbf{C}^2(\mu)$ is a Hilbert space.

Theorem 5.2.5 *Let μ be a complex Daniell measure on $\mathcal{H}(\Omega, \mathbf{C})$. Given a Banach space F, let F' be its normed dual space and γ the bilinear mapping $(z, z') \mapsto z'(z)$ from $F \times F'$ into $\mathbf{C}$. For all mappings f, g from Ω into F and F', respectively, we write fg for $\gamma \circ [f, g]$. Let p, q be two conjugate exponents in $[1, +\infty]$. Denote by B_p (respectively, B_q') the closed ball with center 0 and radius 1 in $\mathcal{L}_F^p(\mu)$ (respectively, in $\mathcal{L}_{F'}^q(\mu)$). Then*

$$N_p(f) = \sup_{g \in B_q'} \left| \int f g d\mu \right| \qquad (1)$$

for all $f \in \mathcal{L}_F^p(\mu)$, and

$$N_q(g) = \sup_{f \in B_p} \left| \int f g d\mu \right| \qquad (2)$$

for all $g \in \mathcal{L}_{F'}^q(\mu)$. Thus, for every $g \in \mathcal{L}_{F'}^q(\mu)$, the mapping $\dot{f} \mapsto \int f g d\mu$ is a continuous linear form on $L_F^p(\mu)$, whose norm is $N_q(g)$.

PROOF: We prove (1).

Suppose that $1 \leq p < +\infty$ and $N_p(f) = 1$. First, assume that f belongs to $St(\hat{\mathcal{R}}, F)$; it can be written $f = \sum_{1 \leq k \leq n} a_k 1_{A_k}$, where the A_k are disjoint integrable sets. By hypothesis,

$$\sum_{1 \leq k \leq n} |a_k|^p \cdot V\mu(A_k) = 1.$$

For fixed $0 < \varepsilon < 1$, partitioning each A_k into integrable sets, if necessary, we may suppose that

$$\sum_{1 \leq k \leq n} |a_k|^p \cdot |\mu(A_k)| > 1 - \varepsilon,$$

and let $0 < \delta < 1$ be such that

$$\delta \cdot \sum_{1 \leq k \leq n} |a_k|^p \cdot |\mu(A_k)| \geq 1 - \varepsilon.$$

Let ζ_k be a complex number of modulus 1 such that $\zeta_k \cdot \mu(A_k) = |\mu(A_k)|$ (we choose ζ_k real whenever μ is real). For every index $1 \leq k \leq n$, there exists $a'_k \in F'$ such that $|a'_k|^q = |a_k|^p$ if $p > 1$ (respectively, $|a'_k| = 1$ if $p = 1$), $a_k \cdot a'_k$ is real, and $a_k \cdot a'_k \geq \delta \cdot |a_k| \cdot |a'_k|$; indeed, when $a_k \neq 0$, there exists $z' \in F'$ such that $|z'| = 1$, $a_k \cdot z'$ is real, and $a_k \cdot z' \geq \delta \cdot |a_k|$; now, it suffices to take $a'_k = |a_k|^{p-1} z'$. Put $g = \sum_{1 \leq k \leq n} \zeta_k a'_k 1_{A_k}$. Then $N_q(g) = 1$. On the other hand,

$$\int fg d\mu = \sum_{1 \leq k \leq n} a_k a'_k \cdot \zeta_k \mu(A_k) \geq \delta \cdot \sum_{1 \leq k \leq n} |a_k| \cdot |a'_k| \cdot |\mu(A_k)|,$$

and, since $|a'_k| = |a_k|^{p-1}$ if $p > 1$ (respectively, $|a'_k| = 1$ if $p = 1$), we have

$$\int fg d\mu \geq \delta \cdot \sum_{1 \leq k \leq n} |a_k|^p \cdot |\mu(A_k)| \geq 1 - \varepsilon.$$

This proves (1) when f belongs to $St(\hat{\mathcal{R}}, F)$.

We now pass to the case in which f is an arbitrary element of $\mathcal{L}^p_F(\mu)$ such that $N_p(f) = 1$. Given $0 < \varepsilon \leq 1$, there exists $\varphi \in St(\hat{\mathcal{R}}, F)$ such that $N_p(f - \varphi) \leq \varepsilon^2$. By what we have just seen, there exists $g \in \mathcal{L}^q_{F'}(\mu)$ such that $N_q(g) = 1$ and $|\int \varphi g d\mu| \geq N_p(\varphi) \cdot (1 - \varepsilon) \geq (1 - \varepsilon)^2$. Then $\int fg d\mu = \int \varphi g d\mu + \int (f - \varphi) g d\mu$ and $|\int (f - \varphi) g d\mu| \leq N_p(f - \varphi) \cdot N_q(g) \leq \varepsilon^2$. Hence $|\int fg d\mu| \geq (1 - \varepsilon)^2 - \varepsilon^2$, which proves (1).

Next, suppose that $p = +\infty$ and $N_\infty(f) > 0$. Let $0 \leq \alpha < N_\infty(f)$. The set $\{x \in \Omega : |f(x)| > \alpha\}$ contains a μ-integrable set E which is not μ-negligible. f/E is the limit, μ-almost everywhere in E, of a sequence $(f_n)_{n \geq 1}$ of $\hat{\mathcal{R}}/E$-simple mappings, and there is a μ-integrable subset Z of E such that $V\mu(Z) < V\mu(E)$ and that $(f_n)_{n \geq 1}$ converges uniformly to f/E on $E - Z$. Therefore, for fixed $\varepsilon > 0$, we can find $n \geq 1$ so that $|f_n - f/E| \leq \varepsilon/2$ on $E - Z$. The mapping f_n takes the values $y_1, \ldots, y_k$ on E, and one of the $f_n^{-1}(y_i) \cap (E - Z)$, say A, is not μ-negligible. Clearly, $|f(x) - y_i| \leq \varepsilon/2$ for all $x \in A$. Given $0 < \delta < 1$, there exists a finite partition $P(A)$ of A into μ-integrable sets such that $\delta \cdot V\mu(A) = \delta \cdot \sum_{B \in P(A)} V\mu(B) < \sum_{B \in P(A)} |\mu(B)|$. Choosing $B \in P(A)$ so that $\delta \cdot V\mu(B) \leq |\mu(B)|$ and letting a be one value of f on B, we have $|a| > \alpha$ and $|f(x) - a| \leq \varepsilon$ for all $x \in B$. There exists $a' \in F'$ such that $|a'| = 1$ and $|aa'| \geq |a| - \varepsilon$. Now the mapping $g = 1_B \cdot a'/V\mu(B)$ is integrable and $N_1(g) = 1$. On the other hand, $\int fg d\mu = (1/V\mu(B)) \cdot \int fa' 1_B d\mu$. Since $\int fa' 1_B d\mu = aa' \mu(B) + \int (f - a)a' 1_B d\mu$ and $|(f - a)a' 1_B| \leq \varepsilon 1_B$, we see that

$$\left| \int fg d\mu \right| \geq \left| aa' \cdot \frac{\mu(B)}{V\mu(B)} \right| - \varepsilon \geq \delta(|a| - \varepsilon) - \varepsilon \geq \delta(\alpha - \varepsilon) - \varepsilon.$$

Since ε and δ are arbitrary, (1) is true.

Arguing as above, we obtain relation (2).   □

If V is a dense vector subspace of $\mathcal{L}^p_{F'}(\mu)$, (1) persists when g extends over $V \cap B'_q$; indeed, the interior B° of $B = B'_q$ is dense in B, and $B^\circ \cap V$ is dense in B°. A similar observation holds for relation (2).

Theorem 5.2.6 *Let $p \geq 1$ be a real number, and assume that $f^p \in \mathcal{H}^+$ for all $f \in \mathcal{H}^+$. Then, for every real Banach space F, $\mathcal{H}(\Omega, \mathbf{R}) \otimes F$ is dense in $\mathcal{L}^p_F(\mu)$.*

PROOF: If A is a μ-integrable set, there exists a sequence $(f_i)_{i \geq 1}$ in $\mathcal{H}^+$ converging to 1_A in the mean. Then $(g_i)_{i \geq 1} = (\inf(1, f_i))_{i \geq 1}$ also converges to 1_A in the mean. Thus, for given $\varepsilon > 0$, there exists an integer $j \geq 1$ such that $\int^* |1_A - g_i| \cdot dV\mu \leq \varepsilon^p$ for every $i \geq j$. But, since $|1_A - g_i| \leq 1$, we see that $N_p(1_A - g_i) \leq \varepsilon$.

Now let $f = \sum_{1 \leq k \leq n} a_k \cdot 1_{A_k}$ be an element of $St(\hat{\mathcal{R}}, F)$. For fixed $\varepsilon > 0$, for every k (with $1 \leq k \leq n$) there exists $g_k \in \mathcal{H}^+$ such that $N_p(1_{A_k} - g_k) \leq \varepsilon/(n \cdot |a_k|)$. Then $N_p(f - \sum_{1 \leq k \leq n} a_k g_k) \leq \varepsilon$. Thus the closure of $\mathcal{H}(\Omega, \mathbf{R}) \otimes F$ in $\mathcal{L}^p_F(\mu)$ contains $St(\hat{\mathcal{R}}, F)$.   □

Observe, however, that $\mathcal{H}(\Omega, \mathbf{R})$ is not a dense subspace of $\mathcal{L}^\infty_{\mathbf{R}}(\mu)$ if μ is Lebesgue measure on $\Omega = [0, 1]$ and $\mathcal{H}(\Omega, \mathbf{R})$ is the space of continuous real-valued functions on Ω.

5.3   Convergence in Measure

Let Ω be a nonempty set and μ a positive Daniell measure on a space $\mathcal{H}(\Omega, \mathbf{C})$.

Definition 5.3.1 *Let A be a μ-measurable subset of Ω. A mapping f from A into a metrizable space F is said to be μ-measurable on A if and only if*

(a) f is measurable $\mathcal{M}/A$;

(b) for every μ-integrable subset E of A, there exists a μ-negligible subset N of E such that $f(E - N)$ is separable.

Thus we see that f is μ-measurable on A if and only if every extension of f which is constant outside A is μ-measurable.

Let F be a metrizable uniform space, and let $\mathcal{L}(A, \mu; F)$ be the set of all μ-measurable mappings from A into F. For every entourage (or vicinity) V of the uniform structure of F, every μ-integrable subset B of A, and every real number $\delta > 0$, define $W(V, B, \delta)$ as the set of pairs (f, g) of functions in $\mathcal{L}(A, \mu; F)$ having the following property: the set M of points $x \in B$ for which $(f(x), g(x))$ does not belong to V satisfies $\mu^*(M) \le \delta$.

Let d be a distance which defines the uniform structure of F. For all μ-integrable subsets B of A and all f, g in $\mathcal{L}(A, \mu; F)$, denote by $d_B(f, g)$ the infimum of those $\varepsilon > 0$ for which $\mu\Big(\big\{x \in B : d(f(x), g(x)) > \varepsilon\big\}\Big)$ is less than ε. If $(\varepsilon_n)_{n \ge 1}$ is a sequence decreasing to $\delta = d_B(f, g)$ in $]\delta, +\infty[$, then, for every $m \ge 1$ and for all $n \ge m$, we have

$$\mu\Big(\big\{x \in B : d(f(x), g(x)) > \varepsilon_n\big\}\Big) \le \varepsilon_m;$$

thus

$$\mu\Big(\big\{x \in B : d(f(x), g(x)) > \delta\big\}\Big) \le \varepsilon_m$$

and

$$\mu\Big(\big\{x \in B : d(f(x), g(x)) > \delta\big\}\Big) \le \delta.$$

For every $\varepsilon > 0$, let V_ε be the set $\{(y, z) \in F \times F : d(y, z) \le \varepsilon\}$. Then, by the above, $d_B(f, g) \le \varepsilon$ if and only if (f, g) belongs to $W(V_\varepsilon, B, \varepsilon)$. The mapping $d_B : (f, g) \mapsto d_B(f, g)$ is a pseudometric on $\mathcal{L}(A, \mu; F)$.

Definition 5.3.2 The uniform structure on $\mathcal{L}(A, \mu; F)$ defined by the pseudometrics d_B (where B runs through the class of μ-integrable subsets of A) is called the uniform structure of convergence in measure on A.

A subset of $\mathcal{L}(A, \mu; F) \times \mathcal{L}(A, \mu; F)$ is an entourage for this uniform structure if and only if it contains a $W(V, B, \delta)$. The topology on $\mathcal{L}(A, \mu; F)$ corresponding to this uniform structure is called the topology of convergence in measure on A. When $A = \Omega$, we often omit reference to A. If f, g belong to $\mathcal{L}(A, \mu; F)$, if B is a μ-integrable subset of A, and if $d_B(f, g) = 0$, then $f = g$ almost everywhere in B. Therefore, the intersection of all the entourages of $\mathcal{L}(A, \mu; F)$ is the set of pairs (f, g) such that $f = g$ locally μ-almost everywhere in A. We shall denote by $L(A, \mu; F)$ the Hausdorff uniform space associated with $\mathcal{L}(A, \mu; F)$.

A sequence $(f_n)_{n \ge 1}$ in $\mathcal{L}(A, \mu; F)$ is a Cauchy sequence in measure (respectively, it converges in measure to $f \in \mathcal{L}(A, \mu; F)$) if and only if, for every entourage V of F and every μ-integrable subset B of A, the number $\mu^*\Big(\big\{x \in B : (f_p(x), f_q(x)) \notin V\big\}\Big)$ goes to 0 as p and q tend to $+\infty$ (respectively, if and only if $\mu^*\Big(\big\{x \in B : (f_n(x), f(x)) \notin V\big\}\Big)$ goes to 0 as $n \to +\infty$).

Every sequence $(f_n)_{n\geq 1}$ in $\mathcal{L}(A, \mu; F)$, converging locally μ-almost everywhere to a mapping f from A into F also converges in measure to f by Egorov's theorem.

Proposition 5.3.1 *Let $(f_n)_{n\geq 1}$ be a Cauchy sequence in $\mathcal{L}(A, \mu; F)$ and let B be a μ-measurable and μ-moderate subset of A. There exists a subsequence $(f_{n_k})_{k\geq 1}$ of $(f_n)_{n\geq 1}$ such that $(f_{n_k}(x))_{k\geq 1}$ is a Cauchy sequence in F for almost all x in B. If $(f_n)_{n\geq 1}$ converges in measure to an element f of $\mathcal{L}(A, \mu; F)$, and if $(f_{n_k})_{n_k\geq 1}$ is as above, then $(f_{n_k})_{k\geq 1}$ converges to f almost everywhere in B.*

PROOF: First, suppose that B is integrable, and let d be a distance compatible with the uniform structure of F. By induction on $m \geq 0$, we can define a double sequence $(f_{m,n})_{m\geq 0, n\geq 1}$ in $\mathcal{L}(A, \mu; F)$ with the following properties:

(a) $f_{0,n} = f_n$ for all $n \geq 1$.

(b) $(f_{m,n})_{n\geq 1}$ is a subsequence of $(f_{m-1,n})_{n\geq 1}$ for all $m > 0$.

(c) For all $m > 0$, the set $E_{m,n}$ of points $x \in B$ for which

$$d(f_{m,n}(x), f_{m,n+1}(x)) > 1/2^{m+n}$$

has measure $\mu(E_{m,n})$ less than $1/2^{m+n}$.

For every $m \geq 1$, put $E_m = \bigcup_{n\geq 1} E_{m,n}$. Then $\mu(E_m) \leq \sum_{n\geq 1} \mu(E_{m,n}) \leq 1/2^m$. For all x of $B - E_m$, we have $d(f_{m,n}(x), f_{m,n+p}(x)) \leq 1/2^{m+n-1}$ for $n \geq 1$ and $p \geq 0$; $(f_{m,n}(x))_{n\geq 1}$ is thus a Cauchy sequence in F. Put $g_n = f_{n,n}$ for all $n \geq 1$, so that, for every $m \geq 1$, $(g_n)_{n\geq m}$ is a subsequence of $(f_{m,n})_{n\geq 1}$. Now $N = \bigcap_{m\geq 1} E_m$ is μ-negligible. If x belongs to $B - N$, there is an index $m \geq 1$ such that x does not lie in E_m, and hence $(g_n(x))_{n\geq 1}$ is a Cauchy sequence in F. Finally, note that $(g_n)_{n\geq 1}$ is a subsequence of $(f_n)_{n\geq 1}$.

We now pass to the case in which B is the union of a sequence $(B_m)_{m\geq 1}$ of μ-integrable sets. By induction on $m \geq 0$, we can define a double sequence $(g_{m,n})_{m\geq 0, n\geq 1}$ in $\mathcal{L}(A, \mu; F)$ with the following properties:

(a) $g_{0,n} = f_n$ for all $n \geq 1$.

(b) $(g_{m,n})_{n\geq 1}$ is a subsequence of $(g_{m-1,n})_{n\geq 1}$ for all $m > 0$.

(c) For every $m > 0$, there exists a μ-negligible subset P_m of B_m such that $(g_{m,n}(x))_{n\geq 1}$ is a Cauchy sequence for all x in $B_m - P_m$.

Set $h_n = g_{n,n}$ for each $n \geq 1$. Then, for every $m \geq 1$, $(h_n)_{n\geq m}$ is a subsequence of $(g_{m,n})_{n\geq 1}$, and so $(h_n(x))_{n\geq 1}$ is a Cauchy sequence in F for all x of $B - P$, where $P = \bigcup_{m\geq 1} P_m$. This proves the first assertion.

Next, suppose that $(f_n)_{n\geq 1}$ converges in measure to $f \in \mathcal{L}(A, \mu; F)$, and that $(f_{n_k})_{k\geq 1}$ is as above. Write $\hat{F}$ for the completion of the uniform space

F. Let N be a μ-negligible subset of B such that $\left(f_{n_k}(x)\right)_{k \geq 1}$ is a Cauchy sequence in F for every $x \in B - N$, and write $f'(x)$ for the limit in $\hat{F}$ of $\left(f_{n_k}(x)\right)_{k \geq 1}$. Clearly, the mapping $f' : x \mapsto f'(x)$ from $B - N$ into $\hat{F}$ is μ-measurable on $B - N$, and $(f_{n_k})_{k \geq 1}$ converges in measure to f' on $B - N$. Therefore, $f' = f$ almost everywhere in $B - N$, whence follows the second assertion. $\qquad\square$

Proposition 5.3.2   *Suppose that we can find in A a family $(A_i)_{i \in I}$ of mutually disjoint, integrable sets with the following property: for every μ-integrable set $E \subset A$, there exists an at most countable subset J of I such that $E - \left(\bigcup_{i \in J} E \cap A_i\right)$ is μ-negligible. Then $f : A \to F$ is μ-measurable if and only if, for all $i \in I$, its restriction f_i to A_i is μ-measurable. The mapping $\dot{f} \mapsto (\dot{f}_i)_{i \in I}$ is an isomorphism of the uniform space $L(A, \mu; F)$ onto the uniform space $\prod_{i \in I} L(A_i, \mu; F)$. When F is complete, $L(A, \mu; F)$ is complete. When A is μ-moderate, $L(A, \mu; F)$ is metrizable.*

PROOF: The first assertion is obvious, and $\psi : \dot{f} \mapsto (\dot{f}_i)_{i \in I}$ is clearly bijective. For every $i \in I$, the mapping $\dot{f} \mapsto \dot{f}_i$ is uniformly continuous from $L(A, \mu; F)$ into $L(A_i, \mu; F)$; hence ψ is uniformly continuous. Now let $W(V, B, \delta)$ be an entourage of $\mathcal{L}(A, \mu; F)$, and let J be a countable subset of I such that $B - \left(\bigcup_{i \in J} B \cap A_i\right)$ is μ-negligible. Since $\mu(B) = \sum_{i \in J} \mu(B \cap A_i)$, there is a finite subset H of J such that $\sum_{i \in J - H} \mu(B \cap A_i) \leq \delta/2$. If $f, g \in \mathcal{L}(A, \mu; F)$ are such that (f_i, g_i) belongs to $W\left(V, B \cap A_i, \delta/(2|H|)\right)$ for all $i \in H$, then $(f, g) \in W(V, B, \delta)$. Thus ψ^{-1} is uniformly continuous, and the second assertion is proved.

When A is μ-moderate, it is the union of a countable family of disjoint integrable sets. By the above, to prove that $L(A, \mu; F)$ is metrizable, there is no restriction in assuming that A is integrable. But, if $(V_n)_{n \geq 1}$ is a basis for the filter of entourages of F, then the $W(V_n, A, 1/n)$ form a basis for the uniformity of $L(A, \mu; F)$, which proves, in fact, that $L(A, \mu; F)$ is metrizable.

Finally, suppose that F is complete. By the second assertion, to prove that $L(A, \mu; F)$ is complete, we may restrict our attention to the case in which A is μ-integrable. Then $L(A, \mu; F)$ is metrizable. Let $(f_n)_{n \geq 1}$ be a Cauchy sequence in $\mathcal{L}(A, \mu; F)$. By Proposition 5.3.1, there is a subsequence $(f_{n_k})_{k \geq 1}$ of $(f_n)_{n \geq 1}$ which converges almost everywhere in A. The limit f of $(f_{n_k})_{k \geq 1}$ (arbitrarily extended to the whole of A) is μ-measurable, and $(f_{n_k})_{k \geq 1}$ converges to f in $\mathcal{L}(A, \mu; F)$. Therefore, the sequence $(f_n)_{n \geq 1}$ itself converges to f in measure. $\qquad\square$

Observe that the condition of Proposition 5.3.2 is satisfied when A is μ-moderate. We shall prove later (Section 19.2) that it is also satisfied when μ is a Radon measure.

Proposition 5.3.3 *Let F be a real Banach space, equipped with the uniform structure defined by its norm.*

(a) For every μ-measurable set A, the topology of convergence in measure is compatible with the vector space structure of $\mathcal{L}(A, \mu; F)$, and the uniform structure associated with the topological vector space so obtained is the uniform structure of convergence in measure.

(b) For every real number $p \geq 1$, the topology of $\mathcal{L}_F^p(\mu)$ is finer than the topology induced on $\mathcal{L}_F^p(\mu)$ by the topology of $\mathcal{L}(\Omega, \mu; F)$.

(c) For $1 \leq p < +\infty$, $\mathcal{L}_F^p(\mu)$ is dense in $\mathcal{L}(\Omega, \mu; F)$.

PROOF: For every μ-integrable subset B of A and for every real number $\delta > 0$, let $T(B, \delta)$ be the set of those $f \in \mathcal{L}(A, \mu; F)$ for which $\mu(\{x \in B : |f(x)| > \delta\})$ is less than δ. Let d be the distance $(y, z) \mapsto \|y - z\|$ on F. Clearly, $d_B(f, g) \leq \delta$ if and only if $f - g$ belongs to $T(B, \delta)$. Thus, to prove assertion (a), it is enough to show that the filter generated by the $T(B, \delta)$ is the set of neighborhoods of 0 for a topology compatible with the vector space structure of $\mathcal{L}(A, \mu; F)$. Now $T(B, \delta)$ contains $\lambda \cdot T(B, \delta)$ for all $\lambda \in \mathbf{C}$ such that $|\lambda| \leq 1$, and it also contains $T(B, \delta/2) + T(B, \delta/2)$. Hence it remains only to be shown that $T(B, \delta)$ is absorbing (N. Bourbaki, *Espaces vectoriels topologiques* Chapter 1, §1.5, Masson). But let f be a μ-measurable mapping from A into F and, for every $n \geq 1$, let E_n be the set $\{x \in B : |f(x)| > n\}$. The E_n decrease to the empty set, so there exists an integer $n \geq 1/\delta$ such that $\mu(E_n) \leq \delta$. Then f/n^2 belongs to $T(B, \delta)$, which gives (a).

Now let $\delta > 0$ be a real number. If $f \in \mathcal{L}_F^p(\mu)$ is such that $\int |f|^p d\mu \leq \delta^{p+1}$ and if $E = \{x \in \Omega : |f(x)| > \delta\}$, then $\delta^p \cdot \mu(E) \leq \int |f|^p d\mu \leq \delta^{p+1}$, and $\mu(E) \leq \delta$, which proves (b).

Finally, let f be an arbitrary element of $\mathcal{L}(\Omega, \mu; F)$ and $T(B, \delta)$ a neighborhood of 0 in this space. Arguing as above, we see that there exists a μ-integrable subset E of B such that $\mu(E) \leq \delta$ and such that f is bounded on $B - E$. Then the mapping g equal to f on $B - E$ and to zero on $\Omega - (B - E)$ lies in $\mathcal{L}_F^p(\mu)$, and, clearly, $f - g$ belongs to $T(B, \delta)$. □

5.4   Uniformly Integrable Sets

Let Ω be a nonempty set and μ a positive Daniell measure on a space $\mathcal{H}(\Omega, \mathbf{C})$. Let F be a real Banach space and $p \geq 1$ a real number.

Definition 5.4.1 Suppose H is a subset of $\mathcal{L}_F^p(\mu)$. H is said to be uniformly integrable of order p if and only if the following conditions hold:

(a) For every $\varepsilon > 0$, there exists $\delta > 0$ such that $\sup_{f \in H} \int |f|^p 1_A d\mu \leq \varepsilon$ for all μ-integrable sets A whose measure $\mu(A)$ is less than δ.

(b) For every $\varepsilon > 0$, there exists a μ-integrable set B such that $\sup_{f \in H} \int |f|^p 1_{\Omega-B} d\mu \leq \varepsilon$.

When $p = 1$, H is said simply to be uniformly integrable.

Proposition 5.4.1 *Let H be a subset of $\mathcal{L}_F^p(\mu)$, uniformly integrable of order p. The uniform structure of convergence in measure, on H, is equal to the uniform structure induced by that of $\mathcal{L}_F^p(\mu)$.*

PROOF: Let $\varepsilon > 0$ be a real number, and let $\delta > 0$ and $B \in \hat{\mathcal{R}}$ be as in Definition 5.4.1. We may assume that $\mu(B) > 0$. Put $\eta = \left(\varepsilon/\mu(B)\right)^{1/p}$ and recall that $\{(y, z) \in F \times F : \|y - z\| \leq \eta\}$ is written V_η. Let (f, g) be an element of $W(V_\eta, B, \delta) \cap (H \times H)$, so that $|f - g|(x) \leq \eta$ for all $x \in B - E$, where E is a μ-integrable subset of B whose measure is smaller than δ. We have $N_p\left((f - g) \cdot 1_{\Omega-B}\right) \leq N_p(f 1_{\Omega-B}) + N_p(g 1_{\Omega-B}) \leq 2\varepsilon^{1/p}$, and likewise $N_p\left((f - g) 1_E\right) \leq 2\varepsilon^{1/p}$; therefore,

$$\int |f - g|^p d\mu = \int_{\Omega-B} |f - g|^p d\mu + \int_E |f - g|^p d\mu + \int_{B-E} |f - g|^p d\mu$$

is less than $2^p \varepsilon + 2^p \varepsilon + \left(\varepsilon/\mu(B)\right) \cdot \mu(B - E) \leq (2^{p+1} + 1)\varepsilon$. $\quad\square$

Proposition 5.4.2 *Let H be a subset of $\mathcal{L}_F^p(\mu)$. H is uniformly integrable of order p if and only if $\sup_{f \in H} \int |f|^p 1_{A_n} d\mu$ converges to 0 as $n \to +\infty$, for every sequence $(A_n)_{n \geq 1}$ of μ-measurable and μ-moderate sets which decreases to $\emptyset$. In this case, $\sup_{f \in H} \int |f|^p 1_{A_n} d\mu$ converges to 0 as $n \to +\infty$, for every sequence $(A_n)_{n \geq 1}$ of μ-measurable sets such that $(1_{A_n})_{n \geq 1}$ converges to 0 almost everywhere.*

PROOF: First, suppose that H is uniformly integrable of order p, and let $(A_n)_{n \geq 1}$ be a sequence of μ-measurable sets such that $(1_{A_n})_{n \geq 1}$ converges to 0 almost everywhere. For fixed $\varepsilon > 0$, let B be a μ-integrable set such that $\sup_{f \in H} \int_{\Omega-B} |f|^p d\mu \leq \varepsilon/2$. By the dominated convergence theorem, $\mu(A_n \cap B)$ goes to 0 as $n \to +\infty$. For $n \geq 1$ large enough, therefore, $\sup_{f \in H} \int_{A_n \cap B} |f|^p d\mu \leq \varepsilon/2$, and $\sup_{f \in H} \int |f|^p 1_{A_n} d\mu \leq \varepsilon$.

Conversely, suppose that $\sup_{f \in H} \int |f|^p 1_{A_n} d\mu$ converges to 0, for every sequence $(A_n)_{n \geq 1}$ of μ-measurable and μ-moderate sets which decreases to $\emptyset$.

Assume, for the moment, that condition (a) of Definition 5.4.1 is not satisfied. There exist $\varepsilon > 0$ and a sequence $(B_n)_{n \geq 1}$ of μ-integrable sets such that $\mu(B_n) \leq 1/2^n$ for every $n \geq 1$ and that $\sup_{f \in H} \int |f|^p 1_{B_n} d\mu > \varepsilon$. Set $A_n = \bigcup_{i \geq n} B_i$ for every integer $n \geq 1$, and put $\bar{B} = \limsup B_n = \bigcap_{n \geq 1} A_n$. Since $\mu(A_n) \leq 1/2^{n-1}$ for each $n \geq 1$, $\bar{B}$ is μ-negligible. Now the sequence $(A_n - \bar{B})_{n \geq 1}$ decreases to the empty set, hence $\sup_{f \in H} \int |f|^p 1_{A_n} d\mu$ converges to 0 as $n \to +\infty$, and we arrive at a contradiction. Condition (a) therefore holds.

Now assume that condition (b) of Definition 5.4.1 is not satisfied. There exists $\varepsilon > 0$ such that $\sup_{f \in H} \int_{\Omega - B} |f|^p d\mu > \varepsilon$ for all integrable sets B. Given $n \in \mathbf{N}$, suppose we have obtained disjoint integrable sets $B_1, \ldots, B_{n-1}$ and elements $f_1, \ldots, f_{n-1}$ of H such that $\int_{B_i} |f_i|^p d\mu > \varepsilon$ for all $1 \leq i \leq n-1$. There exist $f_n \in H$ such that $\int_{\Omega - (B_1 \cup \ldots \cup B_{n-1})} |f_n|^p d\mu > \varepsilon$, and a μ-integrable subset B_n of $\Omega - (B_1 \cup \ldots \cup B_{n-1})$ for which $\int_{B_n} |f_n|^p d\mu > \varepsilon$. We so construct an infinite sequence $(B_n)_{n \geq 1}$ of disjoint integrable sets and a sequence $(f_n)_{n \geq 1}$ in H such that $\int |f_n|^p 1_{B_n} d\mu > \varepsilon$ for all $n \geq 1$. Now, for each $n \geq 1$, put $A_n = \bigcup_{i \geq n} B_i$. The sequence $(A_n)_{n \geq 1}$ decreases to the empty set, hence $\sup_{f \in H} \int |f|^p 1_{A_n} d\mu$ converges to 0 as $n \to +\infty$, and we arrive at a contradiction. Condition (b) therefore holds. $\qquad\square$

Now let H be a subset of $\mathcal{L}_F^p(\mu)$, and consider the following condition

(c) $\sup_{f \in H} \int_{|f| > t} |f|^p d\mu$ goes to 0 as $t \to +\infty$.

Proposition 5.4.3 *Condition (c) implies condition (a) of Definition 5.4.1. They are equivalent when μ is diffuse.*

PROOF: First, suppose that (c) holds. For fixed $\varepsilon > 0$, let $t > 0$ be a real number such that $\sup_{f \in H} \int_{|f| > t} |f|^p d\mu$ is less than $\varepsilon/2$. If E is a μ-integrable set such that $\mu(E) \leq (\varepsilon/2) \cdot t^{-p}$, then $\int_E |f|^p d\mu \leq \int_{|f| > t} |f|^p d\mu + t^p \cdot \mu(E) \leq \varepsilon$, hence condition (a) is satisfied.

Conversely, assume that μ is diffuse and that condition (a) holds. Suppose that $\sup_{f \in H} \mu(|f| > t)$ does not go to 0 as $t \to +\infty$. There exist $\alpha > 0$, a sequence $(t_n)_{n \geq 1}$ of strictly positive numbers converging to $+\infty$, and a sequence $(f_n)_{n \geq 1}$ in H, such that $\mu(|f_n| > t_n) > \alpha$ for every $n \geq 1$. Let $\beta > 0$ be such that $\mu(E) \leq \beta$ implies $\sup_{f \in H} \int_E |f|^p d\mu \leq 1$. For each integer $n \geq 1$, we can find an integrable subset E_n of $(|f_n| > t_n)$ such that $\mu(E_n) = \inf(\alpha, \beta)$. Now $1 \geq \int_{E_n} |f_n|^p d\mu \geq t_n^p \cdot \inf(\alpha, \beta)$, which is absurd because $t_n \to +\infty$. Therefore, $\sup_{f \in H} \mu(|f| > t)$ converges to 0 as $t \to +\infty$. For fixed $\varepsilon > 0$, let $\delta > 0$ be such that $\sup_{f \in H} \int_A |f|^p d\mu \leq \varepsilon$ for every μ-integrable set A whose measure $\mu(A)$ is smaller than δ. Then, for t large enough, $\sup_{f \in H} \mu(|f| > t)$ is smaller than δ, and so we have $\sup_{f \in H} \int_{|f| > t} |f|^p d\mu \leq \varepsilon$. $\qquad\square$

Proposition 5.4.4 *Let H be a subset of $\mathcal{L}_F^p(\mu)$ satisfying condition (b) of Definition 5.4.1. Then (c) holds if and only if H is bounded and uniformly integrable of order p.*

PROOF: First, suppose that (c) holds, and let $\alpha > 0$ be a real number such that $\sup_{f \in H} \int_{|f| > \alpha} |f|^p d\mu \leq 1$. Also, let B be an integrable set such that $\sup_{f \in H} \int |f|^p 1_{\Omega - B} d\mu \leq 1$. Then

$$\int_B |f|^p d\mu \leq \int_{(|f| > \alpha) \cap B} |f|^p d\mu + \alpha^p \mu(B \cap (0 < |f| \leq \alpha)) \leq 1 + \alpha^p \mu(B)$$

for every $f \in H$. Thus $\int |f|^p d\mu \leq 2 + \alpha^p \mu(B)$, and H is bounded in $\mathcal{L}^p_F(\mu)$.

Conversely, suppose that H is bounded and that (a) holds. For fixed $\varepsilon > 0$, let $\delta > 0$ be such that $\sup_{f \in H} \int |f|^p 1_A d\mu$ is smaller than ε for every integrable set A whose measure $\mu(A)$ is less than δ. Choose $t > 0$ so that $t^{-p}(\sup_{f \in H} \int |f|^p d\mu) \leq \delta$. Then $\mu(|f| > t) \leq t^{-p} \int |f|^p d\mu \leq \delta$, and hence $\int_{|f|>t} |f|^p d\mu \leq \varepsilon$, for all $f \in H$. This proves that condition (c) is satisfied. $\square$

Therefore, when μ is diffuse, every uniformly integrable subset H of $\mathcal{L}^p_F(\mu)$ is bounded. On the contrary, when we take for μ the measure on $\mathbf{R}$ defined by the mass 1 at the point 0, and, for each $n \geq 1$, let f_n be the function from $\mathbf{R}$ into $\mathbf{R}$ which is constant, equal to n, then $\{f_n : n \geq 1\}$ is uniformly integrable of order 1, but it is not bounded in $\mathcal{L}^1_F(\mu)$.

Theorem 5.4.1 *If a sequence $(f_n)_{n \geq 1}$ converges in $\mathcal{L}^p_F(\mu)$ to an element f of $\mathcal{L}^p_F(\mu)$, then $H = \{f_n : n \geq 1\}$ is uniformly integrable of order p.*

PROOF: For fixed $\varepsilon > 0$, there exists $\delta > 0$ such that $\int |f|^p 1_A d\mu \leq \varepsilon/2^p$ for every integrable set A whose measure is less than δ. Then $t^p \cdot \mu(|f_n| > t) \leq \int |f_n|^p d\mu \leq \sup_{k \geq 1} N_p(f_k)^p$ for all $n \geq 1$ and all $t > 0$. This implies that we can find $t_0 > 0$ such that $\sup_{k \geq 1} \mu(|f_k| > t)$ is smaller than δ for all $t \geq t_0$. Now let $N \geq 1$ be an integer such that $N_p(f_n - f) \leq \varepsilon^{1/p}/2$ for all $n \geq N$. Then, for every $t \geq t_0$ and every $n \geq N$,

$$\left(\int_{|f_n|>t} |f_n|^p d\mu \right)^{1/p} \leq N_p(f_n - f) + \left(\int_{|f_n|>t} |f|^p d\mu \right)^{1/p} \leq \frac{1}{2}\varepsilon^{1/p} + \frac{1}{2}\varepsilon^{1/p}$$

$$= \varepsilon^{1/p},$$

and $\int_{|f_n|>t} |f_n|^p d\mu \leq \varepsilon$. Now, for every $1 \leq n < N$, $\int_{|f_n|>t} |f_n|^p d\mu$ goes to 0 as $t \to +\infty$, by the dominated convergence theorem. For t large enough, we get $\int_{|f_n|>t} |f_n|^p d\mu \leq \varepsilon$ for all $n \geq 1$. Condition (c) is therefore satisfied.

Next, there exists an integrable set B' such that $\int_{\Omega-B'} |f|^p d\mu \leq \varepsilon/2^p$. Let N be an integer such that $N_p(f_n - f) \leq \varepsilon^{1/p}/2$ for all $n \geq N$. Then, for every $n \geq N$,

$$\left(\int_{\Omega-B'} |f_n|^p d\mu \right)^{1/p} \leq \frac{1}{2} \cdot \varepsilon^{1/p} + \frac{1}{2} \cdot \varepsilon^{1/p} = \varepsilon^{1/p},$$

and $\int_{\Omega-B'} |f_n|^p d\mu \leq \varepsilon$. On the other hand, for every $1 \leq n < N$, there exists an integrable set B'_n such that $\int_{\Omega-B'_n} |f_n|^p d\mu \leq \varepsilon$. If we put $B = B' \cup B'_1 \cup \ldots \cup B'_{N-1}$, then $\int_{\Omega-B} |f_n|^p d\mu \leq \varepsilon$ for all $n \geq 1$, which proves that (b) holds. $\square$

Now we give some sufficient conditions for a sequence which converges in measure to converge in the mean of order p.

Proposition 5.4.5 *Let $(f_k)_{k \geq 1}$ be a sequence in $\mathcal{L}_F^p(\mu)$ converging in measure to a μ-measurable and μ-moderate mapping f. Assume that $(f_k)_{k \geq 1}$ has the following properties:*

(a) *For every $\varepsilon > 0$, there exists $\delta > 0$ such that $\limsup_{k \to +\infty} \int_A |f_k|^p d\mu$ is smaller than ε for every μ-integrable set A whose measure is less than δ.*

(b) *For every $\varepsilon > 0$, there exists a μ-integrable set B such that $\sup_k \int_{\Omega - B} |f_k|^p d\mu$ is smaller than ε.*

Then f belongs to $\mathcal{L}_F^p(\mu)$ and $(f_k)_{k \geq 1}$ converges to f in the mean of order p (i.e., $N_p(f_k - f) \to 0$).

PROOF: First, suppose that $(f_k)_{k \geq 1}$ converges to f almost everywhere. For fixed $\varepsilon > 0$, there exists an integrable set B such that $\sup_{k \geq 1} \int_{\Omega - B} |f_k|^p d\mu$ is smaller than $(\varepsilon/6)^p$. For every integer $n \geq 1$. put

$$B_n = \left\{ x \in B : \sup_{k \geq n} |f_k(x) - f(x)| > \frac{\varepsilon}{6} \cdot \mu(B)^{-1/p} \right\}.$$

The sets B_n decrease and $\mu\left(\bigcap_{n \geq 1} B_n \right) = 0$. There exists $\delta > 0$ such that $\limsup_{k \to +\infty} \int_A |f_k|^p d\mu$ is strictly less than $(\varepsilon/6)^p$ for all integrable sets A satisfying $\mu(A) \leq \delta$.

Let $n_0 \geq 1$ be large enough so that $\mu(B_{n_0}) \leq \delta$, and let $n_1 \geq n_0$ be such that $\sup_{k \geq n_1} \int_{B_{n_0}} |f_k|^p d\mu$ is smaller than $(\varepsilon/6)^p$. Then, for all $r, s \geq n_1$,

$$f_r - f_s = f_r \cdot 1_{\Omega - B} - f_s \cdot 1_{\Omega - B} + (f_r - f) \cdot 1_{B - B_{n_0}}$$
$$+ (f - f_s) \cdot 1_{B - B_{n_0}} + f_r \cdot 1_{B_{n_0}} - f_s \cdot 1_{B_{n_0}};$$

thus $N_p(f_r - f_s)$ is smaller than

$$N_p(f_r \cdot 1_{\Omega - B}) + N_p(f_s \cdot 1_{\Omega - B}) + N_p((f_r - f) \cdot 1_{B - B_{n_0}})$$
$$+ N_p((f - f_s) \cdot 1_{B - B_{n_0}}) + N_p(f_r \cdot 1_{B_{n_0}}) + N_p(f_s \cdot 1_{B_{n_0}}),$$

and so less than ε. Hence $(f_k)_{k \geq 1}$ is a Cauchy sequence in $\mathcal{L}_F^p(\mu)$, and it converges to f in $\mathcal{L}_F^p(\mu)$.

We now pass to the general case, in which $(f_k)_{k \geq 1}$ converges to f in measure. From every subsequence $(f_{k_j})_{j \geq 1}$ of $(f_k)_{k \geq 1}$ we can extract a sequence that converges to f almost everywhere, and that therefore converges to f in $\mathcal{L}_F^p(\mu)$. Thus the sequence $(f_k)_{k \geq 1}$ itself converges to f in $\mathcal{L}_F^p(\mu)$. $\square$

Theorem 5.4.2 (Dominated Convergence in Measure) *Let $(f_n)_{n \geq 1}$ be a sequence in $\mathcal{L}_F^p(\mu)$ converging in measure to a μ-measurable and μ-moderate mapping f. Assume that the $|f_n|$ are almost everywhere dominated by a function g from Ω into $[0, +\infty]$ such that $\int^* g^p d\mu < +\infty$. Then f belongs to $\mathcal{L}_F^p(\mu)$ and $(f_n)_{n \geq 1}$ converges to f in the mean of order p.*

PROOF: This follows from Proposition 5.4.4.   □

Proposition 5.4.6 *Let* $(f_n)_{n\geq 1}$ *be a sequence in* $\mathcal{L}_F^p(\mu)$ *that converges in measure to* $f \in \mathcal{L}_F^p(\mu)$ *and such that* $(N_p(f_n))_{n\geq 1}$ *converges to* $N_p(f)$. *Then* $(f_n)_{n\geq 1}$ *converges to* f *in the mean of order p.*

PROOF: First, suppose that $(f_n)_{n\geq 1}$ converges to f almost everywhere. For fixed $\varepsilon > 0$, let B be an integrable set such that $\int_{\Omega - B} |f|^p d\mu$ is strictly less than $\varepsilon/6$. By Egorov's theorem, we can find a decreasing sequence $(E_r)_{r\geq 1}$ of integrable subsets of B such that $\mu(E_r) \leq 1/r$ for every $r \geq 1$ and such that $(f_n)_{n\geq 1}$ converges uniformly to f on $B - E_r$. If $s \geq 1$ is an integer such that $\int_{E_s} |f|^p d\mu \leq \varepsilon/6$, and if we put $A = B - E_s$, then $\int_{\Omega - A} |f|^p d\mu < \varepsilon/3$. Now $\int_{\Omega - A} |f_n|^p d\mu = \int |f_n|^p d\mu - \int_A |f_n|^p d\mu$ converges to $\int_{\Omega - A} |f|^p d\mu$. Hence, for $k \geq 1$ large enough, each of $\int_{\Omega - A} |f_n|^p d\mu$ and $\int_A |f - f_n|^p d\mu$ is smaller than $\varepsilon/3$ for all $n \geq k$, and then

$$\int |f - f_n|^p d\mu \leq \int_A |f - f_n|^p d\mu + \int_{\Omega - A} |f|^p d\mu + \int_{\Omega - A} |f_n|^p d\mu$$

is less than ε for all $n \geq k$.

Now the general case, in which $(f_n)_{n\geq 1}$ converges to f in measure, is handled by the same argument as that in Proposition 5.4.5, showing that $(f_n)_{n\geq 1}$ converges to f in the mean of order p.   □

Exercises for Chapter 5

1   Let F be a real Banach space and $p \geq 1$ a real number.

   1.   Let a, b be two vectors in F such that $|a| = |b| = 1$. Show that $1 - \rho \leq |a - \rho b|$ and $|a - b| \leq 2|a - \rho b|$ for all $0 \leq \rho \leq 1$. Prove that

$$|a - tb|^p \leq 2^p \cdot |a - t^p b| \qquad (a)$$

     and

$$|a - t^p b| \leq 3p \cdot |a - tb| \qquad (b)$$

     for all $0 \leq t \leq 1$ (assume that $0 \leq t < 1$; observe that

$$a - tb = \lambda(a - t^p b) + (1 - \lambda)(a - b),$$

     where $\lambda = (1 - t)/(1 - t^p)$, and that

$$a - t^p b = \mu(a - tb) + (1 - \mu)(a - b),$$

     where $\mu = (1 - t^p)/(1 - t) = \left(\int_t^1 p u^{p-1} du \right)/(1 - t)$).

2. From part 1 deduce that, for all y, $z \in F$,

$$|y - z|^p \le 2^p \cdot \left| |y|^{p-1} \cdot y - |z|^{p-1} \cdot z \right| \qquad (c)$$

and

$$\left| |y|^{p-1} \cdot y - |z|^{p-1} \cdot z \right| \le 3p \cdot |y - z| \cdot \left(|y| + |z| \right)^{p-1} \qquad (d).$$

3. Henceforth, suppose that μ is a positive Daniell measure on a space $\mathcal{H}(\Omega, \mathbf{C})$. Conclude from inequality (c) of part 2 that the mapping $f \mapsto |f|^{1/p-1} \cdot f$ from $\mathcal{L}_F^1(\mu)$ into $\mathcal{L}_F^p(\mu)$ is uniformly continuous.

4. Suppose that $1 < p < +\infty$, and let q be the exponent conjugate to p. Deduce from inequality (d) of part 2 that

$$N_1\left(|f|^{p-1} \cdot f - |g|^{p-1} \cdot g \right) \le 3p \cdot N_p(f - g) \cdot \left(N_p\left(|f| + |g| \right) \right)^{1/q}$$

for all f and g of $\mathcal{L}_F^p(\mu)$, and therefore that the mapping $f \mapsto |f|^{p-1} \cdot f$ from $\mathcal{L}_F^p(\mu)$ into $\mathcal{L}_F^1(\mu)$ is uniformly continuous on every bounded subset of $\mathcal{L}_F^p(\mu)$.

5. Conclude, for $1 < p < +\infty$, that the mapping $f \mapsto |f|^{p-1} \cdot f$ from $\mathcal{L}_F^p(\mu)$ into $\mathcal{L}_F^1(\mu)$ is a homeomorphism.

2 Let F be a real Banach space, $0 < p < 1$ a real number, and μ a positive Daniell measure on $\mathcal{H}(\Omega, \mathbf{C})$.

1. Define d as the mapping $(f, g) \mapsto \int |f - g|^p d\mu$ from $\mathcal{L}_F^p(\mu) \times \mathcal{L}_F^p(\mu)$ into $[0, +\infty[$. Show that d is a pseudometric and defines a topology compatible with the vector space structure of $\mathcal{L}_F^p(\mu)$ (use Lemma 5.1.2).

2. Using inequality (c) above, show that the mapping $h \mapsto |h|^{p-1} \cdot h$ is uniformly continuous from $\mathcal{L}_F^p(\mu)$ into $\mathcal{L}_F^1(\mu)$.

3. From (d), obtain the inequality

$$\left| |f|^{1/p-1} \cdot f - |g|^{1/p-1} \cdot g \right| \le \frac{3}{p} \cdot |f - g| \cdot \left(|f| + |g| \right)^{1/p-1}$$

for all f, g of $\mathcal{L}_F^1(\mu)$. Deduce that the mapping $f \mapsto |f|^{1/p-1} \cdot f$ from $\mathcal{L}_F^1(\mu)$ into $\mathcal{L}_F^p(\mu)$ is uniformly continuous on every bounded subset of $\mathcal{L}_F^1(\mu)$.

4. Conclude that the mapping $f \mapsto |f|^{p-1} \cdot f$ is a homeomorphism from $\mathcal{L}_F^p(\mu)$ onto $\mathcal{L}_F^1(\mu)$ and that $\mathcal{L}_F^p(\mu)$ is complete. Thus $\mathcal{L}_F^p(\mu)$ is a complete metrizable topological vector space and $St(\hat{\mathcal{R}}, F)$ is dense in $\mathcal{L}_F^p(\mu)$.

3 Let μ be Lebesgue measure on $I = [0, 1]$, p an element of $]0, 1[$, and F a real Banach space $(F \ne \{0\})$.

1. Let $f \in \mathcal{L}_F^p(\mu)$. Show that there exist μ-integrable sets A, B such that $A \cap B = \emptyset$, $A \cup B = I$, and $\int |f|^p 1_A d\mu = \int |f|^p 1_B d\mu = (1/2) \int |f|^p d\mu$.

2. For fixed $a > 0$, put $B_a = \{f \in \mathcal{L}_F^p(\mu) : N_p(f) \le a\}$ and write D for its convex envelope. If $f \in \mathcal{L}_F^p(\mu)$ is such that $N_p(f) \le a \cdot 2^{1/p-1}$, show that f lies in D (in the notation of part 1, consider $f_1 = 2f1_A$ and $f_2 = 2f1_B$). Conclude that $D = \mathcal{L}_F^p(\mu)$.

3. Deduce from part 2 that every continuous linear form on $\mathcal{L}_F^p(\mu)$ is identically zero, and therefore that $\mathcal{L}_F^p(\mu)$ is not locally convex.

4   Let μ be Lebesgue measure on $I = [0,1]$ and let F be a real Banach space ($F \neq \{0\}$). For all real numbers $\delta > 0$, put

$$T(\delta) = \left\{ f \in \mathcal{L}(I, \mu; F) : \mu\big(|f| > \delta\big) \le \delta \right\}.$$

1. Let u be a continuous linear form on $\mathcal{L}(I, \mu; F)$. Choose $\delta > 0$ so that $|u(f)| \le 1$ for every $f \in T(\delta)$. If A is an integrable set whose measure is less than δ, prove that $u(c1_A) = 0$ for every $c \in F$ (consider $\alpha c1_A$ for $\alpha > 0$). Deduce that $u(c1_A) = 0$ for all $c \in F$ and all integrable sets. Conclude that u is identically zero.

2. Deduce from part 1 that $\mathcal{L}(I, \mu; F)$ is not locally convex.

5   Let μ be a positive Daniell measure on a space $\mathcal{H}(\Omega, \mathbf{C})$, A a μ-measurable set, and F a metrizable uniform space.

1. Let $(f_n)_{n \ge 1}$ be a sequence in $\mathcal{L}(A, \mu; F)$, and let f be an element of $\mathcal{L}(A, \mu; F)$. If $(f_n)_{n \ge 1}$ does not converge to f in measure, there exist a $W(V, B, \delta)$ and a subsequence $(f_{n_k})_{k \ge 1}$ of $(f_n)_{n > 1}$ such that none of the (f_{n_k}, f) belongs to $W(V, B, \delta)$. In this case, show that no subsequence of $(f_{n_k})_{k \ge 1}$ can converge to f almost everywhere in B.

2. Let $(f_n)_{n \ge 1}$ be a sequence in $\mathcal{L}(A, \mu; F)$ and f an element of $\mathcal{L}(A, \mu; F)$. Show that $(f_n)_{n \ge 1}$ converges to f in measure if and only if, for every integrable subset B of A, every subsequence of $(f_n)_{n \ge 1}$ contains a further subsequence that converges to f almost everywhere in B. In this case, if g is a continuous function from F into a metrizable uniform space G, show that $(g \circ f_n)_{n \ge 1}$ converges to $g \circ f$ in $\mathcal{L}(A, \mu; G)$.

3. Suppose that A is μ-moderate. Show that the topology of convergence in measure is the finest of those topologies T on $\mathcal{L}(A, \mu; F)$ which have the following property: every sequence $(f_n)_{n \ge 1}$ in $\mathcal{L}(A, \mu; F)$ converging to f almost everywhere in A converges also to f in the topology T.

6   Let μ be a positive Daniell measure on a space $\mathcal{H}(\Omega, \mathbf{C})$, and let A be a μ-measurable set.

1. Suppose that each μ-integrable subset of A which is not μ-negligible contains an atom. Let F be a metrizable uniform space and d a distance compatible with the uniform structure of F. Show that every sequence $(f_n)_{n \ge 1}$ in $\mathcal{L}(A, \mu; F)$ converging in measure to $f \in \mathcal{L}(A, \mu; F)$ converges also to f locally almost everywhere in A.

2. Suppose there exists an integrable subset B of A which is not negligible and which contains no atom. By induction on $h \geq 0$, define μ-integrable sets $B_{h,k}(0 \leq k < 2^h)$ as follows:

(a) $B_{0,0} = B$.

(b) For every $h \geq 1$, $B_{h,2k}$ and $B_{h,2k+1}$ form a partition of $B_{h-1,k}$ and $\mu(B_{h,2k}) = \mu(B_{h,2k+1}) = (1/2)\mu(B_{h-1,k})$ for $0 \leq k < 2^{h-1}$.

Then, for all integer pairs (h, k) such that $h \geq 0$ and $0 \leq k < 2^h$, put $f_{2^h+k} = 1_{B_{h,k}}$. Prove that $(f_n)_{n \geq 1}$ converges to 0 in $\mathcal{L}^1_{\mathbf{R}}(\mu)$, but that it does not converge to 0 locally almost everywhere in A.

7 Let μ be a positive Daniell measure on a space $\mathcal{H}(\Omega, \mathbf{C})$. let $(f_n)_{n \geq 1}$ be a sequence of μ-measurable functions from Ω into $\overline{\mathbf{R}}$, and define $f = \limsup f_n$. For every μ-integrable set B and every $\delta > 0$, construct a μ-integrable subset A of B whose measure $\mu(A)$ is less than δ and for which the following property is satisfied: for each $\eta > 0$, there exists an integer $n_0 \geq 1$ such that $\sup_{n \geq n_0} f_n(x) \leq f(x) + \eta$ for every $x \in B - A$ (for all integers $j \geq 1$ and $k \geq 1$, consider the set $B_j^{(k)} = \{x \in B : f(x) < +\infty \text{ and } f(x) + 1/k < \sup_{n \geq j} f_n(x)\}$, and observe that $\bigcap_{j \geq 1} B_j^{(k)} = \emptyset$).

8 Let μ be a positive Daniell measure on a space $\mathcal{H}(\Omega, \mathbf{C})$, and let $(f_n)_{n \geq 1}$ be a sequence in $\mathcal{L}^1_{\mathbf{R}}(\mu)$ such that $\{f_n : n \geq 1\}$ is uniformly integrable.

1. Prove that $\limsup \int f_n d\mu \leq \int_* (\limsup f_n) d\mu$ (use Exercise 7).

2. Similarly, show that $\int^* (\liminf f_n) d\mu \leq \liminf \int f_n d\mu$. Compare with Proposition 3.2.1.

9 Let $(\Omega, \mathcal{F}, P)$ be a probability space, that is, a nonempty set Ω, a σ-algebra $\mathcal{F}$ in Ω, and a positive measure P on $\mathcal{F}$ such that $P(\Omega) = 1$. For every integer $n \geq 1$, let S_n be a random variable from $(\Omega, \mathcal{F})$ into $\mathbf{R}$, that is, a measurable $\mathcal{F}$ function from Ω into $\mathbf{R}$, and suppose that $P(S_n = k) = (n^k/k!) \cdot e^{-n}$ for all integers $k \geq 0$. In other words, the law of S_n is the Poisson distribution with parameter n. Finally, put $Y_n = ((S_n - n)/\sqrt{n})^- = \sup(0, -(S_n - n)/\sqrt{n})$ for all $n \geq 1$.

1. For every $n \geq 1$, let $u_n = (1/n!)n^{n+1/2}e^{-n}$. From

$$\log\left(1 + \frac{1}{n}\right) = \log\left(1 + \frac{1}{2n+1}\right) - \log\left(1 - \frac{1}{2n+1}\right),$$

deduce that $u_{n+1}/u_n > 1$, and that $(u_n)_{n \geq 1}$ strictly increases to $1/\sqrt{2\pi}$ (use Stirling's formula). Show that $\int Y_n dP = u_n$.

2. Let $\alpha \geq 1$ be a real number. Prove that $\int_{Y_n > \alpha} Y_n dP = 0$ for all $n \leq \alpha^2$, and that $\int_{Y_n > \alpha} Y_n dP = u_n \cdot \prod_{0 \leq k \leq [\alpha\sqrt{n}]}(1 - k/n)$ for all $n > \alpha^2$ (where $[\alpha\sqrt{n}] = p_n$ is the integral part of $\alpha\sqrt{n}$).

3. Let $n > \alpha^2$ be an integer, and apply the Euler–MacLaurin summation formula to the function $x \mapsto \log(1 - x/n)$ from $[0, n - 1]$ into $\mathbf{R}$. Obtain the formula

$$\sum_{0 \le k \le p_n} \log\left(1 - \frac{k}{n}\right) = \int_0^{p_n} \log\left(1 - \frac{x}{n}\right) dx + \frac{1}{2}\log\left(1 - \frac{p_n}{n}\right) + R_0(n),$$

where $R_0(n) \le 0$, and conclude that $\displaystyle\sum_{0 \le k \le p_n} \log\left(1 - \frac{k}{n}\right) + \frac{p_n(p_n + 1)}{2n}$ is negative.

4. Deduce from parts 2 and 3 that $\{Y_n : n \ge 1\}$ is uniformly integrable with respect to P.

Observe, incidentally, that $(S_n - n)/\sqrt{n}$ converges in law to the normal law $N_1(0, 1)$ (by the central limit theorem), and hence that $(Y_n)_{n \ge 1}$ converges in law.

10 Let A be a commutative ring (in the algebraic sense) and J an involutive automorphism $\lambda \mapsto \bar{\lambda}$ of A. Let E be an A-module and Φ a Hermitian form on $E \times E$. Hence $\Phi(y, x) = \overline{\Phi(x, y)}$ for all $x, y \in E$. Finally, let $x_1, \ldots, x_n$ be linearly independent elements of E and F be the submodule of E spanned by $x_1, \ldots, x_n$.

1. Show that $\lambda_1 x_1 + \ldots + \lambda_n x_n \in F$ is orthogonal to F if and only if $(\bar{\lambda}_1, \ldots, \bar{\lambda}_n)$ is a solution of a homogeneous linear system.

2. Show that the restriction of Φ to $F \times F$ is nondegenerate if and only if $\det\left(\Phi(x_i, x_j)\right)_{1 \le i,j \le n}$ is not a zero-divisor in A.

11 If X is a square matrix of order $n \ge 1$ over a commutative ring A and if Y is the matrix of cofactors of X, recall that ${}^t Y \cdot X = (\det X) \cdot I_n$, where I_n is the unit matrix of order n. Let A be a commutative field and $J : \lambda \mapsto \bar{\lambda}$ an involutive automorphism of A. Let E be a vector space over A, Φ a Hermitian form on $E \times E$, and $(x_j)_{1 \le j \le n}$ a finite family of linearly independent vectors of E such that, for each integer $1 \le k \le n$, the subspace $E_k = Ax_1 + \ldots + Ax_k$ is non-isotropic (i.e., the restriction of Φ to $E_k \times E_k$ is nondegenerate). Put $D_{1,1} = 1$ and, for all integers j, k such that $1 \le j \le k \le n$, let $D_{j,k}$ be the cofactor of $\Phi(x_j, x_k)$ in the matrix $\left(\Phi(x_s, x_t)\right)_{1 \le s,t \le k}$.

1. Show that $D_{k,k} \ne 0$ for all $1 \le k \le n$.

2. For every $1 \le k \le n$, put $e_k = \sum_{1 \le j \le k} D_{k,k}^{-1} \cdot D_{j,k} \cdot x_j$. Prove that $(e_1, \ldots, e_k)$ is an orthogonal basis of E_k, for every $1 \le k \le n$, and that

$$\Phi(e_k, e_k) = D_{k,k}^{-1} \cdot \det\left(\Phi(x_i, x_j)\right)_{1 \le i,\, j \le k}$$

(use the remark at the beginning of the exercise). One says that $(e_1, \ldots, e_n)$ is obtained from $(x_1, \ldots x_n)$ by Gram's orthogonalization.

12 Let $(x_i)_{1 \le i \le n}$ be a finite sequence in an inner product space (over $\mathbf{R}$ or $\mathbf{C}$). By definition, Gram's determinant (or the grammian) of this sequence is $G(x_1, \ldots, x_n) = \det(x_i | x_j)_{1 \le i,j \le n}$.

1. Show that $x_1,\ \ldots\ ,\ x_n$ are linearly independent if and only if $G(x_1,\ldots,x_n) \neq 0$ and that, in this case, $G(x_1,\ldots,x_n)$ is strictly positive.

2. Assume that $x_1,\ \ldots\ ,\ x_n$ are linearly independent, and let V be the vector subspace of E generated by $\{x_1,\ldots,x_n\}$. Show that the distance from a point x to V is $\big(G(x_1,\ldots,x_n,x)/G(x_1,\ldots,x_n)\big)^{1/2}$. To see this, consider the sequence $(e_1,\ldots,e_{n+1})$ obtained from $(x_1,\ldots,x_n,x)$ by Gram's orthogonalization.

3. Under the hypotheses of part 2, show that $e_k|x_k$ is strictly positive for all $1 \leq k \leq n$.

4. Under the hypotheses of part 2, let $(\varepsilon_1,\ldots,\varepsilon_n)$ be an orthonormal family in E with the following properties:

    (a) For every $1 \leq k \leq n$, the vector subspace of E generated by $\{\varepsilon_1,\ldots,\varepsilon_k\}$ is identical to the vector subspace of E spanned by $\{x_1,\ldots,x_k\}$.

    (b) For every $1 \leq k \leq n$, $\varepsilon_k|x_k$ is strictly positive.

    Show that $\varepsilon_k = e_k/|e_k|$ for all $1 \leq k \leq n$. We say that $(\varepsilon_1,\ldots,\varepsilon_n)$ is obtained by orthonormalization of the sequence $(x_1,\ldots,x_n)$.

Now, if $(x_i)_{i\geq 1}$ is an infinite sequence in an inner product space E and if the x_i are linearly independent, there is a unique sequence $(\varepsilon_i)_{i\geq 1}$ in E such that, for every integer $n \geq 1$, $(\varepsilon_1,\ldots,\varepsilon_n)$ is the sequence obtained by orthonormalization of $(x_1,\ldots,x_n)$. We also say, in this case, that $(\varepsilon_i)_{i\geq 1}$ is obtained by orthonormalization of the sequence $(x_i)_{i\geq 1}$.

13 If A is a commutative ring, I a nonempty set, P an element of the ring $A[X_k]_{k\in I}$ of polynomials, α and β two distinct elements of I, and if the element of $A[X_k]_{k\in I}$ obtained from P by substituting X_β for X_α is equal to 0, we recall that P is then divisible by $X_\alpha - X_\beta$.

Now let K be a commutative field and, for every $n \in \mathbf{N}$, write $\Delta(X_1,\ldots,X_{2n})$ for the element $\det\left(\big(1/(X_i+X_{n+j})\big)_{1\leq i,j\leq n}\right)$ of the field $K(X_1,\ldots,X_{2n})$ of rational fractions.

1. Show that $\Delta(X_1,\ldots,X_{2n}) = P(X_1,\ldots,X_{2n})/\prod_{1\leq i,j\leq n}(X_i+X_{n+j})$, where $P(X_1,\ldots,X_{2n})$ is a polynomial of degree n^2-n. For integers α, β such that $1 \leq \alpha < \beta \leq n$, substitute X_β for X_α (respectively, $X_{n+\beta}$ for $X_{n+\alpha}$) in $\Delta(X_1,\ldots,X_{2n})$. Show that the element of $K(X_1,\ldots,X_{2n})$ so obtained is zero. Conclude that $P(X_1,\ldots,X_{2n})$ is divisible by $X_\alpha - X_\beta$ (respectively, by $X_{n+\alpha} - X_{n+\beta}$), so that

$$\Delta(X_1,\ldots,X_{2n}) = c_n \cdot \prod_{1\leq i<j\leq n}(X_i-X_j)(X_{n+i}-X_{n+j})\Big/ \prod_{1\leq i,j\leq n}(X_i+X_{n+j}),$$

    where $c_n \in K$.

2. Suppose $n \geq 2$ and c_{n-1} equal 1. For simplicity, denote by y and z the elements $X_1 \cdot \Delta(X_1, \ldots, X_n, 0, X_{n+2}, \ldots, X_{2n})$ and

$$\Delta(X_2, \ldots, X_n, X_{n+2}, \ldots, X_{2n})$$

of $K(X_1, \ldots, X_{2n})$. Expanding $\det\left(1/(X_i + X_{n+j})\right)$ along its first row, show that $y =$

$$z - \sum_{2 \leq j \leq n} (-1)^j \frac{X_1}{X_1 + X_{n+j}} \cdot \Delta(X_2, \ldots, X_n, 0, X_{n+2}, \ldots, \hat{X}_{n+j}, \ldots, X_{2n}),$$

where the $\hat{}$ is a symbol of omission. On the other hand, prove that

$$y = c_n z \prod_{2 \leq j \leq n} (X_1 - X_j)(0 - X_{n+j}) \Big/ \prod_{2 \leq j \leq n} X_j(X_1 + X_{n+j}).$$

Thus we obtain two expressions of y. Substituting 0 for X_1 in each of these expressions, conclude that $c_n = 1$.

3. If $a_1, \ldots, a_n, b_1, \ldots, b_n$ are elements of K such that $a_i + b_j$ differs from 0 for all i, j of $\{1, \ldots, n\}$, show that

$$\det\left(\frac{1}{a_j + b_j}\right) = \prod_{1 \leq i < j \leq n} (a_i - a_j)(b_i - b_j) \Big/ \prod_{1 \leq i,j \leq n} (a_i + b_j)$$

(Cauchy's determinant).

14 Let μ be Lebesgue measure on $]0, 1]$, A a countable subset of $]-1/2, +\infty[$, V the closed vector subspace of $L^2_{\mathbf{C}}(\mu)$ generated by the functions $t^\alpha (\alpha \in A)$. Consider the conditions:

(a) A has no accumulation point in $]-1/2, +\infty[$.

(b) $\sum_{\alpha \leq 0, \alpha \in A}(\alpha + 1/2)$ and $\sum_{\alpha > 0, \alpha \in A} 1/\alpha$ are finite.

We intend to prove that V differs from $L^2_{\mathbf{C}}(\mu)$ if and only if conditions (a) and (b) are simultaneously realized, and that, in this case, t^z does not belong to V, for all z in $C - A$ such that $\operatorname{Re}(z) > -1/2$.

Let z be a point of $]-1/2, +\infty[-A$.

1. If B is a nonempty finite subset of A and if V_B is the vector subspace of $L^2_{\mathbf{C}}(\mu)$ generated by the t^α for $\alpha \in B$, by means of Gram's determinants compute the distance d_B from t^z to V_B. Then use Exercise 13 to show that $d_B = (2z+1)^{-1/2} \cdot \prod_{\alpha \in B} |z - \alpha|/(z + \alpha + 1)$.

2. Prove that $|z - \alpha|/(z + \alpha + 1)$ is strictly less than 1, for every $\alpha \in A$.

3. Suppose that A has an accumulation point in $]-1/2, +\infty[$. Then there exist u, v in $]-1/2, +\infty[$ such that $A' = A \cap [u, v]$ is infinite. Prove that d_B can be made arbitrarily small by taking for B a subset of A' with large enough cardinality.

4. Assume that A has no accumulation point in $]-1/2,+\infty[$. For every α
    in A, $1-|z-\alpha|/(z+\alpha+1)$ is either equal to $(2\alpha+1)/(z+\alpha+1)$ or
    $(2z+1)/(z+\alpha+1)$, as α is smaller or greater than z. Deduce from this
    that the family $1-|z-\alpha|/(z+\alpha+1)$ for $\alpha\in A$ and $\alpha\leq 0$ (respectively, for
    $\alpha\in A$ and $\alpha>0$) is summable if and only if the sum $\sum_{\alpha\in A,\alpha\leq 0}(\alpha+1/2)$
    (respectively, $\sum_{\alpha\in A,\alpha>0}1/\alpha$) is finite. Conclude that

    $$\prod_{\alpha\in A}\frac{|z-\alpha|}{z+\alpha+1}$$

    differs from 0 if and only if condition (b) holds.

5. Prove the statement at the beginning of the exercise.

6

Integrable Functions for Measures on Semirings

In this chapter, we focus our attention on measures defined on semirings and measurable functions.

Summary

6.1 Let μ be a measure on S semiring in Ω. $A \subset \Omega$ is μ-measurable if and only if, for every $E \in S$, the set $A \cap E$ is integrable (Proposition 6.1.3). A function from Ω into some metrizable space is μ-measurable if and only if, for every $E \in S$, the restriction of f to E is the limit a.e. of simple mappings (Proposition 6.1.4). Finally, we give two other important necessary and sufficient conditions for f to be μ-measurable.

6.2 Let F be a Banach space with dual F'. If $f \in \mathcal{L}_F^1(\mu)$, then $N_1(f) = \sup_{g \in B} \left| \int fg d\mu \right|$ where B is the closed unit ball of $St(S, F')$ (Theorem 6.2.1).

6.3 This section generalizes the following example. Let S be the semiring of finite subsets of $\mathbf{N}$. Then $E \mapsto \mathrm{card}\, E$, where $\mathrm{card}\, E$ is the cardinality of E, is a measure on S, called the counting measure. It is defined by the unit mass at each point. Such a measure is atomic (cf. Section 3.8).

6.4 In this section we return to the study of prolongations of a measure (cf. Section 3.9).

6.1 Measurability

Let Ω be a nonempty set, S a semiring in Ω, and μ a complex measure on S.

For all subsets E of Ω, put $V\mu^\circ(E) = \inf_{(A_i)_{i \in I}} \left(\sum_{i \in I} V\mu(A_i) \right)$, where $(A_i)_{i \in I}$ extends over the class of countable families of S-sets such that $\bigcup_{i \in I} A_i$ contains E. If $\mathcal{R}$ is the ring generated by S, $V\mu^\circ(E) = \inf_{(A_i)_{i \in I}} \left(\sum_{i \in I} V\mu(A_i) \right)$, where $(A_i)_{i \in I}$ extends over the class of countable families of $\mathcal{R}$-sets such that $\bigcup_{i \in I} A_i \supset E$. Clearly, $V\mu^\circ(E) \leq V\mu^\circ(F)$ for all subsets E, F such that $E \subset F$, and $V\mu^\circ(E) = V\mu(E)$ for every $E \in \mathcal{R}$.

Theorem 6.1.1 $V\mu^\circ(A) = \int^* 1_A dV\mu$ *for every subset A of Ω.*

PROOF: Clearly, $\int^* 1_A dV\mu \leq V\mu^\circ(A)$. To establish the reverse inequality, suppose that $\int^* 1_A dV\mu < +\infty$. For fixed $\varepsilon > 0$, let $g \in \mathcal{J}^+$ be such that $g \geq 1_A$ and $\int^* g dV\mu \leq \int^* 1_A dV\mu + \varepsilon$, and let $(g_n)_{n \geq 1}$ be an increasing sequence in $St^+(S)$ admitting g as its upper envelope. Choose $\delta \in]0, 1[$ and put $B_n = \{x \in \Omega : g_n(x) \geq 1 - \delta\}$ for all $n \geq 1$. The B_n form an increasing sequence in $\mathcal{R}$, and $A \subset \bigcup_{n \geq 1} B_n$, so

$$V\mu^\circ(A) \leq \sum_{n \geq 1} V\mu\left(B_n \cap \left(\bigcup_{1 \leq i \leq n-1} B_i\right)^c\right) = \sup_{n \geq 1} V\mu(B_n).$$

But

$$V\mu(B_n) = \int 1_{B_n} dV\mu \leq \frac{1}{1-\delta} \int g_n dV\mu \leq \frac{1}{1-\delta} \int^* g dV\mu,$$

whence we see that $V\mu^\circ(A) \leq [(\int^* 1_A dV\mu) + \varepsilon]/(1 - \delta)$. Since ε and δ were arbitrary, we conclude that $V\mu^\circ(A) \leq \int^* 1_A dV\mu$.   □

Proposition 6.1.1 *Let τ be a σ-additive function from $\tilde{S} = \sigma(S)$ into $[0, +\infty]$, which agrees with $V\mu$ on S. Then $\tau(E) = \int^* 1_E \cdot dV\mu$ for all $E \in \tilde{S}$.*

PROOF: Obviously, $\tau(B) \leq V\mu^*(B)$ for all $B \in \tilde{S}$. Now suppose that $\tau(E) < V\mu^*(E)$ for some $E \in \tilde{S}$, and let $(E_n)_{n \geq 1}$ be a sequence of disjoint S-sets whose union contains E. Since

$$\tau(E) = \sum_{n \geq 1} \tau(E_n \cap E) < \sum_{n \geq 1} V\mu^*(E_n \cap E) = V\mu^*(E),$$

there exists an integer $n \geq 1$ such that $\tau(E_n \cap E) < V\mu^*(E_n \cap E)$. Then

$$\begin{aligned}
\tau(E_n \cap E^c) + \tau(E_n \cap E) &= \tau(E_n) \\
&= V\mu(E_n) \\
&= V\mu(E_n \cap E^c) + V\mu(E_n \cap E),
\end{aligned}$$

so

$$\tau(E_n \cap E^c) = V\mu(E_n \cap E^c) + V\mu(E_n \cap E) - \tau(E_n \cap E) > V\mu(E_n \cap E^c),$$

which contradicts the fact that $\tau(B) \leq V\mu^*(B)$ for every $B \in \tilde{S}$.   □

Proposition 6.1.2 *A set $A \subset \Omega$ is μ-measurable if (and only if) $V\mu(E) = V\mu^*(A \cap E) + V\mu^*(A^c \cap E)$ for every S-set E.*

PROOF: Let B be a subset of Ω. We show that $V\mu^*(B) \geq V\mu^*(A \cap B) + V\mu^*(A^c \cap B)$. We may as well assume that $V\mu^*(B) < +\infty$. Fix $\varepsilon > 0$. There exists a sequence $(E_n)_{n\geq 1}$ of S-sets such that $B \subset \bigcup_{n\geq 1} E_n$ and $\sum_{n\geq 1} V\mu(E_n) \leq V\mu^*(B) + \varepsilon$. But

$$V\mu(E_n) = V\mu^*(A \cap E_n) + V\mu^*(A^c \cap E_n)$$

for each n, so

$$V\mu^*(B) \geq \sum_{n\geq 1} V\mu^*(A \cap E_n) + \sum_{n\geq 1} V\mu^*(A^c \cap E_n) - \varepsilon$$

$$\geq V\mu^*\left(\bigcup_{n\geq 1} A \cap E_n\right) + V\mu^*\left(\bigcup_n A^c \cap E_n\right) - \varepsilon,$$

and $V\mu^*(B) \geq V\mu^*(A \cap B) + V\mu^*(A^c \cap B) - \varepsilon$. Since ε was arbitrary, we find that $V\mu^*(B) \geq V\mu^*(A \cap B) + V\mu^*(A^c \cap B)$, and by Theorem 3.7.2 it follows that A is μ-measurable. □

The fact that for a set $A \subset \Omega$ to be μ-measurable it is necessary and sufficient that it split the outer measure $V\mu^*$ can be proved by considering the main prolongation $\hat{\mu}$ of μ, because $V\hat{\mu}^* = V\mu^*$. The proof is left to the reader.

Proposition 6.1.3 *A set $A \subset \Omega$ is μ-measurable if and only if $A \cap E$ is μ-integrable for every $E \in S$.*

PROOF: This follows from Proposition 6.1.2. □

Now we give two characterizations of μ-measurable functions.

Let $\mathcal{R}$ be the ring generated by S and $\hat{\mathcal{R}}$ the ring of μ-integrable sets.

Lemma 6.1.1 *For every $B \in \hat{\mathcal{R}}$ and every $\varepsilon > 0$, there exists $A \in \mathcal{R}$ such that $V\mu^*(A \triangle B) \leq \varepsilon$.*

PROOF: There exists a sequence $(B_n)_{n\geq 1}$ of disjoint $\mathcal{R}$-sets such that $B \subset \bigcup_{n\geq 1} B_n$ and $\sum_{n\geq 1} V\mu(B_n) < V\mu(B) + \varepsilon$. Therefore, letting $p \geq 0$ be an integer such that $V\mu((\bigcup_{n\geq 1} B_n) - B) + \sum_{n\geq p+1} V\mu(B_n) \leq \varepsilon$, we see that $A = \bigcup_{1\leq n\leq p} B_n$ has the desired property. □

Proposition 6.1.4 *Let f be a mapping from Ω into a metrizable space F. Then the following conditions are equivalent:*

(a) f is μ-measurable.

(b) For every $E \in S$, f/E is the limit almost everywhere in E of a sequence of $\mathcal{R}/E$-simple mappings.

(c) For every $E \in S$, f/E is the limit almost everywhere in E of a sequence of $\hat{\mathcal{R}}/E$-simple mappings.

PROOF: Fix $E \in S$. Let g be an $\hat{\mathcal{R}}/E$-simple mapping from E into F and $y_1, \ldots, y_k$ its values. The sets $B_j = g^{-1}(y_j)$ lie in $\hat{\mathcal{R}}$. Given $\varepsilon > 0$, for each $1 \leq j \leq k$ there exists $A_j \in \mathcal{R}$ such that $A_j \subset E$ and $V\mu^*(A_j \triangle B_j) \leq \varepsilon/k$. The sets $\hat{A}_j = A_j - \bigcup_{1 \leq i \leq j-1} A_i$, for $1 \leq j \leq k$, belong to $\mathcal{R}$ and are disjoint. For every $1 \leq j \leq k$, $B_j \cap (\hat{A}_j)^c$ is included in $\left(\bigcup_{1 \leq i \leq j-1}(A_i \cap A_j) \right) \cup (B_j \cap A_j^c)$; indeed, let $x \in B_j \cap (\hat{A}_j)^c$; if x does not belong to A_j, then x lies in $B_j \cap A_j^c$; on the other hand, if x belongs to A_j, then it lies in $\bigcup_{1 \leq i \leq j-1} A_i$. Now, for all $1 \leq i \leq j-1$, $A_i \cap A_j$ is contained in $(A_i - B_i) \cup (A_j - B_j)$, because B_i, B_j are disjoint. So $B_j \cap (\hat{A}_j)^c$ is included in $\bigcup_{1 \leq i \leq k}(A_i \triangle B_i)$. Fix $z \in F$ and let $h : E \to F$ be equal to z outside $\bigcup_{1 \leq j \leq k} \hat{A}_j$, and to y_j on each $\hat{A}_j$. Then, of course, h is $\mathcal{R}/E$-simple. Furthermore, $D = \{x \in E : h(x) \neq g(x)\}$ is integrable and, since it is contained in $\bigcup_{1 \leq i \leq k}(A_i \triangle B_i)$, we have $V\mu(D) \leq \varepsilon$.

Thus, for every $n \geq 1$, we can find an $\mathcal{R}/E$-simple mapping g_n from E into F such that the $V\mu$-measure of $D_n = \{x \in E : g_n(x) \neq g(x)\}$ is less than $1/2^n$. But then $N = \limsup D_n = \bigcap_{p \geq 1} \left(\bigcup_{n \geq p} D_n \right)$ is negligible, and, if x belongs to $E - N$, there exists $p \geq 1$ such that $g_n(x) = g(x)$ for all $n \geq p$. This proves that the sequence $(g_n)_{n \geq 1}$ converges to g on $E - N$, and hence almost everywhere in E.

Now suppose that f is μ-measurable and let $E \in S$. By Proposition 3.5.2, f/E is the limit almost everywhere in E of a sequence $(f_m)_{m \geq 1}$ of $\hat{\mathcal{R}}/E$-simple mappings. By what has just been proven, for every $m \geq 1$, f_m is the limit almost everywhere in E of a sequence $(f_{m,n})_{n \geq 1}$ of $\mathcal{R}/E$-simple mappings. Hence f/E is the limit almost everywhere in E of a sequence of $\mathcal{R}/E$-simple mappings (Proposition 3.5.4). In short, (b) is satisfied if (a) is.

Next, we suppose that (c) is satisfied, and prove that f is μ-measurable. Obviously, for all $E \in \mathcal{R}$, f/E is the limit almost everywhere in E of a sequence of $\hat{\mathcal{R}}/E$-simple mappings. Now let E be an integrable set, and let $(A_m)_{m \geq 1}$ be an increasing sequence of $\mathcal{R}$-sets such that $E \subset A = \bigcup_{m \geq 1} A_m$ and $V\mu^*(A) < +\infty$. Fix $z \in F$ and, for every $m \geq 1$, let f_m be the mapping from A into F equal to z on $A - A_m$ and to f on A_m, so that f/A is the limit of the f_m as $m \to +\infty$.

By hypothesis, for every $m \geq 1$, f/A_m is the limit almost everywhere in A_m of a sequence $(g_{m,n})_{n \geq 1}$ of $\hat{\mathcal{R}}/A_m$-simple mappings. For every $m \geq 1$ and every $n \geq 1$, denote by $f_{m,n}$ the mapping from A into F equal to z on $A - A_m$ and to $g_{m,n}$ on A_m. Then, for every $m \geq 1$, f_m is the limit almost everywhere in A of the sequence $(f_{m,n})_{n \geq 1}$ of $\hat{\mathcal{R}}/A$-simple mappings. But now f/E is the limit almost everywhere in E of a sequence of $\hat{\mathcal{R}}/E$-simple mappings.   □

Define $\tilde{S}$ as the σ-ring generated by S, and recall that the elements of $\tilde{S}$ are "the S Borel sets".

If $E \subset \Omega$ is μ-measurable and μ-moderate, it can be partitioned into μ-integrable sets $E_i (i \geq 1)$. Given $\varepsilon > 0$, for each integer $i \geq 1$ there exists a countable union B_i of S-sets such that $E_i \subset B_i$ and $V\mu^*(B_i - E_i) \leq \varepsilon/2^i$. Then $B = \bigcup_{i \geq 1} B_i$ is a countable union of S-sets, B contains E, and $V\mu^*(B - E) \leq \varepsilon$. It follows immediately that there exists $E'' \in \tilde{S}$, containing E, such that $E'' - E$ is μ-negligible. Now there exists $F \in \tilde{S}$ which contains $E'' - E$ and is μ-negligible. Then $E' = E'' \cap F^c$ belongs to $\tilde{S}$, is included in E, and $E - E'$ is negligible. In short, there are $E' \in \tilde{S}$, $E'' \in \tilde{S}$ such that $E' \subset E \subset E''$ and $E'' - E'$ is negligible.

Observe incidentally that, by the preceding, a set N is μ-negligible if and only if there exists $E \in \tilde{S}$ such that $E \supset N$ and $V\mu^*(E) = 0$.

Theorem 6.1.2 *Let f be a mapping from Ω into a metrizable space F. The following conditions are equivalent:*

(a) f is μ-measurable.

(b) For every $E \in \tilde{S}$, there exists a negligible subset N of E, $N \in \tilde{S}$, such that $f/E - N$ is measurable $\tilde{S}/E - N$ and $f(E - N)$ is separable.

PROOF: First, assume that f is μ-measurable, and let $E \in S$. By Proposition 3.5.2, f/E is the limit of a sequence $(f_n)_{n \geq 1}$ of $\hat{\mathcal{R}}/E$-simple mappings, outside a negligible subset N of E. Now, by the remarks just preceding the present theorem, we may suppose that N lies in $\tilde{S}$ and that each f_n is $\tilde{S}/E$-simple. Fix $z \in F$ and, for every $n \geq 1$, let g_n be the mapping from E into F equal to f_n on $E - N$ and to z on N. The g_n are $\tilde{S}/E$-simple mappings and converge to the mapping g from E into F which agrees with f on $E - N$ and is equal to z on N. This means that g is measurable $\tilde{S}/E$, and that $f/E - N$ is measurable $\tilde{S}/E - N$. Moreover, $f(E - N)$ is separable. Now each $E \subset \tilde{S}$ is contained in a countable union of disjoint S-sets, so (b) is satisfied.

Conversely, f is μ-measurable whenever (b) holds, because, for each integrable set E, there exists $E' \in \tilde{S}$ such that $E' \subset E$ and $E - E'$ is negligible. $\square$

Theorem 6.1.3 *Let f be a mapping from Ω into a real Banach space F. Then f is μ-measurable and μ-moderate if and only if there exists $g : \Omega \to F$ with the following properties:*

(a) g vanishes outside some $\tilde{S}$-set.

(b) $g^{-1}(B) \cap E$ lies in $\tilde{S}$ for every Borel subset B of F and every $E \in \tilde{S}$.

(c) There exists a μ-negligible $\tilde{S}$-set N such that $g(\Omega - N)$ is separable.

(d) $g = f$ almost everywhere.

PROOF: Theorem 6.1.3 follows immediately from Theorem 6.1.2. $\square$

6.2 Complements on the L^p Spaces

Let Ω, S, and μ be as in Section 6.1. Then Theorem 5.2.5 may be refined as follows:

Theorem 6.2.1 *Let F be a Banach space, F' its normed dual, and B the closed ball with center 0 and radius 1 in $St(S, F')$ (considered as a subspace of $\mathcal{L}_{F'}^{\infty}(\mu)$). Then $N_1(f) = \sup_{g \in B} \left| \int fg d\mu \right|$ for all $f \in \mathcal{L}_F^1(\mu)$.*

PROOF: Let $f \in St(S, F)$ be such that $N_1(f) = 1$. Fix $\varepsilon > 0$. By the same argument as that in Theorem 5.2.5, there exists $g \in St(S, F')$ such that $N_\infty(g) = 1$ and $\left| \int fg \, d\mu \right| \geq 1 - \varepsilon$.

Since $St(S, F)$ is dense in $\mathcal{L}_F^1(\mu)$, we conclude, as in Theorem 5.2.5, that $N_1(f) = \sup_{\substack{g \in St(S, F') \\ N_\infty(g) \leq 1}} \left| \int fg d\mu \right|$ for all $f \in \mathcal{L}_F^1(\mu)$. $\square$

Proposition 6.2.1 *Suppose that μ is positive, and let $f : \Omega \to [0, +\infty[$ be μ-measurable. Also, let p be an element of $[1, +\infty[$ and q its conjugate exponent. Then $(\int^{\bullet} f^p d\mu)^{1/p} = \sup \int^* fg d\mu$, where g extends over the set of functions of $St^+(S)$ such that $N_q(g) \leq 1$.*

PROOF: When $N_p(f)$ is finite, the proposition is true, by the remark immediately following Theorem 5.2.5 if $p > 1$, and by Theorem 6.2.1 if $p = 1$.

When $\bar{N}_p(f)$ is finite, f is equal l.a.e. to a positive function φ such that $N_p(\varphi) < +\infty$; thus the proposition is also true in this case.

Finally, suppose that $\int^{\bullet} f^p d\mu = +\infty$, and let A be a μ-integrable set. For all integers $n \geq 1$, put $f_n = \inf(f, n) 1_A$. Then

$$N_p(f_n) = \sup \int^* f_n g d\mu \leq \sup \int^* fg d\mu,$$

so $N_p(f 1_A) \leq \sup \int^* fg d\mu$, which proves that $\sup \int^* fg d\mu = +\infty$. $\square$

Now, we can precise the concept of convergence in measure. Suppose that μ is positive, and let A be a μ-measurable set. If F is a metrizable uniform space, the sets $W(V, E \cap A, \delta)$, where E extends over $\mathcal{R}$, form a basis for the uniformity of $\mathcal{L}(A, \mu; F)$ (notation of Section 5.3). Indeed, if B is a μ-integrable subset of A, there exists $E \in \mathcal{R}$ such that $\mu(E \bigtriangleup B) \leq \delta/2$, and $W(V, B, \delta)$ contains $W(V, E \cap A, \delta/2)$. Similarly, when F is a Banach space, the sets $T(E \cap A, \delta)$ form a basis for the filter of neighborhoods of 0 in $\mathcal{L}(A, \mu; F)$; therefore, a sequence $(f_n)_{n \geq 1}$ in $\mathcal{L}(A, \mu; F)$ converges to $f \in \mathcal{L}(A, \mu; F)$ if and only if, for all $E \in \mathcal{R}$ and all $\delta > 0$, $\mu(\{x \in E \cap A : |f_n - f|(x) > \delta\})$ converges to 0 as $n \to +\infty$.

6.3 Measures Defined by Masses

Let Ω be a nonempty set and S a semiring in Ω. Write $\tilde{S}$ for the σ-ring generated by S.

Proposition 6.3.1 *Let $(\alpha_x)_{x \in X}$ be a family of nonzero complex numbers, whose set of indices is a subset of Ω. Suppose that $\{x\}$ belongs to $\tilde{S}$ for every $x \in X$ and that $\sum_{x \in E \cap X} |\alpha_x|$ is finite for every $E \in S$. Then the function $\mu : E \mapsto \sum_{x \in E \cap X} \alpha_x$ on S is an atomic measure. Moreover, the $\{x\}$ are representatives of the different classes of atoms, as x ranges over X, and $\mu(\{x\}) = \alpha_x$ for every $x \in X$.*

PROOF: Define ν as the measure $E \to \sum_{x \in E \cap X} |\alpha_x|$ on S. By Proposition 6.1.1, $\nu^*(E) = \sum_{x \in E \cap X} |\alpha_x|$ for all $E \in \tilde{S}$; in particular, $\{x\}$ is ν-integrable and $\int 1_{\{x\}} \cdot d\nu = |\alpha_x|$ for each $x \in X$. Clearly, $V\mu \leq \nu$.

Fix $E \in S$. Given $\varepsilon > 0$, let $x_1, \ldots, x_n$ be points of $X \cap E$ such that $\sum_{1 \leq i \leq n} |\alpha_{x_i}| \geq \nu(E) - \varepsilon/2$. For each $1 \leq i \leq n$, there exists an S-set F_i, included in E and containing x_i, such that $\nu(F_i - \{x_i\}) < \inf_{1 \leq j \leq n} |\alpha_{x_j}|$ and $\nu(F_i - \{x_i\}) \leq \varepsilon/(2n)$; the existence of such an F_i follows from the relation $\nu^*(\{x_i\}) = |\alpha_{x_i}|$. Observe that F_i contains no x_j for $j \neq i$. Now we may suppose that the F_i are disjoint. But

$$|\mu(F_i) - \alpha_{x_i}| \leq \nu(F_i) - |\alpha_{x_i}| = \nu(F_i - \{x_i\}) \leq \frac{\varepsilon}{2n}$$

for every $1 \leq i \leq n$, so $|\mu(F_i)| \geq |\alpha_{x_i}| - \varepsilon/(2n)$, and $\sum_{1 \leq i \leq n} |\mu(F_i)| \geq \nu(E) - \varepsilon$, which proves that $V\mu(E) \geq \nu(E)$.

Therefore, $V\mu = \nu$. The set $\Omega - X$ is locally μ-negligible, because

$$\nu(E - E \cap X) = \nu(E) - \sum_{x \in E \cap X} \nu(\{x\}) = 0$$

for all $E \in S$.

Now fix $x \in X$. For every $\varepsilon > 0$, there exists an $E \in S$ containing x such that $\nu(E - \{x\}) \leq \varepsilon/2$. Then

$$|\alpha_x - \mu(\{x\})| = \left| \mu(E) - \mu(\{x\}) - \sum_{y \in E \cap X, y \neq x} \alpha_y \right|$$

$$\leq |\mu(E - \{x\})| + \sum_y |\alpha_y|,$$

so

$$\left| \alpha_x - \mu(\{x\}) \right| \leq \nu(E - \{x\}) + \nu(E) - |\alpha_x| \leq 2\nu(E - \{x\}) \leq \varepsilon.$$

Hence $\mu(\{x\}) = \alpha_x$ for all $x \in X$.

If B is a μ-integrable set such that $V\mu(B) > 0$, then it is not included in the locally μ-negligible set $\Omega - X$. This means that B contains an $x \in X$, which shows that μ is atomic, and the proof is complete. $\square$

Definition 6.3.1 Let $(\alpha_x)_{x \in X}$ be a family of complex numbers, whose set of indices is a subset of Ω. Suppose that $\{x\}$ belongs to $\tilde{S}$, for all $x \in X$ such that $\alpha_x \neq 0$, and suppose that $\sum_{x \in E \cap X} |\alpha_x|$ is finite for every $E \in S$. Then $\mu : E \mapsto \sum_{x \in E \cap X} \alpha_x$ is the measure on S defined by the masses α_x at the points x of X.

For integration with respect to such a measure, see the results of Section 3.8.

Now let μ be an atomic measure on S. If each atom contains an atom of the form $\{x\}$ and if we put $X = \{x \in \Omega : \{x\}$ is integrable and $\mu(\{x\}) \neq 0\}$, then μ is defined by the masses $\alpha_x = \mu(\{x\})$ at the points x of X. Indeed, the $\varrho(\{x\})$, where $x \in X$, are the different classes of atoms, and $\mu(E) = \sum_{x \in X \cap E} \alpha_x$ for all $E \in S$ (Theorem 3.8.1). Moreover, for every $x \in X$, there exists an $\tilde{S}$-set B containing x such that $V\mu(\{x\} - B) = 0$, and, since $V\mu(\{x\}) = |\alpha_x| > 0$, necessarily $B = \{x\}$, which proves that $\{x\}$ lies in $\tilde{S}$.

Even though measures defined by masses are of great importance, there are atomic measures on semirings whose atoms are not points. As an example, take for Ω an uncountable set, for S the semiring in Ω consisting of the countable and the co-countable sets in Ω, and define μ on S by taking $\mu(A) = 0$ if A is countable and $\mu(A) = 1$ if A is co-countable.

6.4   Prolongations of a Measure

Let Ω be a nonempty set, S a semiring in Ω, and μ a complex measure on S.

Integration with respect to the extensions of μ and integration with respect to μ are identical, as shown now.

Proposition 6.4.1 Let Φ be a semiring of μ-integrable sets and μ_Φ the measure $E \to \int 1_E d\mu$ on Φ. Suppose that

(a) for every $A \in S$, there exists $B \in \tilde{\Phi}$ such that $B \subset A$ and $A - B$ is μ-negligible;

(b) $A \cap E$ belongs to $\tilde{\Phi}$ for every $A \in S$ and every $E \in \Phi$.

Then $\bar{\mu}_\Phi = \bar{\mu}$. Moreover, $\hat{\mu}_\Phi = \hat{\mu}$ when $\tilde{\Phi} = \sigma(\Phi)$ contains S.

PROOF: By Propositions 3.9.4 and 3.9.3, $\mathcal{L}_\mathbf{C}^1(\mu)$ contains $\mathcal{L}_\mathbf{C}^1(\mu_\Phi)$, and further $\int f \cdot dV\mu_\Phi = \int f \cdot dV\mu$ and $\int f \cdot d\mu_\Phi = \int f \cdot d\mu$ for all $f \in \mathcal{L}_\mathbf{C}^1(\mu_\Phi)$. It follows that $V\mu_\Phi^*(B) = V\mu^*(B)$ for all $B \in \tilde{\Phi}$ (Proposition 6.1.1).

If F is a locally μ_Φ-negligible set, then $F \cap A$ is μ-negligible for all $A \in S$, by hypothesis (a). This implies that F is locally μ-negligible.

Now, if $B \in \tilde{S}$ is μ-negligible, then $B \cap E$ is a μ_Φ-negligible $\tilde{\Phi}$-set for all $E \in \Phi$. Thus each μ-negligible set meets each $E \in \Phi$ in a μ_Φ-negligible set. Evidently, this remains true for each locally μ-negligible set, which therefore is locally μ_Φ-negligible.

Then define on $St(\Phi, C)$ the seminorm $g \to \int |g| \cdot dV\mu_\Phi = \int |g| \cdot dV\mu$. For f in $\bar{\mathcal{L}}_C^1(\mu)$, there exists a Cauchy sequence $(f_n)_{n \geq 1}$ in $St(\Phi, C)$ which converges to f locally μ-almost everywhere, and so locally μ_Φ-almost everywhere, whence we deduce that f belongs to $\bar{\mathcal{L}}_C^1(\mu_\Phi)$. Now $\bar{\mathcal{L}}_C^1(\mu) = \bar{\mathcal{L}}_C^1(\mu_\Phi)$.

Finally, when $\tilde{\Phi}$ contains S, $\hat{\mu} = \hat{\mu}_\Phi$ because all μ-negligible sets are μ_Φ-negligible. $\qquad\qquad\qquad\qquad\qquad\qquad\qquad\qquad\qquad\qquad\qquad\qquad\qquad\qquad$ $\square$

As an example take for Ω an interval of $\mathbf{R}$, and take for μ Lebesgue measure on Ω. If Φ is a semiring of μ-integrable sets and if $\tilde{\Phi} \supset S$, then $\hat{\mu}_\Phi = \hat{\mu}$.

The following proposition is especially useful to probability theory.

Proposition 6.4.2 *Let ν be a measure on S and C a π-system in Ω. Suppose that*

(a) *each C-set is both essentially μ-integrable and essentially ν-integrable;*

(b) $\int 1_E d\mu = \int 1_E d\nu$ *for all $E \in C$;*

(c) *the σ-ring $\bar{C}$ generated by C contains S.*

Then $\mu = \nu$ on S.

PROOF: For each $A \subset \Omega$, define $\mathcal{L}_A$ as the class of those sets E such that $E \cap A$ is essentially integrable with respect to both μ and ν, and such that $\int 1_{E \cap A} d\mu = \int 1_{E \cap A} d\nu$. For every $A \in C$, $\mathcal{L}_A$ is a λ-system and contains C, so $S \subset \mathcal{L}_A$. Therefore, $\int 1_{E \cap A} d\mu = \int 1_{E \cap A} d\nu$ for all $E \in S$ and all $A \in C$. Now fix $E \in S$. $\mathcal{L}_E$ is a λ-system and contains C, so $S \subset \mathcal{L}_E$ and $\mu(E) = \nu(E)$. $\qquad$ $\square$

Exercises for Chapter 6

1   Let $\alpha_1, \ldots, \alpha_n$ be strictly positive numbers such that $\sum_{1 \leq i \leq n} \alpha_i = 1$. Consider the semiring S in $\Omega = \{1, \ldots, n\}$ consisting of the empty set and the singletons $\{i\}$ $(1 \leq i \leq n)$. Define μ as the measure on S given by the mass α_i at each i. From Jensen's inequality, deduce that

$$u_1^{\alpha_1} \ldots u_n^{\alpha_n} \leq \alpha_1 u_1 + \ldots + \alpha_n u_n$$

for all $u_1 > 0, \ldots, u_n > 0$ (inequality of the means).

2   For every integer $k \geq 1$, every $(p_1, \ldots, p_k) \in]0, 1]^k$ such that $p_1 + \ldots + p_k = 1$, and every integer $r \geq 1$, call the measure μ on the Borel σ-algebra of $\mathbf{R}^k$, defined by the mass

$$\frac{r!}{\alpha_1! \ldots \alpha_k!} p_1^{\alpha_1} \ldots p_k^{\alpha_k}$$

at each point $(\alpha_1, \ldots, \alpha_k)$ of $(\mathbf{Z}^+)^k$ such that $\alpha_1 + \ldots + \alpha_k = r$, the multinomial law with parameters $(p_1, \ldots, p_k)$ and r. On the other hand, for every real number

$0 < p < 1$ and every integer $r \geq 1$, call the measure $B(r,p)$ on the Borel σ-algebra of $\mathbf{R}$, given by the masses $\binom{r}{x}p^x(1-p)^{r-x}$ at the points x of $\{0, 1, \ldots, r\}$, the binomial law with parameters r, p.

1. Let $n \geq 1$, $r \geq 1$ be two integers, G a set with n elements, Ω the set of mappings from $\{1, 2, \ldots, r\}$ into G. Let P be the measure on the class of all subsets of Ω defined by $P(\{\omega\}) = 1/|\Omega| = n^{-r}$ for all $\omega \in \Omega$. Now let $(G_j)_{1 \leq j \leq k}$ be a partition of G such that $n_j \geq 1$ for every $1 \leq j \leq k$. Define X_j as the function $\omega \mapsto |\omega^{-1}(G_j)|$ from Ω into $\mathbf{R}$ for every $1 \leq j \leq k$, and put $X = [X_j]_{1 \leq j \leq k}$.

    Prove that

    $$P\left(X^{-1}(\alpha)\right) = \frac{r!}{\alpha_1! \ldots \alpha_k!} \left(\frac{n_1}{n}\right)^{\alpha_1} \cdots \left(\frac{n_k}{n}\right)^{\alpha_k}$$

    for each $(\alpha_1, \ldots, \alpha_k)$ of $(\mathbf{Z}^+)^k$ such that $\alpha_1 + \ldots + \alpha_k = r$. Show that, for every $1 \leq j \leq k$, $P(X_j = x) = B(r, n_j/n)(\{x\})$ for all integers $x \geq 0$.

2. Consider a box containing n balls of k different colors. For every $1 \leq j \leq k$, there are n_j balls of the kth color. Draw r balls from the box, successively and with replacement. Show that, for every $(\alpha_1, \ldots, \alpha_k) \in (\mathbf{Z}^+)^k$, the probability of drawing α_j balls of the jth color for all $1 \leq j \leq k$ is given by the multinomial law with parameters $(n_1/n, \ldots, n_k/n)$ and r.

3 For each integer $n \geq 1$, let P_n be the measure on the class of all subsets of $\mathbf{N}$ such that $P_n(m) = 1/n$ for every $1 \leq m \leq n$ and $P_n(m) = 0$ for every $m > n$. For each real valued function f on Ω, write $E_n(f)$ for $\int f dP_n$. For integers $m \geq 1$ and primes p, let $\alpha_p(m)$ be the power of p in the prime factorization of m. Also, let $\delta_p(m)$ be 1 or 0 as p divides m or not.

1. $E_n(\alpha_p) = \int_0^{+\infty} P_n(\alpha_p > x)\, dx$ by Proposition 3.5.6, for all integers $n \geq 1$ and all primes p, so

    $$E_n(\alpha_p - \delta_p) = \sum_{k \geq 2} \frac{1}{n}\left[\frac{n}{p^k}\right].$$

    Now, for each integer $m \geq 1$, put $\log^* m = \sum_p \delta_p(m) \log p$. Show that

    $$E_n(\log - \log^*) = E_n\left(\sum_p (\alpha_p - \delta_p)\log p\right)$$

    converges to $\sum_p \left(1/(p(p-1))\right)\log p$ as $n \to +\infty$, and deduce that

    $$E_n(\log^*) = \log n - 1 - \sum_p \frac{1}{p(p-1)}\log p + o(n^\circ),$$

    where $o(n^\circ)$ converges to 0 as $n \to +\infty$ (use Stirling's formula to estimate $E_n(\log)$ as $n \to +\infty$).

2. For each real number $x > 0$, define $\theta(x)$ as $\sum_{p \leq x} \log p$. Prove that

$$0 \leq \sum_{p \leq n} \frac{1}{p} \log p - E_n(\log^*) \leq \frac{\theta(n)}{n}$$

for each integer $n \geq 1$.

3. If $n \geq 1$ is an integer, $\left[2\frac{n}{p}\right] - 2\left[\frac{n}{p}\right]$ is positive for each prime p and equal to 1 when $n < p \leq 2n$, so

$$\frac{1}{2n} \sum_{n < p \leq 2n} \log p \leq E_{2n}(\log^*) - E_n(\log^*) = \log(2) + o(n^\circ).$$

From this last inequality, deduce that $\theta(n)/n$ remains bounded in $\mathbf{N}$ (in fact, it can be shown that $\theta(n)/n$ converges to 1 as $n \to +\infty$). Conclude that

$$\sum_{p \leq x} \frac{1}{p} \log p = \log x + 0(x^\circ),$$

where $0(x^\circ)$ remains bounded as $x \geq 1$ varies.

4. Write dF for the measure on the natural semiring of $]1, +\infty[$ defined by the mass $\log p/p$ at each prime p, and let F be the function $x \mapsto \sum_{p \leq x} \log p/p$ from $]1, +\infty[$ into $\mathbf{R}$. To simplify notation, put $a_p = \log p/p$ for each prime p, and $a_n = 0$ for each integer $n \geq 2$ which is not a prime. For each real number $x \geq 2$, observe that

$$
\begin{aligned}
-\int_2^x \frac{F(u)}{u \cdot \log^2(u)}\,du &= \sum_{2 \leq k < [x]} \int_k^{k+1} (a_2 + \cdots + a_k) \left(\frac{1}{\log u}\right)'\,du \\
&\quad + \int_{[x]}^x (a_2 + \cdots + a_{[x]}) \left(\frac{1}{\log u}\right)'\,du \\
&= \sum_{2 \leq n \leq x} a_n \cdot \int_n^x \left(\frac{1}{\log u}\right)'\,du \\
&= \frac{F(x)}{\log x} - \sum_{p \leq x} \frac{1}{p}.
\end{aligned}
$$

Now deduce from part 3 that

$$\sum_{p \leq x} \frac{1}{p} = \log \log x + \text{constant} + 0\left(\frac{1}{\log x}\right).$$

4   For every $m \geq 1$, let $g(m) = \sum_p \delta_p(m)$ be the number of distinct prime divisors of m.

1. Show that $E_n(g)$ is equivalent to $\log(\log n)$ as $n \to +\infty$ (use the preceding exercise).

2. Prove that

$$E_n\left(\left(\delta_p - \frac{1}{n}\left[\frac{n}{p}\right]\right)^2\right) \le \frac{1}{p} \quad \text{for } n \ge 1 \text{ and } p \le n.$$

Show that

$$E_n\left(\left(\delta_p - \frac{1}{n}\left[\frac{n}{p}\right]\right)\left(\delta_q - \frac{1}{n}\left[\frac{n}{q}\right]\right)\right)$$

is less than $1/(np)+1/(nq)$, for distinct primes p, q such that $p \le n$, $q \le n$, and conclude that

$$\text{Var}_n(g) = \int \left(g - E_n(g)\right)^2 dP_n = \int \left[\sum_{1 \le p \le n}\left(\delta_p - \frac{1}{n}\left[\frac{n}{p}\right]\right)\right]^2 dP_n$$

is smaller than $3\sum_{p \le n} 1/p$.

3. Deduce that, for every $\varepsilon > 0$, $P_n\left(\left|g/(\log\log n) - 1\right| \ge \varepsilon\right)$ converges to 0 as $n \to +\infty$ (Hardy–Ramanujan theorem).

7

Radon Measures

Many of the measures used in analysis and differential geometry belong to a particular class of Daniell measures called Radon measures. Lebesgue measure is an example. These measures are defined in locally compact Hausdorff spaces as continuous linear forms on the space of continuous functions with compact support. By definition, Radon measures are related to the topology of X; as we will see this relationship has many consequences. For example, any compact set is measurable for a Radon measure (in fact integrable cf. Proposition 7.3.1). Also, Radon measures are well-behaved with respect to natural operations on measures (product, images and convolution of measures).

Summary

7.1 For the convenience of the reader, we review some properties of locally compact Hausdorff spaces, such as Urysohn's Lemma.

7.2 Let X be a locally compact Hausdorff space. A linear form μ on the space of complex-valued continuous functions with compact support $\mathcal{H}$ is a Radon measure if, for every compact $K \subset X$, the restriction of μ to the space of continuous functions with support in K is continuous. Every positive linear form on $\mathcal{H}$ is a Radon measure (Theorem 7.2.1) and every Radon measure is a Daniell measure. Then we define the upper integral with respect to a Radon measure and give results analogous to those obtained in Chapter 3.

7.3 The main result of this section (Theorem 7.3.2) is the following. The upper envelope f of an upwardly directed set, H, of integrable, lower semicontinuous functions such that $\sup_{g \in H} \int g \, dV\mu < \infty$ is itself integrable and g converges to f in the mean along the filter of sections of H. Also, a Radon measure is regular in the following sense: if A is measurable and moderate, then, given $\varepsilon > 0$, there is an open set U and a countable union of compact sets F such that $V\mu^*(U - F) \leq \varepsilon$.

7.4 In this section we introduce the notion of Lusin measurable mappings. Intuitively these functions are "almost continuous" on any compact set. A function from X into a metrizable space is Lusin measurable if and only if it is measurable.

7.5 Let $x_0 \in X$. The measure ε_{x_0}, defined by $f \mapsto f(x_0)$ for f continuous with compact support, shows that a Radon measure may have some atoms. However, any atom is essentially a point (Proposition 7.5.1). Counting measures provide us with examples of atomic Radon measures.

7.6 One of the goals of measure theory was to improve the Riemann integral. The reader might be happy to know that what we have done so far has substantially improved Riemann's integral. In this section we prove, in particular, that a bounded function with compact support is Riemann integrable if and only if its set of discontinuity is Lebesgue negligible.

7.7 Various types of convergence, such as vague, weak. and narrow convergence in spaces of measures are studied.

7.8 We continue our investigation of convergence in spaces of measures by looking at tight subsets of $\mathcal{M}^1$. Tightness may be viewed as a relative compactness condition.

7.1 Locally Compact Spaces

We begin by recalling several properties of locally compact Hausdorff spaces, which are especially useful in measure theory.

A Hausdorff space Ω is normal if, for all disjoint closed subsets A, B of Ω, there exist disjoint open sets U, V such that $U \supset A$ and $V \supset B$. Observe that a locally compact Hausdorff space is not necessarily normal (see Exercise 3).

Proposition 7.1.1 (Urysohn's Lemma)   *Let Ω be a normal topological space, F a closed subset of Ω, and U an open set containing F. Then there exists a continuous function f from Ω into $[0, 1]$. equal to 1 on F and to 0 on U^c.*

PROOF: Put $r_1 = 0$, $r_2 = 1$, and let $r_3, r_4, r_5 \ldots$ be an enumeration of the rationals in $]0, 1[$. We can find open sets V_0 and then V_1 such that $F \subset V_1 \subset \bar{V}_1 \subset V_0 \subset \bar{V}_0 \subset U$. Suppose $n \geq 2$ and $V_{r_1}, \ldots, V_{r_n}$ have been chosen so that $r_i < r_j$ implies $\bar{V}_{r_j} \subset V_{r_i}$. Then one of the numbers $r_1, \ldots, r_n$, say r_i, will be the largest one which is smaller than r_{n+1}, and another, say r_j, will be the smallest one larger than r_{n+1}. Then we can find $V_{r_{n+1}}$ so that $\bar{V}_{r_j} \subset V_{r_{n+1}} \subset \bar{V}_{r_{n+1}} \subset V_{r_i}$. Continuing, we obtain a family $(V_r)_{r \in [0,1] \cap \mathbb{Q}}$ of open sets with the following properties: $F \subset V_1 \subset \bar{V}_0 \subset U$, and $s > r$ implies $\bar{V}_s \subset V_r$. Define

$$f_r(x) = \begin{cases} r & \text{if } x \in V_r \\ 0 & \text{otherwise} \end{cases} \qquad g_s(x) = \begin{cases} 1 & \text{if } x \in \bar{V}_s \\ s & \text{otherwise} \end{cases}$$

and $f = \sup_r f_r$, $g = \inf_s g_s$. So f is lower semicontinuous and g is upper semicontinuous. It is clear that $0 \leq f \leq 1$, that $f(x) = 1$ if $x \in F$, and that f vanishes outside V_0. To complete the proof we need only show that $f = g$. The inequality $f_r(x) > g_s(x)$ is possible only if $r > s$, $x \in V_r$, and $x \notin \bar{V}_s$, although $r > s$ implies $\bar{V}_r \subset V_s$. Hence $f_r \leq g_s$ for all r, s, and $f \leq g$. Suppose $f(x) < g(x)$ for some x. Then there are rationals r, s such that $f(x) < r < s < g(x)$. Since $f(x) < r$, $x \notin \bar{V}_r$; since $g(x) > s$, $x \in \bar{V}_s$, and this is a contradiction. Finally, $f = g$. $\qquad\square$

Definition 7.1.1 Let Ω be a topological space and f a mapping from Ω into an **R**-vector space or into $\bar{\mathbf{R}}$. The support of f is the closure $\mathrm{supp}(f)$ of $\{x : f(x) \neq 0\}$. Equivalently, it is the smallest closed subset of Ω outside which f vanishes.

Proposition 7.1.2 *Let Ω be a locally compact Hausdorff space, K a compact subset, and U an open subset of Ω containing K. There exists a continuous function f from Ω into $[0,1]$, with compact support contained in U, such that $f(x) = 1$ for all points x in K.*

PROOF: There exists an open set V, whose closure $\bar{V}$ is compact, such that $K \subset V \subset \bar{V} \subset U$. Since $\bar{V}$ is compact, it is normal (in the relative topology). Hence there exists a continuous function g from $\bar{V}$ into $[0,1]$ which is equal to 1 on K and to 0 on $\bar{V} - V$. Put

$$f(x) = \begin{cases} g(x) & \text{if } x \in \bar{V} \\ 0 & \text{if } x \in \Omega - \bar{V}, \end{cases}$$

and observe that f has the desired properties. $\qquad\square$

Theorem 7.1.1 *Suppose $U_1, \ldots U_n$ are open subsets of a locally compact Hausdorff space Ω, K is compact, and $K \subset \bigcup_{1 \leq i \leq n} U_i$. Then there exist continuous functions f_i $(1 \leq i \leq n)$ from Ω into $[0,1]$ such that*

(a) for each $1 \leq i \leq n$, $\mathrm{supp}(f_i)$ is compact and contained in U_i;

(b) $\sum_{1 \leq i \leq n} f_i(x) \leq 1$ for every $x \in \Omega$, and $\sum_{1 \leq i \leq n} f_i(x) = 1$ for every $x \in K$.

PROOF: Each $x \in K$ has a neighborhood V_x with compact closure $\bar{V}_x \subset U_i$ for some i (depending on x). There are points $x_1, \ldots, x_m$ such that $K \subset V_{x_1} \cup \ldots \cup V_{x_m}$. For $1 \leq i \leq n$, let K_i be the union of those $\bar{V}_{x_j}$ included in U_i; by Proposition 7.1.2, there is a continuous function g_i from Ω into $[0,1]$, with compact support contained in U_i, such that $g_i(x) = 1$ for all $x \in K_i$. Define $f_1 = g_1$, $f_2 = (1 - g_1)g_2, \ldots, f_n = (1 - g_1)(1 - g_2) \ldots (1 - g_{n-1})g_n$. It is easily verified, by induction, that

$$f_1 + f_2 + \ldots + f_n = 1 - (1 - g_1)(1 - g_2) \ldots (1 - g_n).$$

Since $K \subset K_1 \cup \ldots \cup K_n$, for every $x \in K$ $g_i(x) = 1$ for at least one i; so $(f_1 + \ldots + f_n)(x) = 1$. $\qquad\qquad\qquad\qquad\qquad\qquad\qquad\qquad\qquad\qquad\square$

Proposition 7.1.3 *Let Ω be a locally compact Hausdorff space and f a continuous function from Ω into a real Banach space F. Suppose f is supported on some compact set K in Ω. Then, for every $\varepsilon > 0$, there exist continuous functions h_i $(1 \leq i \leq n)$, $h_i \neq 0$, from Ω into $\mathbf{R}^+$ such that*

(a) $\operatorname{supp}(h_i) \subset K$ *for every* $1 \leq i \leq n$;

(b) $\sum_{1 \leq i \leq n} h_i(x) \leq 1$ *for all* $x \in K$;

(c) if x_i is an arbitrary point of $\operatorname{supp}(h_i)$ for $1 \leq i \leq n$, then

$$\left| f(x) - \sum_{1 \leq i \leq n} h_i(x) f(x_i) \right| \leq \varepsilon \quad \text{and} \quad \left| |f(x)| - \sum_{1 \leq i \leq n} h_i(x) |f(x_i)| \right| \leq \varepsilon$$

for all $x \in \Omega$.

PROOF: For any y belonging to the boundary of K, there exists an open neighborhood V_y of y such that $|f(z)| \leq \varepsilon$ for all $z \in V_y$. Let K' be the set of those points in K which belong to no V_y, as y runs through the boundary of K. Now K' is compact and contained in the interior of K. so there exists a finite covering $(U_i)_{1 \leq i \leq n}$ of K', consisting of open subsets of Ω contained in K, such that $|f(x) - f(y)| \leq \varepsilon$ whenever x, y belong to the same U_i. By Theorem 7.1.1, there exist n continuous functions h_i from Ω into $[0,1]$ such that $\operatorname{supp}(h_i) \subset U_i$, $\sum_{1 \leq i \leq n} h_i(x) \leq 1$ on Ω, and $\sum_{1 \leq i \leq n} h_i(x) = 1$ on K'. For $x_i \in \operatorname{supp}(h_i)$ $(1 \leq i \leq n)$, then

$$\left| h_i(x) f(x) - h_i(x) f(x_i) \right| = h_i(x) |f(x) - f(x_i)| \leq \varepsilon h_i(x)$$

for every $x \in U_i$. But this inequality is still true if $x \notin U_i$, because, in this case, $h_i(x) = 0$. Hence, for every $x \in \Omega$,

$$\left| (1 - h_0(x)) \cdot f(x) - \sum_{1 \leq i \leq n} h_i(x) f(x_i) \right| \leq \varepsilon (1 - h_0(x)),$$

where $h_0 = 1 - \sum_{1 \leq i \leq n} h_i$. Therefore,

$$\left| f(x) - \sum_{1 \leq i \leq n} h_i(x) f(x_i) \right| \leq \varepsilon \qquad \text{for every } x \in K'.$$

Next, $|h_0(x) f(x)| \leq \varepsilon \cdot h_0(x)$ for every $x \in K - K'$, so

$$\left| f(x) - \sum_{1 \leq i \leq n} h_i(x) f(x_i) \right| \leq \varepsilon$$

in this case. Evidently, this last inequality remains true when $x \notin K$. Finally, by a similar argument, $\left\|f(x) - \sum_{1 \leq i \leq n} h_i(x)|f(x_i)|\right\| \leq \varepsilon$ for all $x \in \Omega$. $\square$

In other words, f can be approximated by decomposable functions. Likewise, we have the following proposition.

Proposition 7.1.4 *In a locally compact Hausdorff space Ω, let Φ be the semi-ring consisting of the $K \cap L^c$, where K, L range over the class of compact sets. Let F be a real Banach space and $f : \Omega \to F$ a continuous mapping with compact support K. Then there exists a sequence $(g_n)_{n \geq 1}$ in $St(\Phi, F)$, converging uniformly to f, such that $\operatorname{supp}(g_n) \subset K$ for every $n \geq 1$.*

PROOF: For a fixed integer $n \geq 1$, there exists a finite covering $(M_i)_{1 \leq i \leq m}$ of K by compact subsets of K, such that $|f(x) - f(y)| \leq 1/n$ for all $1 \leq i \leq m$ and all x, $y \in M_i$. For every $1 \leq j \leq m$, put $N_j = M_j \cap \left(\bigcup_{1 \leq i \leq j-1} M_i\right)^c$, and let $a_j \in F$ be such that $|f(x) - a_j| \leq 1/n$ for all $x \in N_j$. If we put $g_n = \sum_{1 \leq j \leq m} a_j \cdot 1_{N_j}$, then $|f - g_n| \leq 1/n$. $\square$

In the notation of Proposition 7.1.4, when $F = \mathbf{R}$ and f is positive, we may suppose that $0 \leq g_n \leq f$ (take $a_j = \inf_{x \in N_j} f(x)$ in the above argument).

7.2   Radon Measures

Let Ω be a locally compact Hausdorff space and $\mathcal{H}(\Omega, \mathbf{C})$ the space of continuous, compactly supported functions from Ω into $\mathbf{C}$. For any compact subset K of $\mathbf{C}$, denote by $\mathcal{H}(\Omega, K; \mathbf{C})$ the space of those continuous functions from Ω into $\mathbf{C}$ which vanish outside K. Consider on $\mathcal{H}(\Omega, K; \mathbf{C})$ the norm $f \mapsto \sup_{x \in K} |f(x)| = \sup_{x \in \Omega} |f(x)|$, and, on $\mathcal{C}(K, \mathbf{C})$, consider the norm $g \mapsto \sup_{x \in K} |g(x)|$. The mapping $f \mapsto f/K$ from $\mathcal{H}(\Omega, K; \mathbf{C})$ into $\mathcal{C}(K, \mathbf{C})$ is an isometry of $\mathcal{H}(\Omega, K; \mathbf{C})$ onto the closed subspace of $\mathcal{C}(K, \mathbf{C})$ consisting of those continuous functions which vanish on the boundary of K. Hence $\mathcal{H}(\Omega, K; \mathbf{C})$ is a Banach space.

Definition 7.2.1 A $\mathbf{C}$-linear form μ on $\mathcal{H}(\Omega, \mathbf{C})$ is called a *Radon measure* whenever, for every compact subset K of Ω, its restriction to $\mathcal{H}(\Omega, K; \mathbf{C})$ is continuous.

If $f_1 \in \mathcal{H}^+$, $f_2 \in \mathcal{H}^+$, $g \in \mathcal{H}(\Omega, \mathbf{C})$, and $|g| \leq f_1 + f_2$, the function g_i, equal to $(gf_i)/(f_1 + f_2)$ at points where $(f_1 + f_2)(x) \neq 0$, and to 0 elsewhere, is continuous on Ω (with $i = 1, 2$). Indeed, the inequality $|g_i| \leq |g|$ shows that g_i is continuous at every point where $f_1 + f_2$ vanishes. Moreover, $|g_i| \leq f_i$ and $g = g_1 + g_2$. Thus $\mathcal{H}(\Omega, \mathbf{C})$ has the properties required in Section 1.4.

If μ is a Radon measure on Ω, then $|\mu(g)| \leq \|\mu/\mathcal{H}(\Omega, K; \mathbf{C})\| \cdot \|f\|$ (where $K = \operatorname{supp}(f)$) for every $f \in \mathcal{H}^+$ and for all $g \in \mathcal{H}(\Omega, \mathbf{C})$ such that $|g| \leq f$.

Therefore, μ has finite variation. Moreover, every decreasing sequence $(f_n)_{n \geq 1}$ in $\mathcal{H}^+$ which converges pointwise to 0 also converges uniformly to 0 by Dini's theorem; thus $\mu(f_n)$ tends to 0 as $n \to +\infty$, and μ is a Daniell measure.

Theorem 7.2.1 *Every positive linear form μ on $\mathcal{H}(\Omega, \mathbf{C})$ is a Radon measure.*

PROOF: Let K be a compact subset of Ω. There exists a continuous function f_0 from Ω into $[0, 1]$, with compact support, such that $f_0(x) = 1$ on K. For every $g \in \mathcal{H}(\Omega, K; \mathbf{R})$, we have $-\|g\| \cdot f_0 \leq g \leq \|g\| \cdot f_0$ and so $|\mu(g)| \leq \|g\| \cdot \mu(f_0)$. Finally, for every $g \in \mathcal{H}(\Omega, K; \mathbf{C})$, there exists $\zeta \in \mathbf{T}$ (the unit circle in the complex plane) such that $|\mu(g)| = \zeta \mu(g) = \mu(\zeta g) = \mu(\operatorname{Re}(\zeta g))$, and $|\mu(g)|$ is smaller than $\|\operatorname{Re}(\zeta g)\| \cdot \mu(f_0) \leq \|g\| \cdot \mu(f_0)$. $\qquad\square$

Every $\mathbf{C}$-linear form on $\mathcal{H}(\Omega, \mathbf{C})$ which has a finite variation is a linear combination of four positive linear forms. Thus it is a Radon measure.

In short, the Radon measures on Ω are exactly the $\mathbf{C}$-linear forms of finite variation over $\mathcal{H}(\Omega, \mathbf{C})$, and they are Daniell measures. Therefore, we may apply to Radon measures the results of Section 1.4. In the sequel, we will let $\mathcal{M}(\Omega, \mathbf{C})$ (respectively, $\mathcal{M}(\Omega, \mathbf{R})$) be the space of complex (respectively, real) Radon measures, and $\mathcal{M}^+(\Omega)$ the cone of positive Radon measures on Ω.

If $\mu \in \mathcal{M}(\Omega, \mathbf{C})$, $\nu \in \mathcal{M}^+(\Omega)$ are such that $|\mu(f)| \leq \nu(f)$ for all f in $\mathcal{H}^+$, then $|\mu(g)| \leq \nu(|g|)$ for all g in $\mathcal{H}(\Omega, \mathbf{C})$. Indeed, for $\varepsilon > 0$, let $\delta > 0$ be such that $|\mu(\varphi)| \leq \varepsilon$ and $|\nu(\varphi)| \leq \varepsilon$ for every φ in $\mathcal{H}(\Omega, \operatorname{supp}(g); \mathbf{C})$ satisfying $\|\varphi\| \leq \delta$; by Proposition 7.1.3, we can find a finite family $(h_i)_{i \in I}$ in $\mathcal{H}^+(\Omega, \operatorname{supp}(g))$ and a family $(\alpha_i)_{i \in I}$ of complex numbers such that $|g - \sum_{i \in I} \alpha_i h_i| \leq \delta$ and $||g| - \sum_{i \in I} |\alpha_i| h_i| \leq \delta$; thus

$$
\begin{aligned}
|\mu(g)| \leq \left|\mu\left(\sum_i \alpha_i h_i\right)\right| + \varepsilon \leq \sum_i |\alpha_i| \cdot |\mu(h_i)| + \varepsilon &\leq \sum_i |\alpha_i| \cdot \nu(h_i) + \varepsilon \\
&\leq \nu(|g|) + 2\varepsilon,
\end{aligned}
$$

from which the desired inequality $|\mu(g)| \leq \nu(|g|)$ follows.

Therefore, even the last condition of Section 1.4 holds.

Definition 7.2.2 Let I be a nonempty interval in $\mathbf{R}$, and let λ be Lebesgue measure on the natural semiring of I. The linear form $f \mapsto \int f d\lambda$ on $\mathcal{H}(I, \mathbf{C})$ is also called Lebesgue measure on I.

Observe that $|\int f d\lambda| \leq (\beta - \alpha)\|f\|$ for every $f \in \mathcal{H}(I, \mathbf{C})$, if $[\alpha, \beta]$ is a compact subinterval of I containing $\operatorname{supp}(f)$, so that the linear form $f \mapsto \int f d\lambda$ is actually a Radon measure.

Henceforth, we shall denote by $\mathcal{J}^+$ the set of lower semicontinuous functions from Ω into $[0, +\infty]$, where Ω is a given locally compact Hausdorff space.

Lemma 7.2.1 *Every $f \in \mathcal{J}^+$ is the upper envelope of the set*
$$\{g \in \mathcal{H}^+ : g \leq f\}.$$

PROOF: For every $x \in \Omega$ such that $f(x) > 0$ and every real number a in $]0, f(x)[$, there exists a compact neighborhood V of x for which $f(y) \geq a$ on V. On the other hand, there exists a function $g \in \mathcal{H}^+$ such that $\mathrm{supp}(g) \subset V$, $g(x) = a$, and $g(y) \leq a$ for all $y \in V$. Thus $0 \leq g \leq f$ and $g(x) \geq a$, which proves the lemma. $\square$

For the remainder of this section, μ will denote a positive Radon measure on Ω.

Definition 7.2.3   Let f be a function in $\mathcal{J}^+$. The number $\mu^*(f) = \sup_{g \in \mathcal{H}^+, g \leq f} \mu(g)$ is called the upper integral of f (relative to μ).

Evidently, $\mu^*(f) = \mu(f)$ for every $f \in \mathcal{H}^+$.

Theorem 7.2.2 *Let H be a nonempty, upward-directed subset of $\mathcal{J}^+$. Then $\mu^*(\sup_{g \in H} g) = \sup_{g \in H} \mu^*(g)$.*

PROOF: Put $f = \sup_{g \in H} g$. We first prove the theorem when the functions $g \in H$ and their upper envelope f belong to $\mathcal{H}^+$. Then, by Dini's theorem, the filter of sections of H converges uniformly to f on $\mathrm{supp}(f)$, and, so, on Ω. Since μ is continuous on $\mathcal{H}(\Omega, \mathrm{supp}(f); \mathbf{C})$, the relation $\mu(f) = \sup_{g \in H} \mu(g)$ follows, as desired.

We now pass to the general case. Clearly, $\mu^*(g) \leq \mu^*(f)$ for all $g \in H$. By the definition of $\mu^*(f)$, it suffices to prove that, for every $\psi \in \mathcal{H}^+$ satisfying $\psi \leq f$, $\mu(\psi)$ is less than $\sup_{g \in H} \mu^*(g)$. To this end, let Φ_g be the set $\{\varphi \in \mathcal{H}^+ : \varphi \leq g\}$, for every $y \in H$, and let Φ be the union of the Φ_g when g extends over H. Since H is directed upward, so is Φ, and we have $f = \sup_{\varphi \in \Phi} \varphi$. Since $\psi \leq f$, ψ is the upper envelope of the functions $\inf(\psi, \varphi)$, where φ extends over Φ. But ψ and the function $\inf(\psi, \varphi)$ belong to $\mathcal{H}^+$, so the first part of the argument proves that $\mu(\psi) = \sup_{\varphi \in \Phi} \mu(\inf(\psi, \varphi))$. Now each $\varphi \in \Phi$ lies in a set Φ_g. Thus

$$\mu(\inf(\psi, \varphi)) \leq \mu(\varphi) \leq \mu^*(g) \leq \sup_{g \in H} \mu^*(g),$$

whence we conclude immediately that $\mu(\psi) \leq \sup_{g \in H} \mu^*(g)$. $\square$

Proposition 7.2.1 $\mu^*(f_1 + f_2) = \mu^*(f_1) + \mu^*(f_2)$ *for all $f_1, f_2 \in \mathcal{J}^+$.*

PROOF: When φ_1 (respectively, φ_2) runs through $\{\varphi_1 \in \mathcal{H}^+ : \varphi_1 \leq f_1\}$ (respectively, $\{\varphi_2 \in \mathcal{H}^+ : \varphi_2 \leq f_2\}$), the functions $\varphi_1 + \varphi_2$ form an upward-directed set, whose upper envelope is $f_1 + f_2$. Therefore,

$$\mu^*(f_1 + f_2) \quad = \quad \sup \mu(\varphi_1 + \varphi_2)$$

$$
\begin{aligned}
&= \sup\left(\mu(\varphi_1) + \mu(\varphi_2)\right) \\
&= \sup\mu(\varphi_1) + \sup\mu(\varphi_2) \\
&= \mu^*(f_1) + \mu^*(f_2).
\end{aligned}
$$

$\square$

Proposition 7.2.2 $\mu^*\left(\sum_{i\in I} f_i\right) = \sum_{i\in I}\mu^*(f_i)$ *for every family (finite or infinite) in* $\mathcal{J}^+$.

PROOF: This follows immediately from Proposition 7.2.1 and Theorem 7.2.1.

$\square$

7.3   Regularity of Radon Measures

Let Ω be a locally compact Hausdorff space and $\mathcal{J}^+$ the set of positive, lower semicontinuous functions on Ω. Since the conditions of Section 3.1 hold, we may use freely the results of Chapter 3, and add new ones.

Let μ be a Radon measure on Ω.

Denote by V the union of all μ-negligible open sets. The indicator functions of μ-negligible open sets form an upward-directed set H whose upper envelope is 1_V. Therefore, V is μ-negligible (Theorem 7.2.1).

Definition 7.3.1 The complement in Ω of the greatest μ-negligible open set is called the support of μ, and is written $\mathrm{supp}(\mu)$.

Theorem 7.3.1 $V\mu^*(A) = \inf_U V\mu^*(U)$ *for every subset A of Ω, where U extends over the class of open sets containing A.*

PROOF: We may consider only the case where $V\mu^*(A) < +\infty$. Fix $0 < \varepsilon < 1$. There exists $f \in \mathcal{J}^+$ such that $f \geq 1_A$ and $V\mu^*(A) \leq V\mu^*(f) \leq V\mu^*(A) + \varepsilon$. The open set $G = \{x \in \Omega : f(x) > 1 - \varepsilon\}$ contains A. On the other hand, $f \geq (1-\varepsilon)1_G$, so $V\mu^*(G) \leq \left(1/(1-\varepsilon)\right)V\mu^*(f) \leq \left(1/(1-\varepsilon)\right)(V\mu^*(A) + \varepsilon)$, and the result follows.

$\square$

Theorem 7.3.2   Let $H \neq \emptyset$ be an upward-directed set of μ-integrable, lower semicontinuous functions (respectively, a downward-directed set of μ-integrable, upper semicontinuous functions). If $\sup_{f\in H}\int f\, dV\mu < +\infty$ (respectively, $\inf_{f\in H}\int f\, dV\mu > -\infty$), then the upper (respectively, lower) envelope g of H is μ-integrable, and f converges to g in the mean along the filter of sections of H.

PROOF: We will prove the statement only for lower semicontinuous functions, and we may suppose that H has a smallest element f_0. The functions f^+ (respectively, f^-), where f ranges over H, form an upward-directed set of lower semicontinuous functions whose upper envelope is g^+ (respectively, a downward-directed set of upper semicontinuous functions whose lower envelope is g^-). Moreover, $\int f^+ dV\mu \leq \int f dV\mu + \int f_0^- dV\mu$ for all $f \in H$. Thus we need only prove the two statements of Theorem 7.3.2 when H consists of positive functions.

If H is directed upward and composed of positive, lower semicontinuous functions, then $\int^* g dV\mu = \sup_{f \in H} \int^* f dV\mu = \sup_{f \in H} \int f dV\mu < +\infty$, so the lower semicontinuous function g is integrable and $\int g dV\mu = \sup_{f \in H} \int f dV\mu$. Since $f \leq g$ for every $f \in H$, the filter of sections of H converges to g in the mean.

Next, suppose that H is directed downward and composed of positive, upper semicontinuous functions. We may restrict our attention to the case in which H has a greatest element f_1. There exists an integrable, lower semicontinuous function h such that $f_1 \leq h$. For each $f \in H$, consider the function f' taking the value $h(x) - f(x)$ at points x where $f(x) < +\infty$, and the value zero at points x where $f(x) = +\infty$. When f extends over H, the f' form an upward-directed set of positive functions, integrable and lower semicontinuous. Moreover, $\int f' dV\mu \leq \int h dV\mu$ for all $f \in H$. Thus, along the filter of sections of H, f' converges in the mean to the upper envelope g' of the f'. Now $h - f' = f$ and $h - g' = g$ at points where h is finite, so f converges to g in the mean along the filter of sections of H.   □

Proposition 7.3.1   *Each compact set is μ-integrable, and $V\mu^*(U) = \sup_{K \subset U, K \text{ compact}} V\mu(K)$ for every open set U.*

PROOF: Every compact set K is integrable because $1_K = \inf_{f \in \mathcal{H}^+, f \geq 1_k} f$. Now, given an open set U, let $r < V\mu^*(U)$ be a real number. There exists $f \in \mathcal{H}^+$ such that $f \leq 1_U$ and $V\mu(f) > r$. Put $S = \text{supp}(f)$, and $K_\varepsilon = \{x \in \Omega : f(x) \geq \varepsilon\}$ for every $\varepsilon > 0$. Then $f \leq 1_{K_\varepsilon} + \varepsilon 1_{S - K_\varepsilon}$, so $V\mu(f) \leq V\mu(K_\varepsilon) + \varepsilon V\mu(S)$. Now $V\mu(K_\varepsilon) \geq r$ for ε small enough.   □

Proposition 7.3.2   *A set $A \subset \Omega$ is integrable if and only if, for every $\varepsilon > 0$, there exist a compact set K and an integrable open set U such that $K \subset A \subset U$ and $V\mu(U - K) \leq \varepsilon$. Equivalently, A is μ-integrable if and only if, for every $\varepsilon > 0$, we can find a compact set $K \subset A$ such that $V\mu^*(A - K) \leq \varepsilon$. In this case, there exists a countable union of disjoint compact sets $A_1 \subset A$, such that $A - A_1$ is μ-negligible.*

PROOF: If A is integrable, then, by Theorem 7.3.1, there exists an open set $U \supset A$ such that $V\mu^*(U)$ is arbitrarily close to $V\mu(A)$. Moreover, for every $\varepsilon > 0$, there exists a positive, upper semicontinuous function f, with compact support S, such that $f \leq 1_A$ and $\int (1_A - f) dV\mu \leq \varepsilon/2$ (Proposition 3.7.2). Then $K = \{x \in \Omega : f(x) \geq \delta\}$ is closed and contained in S for every $\delta > 0$. Thus it is compact and, since $f \leq 1_A$, K is included in A. The set $B = A - K$ is integrable, and $f \leq 1_K + \delta 1_B$. Hence

$$\int f dV\mu \leq V\mu(K) + \delta V\mu(B) \leq V\mu(K) + \delta V\mu(A),$$

and finally

$$V\mu(A) \leq \int f dV\mu + \frac{\varepsilon}{2} \leq V\mu(K) + \delta V\mu(A) + \frac{\varepsilon}{2}.$$

But, as δ was arbitrary, $V\mu(A) \leq V\mu(K) + \varepsilon$ for a suitable compact set $K \subset A$.

Conversely, if for every $\varepsilon > 0$ there exists a compact set $K \subset A$ such that $V\mu^*(A - K) \leq \varepsilon$, then 1_A lies in the closure (with respect to $\mathcal{L}_{\mathbb{C}}^1(\mu)$) of the class of the 1_K (K arbitrary compact subset of A), so A is integrable.

Now the proofs of the first two statements are complete. Finally, if A is integrable, define by induction a sequence $(K_n)_{n \geq 1}$ of compact subsets of A as follows: $K_1 \subset A$ and $V\mu(A - K_1) \leq 1$; $K_n \subset (A - \bigcup_{1 \leq i \leq n-1} K_i)$ and $V\mu(A - \bigcup_{1 \leq i \leq n} K_i) \leq 1/n$ for $n > 1$. Then $A - \bigcup_{i \geq 1} K_i$ is negligible. □

By Proposition 7.3.2, a subset A of Ω is locally negligible if and only if $A \cap K$ is negligible for all compact sets K. For every subset A of Ω, $V\mu_*(A) = \sup_{K \subset A,\ K\text{compact}} V\mu(K)$ (Proposition 3.7.9).

A necessary and sufficient condition that a subset A of Ω be μ-measurable is that $A \cap K$ be μ-integrable for all compact sets. Thus closed sets, and hence all Borel sets, are μ-measurable. $A \subset \Omega$ is μ-measurable if and only if it splits the outer measure of compact sets (by Proposition 3.7.8).

Proposition 7.3.3 *Let A be a μ-measurable and μ-moderate set. For every $\varepsilon > 0$, there exist an open set U and a countable union F of compact sets such that $F \subset A \subset U$ and $V\mu^*(U - F) \leq \varepsilon$. When Ω itself is a countable union of compact sets, F may be taken to be closed.*

PROOF: A is the union of disjoint integrable sets A_n ($n \geq 1$). For each integer $n \geq 1$, there exist an open set V_n and a compact set K_n such that $K_n \subset A_n \subset V_n$ and $V\mu(V_n - K_n) \leq \varepsilon/2^n$. Put $U = \bigcup_{n \geq 1} V_n$ and $F = \bigcup_{n \geq 1} K_n$. Then $F \subset A \subset U$ and $U - F \subset \bigcup_{n \geq 1}(V_n - K_n)$, so $V\mu^*(U - F) \leq \varepsilon$.

When Ω is a countable union of compact sets, there exist open sets $U \supset A$, $U' \supset A^c$ such that $V\mu^*(U - A) \leq \varepsilon/2$ and $V\mu^*(U' - A^c) \leq \varepsilon/2$. Then $F = U'^c$ is a closed subset of A, and $U - F$ is included in $(U - A) \cup (U' - A^c)$. Hence $V\mu^*(U - F) \leq \varepsilon$. □

Finally, observe that $\int^{\bullet} f dV\mu = \sup_{K \text{ compact set}} (\int^* f 1_K dV\mu)$ for every $f : \Omega \to [0, +\infty]$.

7.4   Lusin Measurable Mappings

Let Ω and μ be as in Section 7.3.

Proposition 7.4.1 *Let F be a topological space (not necessarily metrizable), and consider the function $f : \Omega \to F$. Then the following conditions are equivalent:*

(a) *For each integrable set A, there exist a negligible subset N of A and a sequence $(K_n)_{n \geq 1}$ of disjoint compact sets whose union is $A - N$, such that each f/K_n is continuous.*

(b) *For each compact set K, there exist a negligible subset N of A and a sequence $(K_n)_{n \geq 1}$ of disjoint compact sets whose union is $A - N$, such that each f/K_n is continuous.*

(c) *For each compact set K and each $\varepsilon > 0$, there exists a compact subset H of K such that $V\mu(K - H) \leq \varepsilon$ and f/H is continuous.*

(d) *For each integrable set A and each $\varepsilon > 0$, there exists a compact subset H of A such that $V\mu(A - H) \leq \varepsilon$ and f/H is continuous.*

PROOF: Clearly, (a) implies (b).

If K is a compact set and $(K_n)_{n \geq 1}$ is a sequence as in condition (b), for every $\varepsilon > 0$ there exists $p \geq 1$ such that $V\mu(K - \bigcup_{1 \leq n \leq p} K_n) \leq \varepsilon$. Then $f/\bigcup_{1 \leq n \leq p} K_n$ is continuous. Thus (b) implies (c).

If (c) holds and A is an integrable set, let $(K_n)_{n \geq 1}$ be a sequence of disjoint compact subsets of A such that $A - \bigcup_{n \geq 1} K_n$ is negligible. For fixed $\varepsilon > 0$ and for every $n \geq 1$, there exists a compact subset H_n of K_n such that $V\mu(K_n - H_n) \leq \varepsilon/2^{n+1}$ and f/H_n is continuous. Then let $p \geq 1$ be an integer such that $V\mu(A - \bigcup_{1 \leq n \leq p} K_n) \leq \varepsilon/2$, and put $H = \bigcup_{1 \leq n \leq p} H_n$. Clearly, $V\mu(A - H) \leq \varepsilon$ and f/H is continuous. Therefore, (c) implies (d).

Finally, (d) implies (a), because, if A is an integrable set, we can inductively construct a sequence $(K_n)_{n \geq 1}$ of disjoint compact subsets of A such that $V\mu(A - \bigcup_{1 \leq n \leq p} K_n) \leq 1/p$ for all $p \geq 1$, and such that f/K_n is continuous. $\qquad\square$

Definition 7.4.1 A mapping f satisfying the conditions of Proposition 7.4.1 is called Lusin μ-measurable.

If f is Lusin μ-measurable, all mappings which agree with f locally almost everywhere are Lusin μ-measurable. A mapping from Ω into F that has a locally μ-negligible set of points of discontinuity is Lusin measurable.

Proposition 7.4.2 *Let $(F_n)_{n \geq 1}$ be a sequence of topological spaces and, for each n, let $f_n : \Omega \to F_n$ be Lusin μ-measurable. For every compact set $K \subset \Omega$ and every $\varepsilon > 0$, there exists a compact subset L of K such that $V\mu(K - L) \leq \varepsilon$ and each f_n/L is continuous.*

PROOF: For each $n \geq 1$, there exists a compact set $K_n \subset K$ such that $V\mu(K - K_n) \leq \varepsilon/2^n$ and f_n/K_n is continuous. Take $L = \bigcap_{n \geq 1} K_n$.   □

Theorem 7.4.1 *Under the hypothesis of Proposition 7.4.2, put $f = [f_n]_{n \geq 1}$ and suppose that u is a continuous mapping from $f(\Omega)$ into a topological space. Then $u \circ f$ is Lusin μ-measurable.*

PROOF: Obvious.   □

Now we compare μ-measurability and Lusin μ-measurability.

Proposition 7.4.3 *Let $\mathcal{M}$ be the σ-algebra of μ-measurable sets. Each $\mathcal{M}$-simple mapping from Ω into a topological space is Lusin μ-measurable.*

PROOF: Let $(A_i)_{i \in I}$ be a finite partition of Ω into $\mathcal{M}$-sets such that f is constant on each A_i. Next, let E be an integrable set. For each $i \in I$, there exists a sequence $(K_{i,n})_{n \geq 1}$ of disjoint compact subsets of $E \cap A_i$ such that $(E \cap A_i) - \bigcup_{n \geq 1} K_{i,n}$ is negligible. Since $f/K_{i.n}$ is continuous for every $i \in I$ and every $n \geq 1$, f is Lusin μ-measurable.   □

Proposition 7.4.4 *Every Lusin μ-measurable mapping f from Ω into a metrizable space F is μ-measurable.*

PROOF: Let d be a compatible distance on F and E an integrable set. There exists a sequence $(K_i)_{i \geq 1}$ of disjoint compact subsets of E such that $E - \bigcup_{i \geq 1} K_i$ is negligible and f/K_i is continuous for every $i \geq 1$. For fixed $n \geq 1$ and each $1 \leq i \leq n$, there exists a partition of K_i into integrable sets $E_{i,j}$ (with $1 \leq j \leq q_i$) such that the oscillation of f over each $E_{i,j}$ is less than $1/n$. Define an $\mathcal{R}/E$-simple mapping g_n from Ω into F as follows: g_n is constant on each $E_{i,j}$ (with $1 \leq i \leq n$, $1 \leq j \leq q_i$), equal to any fixed value of f on this set, and equal to a fixed $z \in F$ for all $x \in E$ which belong to none of the $E_{i,j}$. Then the sequence $(g_n)_{n \geq 1}$ converges to f/E almost everywhere in E.   □

Proposition 7.4.5 *Let F be a metrizable uniform space, and let $(f_n)_{n \geq 1}$ be a sequence of Lusin μ-measurable mappings from Ω into F, converging locally almost everywhere to a mapping f. Then, for every integrable set E and every $\varepsilon > 0$, there exists a compact set $K \subset E$ such that $V\mu(E - K) \leq \varepsilon$, f_n is continuous on K for every $n \in \mathbf{N}$, and f_n converges uniformly to f on K.*

PROOF: By Egorov's theorem, we can find an integrable set $Z \subset E$ such that $V\mu(Z) \leq \varepsilon/2$ and the f_n converge uniformly to f on $E - Z$. By Proposition 7.4.2, there exists a compact subset K of $E - Z$ such that $V\mu((E - Z) - K) \leq \varepsilon/2$ and the functions f_n/K are continuous. Then $V\mu(E - K) \leq \varepsilon$.   □

Theorem 7.4.2 *Let f be a mapping from Ω into a metrizable space. Then f is Lusin μ-measurable if and only if it is μ-measurable.*

PROOF: If f is μ-measurable, then it is Lusin μ-measurable by Propositions 7.4.3 and 7.4.5. Conversely, if f is Lusin μ-measurable, then it is μ-measurable by Proposition 7.4.4.   □

In view of Theorem 7.4.2, a Lusin μ-measurable mapping from Ω into a topological space will now simply be called a μ-measurable mapping.

Now Theorem 5.2.5 may be refined as follows:

Theorem 7.4.3 *Let F be a Banach space, F' its normed dual, and B the closed ball with center 0 and radius 1 in $\mathcal{H}(\Omega, \mathbf{R}) \otimes F'$ (considered as a subspace of $\mathcal{L}_{F'}^\infty(\mu)$). Then $N_1(f) = \sup_{g \in B} \left| \int fg \, d\mu \right|$ for every $f \in \mathcal{L}_F^1(\mu)$.*

PROOF: As in Theorem 5.2.5, we may restrict ourselves to the case in which f belongs to $St(\hat{\mathcal{R}}, F)$ and $N_1(f) = 1$.

Given $\varepsilon > 0$, there exists $g \in St(\hat{\mathcal{R}}, F')$ such that $|g| \leq 1$ and $\left| \int fg \, d\mu \right| \geq 1 - \varepsilon$ (by the same argument as that of Theorem 5.2.5). There exists a finite number of mutually disjoint, compact sets K_i such that g takes the constant value a_i' on each K_i. If K is the union of the K_i, then $\int |f| 1_{K^c} \, dV\mu \leq \varepsilon$. Let U_i be an open neighborhood of K_i such that the U_i are mutually disjoint, and let h_i be a continuous function from Ω into $[0, 1]$ equal to 1 on K_i, whose support is compact and contained in U_i. If we put $h = \sum_i a_i' h_i$, then $h(x) = g(x)$ on K and $|h(x)| \leq 1$ on Ω. Hence $\int |fh| 1_{K^c} \, dV\mu \leq \varepsilon$ and $\left| \int fh \, d\mu \right| \geq 1 - 3\varepsilon$, which proves the theorem.   □

Replacing $St^+(S)$ by $\mathcal{H}^+(\Omega)$, we obtain a result similar to Proposition 6.2.1.

7.5 Atomic Radon Measures

Fix Ω, a locally compact Hausdorff space.

Proposition 7.5.1 *Let μ be a Radon measure on Ω. If A is a μ-atom, there is a point $a \in A$ such that $A - \{a\}$ is negligible.*

PROOF: Let $\mathcal{F}$ be the class of those compact sets $K \subset A$ such that $V\mu(K) = V\mu(A)$. The class of indicator functions 1_K, where K extends over $\mathcal{F}$, is directed downward. Indeed, if K_1, K_2 lie in $\mathcal{F}$, then

$$
\begin{aligned}
V\mu(K_1 \cap K_2) &= V\mu(K_1) + V\mu(K_2) - V\mu(K_1 \cup K_2) \\
&= V\mu(A).
\end{aligned}
$$

Write P for the intersection of all elements of $\mathcal{F}$. Then P lies in $\mathcal{F}$ by Theorem 7.3.2. Now, if $\{a\}$ is negligible for every $a \in P$, there exists a compact neighborhood V_a of a such that $V\mu(V_a) < V\mu(A)$, and then $V_a \cap P$ is negligible. In this case, since P is covered by a finite number of the V_a, $V\mu(P) = 0$, which contradicts the fact that P belongs to $\mathcal{F}$. Thus there is a point a of P such that $V\mu(\{a\}) = V\mu(A)$, and P reduces to the point a.   □

Definition 7.5.1 Let $\alpha : \Omega \to \mathbf{C}$ be such that $\sum_{x \in K} |\alpha(x)|$ is finite for every compact set K. The Radon measure $\mu : f \mapsto \sum_{x \in \Omega} \alpha(x) f(x)$ on Ω is called the Radon measure defined by the point masses $\alpha(x)$ (x in any set X such that α vanishes on X^c).

Proposition 7.5.2 *Let α and μ be as in Definition 7.5.1. Then μ is atomic.*

PROOF: Let $f \in \mathcal{H}^+$ and $\varepsilon > 0$. There exists a finite subset Y of Ω such that $\sum_{x \in \Omega - Y} |\alpha(x)||f(x)| \le \varepsilon$. We can find disjoint neighborhoods V_y of the points $y \in Y$, and, for each $y \in Y$, a function $\varphi_y \in \mathcal{H}(\Omega, \mathbf{C})$ such that $|\varphi_y| \le 1$, $\operatorname{supp}(\varphi_y) \subset V_y$, and $\varphi_y(y)\alpha(y) = |\alpha(y)|$. Then $g = \sum_{y \in Y} \varphi_y f$ lies in $\mathcal{H}(\Omega, \mathbf{C})$, $|g| \le f$, and $\sum_{x \in \Omega - Y} |\alpha(x) \cdot g(x)| \le \varepsilon$. We see that

$$|\mu(g)| \ge \left| \sum_{y \in Y} \alpha(y) g(y) \right| - \varepsilon = \sum_{y \in Y} |\alpha(y)||f(y)| - \varepsilon \ge \sum_{x \in \Omega} |\alpha(x)||f(x)| - 2\varepsilon.$$

Therefore,

$$V\mu(f) \ge \sum_{x \in \Omega} |\alpha(x)||f(x)|$$

for all $f \in \mathcal{H}^+$, and

$$V\mu(f) = \sum_{x \in \Omega} |\alpha(x)||f(x)|.$$

Now, $V\mu^*(h) \le \sum_{x \in \Omega} |\alpha(x)||h(x)|$ for each $h \in \mathcal{J}^+$. Conversely, let $r < \sum_{x \in \Omega} |\alpha(x)||h(x)|$ be a real number, and let Y be a finite subset of Ω such that $r < \sum_{y \in Y} |\alpha(y)||h(y)|$. For each $y \in Y$, choose a real number a_y in $]-\infty, h(y)[$. We can find disjoint neighborhoods V_y of the points $y \in Y$ such that $h(x) \ge a_y$ for all $x \in V_y$, and, for each $y \in Y$, a function $\varphi_y \in \mathcal{H}^+(\Omega)$ such that $\varphi_y \le 1$, $\operatorname{supp}(\varphi_y) \subset V_y$, and $\varphi_y(y) = 1$. Then $g = \sum_{y \in Y} a_y \varphi_y$ lies in $\mathcal{H}^+$, $g \le h$, and $V\mu^*(h) \ge V\mu(g) \ge \sum_{y \in Y} a_y |\alpha(y)|$. For suitable numbers a_y, $\sum_{y \in Y} a_y |\alpha(y)|$ can be made larger than r. Therefore, $V\mu^*(h) \ge \sum_{x \in \Omega} |\alpha(x)||h(x)|$.

Finally, if K is a compact set, there is a function $f \in \mathcal{H}^+$ such that $f \ge 1_K$, and then

$$V\mu(K) = V\mu(f) - V\mu(f - 1_K) = \sum_{x \in K} |\alpha(x)|.$$

Thus μ is atomic.   □

Note that the $\{x\}$ ($x \in \Omega$ such that $\alpha(x) \ne 0$) are representatives of the different classes of atoms.

Conversely, by Proposition 7.5.1 and Theorem 3.8.1, any atomic Radon measure is defined by point masses.

For every atomic Radon measure μ on Ω, all mappings from Ω into a topological space are μ-measurable.

7.6   The Riemann Integral[†]

Let Ω be a locally compact Hausdorff space, μ a Radon measure on Ω, and $\mathcal{H}$ the set of continuous functions from Ω into $\mathbf{R}$ which have compact support.

Definition 7.6.1 Let f be a bounded real-valued function on Ω, with compact support. Then

$$\int^R f dV\mu = \inf_{\varphi \in \mathcal{H},\, \varphi \geq f} V\mu(\varphi) \quad \text{and} \quad \int_R f dV\mu = \sup_{\varphi \in \mathcal{H},\, \varphi \leq f} V\mu(\varphi)$$

are called, respectively, the upper Riemann integral and the lower Riemann integral of f (with respect to $V\mu$).

Clearly, $\int_R f dV\mu = -\int^R (-f) dV\mu$.

Denote by G the set of bounded functions from Ω into $\mathbf{R}$ which have compact support. Then $\int^R (f_1 + f_2) dV\mu \leq \int^R f_1 dV\mu + \int^R f_2 dV\mu$ for all f_1, f_2 in G, $\int^R \alpha f dV\mu = \alpha \int^R f dV\mu$ for all $f \in G$ and $\alpha > 0$, and $\int^* f dV\mu \leq \int^R f dV\mu$. Moreover, by Theorem 7.3.2, $\int^* f dV\mu = \int^R f dV\mu$ for every upper semicontinuous function from Ω into $[0, +\infty[$ which has compact support.

Let F be a real Banach space and G_F the space of bounded mappings with compact support from Ω into F. Define on G_F the seminorm $f \mapsto \int^R |f| dV\mu$.

Definition 7.6.2 Let R_F be the closure of $\mathcal{H}(\Omega, \mathbf{R}) \otimes F$ in G_F. The elements of R_F are called the Riemann μ-integrable functions from Ω into F. The linear mapping $\varphi \mapsto \int \varphi d\mu$ from $\mathcal{H}(\Omega, \mathbf{R}) \otimes F$ into F (F real or complex Banach space, as μ is real or complex) has a unique linear continuous extension to R_F, which we write $f \mapsto \int^R f d\mu$. For every $f \in R_F$, $\int^R f d\mu$ is called the Riemann integral of f (with respect to μ).

Since $\int^* |f| \cdot dV\mu \leq \int^R |f| \cdot dV\mu$ for all $f \in G_F$, we see that any Riemann integrable function is μ-integrable, and, since the linear mapping $f \mapsto \int f d\mu$ from R_F into F is continuous, it is clear that $\int f d\mu = \int^R f d\mu$ for all $f \in R_F$.

Now let $\mathcal{B}$ be a filter basis on G_F, and suppose there is a compact set K such that, for each $M \in \mathcal{B}$, all functions $g \in M$ have their supports included in K. If $\mathcal{B}$ converges uniformly to a function f, then f lies in G_F and $\mathcal{B}$ converges to

---

[†]This section may be omitted.

f in G_F. Proposition 7.1.3 therefore proves that the space $\mathcal{H}(\Omega, F)$, consisting of the continuous mappings with compact support from Ω into F, is contained in R_F.

Next, let f be a bounded function from Ω into $\mathbf{R}$ with compact support. If f is Riemann integrable, for every $\varepsilon > 0$ there exists $g \in \mathcal{H}$ such that $\int^R |f - g| dV\mu < \varepsilon/2$, and $h \in \mathcal{H}^+$ such that $|f - g| \le h$ and $\int h dV\mu \le \varepsilon/2$; then $g - h \le f \le g + h$, so

$$\int g dV\mu - \int h dV\mu \le \int_R f dV\mu \le \int^R f dV\mu \le \int g dV\mu + \int h dV\mu$$

and

$$\int f dV\mu - \varepsilon \le \int_R f dV\mu \le \int^R f dV\mu \le \int f dV\mu + \varepsilon;$$

since ε is arbitrary, we see that $\int_R f \cdot dV\mu = \int^R f \cdot dV\mu = \int f dV\mu$. Conversely, if $\int_R f dV\mu = \int^R f dV\mu$, for every $\varepsilon > 0$ there exist $\varphi, \psi \in \mathcal{H}$ such that $\varphi \le f \le \psi$ and $\int(\psi - \varphi) \cdot dV\mu \le \varepsilon$; then $\int^R (f - \varphi) dV\mu \le \varepsilon$; this proves that f is Riemann integrable. In short, a bounded function f from Ω into $\mathbf{R}$, with compact support, is Riemann integrable if and only if $\int_R f dV\mu = \int^R f dV\mu$, and in this case $\int^R f dV\mu = \int f dV\mu$.

Theorem 7.6.1 *Let f be a bounded mapping, with compact support, from Ω into a real Banach space F. The following conditions are equivalent:*

(a) *The set N of discontinuity points of f is μ-negligible.*

(b) *For every $\varepsilon > 0$, there exist elements $a_1, \ldots, a_n$ of F, elements $g_1, \ldots, g_n$ of $\mathcal{H}(\Omega, \mathbf{R})$, and a function $h \in \mathcal{H}^+$, such that $|f - g_1 a_1 - \cdots - g_n a_n| \le h$ and $\int h \cdot dV\mu \le \varepsilon$.*

(c) *f is Riemann integrable with respect to μ.*

If these conditions are satisfied, we may suppose in (b) that $\left| \sum_{1 \le i \le n} g_i a_i \right| \le M$, where $M = \sup_{x \in \Omega} |f(x)|$.

PROOF: First, suppose that condition (a) is satisfied. f is μ-integrable, so, for all $\varepsilon > 0$, there are elements $a_1, \ldots, a_n$ of F and elements $g_1, \ldots, g_n$ of $\mathcal{H}(\Omega, \mathbf{R})$ such that $\int k dV\mu \le \varepsilon/4$, if $k = |f - g_1 a_1 - \cdots - g_n a_n|$. Let g be equal to $M/|g_1 a_1 + \ldots + g_n a_n|$ on the set $K = \{x \in \Omega : |g_1 a_1 + \ldots + g_n a_n|(x) \ge M\}$ and to 1 on $\Omega - K$. Then g is continuous and

$$\int |f - g(g_1 a_1 + \ldots + g_n a_n)| dV\mu \le$$

$$\int |f - (g_1 a_1 + \ldots + g_n a_n)| dV\mu + \int (1 - g)|g_1 a_1 + \ldots + g_n a_n| dV\mu.$$

But $|g_1a_1 + \ldots + g_na_n| \le M + |f - (g_1a_1 + \ldots + g_na_n)|$, so

$$(1-g)|g_1a_1 + \ldots + g_na_n| \le |f - (g_1a_1 + \ldots + g_na_n)|,$$

which leads to

$$\int |f - g(g_1a_1 + \ldots + g_na_n)|dV\mu \le 2\int |f - (g_1a_1 + \ldots + g_na_n)|dV\mu \le \frac{\varepsilon}{2}.$$

In short, multiplying $g_1, \ldots, g_n$ by the same continuous function g, we may suppose that $|\sum_{1 \le i \le n} g_ia_i| \le M$ and $\int k dV\mu \le \varepsilon/2$. Henceforth, we will assume that the latter two inequalities hold. For every $x \in \Omega$, put $\ell(x) = \inf_{V \in \mathcal{U}(x)} \sup_{y \in V} k(y)$, where $\mathcal{U}(x)$ is the filter of neighborhoods of x. Then ℓ is upper semicontinuous, such that $\ell(x) \le 2M$ for every $x \in \Omega$ and $\ell = k$ outside the negligible set N. Thus $\int^R \ell dV\mu = \int \ell dV\mu = \int k dV\mu \le \varepsilon/2$, and there exists $h \in \mathcal{H}^+$ such that $h \ge \ell$ and $\int h dV\mu \le \varepsilon$. This proves that condition (a) implies (b).

Conversely, suppose that condition (b) is satisfied. For each $x \in \Omega$, write $\delta(x)$ for the oscillation

$$\inf_{V \in \mathcal{U}(x)} \sup_{(y,z) \in V \times V} |f(y) - f(z)|$$

of f at x. The function $\delta : x \mapsto \delta(x)$ is upper semicontinuous. Fix $\varepsilon > 0$, and let $a_1, \ldots, a_n$, $g_1, \ldots, g_n$, h be as in condition (b). Then $\delta(x)$ is the oscillation of $f - g_1a_1 - \cdots - g_na_n$ at x, so $\int \delta dV\mu \le \int 2h dV\mu \le 2\varepsilon$. Since ε is arbitrary, $\delta = 0$ almost everywhere. Therefore, the set of discontinuities of f is negligible.

Finally, conditions (b) and (c) are obviously equivalent.   □

We recall some topological results, which will be used in what follows. Let Ω be a compact Hausdorff space. Every neighborhood in $\Omega \times \Omega$ of the diagonal $\{(x,x) : x \in \Omega\}$ is called an entourage. If V is an entourage of Ω, there exists an entourage W such that

$$W \circ W = \{(x,z) \in \Omega \times \Omega : \exists y \in \Omega, (x,y) \in W, (y,z) \in W\}$$

is included in V. For every entourage V and every subset A of Ω, we put $V(A) = \{y \in \Omega : \exists x \in A, (x,y) \in V\}$. Then, if A is a compact subset of Ω, the $V(A)$, where V extends over the set of entourages, form a basis for the filter of neighborhoods of A. In particular, for all $x \in \Omega$, the $V(x) = V(\{x\})$ form a basis for the filter of neighborhoods of x. If V is an entourage, a subset A of Ω is said to be small of order V whenever $A \times A$ is included in V.

In what follows, Ω will be a compact Hausdorff space and μ a Radon measure on Ω. Let P be a set of finite coverings of Ω by μ-integrable sets, such that $A_{k_1} \cap A_{k_2}$ is μ-negligible for all $(A_k)_{k \in L}$ of P and all distinct k_1, k_2 of L. Suppose that for each entourage V of Ω there exists an element $(A_k)_{k \in L}$ of P such that all the A_k are small of order V.

Proposition 7.6.1 *Let f be a Riemann μ-integrable mapping from Ω into a Banach space. Then, for every neighborhood B of $\int f d\mu$, there exists an entourage W of Ω such that $S(\omega) = \sum_{k \in L} \mu(A_k) \cdot \overline{f(A_k)}$ is included in B for each $\omega = (A_k)_{k \in L}$ of P consisting of sets small of order W.*

PROOF: Fix $\varepsilon > 0$ and let A be the set of points of Ω at which the oscillation of f exceeds ε. Since A is compact and μ-negligible, there exists an open entourage V such that $V \mu^*(V(A)) \leq \varepsilon$.

For every $x \in \Omega - V(A)$, the oscillation of f at x is strictly less than ε, so there exists an entourage V_x of Ω for which

$$\sup_{(y,z) \in [(V_x \circ V_x)(x)]^2} |f(y) - f(z)| \leq \varepsilon.$$

Then let $(x_i)_{i \in I}$ be a finite family in $\Omega - V(A)$ such that $\bigcup_{i \in I} V_{x_i}(x_i)$ contains $\Omega - V(A)$, and put $W = \bigcap_{i \in I} V_{x_i}$. If N is a set small of order W which meets $\Omega - V(A)$, fix $i \in I$ so that $N \cap V_{x_i}(x_i)$ contains at least one point x. Now, since $(x_i, y) = (x_i, x) \circ (x, y)$ lies in $V_{x_i} \circ V_{x_i}$ for all y in N, $\sup_{(y,z) \in N \times N} |f(y) - f(z)|$ is less than ε. Next, let $\omega = (A_k)_{k \in L}$ be an element of P consisting of sets small of order W, and, for every $k \in L$, let θ_k be a point of $\overline{f(A_k)}$. For each $k \in L$ such that A_k meets $\Omega - V(A)$, we have $|f - \theta_k| \leq \varepsilon$ on A_k. Therefore, $\left| \int f d\mu - \sum_{k \in L} \mu(A_k) \theta_k \right| = \left| \sum_{k \in L} \int (f - \theta_k) 1_{A_k} d\mu \right|$ is less than $2\|f\| V \mu(V(A)) + \varepsilon V \mu(\Omega) \leq (2\|f\| + V \mu(\Omega)) \cdot \varepsilon$. $\square$

Now we investigate the properties of real-valued, Riemann integrable functions. We suppose that A_k is μ-quadrable (i.e., A_k has μ-negligible boundary), for all $\omega = (A_k)_{k \in L}$ of P and all $k \in L$.

Definition 7.6.3 *Let $f : \Omega \to \mathbf{R}$ be bounded. For every $\omega = (A_k)_{k \in L}$ of P,*

$$m(\omega) = \sum_{k \in L} V \mu(A_k) \inf_{x \in A_k} f(x) \quad \text{and} \quad M(\omega) = \sum_{k \in L} V \mu(A_k) \sup_{x \in A_k} f(x)$$

are called the Darboux sums relative to f and ω.

Observe that $m(\omega) \leq \int_R f \cdot dV\mu \leq \int^R f \cdot dV\mu \leq M(\omega)$, because 1_{A_k} is Riemann integrable.

Proposition 7.6.2 $\int^R f dV\mu = \inf_{\omega \in P} M(\omega)$. *Precisely, for every $\varepsilon > 0$, there exists an entourage W of Ω such that $M(\omega) \leq \int^R f \cdot dV\mu + \varepsilon$ for each $\omega \in P$ consisting of sets small of order W.*

PROOF: Let $\psi \in \mathcal{H}$ be such that $\psi \geq f$ and $\int \psi dV\mu \leq \int^R f \cdot dV\mu + \varepsilon/2$. Since ψ is uniformly continuous on Ω, there is an entourage W of Ω such that $|\psi(y) - \psi(z)| \leq \varepsilon / (2V\mu(\Omega))$ for every $(y, z) \in W$. If $\omega = (A_k)_{k \in L}$ is an element of P consisting of sets of order W, then $\sup_{x \in A_k} \psi(x) - \psi$ remains smaller than $\varepsilon/(2V\mu(\Omega))$ on A_k, for all $k \in L$. Hence

$$\sum_{k \in L} V\mu(A_k) \sup_{x \in A_k} \psi(x) - \int \psi dV\mu \leq \frac{\varepsilon}{2}.$$

Therefore, $M(\omega) \leq \int^R f \cdot dV\mu + \varepsilon$, and the proof is complete.   □

So, if f is Riemann integrable, then there exists, for every $\varepsilon > 0$, an entourage W of Ω such that $0 \leq M(\omega) - m(\omega) \leq \varepsilon$ for all $\omega \in P$ consisting of sets small of order W. Conversely, we have the following proposition.

Proposition 7.6.3 *Let* $f : \Omega \to \mathbf{R}$ *be bounded. If, for every* $\varepsilon > 0$, *there exists an* $\omega \in P$ *such that* $M(\omega) - m(\omega) \leq \varepsilon$, *then* f *is Riemann integrable.*

PROOF: Clearly, $\displaystyle\int^R \left| f - \sum_{k \in L} 1_{A_k} \cdot \inf_{x \in A_k} f(x) \right| dV\mu \leq \varepsilon.$ □

We now give special attention to measures on intervals.

Theorem 7.6.2 *Let* I *be an interval of* $\mathbf{R}$ *and* μ *a Radon measure on* I. *Finally, let* f *be a Riemann* μ-*integrable mapping from* I *into a Banach space. Thus* f *vanishes outside a compact subinterval* $J = [a, b]$ *of* I. *For every* $\varepsilon > 0$, *there exists* $\delta > 0$ *such that, for all subdivisions* $(a_i)_{0 \leq i \leq n}$ *of* $[a, b]$ *of mesh less than* δ, *we have*

$$\left| \int f d\mu - \mu(\{a\})f(a) - \sum_{1 \leq i \leq n} \mu(]a_{i-1}, a_i]) u_i \right| \leq \varepsilon$$

independent of the choice of u_i *in* $\overline{f(]a_{i-1}, a_i])}$ *(respectively, in* $\overline{f([a_{i-1}, a_i])}$ *when* $\mu(\{a_{i-1}\}) = 0$*).*

PROOF: For each $g \in \mathcal{H}(J, \mathbf{C})$, the function $\tilde{g} : I \to \mathbf{C}$, that agrees with g on J and vanishes outside J, is μ-integrable. Indeed, when g takes its values in $\mathbf{R}^+$, $\tilde{g}$ is upper semicontinuous with compact support. The linear form $g \mapsto \int \tilde{g} d\mu$ on $\mathcal{H}(J, \mathbf{C})$ is a Radon measure μ_J on J. Clearly, for $h \in \mathcal{H}^+(I)$, we have $\int h/J.dV(\mu_J) \leq \int h dV\mu$. Given $\varepsilon > 0$, let $a_1, \ldots, a_n, g_1, \ldots, g_n$, and h be as in Theorem 7.6.1. Then $\left| f - ((g_1 a_1 + \ldots + g_n a_n)/J)^{\sim} \right| \leq h$, so

$$\left| \int f d\mu - \int f/J.d\mu_J \right| \leq \left| \int (f - (g_1 a_1 + \ldots + g_n a_n)/J^{\sim}) d\mu \right|$$

$$+ \left| \int (f/J - (g_1 a_1 + \ldots + g_n a_n)/J) d\mu_J \right| \leq 2\varepsilon,$$

which proves that $\int f d\mu = \int f/J.d\mu_J$. Now, for every $x \in J$, $\mu_J(\{x\}) = \mu(\{x\})$, because $1_{\{x\}} = \inf\{\tilde{g} : g \in \mathcal{H}^+(J), g(x) \geq 1\}$. Therefore, we may replace I by J, and it remains only to apply Proposition 7.6.1.   □

Proposition 7.6.4 *Let I be a nonempty interval in $\mathbf{R}$, and let λ be Lebesgue measure on I. Designate by S the natural semiring of I. Then*

$$\int^R f d\lambda = \inf_{\varphi \in St(S,\mathbf{R}), \varphi \geq f} \int \varphi d\lambda \quad \text{and} \quad \int_R f d\lambda = \sup_{\varphi \in St(S,\mathbf{R}), \varphi \leq f} \int \varphi d\lambda$$

for every bounded function f from I into $\mathbf{R}$ with compact support.

PROOF: All functions $\varphi \in St(S, \mathbf{R})$ are Riemann integrable. Therefore,

$$\int^R f d\lambda \leq r = \inf_{\varphi \in St(S,\mathbf{R}), \varphi \geq f} \int \varphi d\lambda.$$

For fixed $g \in \mathcal{H}(I, \mathbf{R})$ satisfying $g \geq f$, let $[a, b]$ be a compact subinterval of I containing supp(g). For every $\varepsilon > 0$, there exists an integer $n \geq 1$ such that $|g(x) - g(y)| \leq \varepsilon/(b-a)$ for all x, y in $[a, b]$ satisfying $|x - y| \leq (b-a)/n$. Write c_k for the supremum of g on $[a + (k-1)(b-a)/n, a + k(b-a)/n]$, for every $1 \leq k \leq n$, and let φ be the function from I into $\mathbf{R}$ which vanishes outside $[a, b]$, is equal to c_k on $]a + (k-1)(b-a)/n, a + k(b-a)/n]$ for $1 \leq k \leq n$, and is equal to c_1 at a in case $a = \inf(I)$. Then φ lies in $St(S, \mathbf{R})$, $\varphi \geq g$, and $\int (\varphi - g) d\lambda \leq \varepsilon$, so $\int \varphi d\lambda \leq \int g d\lambda + \varepsilon$. This proves that $r \leq \int g d\lambda$, and therefore that $r \leq \int^R f d\lambda$. ☐

Proposition 7.6.5 *Let A be a subset of I with compact closure in I. Then $\int^R 1_A d\lambda = \inf \sum_{i \in I} \lambda(E_i)$, where the infimum is taken over finite families $(E_i)_{i \in I}$ of S-sets such that $A \subset \bigcup_{i \in I} A_i$. Similarly, $\int_R 1_A d\lambda = \sup \sum_{i \in I} \lambda(E_i)$, where the supremum is taken over finite families $(E_i)_{i \in I}$ of disjoint S-sets such that $\bigcup_{i \in I} E_i \subset A$.*

PROOF: If $(E_i)_{i \in I}$ is any finite family of S-sets such that $A \subset \bigcup_{i \in I} E_i$, then $1_A \leq \sum_{i \in I} 1_{E_i}$, so $\int^R 1_A d\lambda \leq \sum_{i \in I} \lambda(E_i)$.

Conversely, for every $\varepsilon > 0$, there exists $\varphi \in St^+(S)$ such that $\varphi \geq 1_A$ and $\int \varphi d\lambda \leq \int^R 1_A d\lambda + \varepsilon$. Write φ in the form $\sum_{i \in I} c_i 1_{E_i}$, where I is finite, the E_i are disjoint S-sets, and the c_i are positive numbers. Then $1_A \leq \sum_{i \in I, E_i \cap A \neq \emptyset} 1_{E_i} \leq \varphi$, so

$$\sum_{i \in I, E_i \cap A \neq \emptyset} \lambda(E_i) \leq \int^R 1_A d\lambda + \varepsilon.$$

This proves the first statement.

The second claim is obvious. ☐

If F is a real Banach space, each $f \in \mathcal{H}(I, F)$ may be uniformly approximated by elements of $St(S, F)$ which vanish outside a fixed compact subinterval of I. Thus R_F is the closure of $St(S, F)$ in G_F. Therefore, Riemann integrable functions (with respect to λ) may be defined through $St(S, F)$. Elementary treatments proceed in this way.

The following examples illustrate the ideas in this section.

Let μ be Lebesgue measure on $I = [0, 1]$. If H is a compact subset of I, nowhere dense in I, such that $0 < \mu(H) < 1$, then 1_H is not Riemann μ-integrable, even though it has a nowhere dense set of discontinuities.

Next, if A is a subset of the Cantor set which is not Borelian (Proposition 4.3.3), then 1_A is Riemann μ-integrable but not a Borel function.

7.7   Weak Convergence

Let Ω be a locally compact Hausdorff space. Write $\mathcal{M}(\Omega, \mathbf{C})$ for the space of Radon measures on Ω, and $\mathcal{M}^+(\Omega)$ for the cone of positive Radon measures on Ω.

Definition 7.7.1   On $\mathcal{M}(\Omega, \mathbf{C})$ the topology of pointwise convergence in $\mathcal{H}(\Omega, \mathbf{C})$ is called the topology of vague convergence. Thus a filter basis $\mathcal{B}$ on $\mathcal{M}(\Omega, \mathbf{C})$ converges vaguely to $\mu \in \mathcal{M}(\Omega, \mathbf{C})$ if, for every $f \in \mathcal{H}(\Omega, \mathbf{C})$, $\nu(f)$ converges to $\mu(f)$ along $\mathcal{B}$.

Theorem 7.7.1   *Let $(\mu_n)_{n \geq 1}$ be a sequence in $\mathcal{M}(\Omega, \mathbf{C})$ such that, for every $f \in \mathcal{H}(\Omega, \mathbf{C})$, the sequence $\big(\mu_n(f)\big)_{n \geq 1}$ converges to a limit $\mu(f) \in \mathbf{C}$. Then $\mu : f \mapsto \mu(f)$ is a Radon measure, and $(\mu_n)_{n \geq 1}$ converges vaguely to μ.*

PROOF: If K is a compact subset of Ω, $\{\mu_n / \mathcal{H}(\Omega, K; \mathbf{C}) : n \geq 1\}$ is pointwise bounded. By the Banach-Steinhaus theorem, there exists a positive number M_K such that $|\mu_n(f)| \leq M_K \|f\|$ for $f \in \mathcal{H}(\Omega, K; \mathbf{C})$ and $n \geq 1$. Then $\mu / \mathcal{H}(\Omega, K; \mathbf{C})$ is linear and $|\mu(f)| \leq M_K \|f\|$ for all $f \in \mathcal{H}(\Omega, K; \mathbf{C})$, which proves that μ is a Radon measure. $\square$

Proposition 7.7.1   *All Cauchy sequences in the vague topology converge vaguely. Every Cauchy filter basis on $\mathcal{M}^+(\Omega)$ converges vaguely.*

PROOF: The first statement follows from Theorem 7.4.1. Since each positive linear form on $\mathcal{H}(\Omega, \mathbf{C})$ is a Radon measure, the second statement is obvious. $\square$

Proposition 7.7.2   *Let $\mathcal{B}$ be a filter basis on $\mathcal{M}^+(\Omega)$ converging vaguely to a Radon measure μ. Let F be a real Banach space, and let f be a bounded mapping, with compact support, from Ω into F. Suppose that f is ν-integrable for every $A \in \mathcal{B}$ and for all $\nu \in A$ and that f has a μ-negligible set of discontinuities. Then $\int f d\nu$ converges to $\int f d\mu$ along $\mathcal{B}$.*

PROOF: By Theorem 7.6.1, for every $\varepsilon > 0$, there exist $a_1, \ldots, a_n \in F$, $g_1, \ldots, g_n \in \mathcal{H}(\Omega, \mathbf{R})$, and $h \in \mathcal{H}^+$, such that $|g_1a_1 + \ldots + g_na_n| \leq \|f\|$, $|f - (g_1a_1 + \ldots + g_na_n)| \leq h$, and $\int h \, d\mu \leq \varepsilon$. We can find $A \in \mathcal{B}$ such that $\int h \, d\nu \leq \int h \, d\mu + \varepsilon$ and $\left| \int (g_1a_1 + \ldots + g_na_n) d\nu - \int (g_1a_1 + \ldots + g_na_n) d\mu \right| \leq \varepsilon$ for all $\nu \in A$. Then $\left| \int f \, d\nu - \int f \, d\mu \right| \leq \int h \, d\nu + \varepsilon + \int h \, d\mu \leq 4\varepsilon$ for all $\nu \in A$, and the proof is complete. $\square$

Let $\mathcal{C}^b(\Omega, \mathbf{C})$ be the space of bounded continuous functions from Ω into $\mathbf{C}$, endowed with the norm $f \mapsto \|f\| = \sup_{x \in \Omega} |f(x)|$. We say that a continuous function f from Ω into $\mathbf{C}$ vanishes at infinity whenever, for every $\varepsilon > 0$, there exists a compact set K such that $|f(x)| \leq \varepsilon$ outside K. The space $\mathcal{C}^0(\Omega, \mathbf{C})$ of continuous functions that vanish at infinity is, in fact, the closure of $\mathcal{H}(\Omega, \mathbf{C})$ in $\mathcal{C}^b(\Omega, \mathbf{C})$. If, on the space $\mathcal{M}^1(\Omega, \mathbf{C})$ of bounded Radon measures, we define the norm $\mu \mapsto \|\mu\| = \int 1 \, dV\mu$, then $\mathcal{M}^1(\Omega, \mathbf{C})$ can be identified with the normed dual of $\mathcal{C}^0(\Omega, \mathbf{C})$, on one hand, and, on the other hand, with a subspace of the normed dual of $\mathcal{C}^b(\Omega, \mathbf{C})$.

Definition 7.7.2 On $\mathcal{M}^1(\Omega, \mathbf{C})$, the topology of pointwise convergence in $\mathcal{C}^0(\Omega, \mathbf{C})$ (respectively, $\mathcal{C}^b(\Omega, \mathbf{C})$) is called the topology of weak convergence (respectively, of narrow convergence). Thus a filter basis $\mathcal{B}$ on $\mathcal{M}^1(\Omega, \mathbf{C})$ converges weakly (respectively, narrowly) to a bounded Radon measure μ if $\nu(f)$ converges to $\mu(f)$ along $\mathcal{B}$, for every f of $\mathcal{C}^0(\Omega, \mathbf{C})$ (respectively, of $\mathcal{C}^b(\Omega, \mathbf{C})$).

Proposition 7.7.3 $P_a = \{ \nu \in \mathcal{M}^1(\Omega, \mathbf{C}) : \|\nu\| \leq a \}$ is weakly compact, for every $a > 0$. Let $\mathcal{B}$ be a filter basis on P_a, and let T be a subset of $\mathcal{C}^0(\Omega, \mathbf{C})$ such that the vector space V generated by T is dense in $\mathcal{C}^0(\Omega, \mathbf{C})$. Then a necessary and sufficient condition that $\mathcal{B}$ converge weakly is that, for all $f \in T$, $\int f \, d\nu$ have a limit in $\mathbf{C}$ along $\mathcal{B}$.

PROOF: P_1 is the closed unit ball with center 0 in the topological dual of $\mathcal{C}^0(\Omega, \mathbf{C})$. Thus it is weakly compact, by the Banach–Aloaglu theorem, and $P_a = aP_1$ is weakly compact as well. Now, for all $f \in V$, $\int f \, d\nu$ has a limit $u(f)$ along $\mathcal{B}$, and $|u(f)| \leq a\|f\|$. Therefore, u has a unique linear continuous extension μ to $\mathcal{C}^0(\Omega, \mathbf{C})$. Fix $h \in \mathcal{C}^0(\Omega, \mathbf{C})$. Given $\varepsilon > 0$, choose f in V so that $\|h - f\| \leq \varepsilon/(3a)$. Then

$$
\begin{aligned}
|\nu(h) - \mu(h)| &\leq |\nu(h - f)| + |\nu(f) - \mu(f)| + |\mu(h - f)| \\
&\leq 2\varepsilon/3 + |\nu(f) - \mu(f)|
\end{aligned}
$$

for all $\nu \in P_a$, and hence $|\nu(h) - \mu(h)| \leq \varepsilon$ when ν lies in a suitable $A \in \mathcal{B}$, which proves that $\mathcal{B}$ converges weakly to μ. $\square$

In particular, a filter basis on P_a converges weakly if and only if it converges vaguely.

Proposition 7.7.4   *A necessary and sufficient condition that a sequence $(\mu_n)_{n\geq 1}$ in $\mathcal{M}^1(\Omega, \mathbf{C})$ converge weakly is that it converge vaguely and that $(\|\mu_n\|)_{n\geq 1}$ be bounded. Any weak Cauchy sequence converges weakly.*

PROOF: If $(\mu_n)_{n\geq 1}$ converges vaguely and $(\|\mu_n\|)_{n\geq 1}$ is bounded, then $(\mu_n)_{n\geq 1}$ converges weakly by Proposition 7.7.3.

Conversely, if $(\mu_n)_{n\geq 1}$ is a weak Cauchy sequence, then $\sup \|\mu_n\| < +\infty$ by the Banach–Steinhaus theorem, and $(\mu_n(f))_{n\geq 1}$ has a limit for every f in $\mathcal{C}^0(\Omega, \mathbf{C})$. So $(\mu_n)_{n\geq 1}$ converges weakly.   □

Proposition 7.7.5 *Suppose the topology of Ω has a countable basis. Then every bounded sequence $(\mu_n)_{n\geq 1}$ in $\mathcal{M}^1(\Omega, \mathbf{C})$ contains a weakly convergent subsequence.*

PROOF: For all $a > 0$, $P_a = \{\mu \in \mathcal{M}^1(\Omega, \mathbf{C}) : \|\mu\| \leq a\}$ is compact and metrizable when it is endowed with the topology of weak convergence. The proposition follows.

Alternatively, we can give a proof that does not use the weak compactness of P_a:

Let Ω' be the compact space obtained by adding to Ω a point ∞ (Alexandrov's compactification). Ω', and hence $\mathcal{C}(\Omega', \mathbf{C})$, is second countable. But $\mathcal{C}^0(\Omega, \mathbf{C})$ may be identified with the subspace of $\mathcal{C}(\Omega', \mathbf{C})$ consisting of the continuous functions which vanish at ∞. Therefore, $\mathcal{C}^0(\Omega, \mathbf{C})$ is separable. Now let $(f_n)_{n\geq 1}$ be a sequence in $\mathcal{H}(\Omega, \mathbf{C})$ such that $H = \{f_n : n \geq 1\}$ is dense in $\mathcal{C}^0(\Omega, \mathbf{C})$. By induction, we construct a sequence $(j_n)_{n\geq 1}$ of strictly increasing functions from $\mathbf{N}$ into $\mathbf{N}$ such that the sequence $(\mu_{(j_k \circ \ldots \circ j_1)(n)}(f_k))_{n\geq 1}$ converges in $\mathbf{C}$ for every $k \geq 1$. Put $\nu_n = \mu_{(j_n \circ \ldots \circ j_1)(n)}$ for all $n \geq 1$. Then the sequence $(\nu_n(f))_{n\geq 1}$ converges, for every $f \in H$, so $(\nu_n)_{n\geq 1}$ converges weakly (Proposition 7.7.3).   □

Now, narrow convergence in the set of positive bounded measures is studied.

Theorem 7.7.2 *Let $\mathcal{B}$ be a filter basis on $\mathcal{M}^1_+(\Omega)$ converging vaguely to an element μ of $\mathcal{M}^1_+(\Omega)$ such that $\|\nu\|$ tends to $\|\mu\|$ along $\mathcal{B}$. Let f be a bounded mapping from Ω into a real Banach space F, and assume that*

(a) the set of discontinuities of f is μ-negligible;

(b) f is ν-measurable for every $A \in \mathcal{B}$ and for all $\nu \in A$.

Then $\int f d\nu$ converges to $\int f d\mu$ along $\mathcal{B}$.

PROOF: Let $\varepsilon > 0$ be a real number. By the same argument as that in Theorem 7.6.1, there exist elements $a_1, \ldots, a_n$ of F, functions $g_1, \ldots, g_n$ in $\mathcal{H}(\Omega, \mathbf{R})$, and a bounded continuous positive function h on Ω such that

$|f - (g_1 a_1 + \ldots + g_n a_n)| \leq h \leq 2\|f\|$ and $\int h d\mu \leq \varepsilon$. Put $M = \|f\|$. There exists a compact set K such that $\mu(\Omega - K) \leq \varepsilon$. Let K' be a compact neighborhood of K in Ω, and let $-h_1$ be a continuous function from Ω into $[0, 2M]$ which agrees with $2M - h$ on K' and vanishes outside K'. Then $h' = 2M + h_1$ is continuous, $h' = h$ on K, and $h' = 2M$ outside K'. Replacing h' by $\sup(h, h')$, we may assume that $h' \geq h$. In this case.

$$0 \leq \int (h' - h) d\mu \leq 2M\mu(\Omega - K) \leq 2M\varepsilon.$$

Now $\int h' d\nu = \int h_1 d\nu + 2M\|\nu\|$ converges to $\int h_1 d\mu + 2M\|\mu\| = \int h' d\mu$ along $\mathcal{B}$. Hence there exists $A \in \mathcal{B}$ such that

$$\left| \int (g_1 a_1 + \ldots + g_n a_n) d\nu - \int (g_1 a_1 + \ldots + g_n a_n) d\mu \right| \leq \varepsilon$$

and

$$\int h d\nu \leq \int h' d\nu \leq \int h' d\mu + \varepsilon \leq \int h d\mu + 2M\varepsilon + \varepsilon \leq 2(M+1)\varepsilon$$

for all $\nu \in A$. These inequalities imply that

$$\left| \int f d\nu - \int f d\mu \right| \leq \int h d\nu$$

$$+ \left| \int (g_1 a_1 + \cdots + g_n a_n) d\nu - \int (g_1 a_1 + \cdots + g_n a_n) d\mu \right| + \int h d\mu$$

is less than $2(M+2)\varepsilon$, which proves the theorem. □

By Theorem 7.7.2, a filter basis on $\mathcal{M}_+^1(\Omega)$ converges narrowly to an element μ of $\mathcal{M}_+^1(\Omega)$ if and only if it converges vaguely to μ and $\|\nu\|$ converges to $\|\mu\|$ along $\mathcal{B}$.

Observe that, if we take for Ω the interval $[0,1]$, for μ_n the Radon measure on Ω defined by the mass $1/n$ at each point k/n $(1 \leq k \leq n)$, and for μ Lebesgue measure on Ω, then $(\mu_n)_{n \geq 1}$ converges narrowly to μ, but $\mu_n(A) = 1$ does not tend to $\mu(A) = 0$ if A is the set of rationals in Ω. Therefore, we see that the condition that f have a μ-negligible set of points of discontinuity in Theorem 7.7.2 is not superfluous.

Proposition 7.7.6 *Let $(\mu_n)_{n \geq 1}$ be a sequence in $\mathcal{M}_+^1(\Omega)$ and μ an element of $\mathcal{M}_+^1(\Omega)$. Then $(\mu_n)_{n \geq 1}$ converges narrowly to μ if and only if $\|\mu_n\|$ tends to $\|\mu\|$ and $\mu(F) \geq \limsup \mu_n(F)$ for every μ-quadrable closed set F (i.e., every closed set with μ-negligible boundary). In this case, the inequality $\mu(F) \geq \limsup \mu_n(F)$ holds for every closed set F.*

PROOF: First, suppose that $(\mu_n)_{n \geq 1}$ converges narrowly to μ. Let U be an open subset of Ω. For every $f \in \mathcal{H}^+$ such that $f \leq 1_U$, the sequence $(\mu_n(f))_{n \geq 1}$ converges to $\mu(f)$; thus

$$\mu(f) = \liminf \mu_n(f) \leq \liminf \mu_n(U),$$

and $\mu(U) \leq \liminf \mu_n(U)$. Now, if F is a closed subset of Ω, we must have

$$\mu(F) = \|\mu\| - \mu(\Omega - F) \geq \lim \|\mu_n\| - \liminf \mu_n(\Omega - F),$$

that is, $\mu(F) \geq \limsup \mu_n(F)$.

Conversely, suppose that $\|\mu_n\|$ converges to $\|\mu\|$ as $n \to +\infty$, and that $\mu(F) \geq \limsup \mu_n(F)$ for all μ-quadrable closed sets. Fix a bounded continuous function $f \geq 0$ on Ω. By Proposition 3.6.5, $\int f d\mu = \int \mu(f^{-1}]t, +\infty[) d\lambda(t)$, where λ is Lebesgue measure on $]0, +\infty[$. The family of the $t\mu(f^{-1}(t))$ $(t \in]0, +\infty[)$ is summable and its sum is smaller than $\int f d\mu$. Hence there exists a countable subset D of $]0, +\infty[$ such that $\mu(f^{-1}(t)) = 0$ for all t in $]0, +\infty[-D$, and for such t, $f^{-1}([t, +\infty[)$ is a μ-quadrable closed set. Then

$$\mu\left(f^{-1}([t, +\infty[)\right) \geq \limsup \mu_n\left(f^{-1}([t, +\infty[)\right)$$

for λ-almost all t, and $\int f d\mu \geq \limsup \int f d\mu_n$ by Proposition 3.2.1. Now, replacing f by $\|f\| - f$, we obtain $\int f d\mu \leq \liminf \int f d\mu_n$, so that $(\mu_n(f))_{n \geq 1}$ converges to $\mu(f)$. $\qquad\square$

7.8   Tight Sequences

To study narrowly convergent sequences in $\mathcal{M}^1(\Omega, \mathbf{C})$, it is useful to introduce the notion of tight set.

Definition 7.8.1 A subset P of $\mathcal{M}^1(\Omega, \mathbf{C})$ is said to be tight whenever, for every $\varepsilon > 0$, there exists a compact set K such that $V\mu(\Omega - K) \leq \varepsilon$ for all $\mu \in P$.

Let $(\mu_n)_{n \geq 1}$ be a sequence in $\mathcal{M}^1(\Omega, \mathbf{C})$, and suppose that $\{\mu_n : n \geq 1\}$ is not tight. If $\varepsilon > 0$ is suitably chosen, for each compact set K we can find $n \geq 1$ such that $V\mu_n(\Omega - K) > \varepsilon$. If $n_0 \geq 1$ is an integer and K a compact set, then we must not have $V\mu_n(\Omega - K) \leq \varepsilon$ for all $n \geq n_0$; otherwise, taking L a compact set such that $K \subset L$ and $V\mu_n(\Omega - L) \leq \varepsilon$ for every $1 \leq n < n_0$, we have $V\mu_n(\Omega - L) \leq \varepsilon$ for all $n \geq 1$, contradicting the choice of ε. Therefore, for each integer $n_0 \geq 1$ and each compact set K, there exists an $n \geq n_0$ such that $V\mu_n(\Omega - K) > \varepsilon$.

Proposition 7.8.1   *Let $(\mu_n)_{n\geq 1}$ be a sequence in $\mathcal{M}^1(\Omega, \mathbf{C})$ converging vaguely to a Radon measure μ, and suppose that $\{\mu_n : n \geq 1\}$ is tight. Then μ is bounded and $(\mu_n)_{n\geq 1}$ converges narrowly to μ.*

PROOF: For every $\varepsilon > 0$, there exists a compact set K such that $V\mu_n(\Omega - K) \leq \varepsilon$ for all $n \geq 1$. If $f \in \mathcal{H}^+$ is such that $f \leq 1_{\Omega - K}$ and $\mathrm{supp}(f) \subset \Omega - K$, then $|\int g d\mu| = |\lim_{n\to +\infty} \int g d\mu_n| \leq \varepsilon$ for each $g \in \mathcal{H}(\Omega, \mathbf{C})$ satisfying $|g| \leq f$; therefore, $\int f dV\mu \leq \varepsilon$. This shows that $V\mu^*(\Omega - K) \leq \varepsilon$. Now fix $h \in \mathcal{C}^b(\Omega, \mathbf{C})$ and $\varepsilon > 0$, and let K be a compact set such that $V\mu_n(\Omega - K) \leq \varepsilon$ for all $n \geq 1$. If $\varphi \in \mathcal{H}^+$ takes its values only in $[0, 1]$ and is equal to 1 on K, then

$$\int |h - h\varphi| dV\mu_n \leq \|h\| \cdot V\mu_n(\Omega - K) \leq \varepsilon\|h\|$$

and

$$\int |h - h\varphi| dV\mu \leq \varepsilon\|h\|.$$

So

$$\left| \int h d\mu_n - \int h d\mu \right| \leq 2\varepsilon\|h\| + \left| \int h\varphi d\mu_n - \int h\varphi d\mu \right|,$$

and

$$\left| \int h d\mu_n - \int h d\mu \right| \leq 3\varepsilon\|h\|$$

for n large enough. ☐

Proposition 7.8.2 *Suppose that Ω is a countable union of compact sets, and let $(\mu_n)_{n\geq 1}$ be a sequence in $\mathcal{M}^1(\Omega, \mathbf{C})$. If, for each subsequence of $(\mu_n)_{n\geq 1}$, there is a further subsequence which converges narrowly, then $\{\mu_n : n \geq 1\}$ is tight.*

PROOF: Suppose that $\{\mu_n : n \geq 1\}$ is not tight. Choose $\varepsilon > 0$ so that, for each compact set K, there is an $n \geq 1$ for which $V\mu_n(\Omega - K) > \varepsilon$. Let $(V_k)_{k\geq 1}$ be an increasing sequence of open sets with compact closures such that $\Omega = \bigcup_{k\geq 1} V_k$. In view of the discussion immediately following Definition 7.8.1, we can construct a strictly increasing sequence $(n_k)_{k\geq 1}$ of integers ≥ 1 such that $V\mu_{n_k}(\Omega - \bar{V}_k) > \varepsilon$ for all $k \geq 1$. Now let $i \mapsto k(i)$ be a strictly increasing function from $\mathbf{N}$ into $\mathbf{N}$. If L is a compact set, it is included in $V_{k(j)}$ for some $j \geq 1$, and $V\mu_{n_{k(j)}}(\Omega - L)$ strictly exceeds ε. Thus $\{\mu_{n_{k(i)}} : i \geq 1\}$ is not tight.

Therefore, if from each subsequence of $(\mu_n)_{n\geq 1}$ we can extract a sequence which is tight, then $\{\mu_n : n \geq 1\}$ is tight. It remains to be shown that a sequence which converges narrowly in $\mathcal{M}^1(\Omega, \mathbf{C})$ is tight. In other words, we suppose that the sequence $(\mu_n)_{n\geq 1}$ converges narrowly, and we have to prove that $\{\mu_n : n \geq 1\}$ is tight.

Write μ for the narrow limit of $(\mu_n)_{n\geq 1}$. Since $\{\mu_n : n \geq 1\}$ is tight when $\{\mu_n - \mu : n \geq 1\}$ is tight, and since $(\mu_n - \mu)_{n\geq 1}$ converges narrowly to 0, we need only consider the case where $\mu = 0$. Now argue by contradiction: suppose that $(\mu_n)_{n \geq 1}$ converges narrowly to 0 and $\{\mu_n : n \geq 1\}$ is not tight. Let $\varepsilon > 0$ and $(V_k)_{k \geq 1}$ be as at the beginning of the proof.

For a fixed integer $k \geq 1$, suppose we have obtained compact sets $L_1, \ldots, L_k$, functions $g_1, \ldots, g_{k-1}$ in $\mathcal{H}(\Omega, \mathbf{C})$, and integers $n_1, \ldots, n_k$, so that

(a) $1 \leq n_1 < \ldots < n_k$;

(b) $L_i^\circ = \text{int}(L_i)$ contains V_i, for every $1 \leq i \leq k$;

(c) $L_{i+1}^\circ \supset L_i$ for every $1 \leq i < k$;

(d) $V\mu_{n_i}(\Omega - L_i) > \varepsilon$ for every $1 \leq i \leq k$;

(e) $V\mu_{n_i}(\Omega - L_j) \leq \varepsilon/3$ for every (i,j) with $i < j \leq k$;

(f) $|g_i| \leq 1$ and $\text{supp}(g_i) \subset L_{i+1}^\circ - L_i$ for every $1 \leq i < k$;

(g) $|\mu_{n_i}(g_i)| \geq \varepsilon$ for every $1 \leq i < k$;

(h) $|\mu_{n_j}(g_i)| \leq (1/2^i)(\varepsilon/3)$ for every (i,j) with $i < j \leq k$.

Since $V\mu_{n_k}(\Omega - L_k) > \varepsilon$, there exists a compact set L_{k+1} such that $L_{k+1}^\circ \supset L_k \cup V_{k+1}$ and $V\mu_{n_k}(L_{k+1}^\circ - L_k) > \varepsilon$. Choosing L_{k+1} large enough, we may assume that $V\mu_{n_k}(\Omega - L_{k+1}) \leq \varepsilon/3$. Then $V\mu_{n_i}(\Omega - L_j) \leq \varepsilon/3$ for all i, j satisfying $i < j \leq k+1$. Let g_k be an element of $\mathcal{H}(\Omega, \mathbf{C})$ such that $|g_k| \leq 1$, $\text{supp}(g_k) \subset L_{k+1}^\circ - L_k$, and $|\mu_{n_k}(g_k)| \geq \varepsilon$. Finally, choose $n_{k+1} > n_k$ so that $|\mu_{n_{k+1}}(g_i)| \leq (1/2^i)(\varepsilon/3)$ for every $i \leq k$ and $V\mu_{n_{k+1}}(\Omega - L_{k+1}) > \varepsilon$ (which is possible by the remark following Definition 7.8.1). Then the construction can continue by induction.

The sets $L_{i+1}^\circ - L_i$ are disjoint. Hence the $\text{supp}(g_i)$ are disjoint, and $|g| \leq 1$ if $g = \sum_{i \geq 1} g_i$. Moreover, $g = g_1 + \ldots + g_{k-1}$ on L_k°, for each $k \geq 1$, so g is continuous on $\Omega = \bigcup_{k \geq 1} L_k^\circ$. Therefore, $\mu_{n_k}(g)$ converges to 0 as $k \to +\infty$. But

$$\left| \mu_{n_k}\left(\sum_{1 \leq i < k} g_i \right) \right| \leq \sum_{1 \leq i < k} |\mu_{n_k}(g_i)| \leq \frac{\varepsilon}{3} \sum_{1 \leq i < k} \frac{1}{2^i} \leq \frac{\varepsilon}{3}$$

and $\left| \mu_{n_k}\left(\sum_{i > k} g_i \right) \right| \leq V\mu_{n_k}(\Omega - L_{k+1}) \leq \varepsilon/3$, whereas $|\mu_{n_k}(g_k)| \geq \varepsilon$. Finally, $|\mu_{n_k}(g)| \geq \varepsilon/3$, whence the desired contradiction. $\square$

Proposition 7.8.3 *Suppose that Ω is a countable union of compact sets. Then every narrow Cauchy sequence $(\mu_n)_{n \geq 1}$ converges narrowly.*

PROOF: Let μ be the weak limit of $(\mu_n)_{n \geq 1}$. Then $(\mu_n - \mu)_{n \geq 1}$ is a narrow Cauchy sequence. Thus it suffices to prove the proposition when $(\mu_n)_{n \geq 1}$ converges weakly to 0.

Suppose that $\{\mu_n : n \geq 1\}$ is not tight and retain the notation of Proposition 7.8.2. Multiplying g_i by a complex number ζ_i with modulus 1, we may assume that $\mu_{n_i}(g_i) = |\mu_{n_i}(g_i)|$ if i is even and that $\mu_{n_i}(g_i) = -|\mu_{n_i}(g_i)|$ if i is odd. Then $\mathrm{Re}\,\mu_{n_k}(g) \geq \varepsilon/3$ if k is even, and $\mathrm{Re}\,\mu_{n_k}(g) \leq -\varepsilon/3$ if k is odd. But now the sequence $(\mu_{n_k}(g))_{k\geq 1}$ cannot be a Cauchy sequence and we arrive at a contradiction, which proves that $\{\mu_n : n \geq 1\}$ is tight and that $(\mu_n)_{n\geq 1}$ converges narrowly to 0. $\square$

Exercises for Chapter 7

1 Let Ω be a topological space and $f : \Omega \to \overline{\mathbf{R}}$ a function. We define the oscillation of f at a point $a \in \Omega$ as the number in $\overline{\mathbf{R}}$

$$\omega(a; f) \equiv \limsup_{x\to a} f(x) - \liminf_{x\to a} f(x),$$

whenever the right-hand side is defined (i.e., when $\limsup_{x\to a} f(x)$ and $\liminf_{x\to a} f(x)$ are neither both equal to $+\infty$, nor both equal to $-\infty$).

1. Show that the function $x \mapsto \omega(x; f)$ is upper semicontinuous on its domain A.

2. If f is finite on Ω, show that $A = \Omega$ and that, for every $a \in \Omega$,

$$\omega(a, f) = \limsup_{(x,y)\to(a,a)} f(x) - f(y).$$

(Let $r < \omega(a; f)$ be a real number; show that $r \leq \sup_{(x,y)\in V\times V} f(x) - f(y)$ for every neighborhood V of a. Conversely, let $r > \omega(a; f)$; prove that $\sup_{(x,y)\in U\times U} f(x) - f(y) \leq r$ for a suitable neighborhood U of a).

3. Let $f : \Omega \to \mathbf{R}$ be a lower semicontinuous function. Show that, if $\omega(a; f)$ is finite at some $a \in \Omega$, then $\liminf_{x\to a} \omega(x; f) = 0$. For this, argue by contradiction, showing that, under the contrary hypothesis, there exist points x arbitrarily close to a such that $f(x)$ is as large as desired.

4. For every rational number $r = p/q$ in lowest terms (with $q > 0$, and $q = 1$ if $r = 0$), put $f(r) = q$. Show that f is lower semicontinuous on $\mathbf{Q}$, but that $\omega(a; f) = +\infty$ for all $a \in \mathbf{Q}$.

2 Let Ω be a locally compact Hausdorff space and f a lower semicontinuous function on Ω.

1. If $\limsup_{x\to a} f(x) = +\infty$ for all $a \in \Omega$, show that the set $f^{-1}(+\infty)$ is dense in Ω (let V be an open subset of Ω; construct a sequence $(U_n)_{n\geq 1}$ of open subsets of V such that each U_n has compact closure contained in V, $U_n \supset \overline{U}_{n+1}$ for every $n \geq 1$ and f is strictly greater than n on U_n).

2. If f takes only finite values, show that the set B, of points x for which $\omega(x; f)$ is finite, is dense in Ω (use part 1).

3. Show that the set of points of continuity of f is dense in Ω (there is no loss of generality in assuming that f is bounded, for we may replace f by $f/(1 + |f|)$; then, observe that the function $x \mapsto 1/\omega(x; f)$ is lower semicontinuous, and use part 1 and Exercise 1, part 3).

3 Let Ω be the subset of the plane $\mathbf{R}^2$ whose elements are the points of the line $D = \{0\} \times \mathbf{R}$ and the points $(1/n, k/n^2)$, where n ranges over the set $\mathbf{N}$ of strictly positive integers and k ranges over the set $\mathbf{Z}$ of positive or negative integers.

   1. For every point $(0, y)$ in D and every integer $n > 0$, let $T_n(y)$ be the set of those points $(u, v) \in \Omega$ such that $u \leq 1/n$ and $|v - y| \leq u$. Take, as basis of the filter of neighborhoods of each point $(0, y) \in D$, the class of the $T_n(y)$, and, as neighborhoods of each of the other points (u, v) in Ω, the subsets of Ω containing (u, v), and prove that this construction actually defines a Hausdorff topology T on Ω. Moreover, prove that, in this topology T, each of the subsets $T_n(y)$ is compact and metrizable.

   2. Deduce that Ω, endowed with the topology T, is locally compact. Prove that the topology induced by T on D is discrete.

   3. Let A be the set $\{0\} \times \mathbf{Q}$, closed in Ω, whose complement B in D is also closed. Given an open subset U of Ω containing B, for every $p \in \mathbf{N}$ denote by D_p the set of those $z \in \mathbf{R} - \mathbf{Q}$ such that $T_p(z) \subset U$. Show that D_p is closed in $\mathbf{R} - \mathbf{Q}$ (with the usual topology on $\mathbf{R}$).

   4. Deduce that there exists an integer $p \geq 1$ and real numbers a, b, with $a < b$, such that the closure $\bar{D}_p$ of D_p in $\mathbf{R}$ (with the usual topology on $\mathbf{R}$) contains $[a, b]$. To this end, observe that $\mathbf{R} - \mathbf{Q} = \bigcup_{p \geq 1} D_p$, and therefore that at least one $\bar{D}_p$ has an interior point.

   5. Conclude that each neighborhood of A in Ω meets U.

   6. Deduce from part 5 that two open subsets of Ω containing the disjoint closed sets A and B, respectively, cannot themselves be disjoint.

4 In the notation of Exercise 3, let α be the function $(1/n, k/n^2) \mapsto 1/n^3$ from $\Omega - D$ into $[0, +\infty[$.

   1. For $y \in \mathbf{R}$, show that $\sum_{\omega \in T_1(y)} \alpha(\omega)$ is smaller than $\sum_{n \geq 1} (2n+1)/n^3$. Deduce that, for every compact set K, the sum $\sum_{\omega \in K} \alpha(\omega)$ is finite.

   2. Let μ be the Radon measure on Ω defined by the point masses $\alpha(\omega)$ (ω in $\Omega - D$). Let U be an open subset of Ω containing D. Recall that we can find an integer $p \geq 1$ and a compact interval $[a, b]$ in $\mathbf{R}$ ($a < b$) such that U contains $T_p(y)$ for all y of $[a, b] \cap (\mathbf{R} - \mathbf{Q})$. Deduce that $\mu^*(U) = +\infty$.

   3. Conclude that the closed subset D of Ω is locally μ-negligible, but that it is not μ-negligible.

5 For every integer $n \geq 1$ and every complex-valued function f on $I = [0, 1]$, the Bernstein polynomial of degree n associated with f is

$$B_{n,f}(x) = \sum_{0 \leq p \leq n} \binom{n}{p} x^p (1 - x)^{n-p} f\left(\frac{p}{n}\right).$$

If $x \in I$ and $n \geq 1$ are given and if h is a function from $\{0, 1, \ldots, n\}$ into $\mathbf{C}$, we denote by $\int h\,d\mu$, or $\int h(p)\,d\mu(p)$, the number $\sum_{0 \leq p \leq n} \binom{n}{p} x^p (1 - x)^{n-p} h(p)$

(integral of h relative to the measure μ on $\{0, 1, \ldots, n\}$ defined by the mass $\binom{n}{p} x^p (1 - x)^{n-p}$ at each point p). For every subset A of $\{0, 1, \ldots, n\}$, we put $\mu(A) = \int 1_A d\mu$.

1. If g is the function $p \mapsto p/n - x$ from $\{0, 1, \ldots, n\}$ into $\mathbf{R}$, show that $\int g d\mu = 0$ and that $\int g^2 d\mu = x(1-x)/n$.

2. For $\alpha > 0$, put $A = \{0 \leq p \leq n : |g(p)| > \alpha\}$, and show that $\mu(A) \leq (1/\alpha^2) \int g^2 d\mu$. Conclude that $\mu(A) \leq 1/(4n\alpha^2)$.

3. Let $f : I \to \mathbf{C}$ be continuous. For every $\varepsilon > 0$, there exists $\alpha > 0$ such that $|f(x) - f(y)| \leq \varepsilon$ for all x, $y \in I$ satisfying $|x - y| \leq \alpha$. Deduce from part 2 that, for every integer n large enough, we have $|f(x) - B_{n,f}(x)| \leq (1 + 2\|f\|)\varepsilon$ for all $x \in I$, where $\|f\| = \sup_{y \in I} |f(y)|$. In other words, the sequence $(B_{n,f})_{n \geq 1}$ converges uniformly to f on $[0, 1]$.

4. Let $f : I \to \mathbf{C}$ be continuously differentiable and let f' be its derivative. Show that $\|f - B_{n,f}\| \leq (\|f\|/2 + \|f'\|) n^{-1/3}$ (take $\alpha = n^{-1/3}$ in part 2).

6 1. For every integer $k \geq 0$, denote by f_k the function $x \to x^k$ from $I = [0, 1]$ into $\mathbf{R}$. Show that $B_{n,f_0}(x) = 1$ and $B_{n,f_1}(x) = x$ for all $n \geq 1$ and all $x \in \mathbf{R}$.

2. Show that, for every $k \geq 1$ and every $n \geq 2$,

$$B_{n,f_k}(x) = x \left[\left(\frac{n-1}{n} \right)^{k-1} B_{n-1,f_{k-1}}(x) + \frac{1}{n} B'_{n,f_{k-1}}(x) \right].$$

3. For every $n \geq 1$, put $a_{0,n} = a_{1,n} = 1$ and $P_{0,n} = P_{1,n} =$ the polynomial identically equal to 0. Finally, for every $k \geq 2$, put $a_{k,1} = 0$ and $P_{k,1}(x) = x$. Then show that, for every $k \geq 0$ and for every $n \geq 1$, $B_{n,f_k}(x)$ has the form $a_{k,n} x^k + P_{k,n}(x)$, where $P_{k,n}$ is a polynomial of degree at most $k - 1$. Moreover, prove that $a_{k,n} = ((n-1)/n)^{k-1} a_{k-1,n-1}$ for all $k \geq 1$ and all $n \geq 2$.

4. With notation as in part 3, show that

$$a_{k,n} = \left(1 - \frac{k-1}{n} \right) \left(1 - \frac{k-2}{n} \right) \cdots \left(1 - \frac{0}{n} \right)$$

for every $k \geq 0$ and every $n \geq 1$.

5. By induction on $\ell \geq 0$, prove the existence of a number $c_\ell \geq 0$ such that, for every $n \geq 1$, the coefficients of $P_{\ell,n}$ are (in absolute value) smaller than c_ℓ/n.

7 Let $(c_n)_{n \geq 0}$ be a sequence of complex numbers. For all integers $k \geq 0$, we put $\Delta^k c_n = \sum_{0 \leq j \leq k} (-1)^j \binom{k}{j} c_{n+j}$. For each $n \geq 0$, define $f_n : [0, 1] \to \mathbf{R}$ by $f_n(x) = x^n$. In this exercise, we show that there exists a Radon measure μ on $I = [0, 1]$ such that $\mu(f_n) = c_n$ for all $n \geq 0$ if and only if there exists a number $A \geq 0$ such that $\sum_{0 \leq p \leq n} \binom{n}{p} |\Delta^{n-p} c_p| \leq A$ for every integer $n \geq 0$ ("Hausdorff's moment problem").

1. Show that the condition is necessary (observe that, for all $n \geq 0$ and all $0 \leq p \leq n$, $\Delta^{n-p}c_p$ must be the value of μ on the polynomial $x^p(1-x)^{n-p}$).

2. Conversely, assume the condition is satisfied. Let V be the vector subspace of $\mathcal{C}(I, \mathbf{C})$ generated by the $f_k(k \geq 0)$, and let v be the linear form on V which sends f_k to c_k for every $k \geq 0$. Using part 5 of Exercise 6, prove that, for every $k \geq 0$, $v(B_{n,f_k})$ converges to c_k as $n \to +\infty$.

3. Now, for each $n \geq 1$, let u_n be the linear form

$$f \mapsto \sum_{0 \leq p \leq n} \binom{n}{p} . \Delta^{n-p} c_p . f\left(\frac{p}{n}\right)$$

on $\mathcal{C}(I, \mathbf{C})$. Show that u_n is continuous and that, for every $k \geq 0$, the sequence $\big(u_n(f_k)\big)_{n \geq 1}$ converges to c_k. Deduce that u_n converges weakly to a Radon measure μ such that $\mu(f_k) = c_k$ for every $k \geq 0$.

4. Show there exists a positive Radon measure μ on I such that $\mu(f_n) = c_n$ for every $n \geq 0$ if and only if $\Delta^k c_n$ is positive for all integers $k \geq 0$ and $n \geq 0$.

8 Let Ω be a locally compact Hausdorff space, E a vector subspace of $\mathcal{C}(\Omega, \mathbf{R})$, and P a convex cone in $\mathcal{C}(\Omega, \mathbf{R})$ (i.e., a subset of $\mathcal{C}(\Omega, \mathbf{R})$ such that $f + g$ and ah belong to P for all f, g, $h \in P$ and all strictly positive real numbers a). Suppose that, for each $h \in \mathcal{H}(\Omega, \mathbf{R})$, there exists $f \in E$ such that $f - h$ belongs to P. Let u be a real linear form on E such that $u(f) \geq 0$ for all $f \in E \cap P$. Let $h \in \mathcal{H}(\Omega, \mathbf{R})$, P_h' the set of those $f \in E$ for which $h - f$ belongs to P, and P_h'' the set of those $f \in E$ for which $f - h$ belongs to P.

1. Show that P_h' is nonempty.

2. If α' is the supremum of the $u(f)$ for $f \in P_h'$, and α'' the infimum of the $u(f)$ for $f \in P_h''$, show that α', α'' are finite and that $\alpha' \leq \alpha''$.

3. Let u_1 be a linear form on $E_1 = E + \mathbf{R}h$ extending u. Show that $u_1(f_1)$ is positive for all $f_1 \in E_1 \cap P$ if and only if $u_1(h)$ lies in the interval $[\alpha', \alpha'']$.

9 Let A be a closed subset of $\mathbf{R}$ and $(c_n)_{n \geq 0}$ a sequence in $\mathbf{R}$. We prove in stages what follows: there exists a positive Radon measure μ on $\mathbf{R}$ with support contained in A, such that x^n is μ-integrable and $\int x^n d\mu(x) = c_n$ for all integers $n \geq 0$, if and only if $\sum_{0 \leq k \leq n} a_k c_k$ is positive for every polynomial $P(X) = \sum_{0 \leq k \leq n} a_k X^k$ in $\mathbf{R}[X]$ such that $P(x) \geq 0$ on A.

The condition is clearly necessary. Assume therefore, in what follows, that it is satisfied.

1. Let E be the vector subspace of $\mathcal{C}(\mathbf{R}, \mathbf{R})$ generated by the functions x^k ($k \geq 0$), v the linear form on E which sends each x^k to c_k, and P the cone in $\mathcal{C}(\mathbf{R}, \mathbf{R})$ consisting of those continuous functions which are positive on A. Prove that there is a linear form u on $E + \mathcal{H}(\mathbf{R}, \mathbf{R})$ extending v such that $u(f)$ is positive for all f of $P \cap (E + \mathcal{H}(\mathbf{R}, \mathbf{R}))$. For this, consider a basis $(h_i)_{i \in I}$ of $\mathcal{H}(\mathbf{R}, \mathbf{R})$, endow I with an order for which every nonempty

subset of I has a smallest element, and construct u by transfinite induction (use Exercise 8).

2. Show that $u(h) \geq 0$ for all $h \in \mathcal{H}^+(\mathbf{R})$. Let μ be the Radon measure on $\mathbf{R}$ such that $\mu(h) = u(h)$ for all $h \in \mathcal{H}(\mathbf{R}, \mathbf{R})$. Prove that A contains $\operatorname{supp}(\mu)$ (observe that $u(h) = 0$ for every $h \in \mathcal{H}(\mathbf{R}, \mathbf{R})$ whose support is disjoint from A).

3. Show that $\int^* x^{2n} d\mu(x)$ is smaller than c_{2n} for all integers $n \geq 0$.

4. Fix an integer $n \geq 0$. For given $\varepsilon > 0$, let $r > 0$ be a real number such that $1 \leq \varepsilon r^{n+2}$, and let $g : \mathbf{R} \to [0, 1]$ be a continuous function, with compact support, equal to 1 on $[-r, r]$. From the relation $|x^n(1 - g)| \leq \varepsilon x^{2n+2}$, deduce that x^n is μ-integrable. Finally, show that $\int x^n d\mu(x) = c_n$.

Incidentally, observe that, if the above condition is satisfied, the moment problem considered may have several solutions when A is not compact (see Patrick Billingsley, *Probability and Measure*, Example 30.2, p. 407).

10   1.   Let E be the subset of $\mathcal{C}(\mathbf{R}, \mathbf{R})$ consisting of the polynomials $P_1^2 + P_2^2$, where P_1, P_2 belong to $\mathbf{R}[X]$. Show that E is stable with respect to the multiplication, and deduce that a polynomial $P \in \mathbf{R}[X]$ lies in E whenever $P(x)$ is positive for all $x \in \mathbf{R}$.

2. When $A = \mathbf{R}$, show that the condition of Exercise 9 holds if and only if the quadratic forms $Q_n : (x_i)_{0 \leq i \leq n} \mapsto \sum_{0 \leq j,k \leq n} c_{j+k} x_j x_k$ are positive for all $n \geq 0$ (use part 1). (This solves Hamburger's moment problem.)

3. Let $P \in \mathbf{R}[X]$. If $P(x)$ is positive for all $x \geq 0$, show that P can be written $P_1^2 + P_2^2 + X(P_3^2 + P_4^2)$, where the P_i lie in $\mathbf{R}[X]$.

4. When $A = [0, +\infty[$, show that the condition of Exercise 9 holds if and only if the quadratic forms $(x_i)_{0 \leq i \leq n} \mapsto \sum_{0 \leq j,k \leq n} c_{j+k+1} \cdot x_j x_k$ are positive for all $n \geq 0$ (use part 3). (This solves Stieltjes' moment problem.)

11   Let Ω be a locally compact Hausdorff space and μ a Radon measure on Ω.

1. Fix a real number $\delta > 0$ and a finite family $(f_i)_{i \in I}$ in $\mathcal{H}(\Omega, \mathbf{C})$, and let h be a continuous function from Ω into $[0, 1]$, which is equal to 1 on $K = \bigcup_{i \in I} \operatorname{supp}(f_i)$ and has compact support. Prove that there exist a finite family $(a_j)_{j \in J}$ of distinct points of K and a family $(g_j)_{j \in J}$ in $\mathcal{H}^+(\Omega)$ such that $\left| \mu(f_i) - \sum_{j \in J} \mu(g_j) f_i(a_j) \right| \leq \delta$ for all $i \in I$.

2. Deduce from part 1 that, when $\mathcal{M}(\Omega, \mathbf{C})$ is equipped with the vague topology, μ belongs to the closure of the space of Radon measures whose supports are finite and contained in $\operatorname{supp}(\mu)$.

3. When μ is bounded, let A be the convex set of Radon measures ν such that $\|\nu\| \leq \|\mu\|$ and whose supports are finite and contained in $\operatorname{supp}(\mu)$. Show that μ belongs to the closure of A in the vague topology.

12   Let Ω be a compact Hausdorff space.

1. If $(f_n)_{n\geq 1}$ converges (uniformly) to 0 in $\mathcal{C}(\Omega, \mathbf{C})$ and $(\mu_n)_{n\geq 1}$ converges vaguely to 0 in $\mathcal{M}(\Omega, \mathbf{C})$, show that the sequence $\big(\mu_n(f_n)\big)_{n\geq 1}$ converges to 0 (observe that $(\|\mu\|)_{n\geq 1}$ is bounded).

2. Assume that Ω is infinite. Let V be a neighborhood of 0 in the vague topology, defined by a finite number of inequalities $|\mu(f_i)| \leq 1$ (f_i in $\mathcal{C}(\Omega, \mathbf{C})$ for each $i \in I$). When f is not a linear combination of the f_i, show that $\{\mu(f) : f \in V\} = \mathbf{C}$ (use the Hahn–Banach theorem). Deduce that the mapping $(\mu, f) \mapsto \mu(f)$ from $\mathcal{M}(\Omega, \mathbf{C}) \times \mathcal{C}(\Omega, \mathbf{C})$ into $\mathbf{C}$ is not continuous, and hence that the vague topology on $\mathcal{M}(\Omega, \mathbf{C})$ is not metrizable.

3. Show that $\mathcal{M}^+(\Omega)$ is locally compact (use Ascoli's theorem).

13 Let Ω be a locally compact Hausdorff space, and equip $\mathcal{M}(\Omega, \mathbf{C})$ with the vague topology.

Assume that Ω is second countable. Let $(V_n)_{n\geq 1}$ be a sequence of open subsets of Ω, with compact closures, such that $\bar{V}_n \subset V_{n+1}$ for all $n \geq 1$ and $\Omega = \bigcup_{n\geq 1} V_n$. For every $n \geq 1$, let f_n be a continuous function from Ω into $[0,1]$, with compact support, equal to 1 on $\bar{V}_n$. Moreover, let $(f_{m,n})_{m\geq 1}$ be a dense sequence in $\mathcal{H}(\Omega, \bar{V}_n; \mathbf{C})$. Then the vague topology is finer than the topology T on $\mathcal{M}(\Omega, \mathbf{C})$ defined by the seminorms $\mu \mapsto |\mu(f_n)|$ and the seminorms $\mu \mapsto |\mu(f_{m,n})|$ (with $m \geq 1$, $n \geq 1$). Show that these two topologies induce the same topology on $\mathcal{M}^+(\Omega)$, and deduce that $\mathcal{M}^+(\Omega)$ is metrizable.

14 Let Ω be a locally compact Hausdorff space, whose topology has a countable basis. Let $(\mu_n)_{n\geq 1}$ be a sequence in $\mathcal{M}^1(\Omega, \mathbf{C})$ converging narrowly to $\mu \in \mathcal{M}^1(\Omega, \mathbf{C})$, and let H be an equicontinuous subset of $\mathcal{C}^b(\Omega, \mathbf{C})$, bounded in $\mathcal{C}^b(\Omega, \mathbf{C})$. We prove that $\sup_{f\in H} |\mu_n(f) - \mu(f)|$ converges to 0 as $n \to +\infty$.

Assume that $\sup_{f\in H} |\mu_n(f) - \mu(f)|$ does not converge to 0 as $n \to +\infty$. Then, for a suitable $\varepsilon > 0$, we can find a strictly increasing sequence $(n_i)_{i\geq 1}$ in $\mathbf{N}$ and a sequence $(f_i)_{i\geq 1}$ in H such that $|\mu_{n_i}(f_i) - \mu(f_i)| > \varepsilon$ for all $i \geq 1$. Put $\mu_{n_i} = \nu_i$.

1. Equip $\mathcal{C}(\Omega, \mathbf{C})$ with the topology of uniform convergence on compact sets. Then, as is well-known, $\mathcal{C}(\Omega, \mathbf{C})$ is metrizable. Deduce from Ascoli's theorem that there exists a subsequence $(f_{i_k})_{k\geq 1}$ of $(f_i)_{i\geq 1}$ which converges in $\mathcal{C}(\Omega, \mathbf{C})$ to a function $f \in \mathcal{C}^b(\Omega, \mathbf{C})$.

2. By Proposition 7.8.2, $\{\nu_i - \mu : i \geq 1\}$ is tight. Deduce from this fact that there exists $\varphi \in \mathcal{H}^+(\Omega)$, with values in $[0,1]$, such that
    $$\big|(\nu_i - \mu)[(f_i - f)(1 - \varphi)]\big| \leq \frac{\varepsilon}{3} \qquad \text{for all} \quad i \geq 1.$$

3. Show that $\big|(\nu_{i_k} - \mu)[(f_{i_k} - f)\varphi]\big| \leq \varepsilon/3$ for k large enough.

4. Conclude that $\big|(\nu_{i_k} - \mu)(f_{i_k})\big|$ is less than ε for k large enough, so that we reach a contradiction.

8
Regularity

As we saw earlier, Radon measures are somehow nicer than general Daniell measures. Also, we introduced two distinct notions of Lebesgue measure; first, we defined Lebesgue measure on the natural semiring of an interval and later, we defined Lebesgue measure as a Radon measure on the interval I. In this chapter we investigate the relationships between measures defined on semirings and Radon measures.

Summary

8.1 A measure defined on a semiring Φ is said to be strictly regular if its main prolongation is a prolongation of a Radon measure. If each open subset of Ω is a countable union of compact sets, if $\tilde{\Phi}$ is the Borel σ-algebra, and if each compact set is μ-integrable, then μ is strictly regular (Theorem 8.1.2).

8.1 Regular Measures

Let Ω be a locally compact Hausdorff space.

Definition 8.1.1 A measure μ_Φ on a semiring Φ in Ω is said to be strictly regular (respectively, regular) if there exists a Radon measure μ on Ω such that $\hat{\mu} = \hat{\mu}_\Phi$ (respectively, such that $\bar{\mu} = \bar{\mu}_\Phi$ and if every compact set is included in a countable union of Φ-sets). Then $\mu : f \mapsto \int f d\mu_\Phi$ is called the Radon measure arising from μ_Φ.

Theorem 8.1.1 Let μ be a Radon measure on Ω and Φ a semiring consisting of μ-integrable sets. Suppose that, for each compact set K, there exist $E_1 \in \Phi$, $E_2 \in \tilde{\Phi}$ such that $E_1 \subset K \subset E_2$ and $E_2 - E_1$ is μ-negligible. Denote by μ_Φ

the measure $E \mapsto \int 1_E d\mu$ on Φ. Then $\bar{\mu} = \bar{\mu}_\Phi$. Moreover, $\hat{\mu} = \hat{\mu}_\Phi$ if and only if every μ-integrable open set is contained in a countable union of Φ-sets.

PROOF: Let S be the semiring in Ω consisting of the $K \cap L^c$ for all compact subsets K, L of Ω. By Proposition 7.1.4, $St(S, \mathbf{C})$ is dense in $\mathcal{L}_\mathbf{C}^1(\mu)$, and, by Proposition 3.9.4, $St(\Phi, \mathbf{C})$ is also dense in $\mathcal{L}_\mathbf{C}^1(\mu)$. Moreover, $\mathcal{L}_\mathbf{C}^1(\mu_\Phi) \subset \mathcal{L}_\mathbf{C}^1(\mu)$ and $\int f dV \mu_\Phi = \int f dV \mu$ for all $f \in \mathcal{L}_\mathbf{C}^1(\mu_\Phi)$, by Proposition 3.9.3.

The restrictions to $\tilde{\Phi}$ of $V\mu^*$ and $V\mu_\Phi^*$ are σ-additive, and they agree on Φ, so they are identical. If K is a compact set and E_1, E_2 are as in the statement of Theorem 8.1.1, then $E_2 - E_1$ is μ_Φ-negligible, hence K is μ_Φ-integrable and $V\mu_\Phi(K) = V\mu(K)$. Now, for each $E \in \Phi$, there is an increasing sequence $(K_n)_{n \geq 1}$ of compact subsets of E such that $E - \bigcup_{n \geq 1} K_n$ is μ-negligible. Then

$$
\begin{aligned}
V\mu_\Phi(E - K_n) &= V\mu_\Phi(E) - V\mu_\Phi(K_n) \\
&= V\mu(E) - V\mu(K_n) \\
&= V\mu(E - K_n)
\end{aligned}
$$

converges to 0 as $n \to \infty$, which implies that $E - \bigcup_{n \geq 1} K_n$ is μ_Φ-negligible. Next,

$$
\begin{aligned}
V\mu_\Phi(E) &= \sup_{n \geq 1} V\mu_\Phi(K_n) \\
&= \sup_{n \geq 1}[V\mu_\Phi(K_n \cap U) + V\mu_\Phi(K_n \cap U^c)] \\
&= \sup_{n \geq 1} V\mu_\Phi(K_n \cap U) + \sup_{n \geq 1} V\mu_\Phi(K_n \cap U^c) \\
&= V\mu_\Phi^*(E \cap U) + V\mu_\Phi^*(E \cap U^c),
\end{aligned}
$$

for each open subset U of Ω.

This proves that each open set U splits the outer measure $V\mu_\Phi^*$ of Φ-sets, and therefore is μ_Φ-measurable. Furthermore,

$$
\begin{aligned}
V\mu^*(U) &= \sup\{V\mu(K) : K \subset U, K \text{ compact}\} \\
&= \sup_K V\mu_\Phi(K) \leq V\mu_\Phi^\bullet(U)
\end{aligned}
$$

and

$$
\begin{aligned}
V\mu_\Phi^\bullet(U) &= \sup\{V\mu_\Phi(B) : B \subset U,\ B\ \mu_\Phi\text{-integrable}\} \\
&= \sup_B V\mu(B) \leq V\mu^*(U),
\end{aligned}
$$

so

$$
V\mu^*(U) = V\mu_\Phi^\bullet(U).
$$

Consequently, each μ-negligible set is locally μ_Φ-negligible.

Each locally μ-negligible set A meets every $E \in \Phi$ in a μ-negligible, and hence μ_Φ-negligible, set $A \cap E$. So A is locally μ_Φ-negligible. Conversely, let A

be a locally μ_Φ-negligible set. A meets every compact set K in a μ_Φ-negligible, and hence μ-negligible, set $A \cap K$, which proves that A is locally μ-negligible. What we have shown is that the locally μ_Φ-negligible sets are exactly the locally μ-negligible sets, and then $\bar\mu_\Phi = \bar\mu$ by Proposition 3.9.2.

Finally, if each μ-integrable open set, U, is included in a countable union of Φ-sets, then $V\mu^*(U) = V\mu_\Phi^\bullet(U) = V\mu_\Phi^*(U)$. Therefore, each μ-negligible set is μ_Φ-negligible, and $\hat\mu_\Phi = \hat\mu$ by Proposition 3.9.3.   □

So, given a Radon measure μ on Ω, we see that some measures μ_Φ on semirings satisfy $\hat\mu = \hat\mu_\Phi$ (or $\bar\mu = \bar\mu_\Phi$). Now, the following proposition goes the other way.

Proposition 8.1.1 *Let Φ be a semiring in Ω and μ_Φ a measure on Φ. Then μ_Φ is strictly regular if and only if the following conditions are satisfied:*

(a) *For every compact set K, there exists $B \in \check\Phi$ such that $B \triangle K$ is μ_Φ-negligible.*

(b) *$V\mu_\Phi^*(K) < +\infty$ for every compact set K.*

(c) *$V\mu_\Phi^*(U) = \sup\{V\mu_\Phi(K) : K \subset U, K \text{ compact}\}$, for every open set U.*

(d) *$V\mu_\Phi(E) = \inf\{V\mu_\Phi^*(U) : U \supset E, U \text{ open}\}$, for every $E \in \Phi$.*

PROOF: Evidently, the conditions are necessary. Conversely, suppose that each of them is satisfied. By (a), each compact set is μ_Φ-measurable, and it is μ_Φ-integrable by (b). Each $f \in \mathcal{H}(\Omega, \mathbf{C})$ is μ_Φ-integrable by Proposition 7.1.4. Denote the linear forms $f \mapsto \int f d\mu_\Phi$ and $f \mapsto \int f dV\mu_\Phi$ on $\mathcal{H}(\Omega, \mathbf{C})$, which are clearly Radon measures, by μ and ν, so that $V\mu \le \nu$. Then

$$\int 1_K \, d\nu = \inf_{f \in \mathcal{H}^+, f \ge 1_K} \int f dV\mu_\Phi \ge V\mu_\Phi(K)$$

for all compact sets K, so $V\mu_\Phi^*(U) \le \nu^*(U)$ for each open set U. On the other hand, $\nu^*(U) = \sup\{\int f \, d\nu : f \in \mathcal{H}^+, f \le 1_U\} \le V\mu_\Phi^*(U)$. Thus $V\mu_\Phi^*(U) = \nu^*(U)$. It follows that $V\mu_\Phi^*(X) \le \nu^*(X)$ for all subsets X of Ω.

Now $V\mu_\Phi(E) = \inf\{V\mu_\Phi^*(U) : U \supset E, U \text{ open}\} = \inf_U \nu^*(U) = \nu^*(E)$, for each $E \in \Phi$, so $\nu^*(X) \le V\mu_\Phi^*(X)$ for each subset X of Ω, by definition of $V\mu_\Phi^*(X)$. Finally, $\nu^*(X) = V\mu_\Phi^*(X)$. The main prolongations of ν and $V\mu_\Phi$ are therefore identical. In particular, each $E \in \Phi$ is ν-integrable. Since $V\mu \le \nu$, every complex-valued ν-integrable function is μ-integrable and the linear form $f \mapsto \int f d\mu$ is continuous on $\mathcal{L}_{\mathbf{C}}^1(\nu) = \mathcal{L}_{\mathbf{C}}^1(\mu_\Phi)$. But, since it agrees with the linear form $f \mapsto \int f d\mu_\Phi$ on $\mathcal{H}(\Omega, \mathbf{C})$, $\int f d\mu = \int f d\mu_\Phi$ for all $f \in \mathcal{L}_{\mathbf{C}}^1(\nu)$. In particular, $\mu_\Phi(E) = \int 1_E d\mu$ for all $E \in \Phi$.

Next, $St(\Phi, \mathbf{C})$ is dense in $L_{\mathbf{C}}^1(\mu)$, and now $\int g \, dV\mu = \int g \, dV\mu_\Phi$ for all $g \in \mathcal{H}(\Omega, \mathbf{C})$ by Proposition 3.9.3. The foregoing shows that $\mathcal{L}_{\mathbf{C}}^1(\mu) = \mathcal{L}_{\mathbf{C}}^1(\nu) = \mathcal{L}_{\mathbf{C}}^1(\mu_\Phi)$, and so $\int f d\mu = \int f d\mu_\Phi$ for all $f \in \mathcal{L}_{\mathbf{C}}^1(\mu)$.   □

The following result is particularly important.

Theorem 8.1.2 *Suppose that each open subset of Ω is a countable union of compact sets. Let Φ be a semiring in Ω such that $\tilde{\Phi}$ is the Borel σ-algebra. Then every measure μ_Φ on Φ such that $V\mu_\Phi^*(K) < +\infty$ for all compact sets K is strictly regular.*

PROOF: Let ν be the Radon measure $\mathcal{H}(\Omega, \mathbf{C}) \ni f \mapsto \int f dV\mu_\Phi$. If U is an open subset of Ω, there exists an increasing sequence $(f_n)_{n\geq 1}$ in $\mathcal{H}^+$ such that $1_U = \sup_{n\geq 1} f_n$ and that $K_n = \text{supp}(f_n) \subset U$ for every $n \geq 1$. Then

$$V\mu_\Phi^*(U) = \sup_n \int f_n dV\mu_\Phi = \sup_n \int f_n d\nu = \nu^*(U).$$

Since $f_n \leq 1_{K_n}$,

$$\int f_n dV\mu_\Phi \leq V\mu_\Phi(K_n) \leq V\mu_\Phi^*(U) \qquad \text{for each integer } n \geq 1.$$

Therefore,

$$V\mu_\Phi^*(U) = \sup\{V\mu_\Phi(K) : K \subset U, K \text{ compact}\}.$$

Now fix $E \in \Phi$. By Proposition 7.3.3, there exist, for all $\varepsilon > 0$, an open set U and a closed set F such that $F \subset E \subset U$ and $\nu^*(U - F) \leq \varepsilon$. Then $V\mu_\Phi^*(U - E) \leq V\mu_\Phi^*(U - F) = \nu^*(U - F) \leq \varepsilon$, and $V\mu_\Phi^*(U) \leq V\mu_\Phi(E) + \varepsilon$. By Proposition 8.1.1, μ_Φ is strictly regular.   □

For example, take the measures on the natural semiring of an interval. These are all strictly regular.

Proposition 8.1.2 may be simplified when Φ contains the compact sets.

In view of probability theory, we now fix some notation. Let Φ be a semiring in Ω such that each compact set is included in a countable union of Φ-sets. Write $\mathcal{M}_{reg}(\Phi, \mathbf{C})$ (respectively, $\mathcal{M}_{reg}^1(\Phi, \mathbf{C})$) for the set of regular measures (respectively, bounded regular measures) on Φ. Then $\mathcal{M}_{reg}(\Phi, \mathbf{C})$ may be identified with a subset (in fact, a vector subspace, as we shall see in Chapter 22) of $\mathcal{M}(\Omega, \mathbf{C})$. Thus $\mathcal{M}_{reg}(\Phi, \mathbf{C})$ may be equipped with a topology of vague convergence and $\mathcal{M}_{reg}^1(\Phi, \mathbf{C})$ with topologies of weak convergence and of narrow convergence. For instance, a filter basis $\mathcal{B}$ on $\mathcal{M}_{reg}^1(\Phi, \mathbf{C})$ converges narrowly to $\mu \in \mathcal{M}_{reg}^1(\Phi, \mathbf{C})$ if, for all $f \in \mathcal{C}^b(\Omega, \mathbf{C})$, $\int f d\nu$ converges to $\int f d\mu$ along $\mathcal{B}$. In the particular case where Φ is a semiring of Borel sets such that $\tilde{\Phi}$ contains the compact sets, then $\mathcal{M}_{reg}^1(\Phi, \mathbf{C})$ may be identified with the whole of $\mathcal{M}^1(\Omega, \mathbf{C})$ (Theorem 8.1.1), and the study of weak convergence is simplified.

Exercises for Chapter 8

1   Let Ω be a locally compact Hausdorff space and let S be a semiring in Ω such that each compact set is contained in a countable union of S-sets. Let μ be a

measure on S defined by point masses α_x $(x \in X)$. Prove that μ is regular if and only if $\sum_{x \in X \cap K} |\alpha_x|$ is finite for every compact set K and that, in this case, the Radon measure arising from μ is defined by the point masses α_x.

2 Let $f : \mathbf{R} \to \mathbf{R}$ be such that $f(x + y) = f(x) + f(y)$ for all x, y in $\mathbf{R}$.

1. Prove that $f\big((p/q)x\big) = (p/q)f(x)$ for all $x \in \mathbf{R}$, $q \in \mathbf{N}$, and $p \in \mathbf{Z}$.

2. Suppose that the graph of f is not dense in $\mathbf{R}^2$. Then show that f is continuous at 0. Conclude that $f(x) = xf(1)$ for all $x \in \mathbf{R}$.

3. Let λ be Lebesgue measure on $\mathbf{R}$, and suppose that f is λ-measurable. Show that f is continuous at 0 (use Proposition 4.3.1). Conclude that $f(x) = xf(1)$ for all $x \in \mathbf{R}$.

3 Let B be a basis of $\mathbf{R}$ considered as a vector space over the field $\mathbf{Q}$ of rational numbers (Hamel basis).

1. Show that B is uncountable.

2. Let φ be a bijection from a proper subset C of B onto B. Define a function f from $\mathbf{R}$ into $\mathbf{R}$ by $f(x) = \sum_{u \in C} \lambda(u)\varphi(u)$ for each $x = \sum_{u \in B} \lambda(u)u$ of $\mathbf{R}$, where $\lambda(u) \in \mathbf{Q}$ and $\lambda(u) = 0$ except for finitely many $u \in B$.

    Show that $f(x + y) = f(x) + f(y)$ for all x, y in $\mathbf{R}$, but that, for all $z \in \mathbf{R}$, $f^{-1}(z)$ is a dense subset of $\mathbf{R}$. Prove that f takes all real values on every interval of $\mathbf{R}$ which is not a point.

3. Refer to part 2 and show that f is not Lebesgue measurable.

4 Let μ be Lebesgue measure on $I = [0, 1]$. Consider, in I, the family $(I_{n,p})_{n,p}$ of disjoint open intervals defined by induction on n as follows. The integer n takes all positive values; for each $n \geq 0$, p takes the values 1, 2, 3, ..., 2^n, and $I_{0,1} =]1/3, 2/3[$. If J_n is the union of the $I_{m,p}$ corresponding to the numbers $m \leq n$, the complement of J_n in I is the union of 2^{n+1} disjoint compact intervals $K_{n,p}$ $(1 \leq p \leq 2^{n+1})$ such that $\alpha(K_{n,1}) < \alpha(K_{n,2}) < \cdots < \alpha(K_{n,2^{n+1}})$. If $K_{n,p} = [a, b]$, then we take for $I_{n+1,p}$ the interval $]b - (1 + 2^{-n})\big((b-a)/3\big), b - 2^{-n}\big((b-a)/3\big)[$. Now write E for the complement in I of the union of all the $I_{n,p}$.

1. Show that $\mu(I - J_n) = (2/3)^{n+1}$ for all integers $n \geq 0$ and deduce that E is μ-negligible.

2. Define a function f on $[0, 1]$ as follows:

    f vanishes on E and on $]1/3, 2/3[$;

    for all integers $n \geq 0$ and $1 \leq p \leq 2^{n+1}$, if we put $K_{n,p} = [a, b]$, f takes the value $(b - a)/2^{n+1}$ at $b - (b - a)/2^{n+1}$ and is affine on each of the intervals $[\alpha(I_{n+1,p}), b - (b - a)/2^{n+1}]$ and $[b - (b - a)/2^{n+1}, \beta(I_{n+1,p})]$. Let x be a point of $E - \{1\}$ which is the origin of no interval contiguous to E. Show that the right-hand derivative of f at x is 0.

3. Let x be a point of $E - \{0\}$ which is the right endpoint of no interval contiguous to E. Show that f has no left-hand derivative (finite or infinite) at x. Since E is metrizable, compact, totally disconnected, and has no isolated point, it is homeomorphic to the Cantor set by a classical topological result. Therefore, f has no left-hand derivative at uncountably many points of $]0, 1]$.

4. Prove the existence of a function g from I into $\mathbf{R}$ such that $|g| \leq 3$, g has a μ-negligible set of points of discontinuity, and $f(x) = \int_0^x g(t)dt = \int g \cdot 1_{[0,x]}d\mu$ for every $x \in I$ (use the dominated convergence theorem). Conclude that g is Riemann integrable, even though its indefinite integral $x \mapsto \int_0^x g(t)dt$ has no left-hand derivative at uncountably many points.

5 Let μ be Lebesgue measure on $I = [0, 1]$, and, for all $x \in I$, denote by ε_x the measure on the natural semiring of I defined by the unit mass at x. Let $(x_k)_{k \geq 1}$ be a sequence in I. For all $a, b \in I$ satisfying $a \leq b$ and all integers $n \geq 1$, define $\nu(n, a, b)$ as the number of integers $1 \leq k \leq n$ such that $a \leq x_k \leq b$. Clearly, $(1/n)\sum_{1 \leq k \leq n} 1_{[a,b]}(x_k)$ converges to $\int 1_{[a,b]} d\mu$ as $n \to +\infty$, for all compact subintervals $[a, b]$ of I, if and only if $(1/n)\nu(n, a, b)$ tends to $b - a$. In this case, $(x_k)_{k \geq 1}$ is said to be uniformly distributed modulo 1.

1. Show that $(x_k)_{k \geq 1}$ is uniformly distributed modulo 1 if and only if $\left((1/n)\sum_{1 \leq k \leq n} \varepsilon_{x_k}\right)_{n \geq 1}$ converges vaguely to μ.

2. Suppose that $(1/n)\sum_{1 \leq k \leq n} \exp(2i\pi p x_k)$ converges to 0 as $n \to +\infty$, for all integers $p \geq 1$. Show that $(1/n)\sum_{1 \leq k \leq n} f(x_k)$ tends to $\int f d\mu$ for all continuous functions f from I into $\mathbf{C}$ such that $f(0) = f(1)$, and prove that this result persists even when $f(0) \neq f(1)$.

3. Deduce from parts 1 and 2 that $(x_k)_{k \geq 1}$ is uniformly distributed modulo 1 if and only if $(1/n)\sum_{1 \leq k \leq n} \exp(2i\pi p x_k)$ converges to 0 as $n \to +\infty$, for all integers $p \geq 1$.

6 Retain the notation of Exercise 5. A sequence $(x_k)_{k \geq 1}$ in I is said to have a limit distribution μ, where μ is a measure on the natural semiring of I, if $(1/n)\sum_{1 \leq k \leq n} \varepsilon_{x_k}$ converges vaguely to μ. Let θ be a real number and put $x_k = k\theta - [k\theta]$ for all integers $k \geq 1$, where $[k\theta]$ is the integral part of $k\theta$.

1. When θ is irrational, prove that $(x_k)_{k \geq 1}$ is uniformly distributed modulo 1 (Bohl's theorem).

2. When θ is rational, find the limit distribution of $(x_k)_{k \geq 1}$.

Part II

Operations on Measures Defined on Semirings

9

Induced Measures and Product Measures

Throughout Chapters 9–12 we shall be concerned with measures on semirings and their related operations. We shall also see how these operations combine with each other.

Summary

9.1   Let μ be a measure on a semiring S whose underlying set is X. Let Y be a μ-measurable subset of X. Let T be a semiring whose underlying set is Y. Assume that μ, Y, and T are endowed with adequate properties. We define the measure $\mu_{/T}$ induced by μ on T and see how to integrate with respect to this induced measure (Theorem 9.1.1).

9.2   Let μ', μ'' be two measures on semirings S', S'' (with Ω', Ω'' as their underlying sets, respectively). The measure $\mu : A' \times A'' \longmapsto \mu'(A')\mu''(A'')$ on the semiring $S = \{A' \times A'' : A' \in S', A'' \in S''\}$ (whose underlying set is $\Omega = \Omega' \times \Omega''$) is called the product of μ' and μ''. $V(\mu' \otimes \mu'') = V\mu' \otimes V\mu''$ (Theorem 9.2.1). If $f : \Omega \to [0, +\infty]$ is μ-measurable and μ-moderate, then $\int^* f \, dV\mu = \int^* dV\mu'(x') \int^* f(x', x'') \, dV\mu''(x'')$ can be computed by means of iterated integrals (Theorem 9.2.5). Likewise, we can compute $\int f \, d\mu$ for every μ-integrable mapping from Ω into a Banach space (Theorem 9.2.4, Fubini's).

9.3   We define Lebesgue measure on an open subset Ω of $\mathbf{R}^k$ (with $k \geq 1$).

9.1   Measure Induced on a Measurable Set

Let X be a nonempty set, Y a nonempty subset of X, S a semiring in X, T a semiring in Y, and μ a complex measure on S. We shall say that $\mu_{/T}$ exists if

(a) Y is μ-measurable and the T-sets are μ-integrable;

(b) for each A in S, there exists $B \in \tilde{T}$ such that $B \subset A$ and $A \cap Y - B$ is μ-negligible;

(c) $A \cap B$ belongs to $\tilde{T}$, for all $A \in S$ and for all $B \in T$.

Definition 9.1.1 The measure $\mu_{/T} : B \mapsto \int 1_B d\mu$ on T is called the measure induced by μ on T.

The rules of integration with respect to $\mu_{/T}$ are given by the following:

Theorem 9.1.1 $|\mu_{/T}| = |\mu|_{/T}$. *For every mapping f from Y into $\overline{\mathbf{R}}$ (respectively, into a Banach space F), write $\tilde{f}$ for the mapping from X into $\overline{\mathbf{R}}$ (respectively, into F) which agrees with f on Y and vanishes outside Y. Then $\int^\bullet f d|\mu_{/T}| = \int^\bullet \tilde{f} d|\mu|$ for all $f : Y \to [0, +\infty]$. A mapping f from Y into F is essentially $\mu_{/T}$-integrable if and only if $\tilde{f}$ is essentially μ-integrable, and then $\int f d\mu_{/T} = \int \tilde{f} d\mu$. A mapping g from Y into a metrizable space is $\mu_{/T}$-measurable if and only if g is μ-measurable on Y.*

PROOF: First, suppose that $T = \{A \cap Y : A \in S\}$ and write ν instead of $\mu_{/T}$. The fact that $|\nu(B)| = |\int 1_B d\mu| \le \int 1_B d|\mu| = |\mu|_{/T}(B)$ for all $B \in T$ leads to $|\nu| \le |\mu|_{/T}$. Conversely, let $A \in S$ and $B = A \cap Y$. Then

$$|\mu|_{/T}(B) = \int 1_B d|\mu| = \sup_{\alpha \in St(S,\mathbf{C}), |\alpha| \le 1} \left| \int \alpha \cdot 1_B d\mu \right|.$$

Every $\alpha \in St(S, \mathbf{C})$ such that $|\alpha| \le 1$ can be written $\alpha = \sum_{i \in I} 1_{A_i} \cdot y_i$, where I is finite, the A_i are disjoint S-sets, and $|y_i| \le 1$ for all $i \in I$. But each $A_i \cap A$ can be partitioned into S-sets $E_{i,j}$ ($j \in J_i$, $|J_i| < +\infty$), so

$$\left| \int \alpha \cdot 1_B d\mu \right| = \left| \sum_{i \in I} \int 1_{A_i \cap B} \cdot y_i d\mu \right| \le \sum_{i \in I} |\mu(A_i \cap A \cap Y)|$$

$$\le \sum_{i \in I} \sum_{j \in J_i} |\mu(E_{i,j} \cap Y)| \le \sum_{i \in I} \sum_{j \in J_i} |\nu(E_{i,j} \cap Y)| \le |\nu|(B).$$

Thus $|\mu|_{/T}(B) \le |\nu|(B)$ and, finally, $|\nu| = |\mu|_{/T}$.

Clearly, $\int g d\nu = \int \tilde{g} d\mu$ and $\int |g| d|\nu| = \int |\tilde{g}| d|\mu|$ for all $g \in St(T, F)$ (F is a Banach space).

If g is the upper envelope of an increasing sequence $(g_n)_{n \ge 1}$ of elements of $St^+(T)$, then $\int^* g d|\nu| = \sup_{n \ge 1} \int \tilde{g}_n d|\mu| = \int^* \tilde{g} d|\mu|$. so $\int^* \tilde{f} d|\mu| \le \int^* f d|\nu|$ for all $f : Y \to [0, +\infty]$. Conversely, if $h \ge \tilde{f}$ is the upper envelope of an increasing sequence $(h_n)_{n \ge 1}$ in $St^+(S, \mathbf{R})$,

$$\int^* h d|\mu| = \sup_{n \ge 1} \int h_n d|\mu| \ge \sup_n \int h_n \cdot 1_Y d|\mu| = \sup_n \int h_{n/Y} d|\nu|$$

$$= \int^* h_{/Y} d|\nu| \ge \int^* f d|\nu|,$$

whence follows $\int^* \tilde{f} d|\mu| \geq \int^* f d|\nu|$ and finally $\int^* f d|\nu| = \int^* \tilde{f} d|\mu|$.

Let $\mathcal{A}$ (respectively, $\mathcal{B}$) be the ring in X (respectively, Y) generated by S (respectively, T). Then

$$\int^\bullet \tilde{f} d|\mu| = \sup_{E \in \mathcal{A}} \int^* \tilde{f} \cdot 1_E d|\mu|$$

$$= \sup_{E \in \mathcal{A}} \int^* f \cdot 1_{E \cap Y} d|\nu| \leq \int^\bullet f d|\nu|$$

and

$$\int^\bullet f d|\nu| = \sup_{F \in \mathcal{B}} \int^* f \cdot 1_F d|\nu|$$

$$= \sup_{F \in \mathcal{B}} \int^* \tilde{f} \cdot 1_F d|\mu| \leq \int^\bullet \tilde{f} d|\mu|,$$

that is,

$$\int^\bullet f d|\nu| = \int^\bullet \tilde{f} d|\mu|$$

for all $f : Y \to [0, +\infty]$.

Next, let f be a mapping from Y into a Banach space F. If f is essentially ν-integrable, it is the limit locally ν-almost everywhere of a Cauchy sequence $(g_n)_{n \geq 1}$ in $St(T, F)$ (considered as a subspace of $\mathcal{L}_F^1(\nu)$); in this case, it is obvious that $\tilde{f}$ is essentially μ-integrable and that $\int f d\nu = \int \tilde{f} d\mu$. Conversely, if $\tilde{f}$ is essentially μ-integrable, it is the limit locally μ-almost everywhere of a Cauchy sequence $(h_n)_{n \geq 1}$ in $St(S, F)$ (considered as a subspace of $\mathcal{L}_F^1(\mu)$), and, since $\int |h_{p/Y} - h_{q/Y}| d|\nu| \leq \int |h_p - h_q| d|\mu|$ for all $p \geq 1$, $q \geq 1$, we see that f is essentially ν-integrable.

In short, f is essentially ν-integrable if and only if $\tilde{f}$ is essentially μ-integrable. Moreover, f is ν-integrable if and only if $\tilde{f}$ is μ-integrable. Now this implies that a mapping g from Y into a metrizable space is ν-measurable if and only if g is μ-measurable on Y.

Finally, we go to the general case, in which T is not necessarily the semiring $S_Y = \{A \cap Y : A \in S\}$, and write ν for the measure $A \cap Y \mapsto \int 1_{A \cap Y} d\mu$ on S_Y. Every $B \in T$ is ν-integrable, and $\mu_{/T}(B) = \int 1_B d\nu$. For all $A \cap Y$ of S_Y, there exists $B \in \tilde{T}$ such that $B \subset A \cap Y$ and $A \cap Y - B$ is ν-negligible. Moreover, $(A \cap Y) \cap B = A \cap B$ belongs to $\tilde{T}$ for all $A \in S$ and $B \in T$. Hence $\overline{\mu_{/T}} = \bar{\nu}$ by Propositions 3.9.2 and 3.9.4, and the desired results follow. $\square$

Proposition 9.1.1 *Let μ, Y, and T be as in Theorem 9.1.1. Also, let Z be a subset of Y and let U be a semiring in Z. Then $(\mu_{/T})_{/U}$ exists if and only if the following conditions hold:*

(a) $\mu_{/U}$ *exists.*

(b) $B \cap C$ belongs to $\tilde{U}$ for all $B \in T$ and $C \in U$.

(c) Each $C \in U$ is contained in a countable union of T-sets.

In this case, $(\mu_{/T})_{/U} = \mu_{/U}$ (transitivity of induced measures).

PROOF: Observe that $B \cap C$ belongs to $\tilde{U}$ for all $B \in \tilde{T}$ and $C \in \tilde{U}$ if it does for all $B \in T$ and $C \in U$.

First, suppose that $(\mu_{/T})_{/U}$ exists. Conditions (b) and (c) are clearly satisfied. Z is μ-measurable and each $C \in U$ is μ-integrable. Next, for fixed $A \in S$, choose $B \in \tilde{T}$, $B \subset A$, so that $A \cap Y - B$ is μ-negligible, and let $(B_n)_{n \geq 1}$ be a sequence of disjoint T-sets whose union contains B; for all $n \geq 1$, there exists $C_n \in \tilde{U}$, $C_n \subset B_n$, such that $B_n \cap Z - C_n$ is μ-negligible; then $B \cap (\bigcup_{n \geq 1} C_n)$ lies in $\tilde{U}$, and $A \cap Z - B \cap (\bigcup_{n \geq 1} C_n)$, as the union of $(A \cap Y - B) \cap Z$ and the $B \cap (B_n \cap Z - C_n)$, is μ-negligible. Last, each $C \in U$ is contained in the union of T-sets B_n ($n \geq 1$), so $A \cap C = \bigcup_{n \geq 1} (A \cap B_n) \cap C$ lies in $\tilde{U}$. Hence $\mu_{/U}$ exists.

Conversely, assume that conditions (a), (b), and (c) hold. Z is $\mu_{/T}$-measurable. Each $C \in U$ is μ-integrable and contained in a countable union of T-sets, and hence is $\mu_{/T}$-integrable. For a given $B \in T$, let $(A_n)_{n \geq 1}$ be a sequence of disjoint S-sets whose union contains B. For every $n \geq 1$, there exists $C_n \in \tilde{U}$, $C_n \subset A_n$, such that $A_n \cap Z - C_n$ is μ-negligible. Then $B \cap (\bigcup_{n \geq 1} C_n)$ belongs to $\tilde{U}$, and

$$B \cap Z - B \cap \left(\bigcup_{n \geq 1} C_n \right) = B \cap \bigcup_{n \geq 1} (A_n \cap Z - C_n)$$

is $\mu_{/T}$-negligible, which proves that $(\mu_{/T})_{/U}$ exists. $\square$

For example, if μ is Lebesgue measure on the natural semiring S of an interval I, and if T is the natural semiring of a subinterval J of I, then $\mu_{/T}$ exists and is Lebesgue measure on T.

9.2 Fubini's Theorem

Let Ω', Ω'' be two nonempty sets and S', S'' two semirings in Ω', Ω'', respectively. Consider the class $S = \{A' \times A'' : A' \in S', A'' \in S''\}$. For any two S-sets, $A = A' \times A''$ and $B = B' \times B''$, the set $A \cap B$, if it is nonempty, can be partitioned into a finite number of S-sets. Similarly,

$$A \cap B^c = \left((A' \cap B'^c) \times A'' \right) \cup \left((A' \cap B') \times (A'' \cap B''^c) \right),$$

if it is nonempty, can be partitioned into a finite number of S-sets. Therefore, S is a semiring, called the product of S' and S'', denoted by $S' \times S''$.

Theorem 9.2.1 *Let μ' and μ'' be two measures on S' and S'', respectively, and let μ be the function defined on S by $A = A' \times A'' \mapsto \mu'(A') \cdot \mu''(A'')$. Then μ is a measure on S, written $\mu' \otimes \mu''$, which satisfies*

$$V\mu = (V\mu') \otimes (V\mu'').$$

PROOF: Let $(A_i)_{i \geq 1} = (A_i' \times A_i'')_{i \geq 1}$ be a sequence of disjoint nonempty S-sets whose union is an S-set $A = A' \times A''$. Then

$$1_A(x', x'') = \lim_{n \to +\infty} \sum_{1 \leq i \leq n} 1_{A_i}(x', x'')$$

for all (x', x'') in $\Omega = \Omega' \times \Omega''$. For a given x', the functions $x'' \longmapsto \sum_{1 \leq i \leq n} 1_{A_i' \times A_i''}(x', x'')$ are μ''-integrable and dominated by the μ''-integrable function $x'' \mapsto 1_{A' \times A''}(x', x'')$. The dominated convergence theorem shows that

$$\int 1_{A' \times A''}(x', x'') d\mu''(x'') = \lim_{n \to +\infty} \int \left[\sum_{1 \leq i \leq n} 1_{A_i' \times A_i''}(x', x'') \right] d\mu''(x'').$$

Now the functions

$$x' \mapsto \int \left[\sum_{1 \leq i \leq n} 1_{A_i' \times A_i''}(x', x'') \right] d\mu''(x'') = \sum_{1 \leq i \leq n} 1_{A_i'}(x') \mu''(A_i'')$$

are μ'-integrable, and their absolute values are dominated by the μ'-integrable function $x' \mapsto 1_{A'}(x') V\mu''(A'')$ because

$$\left| \sum_{1 \leq i \leq n} 1_{A_i'}(x') \mu''(A_i'') \right| \leq \int \left[\sum_{1 \leq i \leq n} 1_{A_i' \times A_i''}(x', x'') \right] dV\mu''(x'')$$

$$\leq \int 1_{A' \times A''}(x', x'') dV\mu''(x'').$$

Resorting once again to the dominated convergence theorem,

$$\lim_{n \to +\infty} \int \left[\sum_{1 \leq i \leq n} 1_{A_i'}(x') \mu''(A_i'') \right] d\mu'(x') =$$

$$\int d\mu'(x') \int 1_{A' \times A''}(x', x'') d\mu''(x''),$$

which is equivalent to

$$\sum_{i \geq 1} (\mu' \otimes \mu'')(A_i' \times A_i'') = (\mu' \otimes \mu'')(A' \times A'').$$

Thus μ is countably additive.

This argument also applies to the function $(V\mu') \otimes (V\mu'')$. which is therefore a measure on S.

If $(A_i)_{i\geq 1} = (A_i' \times A_i'')_{i\geq 1}$ is a sequence of disjoint S-sets all contained in an S-set $A = A' \times A''$, then

$$\sum_{i\geq 1} |(\mu' \otimes \mu'')(A_i' \times A_i'')| \leq \sum_{i\geq 1} ((V\mu') \otimes (V\mu''))(A_i' \times A_i'')$$

$$\leq ((V\mu') \otimes (V\mu''))(A' \times A'') < +\infty.$$

Hence μ is a measure and $V(\mu' \otimes \mu'') \leq (V\mu') \otimes (V\mu'')$. It remains to be shown that $V(\mu' \otimes \mu'') \geq (V\mu') \otimes (V\mu'')$.

But, for all finite partitions $P(A')$ and $P(A'')$ of A' and A'' into S'-sets and S''-sets, respectively,

$$\left(\sum_{B' \in P(A')} |\mu'(B')| \right) \left(\sum_{B'' \in P(A'')} |\mu''(B'')| \right) =$$

$$\sum_{\substack{B' \in P(A') \\ B'' \in P(A'')}} |(\mu' \otimes \mu'')(B' \times B'')| \leq V(\mu' \otimes \mu'')(A' \times A''),$$

and the conclusion follows. $\square$

Definition 9.2.1 The measure $\mu = \mu' \otimes \mu''$ is called the product measure of μ' and μ''.

If f is a function from Ω into $[0, +\infty]$, we shall often write

$$\iint^* f(x', x'')dV\mu'(x')dV\mu''(x'')$$

for $\int^* fdV\mu$, and

$$\iint_* f(x', x'')dV\mu'(x')dV\mu''(x'')$$

for $\int_* fdV\mu$. Also, for a μ-integrable mapping, f, from Ω into a Banach space, we shall often write $\int \int f(x', x'')d\mu'(x')d\mu''(x'')$ instead of $\int fd\mu$.

If F is a Banach space and α belongs to $St(S, F)$, observe that $\alpha(x', \cdot)$ lies in $St(S'', F)$ for all $x' \in X'$ and that $x' \mapsto \int \alpha(x', \cdot)d\mu''$ lies in $St(S', F)$. Moreover, $\int d\mu' \int \alpha(x', \cdot)d\mu'' = \int \alpha d\mu$.

Henceforth, unless otherwise stated, assume that μ' and μ'' are positive and put $\mu = \mu' \otimes \mu''$. Denote by $\mathcal{J}^+$ ($\mathcal{J}'^+$, $\mathcal{J}''^+$, respectively) the set of those functions from Ω (Ω', Ω'', respectively) into $[0, +\infty]$ which are upper envelopes of increasing sequences in $St^+(S)$ ($St^+(S')$, $St^-(S'')$, respectively).

Proposition 9.2.1 $\int^* hd\mu = \int^* d\mu' \int^* h(x'. \cdot)d\mu''$ for all $h \in \mathcal{J}^+$, and $\int^* fd\mu \geq \int^* d\mu' \int^* f(x', \cdot)d\mu''$ for all functions f from Ω into $[0, +\infty]$.

PROOF: Let $h \in \mathcal{J}^+$, and let $(h_n)_{n \geq 1}$ be an increasing sequence in $St^+(S)$ whose upper envelope is h. The function $x' \mapsto \int^* h(x', \cdot)d\mu''$ is the upper envelope of the increasing sequence of the functions $x' \mapsto \int h_n(x', \cdot)d\mu''$, so

$$\int^* d\mu' \int^* h(x', \cdot)d\mu'' = \sup_n \int d\mu' \int h_n(x', \cdot)d\mu''$$

$$= \sup_n \int h_n d\mu$$

$$= \int^* h d\mu.$$

Now the second claim is clear. □

Proposition 9.2.2 Let $f' : \Omega' \to [0, +\infty]$ and $f'' : \Omega'' \to [0, +\infty]$ be two functions. Denote by $f' \otimes f''$ the function $(x', x'') \mapsto f'(x')f''(x'')$ from Ω into $[0, +\infty]$. Then $\int^* f' \otimes f'' d\mu = (\int^* f' d\mu')(\int^* f'' d\mu'')$ unless one factor of the right-hand side is 0 and the other is $+\infty$.

PROOF: By Proposition 9.2.1,

$$\int^* (f' \otimes f'')d\mu \geq \int^* d\mu'(x') \int^* f'(x')f''(x'')d\mu''(x'').$$

But

$$\int^* f'(x')f''(x'')d\mu''(x'') = f'(x') \int^* f''(x'')d\mu''(x'')$$

for all x' because, by convention, $0 \times (\infty) = 0$. Therefore,

$$\int^* d\mu'(x') \int^* f'(x')f''(x'')d\mu''(x'') = \left(\int^* f' d\mu' \right)\left(\int^* f'' d\mu'' \right).$$

Thus it suffices to show that $\int^*(f' \otimes f'')d\mu \leq (\int^* f' d\mu')(\int^* f'' d\mu'')$. This inequality is clear when the right-hand side is $+\infty$. We exclude the case in which one factor of the right-hand side is 0 and the other $+\infty$, and so it remains only to be shown that the inequality is true when both $\int^* f' d\mu'$ and $\int^* f'' d\mu''$ are finite.

Let $g' \in \mathcal{J}'^+$ be such that $f' \leq g'$ and $\int^* g' d\mu' < +\infty$. Similarly, let g'' in $\mathcal{J}''^+$ be such that $f'' \leq g''$ and $\int^* g'' d\mu'' < +\infty$. The convention $0 \times (+\infty) = 0$ implies that $g' \otimes g''$ belongs to $\mathcal{J}^+$, so that $\int^* g' \otimes g'' d\mu = (\int^* g' d\mu')(\int^* g'' d\mu'')$, and the desired inequality follows. □

Proposition 9.2.3 $A' \times A''$ is μ-negligible for every μ'-negligible set A' and every μ''-moderate set A''. Moreover, $A' \times \Omega''$ is locally μ-negligible for every locally μ'-negligible set A'.

PROOF: To prove the first assertion, we may assume A'' μ''-integrable; the result is then a consequence of Proposition 9.2.2. $\qquad\square$

Theorem 9.2.2 *Let f' be a μ'-measurable mapping from Ω' into a metrizable space F. Then $f : (x', x'') \mapsto f'(x')$ is μ-measurable.*

PROOF: Denote by $\mathcal{R}'$ ($\mathcal{R}''$ and $\mathcal{R}$,respectively) the ring generated by S' (S'' and S, respectively). Let $A = A' \times A''$ be a nonempty S-set. If α' is an $\mathcal{R}'|A'$-simple mapping from A' into F, then $\alpha : (x', x'') \mapsto \alpha'(x')$ is $\mathcal{R}|A$-simple. There exists a μ'-negligible subset N' of A' such that $f'|A'$ is the limit, on $A' - N'$, of a sequence $(\alpha'_n)_{n \geq 1}$ of $\mathcal{R}'|A'$-simple mappings (Proposition 6.1.4). Then $f|A$ is the limit, on $A - (N' \times A'')$, of the sequence $(\alpha_n)_{n \geq 1}$ of $\mathcal{R}|A$-simple mappings, which proves that f is μ-measurable. $\qquad\square$

In particular, if A' is μ'-measurable and A'' is μ'''-measurable, then $A' \times A'' = (A' \times \Omega'') \cap (\Omega' \times A'')$ is μ-measurable.

Proposition 9.2.4 $\int^{\bullet}(f' \otimes f'')d\mu = \left(\int^{\bullet} f'd\mu'\right)\left(\int^{\bullet} f''d\mu''\right)$ *for all functions f' and f'' from Ω', Ω'' into $[0, +\infty]$.*

PROOF: If A' is a μ'-measurable and μ'-moderate subset of Ω' and if A'' is a subset of Ω'' which is μ''-measurable and μ''-moderate, then

$$\left(\int^* f'1_{A'}d\mu'\right)\left(\int^* f''1_{A''}d\mu''\right) \leq \int^* (f' \otimes f'')1_{A' \times A''}d\mu \leq \int^{\bullet} (f' \otimes f'')d\mu.$$

Therefore, $\left(\int^{\bullet} f'd\mu'\right)\left(\int^{\bullet} f''d\mu''\right) \leq \int^{\bullet}(f' \otimes f'')d\mu$. It remains to be proven that $\int^{\bullet}(f' \otimes f'')d\mu \leq \left(\int^{\bullet} f'd\mu'\right)\left(\int^{\bullet} f''d\mu''\right)$. This inequality is clear when $\int^{\bullet} f'd\mu' = 0$ or $\int^{\bullet} f''d\mu'' = 0$ (Proposition 9.2.3), and we may assume $\int^{\bullet} f'd\mu' < +\infty$ and $\int^{\bullet} f''d\mu'' < +\infty$. Then there exists a μ'-moderate function $g' \geq 0$ such that $f' = g'$ locally μ'-a.e. Similarly, there exists a μ''-moderate function $g'' \geq 0$ such that $f'' = g''$ locally μ''-a.e., and now $f' \otimes f'' = g' \otimes g''$ locally μ-a.e. (Proposition 9.2.3.). Therefore,

$$\begin{aligned}
\int^{\bullet} f' \otimes f''d\mu &= \int^* g' \otimes g''d\mu \\
&= \left(\int^* g'd\mu'\right)\left(\int^* g''d\mu''\right) \\
&= \left(\int^{\bullet} f'd\mu'\right)\left(\int^{\bullet} f''d\mu''\right).
\end{aligned}$$

$\qquad\square$

Proposition 9.2.5 *Let N be a subset of Ω. Denote by N' the set of those $x' \in \Omega'$ such that the section $N(x',\cdot) = \{x'' : (x',x'') \in N\}$ of N determined by x' is not μ''-negligible. The set N' is μ'-negligible when N is μ-negligible, and N' is locally μ'-negligible when N is locally μ-negligible and μ'' is moderate.*

PROOF: Because $\int^* 1_N d\mu \geq \int^* d\mu'(x') \int^* 1_{N(x',\cdot)} d\mu''$, the first assertion is obvious. The second follows directly from the first one.   □

Theorem 9.2.3 *Let f be a μ-measurable mapping from Ω into a metrizable space, and let N' be the set of those $x' \in \Omega'$ for which $f(x',\cdot)$ is not μ''-measurable. The set N' is μ'-negligible when f is constant outside a μ-moderate set. N' is locally μ'-negligible when μ'' is moderate.*

PROOF: Suppose that f takes the constant value z outside $\bigcup_{i \in I} A_i$, where $(A_i)_{i \in I}$ is a finite or countable family of disjoint nonempty S-sets. For every $i \in I$, there exists a μ-negligible subset N_i of A_i such that $f|A_i$ is the limit, on $A_i - N_i$, of a sequence of $\mathcal{R}|A_i$-simple mappings. There also exists a μ'-negligible subset N'_i of A'_i such that $N_i(x',\cdot)$ is μ''-negligible for all x' in $A'_i - N'_i$.

If $x' \notin \bigcup_{i \in I} A'_i$, then $f(x',\cdot) = z$, so that $f(x',\cdot)$ is μ''-measurable. Now let $x' \in (\bigcup_{i \in I} A'_i) - (\bigcup_{i \in I} N'_i)$, and define $J = \{i \in I : x' \in A'_i\}$. The A''_i, for i in J, are disjoint and, for each i in J, $f(x',\cdot)|A''_i$ is the limit, on $A''_i - N_i(x',\cdot)$, of a sequence of $\mathcal{R}''|A''_i$-simple mappings. It follows that $f(x',\cdot)|A''_i$ is measurable $\mathcal{M}''/A''_i$ (where $\mathcal{M}''$ denotes the σ-algebra of μ''-measurable sets) and that $f(x',\cdot)(A''_i - N_i(x',\cdot))$ is separable. $f(x',\cdot) = z$ outside $\bigcup_{i \in J} A''_i$. Therefore, $f(x',\cdot)$ is measurable $\mathcal{M}''$ and $f(x',\cdot)(\Omega'' - \bigcup_{i \in J} N_i(x',\cdot))$ is separable, so $f(x',\cdot)$ is μ''-measurable.

To prove the second assertion, let $A' \in S'$. The mapping g equal to f on $A' \times \Omega''$ and equal to z outside $A' \times \Omega''$ is μ-measurable, which shows that $N' \cap A'$ is μ'-negligible. Hence N' is locally μ'-negligible and the proof is complete.   □

We now return to the case where μ' and μ'' are complex measures.

Theorem 9.2.4 (Fubini) *Let f be a mapping from Ω into a Banach space F, and let N' be the set of those $x' \in \Omega'$ for which $f(x',\cdot)$ is not μ''-integrable. When f is μ-integrable, N' is μ'-negligible and $x' \mapsto \int f(x',\cdot) d\mu''$ (which is defined for $x' \notin N'$) is μ'-integrable; moreover,*

$$\int f d\mu = \int d\mu'(x') \int f(x',\cdot) d\mu''.$$

When μ'' is moderate and f is essentially μ-integrable, then N' is locally μ'-negligible, the mapping $x' \mapsto \int f(x',\cdot) d\mu''$ is essentially μ'-integrable, and $\int f d\mu = \int d\mu'(x') \int f(x',\cdot) d\mu''$.

PROOF: Assume that f is μ-integrable. Let $(\varphi_n)_{n\geq 1}$, a sequence in $St(S, F)$, and let $h \in \mathcal{J}^+$ be such that

(a) $(\varphi_n)_{n\geq 1}$ converges to f on the complement of a μ-negligible set N,

(b) $|\varphi_n| \leq h$ for all $n \geq 1$ and $\int^* h dV\mu < +\infty$.

There exists a μ'-negligible subset A' of Ω' such that $N(x', \cdot)$ is μ''-negligible for all $x' \notin A'$. Therefore, $\lim_{n\to\infty} \varphi_n(x', x'') = f(x', x'')$ μ''-a.e. for all $x' \notin A'$. On the other hand, because $\int^* dV\mu' \int^* h(x', \cdot)dV\mu''$ is finite, there exists a μ'-negligible subset B' of Ω' such that $\int^* h(x', \cdot)dV\mu'' < +\infty$ for all $x' \notin B'$. If $x' \notin A' \cup B'$, the sequence $(\varphi_n(x', \cdot))_{n\geq 1}$ converges to $f(x', \cdot)$ μ''-a.e. and, for $n \geq 1$, $|\varphi_n(x', \cdot)|$ is dominated by $h(x', \cdot)$. It follows that $f(x', \cdot)$ is μ''-integrable and $\int f(x', \cdot)d\mu'' = \lim_{n\to\infty} \int \varphi_n(x', \cdot)d\mu''$. Denote by ψ_n the mapping $x' \mapsto \int \varphi_n(x', \cdot)d\mu''$ from Ω' into F. By what we have just shown, $\psi_n(x')$ converges to $\int f(x', \cdot)d\mu''$ as $n \to \infty$, for all $x' \notin A' \cup B'$. Moreover,

$$|\psi_n(x')| \leq \int |\varphi_n(x', \cdot)|dV\mu'' \leq \int^* h(x', \cdot)dV\mu''$$

for all $x' \in \Omega'$, and $\int^* dV\mu' \int^* h(x', \cdot)dV\mu''$ is finite. As a consequence, $x' \mapsto \int f(x', \cdot)d\mu''$, defined on $(A' \cup B')^c$, is μ'-integrable and

$$\int d\mu' \int f(x', \cdot)d\mu'' = \lim_{n\to\infty} \int \psi_n d\mu'$$
$$= \lim_{n\to\infty} \int \varphi_n d\mu$$
$$= \int f d\mu.$$

Next, assume that μ'' is moderate and that f is essentially μ-integrable. Let g be a μ-integrable mapping from Ω into F equal to f outside a locally μ-negligible set N. By Proposition 9.2.5 there is a locally μ'-negligible subset E' of Ω' such that $N(x', \cdot)$ is μ''-negligible for all $x' \notin E'$. Put $F' = \{x' \in \Omega' : g(x', \cdot)$ is not μ''-integrable $\}$. Then $f(x', \cdot)$ is μ''-integrable for all $x' \notin E' \cup F'$ and $\int f(x', \cdot)d\mu'' = \int g(x', \cdot)d\mu''$. The second assertion follows. $\quad\square$

Theorem 9.2.5 *Let f be a μ-measurable function from Ω into $[0, +\infty]$. If f is μ-moderate, the functions*

$$x' \mapsto \int^* f(x', \cdot)dV\mu'' \quad and \quad x'' \mapsto \int^* f(\cdot, x'')dV\mu'$$

are measurable and moderate for μ' and μ'', respectively; furthermore,

$$\int^* f dV\mu = \int^* dV\mu'(x') \int^* f(x', \cdot)dV\mu''$$
$$= \int^* dV\mu''(x'') \int^* f(\cdot, x'')dV\mu'.$$

If μ'' is moderate, the function $x' \mapsto \int^\bullet f(x', \cdot)dV\mu''$ is μ'-measurable and

$$\int^\bullet fdV\mu = \int^\bullet dV\mu'(x') \int^\bullet f(x', \cdot)dV\mu''.$$

PROOF: Assume that f is μ-moderate and let $(A_n)_{n \geq 1}$ be an increasing sequence of μ-integrable sets such that f vanishes on $(\bigcup_n A_n)^c$. For each integer $n \geq 1$, define $f_n = \inf(f, n)1_{A_n}$. By Fubini's theorem, the function $x' \mapsto \int f_n(x', \cdot)dV\mu''$ is defined μ'-a.e. and is μ'-integrable. Therefore, $g_n :$ $x' \mapsto \int^* f_n(x', \cdot)dV\mu''$ is μ'-measurable. Moreover, $\int^* g_n dV\mu' = \int^* f_n dV\mu$. The first assertion follows.

Suppose now that μ'' is moderate and let A' be a μ'-measurable and μ'-moderate set. The mapping $(x', x'') \mapsto f(x', x'')1_{A'}(x')$ is μ-measurable and μ-moderate, so $x' \mapsto 1_{A'}(x')\int^* f(x', \cdot)dV\mu''$ is μ'-measurable and

$$\int^* dV\mu'(x')1_{A'}(x') \int^* f(x', \cdot)dV\mu'' = \int^* f \cdot 1_{A' \times \Omega''}dV\mu \leq \int^\bullet fdV\mu.$$

This proves that $x' \mapsto \int^* f(x', \cdot)dV\mu''$ is μ'-measurable and that

$$\int^\bullet dV\mu' \int^* f(x', \cdot)dV\mu'' \leq \int^\bullet fdV\mu.$$

Now let A be a μ-measurable and μ-moderate subset of Ω. The function $x' \mapsto \int^* 1_A(x', \cdot)f(x', \cdot) \, dV\mu''$ is μ'-moderate and

$$\int^* f \cdot 1_A dV\mu = \int^\bullet dV\mu' \int^* 1_A(x', \cdot)f(x', \cdot)dV\mu''$$

$$\leq \int^\bullet dV\mu' \int^* f(x', \cdot)dV\mu''.$$

Therefore,

$$\int^\bullet fdV\mu \leq \int^\bullet dV\mu'(x') \int^\bullet f(x', \cdot)dV\mu''$$

which completes the proof. □

Theorem 9.2.6 (Tonelli) *Let f be a μ-measurable and μ-moderate mapping from Ω into a real Banach space. Then f is μ-integrable if and only if either of $\int^* dV\mu'(x') \int^* |f|(x', \cdot)dV\mu''$ or $\int^* dV\mu''(x'') \int^* |f|(\cdot, x'')dV\mu'$ is finite.*

PROOF: This is a direct consequence of Theorem 9.2.5. □

Proposition 9.2.6   *Let F', F'', and F be three Banach spaces, and let $(u', u'') \mapsto u' \cdot u''$ be a continuous bilinear mapping from $F' \times F''$ into F. Let f' (respectively, f'') be an essentially μ' (respectively, μ'') -integrable mapping from Ω' into F' (respectively, from Ω'' into F''). Then $f : (x', x'') \longmapsto f'(x') \cdot f''(x'')$ is essentially μ-integrable and $\int f d\mu = (\int f' d\mu')(\int f'' d\mu'')$. If f' and f'' are integrable, then f is μ-integrable.*

PROOF: By Theorem 9.2.2, f is μ-measurable. On the other hand, if b is the norm of the bilinear mapping $(u', u'') \mapsto u' \cdot u''$,

$$\int^\bullet |f| dV\mu \le b \int^\bullet |f'| \otimes |f''| dV\mu = b \left(\int^\bullet |f'| dV\mu' \right) \left(\int^\bullet |f''| dV\mu'' \right)$$

by Proposition 9.2.4. This shows that f is essentially μ-integrable. Suppose that f' and f'' are integrable. Then f is μ-moderate, and therefore μ-integrable, and Fubini's theorem shows that $\int f d\mu = \left(\int f' d\mu' \right) \left(\int f'' d\mu'' \right)$. Finally, we return to the case where f' and f'' are essentially integrable. Let g' (respectively, g'') be a mapping from Ω' into F' (respectively, from Ω'' into F'') equal to f' (respectively, f'') locally a.e. Then $g : (x', x'') \mapsto g'(x') \cdot g''(x'')$ is equal to f locally a.e. (Proposition 9.2.3) and

$$
\begin{aligned}
\int f d\mu &= \int g d\mu \\
&= \left(\int g' d\mu' \right) \left(\int g'' d\mu'' \right) \\
&= \left(\int f' d\mu' \right) \left(\int f'' d\mu'' \right).
\end{aligned}
$$

$\square$

Let Y' be a μ'-measurable subset of Ω', let Y'' be a μ''-measurable subset of Ω'', and let T' (respectively, T'') be a semiring in Y' (respectively, Y''). If $\mu'_{/T'}$ and $\mu''_{/T''}$ exist, then $(\mu' \otimes \mu'')_{/(T' \times T'')}$ exists and is equal to $\mu'_{/T'} \otimes \mu''_{/T''}$.

As we shall see later, all the preceding results are true even for Radon measures. This, however, is not the case for Proposition 9.2.7. which is especially useful in probability theory.

Proposition 9.2.7 *Let Ω' and Ω'' be two nonempty sets. Let μ'' be a positive measure on a semiring S'' in Ω''. Let S' be a semiring in Ω', and put $S = S' \times S''$. Suppose that $\Omega' \in \sigma(S')$ and $\Omega'' \in \sigma(S'')$. If $f : \Omega \to [0, +\infty]$ is S-Borelian (i.e., measurable $\sigma(S)$), then $f(x', \cdot)$ is S''-Borelian and the function $x' \mapsto \int^* f(x', \cdot) d\mu''$ is S'-Borelian.*

PROOF: Let $A = A' \times A'' \in S$ be given. Denote by $\mathcal{L}$ the class of those $E \subset \Omega$ such that $(E \cap A)(x', \cdot)$ is an $\widetilde{S''}$-set for all $x' \in X'$, and such that the function $x' \mapsto \int 1_{E \cap A}(x', \cdot) d\mu''$ is S'-Borelian. Then $\mathcal{L}$ is a λ-system and contains S, so $\mathcal{L} \supset \widetilde{S}$.

Fix $E \in \widetilde{S}$, and let $(A'_n \times A''_n)_{n \ge 1}$ be a sequence of disjoint S-sets whose union contains E. Put $E_n = E \cap (A'_n \times A''_n)$ for each $n \ge 1$. Then $E(x', \cdot) = \bigcup_{n \ge 1} E_n(x', \cdot)$ is an $\widetilde{S''}$-set for all $x' \in \Omega'$, and the function

$$x' \mapsto \int^* 1_E(x', \cdot) d\mu'' = \sum_{n \ge 1} \int^* 1_{E_n}(x', \cdot) d\mu''$$

is S'-Borelian. Thus, if g belongs to $St^+(\widetilde{S})$, then $g(x', \cdot)$ is S''-Borelian and the function $x' \mapsto \int^* g(x', \cdot) d\mu''$ is S'-Borelian.

Finally, if $f : \Omega \to [0, +\infty]$ is S-Borelian, there is an increasing sequence $(f_n)_{n \geq 1}$ in $St^+(\widetilde{S})$ admitting f as its upper envelope. The result follows. $\square$

What we have done so far for the product of two measures extends easily to finite products of measures. The details are left to the reader.

If $\mu_1, \ldots, \mu_n$ are complex measures on semirings $S_1, \ldots, S_n$ in $\Omega_1, \ldots, \Omega_n$, with product measure μ, and if f is a μ-integrable mapping from $\Omega = \Omega_1 \times \ldots \times \Omega_n$ into a Banach space, we write

$$\iint \ldots \int f d\mu_1 d\mu_2 \ldots d\mu_n \quad \text{or} \quad \iint \ldots \int f(x_1, \ldots, x_n) d\mu_1(x_1) \ldots d\mu_n(x_n)$$

instead of $\int f d\mu$, and retain analogous notations for upper integrals with respect to $V\mu$.

9.3   Lebesgue Measure on $\mathbf{R}^k$

In the Euclidean space $\mathbf{R}^k$, we consider the order:

$$(a_1, \ldots, a_k) \leq (b_1, \ldots, b_k) \qquad \text{if } a_i \leq b_i \text{ for every } 1 \leq i \leq k.$$

For all $a, b \in \mathbf{R}^k$ satisfying $a < b$, we define $]a, b]$ as the k-dimensional rectangle $\prod_{1 \leq i \leq k}]a_i, b_i]$. By what has been shown at the beginning of Section 9.2, the empty set and these rectangles $]a, b]$ form a semiring.

Definition 9.3.1 Let Ω be a nonempty open subset of $\mathbf{R}^k$, and let S be the class consisting of the empty set and the rectangles A whose closure $\bar{A}$ is contained in Ω. S is called the natural semiring in Ω.

Proposition 9.3.1 *In the notation of Definition 9.3.1, Ω is a countable union of disjoint S-rectangles, each of which may be assumed to be of the form $\prod_{1 \leq i \leq k}]p_i / 2^m, (p_i + 1) / 2^m]$ $(m \geq 0, p_i \in \mathbf{Z})$.*

PROOF: For each integer $m \geq 0$, let Q_m be the class consisting of all rectangles of the form $\prod_{1 \leq i \leq k}]p_i / 2^m, (p_i + 1) / 2^m]$ $(p_i \in \mathbf{Z}$ for all $1 \leq i \leq k)$. Put

$$P_0 = \{A \in Q_0 : \bar{A} \subset \Omega\}$$
$$P_1 = \{A \in Q_1 : \bar{A} \subset \Omega, \ A \cap B = \emptyset \ \text{ for all } B \in P_0\}$$
$$\vdots$$
$$P_m = \{A \in Q_m : \bar{A} \subset \Omega, \ A \cap B = \emptyset \ \text{ for all } B \in \textstyle\bigcup_{0 \leq j \leq m-1} P_j\}$$
$$\vdots$$

The elements of $P = \bigcup_{m \geq 0} P_m$ are clearly disjoint S-rectangles. Let $x \in \mathbf{R}^k$, $x \notin \bigcup_{A \in P} A$, and, for each $m \geq 0$, let A_m be the element of Q_m containing x.

Suppose that $\overline{A_m}$ is included in Ω for some integer $m \geq 0$, and denote by r the smallest integer in $\{0, \ldots, m\}$ such that $\overline{A_r} \subset \Omega$. Then, for every $0 \leq s \leq r - 1$, $A_s \cap B = \emptyset$ for all $B \in P_s$, because A_s does not belong to P_s and because the elements of Q_s are disjoint. Now $A_r \subset A_s$ for all $0 \leq s \leq r-1$, so A_r lies in P_r, which contradicts the fact that x does not belong to $\bigcup_{A \in P} A$.

Therefore, $\overline{A_m} \cap \Omega^c \neq \emptyset$ for each $m \geq 0$, and we can choose a point x^m in $\overline{A_m} \cap \Omega^c$. Since $|x_i^m - x_i| \leq 1/2^m$ for all $1 \leq i \leq k$, the sequence $(x^m)_{m \geq 0}$ converges to x, and x belongs to the closed set Ω^c. This shows that $\Omega = \bigcup_{A \in P} A$, and the proof is complete. $\square$

By Proposition 9.3.1, each nonempty open subset of Ω is a countable union of disjoint S-rectangles, hence the σ-ring generated by S contains the open subsets of Ω, and it is the Borel σ-algebra in Ω.

Now let λ_1 be Lebesgue measure on the natural semiring of $\mathbf{R}$, and put $\lambda_k = \lambda_1 \otimes \cdots \otimes \lambda_1$.

Definition 9.3.2 Given Ω, a nonempty open subset of $\mathbf{R}^k$, we call Lebesgue measure on Ω the measure μ induced by λ_k on the natural semiring S of Ω.

So $\mu(A) = \prod_{1 \leq i \leq k}(b_i - a_i)$ is the volume of $A = \prod_{1 \leq i \leq k}]a_i, b_i]$, for every S-rectangle A.

Exercises for Chapter 9

1 Write λ for Lebesgue measure on $\mathbf{R}$. Let α be a nonzero complex number whose real part is negative.

1. Prove that $x \mapsto e^{\alpha x}/\sqrt{x}$ is $\lambda_{/]0,+\infty[}$-integrable when $\mathrm{Re}(\alpha)$ is strictly negative, but not when $\alpha = i\beta$ lies on the imaginary axis. However, in this last case, show that $\int_y^z e^{i\beta x}/\sqrt{x}\,dx$ has a limit. denoted by $\int_0^{+\infty} e^{i\beta x}/\sqrt{x}\,dx$, as $y \to 0$ in $]0, +\infty[$ and $z \to +\infty$ (use the second mean value formula). In what follows, $F(\alpha)$ stands for $\int_0^{+\infty} e^{\alpha x}/\sqrt{x}\,dx$.

2. For all integers $n \geq 1$ and $p \geq 1$, define the functions f_n and g_p on $]0, +\infty[$ by

$$f_n(t) = \int_{1/n}^n \frac{1}{\sqrt{t}} e^{x(\alpha - t)}\,dx \qquad \text{and} \qquad g_p(x) = \int_{1/p}^{+\infty} \frac{1}{\sqrt{t}} e^{x(\alpha - t)}\,dt.$$

Show that $\int_{1/p}^{+\infty} f_n(t)\,dt = \int_{1/n}^n g_p(x)\,dx$.

3. Let g be the function

$$x \mapsto \int_0^{+\infty} \frac{1}{\sqrt{t}} e^{x(\alpha - t)}\,dt = F(-1)\frac{1}{\sqrt{x}} e^{\alpha x}$$

from $]0, +\infty[$ into $\mathbf{C}$. Given $n \in \mathbf{N}$, show that $(g_p)_{p \geq 1}$ converges uniformly to g on $[1/n, n]$, and deduce that $\int_{1/n}^n g(x)\,dx = \int_0^{+\infty} f_n(t)\,dt$.

4. Show that $|f_n(t)| \leq 2/(\sqrt{t} \cdot |t - \alpha|)$ for all $n \in \mathbf{N}$ and all $t \in]0, +\infty[$. Deduce that $F(-1)F(\alpha) = \int_0^{+\infty} 1/(\sqrt{t} \cdot (t - \alpha))\,dt$.

5. Deduce from part 4 that $F(-1) = \sqrt{\pi}$ and $F(i) = \sqrt{\pi} e^{i\pi/4}$.

2

1. Let h be the function $(x, y) \mapsto e^{-ix} - y$ from $]0, 2\pi[\times] - 1, +\infty[$ into $\mathbf{C}$. Show that, for every integer $m \in \mathbf{Z}^+$ and every compact subset K of $]0, 2\pi[$, there exists a constant $C_{K,m}$ such that $\left| \dfrac{\partial^m}{\partial x^m}\left(\dfrac{1}{h}\right)(x, y) \right| \leq C_{K,m}$ for all (x, y) in $K \times [0, 1]$.

2. For each $n \in \mathbf{N}$, let S_n be the function $x \mapsto \displaystyle\sum_{1 \leq k \leq n} \dfrac{e^{ikx}}{\sqrt{k}}$ from $]0, 2\pi[$ into $\mathbf{C}$.
    Show that $S_n(x) = \dfrac{1}{\sqrt{\pi}} \displaystyle\int_0^{+\infty} \dfrac{e^{ix-t} - e^{(n+1)(ix-t)}}{1 - e^{ix-t}} \cdot \dfrac{1}{\sqrt{t}}\,dt$ for all $x \in]0, 2\pi[$.

3. Given x in a compact subset K of $]0, 2\pi[$, prove that
    $$\left| \dfrac{1}{\sqrt{\pi}} \int_0^{+\infty} \dfrac{e^{(n+1)(ix-t)}}{1 - e^{ix-t}} \cdot \dfrac{1}{\sqrt{t}}\,dt \right| \leq \dfrac{C_{K,0}}{\sqrt{n+1}}$$

    (notation of part 1). Deduce that S_n converges uniformly on the compact subsets of $]0, 2\pi[$ to the function $S : x \mapsto \dfrac{1}{\sqrt{\pi}} \displaystyle\int_0^{+\infty} \dfrac{e^{ix-t}}{1 - e^{ix-t}} \cdot \dfrac{1}{\sqrt{t}}\,dt$. Prove that S is infinitely differentiable.

4. Let $x \in]0, 2\pi[$. Prove that $\left| \dfrac{e^{ix} - 1}{ix} \cdot \dfrac{e^{ikx}}{\sqrt{k}} - \displaystyle\int_k^{k+1} \dfrac{e^{itx}}{\sqrt{t}}\,dt \right| \leq \dfrac{1}{2}k^{-3/2}$ for all integers $k \geq 1$, and deduce that $\left| \sqrt{x} \cdot \dfrac{e^{ix} - 1}{ix} \cdot S(x) - \displaystyle\int_x^{+\infty} \dfrac{e^{iu}}{\sqrt{u}}\,du \right| \leq$
    $\dfrac{\sqrt{x}}{2} \cdot \zeta\left(\dfrac{3}{2}\right)$ (where $\zeta(3/2) = \sum_{k \geq 1} k^{-3/2}$)

5. Deduce from part 4 that $S(x)$ is asymptotic to $\sqrt{\pi} \cdot e^{i\pi/4} \dfrac{1}{\sqrt{x}}$ as x tends to 0 in $]0, 2\pi[$. Now, show that $S(2\pi - x)$ is asymptotic to $\sqrt{\pi} \cdot e^{-i\pi/4} \dfrac{1}{\sqrt{x}}$, and conclude that S is μ-integrable ($\mu = \dfrac{1}{2\pi} \cdot \lambda_{]0,2\pi[}$).

6. Put $M = \sup \left\{ \left| \displaystyle\int_y^z \dfrac{e^{iu}}{\sqrt{u}}\,du \right| : y, z \text{ in }]0, +\infty[\right\}$. Given $n \in \mathbf{N}$, show that $\left| \dfrac{e^{ix} - 1}{ix} \cdot S_n(x) \right| \leq \dfrac{M}{\sqrt{x}} + \dfrac{1}{2}\zeta\left(\dfrac{3}{2}\right)$ for all $x \in]0, 2\pi[$ (use the first inequality in part 4).

7. Deduce from part 6 that $S_n \cdot 1_{]0,\pi]}$ converges to $S \cdot 1_{]0,\pi]}$ in $L^1_{\mathbf{C}}(\mu)$ (i.e., in the mean), and also that $S_n \cdot 1_{[\pi,2\pi[}$ converges to $S \cdot 1_{[\pi,2\pi[}$ in the mean.

Conclude that S_n converges to S in the mean and that

$$\frac{1}{2\pi} \int_0^{2\pi} S(x) e^{-\imath kx} \, dx = \begin{cases} 0 & \text{for } k < 0 \\ 1/\sqrt{k} & \text{for } k \geq 1 \end{cases} \quad (k \in \mathbf{Z}).$$

8. Prove that the Fourier series of S diverges at all the points $2k\pi$ $(k \in \mathbf{Z})$.

3 Let μ be Lebesgue measure on $]0,1]$ and f the function $(x,y) \longmapsto$ $(x^2 - y^2)/(x^2 + y^2)^2$ from $\Omega =]0,1] \times]0,1]$ into $\mathbf{R}$.

1. Show that $\int^* |f| \, d(\mu \otimes \mu) = +\infty$.

2. Prove that the functions $x \mapsto \int f(x,y) \, d\mu(y)$ and $y \mapsto \int f(x,y) \, d\mu(x)$ are defined on $]0,1]$ and are integrable. Show that $\int d\mu(x) \int f(x,y) \, d\mu(y) \neq \int d\mu(y) \int f(x,y) \, d\mu(x)$.

4 Let Ω be an uncountable set and $\mathcal{F}$ the σ-algebra in Ω consisting of the countable and co-countable sets. Define a measure μ on $\mathcal{F}$ by $\mu(A) = 0$ if A is countable, and $\mu(A) = 1$ if A is co-countable.

1. Prove that $\Delta = \big\{ (x,y) \in \Omega \times \Omega : x = y \big\}$ is not $\mu \otimes \mu$-negligible.

2. Conclude that Δ is not $\mu \otimes \mu$-measurable, even though the sections $\Delta(x, \cdot)$ and $\Delta(\cdot, y)$ are μ-measurable for $x \in \Omega$ and $y \in \Omega$.

5 Denote by λ_2 Lebesgue measure on $\mathbf{R}^2$.

We construct a continuous and injective function g from $[0.7]$ into $\mathbf{R}^2$ such that $g\big([0,7] \big)$ is not λ_2-negligible.

For each real number $0 < \alpha < 1$, let $Q(\alpha)$ be the subset of $\mathbf{R}^2$ which is the complement of the union of $]-1, 1[\times]-\alpha, \alpha[$ and $]-\alpha, \alpha[\times]-1, 1[$ in the square $Q = [-1,1] \times [-1,1]$. The set $Q(\alpha)$ is the union of its four connected components

$$Q_0(\alpha) = [-1, -\alpha] \times [-1, -\alpha] \qquad Q_1(\alpha) = [-1, -\alpha] \times [\alpha, 1]$$
$$Q_2(\alpha) = [\alpha, 1] \times [\alpha, 1] \qquad Q_3(\alpha) = [\alpha. 1] \times [-1, -\alpha].$$

Let $H_{0,\alpha}, h_{1,\alpha}, h_{2,\alpha}, H_{3,\alpha}$ be the homotheties of $\mathbf{R}^2$, with ratio $(1-\alpha)/2$, whose respective centers are $(-1,-1)$, $(-1,1)$, $(1,1)$, $(1,-1)$. So these homotheties map Q onto $Q_0(\alpha)$, $Q_1(\alpha)$, $Q_2(\alpha)$, $Q_3(\alpha)$, respectively. On the other hand, let $R_{0,\alpha}$ be the symmetry of $\mathbf{R}^2$ with respect to the line through $(-1,-1)$ and $(1,1)$, and similarly let $R_{3,\alpha}$ be the symmetry of $\mathbf{R}^2$ with respect to the line through $(1,-1)$ and $(-1,1)$. Finally, put $h_{0,\alpha} = R_{0,\alpha} \circ H_{0,\alpha}$ and $h_{3,\alpha} = R_{3,\alpha} \circ H_{3,\alpha}$. For each integer $0 \leq k \leq 6$, define u_k the function from $[0,7]$ into $[0,7]$, the restriction of an affine function from $\mathbf{R}$ into $\mathbf{R}$, which sends 0 to k and 7 to $k+1$, so that u_k maps $[0,7]$ onto $I_k = [k, k+1]$.

1. Let f_α be the function from $[0,7]$ into Q which is affine on each of the intervals I_k $(0 \leq k \leq 6)$ and is such that the images of $0, 1, 2, \ldots, 7$ are, respectively, $(-1,-1)$, $(-1,-\alpha)$, $(-1,\alpha)$, $(-\alpha,\alpha)$, (α,α), $(1,\alpha)$, $(1,-\alpha)$, $(1,-1)$. Prove that $f_\alpha(2j) = h_{j,\alpha}(-1,-1)$ and $f_\alpha(2j+1) = h_{j,\alpha}(1,-1)$ for all $0 \leq j \leq 3$.

2. For each integer $n \geq 1$ and each sequence $s = (i_1, \ldots, i_n)$ of n terms in $\{0, 1, 2, 3, 4, 5, 6\}$, put $v_s = u_{i_1} \circ u_{i_2} \circ \cdots \circ u_{i_n}$. Show that $v_s(x) = i_1 + i_2/7 + \cdots + i_n/7^{n-1} + x/7^n$ for every $x \in [0, 7]$. Prove that a point $y \in [0, 7[$ lies in $v_s([0, 7[)$ if and only if its proper base-7 expansion has the form $\sum_{p \geq 1} k_p/7^{p-1}$, with $k_1 = i_1, \ldots, k_n = i_n$. Conclude, for fixed $n \geq 1$, that the $v_s([0, 7[)$ are disjoint and their union is $[0, 7[$.

3. Fix $n \geq 1$. Show that $y \in]0, 7[$ lies in two of the $v_s([0, 7])$ if and only if it has the form $r/7^{n-1}$ for some integer $0 < r < 7^n$. In this case, if $y = \sum_{p \geq 1}(k_p/7^{p-1})$ and $y = \sum_{p \geq 1}(l_p/7^{p-1})$ are respectively the proper expansion of y and the improper expansion of y, prove that the unique elements $s = (i_1, \ldots, i_n)$ such that y lies in $v_s([0, 7])$ are $(l_1, \ldots, l_n)$ and $(k_1, \ldots, k_n)$.

4. Let $(\alpha_n)_{n \geq 1}$ be a sequence in $]0, 1[$ such that $\sum_{n \geq 1} \alpha_n = +\infty$. Thus $\prod_{n \geq 1}(1 - \alpha_n) = 0$. Define a sequence $(g_n)_{n \geq 1}$ of continuous functions from $[0, 7]$ into Q as follows: $g_1 = f_{\alpha_1}$; for $n \geq 1$, suppose $g_1, \ldots, g_n$ have been obtained, and define $g_{n+1}(v_s(t))$ for $0 \leq t \leq 7$ and for each of the 7^n sequences $s = (i_1, \ldots, i_n)$; if one at least of the i_r is odd, put $g_{n+1}(v_s(t)) = g_n(v_s(t))$; on the other hand, if $i_1, \ldots, i_n$ are all even integers and if $j_r = (1/2)i_r$ for all $1 \leq r \leq n$, then put $g_{n+1}(v_s(t)) = w_s(f_{\alpha_{n+1}}(t))$, where $w_s = h_{j_1, \alpha_1} \circ \cdots \circ h_{j_n, \alpha_n}$. Prove, for fixed $n \geq 1$, that, if $y \in]0, 7[$ lies in two of the $v_s([0, 7])$ and $(k_1, \ldots, k_n)$, $(l_1, \ldots, l_n)$ are as in part 3, the definitions of g_{n+1} on $v_{(l_1, \ldots, l_n)}([0, 7])$ and on $v_{(k_1, \ldots, k_n)}([0, 7])$ give $g_{n+1}(y)$ the same value, so that the definition of g_{n+1} is consistent.

5. Let $y = \sum_{r \geq 1} i_r/7^{r-1}$ be a point of $[0, 7]$, and let m be the greatest integer of $\{0, 1, \ldots, n\}$ such that $i_1, \ldots, i_m$ are all even. Prove that $g_{n+1}(y) = (h_{j_1, \alpha_1} \circ \cdots \circ h_{j_m, \alpha_m}) \cdot f_{\alpha_{m+1}}\left(\sum_{r \geq m+1} i_r/7^{r-(m+1)}\right)$, where $j_r = i_r/2$ for all $1 \leq r \leq m$.

6. Let d be the distance on $\mathbf{R}^2$ associated with the norm $\|(a, b)\| = \sup(|a|, |b|)$. Show that $d(g_n(y), g_{n+p}(y)) \leq (1 - \alpha_1) \cdots (1 - \alpha_n)$ for all integers $n \geq 1$ and $p \geq 1$ and every $y \in [0, 7]$. Deduce that the sequence $(g_n)_{n \geq 1}$ converges uniformly on $[0, 7]$ to a continuous function g.

7. Prove that g is injective.

8. Prove that $g([0, 7])$ is λ_2-negligible if $2^n \cdot \prod_{1 \leq k \leq n}(1 - \alpha_k)$ converges to 0 as $n \to +\infty$ (argue by contradiction).

6   In this exercise, we prove that the hypotheses of a classical Sard's theorem are optimal.

1. In the notation of exercise 5, for each integer $n \geq 0$, for each $s = (i_1, \ldots, i_n)$ of $\{0, 2, 4, 6\}^n$, and for each integer $1 \leq k \leq 3$, put

$$K_{s,k} = g_{n+1}\left(v_s([2k - 1, 2k])\right).$$

So $K_{s,k} = (h_{j_1, \alpha_1} \circ \cdots \circ h_{j_n, \alpha_n}) \cdot f_{\alpha_{n+1}}([2k - 1, 2k])$ where it is understood that $j_1 = i_1/2, \ldots, j_n = i_n/2$. Prove that the sets $K_{s,k}$ (for all $n \geq 0$ and all s and k) are disjoint.

2. Define a sequence $(K_n)_{n\geq 1}$ of compact sets in $\mathbf{R}^2$ as follows: K_1 is the union of $\{(-1,-1)\}$, $\{(1,-1)\}$ and $g_1([1,2])$, $g_1([3,4])$, $g_1([5,6])$; for each integer $n \geq 1$, K_{n+1} is the union of K_n and the $K_{s,k}$ (where $s \in \{0, 2, 4, 6\}^n$ and $1 \leq k \leq 3$).

   Next, define a sequence $(F_n)_{n\geq 1}$ of functions by putting $F_1((-1,-1)) = 0$, $F_1((1,-1)) = 1$, and putting F_1 equal to $k/4$ on $g_1([2k-1, 2k])$ for $1 \leq k \leq 3$; for all $n \geq 1$, $F_{n+1} = F_n$ on K_n, and F_{n+1} is equal to $(1 - k/4) \cdot F_n(g_{n+1} \cdot v_s(0)) + (k/4) \cdot F_n(g_{n+1} \cdot v_s(7))$ on each $K_{s,k}$ (with $s \in \{0, 2, 4, 6\}^n$ and $1 \leq k \leq 3$).

   Let G be the function, defined on $\bigcup_{n\geq 1} K_n$, equal to F_n on each K_n. Prove that, for all $n \geq 0$, for all $s = (i_1, \ldots i_n) \in \{0, 2, 4, 6\}^n$, and for all $1 \leq k \leq 3$, G takes the value $j_1/4 + \cdots + j_n/4^n + k/4^{n+1}$ on $K_{s,k}$.

3. Let $y = \sum_{r\geq 1}(i_r/7^{r-1})$ be a point of $[0, 7]$ ($0 \leq i_r \leq 6$ for all $r \geq 1$). Set $j_r = i_r/2$ for each integer $r \geq 1$ such that i_r is even. If one of the i_r is odd, and if n is the smallest positive integer such that $i_{n+1} = 2k - 1$ is odd, put $F(gy) = j_1/4 + \cdots + j_n/4^n + k/4^{n+1}$. On the other hand, if all the i_r are even integers, put $F(gy) = \sum_{r\geq 1}(j_r/4^r)$. Show that $F(gy)$ does not depend on any particular expansion of y when y has the form $p/7^q$ ($q \geq 0$, $0 < p < 7^{q+1}$). Thus we have defined a function F from $g([0, 7])$ into $[0, 1]$. Prove that F extends G and that $F(g([0,7])) = [0, 1]$.

4. Fix $z \in g([0, 7])$. Show that z lies in the closure of $D = \bigcup_{n\geq 1} K_n$. Moreover, if $\mathcal{V}$ designates the filter induced on D by the filter of neighborhoods of z in $\mathbf{R}^2$, prove that $G(\mathcal{V})$ converges to $F(z)$, so that F is continuous on $g([0,7])$.

5. Suppose that $\alpha_n \leq 1/3$ for all $n \geq 1$ (and $\sum_{n\geq 1} \alpha_n = +\infty$ as before). Prove that, for a suitably chosen sequence $(\alpha_n)_{n\geq 1}$, there exists a real number $c > 0$ such that, for arbitrary points z, z' of $g([0,7])$, $|F(z) - F(z')| \leq c \cdot \|z' - z\|^{1+\epsilon}$.

6. Suppose that $(\alpha_n)_{n\geq 1}$ has been chosen as in part 5. By Whitney's extension theorem, there exists a continuously differentiable function f from $\mathbf{R}^2$ into $\mathbf{R}$ which extends F and is such that $Df(z) = 0$ for all z of $g([0,7])$. Then all points of $g([0,7])$ are critical points of f, but $f(g([0,7])) = [0, 1]$ is not λ_1-negligible. Compare this result with Sard's theorem.

10

Radon-Nikodym Derivatives

In this chapter we consider the measures $g\mu$ that have a numerical density g with respect to a given measure μ, and we define mutually singular measures. (The case where μ is Lebesgue measure on an interval shall be dealt with in Chapter 13.) We show then that $L^q_{\mathbf{C}}(\mu)$ may be regarded as the dual of $L^p_{\mathbf{C}}(\mu)$ (with $1 \leq p < +\infty$, q exponent conjugate to p).

Summary

10.1 We define summable families of measures on a same semiring S.

10.2 Let μ be a measure on a semiring S whose underlying set is Ω. A function $g : \Omega \to \mathbf{C}$ is said to be locally μ-integrable whenever $g1_A$ is μ-integrable for every A in S; then the measure $g\mu : A \mapsto \int g1_A \, d\mu$ is called the measure with density g relative to μ. $V(g\mu) = |g|V\mu$ (Proposition 10.2.1), and $\int^\bullet f \, dV(g\mu) = \int^\bullet f|g| \, dV\mu$ for all functions $f : \Omega \to [0, +\infty]$ (Theorem 10.2.2). A mapping f from Ω into a Banach space is $g\mu$-measurable (respectively, essentially $g\mu$-integrable) if and only if fg is μ-measurable (respectively, essentially μ-integrable) (Theorem 10.2.1).

When μ is Lebesgue measure on an interval, let us observe that every function continuous on that interval is locally μ-integrable.

10.3 Let μ be a measure on a semiring S. Let $\mathcal{R}$ be the ring generated by S. A measure ν on S is said to be absolutely continuous with respect to μ if every μ-negligible $\sigma(S)$-set is ν-negligible. Another equivalent condition is the following condition. For every E in S and for every $\varepsilon > 0$, there exists $\delta > 0$ such that for all F in $\mathcal{R}$ contained in E and satisfying the inequality $|\mu|(F) \leq \delta$, we have $|\nu|(F) \leq \varepsilon$ (Theorem 10.3.2). This also amounts to saying that ν has a density g with respect to μ (Theorem 10.3.1, Radon–Nikodym). Measures μ and ν on S are said to be mutually singular whenever $\inf\big(|\mu|, |\nu|\big) = 0$. This means that μ and ν are concentrated on disjoint sets (Propositions 10.3.3 and 10.3.4). Every measure on S can be written, uniquely, as the sum of a measure $g\mu$ absolutely continuous with respect to μ and a measure ν such that μ and ν are mutually singular (Theorem 10.3.3).

10.4   We combine different operations on the measures.

10.5   We show that $L^q_{\mathbb{C}}(\mu)$ may be regarded as the dual of $L^p_{\mathbb{C}}(\mu)$ (with $1 \le p < +\infty$, q exponent conjugate to p) (Theorems 10.5.1 and 10.5.2). We also characterize those continuous linear functionals on $L^\infty_{\mathbb{C}}(\mu)$ that can be written $f \mapsto \int fg\,d\mu$ for g in $L^1_{\mathbb{C}}(\mu)$ (Proposition 10.5.1).

10.6   We describe the dual of $L^\infty_{\mathbb{C}}(\mu)$ (Proposition 10.6.1). A necessary and sufficient condition that this dual be equal to $L^1_{\mathbb{C}}(\mu)$ is that Ω be a union of a finite number of atoms and a locally μ-negligible set (Proposition 10.6.2).

10.1   Sums of Measures

Let Ω be a nonempty set, S a semiring in Ω, and $\mathcal{R}$ the ring generated by S.

Let μ_1, μ_2 be two positive measures on S, and put $\mu = \mu_1 + \mu_2$. Then $\int^* f\,d\mu = \int^* f\,d\mu_1 + \int^* f\,d\mu_2$ and $\int^\bullet f\,d\mu = \sup_{B \in \mathcal{R}} \int^* f \cdot 1_B\,d\mu = \int^\bullet f\,d\mu_1 + \int^\bullet f\,d\mu_2$ for any function f from Ω into $[0, +\infty]$. Next, let f be a mapping from Ω into a metrizable space. By Theorem 6.1.2, f is μ-measurable if and only if it is measurable with respect to both μ_1 and μ_2.

Now, for complex measures μ_1, μ_2 on S, put $\mu = \mu_1 + \mu_2$. By the preceding, a mapping f from Ω into a Banach space is $(V\mu_1 + V\mu_2)$-integrable (respectively, essentially integrable) if and only if it is integrable (respectively, essentially integrable) with respect both to μ_1 and μ_2. In this case, $\int f\,d(\mu_1 + \mu_2) = \int f\,d\mu_1 + \int f\,d\mu_2$.

Proposition 10.1.1 *Let H be an upward-directed set of positive measures on S, bounded above in $\mathcal{M}^+(S)$, and suppose μ is its supremum in $\mathcal{M}^+(S)$. Then $\mu^\bullet(f) = \sup_{\nu \in H} \nu^\bullet(f)$ for all functions f from Ω into $[0, +\infty]$.*

PROOF: First, assume that f is bounded and vanishes outside an $\mathcal{R}$-set E. Given $\varepsilon > 0$, we can find $\nu \in H$ so that $\nu(E) \ge \mu(E) - \varepsilon/\|f\|$ (Lemma 1.3.1); then $(\mu - \nu)^*(f) \le (\mu - \nu)^*(\|f\| \cdot 1_E) \le \varepsilon$ and $\nu^*(f) \ge \mu^*(f) - \varepsilon$, and the result follows.

Now suppose that f vanishes outside an $\mathcal{R}$-set E, but is not necessarily bounded. Since $f = \sup_{n \ge 0} \inf(f, n)$, we still have $\mu^*(f) = \sup_{\nu \in H} \nu^*(f)$.

Finally, since $\mu^\bullet(f) = \sup_{B \in \mathcal{R}} \mu^*(f \cdot 1_B)$, the result follows in the general case. □

Proposition 10.1.2 *Let H and μ be as in Proposition 10.1.1, and let f be a mapping from Ω into a metrizable space. Then f is μ-measurable if (and only if) it is ν-measurable for all $\nu \in H$.*

PROOF: Denote by $\mathcal{C}$ the class of those $\tilde{S}$-sets E such that $f_{/E}$ is measurable $\tilde{S}_{/E}$ and $f(E)$ is separable. Clearly, the union of a sequence of $\mathcal{C}$-sets is a $\mathcal{C}$-set.

Next, let K be a μ-integrable $\tilde{S}$-set and put $r = \sup_{E \in \mathcal{C}, E \subset K} \mu(E)$. By the preceding, there exists a $\mathcal{C}$-set L, included in K, such that $\mu(L) = r$. Suppose, to get a contradiction, that $\mu(K - L)$ is strictly positive. Then there exists $\nu \in H$ such that $\nu(K - L) > 0$. Since f is ν-measurable, we can find $M \in \mathcal{C}$ such that $M \subset K - L$ and $\nu(M) = \nu(K - L)$. Then $\mu(M) \geq \nu(M) > 0$ and $\mu(L \cup M) > r$, which contradicts the definition of r. Therefore, $K - L$ is μ-negligible and f is μ-measurable. $\square$

Definition 10.1.1 A family $(\mu_\alpha)_{\alpha \in A}$ of positive measures on S is said to be summable whenever $\big(\mu_\alpha(E)\big)_{\alpha \in A}$ is summable for all $E \in S$.

$(\mu_\alpha)_{\alpha \in A}$ is summable if and only if the upward-directed set of the $\sum_{\alpha \in B} \mu_\alpha$, where B extends over the class of finite subsets of A, has a supremum. In this case, $\mu : E \mapsto \sum_{\alpha \in A} \mu_\alpha(E)$ is a positive measure on S, called the sum of the μ_α and written $\sum_{\alpha \in A} \mu_\alpha$.

Suppose $(\mu_\alpha)_{\alpha \in A}$ is a summable family of positive measures on S. Then $\mu^\bullet(f) = \sum_{\alpha \in A} \mu_\alpha^\bullet(f)$ for all $f : \Omega \to [0, +\infty]$; a mapping g from Ω into a metrizable space is μ-measurable if and only if it is μ_α-measurable for all $\alpha \in A$; a mapping f from Ω into a real Banach space is essentially μ-integrable if and only if it is essentially μ_α-integrable for all $\alpha \in A$ and the sum $\sum_{\alpha \in A} \int^\bullet |f| \, d\mu_\alpha$ is finite, in which case $\int f \, d\mu = \sum_{\alpha \in A} \int f \, d\mu_\alpha$.

10.2 Locally Integrable Functions

Let Ω be a nonempty set, S a semiring in Ω, and μ a complex measure on S.

A mapping g from Ω into a real Banach space is said to be locally μ-integrable if $g \cdot 1_E$ is μ-integrable for all $E \in S$; in this case, g is μ-measurable.

Proposition 10.2.1 If $g : \Omega \to \mathbf{C}$ is locally μ-integrable, then the mapping $E \mapsto \int g \cdot 1_E \, d\mu$ from S into $\mathbf{C}$ is a measure $g\mu$ (or $g \cdot \mu$), called the measure with density g relative to μ, and $V(g\mu) = |g| \cdot V\mu$.

PROOF: $g\mu$ is obviously a measure. Since $\big|(g\mu)(E)\big| \leq \big(|g| \cdot V\mu\big)(E)$ for all $E \in S$, we have $V(g\mu) \leq |g| \cdot V\mu$, and it remains to be shown that $|g| \cdot V\mu \leq V(g\mu)$.

First, suppose that $g = \sum_{i \in I} c_i \cdot 1_{A_i}$, where I is finite, the c_i are complex numbers, and the A_i are disjoint S-sets. Given $E \in S$, for every $i \in I$ there exists a finite partition $(B_{i,j})_{j \in J_i}$ of $E \cap A_i$ into S-sets. Now, for all $i \in I$ and for all $j \in J_i$, let $P(B_{i,j})$ run through the class of finite partitions of $B_{i,j}$ into S-sets. Then

$$\big(|g| \cdot V\mu\big)(E) = \sum_{i \in I} |c_i| \cdot V\mu(E \cap A_i)$$

$$= \sum_{i \in I} |c_i| \cdot \sum_{j \in J_i} \sup_{P(B_{i,j})} \sum_{F \in P(B_{i,j})} |\mu(F)|$$

$$= \sum_{i \in I} \sum_{j \in J_i} \sup_{P(B_{i,j})} \sum_{F \in P(B_{i,j})} |g\mu(F)|$$

$$= \sum_{i \in I} \sum_{j \in J_i} V(g\mu)(B_{i,j}) \le V(g\mu)(E).$$

In the general case, let E be an S-set. Given $\varepsilon > 0$, there exists α in $St(S, \mathbf{C})$ such that $\int |\alpha - g \cdot 1_E| \, dV\mu \le \varepsilon/2$. For all finite partitions $(E_i)_{i \in I}$ of E into S-sets,

$$\sum_{i \in I} \left| \int \alpha \cdot 1_{E_i} \, d\mu \right| \le \sum_{i \in I} \left| \int g \cdot 1_{E_i} \, d\mu \right| + \sum_{i \in I} \int |\alpha - g| \cdot 1_{E_i} \, dV\mu$$

$$\le V(g\mu)(E) + \int |\alpha - g| \cdot 1_E \, dV\mu.$$

So

$$V(\alpha\mu)(E) \le V(g\mu)(E) + \varepsilon/2.$$

Therefore,

$$(|g| \cdot V\mu)(E) = \int |g| \cdot 1_E \, dV\mu \le \int |\alpha| \cdot 1_E \, dV\mu + \int |\alpha - g| \cdot 1_E \, dV\mu$$

$$\le V(\alpha\mu)(E) + \varepsilon/2 \le V(g\mu)(E) + \varepsilon,$$

and, since ε was arbitrary, we conclude that $(|g| \cdot V\mu)(E) \le V(g\mu)(E)$. □

The set $\mathcal{L}^1_{\mathrm{loc}}(\mu)$ of locally μ-integrable functions from Ω into $\mathbf{C}$ is a complex vector space, and the mapping $g \mapsto g\mu$ from $\mathcal{L}^1_{\mathrm{loc}}(\mu)$ into $\mathcal{M}(S, \mathbf{C})$ is linear. $g_1\mu = g_2\mu$ if and only if $|g_1 - g_2| \cdot V\mu = 0$. Equivalently, $g_1\mu = g_2\mu$ if and only if $g_1 = g_2$ locally μ-almost everywhere.

When μ is positive,

$$\sup(g_1\mu, g_2\mu) = \sup(g_1, g_2) \cdot \mu$$

and

$$\inf(g_1\mu, g_2\mu) = \inf(g_1, g_2) \cdot \mu$$

for all locally μ-integrable functions from Ω into $\mathbf{R}$: indeed, $\sup(g_1, g_2)$ is locally μ-integrable and

$$\sup(g_1\mu, g_2\mu) = \frac{1}{2}(g_1\mu + g_2\mu) + \frac{1}{2}V(g_1\mu - g_2\mu)$$

$$= \frac{1}{2}(g_1 + g_2)\mu + \frac{1}{2}|g_1 - g_2| \cdot \mu$$

$$= \frac{1}{2}(g_1 + g_2 + |g_1 - g_2|) \cdot \mu$$

$$= \sup(g_1, g_2) \cdot \mu.$$

Definition 10.2.1 A measure ν on S is said to be a measure with base μ if there is a locally μ-integrable function g from Ω into $\mathbf{C}$ such that $\nu = g\mu$. Then g, which is defined up to a locally μ-negligible set, is called a density of ν relative to μ, or a Radon–Nikodym derivative of ν with respect to μ, sometimes written $d\nu/d\mu$.

Theorem 10.2.1 *Let $g : \Omega \to \mathbf{C}$ be locally μ-integrable, and define X as the set of points $x \in \Omega$ such that $g(x) \neq 0$.*

(a) *A subset E of Ω is locally $g\mu$-negligible if and only if $E \cap X$ is locally μ-negligible.*

(b) *A mapping f from Ω into a metrizable space is $g\mu$-measurable if and only if $f_{/X}$ is μ-measurable on X.*

(c) *A mapping f from Ω into a Banach space is $g\mu$-measurable if and only if fg is μ-measurable; it is essentially $g\mu$-integrable if and only if fg is essentially μ-integrable, and then $\int f \, d(g\mu) = \int fg \, d\mu$.*

PROOF: Let E be a locally μ-negligible subset of Ω. Fix $A \in S$. Given $\varepsilon > 0$, there exists $\delta > 0$ such that $\int_B |g| \, dV\mu \leq \varepsilon$ for every μ-integrable subset B of A whose measure $V\mu(B)$ is less than δ (Section 5.4). Let $(A_n)_{n \geq 1}$ be a sequence of disjoint S-sets included in A such that $E \cap A \subset \bigcup_{n \geq 1} A_n$ and $\sum_{n \geq 1} V\mu(A_n) \leq \delta$. Then

$$
\sum_{n \geq 1} V(g\mu)(A_n) = \sum_{n \geq 1} \int |g| \cdot 1_{A_n} \, dV\mu
$$
$$
= \int |g| \cdot 1_{\bigcup_n A_n} \, dV\mu \leq \varepsilon,
$$

and hence $V(g\mu)^*(E \cap A) \leq \varepsilon$. This proves that E is locally $g\mu$-negligible.

Now, if f is a μ-measurable mapping from Ω into a metrizable space, it is $g\mu$-measurable by Proposition 6.1.4.

Next, let E be a locally $g\mu$-negligible set. We show that

$$
E \cap A \cap \{x : |g(x)| \geq \varepsilon\}
$$

is μ-negligible for all $A \in S$ and all $\varepsilon > 0$. For each integer $m \geq 1/\varepsilon$, let $(A_{m,n})_{n \geq 1}$ be a sequence of S-sets contained in A such that

$$
\bigcup_{n \geq 1} A_{m,n} \supset E \cap A \quad \text{and} \quad \sum_{n \geq 1} \int_{A_{m,n}} |g| \, dV\mu \leq 1/m^2.
$$

Then

$$
E_m = \left(\bigcup_{n \geq 1} A_{m,n} \right) \cap \{x : |g(x)| \geq \varepsilon\} \supset E \cap A \cap \{x : |g(x)| \geq \varepsilon\}
$$

and

$$V\mu(E_m) \leq \sum_{n \geq 1} V\mu\left(A_{m,n} \cap \{x : |g(x)| \geq \varepsilon\}\right)$$

$$\leq \frac{1}{\varepsilon} \sum_{n \geq 1} \int |g| \cdot 1_{A_{m,n}} \, dV\mu$$

$$\leq \frac{1}{m},$$

which proves that $E \cap A \cap \{x : |g(x)| \geq \varepsilon\}$ is μ-negligible. Therefore, $E \cap X$ is locally μ-negligible.

$g^{-1}(0)$ is μ-measurable and so $g\mu$-measurable. Let $A \in S$. Since $A \cap g^{-1}(0)$ is $g\mu$-integrable, we can find a $g\mu$-Cauchy sequence $(\alpha_n)_{n \geq 1}$ in $St(S, \mathbf{C})$ which converges to $1_{\{x \in A : g(x) = 0\}}$ outside a $g\mu$-negligible set B. Then $(\alpha_n g)_{n \geq 1}$ converges to $1_{\{x \in A : g(x) = 0\}} \cdot g = 0$ outside the set $B \cap \{x : g(x) \neq 0\}$, which is μ-negligible. Because $\int |\alpha_p g - \alpha_q g| \, dV\mu = \int |\alpha_p - \alpha_q| \, d(|g|V\mu)$ for all $p \geq 1$ and $q \geq 1$, $(\alpha_n g)_{n \geq 1}$ is a μ-Cauchy sequence. Thus

$$\int 1_{\{x \in A : g(x) = 0\}} \, d(|g|V\mu) = \lim_n \int |\alpha_n| \, d(|g|V\mu)$$

$$= \lim_{n \to +\infty} \int |\alpha_n g| \, dV\mu$$

$$= 0,$$

which says that $A \cap g^{-1}(0)$ is $g\mu$-negligible, and therefore that $g^{-1}(0)$ is locally $g\mu$-negligible. So a set E is locally $g\mu$-negligible whenever $E \cap X$ is locally μ-negligible.

Now let f be a mapping from Ω into a metrizable space. First, suppose that $f_{/X}$ is μ-measurable. For every $B \in \tilde{S}$, we can find a subset $A \in \tilde{S}$ of $B \cap X$ so that $B \cap X - A$ is μ-negligible. There exists a μ-negligible subset $N \in \tilde{S}$ of A such that $f_{/A-N}$ is measurable $\tilde{S}_{/A-N}$ and $f(A - N)$ is separable (Theorem 6.1.2). Since $N \cup (B - A)$ is $g\mu$-negligible, we see that f is $g\mu$-measurable.

Conversely, suppose that f is $g\mu$-measurable. Given a μ-integrable subset B of X, let $A \in \tilde{S}$ be a subset of B such that $B - A$ is μ-negligible. There exists a $g\mu$-negligible subset $N \in \tilde{S}$ of A such that $f_{/A-N}$ is measurable $\tilde{S}_{/A-N}$ and $f(A - N)$ is separable. Then, since $N \cup (B - A)$ is μ-negligible, $f_{/B}$ is measurable $\widehat{\mathcal{R}}_{/B}$, and so $f_{/X}$ is μ-measurable.

In short, f is $g\mu$-measurable if and only if $f_{/X}$ is μ-measurable. In particular, a mapping f from Ω into a Banach space is $g\mu$-measurable if and only if fg is μ-measurable.

Now let f be a mapping from Ω into a Banach space F. First, suppose that f is essentially $g\mu$-integrable. There exists a $g\mu$-Cauchy sequence $(\alpha_n)_{n \geq 1}$ in $St(S, \mathbf{C})$ which converges to f locally $g\mu$-almost everywhere. Then $(\alpha_n g)_{n \geq 1}$

is a Cauchy sequence in $\mathcal{L}_C^1(\mu)$ and it converges to fg locally μ-almost every-where. Therefore, fg is essentially μ-integrable and

$$
\begin{aligned}
\int fg \, d\mu &= \lim_{n \to +\infty} \int \alpha_n g \, d\mu \\
&= \lim_{n \to +\infty} \int \alpha_n \, d(g\mu) \\
&= \int f \, d(g\mu).
\end{aligned}
$$

Conversely, suppose that fg is essentially μ-integrable. There exists a μ-measurable and μ-moderate set A such that fg vanishes locally μ-almost everywhere in $\Omega - A$. If $\hat{f}$ is the mapping from Ω into F which agrees with f on A and vanishes outside A, then $\hat{f}g = fg$ except on $A^c \cap \{x \in \Omega : fg(x) \neq 0\}$. Thus $\hat{f}g = fg$ locally μ-almost everywhere, and $\hat{f} = f$ locally $g\mu$-almost everywhere. Next, let $(E_n)_{n \geq 1}$ be an increasing sequence of $\mathcal{R}$-sets such that $\hat{f}$ vanishes outside $\bigcup_{n \geq 1} E_n$, and, for all $n \geq 1$, put $F_n = \{x \in E_n : |\hat{f}(x)| \leq n\}$. Because the $\hat{f} \cdot 1_{F_n}$ are $g\mu$-integrable, the mappings $\hat{f} \cdot 1_{F_n} \cdot g$ are μ-integrable. Moreover,

$$
\begin{aligned}
\int |\hat{f}| \cdot 1_{F_n} \, d(|g|V\mu) &= \int |\hat{f}| \cdot 1_{F_n} \cdot |g| \, dV\mu \\
&\leq \int |\hat{f}g| \, dV\mu,
\end{aligned}
$$

and since $|\hat{f}|1_{F_n}$ increases to $|\hat{f}|$ as n tends to $+\infty$, we see that $\int^* |\hat{f}| \, d(|g|V\mu)$ is finite. Therefore, $\hat{f}$ is $g\mu$-integrable, and f is essentially $g\mu$-integrable. $\square$

Theorem 10.2.2 *Let $g : \Omega \to \mathbf{C}$ be locally μ-integrable. Then*

$$
\int^\bullet f \, dV(g\mu) = \int^\bullet f|g| \, dV\mu
$$

for all functions f from Ω into $[0, +\infty]$.

PROOF: We may suppose that μ and g are positive. For simplicity, put $\nu = g\mu$.

If h is the upper envelope of an increasing sequence $(h_n)_{n \geq 1}$ in $St^+(S)$, then

$$
\int^* h \, d\nu = \sup_{n \geq 1} \int h_n \, d\nu = \sup_{n \geq 1} \int h_n g \, d\mu = \int^* hg \, d\mu.
$$

It follows that $\int^* fg \, d\mu \leq \int^* f \, d\nu$ for all functions f from Ω into $[0, +\infty]$. If $\int^\bullet f \, d\nu$ is finite, there exists a countable union E of S-sets such that f vanishes locally ν-almost everywhere outside E; then $fg = 0$ locally μ-almost everywhere on $\Omega - E$, and

$$
\int^\bullet fg \, d\mu = \int^* fg \cdot 1_E \, d\mu \leq \int^* f \cdot 1_E \, d\nu = \int^\bullet f \, d\nu.
$$

Conversely, let f be a function from Ω into $[0, +\infty]$. We prove that $\int^\bullet f \, d\nu \le \int^\bullet f g \, d\mu$.

We may restrict our attention to the case $\int^\bullet f g \, d\mu < +\infty$. Then there exists a countable union E of S-sets such that fg vanishes locally μ-almost everywhere on $\Omega - E$, and therefore such that f vanishes locally ν-almost everywhere on $\Omega - E$. Now $\int^\bullet f \, d\nu = \int^* f \cdot 1_E \, d\nu$ and $\int^\bullet f g \, d\mu = \int^* f 1_E g \, d\mu$.

In view of the above, we may suppose that f vanishes outside a countable union of S-sets. Define φ to be the function which is equal to $1/g$ on X and vanishes on $\Omega - X$. Finally, let h be a μ-integrable function from Ω into $[0, +\infty]$ such that $h \ge fg$. Then $h \cdot 1_X = h\varphi g$ is μ-integrable, so $h\varphi$ is ν-integrable and $\int h\varphi \, d\nu = \int h \cdot 1_X \, d\mu$. But $f \cdot 1_X \le h\varphi$, whence $\int^* f \cdot 1_X \, d\nu \le \int h\varphi \, d\nu = \int h \cdot 1_X \, d\mu \le \int h \, d\mu$, and $\int^* f \, d\nu \le \int h \, d\mu$ (because $\Omega - X$ is locally ν-negligible). We conclude that $\int^* f \, d\nu \le \int^* f g \, d\mu$, as desired.   □

Theorem 10.2.3 *Let* $g_2 : \Omega \to \mathbf{C}$ *be locally μ-integrable. A complex-valued function g_1 on Ω is locally $g_2\mu$-integrable if and only if $g_1 g_2$ is locally μ-integrable. In this case, $g_1(g_2\mu) = (g_1 g_2)\mu$.*

PROOF: This follows from Theorem 10.2.1.   □

10.3   The Radon–Nikodym Theorem

Let Ω be a nonempty set and S a semiring in Ω.

Definition 10.3.1 *Let μ be a measure on S. A measure ν on S is said to be absolutely continuous with respect to μ (and we write then $\nu \ll \mu$) if every locally μ-negligible set is locally ν-negligible.*

Equivalently, $\nu \ll \mu$ if every μ-negligible $\tilde{S}$-set is ν-negligible. Any measure ν with base μ is absolutely continuous with respect to μ, by Theorem 10.2.2. Conversely, we have the following result.

Theorem 10.3.1 (Radon–Nikodym Theorem) *Let μ. ν be two complex measures on S such that $\nu \ll \mu$. Assume that there exists a class C of subsets of Ω with the following properties:*

(a) *The C-sets are mutually disjoint and essentially ν-integrable.*

(b) *Each $E \in S$ is contained in a union of countably many C-sets and of a μ-negligible set.*

Then ν is a measure with base μ.

PROOF: To begin, suppose that μ and ν are positive and that ν is bounded. Let $\mathcal{F}$ be the class of all μ-integrable functions f from Ω into $[0, +\infty[$ such that $f\mu \leq \nu$. Then $M = \sup_{f \in \mathcal{F}} \int f \, d\mu$ is smaller than $\nu(\Omega)$, and there is an increasing sequence $(f_n)_{n \geq 1}$ in $\mathcal{F}$ such that $\int f_n \, d\mu$ converges to M as $n \to +\infty$. Let f be a function from Ω into $[0, +\infty[$ that is equal to $\sup_{n \geq 1} f_n$ μ-almost everywhere. By the monotone convergence theorem, f is μ-integrable and $(f\mu)(E) = \sup_n (f_n\mu)(E) \leq \nu(E)$ for all $E \in S$. Thus $f\mu \leq \nu$.

Now we argue by contradiction. Suppose that $(f\mu)(E) < \nu(E)$ for some $E \in S$, and let $\varepsilon > 0$ be such that $\nu(E) - (f\mu)(E) - \varepsilon\mu(E) > 0$. In Proposition 3.3.1, replace μ by $\nu - f\mu - \varepsilon\mu$ and assume the F_m are $\tilde{S}$-sets. We conclude that there is an $\tilde{S}$-set $B \subset E$ such that

$$\nu(B) - (f\mu)(B) - \varepsilon\mu(B) = \sup_{A \in \tilde{S}, A \subset E} \left(\nu(A) - (f\mu)(A) - \varepsilon\mu(A) \right) > 0.$$

Obviously, B is not μ-negligible and is such that

$$\nu(A) - (f\mu)(A) - \varepsilon\mu(A) \geq 0$$

for all $\tilde{S}$-subsets A of B. Put $g = f + \varepsilon \cdot 1_B$. Then, for all $F \in S$,

$$\int_F g \, d\mu = \int_F f \, d\mu + \varepsilon\mu(B \cap F) \leq \int_F f \, d\mu + \nu(B \cap F) - \int_{B \cap F} f \, d\mu$$

$$= \int_{F-B} f \, d\mu + \nu(B \cap F)$$

is less than $\nu(F - B) + \nu(B \cap F) = \nu(F)$, which shows that g lies in $\mathcal{F}$. But

$$\int g \, d\mu = \int f \, d\mu + \varepsilon\mu(B) = M + \varepsilon\mu(B) > M,$$

and so we have a contradiction. Therefore, $f\mu = \nu$ and the theorem is proved when μ and ν are positive and ν is bounded.

Now, suppose that ν is bounded, positive, and let μ be arbitrary. If g is an essentially μ-integrable function from Ω into $[0, +\infty[$ such that $\nu = g \cdot |\mu|$, there exists $k : \Omega \to \mathbf{C}$, essentially ν-integrable, such that $g\mu = k(g \cdot |\mu|)$. We may suppose that $|k| = 1$ everywhere, in which case $\nu = (1/k)(g\mu)$. Thus ν is a measure with base μ.

We now pass to the general case. For all $A \in \mathcal{C}$, $1_A \cdot \nu$ is bounded, and so there exists an essentially μ-integrable function g_A such that $1_A\nu = g_A\mu$. Since 1_{A^c} is locally $g_A\mu$-negligible, $g_A \cdot 1_{A^c}$ is locally μ-negligible, and we may suppose that g_A vanishes outside A. Then put $g = \sum_{A \in \mathcal{C}} g_A$, and fix $E \in S$. By hypothesis, E is contained in the union of a sequence $(A_n)_{n \geq 1}$ of $\mathcal{C}$-sets and of a μ-negligible set N. To save notation, put $g_n = g_{A_n}$ for all $n \geq 1$. Then

$$\int |g_n \cdot 1_E| \, d|\mu| = \int 1_E \, d|g_n\mu| = \int 1_E \, d(1_{A_n}|\nu|) = |\nu|(A_n \cap E)$$

for all $n \geq 1$. Thus

$$\int^* |g \cdot 1_E| \, d|\mu| = \sum_{n \geq 1} \int |g_n \cdot 1_E| \, d|\mu| \leq |\nu|(E).$$

The dominated convergence theorem shows that $g \cdot 1_E$ is μ-integrable and that $\int g \cdot 1_E \, d\mu = \sum_{n \geq 1} \int g_n \cdot 1_E \, d\mu = \sum_{n \geq 1} \nu(A_n \cap E)$. Since N is ν-negligible, $\sum_{n \geq 1} \nu(A_n \cap E) = \nu(E)$. Hence $g\mu = \nu$.   □

Notice that the conditions of Theorem 10.3.1 are satisfied if Ω belongs to $\tilde{S}$ or, as we shall see later, if μ is regular.

Theorem 10.3.2 *Let μ be a positive measure on S, and let ν be a real measure on S. The following conditions are equivalent:*

(a) *For every essentially ν-integrable function f from Ω into $[0, +\infty[$ and for every $\varepsilon > 0$, there exists $\delta > 0$ such that the relations $0 \leq h \leq f$ and $\int^\bullet h \, d\mu \leq \delta$ imply $\int^\bullet h \, d|\nu| \leq \varepsilon$.*

(b) *ν belongs to the band generated by μ in the Dedekind complete Riesz space $\mathcal{M}(S, \mathbf{R})$.*

(c) *For every $E \in S$ and every $\varepsilon > 0$, there exists $\delta > 0$ such that, for all $F \in \mathcal{R}$ contained in E and with $\mu(F) \leq \delta$, we have $|\nu|(F) \leq \varepsilon$.*

(d) *ν is absolutely continuous with respect to μ.*

PROOF: We can restrict our attention to the case in which ν is positive. By Proposition 1.3.5, ν belongs to the band generated by μ if and only if, for every $f \in St^+(S)$ and every $\varepsilon > 0$, there exists $\delta > 0$ such that the relations $h \in St^+(S)$, $0 \leq h \leq f$, and $\int h \, d\mu \leq \delta$ imply $\int h \, d\nu \leq \varepsilon$. Thus condition (a) implies condition (b), which in turn implies condition (c).

Suppose now that condition (c) holds, and let B be a μ-negligible $\tilde{S}$-set. Given $E \in S$ and $\varepsilon > 0$, let δ be as in condition (c). There is an increasing sequence $(B_n)_{n \geq 1}$ of $\mathcal{R}$-sets such that $E \cap B \subset \bigcup_{n \geq 1} B_n \subset E$ and $\mu(\bigcup_{n \geq 1} B_n) \leq \delta$. Then $\nu(\bigcup_{n \geq 1} B_n) \leq \varepsilon$, and so $\nu(E \cap B) \leq \varepsilon$. Since ε was arbitrary, $E \cap B$ is ν-negligible for all $E \in S$. Hence B is ν-negligible and condition (d) is satisfied.

Finally, suppose that ν is absolutely continuous with respect to μ, and let $f : \Omega \to [0, +\infty[$ be essentially ν-integrable. Put $X = \{x \in \Omega : f(x) > 0\}$ and write $1/f$ for the function that is equal to $1/f(x)$ at every $x \in X$ and vanishes on $\Omega - X$. There exists $g : \Omega \to [0, +\infty[$, essentially μ-integrable, such that $f\nu = g\mu$. Since 1_X is locally ν-integrable, $1/f$ is locally $f\nu$-integrable. Thus $k = (1/f)g$ is locally μ-integrable and

$$\left(\frac{1}{f}g\right)\mu = \left(\frac{1}{f}\right)(g\mu) = \left(\frac{1}{f}\right)(f\nu) = 1_X \cdot \nu.$$

Now, for every integer $n \geq 1$, put $A_n = \{x \in \Omega : k(x) \geq n\}$. Since $k \cdot 1_{A_n}$ is μ-measurable, 1_{A_n} is $k\mu$-measurable and so A_n is ν-measurable. The functions $f \cdot 1_{A_n}$ decrease to 0 as $n \to +\infty$, and they remain bounded by f; for fixed $\varepsilon > 0$, we can therefore find an integer $N \geq 1$ such that $\int f \cdot 1_{A_N} \, d\nu \leq \varepsilon/2$. Then, if $h : \Omega \to [0, +\infty[$ satisfies $0 \leq h \leq f$ and $\int^\bullet h \, d\mu \leq \varepsilon/(2N)$, we have

$$
\begin{aligned}
\nu^\bullet(h) &\leq \nu^\bullet(h \cdot 1_{A_N}) + \nu^\bullet(h(1 - 1_{A_N})) \\
&\leq \nu^\bullet(f \cdot 1_{A_N}) + \nu^\bullet(1_X \cdot h(1 - 1_{A_N})) \\
&\leq \frac{\varepsilon}{2} + (1_X \cdot \nu)^\bullet(h(1 - 1_{A_N})) \\
&\leq \frac{\varepsilon}{2} + \mu^\bullet(h(1 - 1_{A_N})k) \\
&\leq \frac{\varepsilon}{2} + N\mu^\bullet(h) \\
&\leq \varepsilon,
\end{aligned}
$$

which proves that condition (a) holds. $\square$

Definition 10.3.2 Two complex measures μ, ν on S are said to be disjoint (or mutually singular) if $\inf(|\mu|, |\nu|) = 0$, and, in this case, we write $\mu \perp \nu$.

Proposition 10.3.1 *If μ and ν are disjoint, then $V(\mu + \nu) = V\mu + V\nu$.*

PROOF: The relation

$$
0 = \inf(V\mu, V\nu) = \frac{V\mu + V\nu}{2} - \frac{|V\mu - V\nu|}{2}
$$

implies

$$
|V\mu - V\nu| = V\mu + V\nu.
$$

On the other hand,

$$
V\mu = V(\mu + \nu - \nu) \leq V(\mu + \nu) + V\nu
$$

implies

$$
V\mu - V\nu \leq V(\mu + \nu);
$$

similarly,

$$
V\nu - V\mu \leq V(\mu + \nu).
$$

Hence

$$
|V\mu - V\nu| \leq V(\mu + \nu) \leq V\mu + V\nu.
$$

$\square$

Proposition 10.3.2 *μ and ν are disjoint if and only if 0 is the unique measure that is absolutely continuous with respect to both μ and ν.*

PROOF: We may suppose that μ and ν are positive. μ and ν are disjoint if and only if $B(\mu) \cap B(\nu) = \{0\}$, where $B(\mu)$ and $B(\nu)$ are the bands in $\mathcal{M}(S, \mathbf{R})$ respectively generated by μ and by ν. The result follows on account of Theorem 10.3.2. $\square$

Theorem 10.3.3 (Lebesgue's Decomposition Theorem)   *Let μ and ν be two complex measures on S. Then ν can be written $\nu_a + \nu_s$, where $\nu_a \ll \mu$ and $\nu_s \perp \mu$. Moreover, ν_a and ν_s are uniquely determined by these conditions.*

PROOF: Assume that μ is positive. By the Riesz decomposition theorem, $\mathrm{Re}\,\nu = \nu_a' + \nu_s'$, where $\nu_a' \in \mathcal{M}(S, \mathbf{R})$, $\nu_s' \in \mathcal{M}(S, \mathbf{R})$, $\nu_a' \ll \mu$ and $\nu_s' \perp \mu$. Similarly, $\mathrm{Im}\,\nu = \nu_a'' + \nu_s''$. Then $\nu = \nu_a + \nu_s$, where $\nu_a = \nu_a' + i\nu_a''$ is absolutely continuous with respect to μ and $\nu_s = \nu_s' + i\nu_s''$ is disjoint from μ. Now let $\nu = \theta_a + \theta_s$ be another decomposition of ν with respect to μ. Then $\nu_s - \theta_s = \theta_a - \nu_a$ is absolutely continuous with respect to μ and to $\nu_s - \theta_s$. Thus $\nu_s - \theta_s = 0$ by Proposition 10.3.2, whence $\theta_s = \nu_s$ and $\theta_a = \nu_a$. $\square$

A necessary and sufficient condition that two measures μ and ν on S be disjoint is that they be concentrated on disjoint sets, as we show now.

Definition 10.3.3 A measure μ on S is said to be concentrated on $E \subset \Omega$ (or carried by E) whenever $\Omega - E$ is locally μ-negligible.

Proposition 10.3.3 *Let μ and ν be two measures on S. If, for every $A \in S$, $1_A \cdot \mu$ and $1_A \cdot \nu$ are concentrated on disjoint sets E and F, then μ and ν are disjoint.*

PROOF: Assume that μ and ν are positive. For every $A \in S$, $1_A \cdot \inf(\mu, \nu) = \inf(1_A \cdot \mu, 1_A \cdot \nu)$ is concentrated on $E \cap F$, which proves that $1_A \cdot \inf(\mu, \nu) = 0$. Therefore, $\inf(\mu, \nu) = 0$. $\square$

Conversely, we have the following proposition.

Proposition 10.3.4 *Let $(\mu_n)_{n \geq 1}$ be a sequence of measures on S such that μ_p and μ_q are mutually singular for all distinct $p \in \mathbf{N}$ and $q \in \mathbf{N}$. Then, for every $A \in S$, there exists a sequence $(E_n)_{n \geq 1}$ of disjoint $\tilde{S}$-sets contained in A, such that each $1_A \cdot \mu_n$ is concentrated on E_n.*

PROOF: Assume that μ_n is positive.

Put $\mu = \mu_1 + \mu_2$. By the Radon–Nikodym theorem, there exist μ-integrable functions f_1 and f_2 from Ω into $[0, +\infty[$ such that $1_A \cdot \mu_i = f_i \mu$ for all $1 \leq i \leq 2$. We may suppose that f_i vanishes on $\Omega - A$ and that $f_{i/A}$ is measurable

$\tilde{S}_{/A}$ (Theorem 6.1.2). Now $\inf(f_1, f_2) = 0$ μ-almost everywhere, so $1_A \cdot \mu_1$ is concentrated on

$$E_1^2 = \{x \in \Omega : f_1(x) > 0,\ f_2(x) = 0\}$$

and $1_A \cdot \mu_2$ is concentrated on

$$F_1^2 = \{x \in \Omega : f_1(x) = 0,\ f_2(x) > 0\}.$$

For each integer $j > 1$, there exist disjoint $\tilde{S}$-sets E_1^j, F_1^j, included in A, such that $1_A \cdot \mu_1$ is carried by E_1^j and $1_A \cdot \mu_j$ by F_1^j. Then $1_A \cdot \mu_1$ is carried by $E_1 = \bigcap_{j \geq 1} E_1^j$ and the $1_A \mu_j$ are concentrated on $F_1 = \bigcup_{j \geq 1} F_1^j$.

In the same way, we can find $E_2 \subset F_1$ and $F_2 \subset F_1$ such that $1_A \cdot \mu_2$ is carried by E_2 and $1_A \cdot \mu_j$ is carried by F_2 for all $j > 2$.

Proceeding with the construction step-by-step, we obtain the sequence $(E_n)_{n \geq 1}$, which has the desired property.   □

In particular, if Ω belongs to $\tilde{S}$, the measures μ_n themselves are concentrated on disjoint $\tilde{S}$-sets.

10.4 Combination of Operations on Measures

The following facts are immediate, and require little comment.

Let X be a nonempty set, S a semiring in X, Y a nonempty subset of X, and T a semiring in Y.

If μ_1 and μ_2 are two measures on S such that $\mu_{1/T}$ and $\mu_{2/T}$ exist, then $(\mu_1 + \mu_2)_{/T}$ also exists and is equal to $(\mu_{1/T}) + (\mu_{2/T})$.

Proposition 10.4.1 *Let H be an upward-directed subset of $\mathcal{M}^+(S)$, bounded above, and let μ be its supremum in $\mathcal{M}(S, \mathbf{R})$. Then a necessary and sufficient condition that $\mu_{/T}$ exist is that $\nu_{/T}$ exist for all $\nu \in H$ and that $\sup_{\nu \in H} \nu(B)$ be finite for each $B \in T$. In this case, $\mu_{/T} = \sup_{\nu \in H} \nu_{/T}$.*

PROOF: Assume that $\nu_{/T}$ exists for all $\nu \in H$ and that $\sup_{\nu \in H} \nu(B)$ is finite for each $B \in T$. Then Y is μ-measurable, the T-sets are μ-integrable, and $\mu(B) = \sup_{\nu \in H} \nu(B)$ for all $B \in T$. Fix $A \in S$. For every $n \geq 1$, there exist $\nu_n \in H$ such that $\mu(A \cap Y) - \nu_n(A \cap Y) \leq 1/n$, and $B_n \in \tilde{T}$ such that $B_n \subset A \cap Y$ and $A \cap Y - B_n$ is ν_n-negligible. Now

$$\mu\left(A \cap Y - \bigcup_{n \geq 1} B_n\right) - \nu_p\left(A \cap Y - \bigcup_{n \geq 1} B_n\right) \leq (\mu - \nu_p)(A \cap Y) \leq 1/p,$$

so $\mu(A \cap Y - \bigcup_{n \geq 1} B_n) \leq 1/p$, for all $p \geq 1$. This proves that $A \cap Y - \bigcup_{n \geq 1} B_n$ is μ-negligible and that $\mu_{/T}$ exists.   □

Proposition 10.4.2 *Let H be a nonempty subset of $\mathcal{M}^+(S)$, bounded above, and let μ be its supremum in $\mathcal{M}(S, \mathbf{R})$. If $\mu_{/T}$ exists, then $\mu_{/T} = \sup_{\nu \in H} \nu_{/T}$.*

PROOF: For all ν_1, ν_2 in H, $\sup(\nu_1, \nu_2) = \left(\nu_1 + \nu_2 + |\nu_1 - \nu_2|\right)/2$, hence

$$\sup(\nu_1, \nu_2)_{/T} = \sup(\nu_{1/T}, \nu_{2/T}).$$

It follows that $(\sup_{\nu \in J} \nu)_{/T} = \sup_{\nu \in J}(\nu_{/T})$ for every nonempty finite subset J of H. Since μ is the supremum of $\sup_{\nu \in J} \nu$, where J extends over the class of nonempty finite subsets of H, the proof is complete. $\qquad\square$

Next, let $(\mu_i)_{i \in I}$ be a summable family of positive measures on S with sum μ. Then $\mu_{/T}$ exists if and only if $\mu_{i/T}$ exists for all $i \in I$, and the family $(\mu_i(B))_{i \in I}$ is summable for each $B \in T$. In this case, $\mu_{/T} = \sum_{i \in I} \mu_{i/T}$.

Now let Ω be a nonempty set and S a semiring in Ω. If μ_1 and μ_2 are two measures on S and if $g : \Omega \to \mathbf{C}$ is locally integrable, for both μ_1 and μ_2, then g is locally $(\mu_1 + \mu_2)$-integrable and $g(\mu_1 + \mu_2) = g\mu_1 + g\mu_2$.

Proposition 10.4.3 *Let H be a nonempty subset of $\mathcal{M}^+(S)$, bounded above, and let μ be its supremum. Then $g : \Omega \to [0, +\infty[$ is locally μ-integrable if (and only if) it is locally ν-integrable for all $\nu \in H$ and $\{g\nu : \nu \in H\}$ is bounded above. In this case, $g\mu = \sup_{\nu \in H} g\nu$.*

PROOF: As in Proposition 10.4.2, we may suppose that H is directed upward. By Proposition 10.1.2, g is μ-measurable. Moreover,

$$\int^\bullet g \cdot 1_E \, d\mu = \sup_{\nu \in H} \int^\bullet g \cdot 1_E \, d\nu = \sup_\nu (g\nu)(E)$$

for all $E \in S$. Thus g is locally μ-integrable and $g\mu = \sup_{\nu \in H} g\nu$. $\qquad\square$

Next, let $(\mu_i)_{i \in I}$ be a summable family of positive measures on S with sum μ. Then $g : \Omega \to [0, +\infty[$ is locally μ-integrable if and only if it is locally μ_i-integrable for all $i \in I$ and the family $(\int g \cdot 1_E \, d\mu_i)_{i \in I}$ is summable for each $E \in S$. In this case, $g\mu = \sum_{i \in I} g\mu_i$.

Propositions 10.4.2 and 10.4.3 are particularly useful in the study of vector measures; we shall, however, not pursue this topic.

The following two easy theorems are very useful.

Theorem 10.4.1 *Let X be a nonempty set, S a semiring in X, and μ a measure on S. Let Y be a subset of X and T a semiring in Y, such that $\mu_{/T}$ exists. Finally, let $g : X \to \mathbf{C}$ be locally μ-integrable. Then a necessary and sufficient condition that $(g\mu)_{/T}$ exist is that $g \cdot 1_B$ be μ-integrable for all $B \in T$. In this case, $g_{/Y}$ is locally $\mu_{/T}$-integrable and $(g\mu)_{/T} = (g_{/Y}) \cdot (\mu_{/T})$.*

PROOF: Each $B \in T$ is μ-integrable, and hence is contained in a countable union of S-sets. Assume that $g \cdot 1_B$ is μ-integrable for all $B \in T$. Then 1_B is $g\mu$-integrable. Moreover, for every $A \in S$, there exists $B \in \tilde{T}$ such that $A \cap Y - B$

is μ-negligible, and therefore $g\mu$-negligible, which proves that $(g\mu)_{/T}$ exists. Finally, for all $B \in T$, $(g_{/Y}) \cdot 1_B$ is $\mu_{/T}$-integrable and

$$\int 1_B \cdot g_{/Y} \, d(\mu_{/T}) = \int g \cdot 1_B \, d\mu = (g\mu)_{/T}(B).$$

$\square$

Theorem 10.4.2 *Let Ω', Ω'' be two nonempty sets and μ', μ'' two measures on semirings S' and S'' in Ω' and Ω'', respectively. If $g' : \Omega' \to \mathbf{C}$ is locally μ'-integrable and $g'' : \Omega'' \to \mathbf{C}$ is locally μ''-integrable, then $g' \otimes g''$ is locally $\mu' \otimes \mu''$-integrable and $(g' \otimes g'') \cdot (\mu' \otimes \mu'') = (g'\mu') \otimes (g''\mu'')$.*

PROOF: Obvious.

$\square$

10.5 Duality of L^p Spaces

Let Ω be a nonempty set, S a semiring in Ω, and μ a complex measure on S. Let $p \in [1, +\infty]$ be given, and let q be its conjugate exponent. For all $g \in L^q_\mathbf{C}(\mu)$, write θ_g for the continuous linear form $f \mapsto \int fg \, d\mu$ on $L^p_\mathbf{C}(\mu)$. Then $\theta : g \mapsto \theta_g$ is an isometry from $L^q_\mathbf{C}(\mu)$ onto a subspace of the normed dual $(L^p_\mathbf{C}(\mu))'$ of $L^p_\mathbf{C}(\mu)$ (Theorem 5.2.5).

Theorem 10.5.1 *When $1 < p < +\infty$, θ maps $L^q_\mathbf{C}(\mu)$ onto $(L^p_\mathbf{C}(\mu))'$.*

PROOF: Let T be a continuous linear form on $L^p_\mathbf{C}(\mu)$, and put $\nu(A) = T(1_A)$ for each $A \in S$. If $(A_i)_{i \geq 1}$ is a sequence of disjoint S-sets contained in an $A \in S$, the series $\sum_{i \geq 1} 1_{A_i}$ converges to $1_{\bigcup_{i \geq 1} A_i}$ in $L^p_\mathbf{C}(\mu)$; hence

$$\sum_{i \geq 1} \nu(A_i) = \sum_{i \geq 1} T(1_{A_i}) = T\Big(1_{\bigcup_{i \geq 1} A_i}\Big),$$

which proves that ν is a measure on S.

Given $f \in St(S, \mathbf{C})$, there exist disjoint S-sets $A_1, \ldots, A_n$ and complex numbers $y_1, \ldots, y_n$ such that $f = \sum_{1 \leq i \leq n} 1_{A_i} \cdot y_i$. For each $1 \leq i \leq n$, let $(B_{i,j})_{1 \leq j \leq j_i}$ be an S-partition of A_i and, for every $1 \leq j \leq j_i$, let $c_{i,j}$ be a complex number such that $|c_{i,j}| = 1$ and $c_{i,j} \cdot \nu(B_{i,j}) \cdot y_i = |\nu(B_{i,j}) \cdot y_i|$. Then

$$\sum_{1 \leq i \leq n} \sum_{1 \leq j \leq j_i} |\nu(B_{i,j})| \cdot |y_i| = T\left(\sum_{1 \leq i \leq n} \sum_{1 \leq j \leq j_i} c_{i,j} \cdot 1_{B_{i,j}} \cdot y_i\right)$$

is less than

$$\|T\| \cdot \left(\int \Big| \sum_{1 \le i \le n} \sum_{1 \le j \le j_i} c_{i,j} \cdot 1_{B_{i,j}} \cdot y_i \Big|^p d|\mu| \right)^{1/p}$$

$$= \|T\| \cdot \left(\sum_{1 \le i \le n} |\mu|(A_i) \cdot |y_i|^p \right)^{1/p}$$

$$= \|T\| \cdot N_p(f).$$

Thus $\sum_{1 \le i \le n} |\nu|(A_i) \cdot |y_i| = \int |f| \, d|\nu|$ is smaller than $\|T\| \cdot N_p(f)$.

In particular, for all $\varepsilon > 0$, we have $|\nu|(A) \le \varepsilon$ for all $A \in \mathcal{R}$ such that $|\mu|(A) \le (\varepsilon/\|T\|)^p$, which proves that ν is absolutely continuous with respect to μ (Theorem 10.3.2).

Now let $\mathcal{F}$ be the class of those positive functions $g \in \mathcal{L}_{\mathbf{C}}^q(\mu)$ for which $g|\mu| \le |\nu|$. Then

$$N_q(g) = \sup_{\substack{f \in St(S,\mathbf{C}) \\ N_p(f) \le 1}} \left| \int f g \, d\mu \right| \le \sup_f \int |f| \, d(g|\mu|) \le \sup_f \int |f| \, d|\nu| \le \|T\|$$

for all $g \in \mathcal{F}$. Put $M = \sup_{g \in \mathcal{F}} N_q(g)$, and let $(g_n)_{n \ge 1}$ be an increasing sequence in $\mathcal{F}$ such that $N_q(g_n)$ converges to M as $n \to +\infty$. There is a positive function $g \in \mathcal{L}_{\mathbf{C}}^q(\mu)$ such that $(g_n)_{n \ge 1}$ converges to g μ-almost everywhere. Obviously, $g|\mu| \le |\nu|$. Assume that $(g|\mu|)(E) < |\nu|(E)$ for some $E \in S$, and choose $\varepsilon > 0$ so that $|\nu|(E) - (g|\mu|)(E) - \varepsilon|\mu|(E) > 0$. By the same argument as that in Theorem 10.3.1, we can find an $\bar{S}$-set B included in E for which $|\mu|(B) > 0$ and $h = g + \varepsilon \cdot 1_B$ lies in $\mathcal{F}$. But $g^q + (\varepsilon \cdot 1_B)^q \le h^q$ (Lemma 5.1.2) yields $\int g^q \, d|\mu| + \varepsilon^q \cdot |\mu|(B) \le \int h^q \, d|\mu|$ and $N_q(g) < N_q(h)$. Thus we arrive at a contradiction; we conclude that $|\nu| = g \cdot |\mu|$.

For this g, we can find a sequence $(E_n)_{n \ge 1}$ of disjoint S-sets such that g vanishes on $\Omega - \bigcup_{n \ge 1} E_n$. Then $\Omega - \bigcup_{n \ge 1} E_n$ is locally ν-negligible. By the Radon–Nikodym theorem, ν is a measure with base $g\mu$, and so there exists $h \in \mathcal{L}_{\mathbf{C}}^q(\mu)$ such that $\nu = h\mu$. Since T and the linear form $f \mapsto \int fh \, d\mu$ are continuous on $L_{\mathbf{C}}^p(\mu)$ and agree on $St(S, \mathbf{C})$, they are identical. $\square$

Theorem 10.5.2 *Suppose we can find a class $\mathcal{C}$ of disjoint essentially μ-integrable sets such that each $E \in S$ is contained in a union of countably many $\mathcal{C}$-sets and of a μ-negligible set. Then (taking $p = 1$), θ maps $L_{\mathbf{C}}^\infty(\mu)$ onto $\left(L_{\mathbf{C}}^1(\mu) \right)'$.*

PROOF: Let T be a continuous linear form on $L_{\mathbf{C}}^1(\mu)$, and let ν be the function $A \mapsto T(1_A)$ on S. Given $A \in S$, if $(A_n)_{n \ge 1}$ is a sequence of disjoint S-sets whose union is A, then,

$$\left| T(1_A) - \sum_{1 \le i \le n} T(1_{A_i}) \right| \le \|T\| \cdot N_1 \left(1_A - \sum_{1 \le i \le n} 1_{A_i} \right)$$

converges to 0 as $n \to +\infty$, and so ν is σ-additive. Moreover, since $|\nu(A)| \leq \|T\| \cdot |\mu|(A)$ for all $A \in S$, ν is a measure.

Since $|\nu| \leq \|T\| \cdot |\mu|$, each essentially μ-integrable set is essentially ν-integrable. By the Radon–Nikodym theorem, there exists a locally μ-integrable function g such that $\nu = g\mu$. We have $|g| \leq \|T\|$ locally μ-almost everywhere.

As the linear forms T and θ_g on $L^1_{\mathbf{C}}(\mu)$ are continuous and agree on $St(S, \mathbf{C})$, they are identical. □

Now suppose that $p = +\infty$. Write $\mathcal{M}$ for the class of μ-measurable sets.

Lemma 10.5.1 $St(\mathcal{M}, \mathbf{C})$ is dense in $L^\infty_{\mathbf{C}}(\mu)$.

PROOF: Let $f \in L^\infty_{\mathbf{C}}(\mu)$ and $N = \{x \in \Omega : |f(x)| > N_\infty(f)\}$. Given an integer $n \geq 0$, let I be the set of those $i = (p, q) \in \mathbf{Z} \times \mathbf{Z}$ such that the square $R_i = \,]p/2^n\,,\,(p+1)/2^n] \times \,]q/2^n\,,\,(q+1)/2^n]$ intersects $f(\Omega - N)$. For each $i \in I$, put $E_i = f^{-1}(R_i)$ and choose any $c_i \in R_i$. Then $\alpha = \sum_{i \in I} 1_{E_i} \cdot c_i$ belongs to $St(\mathcal{M}, \mathbf{C})$ and $N_\infty(f - \alpha) \leq 2^{-n} \cdot \sqrt{2}$, which proves the lemma. □

Proposition 10.5.1 Let T be a linear form on $St(\mathcal{M}, \mathbf{C})$. In order that there exist $g \in L^1_{\mathbf{C}}(\mu)$ such that $T(h) - \int gh \, d\mu$ for all $h \in St(\mathcal{M}, \mathbf{C})$, it is necessary and sufficient that the following conditions hold:

(a) $T(1_E) = 0$ for every μ-negligible $\tilde{S}$-set E.

(b) $T(1_{E_n})$ converges to 0 for every sequence $(E_n)_{n \geq 1}$ in $\tilde{S}$ which decreases to the empty set.

(c) If $E \in \mathcal{M}$ and $T(1_F) = 0$ for all $\tilde{S}$-sets F included in E, then $T(1_E) = 0$.

PROOF: First, suppose that $T(1_E) = \int g \cdot 1_E \, d\mu$ for all $E \in \mathcal{M}$, for a suitably chosen $g \in L^1_{\mathbf{C}}(\mu)$. Condition (b) holds by the dominated convergence theorem. On the other hand, let $E \in \mathcal{M}$ be such that $T(1_F) = 0$ for all $\tilde{S}$-sets F included in E; if F is a Borel subset of $E \cap \{x : g(x) \neq 0\}$ such that $E \cap \{x : g(x) \neq 0\} \cap F^c$ is μ-negligible, then $g1_E = g1_F$ μ-almost everywhere; thus $T(1_E) = \int g1_E \, d\mu = \int g1_F \, d\mu = T(1_F) = 0$.

Conversely, assume that conditions (a), (b), and (c) are satisfied. Then $T(1_N) = 0$ for every locally μ-negligible set N. The function $\tilde{\nu} : E \mapsto T(1_E)$ from $\tilde{S}$ into $\mathbf{C}$ is a measure, which is bounded (Proposition 2.2.3). If ν is its restriction to S, then $\hat{\nu} = \tilde{\nu}$ (Proposition 6.4.1). Since ν is bounded, there exists $g \in L^1_{\mathbf{C}}(\mu)$ such that $\nu = g\mu$. For all $E \in \tilde{S}$, $T(1_E) = \tilde{\nu}(E) = \int 1_E \, d\nu = \int g \cdot 1_E \, d\mu$.

Now let $E \in \mathcal{M}$. Then $T(1_{E \cap \{x : g(x) = 0\}}) = 0$ by condition (c). On the other hand, if F is a Borel subset of $E \cap \{x : g(x) \neq 0\}$ such that $E \cap \{x : g(x) \neq 0\} \cap F^c$ is μ-negligible, then

$$T(1_{E \cap \{x : g(x) \neq 0\}}) = T(1_F) = \int g \cdot 1_F \, d\mu = \int g \cdot 1_E \, d\mu.$$

Hence $T(1_E) = \int g \cdot 1_E \, d\mu.$ □

Corollary *Let T be a continuous linear form on $L_{\mathbf{C}}^\infty(\mu)$. There exists g in $\mathcal{L}_{\mathbf{C}}^1(\mu)$ such that $T = \theta_g$ if and only if conditions (b) and (c) of Proposition 10.5.1 hold.*

PROOF: If these conditions hold, there exists $g \in \mathcal{L}_{\mathbf{C}}^1(\mu)$ such that $T(h) = \int gh \, d\mu$ for all $h \in St(\mathcal{M}, \mathbf{C})$. Since T and θ_g agree on $St(\mathcal{M}, \mathbf{C})$, they are identical. □

On $L_{\mathbf{R}}^\infty(V\mu)$, consider the order relation: $\dot{f}_2 \leq \dot{f}_1$ whenever $f_2 \leq f_1$ locally almost everywhere. Thus a sequence $(\dot{f}_n)_{n \geq 1}$ in $L_{\mathbf{R}}^\infty(V\mu)$ has $\dot{0}$ as its infimum if and only if $\inf_{n \geq 1} f_n = 0$ locally μ-almost everywhere.

Proposition 10.5.2 *Let T be a linear form on $L_{\mathbf{C}}^\infty(\mu)$. There exists $g \in \mathcal{L}_{\mathbf{C}}^1(\mu)$ such that $T = \theta_g$ if and only if*

(a) *$T(f_n)$ converges to 0 for every decreasing sequence $(\dot{f}_n)_{n \geq 1}$ in $L_{\mathbf{R}}^\infty(V\mu)$ admitting $\dot{0}$ as its infimum;*

(b) *for every $E \in \mathcal{M}$ such that $T(1_F) = 0$ for all $\tilde{S}$-sets $F \subset E$, we have $T(1_E) = 0$.*

PROOF: If g exists, condition (a) holds by the dominated convergence theorem.

Conversely, suppose that conditions (a) and (b) hold, and let $g \in \mathcal{L}_{\mathbf{C}}^1(\mu)$ be such that $T(h) = \int gh \, d\mu$ for all $h \in St(\mathcal{M}, \mathbf{C})$ (Proposition 10.5.1). If f is positive and belongs to $\mathcal{L}_{\mathbf{R}}^\infty(\mu)$, it is the upper envelope of an increasing sequence $(f_n)_{n \geq 1}$ of elements of $St^+(\mathcal{M})$ (Proposition 3.3.3), and $T(f - f_n)$ converges to 0 as $n \to +\infty$. Therefore,

$$T(f) = \lim_{n \to +\infty} T(f_n) = \lim \int f_n g \, d\mu = \int fg \, d\mu.$$

and $T = \theta_g$. □

10.6 The Yosida–Hewitt Decomposition Theorem[†]

We now consider the normed dual of $L_{\mathbf{C}}^\infty(\mu)$. Observe that

$$\{\theta_g : g \in \mathcal{L}_{\mathbf{C}}^1(\mu)\} = \left\{ f \mapsto \int fh \, dV\mu : h \in \mathcal{L}_{\mathbf{C}}^1(\mu) \right\}.$$

[†]This section may be omitted.

Therefore, in what follows, we may suppose that μ is positive.

Write $L_{\mathbf{R}}^{\infty}(\mu)'$ for the space of all continuous linear forms on the real Banach space $L_{\mathbf{R}}^{\infty}(\mu)$. It is the order dual of the Riesz space $L_{\mathbf{R}}^{\infty}(\mu)$. Similarly, let $L_{\mathbf{C}}^{\infty}(\mu)'$ be the normed dual of the Banach space $L_{\mathbf{C}}^{\infty}(\mu)$. Then $L_{\mathbf{R}}^{\infty}(\mu)'$ can be identified with the real vector subspace of $L_{\mathbf{C}}^{\infty}(\mu)'$ consisting of those T in $L_{\mathbf{C}}^{\infty}(\mu)'$ such that $T(f)$ is real for all $f \in L_{\mathbf{R}}^{\infty}(\mu)$. For every $T \in L_{\mathbf{C}}^{\infty}(\mu)'$, the mapping $\bar{T} : f \mapsto \overline{T(\bar{f})}$ from $L_{\mathbf{C}}^{\infty}(\mu)$ into $\mathbf{C}$ belongs to $L_{\mathbf{C}}^{\infty}(\mu)'$. Since $\operatorname{Re} T = (T + \bar{T})/2$ and $\operatorname{Im} T = (T - \bar{T})/2i$ are real, they lie in $L_{\mathbf{R}}^{\infty}(\mu)'$; moreover, $T = \operatorname{Re} T + i \operatorname{Im} T$.

Using arguments similar to those in Section 1.4, we find in $L_{\mathbf{R}}^{\infty}(\mu)'$ a smallest positive element L such that $|T(f)| \le L(|f|)$ for all $f \in \mathcal{L}_{\mathbf{C}}^{\infty}(\mu)$. This element is written $|T|$, and

$$|T|(f) = \sup_{h \in \mathcal{L}_{\mathbf{C}}^{\infty}(\mu),\, |h| \le f} |T(h)|$$

for all positive $f \in \mathcal{L}_{\mathbf{R}}^{\infty}(\mu)$. Moreover, $|T_1 + T_2| \le |T_1| + |T_2|$ for all T_1, T_2 of $L_{\mathbf{C}}^{\infty}(\mu)'$, $|\alpha T| = |\alpha| \cdot |T|$ for $T \in L_{\mathbf{C}}^{\infty}(\mu)'$ and $\alpha \in \mathbf{C}$, and $|\operatorname{Re} T| \le |T|$, $|\operatorname{Im} T| \le |T|$. If T is real, then $|T|(f) = \sup_{h \in \mathcal{L}_{\mathbf{R}}^{\infty}(\mu),\, |h| \le f} |T(h)|$ for all positive $f \in \mathcal{L}_{\mathbf{R}}^{\infty}(\mu)$, so $|T|$ is the variation of $T/_{L_{\mathbf{R}}^{\infty}(\mu)}$.

Fix $T \in L_{\mathbf{C}}^{\infty}(\mu)'$ and let f be a positive element of $\mathcal{L}_{\mathbf{R}}^{\infty}(\mu)$. If $h \in \mathcal{L}_{\mathbf{C}}^{\infty}(\mu)$ is such that $|h| \le f$, let k be the function on Ω equal to h/f at points where $f > 0$, to 0 elsewhere. Then k is μ-measurable, $|k| \le 1$, and $h = fk$. Hence $|T|(f) = \sup_{k \in \mathcal{L}_{\mathbf{C}}^{\infty}(\mu),\, N_{\infty}(k) \le 1} |T(fk)|$. In particular, when there exists $g \in \mathcal{L}_{\mathbf{C}}^{1}(\mu)$ such that $T = \theta_g$, we see that $|T|(f) = \int f|g|\, d\mu$ for all $f \in L_{\mathbf{C}}^{\infty}(\mu)$.

Definition 10.6.1 T_1, T_2 in $L_{\mathbf{C}}^{\infty}(\mu)'$ are said to be disjoint, or mutually singular, if $\inf(|T_1|, |T_2|) = 0$ in $L_{\mathbf{R}}^{\infty}(\mu)'$.

Then, from the relation

$$\inf\left(|T_1|, |T_2|\right) = \frac{|T_1| + |T_2|}{2} - \frac{\big||T_1| - |T_2|\big|}{2},$$

it follows that $\big||T_1| - |T_2|\big| = |T_1| + |T_2|$. But, since

$$|T_1| = |T_1 + T_2 - T_2| \le |T_1 + T_2| + |T_2|$$

and

$$|T_2| \le |T_1 + T_2| + |T_1|,$$

we have

$$\big||T_1| - |T_2|\big| \le |T_1 + T_2| \le |T_1| + |T_2|,$$

and we conclude that $|T_1 + T_2| = |T_1| + |T_2|$.

Proposition 10.6.1 (Yosida–Hewitt) *Let G be the space of all θ_g, for g in $\mathcal{L}_{\mathbf{C}}^{1}(\mu)$, and let H be the vector subspace of $L_{\mathbf{C}}^{\infty}(\mu)'$ consisting of those $T \in L_{\mathbf{C}}^{\infty}(\mu)'$ which are disjoint from G (i.e., from all elements of G). Then* *every T in $L_{\mathbf{C}}^{\infty}(\mu)'$ can be written $T_1 + T_2$, where T_1 belongs to G and T_2 belongs to H; the decomposition is unique and $\|T\| = \|T_1\| + \|T_2\|$. Finally, T_1 and T_2 are positive if T is positive.*

PROOF: For all $g \in \mathcal{L}_{\mathbf{R}}^1(\mu)$, define by φ_g the mapping $f \mapsto \int fg\,d\mu$ from $L_{\mathbf{R}}^{\infty}(\mu)$ into $\mathbf{R}$. Thus $G_{\mathbf{R}} = \{\varphi_g : g \in \mathcal{L}_{\mathbf{R}}^1(\mu)\}$ is a vector subspace of $L_{\mathbf{R}}^{\infty}(\mu)'$, and $|\varphi_g| = \varphi_{|g|}$ for every $g \in \mathcal{L}_{\mathbf{R}}^1(\mu)$ by the remarks preceding Definition 10.6.1. We prove that $G_{\mathbf{R}}$ is a band in $L_{\mathbf{R}}^{\infty}(\mu)'$.

For this, let $g \in \mathcal{L}_{\mathbf{R}}^1(\mu)$ and $T \in L_{\mathbf{R}}^{\infty}(\mu)'$ be such that $|T| \leq |\varphi_g|$ in $L_{\mathbf{R}}^{\infty}(\mu)'$. The set function $\tilde{\rho} : E \mapsto T(1_E)$ is finitely additive on $\tilde{S}$, and

$$|\tilde{\rho}(E)| = |T(1_E)| \leq |T|(1_E) \leq |\varphi_g|(1_E) = \int |g| \cdot 1_E\,d\mu$$

for all $E \in \tilde{S}$. The restriction ρ of $\tilde{\rho}$ to S is a bounded measure, and $\tilde{\rho}(E) = \int 1_E\,d\rho$ for all $E \in \tilde{S}$. Since ρ is bounded, there exists a $|g|\mu$-integrable function k from Ω into $\mathbf{R}$ such that $\rho = k(|g|\mu)$. Now, for all $E \in \tilde{S}$ we have $T(1_E) = \int k|g| \cdot 1_E\,d\mu$, and, for all $E \in \mathcal{M}$, we have $T(1_{E \cap \{x:g(x)=0\}}) = 0$ because $|T| \leq |\varphi_g|$; on the other hand, if F is a Borel subset of $E \cap \{x : g(x) \neq 0\}$ such that $E \cap \{x : g(x) \neq 0\} \cap F^c$ is μ-negligible, then

$$T\left(1_{E \cap \{x:g(x)\neq 0\}}\right) = T(1_F) = \int k|g| \cdot 1_F\,d\mu = \int k\,g \cdot 1_E\,d\mu,$$

and $T(1_E) = \int k|g| \cdot 1_E\,d\mu$. Hence T and $\varphi_{k|g|}$, which agree on $St(\mathcal{M}, \mathbf{R})$, are identical. This proves that $G_{\mathbf{R}}$ is an ideal in $L_{\mathbf{R}}^{\infty}(\mu)'$.

Next, let D be an upward-directed subset of $G_{\mathbf{R}}^+$, bounded above in $L_{\mathbf{R}}^{\infty}(\mu)'$, and suppose T is its supremum in $L_{\mathbf{R}}^{\infty}(\mu)'$. If $(E_n)_{n \geq 1}$ is a sequence in $\tilde{S}$ which decreases to the empty set, then, for every $\varepsilon > 0$, there exists $\varphi_g \in D$ such that $\varphi_g(1_{E_1}) \geq T(1_{E_1}) - \varepsilon/2$. For large enough n, we have $\varphi_g(1_{E_n}) \leq \varepsilon/2$; then

$$T(1_{E_n}) = (T - \varphi_g)(1_{E_n}) + \varphi_g(1_{E_n}) \leq (T - \varphi_g)(1_{E_1}) + \varepsilon/2$$

is less than ε, and so $T(1_{E_n})$ converges to 0 as $n \to +\infty$. On the other hand, if $E \in \mathcal{M}$ is such that $T(1_F) = 0$ for all $\tilde{S}$-sets $F \subset E$, then $\varphi_g(1_F) = 0$ for all such F, so $\varphi_g(1_E) = 0$, for all $\varphi_g \in D$, and $T(1_E) = \sup_{\varphi_g \in D} \varphi_g(1_E) = 0$. By the corollary to Proposition 10.5.1, T belongs to $G_{\mathbf{R}}$, and we conclude that $G_{\mathbf{R}}$ is actually a band in $L_{\mathbf{R}}^{\infty}(\mu)'$.

By the Riesz decomposition theorem, for every $T \in L_{\mathbf{R}}^{\infty}(\mu)'$, there exist $T_1 \in G_{\mathbf{R}}$ and T_2 disjoint from $G_{\mathbf{R}}$ in $L_{\mathbf{R}}^{\infty}(\mu)'$, uniquely determined, such that $T = T_1 + T_2$. Now $T \in L_{\mathbf{C}}^{\infty}(\mu)'$ is disjoint from G in $L_{\mathbf{C}}^{\infty}(\mu)'$ if and only if $|T|$ is disjoint from $G_{\mathbf{R}}$ in $L_{\mathbf{R}}^{\infty}(\mu)'$, that is, if and only if $\mathrm{Re}\,T$ and $\mathrm{Im}\,T$ are disjoint from $G_{\mathbf{R}}$ in $L_{\mathbf{R}}^{\infty}(\mu)'$. Each $T = \mathrm{Re}\,T + i\,\mathrm{Im}\,T$ of $L_{\mathbf{C}}^{\infty}(\mu)'$ can therefore be written $T_1 + T_2$, where $T_1 \in G$, T_2 is disjoint from G, and T_1, T_2 are uniquely determined. Moreover,

$$\|T\| = |T|(1_\Omega) = |T_1|(1) + |T_2|(1) = \|T_1\| + \|T_2\|,$$

and, if T is positive, then so are T_1 and T_2.   □

Let $L^\infty_{\mathbf{R}}(\mu)'_c$ be the set of those $T \in L^\infty_{\mathbf{R}}(\mu)'$ such that $T(\dot{f}_n)$ converges to 0 as $n \to +\infty$, for every decreasing sequence $(\dot{f}_n)_{n\geq 1}$ in $L^\infty_{\mathbf{R}}(\mu)$ admitting $\dot{0}$ as infimum. By Proposition 10.5.2, $L^\infty_{\mathbf{R}}(\mu)'_c$ contains $G_{\mathbf{R}}$, and $L^\infty_{\mathbf{R}}(\mu)'_c = G_{\mathbf{R}}$ when Ω is a countable union of essentially μ-integrable sets.

From the general theory of Riesz spaces, it follows that $L^\infty_{\mathbf{R}}(\mu)'_c$ is a band in $L^\infty_{\mathbf{R}}(\mu)'$ (see *Introduction to Riesz Spaces*, Jonge and Van Rooij, Mathematical Centre Tracts 78, Amsterdam, 1981, Theorem 7.3 p. 50). Denote by $L^\infty_{\mathbf{R}}(\mu)'_s$ the disjoint complement of $L^\infty_{\mathbf{R}}(\mu)'_c$ in $L^\infty_{\mathbf{R}}(\mu)'$. The annihilator of $L^\infty_{\mathbf{R}}(\mu)'_s$ is by definition the set $\left\{ \dot{f} \in L^\infty_{\mathbf{R}}(\mu) : T(\dot{f}) = 0 \text{ for all } T \in L^\infty_{\mathbf{R}}(\mu)'_s \right\}$. An element $\dot{f}$ of $L^\infty_{\mathbf{R}}(\mu)$ belongs to this annihilator if and only if $N_\infty(\dot{f}_n)$ converges to 0 as $n \to +\infty$, for every decreasing sequence $(\dot{f}_n)_{n\geq 1}$ in $L^\infty_{\mathbf{R}}(\mu)$ admitting $\dot{0}$ as infimum and such that $\dot{f}_n \leq |\dot{f}|$ for all $n \geq 1$ (*Ibid.*, Theorem 11.4, p. 70). From this, we deduce the following:

Proposition 10.6.2 θ *maps* $L^1_{\mathbf{C}}(\mu)$ *onto* $L^\infty_{\mathbf{C}}(\mu)'$ *if and only if* Ω *is a union of a finite number of atoms and of a locally* μ-*negligible set.*

PROOF: Suppose that $G = L^\infty_{\mathbf{C}}(\mu)'$. Then $G_{\mathbf{R}} = L^\infty_{\mathbf{R}}(\mu)'$, so $L^\infty_{\mathbf{R}}(\mu)'_c = L^\infty_{\mathbf{R}}(\mu)'$ and $L^\infty_{\mathbf{R}}(\mu)'_s = \{0\}$. Then $\left(N_\infty(\dot{f}_n)\right)_{n\geq 1}$ converges to 0 for every decreasing sequence $(\dot{f}_n)_{n\geq 1}$ in $L^\infty_{\mathbf{R}}(\mu)'$ admitting $\dot{0}$ as infimum, because the annihilator of $L^\infty_{\mathbf{R}}(\mu)'_s$ is the whole of $L^\infty_{\mathbf{R}}(\mu)$.

Suppose there is a sequence $(A_n)_{n\geq 1}$ of inequivalent atoms. Replacing A_n by $A_n \cap (A_1 \cup \cdots \cup A_{n-1})^c$, if necessary, we may assume that the A_n are disjoint. For all $n \geq 1$, put $E_n = \bigcup_{p \geq n} A_p$. Then $\bigcap_{n\geq 1} E_n = \emptyset$. But $\mu^*(E_n) > 0$ for each $n \geq 1$, hence $N_\infty(1_{E_n}) \geq 1$ and we arrive at a contradiction. Thus there are finitely many classes of atoms. Let $A_1, \ldots, A_k$ be representatives of these different classes, which we may assume to be disjoint. Now let E be a μ-integrable subset of $\Omega - (A_1 \cup \cdots \cup A_k)$; E contains no atom. If $(E_n)_{n\geq 1}$ is a decreasing sequence of μ-integrable subsets of E such that $\mu(E_n) = (1/n)\mu(E)$ for all $n \geq 1$, then $\inf_n(1_{E_n}) = \dot{0}$, and $N_\infty(1_{E_n})$ converges to 0 as $n \to +\infty$. Therefore, $\mu(E) = 0$, and $\Omega - (A_1 \cup \cdots \cup A_k)$ is locally μ-negligible.

Conversely, assume that Ω is the union of disjoint atoms $A_1, \ldots, A_k$ and of a locally μ-negligible set. Let $(\dot{f}_n)_{n\geq 1}$ be a decreasing sequence in $L^\infty_{\mathbf{R}}(\mu)$ admitting $\dot{0}$ as infimum. For every $1 \leq l \leq k$ and for every $n \geq 1$, there exists a real number $c_{l,n}$ such that $f_n = c_{l,n}$ almost everywhere in A_l, and, for each $1 \leq l \leq k$, the sequence $(c_{l,n})_{n\geq 1}$ decreases to 0. For every $\varepsilon > 0$, there exists $n \geq 1$ such that $c_{l,n} \leq \varepsilon$ for all $1 \leq l \leq k$, and then $N_\infty(\dot{f}_n) \leq \varepsilon$. Therefore, $N_\infty(\dot{f}_n)$ converges to 0 as $n \to +\infty$. Finally, since $L^\infty_{\mathbf{R}}(\mu)'_s = \{0\}$ and $G_{\mathbf{R}} = L^\infty_{\mathbf{R}}(\mu)'_c = L^\infty_{\mathbf{R}}(\mu)'$, we see that $L^\infty_{\mathbf{C}}(\mu)' = G$.   □

For another proof of Proposition 10.6.2, see Exercise 13.

Exercises for Chapter 10

1 Given a and b in $\overline{\mathbf{R}}$ such that $a < b$, denote by dx Lebesgue measure on $I =]a, b[$. Let w be a "weight function" on I, that is, a continuous function from I into $]0, +\infty[$ such that x^n is $w\,dx$-integrable for all integers $n \geq 0$. Denote by E the Hilbert space $L^2_{\mathbf{C}}(w\,dx)$ with inner product $(g, h) = \int g\bar{h}w\,dx$ and by $|f| = (f, f)^{1/2}$ the norm of $f \in E$. We will identify a polynomial (with complex coefficients) with its restriction to I.

Let $(p_n)_{n \geq 0}$ be the sequence obtained from $(x^n)_{n \geq 0}$ by Gram's orthogonalization (Chapter 5, Exercise 11).

1. Prove that p_n is a unitary polynomial of degree n with real coefficients.

2. For each integer $n \geq 2$, show that $p_n = (x - \lambda_n)p_{n-1} - \mu_n p_{n-2}$, where $\lambda_n = |p_{n-1}|^{-2} \cdot (xp_{n-1}, p_{n-1})$ and $\mu_n = |p_{n-1}|^2 \cdot |p_{n-2}|^{-2}$ (observe that $p_n - xp_{n-1}$ is a linear combination of $p_0, \ldots, p_{n-1}$).

3. For each integer $n \geq 1$, let s_n be the coefficient of x^{n-1} in p_n. Show that $|p_n|^{-2} \cdot (xp_n, p_n) = s_n - s_{n+1}$. (Observe that xp_n may be written $\alpha_0 p_0 + \cdots + \alpha_{n+1}p_{n+1}$, and compute α_{n+1}, α_n.)

4. Denote by $x_1, \ldots, x_k$ (with $x_1 < x_2 < \ldots < x_k$) the real roots of p_n which lie in $]a, b[$ and whose orders of multiplicity are odd. Assume that $k < n$. Observe that $q = (x - x_1) \cdots (x - x_k)$ is orthogonal to p_n, and obtain a contradiction.

5. Let $n \in \mathbf{N}$. For every integer $1 \leq k \leq n$, denote by q_k the polynomial $\prod_{1 \leq i \leq n, i \neq k}(x - x_i)/(x_k - x_i)$ and write $c_k = \int q_k w\,dx$. If Q is a polynomial with complex coefficients, whose degree is less than $2n - 1$, show that $\int Q \cdot w\,dx = \sum_{1 \leq k \leq n} c_k \cdot Q(x_k)$. (Observe that $Q = R + p_n S$, where S is a polynomial with degree less than $n - 1$, hence orthogonal to p_n, and where $R = \sum_{1 \leq k \leq n} q_k \cdot Q(x_k)$ is the Lagrange interpolation polynomial associated with the x_k and the $Q(x_k)$.)

6. From the relations $q_k(x_j) = 0$ for $j \neq k$ and $q_k(x_k) = 1$, and from part 5, deduce that $\int q_k^2 w\,dx = c_k$; therefore, c_k is strictly positive.

2 Let f be a real-valued function, defined on an interval $I \subset \mathbf{R}$. Let $x_1, \ldots, x_p$ be distinct elements of I ($p \geq 2$), and let $n_1, \ldots, n_p$ be strictly positive integers. Write n for $n_1 + \cdots + n_p$. Assume that f has $n - 1$ derivatives on I and that f and its first $(n_i - 1)$ derivatives vanish at x_i. Writing J for the smallest compact interval containing all the x_i, we prove the existence of a point ξ in the interior of J such that $f^{(n-1)}(\xi) = 0$.

1. For all $1 \leq q \leq n - 1$, let s_q be the number of indices $1 \leq i \leq p$ for which n_i is greater than q. Computing in two ways the cardinality of $\Omega = \{(i, q) : 1 \leq i \leq p, 1 \leq q \leq n - 1, n_i \geq q\}$, prove that $s_1 + \cdots + s_{n-1} = n$.

2. Given an integer $1 \leq q \leq n - 2$, suppose it has been established that $f^{(q-1)}$ vanishes at $s_1 + \cdots + s_q - (q - 1)$ points of J at least. Then prove that $f^{(q)}$ vanishes at $s_1 + \cdots + s_{q+1} - q$ points of J at least. (Use Rolle's theorem.)

3. Show that $f^{(n-1)}$ vanishes at one point at least in the interior of J.

3 Let $f, x_1, \ldots, x_p$ and $n_1, \ldots, n_p$ be as in Exercise 2, except that f and its first $(n_i - 1)$ derivatives may now be arbitrary at x_i.

1. Show that the unique polynomial of degree less than $n - 1$ that vanishes at x_i, as well as its first $(n_i - 1)$ derivatives (for all $1 \leq i \leq p$), is the polynomial 0.

2. Let $g(X) = \sum_{0 \leq k \leq n-1}(1/k!)a_k X^k$ be a polynomial of degree less than $n - 1$. Show that g and its first $(n_i - 1)$ derivatives are equal at x_i for each $1 \leq i \leq p$ to f and its first $(n_i - 1)$ derivatives, respectively, if and only if $(a_0, \ldots, a_{n-1})$ is a solution of a system of n linear equations with n variables. Conclude that there exists one and only one $g(X)$ of degree less than $n - 1$ which fulfills these requirements.

3. Assume $p \geq 1$ (instead of $p \geq 2$) and suppose that f has n derivatives on I. Denote by g the unique polynomial of degree less than $n - 1$ such that $g^{(k)}(x_i) = f^{(k)}(x_i)$ for all $1 \leq i \leq p$ and all $0 \leq k \leq n_i - 1$. Given $x \in I$, different from the x_i, let a be a real number such that the function defined by

$$t \mapsto f(t) - g(t) - \frac{a}{n!}(t - x_1)^{n_1} \cdots (t - x_p)^{n_p}$$

vanishes at x. Show that there exists ξ (depending on x), in the interior of the smallest compact interval containing x and the x_i, for which

$$f(x) = g(x) + \frac{f^{(n)}(\xi)}{n!}(x - x_1)^{n_1} \cdots (x - x_p)^{n_p}.$$

4 Assume the hypothesis and use the notation of Exercise 1.

Given an integer $n \geq 1$, let f be a real-valued function on I, which has $2n$ continuous derivatives. Assume that f is $w\,dx$-integrable. Let $x_1, \ldots, x_n$ be the roots of the polynomial p_n (with $x_1 < \cdots < x_n$).

1. Denote by g the unique polynomial of degree less than $2n - 1$ such that $g^{(k)}(x_i) = f^{(k)}(x_i)$ for all $1 \leq i \leq n$ and all $0 \leq k \leq 1$. Show that there is a continuous function h on I which is equal to $(f-g)/p_n^2$ on $I - \{x_1, \ldots, x_n\}$. Then prove the existence of a point $t \in I$ such that

$$\int (f - g)w\,dx = \int hp_n^2 w\,dx = h(t)\int p_n^2 w\,dx.$$

2. Deduce from part 3 of Exercise 3 that there exists $\xi \in I$ for which

$$\int (f - g)w\,dx = |p_n|^2 \cdot \frac{1}{(2n)!} \cdot f^{(2n)}(\xi).$$

Conclude that

$$\int fw\,dx = \sum_{1 \leq k \leq n} c_k \cdot f(x_k) + |p_n|^2 \cdot \frac{1}{(2n)!} \cdot f^{(2n)}(\xi)$$

(Markov's equality).

Exercise 4 gives a method to approximate the integral $\int fw\,dx$. This is often used in numerical analysis.

5   Let w be the weight function $\exp(-t^2/2)$ on $\mathbf{R}$, and denote by E the Hilbert space $L^2_{\mathbf{C}}(w\,dt)$. For every integer $n \geq 0$ and all $t \in \mathbf{R}$, put
$$H_n(t) = (-1)^n \cdot \exp\left(t^2/2\right) \cdot D^n\left(\exp\left(-t^2/2\right)\right).$$

1.   Show that $H_n(t)$ is a unitary polynomial, called the Hermite polynomial of degree n. In fact, it can be shown that
$$H_n(t) = \sum_{0 \leq k \leq [n/2]} (-1)^k \cdot \frac{1}{2^k} \cdot \frac{n!}{k!(n-2k)!} \cdot t^{n-2k},$$
where $[n/2]$ is the integral part of $n/2$.

2.   Given $n \in \mathbf{N}$, show that $\int H_n(t)t^k w(t)\,dt = 0$ for all integers $0 \leq k < n$ and that $\int H_n(t)t^n w(t)\,dt = (2\pi)^{1/2}\cdot n!$. (Use integration by parts.) Deduce that $(H_n)_{n \geq 0}$ is obtained in E from $(t^n)_{n \geq 0}$ by Gram's orthogonalization.

3.   Prove that $H_n(t) = tH_{n-1}(t) - (n-1)H_{n-2}(t)$ for all integers $n \geq 2$. (Use part 3 of Exercise 1.)

4.   Show that $H'_n(t) = nH_{n-1}(t)$ for $n \in \mathbf{N}$.

5.   Given $n \in \mathbf{N}$, show that $\int D\left(\exp\left(-t^2/2\right) \cdot H'_n(t)\right) \cdot t^k\,dt = 0$ for all integers $0 \leq k < n$. Deduce that $H''_n(t) - tH'_n(t) = aH_n(t)$ where a is a suitable real number, and finally that $H''_n(t) - tH'_n(t) + nH_n(t) = 0$.

6.   Given $x \in \mathbf{R}$, the function $f : u \mapsto \exp(-u^2/2) \cdot \exp(ux)$ has on $\mathbf{R}$ an expansion in power series of the form $\sum_{n \geq 0} Q_n(x) \cdot (u^n/n!)$. Show that $Q_{n+1}(x) = xQ_n(x) - nQ_{n-1}(x)$ for all integers $n \geq 1$. Conclude that $\exp(-u^2/2) \cdot \exp(ux) = \sum_{n \geq 0} H_n(x) \cdot (u^n/n!)$ for all x and u in $\mathbf{R}$.

7.   Let $a > 0$ and $B = \{z \in \mathbf{C} : |\operatorname{Im} z| \leq a/2\}$. Given an integer $n \geq 1$, the function $t \mapsto t^n \cdot e^{itz}$ from $\mathbf{R}$ into $\mathbf{C}$ is bounded (in absolute value) by $g : t \mapsto |t|^n \exp(a \cdot |t|/2)$, as z runs over B. Prove that g belongs to $L^2_{\mathbf{C}}(w\,dt)$. (Observe that $a|t| - t^2/2 \leq 2a^2 - a|t|$.)

8.   Let $f \in E$ be orthogonal, in E, to the functions t^n $(n \geq 0)$. Prove that $z \mapsto \int f(t) \cdot \exp(itz) \cdot \exp(-t^2/2)\,dt$ is holomorphic in $\mathbf{C}$, and that all its derivatives vanish at 0. (Differentiate under the integral.) Deduce that the t^n, for $n \geq 0$, form a total system in E, that is, that the closed vector subspace of E spanned by the t^n is E itself. (Use Theorem 16.4.1.)

9.   For every $f \in E$, let g_f be the function
$$u \mapsto 2^{1/2} \cdot \pi^{1/4} \cdot f(2\pi^{1/2} \cdot u) \cdot \exp(-\pi u^2)$$
on $\mathbf{R}$. Show that $f \mapsto g_f$ is a linear isometry of $L^2_{\mathbf{C}}(w\,dt)$ onto $L^2_{\mathbf{C}}(dt)$.

10.   Conclude that the Hermite functions
$$h_n : t \mapsto (-1)^n \cdot 2^{1/4} \cdot (n!)^{-1/2} \cdot 2^{-n} \cdot \pi^{-n/2} \cdot \exp(\pi t^2) \cdot D^n\left(\exp(-2\pi t^2)\right)$$
form a total orthonormal system in $L^2_{\mathbf{C}}(dt)$, obtained by orthonormalization in $L^2_{\mathbf{C}}(dt)$ of the sequence $(t^n \cdot \exp(-\pi t^2))_{n \geq 0}$.

11. For every integer $n \geq 0$, let f_n and g_n be the functions

$$x \mapsto H_n(x\sqrt{2}) \cdot \exp(-x^2/2) \qquad \text{and} \qquad x \mapsto i^n \cdot (2\pi)^{-1/2} \cdot \int e^{-ixu} \cdot f_n(u)\, du$$

on $\mathbf{R}$. Show that $\sqrt{2} \cdot f_{n+1}(x) = x f_n(x) - f'_n(x)$ and $\sqrt{2} \cdot g_{n+1}(x) = x g_n(x) - g'_n(x)$. Deduce that $f_n = g_n$.

12. Define the Fourier transform $\mathcal{F}h_n$ of h_n by

$$\mathcal{F}h_n(t) = \int e^{-2i\pi tx} \cdot h_n(x)\, dx.$$

Take $x = t\sqrt{2\pi}$ in $f_n(x) = g_n(x)$, and conclude that $\mathcal{F}h_n(t) = (-i)^n \cdot h_n(t)$ for $t \in \mathbf{R}$.

6 Given $\alpha > -1$ and $\beta > -1$, we consider w the function given by

$$t \mapsto (1-t)^\alpha (1+t)^\beta \qquad \text{on } I =]-1, 1[.$$

1. Prove that, for each integer $n \geq 0$,

$$\frac{1}{w(t)} \cdot D^n\big(w(t) \cdot (t^2 - 1)^n\big) =$$

$$\sum_{0 \leq k \leq n} (-1)^k \cdot n! \cdot \binom{\alpha + n}{n - k} \cdot \binom{\beta + n}{k} \cdot (1-t)^k \cdot (1+t)^{n-k},$$

where $D^n\big(w(t) \cdot (t^2 - 1)^n\big)$ is the nth derivative of $w(t) \cdot (t^2 - 1)^n$.

2. Define the Jacobi polynomial $P_n^{\alpha,\beta}$ of degree n (with parameters α, β) by

$$P_n^{\alpha,\beta}(t) = \frac{1}{2^n \cdot n!} \cdot \frac{1}{w(t)} \cdot D^n\big(w(t) \cdot (t^2 - 1)^n\big)$$

for $t \in]-1, 1[$. Prove that the leading coefficient of $P_n = P_n^{\alpha,\beta}$ is

$$\frac{1}{2^n} \cdot \binom{\alpha + \beta + 2n}{n}.$$

3. Let V_n be the vector subspace of $E = L_{\mathbf{C}}^2(I, w\, dt)$ generated by the functions $t^0, t^1, \ldots, t^n$. Show that $\int_{-1}^{1} P_n(t) \cdot t^k \cdot w(t)\, dt = 0$ for all integers $0 \leq k < n$. (Use integration by parts of order n:

$$\int_a^b f^{(n)}(x) g(x)\, dx =$$

$$\sum_{0 \leq p \leq n-1} (-1)^p \cdot f^{(n-p-1)}(x) \cdot g^{(p)}(x)\Big|_a^b + (-1)^n \cdot \int_a^b f(x) \cdot g^{(n)}(x)\, dx$$

if f, g are functions on a compact interval $[a, b]$ which have n continuous derivatives.) Deduce that $P_0, P_1, \ldots, P_n$ form an orthogonal basis of V_n.

4. Show that

$$\int_{-1}^{1} P_n(t) \cdot t^n \cdot w(t)\, dt = 2^{\alpha+\beta+n+1} \cdot \frac{\Gamma(\alpha+n+1) \cdot \Gamma(\beta+n+1)}{\Gamma(\alpha+\beta+2n+2)},$$

where Γ is the Euler gamma function. Conclude that, for all $n \geq 0$,

$$\Pi_n = \left[\frac{2^{\alpha+\beta+1}}{\alpha+\beta+2n+1} \cdot \frac{\Gamma(\alpha+n+1) \cdot \Gamma(\beta+n+1)}{\Gamma(n+1) \cdot \Gamma(\alpha+\beta+n+1)} \right]^{-1/2} \cdot P_n^{\alpha,\beta}$$

is the polynomial of degree n in the sequence obtained by orthonormalization of $(t^n)_{n \geq 0}$.

5. Let $n \geq 1$ be an integer. Show that

$$\frac{1}{2^n} \cdot (\alpha - \beta) \cdot \binom{\alpha+\beta+2n-1}{n-1} \cdot t^{n-1}$$

is the monomial of $P_n^{\alpha,\beta}$ of degree $n-1$.

6. Let $n \geq 2$ be an integer. Show that

$$n \cdot P_n(t) -$$

$$\frac{1}{2} \cdot \frac{\alpha+\beta+2n-1}{\alpha+\beta+n} \left[(\alpha+\beta+2n)t - \frac{\alpha-\beta}{n(n-1)} \cdot \frac{\alpha+\beta+4n-2}{\alpha+\beta+2n-2} \right] P_{n-1}(t)$$

$$+ \frac{(\alpha+\beta+2n)(\alpha+n-1)(\beta+n-1)}{(\alpha+\beta+n)(\alpha+\beta+2n-2)} P_{n-2}(t)$$

is identically zero. (Use Exercise 1, part 2.)

7. Given $n \in \mathbf{Z}^+$, there exist $\alpha_0 \in \mathbf{R}, \ldots, \alpha_n \in \mathbf{R}$ such that

$$\frac{1}{w(t)} \cdot \frac{d}{dt} \left(w(t) \cdot (1-t^2) P_n'(t) \right) = \sum_{0 \leq j \leq n} \alpha_j P_j(t)$$

for all $t \in\ ]-1, 1[$. Show that, for all $0 \leq k \leq n$, $\alpha_k \int P_k^2 w\, dt$ is the scalar product in E of

$$\frac{1}{w(t)} \cdot \frac{d}{dt} \left(w(t) \cdot (1-t^2) P_k'(t) \right) \qquad \text{and} \qquad P_n.$$

Deduce that $\alpha_k = 0$ for $0 \leq k < n$ and $\alpha_n = -n(\alpha+\beta+n+1)$. Conclude that

$$(1-t^2) \cdot P_n''(t) + \big(\beta - \alpha - (\alpha+\beta+2)t\big) \cdot P_n'(t) + n(\alpha+\beta+n+1) \cdot P_n(t) = 0.$$

8. Show that every polynomial of degree less than n which satisfies the differential equation of part 7 is necessarily proportional to $P_n^{\alpha,\beta}$.

7 Fix $\gamma > -1/2$. For each integer $n \geq 0$, define the Gebenbauer polynomial of degree n by

$$C_n^\gamma(t) = \frac{\binom{2\gamma-1+n}{n}}{\binom{\gamma-1/2+n}{n}} \cdot P_n^{\gamma-1/2,\gamma-1/2}(t).$$

(Notation of Exercise 6.)

1. Show that, for each integer $n \geq 2$,

$$n \cdot C_n^\gamma(t) - 2(\gamma + n - 1) \cdot t \cdot C_{n-1}^\gamma(t) + (2\gamma + n - 2) \cdot C_{n-2}^\gamma(t)$$

   is identically zero.

2. Fix $\theta \in \mathbf{R}$. Show that the function $h : u \mapsto (1 - 2u\cos\theta + u^2)^{-\gamma}$ on $I =]-1, 1[$ can be expanded in a power series. (Consider $(1 - ue^{i\theta})^{-\gamma}$ and $(1 - ue^{-i\theta})^{-\gamma}$.) Prove that $h^{(n)}(0)/n! = Q_n(\cos\theta)$, where Q_n is the polynomial

$$x \mapsto \sum_{0 \leq k \leq [n/2]} (-1)^k \cdot \frac{2^{n-2k}}{k!(n-2k)!} \gamma(\gamma+1)\cdots(\gamma+n-k-1) \cdot x^{n-2k}.$$

3. Let $x \in [-1, 1]$ and let φ be the function $u \mapsto (1 - 2ux + u^2)^{-\gamma}$ on I. By the preceding, $\varphi(u) = \sum_{n \geq 0} Q_n(x)u^n$ for every $u \in I$. Show that

$$(1 - 2ux + u^2)\varphi'(u) + 2\gamma(u - x)\varphi(u) = 0,$$

   and that

$$nQ_n(x) - 2(\gamma + n - 1)xQ_{n-1}(x) + (2\gamma + n - 2)Q_{n-2}(x) = 0$$

   for all integers $n \geq 2$. Conclude that $Q_n(x) = C_n^\gamma(x)$ for all $n \geq 0$. Therefore, $(1 - 2ux + u^2)^{-\gamma} = \sum_{n \geq 0} C_n^\gamma(x)u^n$ for every $u \in]-1, 1[$ (Gebenbauer's expansion of $(1 - 2ux + u^2)^{-\gamma}$).

4. Let $\theta \in \mathbf{R}$. Show that

$$(1 - 2u\cos\theta + u^2)^{-\gamma} = \sum_{n \geq 0} (-1)^n e^{-in\theta} u^n \cdot \left(\sum_{0 \leq k \leq n} \binom{-\gamma}{k}\binom{-\gamma}{n-k} e^{2ik\theta} \right)$$

   for all $u \in]-1, 1[$.

5. Suppose $\gamma > 0$. Deduce from part 4 that $|C_n^\gamma(x)| < C_n^\gamma(1)$ for every integer $n \geq 1$ and every $x \in]-1, 1[$.

8 Let S be a semiring in a nonempty set Ω, μ a complex measure on S, and $1 \leq p < +\infty$ a real number whose conjugate exponent is q. Let $g : \Omega \to \mathbf{C}$ be such that fg is μ-integrable for all $f \in \mathcal{L}_\mathbf{C}^p(\mu)$, so that, in fact, g is locally μ-integrable.

1. Show that $\theta : f \mapsto \int fg \, d\mu$ is continuous on $L_\mathbf{C}^p(\mu)$. (Use Theorem 5.2.1 and the closed graph theorem.)

2. Deduce from part 1 that, if $1 < p < +\infty$, there is $h \in \mathcal{L}_\mathbf{C}^q(\mu)$ such that $g = h$ locally a.e.

3. If $p = 1$, show that g lies in $\mathcal{L}_\mathbf{C}^\infty(\mu)$.

9 Let $\Omega = \{0, 1, 2, \ldots\}$ be the set of positive integers, and let S be the semiring in Ω consisting of the empty set and the singletons. Finally, let μ be the measure on S such that $\mu(\{n\}) = 1$ for every $n \geq 0$.

1. Denote by l^1 (respectively, l^∞) the normed vector space consisting of those sequences $x = (x_n)_{n \geq 0}$ of complex numbers such that $\|x\| = \sum_{n \geq 0} |x_n| < +\infty$ (respectively, $\|x\| = \sup_{n \geq 0} |x_n| < +\infty$). Show that the mapping $f \mapsto \big(f(n)\big)_{n \geq 0}$ is a linear isometry of $L^1_{\mathbf{C}}(\mu)$ onto l^1 (respectively, of $L^\infty_{\mathbf{C}}(\mu)$ onto l^∞).

2. For every integer $n \geq 0$, put $e_n = (\delta_{m,n})_{m \geq 0}$, where $\delta_{m,n} = 1$ or 0 as $m = n$ or $m \neq n$. Show that each $x = (x_n)_{n \geq 0} \in l^1$ is the limit in l^1 of $\sum_{0 \leq n \leq p} x_n e_n$ as $p \to +\infty$.

3. For each $y = (y_n)_{n \geq 0} \in l^\infty$, let u_y be the continuous linear form $(x_n)_{n \geq 0} \mapsto \sum_{n \geq 0} x_n y_n$ on l^1. Show that the mapping $y \mapsto u_y$ is a linear isometry of l^∞ onto the normed dual $(l^1)'$ of l^1.

4. Denote by c^0 the closed vector subspace of l^∞ consisting of those sequences $(x_n)_{n \geq 0}$ of complex numbers which converge to 0. For each $x \in c^0$, show that the series $\sum_{n \geq 0} x_n e_n$ converges commutatively to x in c^0 (i.e., for every bijection from $\mathbf{Z}^+$ onto $\mathbf{Z}^+$, $\sum_{n \geq 0} x_{\sigma(n)} e_{\sigma(n)}$ converges to x).

5. For each $y \in l^1$, let v_y be the continuous linear form $(x_n)_{n \geq 0} \mapsto \sum_{n \geq 0} x_n y_n$ on c^0. Show that the mapping $y \mapsto v_y$ is a linear isometry of l^1 onto the normed dual $(c^0)'$ of c^0.

6. Denote by $l^\infty_{\mathbf{R}}$ (respectively, $c^0_{\mathbf{R}}$) the space of those sequences $(x_n)_{n \geq 0}$ of real numbers which are bounded (respectively, which converge to 0). Write $(l^\infty_{\mathbf{R}})'$ for the order dual of the Riesz space $l^\infty_{\mathbf{R}}$. Show that the set of those $T \in (l^\infty_{\mathbf{R}})'$ which vanish on $c^0_{\mathbf{R}}$ is an ideal in $(l^\infty_{\mathbf{R}})'$. For every $y \in l^1_{\mathbf{R}}$, let θ_y be the linear form $(x_n)_{n \geq 0} \mapsto \sum_{n \geq 0} x_n y_n$ on $l^\infty_{\mathbf{R}}$, and recall that $\{\theta_y : y \in l^1_{\mathbf{R}}\}$ is a band B in $(l^\infty_{\mathbf{R}})'$. If $T \in (l^\infty_{\mathbf{R}})'$ is positive, smaller than one θ_y ($y \in l^1_{\mathbf{R}}$, $y \geq 0$), and if T vanishes on $c^0_{\mathbf{R}}$, show that $T = 0$. Conclude that every $T \in (l^\infty_{\mathbf{R}})'$ which vanishes on $c^0_{\mathbf{R}}$ is disjoint from B.

7. If T is a positive element of $(l^\infty_{\mathbf{R}})'$, disjoint from all θ_{e_n}, prove that T vanishes on $c^0_{\mathbf{R}}$.

8. Conclude that $T \in (l^\infty)'$ is disjoint from the subspace l^1 of $(l^\infty)'$ if and only if T vanishes on c^0.

10 Let μ and ν be two disjoint measures on a semiring S in Ω. Assume that $L^\infty_{\mathbf{C}}(\mu + \nu)' = L^1_{\mathbf{C}}(\mu + \nu)$.

Let u be a continuous linear form on $L^\infty_{\mathbf{C}}(\mu)$, and write p for the mapping $\dot\varphi \mapsto \dot\varphi$ from $L^\infty_{\mathbf{C}}(\mu + \nu)$ into $L^\infty_{\mathbf{C}}(\mu)$. By hypothesis, there exists $f \in \mathcal{L}^1_{\mathbf{C}}(\mu + \nu)$ such that $(u \circ p)(\dot\varphi) = \int f\varphi \, d(\mu + \nu)$ for all $\varphi \in \mathcal{L}^\infty_{\mathbf{C}}(\mu + \nu)$. Let B be a countable union of S-sets such that f vanishes outside B, and let E, F be disjoint $\tilde{S}$-sets which carry $1_B\mu$ and $1_B\nu$, respectively. Show that $u(\dot\psi) = \int f \cdot 1_E \cdot \psi \, d\mu$ for all $\psi \in \mathcal{L}^\infty_{\mathbf{C}}(\mu)$, and conclude that $L^\infty_{\mathbf{C}}(\mu)' = L^1_{\mathbf{C}}(\mu)$.

11 Let S be a semiring in a nonempty set Ω and $\mu \neq 0$ a positive diffuse measure on S. Assume that $L^\infty_{\mathbf{C}}(\mu)' = L^1_{\mathbf{C}}(\mu)$, so that the closed unit ball with center 0 in $L^1_{\mathbf{C}}(\mu)$ is weakly compact.

Let E be a μ-integrable and nonnegligible set, and let $(E_n)_{n \geq 1}$ be a decreasing sequence of μ-integrable subsets of E such that $\mu(E_n) = (1/n)\mu(E)$ for each $n \geq 1$. By Smulian's theorem, from $\big((n/\mu(E))1_{E_n}\big)_{n \geq 1}$ we can extract a sequence which converges weakly to $g \in L^1_{\mathbf{C}}(\mu)$. Show that $\int g \cdot 1_{E_p}\, d\mu = 1$ for all $p \geq 1$, and obtain a contradiction.

12 Let S be a semiring in a nonempty set Ω and μ a positive atomic measure on S. Assume that there exists a sequence $(A_n)_{n \geq 0}$ of inequivalent atoms, and put $B = \bigcup_{n \geq 0} A_n$.

1. For each $1_B\mu$-measurable complex-valued function g on Ω and each integer $n \geq 0$, denote by $c_n(g)$ the unique complex number such that $g = c_n(g)$ μ-almost everywhere in A_n. Show that the mapping $\dot{g} \mapsto \big(c_n(g)\mu(A_n)\big)_{n \geq 0}$ is an isometry of $L^1_{\mathbf{C}}(1_B\mu)$ onto l^1 and that the mapping $\dot{f} \mapsto \big(c_n(f)\big)_{n \geq 0}$ is an isometry of $L^\infty_{\mathbf{C}}(1_B\mu)$ onto l^∞. Conclude that $L^1_{\mathbf{C}}(1_B\mu) \neq L^\infty_{\mathbf{C}}(1_B\mu)'$.

2. Prove that $L^1_{\mathbf{C}}(\mu) \neq L^\infty_{\mathbf{C}}(\mu)'$ (use Exercise 10).

13 Let S be a semiring in a nonempty set Ω and μ a measure on S.

1. If $L^\infty_{\mathbf{C}}(\mu)' = L^1_{\mathbf{C}}(\mu)$, deduce from Exercises 10, 11, and 12 that Ω is a union of a finite number of atoms and of a locally μ-negligible set.

2. Conversely, if Ω is a union of a finite number of atoms and of a locally μ-negligible set, show that $L^\infty_{\mathbf{C}}(\mu)' = L^1_{\mathbf{C}}(\mu)$.

14 In this exercise, we prove that, if Ω is a nonempty set, $\mathcal{F}$ a σ-algebra in Ω, and if $\mu_1, \ldots, \mu_n$ are positive diffuse measures on $\mathcal{F}$, then

$$\big\{\big(\mu_1(B), \ldots, \mu_n(B)\big) : B \in \mathcal{F}\big\}$$

is a compact convex subset of $\mathbf{R}^n$ (Lyapounov's theorem).

Given an integer $n \geq 2$, suppose the theorem has been proved for the integer $n - 1$. We show that it is true for n itself. Put $\mu = \mu_1 + \cdots + \mu_n$.

1. Endow the dual space $L^\infty_{\mathbf{R}}(\mu)$ of $L^1_{\mathbf{R}}(\mu)$ with its weak topology. Prove that $W = \{g \in L^\infty_{\mathbf{R}}(\mu) : 0 \leq g\mu \leq \mu\}$ is compact and convex in $L^\infty_{\mathbf{R}}(\mu)$. Let u be the mapping $f \mapsto \big(\int f\, d\mu_i\big)_{1 \leq i \leq n}$ from $L^\infty_{\mathbf{R}}(\mu)$ into $\mathbf{R}^n$, and let $\alpha = (\alpha_1, \ldots, \alpha_n)$ be an element of $u(W)$. Show that $V = W \cap u^{-1}(\alpha)$ is a compact convex subset of $L^\infty_{\mathbf{R}}(\mu)$, hence has an extremal point g (which we may choose measurable $\mathcal{F}$ with values in $[0, 1]$).

2. Assume $\mu\big(\{x : g(x) \in]0, 1[\}\big) > 0$. Then $\mu_i\big(g \in]0, 1[\big) > 0$ for at least one $1 \leq i \leq n$, and, for simplicity, we suppose that $\mu_n\big(g \in]0, 1[\big) > 0$. Choose $0 < \varepsilon < 1/2$ so that $\mu_n(Z) > 0$, where $Z = g^{-1}\big(]\varepsilon, 1 - \varepsilon[\big)$, and let $A \in \mathcal{F}$ be such that $A \subset Z$ and $0 < \mu_n(A) < \mu_n(Z)$. By the induction hypothesis, there exist $\mathcal{F}$-sets $B \subset A$, $C \subset Z - A$ such that $\mu_i(B) = (1/2)\mu_i(A)$ and $\mu_i(C) = (1/2)\mu_i(Z - A)$ for $1 \leq i \leq n - 1$. For suitable real numbers s, t, show that $g - h$ and $g + h$ belong to V, where $h = s(2 \cdot 1_B - 1_A) + t(1_{Z - A} - 2 \cdot 1_C)$, and obtain a contradiction.

3. From part 2, deduce that $\alpha = \Big(\mu_1\big(g^{-1}(1)\big), \ldots, \mu_n\big(g^{-1}(1)\big) \Big)$.

4. Prove that Lyapounov's theorem remains valid for real diffuse measures.

15 Let S be a semiring in a nonempty set Ω, and let $(\mu_n)_{n \geq 1}$ be a sequence of positive bounded measures on Ω.

1. Prove that there is a positive bounded measure μ on S such that the relation $\mu^\bullet(N) = 0$ is equivalent to $\mu_n^\bullet(N) = 0$ for all $n \geq 1$. (Consider $\mu = \sum_{n \geq 1} 2^{-n} \cdot \dfrac{\mu_n}{1 + \|\mu_n\|}$.)

2. If μ, μ' are as in part 1, show that there exists an essentially μ-integrable function k from Ω into $]0, +\infty[$ such that $\mu' = k\mu$.

16 Let E be a Riesz space and u a positive linear form on E such that $u(|x|) = 0$ implies $x = 0$. The function $x \mapsto u(|x|)$ defines a norm on E. Suppose E is complete in this norm. Then, we know that E is Dedekind complete (Chapter 1, Exercise 13).

In the notation of Exercise 12, Chapter 1,

1. Show that, for every $x \in E^+$, the filter $\mathcal{G}$ considered converges to x in the topological space E (use Exercise 13 of Chapter 1), so that $\Phi^{-1}(\mathcal{H}^+)$ is dense in E^+.

2. Write μ for the Radon measure $\mathcal{H} \ni f \mapsto u\big(\Phi^{-1}(f)\big)$ on X. For each $x \in E^+$, recall that $\Phi_x = \Phi(x)$ is a continuous function from X into $[0, +\infty]$. Prove that $\int^* \Phi_x \, d\mu \leq u(x)$, so that Φ_x is μ-integrable. Show, in fact, that $\int \Phi_x \, d\mu = u(x)$ for all $x \in E$.

3. Prove that $x \mapsto \widehat{\Phi}(x)$ is an isometry of E onto $L^1_{\mathbf{R}}(\mu)$ and an isomorphism of ordered vector spaces (Kakutani's representation theorem).

4. Let H be an upward-directed subset of $\mathcal{H}(X, \mathbf{R})$, bounded above in $\Phi(E)$. Write g for the supremum of H in $\Phi(E)$ and f for the upper envelope of H. Recall that g is the upper semicontinuous regularization of f. Show that $u\big(\Phi^{-1}(g)\big) = \sup_{h \in H} u\big(\Phi^{-1}(h)\big)$, and conclude that $g = f$ μ-almost everywhere.

5. Fix $\alpha \in A$, and let F be a closed and nowhere dense subset of K_α. Denote by $\mathcal{C}$ the class of all clopen subsets of K_α containing F and, for all $U \in \mathcal{C}$, write 1_U for the indicator function of U on X. Thus 1_F is the lower envelope of the 1_U (where $U \in \mathcal{C}$). Show that 0 is the infimum of the 1_U in $\Phi(E)$, and conclude that F is μ-negligible.

6. Fix $\alpha \in A$, and let M be a μ-negligible subset of K_α. Denote by $\mathcal{C}$ the class of all clopen subsets of K_α containing M. Show that $\hat{0} = \hat{1}_M$ is the infimum in $L^1_{\mathbf{R}}(\mu)$ of the $\hat{1}_U$ when U runs over $\mathcal{C}$ (observe that $\mu(\bar{U}) = \mu(U)$ for all open subsets of K_α containing M). Prove that 0 is the infimum of the $\hat{1}_U$ in $\Phi(E)$ ($U \in \mathcal{C}$) and that $\bar{M}$ is nowhere dense.

7. From parts 5 and 6, deduce that the locally μ-negligible sets are exactly the nowhere dense sets. Show that $\text{supp}(\mu) = X$.

17 Let S be a semiring in a nonempty set Ω. Equip the space $\mathcal{M}^1(S, \mathbf{C})$ of bounded measures on Ω with the norm $\mu \mapsto \|\mu\| = \int 1 \, dV\mu$. Note that $\mathcal{M}^1(S, \mathbf{R})$ is an order ideal of the Riesz space $\mathcal{M}(S, \mathbf{R})$.

Let $\mu \in \mathcal{M}^1(S, \mathbf{R})$. Prove that the band in $\mathcal{M}^1(S, \mathbf{R})$ generated by μ consists of those $\nu \in \mathcal{M}^1(S, \mathbf{R})$ which are absolutely continuous with respect to μ.

18 Let S be a semiring in a nonempty set Ω. Let X, μ, and Φ be as in Exercise 16, such that Φ is an isometry and an isomorphism of ordered vector spaces from $\mathcal{M}^1(S, \mathbf{R})$ onto $L^1_{\mathbf{R}}(\mu)$.

1. Prove that Φ has a unique $\mathbf{C}$-linear extension, still denoted by Φ, from $\mathcal{M}^1(S, \mathbf{C})$ into $L^1_{\mathbf{C}}(\mu)$ and that this extension is a linear isometry of $\mathcal{M}^1(S, \mathbf{C})$ onto $L^1_{\mathbf{C}}(\mu)$.

   (Observe that, if $\mathbf{T} = \{\alpha \in \mathbf{C} : |\alpha| = 1\}$, then $|\theta| = \sup_{\alpha \in \mathbf{T}} \text{Re}(\alpha\theta)$ for each $\theta \in \mathcal{M}^1(S, \mathbf{C})$.)

2. Henceforth, H will denote a weakly compact subset of $\mathcal{M}^1(S, \mathbf{C})$. A classical proposition says that a subset H' of $L^1_{\mathbf{C}}(\mu)$ has weakly compact closure if and only if it is bounded and uniformly integrable. Also, it says that $\{|g| : g \in H'\}$ has then weakly compact closure. Thus, given $\varepsilon > 0$, there is a compact subset L or X such that $\int_{X-L} \Phi(|\theta|) \, d\mu \leq \varepsilon$ for all $\theta \in H$.

   Now put $\rho_\varepsilon = \Phi^{-1}(1_L)$. Let $\theta \in H$, $\hat{f} = \Phi(\theta)$, $\theta_1 = \Phi^{-1}(\hat{f} \cdot 1_L)$, and $\theta_2 = \Phi^{-1}(\hat{f} \cdot 1_{X-L})$. Show that $\|\theta_1\| \leq \varepsilon$, that $|\theta_1|$ belongs to the band generated by ρ_ε in $\mathcal{M}^1(S, \mathbf{R})$, and that $|\theta_2|$ is disjoint from ρ_ε in $\mathcal{M}^1(S, \mathbf{R})$.

3. Let ρ be a positive bounded measure on Ω, such that the $\rho_{1/n}$ are all measures with base ρ (Exercise 15). By part 2, each $\theta \in H$ can be written $f_n\rho + \theta_n$, where f_n is ρ-integrable and $\|\theta - f_n\rho\|$ is less than $1/n$. We have $\big| \|\theta\| - |f_n\rho| \big| \leq 1/n$. By Exercise 13, part 2, of Chapter 1, the band generated by ρ in $\mathcal{M}^1(S, \mathbf{R})$ is closed in $\mathcal{M}^1(S, \mathbf{R})$. Prove that ν is a measure with base ρ.

4. Conclude that the $\theta \in H$ have a common base.

11

Images of Measures

Let f be $\mathbf{R}^2 \ni (x, y) \mapsto (r = \sqrt{x^2 + y^2}, \theta)$, where θ is a measure of the argument of $x + iy$ in $[0, 2\pi[$. We have $\int (g \circ f) \, dx \, dy = \int_{[0, +\infty[\times [0, 2\pi[} g(r, \theta) \, d\mu$ where $\mu = r \, dr d\theta$. For any $r \, dr d\theta$-integrable set A in $\mathbf{R}^2$, $f^{-1}(A)$ is Lebesgue integrable, and μ is said to be the image of $dx \, dy$ under f. However, this example could be misleading. As we shall see, the fact that $\mu(A) = \lambda\big(f^{-1}(A)\big)$ defines a measure does not depend on the differentiability of f but rather on some properties of measurability.

A generalization of this idea leads to the definition of images of measures. Images of measures are important in many areas of mathematics, in particular in differential geometry, ergodic theory, and the theory of probability.

Summary

11.1   Given a measure μ on S in Ω and a mapping π from Ω into Ω', we can try to define a measure on a semiring S' in Ω' by $\mu'(A) = \mu\big(\pi^{-1}(A)\big)$. This will define a measure if certain conditions are satisfied, in which case we say that the pair (π, S') is μ-suited. Then $\int f \, d\mu' = \int (f \circ \pi) \, d\mu$ (Theorem 11.1.1).

11.2   In this section, we define compact classes, projective systems of measures, and prove Kolmogorov's theorem, which gives a sufficient condition for a projective system of measures to define a measure called the projective limit of the system (Theorem 11.2.1). In particular, if for each $i \in I$ μ_i is a positive measure with total mass 1 on S_i, and if $\mu_J = \bigotimes_J \mu_i$ for all finite subsets $J \subset I$, then $\{\mu_J\}$ has a projective limit (Theorem 11.2.2).

11.3   We apply the notion of image of a measure to Lebesgue measure on $\mathbf{R}$.

11.4   This is a short introduction to ergodic theory. The main result of this section is Birkhoff's ergodic theorem: If $f : \Omega \to \mathbf{R}$ is essentially μ-integrable and $f_k = f \circ u^k$ (where $u : \Omega \to \Omega$ satisfies $u(\mu) = \mu$), then $(1/n) \sum_{0 \le k \le n-1} f_k$ converges to some essentially μ-integrable function f^* locally μ-a.e. and $\tilde{f^*} = f^* \circ u$ locally μ-a.e.

11.1 μ-Suited Pairs

Definition 11.1.1 Let Ω be a nonempty set, μ a complex measure on a semiring S in Ω, π a mapping from Ω into a set Ω', and S' a semiring in Ω'. The pair (π, S') is said to be μ-suited if

(a) $\pi^{-1}(A')$ is essentially μ-integrable for all $A' \in S'$;

(b) for every $A \in S$, there exist a sequence $(A'_n)_{n \geq 1}$ in S' and a μ-negligible set N such that $N \cup [\bigcup_{n \geq 1} \pi^{-1}(A'_n)]$ contains A.

In this case, the function $A' \mapsto \mu(\pi^{-1}(A'))$ is a measure on S' called the image of μ under π, denoted by $\pi(\mu)$.

In the following theorem, we assume that μ is positive.

Theorem 11.1.1 *Assume that (π, S') is μ-suited, and put $\mu' = \pi(\mu)$. Then*

(a) $\int^{\bullet}(f \circ \pi) \, d\mu \leq \int^{\bullet} f \, d\mu'$ *for all* $f : \Omega' \to [0, +\infty]$, *and* $\int^{\bullet}(f \circ \pi) \, d\mu = \int^{\bullet} f \, d\mu'$ *whenever f is μ'-measurable;*

(b) $g \circ \pi$ *is μ-measurable, for every μ'-measurable mapping g from Ω' into a metrizable space;*

(c) $f \circ \pi$ *is essentially μ-integrable, for every essentially μ'-integrable mapping f from Ω' into a real Banach space; moreover, $\int(f \circ \pi) \, d\mu = \int f \, d\mu'$.*

PROOF: Every $g \in St^+(S')$ can be written $\sum_{i \in I} \alpha_i \cdot 1_{A'_i}$, where I is finite, the A'_i are disjoint S'-sets, and the α_i are positive numbers. Now

$$\int g \, d\mu' = \sum_{i \in I} \alpha_i \cdot \mu'(A'_i) = \sum_{i \in I} \alpha_i \cdot \mu(\pi^{-1}(A'_i)) = \int^{\bullet} (g \circ \pi) \, d\mu.$$

If h is the upper envelope of an increasing sequence $(g_n)_{n \geq 1}$ of elements of $St^+(S')$, then

$$\int^* h \, d\mu' = \sup_n \int g_n \, d\mu' = \sup_n \int^{\bullet} (g_n \circ \pi) \, d\mu = \int^{\bullet} (h \circ \pi) \, d\mu.$$

Hence $\int^{\bullet}(f \circ \pi) \, d\mu \leq \int^* f \, d\mu'$ for all functions f from Ω' into $[0, +\infty]$. Now let $\mathcal{R}$ be the ring generated by S. For every $A \in \mathcal{R}$, there exist a sequence $(A'_n)_{n \geq 1}$ of S'-sets and a μ-negligible set N such that $N \cup (\bigcup_{n \geq 1} \pi^{-1}(A'_n))$ contains A; so

$$\int^{\bullet} (f \circ \pi) \cdot 1_A \, d\mu \leq \int^{\bullet} (f \circ \pi) \cdot 1_{\pi^{-1}(\bigcup A'_n)} \, d\mu$$

$$\leq \int^* f \cdot 1_{\bigcup A'_n} \, d\mu'$$

$$\leq \int^{\bullet} f \, d\mu'.$$

Finally,

$$\int^\bullet (f \circ \pi)\, d\mu \le \int^\bullet f\, d\mu'.$$

If E' is a μ'-integrable set, there exists $A' \in \tilde{S}'$ such that $A' \subset E'$ and $E' - A'$ is μ'-negligible. Since $\pi^{-1}(E' - A')$ is locally μ-negligible, the set $\pi^{-1}(E')$ is μ-measurable.

Now fix a μ'-measurable set E'. The sets $A \in \tilde{S}$ for which $A \cap \pi^{-1}(E')$ is μ-measurable form a σ-ring C. If A lies in S, there exist a sequence $(A'_n)_{n\ge 1}$ of S'-sets and a μ-negligible set N such that $N \cup (\bigcup_{n\ge 1} \pi^{-1}(A'_n))$ contains A, and then $A \cap \pi^{-1}(E')$ is μ-measurable because it is the union of $A \cap N \cap \pi^{-1}(E')$ and the $A \cap \pi^{-1}(A'_n \cap E')$. Hence $C = \tilde{S}$. We conclude that $E \cap \pi^{-1}(E')$ is μ-integrable for all μ-integrable sets E. So $\pi^{-1}(E')$ is μ-measurable.

Next, let g be a μ'-measurable mapping from Ω' into a metrizable space. Given a μ-integrable set E, there exist a sequence $(A'_n)_{n\ge 1}$ of S'-sets and a μ-negligible set N such that $N \cup (\bigcup_{n\ge 1} \pi^{-1}(A'_n))$ contains E. For every $n \ge 1$, let N'_n be a μ'-negligible subset of A'_n such that $g(A'_n - N'_n)$ is separable. Then $P = E \cap (N \cup \bigcup_{n\ge 1} \pi^{-1}(N'_n))$ is μ-negligible, and $(g \circ \pi)(E - P)$, which is contained in $\bigcup_{n\ge 1} g(A'_n - N'_n)$, is separable. This proves that $g \circ \pi$ is μ-measurable.

Now the function $A' \longmapsto \mu^\bullet(\pi^{-1}(A'))$ is σ-additive on $\tilde{S}'$, and $\mu^\bullet(\pi^{-1}(A')) = \mu'(A')$ for all $A' \in S'$. Hence $\mu^\bullet(\pi^{-1}(A')) = \mu'^*(A')$ for all $A' \in \tilde{S}'$ (Proposition 6.1.1). Given the μ'-measurable set E', we have $\mu'^\bullet(E') = \sup_{A' \in \mathcal{R}'} \mu'^*(A' \cap E')$; but $A' \cap E'$ is the union of an $\tilde{S}'$-set and a μ'-negligible set, so

$$\mu'^\bullet(E') = \sup_{B' \in \tilde{S}', B' \subset E'} \mu'^*(B') = \sup_{B'} \mu^\bullet(\pi^{-1}(B')) \le \mu^\bullet(\pi^{-1}(E'))$$

and finally,

$$\mu'^\bullet(E') = \mu^\bullet(\pi^{-1}(E')).$$

Each μ'-measurable function f from Ω' into $[0, +\infty]$ is the upper envelope of an increasing sequence of functions in $St^+(\mathcal{M}')$, so

$$\int^\bullet f\, d\mu' = \int^\bullet (f \circ \pi)\, d\mu.$$

Finally, let F be a real Banach space. The functionals $f \mapsto \int f\, d\mu'$ and $f \mapsto \int (f \circ \pi)\, d\mu$ from $\bar{\mathcal{L}}^1_F(\mu')$ into F are continuous and they agree on $St(S', F)$. Hence they are identical. $\qquad \square$

If (π, S') is μ-suited and g is a complex-valued function on Ω' for which $g \circ \pi$ is μ-measurable, there is no reason to suppose that g is μ'-measurable. We shall pursue this in Section 22.1.

When μ is complex, $\int f\, d\mu' = \int f \circ \pi\, d\mu$ for each f in $\mathcal{L}^1_F(\pi(|\mu|))$.

Proposition 11.1.1 (Transitivity of Images of Measures)   *If* (π, S') *is μ-suited and (π', S'') is $\pi(|\mu|)$-suited, then $(\pi' \circ \pi, S'')$ is μ-suited and* $(\pi' \circ \pi)(\mu) = \pi'(\pi(\mu))$.

PROOF: Obvious.   □

If μ_1 and μ_2 are two complex measures on the same semiring S, and if (π, S') is both μ_1-suited and μ_2-suited, then (π, S') is $\mu_1 + \mu_2$-suited and $\pi(\mu_1 + \mu_2) = \pi(\mu_1) + \pi(\mu_2)$.

Proposition 11.1.2 *Let S be a semiring in a nonempty set Ω, and let H be an upward-directed subset of $\mathcal{M}^+(S)$ having a supremum μ. Finally, let π be a mapping from Ω into a set Ω', and let S' be a semiring in Ω'. Then a necessary and sufficient condition that (π, S') be μ-suited is that it be ν-suited for all $\nu \in H$ and that $\sup_{\nu \in H} \nu(\pi^{-1}(A'))$ be finite for all $A' \in S'$. In this case, $\pi(\mu) = \sup_{\nu \in H} \pi(\nu)$.*

PROOF: Assume that (π, S') is ν-suited for all $\nu \in H$, and that $\sup_{\nu \in H} \nu(\pi^{-1}(A'))$ is finite for all $A' \in S'$. Then $\pi^{-1}(A')$ is essentially μ-integrable for all $A' \in S'$. Fix $A \in S$ and, for each $n \geq 1$, choose $\nu_n \in H$ so that $\mu(A) - \nu_n(A) \leq 1/n$. There exists a sequence $(A'_{m,n})_{m \geq 1}$ of S'-sets such that $A \cap \left(\bigcup_{m \geq 1} \pi^{-1}(A'_{m,n}) \right)^c$ is ν_n-negligible. If we put

$$B = \bigcup_{\substack{m \geq 1, \\ n \geq 1}} \pi^{-1}(A'_{m,n}),$$

then, for all $p \geq 1$, $A \cap B^c$ is ν_p-negligible. So $\mu(A \cap B^c) = (\mu - \nu_p)(A \cap B^c) \leq (\mu - \nu_p)(A)$ is less than $1/p$, which proves that $A \cap B^c$ is μ-negligible.   □

Now let $(\mu_i)_{i \in I}$ be a summable family of positive measures on S (S semiring in Ω). Define $\mu = \sum_{i \in I} \mu_i$. Let π be a mapping from Ω into a set Ω', and let S' be a semiring in Ω'. Then (π, S') is μ-suited if and only if (π, S') is μ_i-suited for all $i \in I$ and the family $(\pi(\mu_i))_{i \in I}$ is summable. In this case, $\pi(\mu) = \sum_{i \in I} \pi(\mu_i)$.

Proposition 11.1.3 *Let Ω be a nonempty set, S a semiring in Ω, and μ a complex measure on S. Let Y be a subset of Ω and let T be a semiring in Y. Finally, let (π, S') be a μ-suited pair. If $\mu_{/T}$ exists, then $(\pi_{/Y}, S')$ is $\mu_{/T}$-suited and $(\pi_{/Y})(\mu_{/T}) = \pi(1_Y \cdot \mu)$.*

PROOF: Obvious.   □

Proposition 11.1.4 *Let μ, π, and S' be as in Proposition 11.1.3. Assume that (π, S') is μ-suited and put $\mu' = \pi(|\mu|)$. Let $g : \Omega' \to \mathbf{C}$ be μ'-measurable, such that $g \circ \pi$ is locally μ-integrable. Then g is locally μ'-integrable if and only if (π, S') is $(g \circ \pi)\mu$-suited. In this case, $\pi((g \circ \pi)\mu) = g \cdot \pi(\mu)$.*

PROOF: Assume that g is locally μ'-integrable. Then, for all $A' \in S'$, $g \cdot 1_{A'}$ is μ'-integrable, so $(g \cdot 1_{A'}) \circ \pi$ is essentially μ-integrable and $\pi^{-1}(A')$ is essentially $(g \circ \pi)\mu$-integrable. Moreover,

$$(g \cdot \pi(\mu))(A') = \int 1_{\pi^{-1}(A')} \, d((g \circ \pi)\mu).$$

Conversely, assume that (π, S') is $(g \circ \pi)\mu$-suited. For each $A' \in S'$, $(g \cdot 1_{A'}) \circ \pi$ is essentially μ-integrable. Now, since $g \cdot 1_{A'}$ is μ'-measurable, we see that $g \cdot 1_{A'}$ is essentially μ'-integrable. Hence g is locally μ'-integrable. $\square$

Proposition 11.1.5 *For all $i \in \{1, 2\}$, let Ω_i be a nonempty set, S_i a semiring in Ω_i, and μ_i a measure on S_i. Let π_i be a mapping from Ω_i into a set Ω'_i, and let S'_i be a semiring in Ω'_i. If (π_i, S'_i) is μ_i-suited, then $(\pi_1 \times \pi_2, S'_1 \times S'_2)$ is $\mu_1 \otimes \mu_2$-suited and $(\pi_1 \times \pi_2)(\mu_1 \otimes \mu_2) = \pi_1(\mu_1) \otimes \pi_2(\mu_2)$.*

PROOF: Obvious. $\square$

11.2 Infinite Product of Measures

This section is particularly useful in probability theory.

Definition 11.2.1 Let Ω be a nonempty set. A class $\mathcal{C} \ni \emptyset$ of subsets of Ω is said to be compact if $\bigcap_{i \geq 1} K_i \neq \emptyset$ for every sequence $(K_i)_{i \geq 1}$ of $\mathcal{C}$-sets satisfying $\bigcap_{1 \leq i \leq n} K_i \neq \emptyset$ for all $n \geq 1$.

Proposition 11.2.1 *Let $\mathcal{C}$ be a compact class of subsets of Ω. The class $\mathcal{C}'$ of all finite unions of $\mathcal{C}$-sets is a compact class.*

PROOF: Let $(H_i)_{i \geq 1}$ be a sequence of $\mathcal{C}'$-sets such that $\bigcap_{1 \leq i \leq p} H_i$ is nonempty for all integers $p \geq 1$. For each $i \geq 1$, H_i is a union of finitely many $\mathcal{C}$-sets $K_{i,j}$ (where $1 \leq j \leq n_i$). Denote by Z the set $\prod_{i \geq 1}\{1, 2, \ldots, n_i\}$, so that $\bigcap_{i \geq 1} H_i = \bigcup_{\alpha \in Z}(\bigcap_{i \geq 1} K_{i,\alpha_i})$.

For each integer $p \geq 1$, the set $Z_p = \{\alpha \in Z : \bigcap_{1 \leq i \leq p} K_{i,\alpha_i} \neq \emptyset\}$ is nonempty because $\bigcap_{1 \leq i \leq p} H_i = \bigcup_{\alpha \in Z}(\bigcap_{1 \leq i \leq p} K_{i,\alpha_i})$ is nonempty.

Let $\alpha^q = (\alpha_1^q, \ldots, \alpha_i^q, \ldots) \in Z_q$, for each $q \geq 1$, and consider the sequence $(\alpha_1^q)_{q \geq 1}$. It is a sequence of elements of $\{1, \ldots, n_1\}$. Thus there exists α_1 in $\{1, \ldots, n_1\}$ such that $\alpha_1^q = \alpha_1$ for infinitely many values of q. Similarly, there exists $\alpha_2 \in \{1, \ldots, n_2\}$ such that, among the values of q for which $\alpha_1^q = \alpha_1$, there are infinitely many for which $\alpha_2^q = \alpha_2$. By induction we can define a sequence $\alpha = (\alpha_1, \ldots, \alpha_n, \ldots)$, which belongs to Z and which satisfies the following property: for every integer $p \geq 1$, there exist infinitely many integers

$q \geq 1$ such that $\alpha_1^q = \alpha_1, \alpha_2^q = \alpha_2, \ldots, \alpha_p^q = \alpha_p$. Since some of these integers are larger than p, $\alpha \in Z_p$.

Now $\bigcap_{1 \leq i \leq p} K_{i,\alpha_i}$ is nonempty for all $p \geq 1$, whereas C is a compact class. Therefore, $\bigcap_{i \geq 1} K_{i,\alpha_i} \neq \emptyset$, and a fortiori $\bigcap_{i \geq 1} H_i \neq \emptyset$. □

Proposition 11.2.2 *Let Ω be a nonempty set, S a semiring in Ω, and μ an additive mapping from S into $[0, +\infty[$. If there is a compact class C of subsets of Ω contained in S such that $\mu(A) = \sup\{\mu(K) : K \in C, K \subset A\}$ for all $A \in S$, then μ is σ-additive.*

PROOF: Let S' be the ring generated by S, and let C' be the class of all finite unions of C-sets. The function μ has a unique additive extension to S', still denoted by μ.

Let $A \in S'$. There are mutually disjoint S-sets $B_1, \ldots, B_n$ whose union is A. Given $\varepsilon > 0$, we can find $K_i \in C$ with $K_i \subset B_i$ such that $\mu(B_i - K_i) \leq \varepsilon/n$. Hence $\mu(A - \bigcup_{1 \leq i \leq n} K_i) \leq \varepsilon$, and

$$\mu(A) = \sup\{\mu(K) : K \in C', K \subset A\}.$$

Now it suffices to show that, if $(A_i)_{i \geq 1}$ is a sequence of S'-sets which decreases to $\emptyset$, $\mu(A_i)$ converges to 0 as $i \to +\infty$. If not, since $(\mu(A_i))_{i \geq 1}$ decreases, $\varepsilon = \inf \mu(A_i)$ is strictly positive. Then, for every $i \geq 1$, there exists $K_i \in C'$ with $K_i \subset A_i$ such that $\mu(A_i - K_i) \leq 2^{-(i+1)}\varepsilon$. We have

$$K_n \subset \left(\bigcap_{1 \leq i \leq n} K_i \right) \cup \left(\bigcup_{1 \leq i \leq n-1} (A_i - K_i) \right),$$

80

$$\varepsilon - \frac{\varepsilon}{2^{n+1}} \leq \mu(A_n) - \frac{\varepsilon}{2^{n+1}} \leq \mu(K_n) \leq \mu\left(\bigcap_{1 \leq i \leq n} K_i \right) + \sum_{1 \leq j \leq n-1} \frac{\varepsilon}{2^{j+1}}.$$

This implies that $\mu(\bigcap_{1 \leq i \leq n} K_i) \geq \varepsilon/2$ for all $n \geq 1$. Now it follows that $\bigcap_{1 \leq i \leq n} K_i \neq \emptyset$ for all $n \geq 1$, and that $\bigcap_{i \geq 1} K_i \neq \emptyset$ because C' is a compact class. This contradicts the hypothesis $\bigcap_{i \geq 1} A_i = \emptyset$. □

Let $(\Omega_i)_{i \in I}$ be an infinite family of nonempty sets. For each $i \in I$, let S_i be a semialgebra in Ω_i, that is, a semiring in Ω_i containing Ω_i. Denote by $\mathcal{F}$ the class of all finite nonempty subsets of I. For each $J \in \mathcal{F}$, write S_J for the semiring $\prod_{i \in J} S_i$ in $\Omega_J = \prod_{i \in J} \Omega_i$ and denote by q_J the projection $(x_i)_{i \in I} \mapsto (x_i)_{i \in J}$ from $\Omega = \prod_{i \in I} \Omega_i$ onto $\Omega_J = \prod_{i \in J} \Omega_i$. Finally, for all J_1, J_2 in $\mathcal{F}$ such that $J_2 \subset J_1$, denote by π_{J_2, J_1} the canonical projection from Ω_{J_1} onto Ω_{J_2}. The class S of all the $q_J^{-1}(A_J)$, for all $J \in \mathcal{F}$ and all $A_J \in S_J$, is a semialgebra of subsets of Ω. Indeed, if $q_J^{-1}(A_J)$ and $q_K^{-1}(B_K)$ belong to S, then $q_J^{-1}(A_J) = q_{J \cup K}^{-1}(A_{J \cup K})$ and $q_K^{-1}(B_K) = q_{J \cup K}^{-1}(B_{J \cup K})$, where $A_{J \cup K} = \pi_{J, J \cup K}^{-1}(A_J)$ and $B_{J \cup K} = \pi_{K, J \cup K}^{-1}(B_K)$ belong to $S_{J \cup K}$.

Definition 11.2.2 S is called the product of the semialgebras S_i and is denoted by $\prod_{i \in I} S_i$.

Notice that the $q_J^{-1}(A_J)$, for $J \in \mathcal{F}$ and $A_J \in \tilde{S}_J = \bigotimes_{i \in J} \tilde{S}_i$, form an algebra $\mathcal{A}$ and that $\bigotimes_{i \in I} \tilde{S}_i$ is the σ-ring generated by S (and by $\mathcal{A}$).

Definition 11.2.3 For each $J \in \mathcal{F}$, let μ_J be a positive measure on S_J with mass 1. If $\mu_{J_2} = \pi_{J_2, J_1}(\mu_{J_1})$ for all J_1 and $J_2 \in \mathcal{F}$ such that $J_2 \subset J_1$, then $(\mu_J)_{J \in \mathcal{F}}$ is said to be a projective system of measures.

Theorem 11.2.1 (Kolmogorov) *Let $(\mu_J)_{J \in \mathcal{F}}$ be a projective system of measures.*

(a) *There is an additive function ν on $\mathcal{A}$ such that $\nu(q_J^{-1}(A_J)) = \mu_J(A_J)$ for all $J \in \mathcal{F}$ and all $A_J \in \tilde{S}_J$.*

(b) *Write $\mu_i = \mu_{\{i\}}$ for all $i \in I$. If, for every $i \in I$, there is a compact class C_i of $\tilde{S}_i$-sets such that $\mu_i(A_i) = \sup\{\mu_i(K_i) : K_i \in C_i, K_i \subset A_i\}$ for all $A_i \in \tilde{S}_i$, then $\mu = \nu_{/S}$ is a measure.*

PROOF: Let E, F be two $\mathcal{A}$-sets and let $J, K \in \mathcal{F}$, $A_J \in \tilde{S}_J$, and $B_K \in \tilde{S}_K$ be such that $E = q_J^{-1}(A_J)$ and $F = q_K^{-1}(B_K)$. Then, in our previous notation, $q_J^{-1}(A_J) = q_{J \cup K}^{-1}(A_{J \cup K})$, so $A_{J \cup K} = q_{J \cup K}(E)$ and, similarly, $B_{J \cup K} = q_{J \cup K}(F)$.

Moreover,

$$\mu_J(A_J) = (\pi_{J, J \cup K}(\mu_{J \cup K}))(A_J) = \mu_{J \cup K}(A_{J \cup K})$$

and

$$\mu_K(B_K) = \mu_{J \cup K}(B_{J \cup K}).$$

If $E = F$, then $A_{J \cup K} = B_{J \cup K}$. This says that $\mu_J(A_J) = \mu_K(B_K)$, and proves that ν is well defined. When E and F are disjoint, so are $A_{J \cup K}$ and $B_{J \cup K}$, and

$$\nu(E) + \nu(F) = \mu_{J \cup K}(A_{J \cup K}) + \mu_{J \cup K}(B_{J \cup K}) = \mu_{J \cup K}(A_{J \cup K} \cup B_{J \cup K})$$

is equal to

$$\nu\left(q_{J \cup K}^{-1}(A_{J \cup K} \cup B_{J \cup K})\right) = \nu(E \cup F).$$

Thus ν is additive on $\mathcal{A}$.

Now assume that, for every $i \in I$, there exists a compact class C_i of $\tilde{S}_i$-sets such that $\mu_i(A_i) = \sup\{\mu_i(K_i) : K_i \in C_i, K_i \subset A_i\}$. Denote by C the class consisting of the sets $q_J^{-1}(\prod_{i \in J} A_i)$ where $J \in \mathcal{F}$ and $A_i \in C_i$ for all $i \in J$, and let $(K^{(n)})_{n \geq 1}$ be a sequence of C-sets. For every $n \geq 1$. $K^{(n)} = \prod_{i \in I} K_i^{(n)}$ where $K_i^{(n)} \in C_i$ for all i in some subset $J_n \in \mathcal{F}$ and $K_i^{(n)} = \Omega_i$ for all $i \in (I - J_n)$. If $\bigcap_{1 \leq n \leq p} K^{(n)}$ is nonempty for all $p \geq 1$, then $\bigcap_{1 \leq n \leq p} K_i^{(n)}$ is nonempty for all $p \geq 1$ and all $i \in I$; so $\bigcap_{n \geq 1} K_i^{(n)}$ is nonempty and, finally, $\bigcap_{n \geq 1} K^{(n)} \neq \emptyset$. This shows that C is a compact class.

Let $q_J^{-1}(A_J) \in \prod_{i \in I} \tilde{S}_i$, and put $n = \mathrm{card}(J)$. Given $\varepsilon > 0$, there exists, for every $i \in J$, $K_i \in C_i$ such that $K_i \subset A_i$ and $\mu_i(A_i - K_i) \leq \varepsilon/n$. Put $K_J = \prod_{i \in J} K_i$. Since $\bigcup_{j \in J} \left((A_j - K_j) \times \prod_{i \in I - \{j\}} \Omega_i \right)$ contains $q_J^{-1}(A_J) - q_J^{-1}(K_J)$, we have $\nu(q_J^{-1}(A_J) - q_J^{-1}(K_J)) \leq \sum_{j \in J} \mu_j(A_j - K_j) \leq \varepsilon$, which shows that

$$\nu(q_J^{-1}(A_J)) = \sup \{ \nu(K) : K \in C, K \subset q_J^{-1}(A_J) \}.$$

Furthermore, by Proposition 11.2.2, ν is σ-additive on $\prod_{i \in I} \tilde{S}_i$.   □

Definition 11.2.4 In the previous notation, if $\mu = \nu_{/S}$ is σ-additive, it is called the projective limit of the μ_J. Then $\mu_J = q_J(\mu)$ for all $J \in \mathcal{F}$, $\nu(E) = \int 1_E \, d\mu$ for all $E \in \mathcal{A}$, and ν is σ-additive.

Theorem 11.2.1 involved compact classes. This is not the case with the following result.

Theorem 11.2.2 *For every $i \in I$, let μ_i be a positive measure with mass 1 on S_i. For each $J \in \mathcal{F}$, put $\mu_J = \bigotimes_{i \in J} \mu_i$. Then $(\mu_J)_{J \in \mathcal{F}}$ has a projective limit.*

PROOF: Let $(E_n)_{n \geq 1}$ be a decreasing sequence in $\mathcal{A}$ such that $(\nu(E_n))_{n \geq 1}$ does not converge to 0. It suffices to show that $\bigcap_{n \geq 1} E_n$ is nonempty.

For each $n \geq 1$, there exists $J_n \in \mathcal{F}$ and $A_{J_n} \in \tilde{S}_{J_n}$ such that $q_{J_n}^{-1}(A_{J_n}) = E_n$. Let $(i_p)_{p \geq 1}$ be a sequence of distinct elements of I such that $\{i_p : p \geq 1\} \supset \bigcup_{n \geq 1} J_n$.

First, we introduce some notation. For every integer $p \geq 1$, what we have done so far for Ω can be done for $\Omega^{(p)} = \prod_{i \in (I - \{i_1, \ldots, i_p\})} \Omega_i$. Let $\mathcal{F}^{(p)}$ be the class of all nonempty finite subsets of $I - \{i_1, \ldots, i_p\}$. For every $J \in \mathcal{F}^{(p)}$, denote by $q_J^{(p)}$ the canonical projection from $\Omega^{(p)}$ onto Ω_J. There exists an additive function $\nu^{(p)}$ on the algebra $\mathcal{A}^{(p)}$ consisting of the $(q_J^{(p)})^{-1}(A_J)$ for $J \in \mathcal{F}^{(p)}$ and $A_J \in \tilde{S}_J$. Moreover, $\nu^{(p)}$ satisfies $\nu^{(p)}((q_J^{(p)})^{-1}(A_J)) = \mu_J(A_J)$.

For every integer $p \geq 1$, every $(x_{i_1}, \ldots, x_{i_p})$ in $\Omega_{i_1} \times \cdots \times \Omega_{i_p}$, and every integer $n \geq 1$, denote by $E_n(x_{i_1}, \ldots, x_{i_p})$ the section of E_n determined by $(x_{i_1}, \ldots, x_{i_p})$. By definition, $E_n(x_{i_1}, \ldots, x_{i_p})$ consists of those $(x_i)_{i \in (I - \{i_1, \ldots, i_p\})}$ in $\Omega^{(p)}$ such that $(x_i)_{i \in I}$ belongs to E_n. If $\{i_1, \ldots, i_p\}$ does not contain J_n, let $A_{J_n}((x_i)_{i \in J_n \cap \{i_1, \ldots, i_p\}})$ be the section of A_{J_n} determined by $(x_i)_{i \in J_n \cap \{i_1, \ldots, i_p\}}$. Then

$$E_n(x_{i_1}, \ldots, x_{i_p}) = \left(q_{J_n - (J_n \cap \{i_1, \ldots, i_p\})}^{(p)} \right)^{-1} A_{J_n}((x_i)_{i \in J_n \cap \{i_1, \ldots, i_p\}})$$

belongs to $\mathcal{A}^{(p)}$. In case $\{i_1, \ldots, i_p\}$ contains J_n, $E_n(x_{i_1}, \ldots, x_{i_p})$ is equal to $\Omega^{(p)}$ if $(x_i)_{i \in J_n}$ belongs to A_{J_n} and is empty otherwise. Set $\varepsilon = \inf_{n \geq 1} \nu(E_n)$ and, for every $n \geq 1$, define

$$B_n = \left\{ x_{i_1} \in \Omega_{i_1} : \nu^{(1)}(E_n(x_{i_1})) \geq \varepsilon/2 \right\}.$$

Suppose $i_1 \notin J_n$; then

$$E_n(x_{i_1}) = \left(q_{J_n}^{(1)}\right)^{-1}(A_{J_n})$$

and

$$\nu^{(1)}(E_n(x_{i_1})) = \mu_{J_n}(A_{J_n}) = \nu(E_n) \geq \varepsilon$$

for all $x_{i_1} \in \Omega_{i_1}$, whence $B_n = \Omega_{i_1}$. Now, if $J_n = \{i_1\}$. then $B_n = A_{J_n}$ and $\mu_{i_1}(B_n) = \mu_{J_n}(A_{J_n}) = \nu(E_n) \geq \varepsilon$. Finally, if $i_1 \in J_n$ and $J_n \neq \{i_1\}$, Proposition 9.2.7, applied to $\mu_{i_1} \otimes \mu_{J_n - \{i_1\}}$, shows that the function

$$x_{i_1} \mapsto \mu_{J_n - \{i_1\}}(A_{J_n}(x_{i_1})) = \nu^{(1)}(E_n(x_{i_1}))$$

is S_{i_1}-Borelian; thus $B_n \in \tilde{S}_{i_1}$ and

$$\varepsilon \leq \nu(E_n) = \mu_{J_n}(A_{J_n}) = \int \mu_{J_n - \{i_1\}}(A_{J_n}(x_{i_1})) \, d\mu_{i_1}(x_{i_1})$$

$$= \int \nu^{(1)}(E_n(x_{i_1})) \, d\mu_{i_1}(x_{i_1}),$$

which yields

$$\varepsilon \leq \int_{B_n} \nu^{(1)}(E_n(x_{i_1})) \, d\mu_{i_1}(x_{i_1}) + \int_{\Omega_{i_1} - B_n} \nu^{(1)}(E_n(x_{i_1})) \, d\mu_{i_1}(x_{i_1})$$

$$\leq \mu_{i_1}(B_n) + \varepsilon/2.$$

Therefore, $(B_n)_{n \geq 1}$ is a decreasing sequence in $\tilde{S}_{i_1}$ such that $\mu_{i_1}(B_n) \geq \varepsilon/2$ for all $n \geq 1$. Since μ_{i_1} is σ-additive, $\bigcap_{n \geq 1} B_n$ is nonempty, and there exists $x_{i_1} \in \Omega_{i_1}$ satisfying $\nu^{(1)}(E_n(x_{i_1})) \geq \varepsilon/2$ for every $n \geq 1$. But, as $(E_n(x_{i_1}))_{n \geq 1}$ is decreasing in $\mathcal{A}^{(1)}$, we may apply the argument used above for E_n to find $x_{i_2} \in \Omega_{i_2}$ with $\nu^{(2)}(E_n(x_{i_1}, x_{i_2})) \geq \varepsilon/4$. By induction, w : define a sequence of points $(x_{i_k})_{k \geq 1} \in \prod_{k \geq 1} \Omega_{i_k}$ so that $\nu^{(p)}(E_n(x_{i_1}, \ldots, x_{i_p})) \geq 2^{-p}\varepsilon$ for all $p \geq 1$ and all $n \geq 1$. Let $y \in \Omega$ be a point whose canonical projection on $\prod_{p \geq 1} \Omega_{i_p}$ is $(x_{i_p})_{p \geq 1}$. Given an integer $n \geq 1$, let $p \geq 1$ be such that $\{i_1, \ldots, i_p\}$ contains J_n. As $E_n(x_{i_1}, \ldots, x_{i_p})$ is nonempty, $(x_i)_{i \in J_n}$ belongs to A_{J_n}. So $y \in E_n$, and $\bigcap_{n \geq 1} E_n \neq \emptyset$. $\quad\square$

Definition 11.2.5 Under the hypotheses of Theorem 11.2.2, $\mu = \nu_{/S}$ is called the infinite product of the μ_i and is denoted by $\bigotimes_{i \in I} \mu_i$.

For every $i \in I$, let μ_i be a positive measure with mass 1 on a semialgebra S_i, and suppose $(I_j)_{j \in J}$ is a partition of I into nonempty subsets. Put $\mu = \bigotimes_{j \in J}(\bigotimes_{i \in I_j} \mu_i)$. For all i in some finite subset K of I let $A_i \in S_i$, and let $A_i = \Omega_i$ if $i \in (I - K)$. Denote by J_K the finite set $\{j \in J : I_j \cap K \neq \emptyset\}$. Then

$$\mu\left(\prod_{i \in I} A_i\right) = \prod_{j \in J_K}\left(\prod_{i \in I_j \cap K} \mu_i(A_i)\right)$$
$$= \prod_{i \in K} \mu_i(A_i)$$
$$= \prod_{i \in I} \mu_i(A_i).$$

Therefore, $\mu = \bigotimes_{i \in I} \mu_i$ (associativity of product measures). This last result persists for finite I when the μ_i are complex measures defined on semirings.

11.3   Change of Variable

Let $I = \langle a, b \rangle$ be a nonempty interval of $\mathbf{R}$, with left endpoint a and right endpoint b in $\bar{\mathbf{R}}$. Let λ be Lebesgue measure on I and let F be a real Banach space.

Now, given a locally λ-integrable function g from I into $\mathbf{R}$, let G be an indefinite integral of $\mu = g\lambda$. So $G(\beta) - G(\alpha) = \int_{\alpha}^{\beta} g(t)\,dt$ for all $\alpha, \beta \in I$. Since G is continuous (Corollary to Theorem 3.2.2), $G(I)$ is an interval of $\mathbf{R}$, and we let ν be Lebesgue measure on the natural semiring S' of $G(I)$.

Suppose for the moment that g is continuous. If $f : G(I) \to F$ is continuous, then $\int_{\alpha}^{\beta}(f \circ G)g\,dt = \int_{G(\alpha)}^{G(\beta)} f(u)\,du$ for all α, β in I. Indeed, for every $\alpha \in I$, the function $h : y \mapsto \int_{G(\alpha)}^{y} f(u)\,du$ from $G(I)$ into F is a primitive of f; so $(h \circ G)'(x) = f(Gx)g(x)$ for all $x \in I$, from which we obtain

$$(h \circ G)(\beta) - (h \circ G)(\alpha) = \int_{\alpha}^{\beta}(f \circ G)g\,dt.$$

The formula $\int_{\alpha}^{\beta}(f \circ G)g\,dt = \int_{G(\alpha)}^{G(\beta)} f(u)\,du$ can also be written

$$\int_{\alpha}^{\beta}(f \circ G)(t)G'(t)\,dt = \int_{G(\alpha)}^{G(\beta)} f(u)\,du.$$

The latter form is recognized as a change of variable formula. In fact, we can prove much more general change of variable formulas. For this, we return to the case in which g is locally λ-integrable.

Proposition 11.3.1 *Assume that g is λ-integrable. Then (G, S') is μ-suited. The limits $G(a^+) = \lim_{\substack{x \to a \\ x > a}} G(x)$ and $G(b_-)$ exist, and if J is an interval* *included in $G(I)$ with endpoints $G(a^+)$ and $G(b_-)$, then $G(\mu) = 1_J \cdot \nu$ or $G(\mu) = -1_J \cdot \nu$, as $G(a^+) \leq G(b_-)$ or $G(a^+) \geq G(b_-)$.*

PROOF: Since μ is bounded, (G, S') is μ-suited. By the dominated convergence theorem, $G(a^+)$ and $G(b_-)$ exist in $\mathbf{R}$, and it remains to be proven that $G(\mu) = \pm 1_J \cdot \nu$. Let $f \in \mathcal{H}(G(I), \mathbf{R})$. It suffices to show that $\int (f \circ G) g \, d\lambda = \pm \int f \cdot 1_J \, d\nu$ (Theorem 8.1.2).

Let φ be a bounded continuous function from $\mathbf{R}$ into $\mathbf{R}$ that extends f. By Theorem 3.1.2, we can find a λ-integrable function h from I into $[0, +\infty]$ and a sequence $(g_n)_{n \geq 1}$ in $\mathcal{H}(I, \mathbf{R})$ with the following properties:

(a) $(g_n)_{n \geq 1}$ converges to g λ-almost everywhere;

(b) $|g_n| \leq h$ for all $n \geq 1$.

$(g_n)_{n \geq 1}$ converges to g in $\mathcal{L}^1_{\mathbf{R}}(\lambda)$. If we put $G_n(x) = G(x_0) + \int_{x_0}^x g_n(t) \, dt$, where x_0 is a fixed point of I, the functions G_n converge uniformly to G, so $G_n(a^+), G_n(b_-)$ converge to $G(a^+), G(b_-)$, respectively. Now $|(\varphi \circ G_n) g_n|$ is smaller than $\|\varphi\| \cdot h$, for all $n \in \mathbf{N}$. Thus $\int (\varphi \circ G_n) g_n \, d\lambda$ converges to $\int (\varphi \circ G) g \, d\lambda$ as $n \to +\infty$. For every $n \in \mathbf{N}$, let $[c_n, d_n]$ (with $c_n \leq d_n$) be a compact subinterval of I containing the support of g_n. Then

$$\int (\varphi \circ G_n) g_n \, d\lambda = \int_{c_n}^{d_n} (\varphi \circ G_n)(t) g_n(t) \, dt$$
$$= \int_{G_n(c_n)}^{G_n(d_n)} \varphi(\xi) \, d\xi,$$

where $G_n(c_n) = G_n(a^+)$ and $G_n(d_n) = G_n(b_-)$ because G_n is constant on each of the intervals $\langle a, c_n]$ and $[d_n, b \rangle$. Letting n converge to $+\infty$, we see that $\int (f \circ G) g \, d\lambda = \int_{G(a^+)}^{G(b_-)} \varphi(\xi) \, d\xi$.

Next, suppose $G(a^+) \neq G(b_-)$ and let $[u, v]$ be a compact subinterval of $G(I)$, included in $[G(a^+), G(b_-)]$, outside which $f \cdot 1_J$ vanishes. Then $\int f \cdot 1_J \, d\nu = \int_u^v \varphi(\xi) \, d\xi$. But, since φ vanishes on the complement of $[u, v]$ with respect to $[G(a^+), G(b_-)]$, we see finally that $\int f \cdot 1_J d\nu$ is equal to $\int_{G(a^+)}^{G(b_-)} \varphi(\xi) \, d\xi$ or $-\int_{G(a^+)}^{G(b_-)} \varphi(\xi) \, d\xi$, as $G(a^+) < G(b_-)$ or $G(a^+) > G(b_-)$.

□

Proposition 11.3.2 *Let $x_0 \in I$. Then, (G, S') is μ-suited if and only if*

(a) *the limits $G(a^+)$ and $G(b_-)$ exist in $\bar{\mathbf{R}}$;*

(b) *$g \cdot 1_{\langle a, x_0]}$ is λ-integrable whenever $G(a^+)$ belongs to $G(I)$;*

(c) *$g \cdot 1_{[x_0, b \rangle}$ is λ-integrable whenever $G(b_-)$ belongs to $G(I)$.*

If (G, S') is μ-suited and if J is a subinterval of $G(I)$ with endpoints $G(a^+)$ and $G(b_-)$, then $G(\mu) = 1_J \cdot \nu$ or $G(\mu) = -1_J \cdot \nu$ as $G(a^+) \leq G(b_-)$ or $G(a^+) \geq G(b_-)$.

PROOF: Assume that these conditions are satisfied, and let K be a compact subset of $G(I)$. Then $B = G^{-1}(K) \cap [x_0, b\rangle$ is closed in $[x_0, b\rangle$. If $G(b_-)$ belongs to K, $[x_0, b\rangle$ is μ-integrable, and the same is true of B. On the other hand, if $G(b_-)$ does not belong to K, b does not lie in the closure of B (with respect to $\bar{\mathbf{R}}$), hence B is compact. Similarly, $A = \langle a, x_0] \cap G^{-1}(K)$ is μ-integrable. Thus (G, S') is μ-suited. Conversely, assume that (G, S') is μ-suited. We argue by contradiction. Suppose that the supremum, M, for $G_{/[x_0,b\rangle}$ is not attained and that $G(x)$ does not converge to M as x converges to b in $[x_0, b\rangle$. We can find $r < M$ with the following property: for all $x \in [x_0, b\rangle$, there exists $y \in [x, b\rangle$ such that $G(y) \leq r$. Letting s be an arbitrary real number in $]r, M[$, there exists a strictly increasing sequence $(x_n)_{n \geq 1}$ in $[x_0, b\rangle$ such that $G([x_{2n-1}, x_{2n}])$ is included in $[r, s]$, $G(x_{2n-1}) = r$, and $G(x_{2n}) = s$. Since the set $G^{-1}([r, s])$ is μ-integrable by hypothesis, $E = \bigcup_{n \geq 1}[x_{2n-1}, x_{2n}]$ is μ-integrable as well, and $g \cdot 1_E$ is λ-integrable. Then $\int^* |g| \cdot 1_E \, d\lambda = \sum_{n \geq 1} \int^* |g| \cdot 1_{[x_{2n-1}, x_{2n}]} \, d\lambda$ is finite. But this is absurd, because $s - r = G(x_{2n}) - G(x_{2n-1})$ is less than $\int |g| \cdot 1_{[x_{2n-1}, x_{2n}]} \, d\lambda$ for all $n \geq 1$.

In short, if the supremum, M, for $G_{/[x_0,b\rangle}$ is not attained, then $G(x)$ converges to M as x tends to b in $[x_0, b\rangle$. Likewise, if the infimum, m, for $G_{/[x_0,b\rangle}$ is not attained, then $G(x)$ converges to m as x tends to b in $[x_0, b\rangle$. In each of these two cases and if $G(b_-)$ belongs to $G(I)$, then $[x_0, b\rangle$ is included in $G^{-1}([m, M])$; thus it is μ-integrable.

Finally, suppose that M and m are attained by $G_{/[x_0,b\rangle}$. Then $[x_0, b\rangle$ is μ-integrable, because it is included in $G^{-1}([m, M])$. So $g \cdot 1_{[x_0,b\rangle}$ is λ-integrable and $G(b_-)$ exists.

Next, let $f \in \mathcal{H}(G(I), \mathbf{C})$. If $G(b_-)$ belongs to $G(I)$, then $g \cdot 1_{[x_0,b\rangle}$ is λ-integrable and $\int_{G(x_0)}^{G(b_-)} f \, d\nu = \int (f \circ G) g \cdot 1_{[x_0,b\rangle} \, d\lambda$ (Proposition 11.3.1). On the other hand, if $G(b_-)$ does not belong to $G(I)$, there exists $x_0 \leq \beta < b$ such that $]\beta, b\rangle$ does not meet $B = G^{-1}(\text{supp } f) \cap [x_0, b\rangle$; then

$$\int_{G(x_0)}^{G(\beta)} f \, d\nu = \int (f \circ G) g \cdot 1_{[x_0, \beta]} \, d\lambda$$

(Proposition 11.3.1), that is,

$$\int_{G(x_0)}^{G(b_-)} f \, d\nu = \int (f \circ G) g \cdot 1_{[x_0,b\rangle} \, d\lambda.$$

In both cases, the last equality is true. Likewise,

$$\int_{G(a^+)}^{G(x_0)} f \, d\nu = \int (f \circ G) g \cdot 1_{\langle a, x_0]} \, d\lambda.$$

We conclude that

$$\int_{G(a^+)}^{G(b_-)} f\, d\nu = \int (f \circ G)g\, d\lambda$$

and $G(\mu) = \pm 1_J \nu$.   □

Theorem 11.3.1 *Suppose g is positive. Then (G, S') is μ-suited and $G(\mu) = \nu$. If $f : G(I) \to F$ is ν-integrable, then $(f \circ G)g$ is λ-integrable and $\int f\, d\nu = \int (f \circ G)g\, d\mu$.*

PROOF: G increases, so $G(a^+)$ and $G(b_-)$ exist in $\bar{\mathbf{R}}$. Let $x_0 \in I$. If $G(b_-)$ belongs to $G(I)$, there exists c in $[x_0, b\rangle$ such that $G(c) = G(b_-)$. Then g vanishes λ-almost everywhere in $[c, b\rangle$, hence $g \cdot 1_{[x_0,b\rangle}$ is λ-integrable. Similarly, $g \cdot 1_{\langle a,x_0]}$ is λ-integrable if $G(a^+)$ belongs to $G(I)$. Therefore, (G, S') is μ-suited. Now the last assertion follows from Section 11.1.   □

As we shall see in Chapter 22, these results can be refined.

11.4   Elements of Ergodic Theory

In this section, we prove Birkhoff's ergodic theorem.

Lemma 11.4.1 *Let $(t_i)_{0 \leq i < n}$ be a finite sequence of real numbers, and let m be an integer such that $0 \leq m \leq n$. Denote by L_m the set of indices i with the following property: there exists an integer $0 \leq p \leq m$ such that $t_i + \cdots + t_{i+p} \geq 0$. Then $\sum_{i \in L_m} t_i$ is positive.*

PROOF: Let $i \in L_m$, and let p be the smallest integer in $\{0, \ldots, m\}$ such that $t_i + \cdots + t_{i+p} \geq 0$. For $1 \leq q \leq p$, we have $t_i + \cdots + t_{i+q-1} < 0$ and $t_i + \cdots + t_{i+p} \geq 0$, hence $t_{i+q} + \cdots + t_{i+p} \geq 0$, which proves that $i + q$ lies in L_m. Thus $i, i+1, \ldots, i+p$ all belong to L_m.

If L_m is nonempty, we can construct a sequence $J_1, \ldots, J_r$ ($r \geq 1$) of nonempty intervals of $[0, n] \cap \mathbf{Z}^+$ such that

(a) $\beta(J_{k-1}) < \alpha(J_k)$ for all $2 \leq k \leq r$;

(b) $J_1 \cup \cdots \cup J_k = [0, \beta(J_k)] \cap L_m$ for all $1 \leq k \leq r$;

(c) $\sum_{i \in J_k} t_i \geq 0$ for each $1 \leq k \leq r$;

(d) $J_1 \cup \cdots \cup J_r = L_m$.

Indeed, suppose we have obtained nonempty intervals $J_1, \ldots, J_{s-1}$ (with $s \geq 1$) of $[0, n] \cap \mathbf{Z}^+$ so that the above properties (a), (b), and (c) hold for all $k \leq s - 1$, and that $J_1 \cup \cdots \cup J_{s-1} \neq L_m$. If i_s is the smallest integer of $L_m - (J_1 \cup \cdots \cup J_{s-1})$ and p_s is the smallest of those integers $0 \leq p \leq m$ such that $t_{i_s} + \cdots + t_{i_s+p} \geq 0$, we take $J_s = \{i_s, \ldots, i_s + p_s\}$. Thus the induction may proceed.

Now we see that $\sum_{i \in L_m} t_i = \sum_{1 \leq k \leq r} \left(\sum_{i \in J_k} t_i \right)$ is positive.   □

In what follows, let Ω be a nonempty set, S a semiring in Ω, $\mu \neq 0$ a positive measure on S, and u a mapping from Ω into Ω. Assume that (u, S) is μ-suited and $u(\mu) = \mu$. Then (u^k, S) is μ-suited and $u^k(\mu) = \mu$, for all $k \in \mathbf{Z}^+$. For each mapping f from Ω into a set Ω', put $f_k = f \circ u^k$ for all $k \in \mathbf{Z}^+$.

Proposition 11.4.1 (Maximal Ergodic Theorem) *Let $f : \Omega \to \mathbf{R}$ be essentially μ-integrable, and put*

$$A = \{x \in \Omega : f_0(x) + \cdots + f_p(x) \geq 0 \text{ for at least one } p \geq 0\}.$$

Then $\int_A f \, d\mu \geq 0$.

PROOF: For all $k \in \mathbf{Z}^+$, f_k is essentially μ-integrable. Fix $m \in \mathbf{Z}^+$, and let A_m be the set of those $x \in \Omega$ such that $f_0(x) + \cdots + f_p(x) \geq 0$ for at least one $0 \leq p \leq m$. Let an integer $n \geq 0$ be given. For each $0 \leq k \leq m + n$ the set B_k, of those $x \in \Omega$ such that $(f_k + \cdots + f_{k+p})(x) \geq 0$ for at least one $0 \leq p \leq \inf(m, n + m - k)$, is μ-measurable. Moreover, $B_k = u^{-k}(A_m)$ for $0 \leq k \leq n$. Now, for every $x \in \Omega$, let $L_m(x)$ be the set of those indices $0 \leq k \leq m + n$ for which there exists $0 \leq p \leq \inf(m, m + n - k)$ such that $(f_k + \cdots + f_{k+p})(x) \geq 0$. Then, for all $x \in \Omega$ and for all $0 \leq k \leq m + n$, we have $1_{B_k}(x) = 1$ or $1_{B_k}(x) = 0$ as k belongs to $L_m(x)$ or not. Hence $\sum_{0 \leq k \leq m+n} f_k(x) \cdot 1_{B_k}(x) = \sum_{k \in L_m(x)} f_k(x)$ for all $x \in \Omega$; however, the right-hand side is positive by Lemma 11.4.1, and so $\sum_{0 \leq k \leq m+n} \int f_k \cdot 1_{B_k} \, d\mu$ is positive.

Next,

$$\sum_{0 \leq k \leq m+n} \int f_k \cdot 1_{B_k} \, d\mu = (n+1) \int f \cdot 1_{A_m} \, d\mu + \sum_{n+1 \leq k \leq n+m} \int f_k \cdot 1_{B_k} \, d\mu.$$

Since $\sum_{n+1 \leq k \leq n+m} \int f_k \cdot 1_{B_k} \, d\mu$ is less than $m \int |f| \, d\mu$, we see, finally, that $(n+1) \int f \cdot 1_{A_m} \, d\mu + m \int |f| \, d\mu$ is positive. Letting n tend to $+\infty$, we conclude that $\int f \cdot 1_{A_m} \, d\mu$ is positive.

Now A_m is included in A_{m+1}, for all $m \geq 0$, and the dominated convergence theorem shows that $\int f \cdot 1_{A_m} \, d\mu$ converges to $\int f \cdot 1_A \, d\mu$ as $m \to +\infty$. Therefore, $\int f \cdot 1_A \, d\mu$ is positive.   □

Theorem 11.4.1 (G.D. Birkhoff's Ergodic Theorem) *Let $f : \Omega \to \mathbf{R}$ be essentially μ-integrable. There exists an essentially μ-integrable function f^* from Ω into $\mathbf{R}$ such that $(1/n) \sum_{0 \leq k \leq n-1} f_k$ converges to f^* locally μ-almost everywhere as $n \to +\infty$. Moreover, $f^* = f^* \circ u$ locally μ-almost everywhere.*

PROOF: Let b be a real number and let C an essentially μ-integrable set such that $b < \limsup_{n \to +\infty}(1/n)\sum_{0 \le k \le n-1} f_k(x)$ for all $x \in C$. Denote by A the set of those $x \in \Omega$ for which there exists $p \ge 0$ such that $(f_0 + \cdots + f_p)(x)$ exceeds $b \cdot \mathrm{card}\{0 \le k \le p : u^k(x) \in C\}$. For each $x \in C$ and each $n_0 \ge 1$, we can find $n \ge n_0$ so that $(1/n)\sum_{0 \le k \le n-1} f_k(x) \ge b$. Thus A contains C. By the maximal ergodic theorem, $\int_A (f - b \cdot 1_C)\, d\mu = \int_A f\, d\mu - b\mu(C)$ is positive, which leads to $b\mu(C) \le \int |f|\, d\mu$.

Now let a, $b \in \mathbf{R}$ be such that $a < b$, and denote by E the set of those $x \in \Omega$ such that

$$\liminf_{n \to +\infty} \frac{1}{n} \sum_{0 \le k \le n-1} f_k(x) < a < b < \limsup_{n \to +\infty} \frac{1}{n} \sum_{0 \le k \le n-1} f_k(x).$$

For each μ-integrable subset C of E,

$$- \int |f|\, d\mu \le a\mu(C) \le b\mu(C) \le \int |f|\, d\mu$$

(by what we have just shown), and hence $(b-a)\mu(C) \le 2\int |f|\, d\mu$. This implies that $\mu^\bullet(E)$ is finite and E is essentially μ-integrable.

Let $x \in \Omega$. The following conditions are clearly equivalent.

(a) $b < \limsup_{n \to +\infty}(1/n)\sum_{0 \le k \le n-1} f_k(ux)$.

(b) $b < \limsup_{n \to +\infty}(1/n)\sum_{0 \le k \le n} f_k(x)$.

(c) There exists $b' > b$ such that $b'\big(n/(n+1)\big) \le \big(1/(n+1)\big)\sum_{0 \le k \le n} f_k(x)$ for infinitely many indices n.

(d) There exists $b'' > b$ such that $b'' \le \big(1/(n+1)\big)\sum_{0 \le k \le n} f_k(x)$ for infinitely many indices n.

Thus

$$b < \limsup_{n \to +\infty} \frac{1}{n} \sum_{0 \le k \le n-1} f_k(ux) \quad \text{if} \quad b < \limsup_{n \to +\infty} \frac{1}{n} \sum_{0 \le k \le n-1} f_k(x),$$

and $u(E)$ is included in E. In fact, $u^k(E) \subset E$ for all $k \ge 0$.

Write A for the set of those $x \in \Omega$ such that there exists $p \ge 0$ with $[(f-b) \cdot 1_E]_0(x) + \cdots + [(f-b) \cdot 1_E]_p(x) \ge 0$. Then $A \cap E$ is the set of those $x \in E$ such that $f_0(x) + \cdots + f_p(x) \ge (p+1)b$ for at least one $p \ge 0$. Thus $A \cap E = E$ in view of the definition of E. Now, applying the maximal ergodic theorem to $(f - b) \cdot 1_E$, we get $b\mu(E) \le \int f \cdot 1_E\, d\mu$. Similarly, $\int f \cdot 1_E\, d\mu \le a\mu(E)$. So $\mu(E) = 0$ and E is locally μ-negligible.

For each $(a, b) \in \mathbf{Q} \times \mathbf{Q}$ such that $a < b$, put

$$F_{(a,b)} = \left\{ x \in \Omega : \liminf \frac{1}{n} \sum_{0 \le k \le n-1} f_k(x) < a < b < \limsup \frac{1}{n} \sum_{0 \le k \le n-1} f_k(x) \right\}.$$

Then

$$F = \left\{ x \in \Omega : \liminf \frac{1}{n} \sum_{0 \le k \le n-1} f_k(x) < \limsup \frac{1}{n} \sum_{0 \le k \le n-1} f_k(x) \right\},$$

as the union of the $F_{(a,b)}$, is locally μ-negligible.

When f is positive, we have $\int^{\bullet} \left(\liminf (1/n) \sum_{0 \le k \le n-1} f_k \right) d\mu \le \int f \, d\mu$ by Fatou's lemma, so $\liminf (1/n) \sum_{0 \le k \le n-1} f_k$ is essentially μ-integrable. Therefore, there exists an essentially μ-integrable function f^* from Ω into $\mathbf{R}$ such that $\left((1/n) \sum_{0 \le k \le n-1} f_k \right)_{n \ge 1}$ converges to f^* locally μ-almost everywhere. Writing $f = f^+ - f^-$, we see that the preceding is true even when f is not positive.

The last assertion is obvious. □

This theorem extends to complex-valued functions.

Theorem 11.4.2 *Suppose μ is bounded. Let $f : \Omega \to \mathbf{C}$ be essentially μ-integrable, and let $f^* : \Omega \to \mathbf{C}$ be such that $\left((1/n) \sum_{0 \le k \le n-1} f_k \right)_{n \ge 1}$ converges to f^* locally μ-almost everywhere. Then $\int f^* \, d\mu = \int f \, d\mu$.*

PROOF: Suppose that f and f^* are positive. Since $(1/n) \sum_{0 \le k \le n-1} f_k \le \|f\|$ for all $n \ge 1$ if f is bounded (and positive), the dominated convergence theorem then shows that $\int f^* \, d\mu = \lim_{n \to +\infty} (1/n) \sum_{0 \le k \le n-1} \int f_k \, d\mu = \int f \, d\mu$. We now pass to the general case. For all $r < \int f \, d\mu$, there exists an integer $p \ge 1$ such that $r \le \int \inf(f, p) \, d\mu$. If $\inf(f, p)^*$ is positive and associated with $\inf(f, p)$ in the same way that f^* is associated with f, then $\inf(f, p)^* \le f^*$ locally almost everywhere. So $\int f^* \, d\mu \ge \int \inf(f, p)^* \, d\mu = \int \inf(f, p) \, d\mu > r$. This yields $\int f^* \, d\mu \ge \int f \, d\mu$. The reverse inequality follows from Fatou's lemma. □

If μ is bounded, observe that $\left((1/n) \sum_{0 \le k \le n-1} f_k \right)_{n \ge 1}$ converges to f^* in $\bar{\mathcal{L}}_{\mathbf{C}}^1(\mu)$ by Proposition 5.4.6.

Now we introduce the concept of μ-ergodic function.

Definition 11.4.1 $f : \Omega \to \mathbf{C}$ is said to be μ-invariant if $f = f \circ u$ locally μ-almost everywhere. u is said to be μ-ergodic whenever each μ-measurable and μ-invariant function from Ω into $\mathbf{C}$ is constant locally μ-a.e.

Proposition 11.4.2 *u is μ-ergodic if (and only if) $\mu^{\bullet}(A) = 0$ or $\mu^{\bullet}(A^c) = 0$ for each μ-measurable set A such that $1_A = 1_A \circ u$ locally μ-a.e.*

PROOF: Let $f : \Omega \to [0, +\infty[$ be μ-measurable and μ-invariant. Define f_n as in Proposition 3.3.3. Then f_n is constant locally μ-a.e. As $f_n \to f$, we see that the same is true of f, and the proposition follows. □

Exercises for Chapter 11

1 Let μ be Lebesgue measure on $[0,1]$, and let φ be the function $x \mapsto \exp(2i\pi x)$ from $[0,1]$ onto the unit circle in the plane $\mathbf{R}^2$. Write $\mathcal{B}$ for the Borel σ-algebra of $\mathbf{T} = \{z \in \mathbf{C} : |z| = 1\}$, and ν for the measure on $\mathcal{B}$ which is the image of μ under φ.

1. Show that ν remains invariant under any translation $z \mapsto zz_0$ ($z_0 \in \mathbf{T}$).

2. Show that the image of ν under the mapping $z \mapsto z^{-1}$ is ν itself.

2 Let $\mathbf{T} = \{z \in \mathbf{C} : |z| = 1\}$. Given an irrational number θ, let u be the mapping $z \mapsto ze^{2i\pi\theta}$ from $\mathbf{T}$ onto $\mathbf{T}$.

Let $f : \mathbf{T} \to \mathbf{R}$ be ν-integrable, and put $f_k = f \circ u^k$ for all integers $k \geq 0$. By Birkhoff's ergodic theorem, there exists a ν-integrable real-valued function f^* on $\mathbf{T}$ such that $\left((1/n)\sum_{0 \leq k \leq n-1} f_k\right)_{n \geq 1}$ converges to f^* almost everywhere. Moreover, $\int f^* \, d\nu = \int f \, d\nu$.

1. Suppose first that f is continuous. Show that, for all $z \in \mathbf{T}$, the sequence $\left((1/n) \cdot \sum_{0 \leq k \leq n-1} f_k(z)\right)_{n \geq 1}$ converges to $\int f \, d\nu$ as $n \to +\infty$ (use Exercise 6 of Chapter 8).

2. Assume that f is upper semicontinuous. Prove that $f^*(z) \leq \int f \, d\nu$ for ν-almost all z, so that $f^*(z) = \int f \, d\nu$ ν-a.e.

3. In the general case, when f is assumed only to be ν-integrable, prove that $f^*(z) \geq \int f \, d\nu$ for ν-almost all z (use Proposition 3.7.3), so that $f^*(z) = \int f \, d\nu$ ν-a.e.

4. Conclude that u is ν-ergodic.

3 Write λ for Lebesgue measure on $\mathbf{R}$. Let $f \in \mathcal{L}_F^1(\lambda)$, where F is a Banach space, and let $g \in \mathcal{L}_{\mathbf{C}}^\infty(\lambda)$ be periodic with period 1. Finally, let $(a_n)_{n \geq 1}$ be any sequence of real numbers. Prove Fejer's formula:

$$\lim_{n \to +\infty} \int f(x)g(nx + a_n) \, dx = \left(\int f(x) \, dx\right) \cdot \left(\int_0^1 g(x) \, dx\right).$$

(First, suppose that f belongs to $St(S, F)$, where S is the natural semiring of $\mathbf{R}$.)

4 Let ν be as in Exercise 1. Given $k \in \mathbf{Z} - \{0\}$, write u for $\mathbf{T} \ni z \mapsto z^k$.

1. Show that $u(\nu) = \nu$.

2. Let $f \in \mathcal{L}_{\mathbf{C}}^1(\nu)$ and $g \in \mathcal{L}_{\mathbf{C}}^\infty(\nu)$. Show that $\int f(z)g(z^n) \, d\nu(z)$ converges to $(\int f \, d\nu) \cdot (\int g \, d\nu)$ as $n \in \mathbf{N}$ tends to $+\infty$. (Apply Fejer's formula.)

3. Let $g \in \mathcal{L}_{\mathbf{R}}^\infty(\nu)$ be such that $g \circ u = g$ ν-a.e., and put $c = \int g \, d\nu$. Let f be the indicator function of $\{z \in \mathbf{T} : g(z) \geq c\}$ on $\mathbf{T}$. By part 2, $\int f(g \circ u^p) \, d\nu$ converges to $(\int f \, d\nu) \cdot (\int g \, d\nu)$ as $p \to +\infty$. Deduce that $\int f(g - c) \, d\nu = 0$, and therefore that $g = c$ almost everywhere in $\{z \in \mathbf{T} : g(z) \geq c\}$.

4. Conclude that u is ν-ergodic.

5 1. Show that $\displaystyle\sum_{-n\leq k\leq n}\left(e^{2i\pi t}\right)^k = \frac{\sin\left((2n+1)\pi t\right)}{\sin(\pi t)}$ for every integer $n\geq 0$ and

all $t\in\mathbf{R}-\mathbf{Z}$. Deduce that $\displaystyle\int_0^{1/2}\frac{\sin\left((2n+1)\pi t\right)}{\sin(\pi t)}\,dt=\frac{1}{2}$.

2. Write λ for Lebesgue measure on $\mathbf{R}$. Let f be a locally λ-integrable function from $\mathbf{R}$ into a Banach space F, periodic with period 1. For each $k\in\mathbf{Z}$, set $\hat{f}(k)=\int_0^1 f(t)e^{-2i\pi kt}\,dt$. Prove that

$$\int_0^{1/2}\left(f(x+t)-f(x-t)\right)\frac{\sin\left((2n+1)\pi t\right)}{\sin(\pi t)}\,dt=\sum_{-n\leq k\leq n}\hat{f}(k)e^{2i\pi kx}=S_n(x)$$

for all integers $n\geq 0$ and all $x\in\mathbf{R}$.

3. Given $x\in\mathbf{R}$ and α, β in F, suppose that the functions $t\mapsto\dfrac{f(x+t)-\alpha}{t}$

and $t\mapsto\dfrac{f(x-t)-\beta}{t}$ from $]0,1/2]$ into F are $\lambda_{/]0,1/2]}$-integrable. Deduce

from Fejer's formula that $\displaystyle\int_0^{1/2}\left(f(x+t)-\alpha\right)\frac{\sin\left((2n+1)\pi t\right)}{\sin(\pi t)}\,dt$ (and

similarly $\displaystyle\int_0^{1/2}\left(f(x-t)-\beta\right)\frac{\sin\left((2n+1)\pi t\right)}{\sin(\pi t)}\,dt$) goes to 0 as $n\to+\infty$,

and conclude that $S_n(x)$ approaches $(\alpha+\beta)/2$.

4. Given $x\in\mathbf{R}$, suppose that f has at x a right-hand limit $f(x_+)$ and a left-hand limit $f(x_-)$. Also, suppose that $\left(f(x+t)-f(x_+)\right)/t$ and $\left(f(x-t)-f(x_-)\right)/(-t)$ converge to some limits as $t>0$ goes to 0. Show that $S_n(x)$ approaches $\dfrac{1}{2}\left(f(x_+)+f(x_-)\right)$ as $n\to+\infty$.

12

Change of Variables

In this chapter we study the image of Lebesgue measure in an open subset V of $\mathbf{R}^k$ under a function $T : V \to \mathbf{R}^k$. By making some assumptions about the differentiability of T, it becomes possible to find the image explicitly in terms of Lebesgue measure and the determinant of the differential of T.

Summary

12.1 In this section we define derivatives $D\mu$, $\underline{D}\mu$. and $\overline{D}\mu$ of a measure μ with respect to Lebesgue measure in V. Next, we analyze the relationships between the Lebesgue decomposition of μ and the properties of its derivatives; for example, if $\overline{D}\mu$ is finite everywhere, μ is absolutely continuous with respect to Lebesgue measure.

12.2 First, we study the image of λ under a linear automorphism. The modulus of an automorphism of $\mathbf{R}^k$ is the absolute value of its determinant (Theorem 12.2.1).

12.3 In this section we prove the change of variables formula: if T is a differentiable homeomorphism from V onto W, then λ_W is the image of $|J(T)|\lambda_V$ under T where $J(T)$ is the Jacobian of T (Theorem 12.3.1).

12.4 This section is devoted to polar coordinates in $\mathbf{R}^n$.

12.1 Differentiation in $\mathbf{R}^k$

Given an integer $k \geq 1$ and a nonempty open subset Ω of $\mathbf{R}^k$, let λ be Lebesgue measure on the natural semiring S of Ω.

Definition 12.1.1 Equip $\mathbf{R}^k$ with the Euclidean metric. A sequence $(E_i)_{i \geq 1}$ of Borel sets in Ω is said to shrink to $x \in \mathbf{R}^k$ nicely if there is a number $\alpha > 0$

and a sequence $(r_i)_{i\geq 1}$ of strictly positive numbers converging to 0, which satisfy the following conditions:

(a) For all $i \geq 1$, Ω contains the closed ball $B'(x, r_i)$ of radius r_i centered at x.

(b) For all $i \geq 1$, E_i is included in the closed ball $B'(x, r_i)$ and $\lambda(E_i) \geq \alpha \cdot \lambda\big(B'(x, r_i)\big)$.

Definition 12.1.2 Let μ be a complex measure on the natural semiring of Ω, and let $x \in \mathbf{R}^k$. If $A \in \mathbf{C}$ is such that $\lim_{i\to+\infty} \mu(E_i)/\lambda(E_i) = A$ for every sequence $(E_i)_{i\geq 1}$ that shrinks to x nicely, we call A the derivative of μ at x, and we write $(D\mu)(x) = A$.

The principal result of this section is Theorem 12.1.1. The following Lemmas 12.1.1 and 12.1.2 will be needed for the proof.

Lemma 12.1.1 *If C is a collection of open balls in $\mathbf{R}^k$, whose union W is contained in Ω, and if $t < \lambda^*(W)$, then there is a finite disjoint subcollection $\{B_1, \ldots, B_n\}$ of C such that $\sum_{1\leq i\leq n} \lambda(B_i) > 3^{-k}t$.*

PROOF: Choose a compact set K so that $K \subset W$ and $\lambda(K) > t$. Since K is compact, it is covered by finitely many elements of C, say $U_1, \ldots, U_p$, which we may order so that their radii $r(U_j)$ satisfy $r(U_j) \geq r(U_{j+1})$ for $1 \leq j \leq p-1$.

Put $B_1 = U_1$. Discard all U_j such that $j > 1$ and U_j intersects B_1. Let $B_2 = U_{i_2}$ be the first of the remaining U_j (if there are any). Discard all U_j such that $j > i_2$ and U_j intersects B_2; let $B_3 = U_{i_3}$ be the first of the remaining U_j. Repeat this process as often as possible. This gives the disjoint collection $\{B_1, \ldots, B_n\}$.

Now each discarded U_j is a subset of a ball C_i concentric with some B_i, whose radius is three times that of B_i. Thus $K \subset \bigcup_{1\leq i\leq n} C_i$. It follows that

$$t < \lambda(K) \leq 3^k \cdot \sum_{1\leq i\leq n} \lambda(B_i). \qquad \square$$

Lemma 12.1.2 *Suppose μ is a positive measure on S, and let A be a μ-negligible subset of Ω. Then there is a set $A' \subset A$ such that $\lambda(A - A') = 0$ and $(D\mu)(x) = 0$ for every $x \in A'$.*

PROOF: Define A' as the set of all $x \in A$ for which

$$\lim_{r\to 0} \frac{\mu\big(B'(x, r)\big)}{\lambda\big(B'(x, r)\big)} = 0,$$

or, equivalently,

$$\lim_{r\to 0} \frac{\mu\big(B(x, r)\big)}{\lambda\big(B(x, r)\big)} = 0.$$

For every $j \in \mathbf{N}$, let P_j be the set of all $x \in A$ for which

$$\limsup_{r \to 0} \frac{\mu(B(x,r))}{\lambda(B(x,r))} > \frac{1}{j}.$$

We show that $\lambda(P_j) = 0$ for all $j \in \mathbf{N}$. Since $A - A' = \bigcup_{j \geq 1} P_j$, it will follow that $\lambda(A - A') = 0$.

Fix j, and let $\varepsilon > 0$ be given. Since $\mu(A) = 0$, there is an open subset V of Ω containing A, with $\mu(V) \leq \varepsilon$. Every $x \in P_j$ is the center of an open ball $B_x \subset V$ such that $\lambda(B_x) \leq j \cdot \mu(B_x)$. Let W be the union of these B_x. For all $t < \lambda^*(W)$, Lemma 12.1.1 shows that there is a disjoint subcollection $\{B_{x_1}, \ldots, B_{x_n}\}$ of $\{B_x : x \in P_j\}$ such that

$$t < 3^k \cdot \sum_{1 \leq i \leq n} \lambda(B_{x_i}) \leq 3^k j \cdot \sum_{1 \leq i \leq n} \mu(B_{x_i}) = 3^k j \cdot \mu(B_{x_1} \cup \cdots \cup B_{x_n})$$

$$\leq 3^k j \cdot \mu(V) \leq 3^k j \varepsilon.$$

Thus $\lambda(W) \leq 3^k j \varepsilon$. Since $P_j \subset W$ and ε is arbitrary, we see that $\lambda(P_j) = 0$.

Now fix $x \in A'$ and let $(E_i)_{i \geq 1}$ be a sequence that shrinks to x nicely. Let $\alpha > 0$ and $(r_i)_{i \geq 1}$ be as in Definition 12.1.1. Then

$$\frac{\mu(E_i)}{\lambda(E_i)} \leq \frac{1}{\alpha} \cdot \frac{\mu(B'(x, r_i))}{\lambda(B'(x, r_i))} \qquad \text{for every } i \geq 1.$$

Therefore, $\mu(E_i)/\lambda(E_i)$ converges to 0 as i tends to $+\infty$, which proves that $D\mu(x) = 0$. □

Theorem 12.1.1 *Let μ be a complex measure on S. $D\mu(x)$ is defined for λ-almost all x of Ω, and the function $x \mapsto D\mu(x)$ (defined λ-almost everywhere) is locally λ-integrable. Moreover, the Lebesgue decomposition of μ relative to λ is $\mu = (D\mu)\lambda + \mu_s$, with $D\mu_s(x) = 0$ λ-almost everywhere.*

PROOF: It suffices to prove the theorem separately for $\mu \perp \lambda$ and for $\mu \ll \lambda$. Also, we need only obtain the result for real μ.

If $\mu \perp \lambda$, then $\mu^+ \perp \lambda$, and there is a set $A \subset \Omega$ with $\mu^+(A) = 0$ and $\lambda(A^c) = 0$ (Proposition 10.3.4). By Lemma 12.1.2, $(D\mu^+)(x) = 0$ λ-almost everywhere. The same argument also shows that $(D\mu^-)(x) = 0$ λ-a.e. Hence $(D\mu)(x) = 0$ λ-a.e.

Next, we assume that $\mu \ll \lambda$ and μ is real.

There is a locally λ-integrable function f from Ω into $\mathbf{R}$ such that $\mu = f\lambda$. The theorem will follow once we show that $D\mu(x) = f(x)$ (λ-almost everywhere).

Associate with each rational number r the sets $A_r = \{x \in \Omega : f(x) < r\}$, $B_r = \{x \in \Omega : f(x) \geq r\}$, and define measures θ_r by $\theta_r = 1_{B_r}(f - r)\lambda$. Since $\theta_r(A_r) = 0$, Lemma 12.1.2 shows that there are sets $A'_r \subset A_r$ such that $\lambda(A_r - A'_r) = 0$ and $(D\theta_r)(x) = 0$ if $x \in A'_r$.

Put $Y = \bigcup_r (A_r - A'_r)$. Then $\lambda(Y) = 0$. Pick some $x \in \Omega$ outside Y, let $(E_i)_i$ shrink to x nicely, and choose $r > f(x)$. Then $x \in A_r$. Since $\mu - r\lambda = (f - r)\lambda \leq \theta_r$, we have

$$\frac{\mu(E_i)}{\lambda(E_i)} \leq \frac{\theta_r(E_i)}{\lambda(E_i)} + r \qquad \text{for every} \quad i \geq 1.$$

Since $x \notin Y$, we have $x \in A'_r$. Hence $\limsup_{i \to +\infty} (\mu(E_i)/\lambda(E_i)) \leq r$. This is true for every rational $r > f(x)$. Therefore,

$$\limsup_{i \to +\infty} \frac{\mu(E_i)}{\lambda(E_i)} \leq f(x).$$

We have now proved the following: almost every $x \in \Omega$ satisfies

$$\limsup_{i \to +\infty} \frac{\mu(E_i)}{\lambda(E_i)} \leq f(x)$$

for every sequence $(E_i)_{i \geq 1}$ that shrinks to x nicely.

If we replace μ by $-\mu$, and f by $-f$, it follows that almost every $x \in \Omega$ satisfies

$$\liminf_{i \to +\infty} \frac{\mu(E_i)}{\lambda(E_i)} \geq f(x)$$

for every sequence $(E_i)_{i \geq 1}$ that shrinks to x nicely.

The proof of the theorem is thus complete.   □

Another important result is the following theorem.

Theorem 12.1.2 *Let f be a locally λ-integrable mapping from Ω into a real Banach space F. Then, for λ-almost all x_0 of Ω,*

$$\lim_{i \to +\infty} \frac{1}{\lambda(E_i)} \cdot \int_{E_i} |f(x) - f(x_0)| \, d\lambda = 0 \qquad (1)$$

for every sequence $(E_i)_{i \geq 1}$ that shrinks to x_0 nicely.

PROOF: There are a λ-negligible subset N of Ω and a countable subset D of F such that $f(\Omega - N)$ is included in the closure of D. For $r \in D$, put $\mu_r = |f - r| \cdot \lambda$. Theorem 12.1.1 shows that $(D\mu_r)(x_0) = |f(x_0) - r|$ for almost all x_0. Let Y_r be the exceptional set, and put $Y = N \cup (\bigcup_r Y_r)$. Then $\lambda(Y) = 0$.

If $x_0 \notin Y$, if $(E_i)_{i \geq 1}$ shrinks to x_0 nicely, and if $\varepsilon > 0$, there exists $r \in D$ such that $|f(x_0) - r| < \varepsilon$. Hence, $|f(x) - f(x_0)| < |f(x) - r| + \varepsilon$ for all $x \in \Omega$, and

$$\frac{1}{\lambda(E_i)} \cdot \int_{E_i} |f(x) - f(x_0)| \, d\lambda(x) < \frac{\mu_r(E_i)}{\lambda(E_i)} + \varepsilon. \qquad (2)$$

Since $x_0 \notin Y$, $D\mu_r(x_0) = |f(x_0) - r|$, and therefore the left-hand side of (2) is less than 2ε for all sufficiently large i. This proves (1) for every $x_0 \notin Y$, and completes the proof. $\square$

We now restrict our attention to special types of sequences that shrink nicely.

Lemma 12.1.3 *Let $E = \prod_{1 \leq i \leq k}]\alpha_i, \beta_i]$ be a nonempty rectangle in $\mathbf{R}^k$. Then E is a disjoint union of rectangles* $\prod_{1 \leq i \leq k}\left]\beta_i - \dfrac{p_i + 1}{2^n}, \beta_i - \dfrac{p_i}{2^n}\right]$ $(n \in \mathbf{Z}^+,$ $p_i \in \mathbf{Z}^+$ *for every* $1 \leq i \leq k)$.

PROOF: For each $n \in \mathbf{Z}^+$, let Q_n be the class consisting of all the rectangles

$$\prod_{1 \leq i \leq k}\left]\beta_i - \frac{p_i + 1}{2^n}, \beta_i - \frac{p_i}{2^n}\right] \qquad (p_i \in \mathbf{Z}^+ \quad \text{for all } 1 \leq i \leq k).$$

Put

$$P_0 = \{A \in Q_0 : A \subset E\}$$
$$P_1 = \{A \in Q_1 : A \subset E, A \cap B = \emptyset \quad \text{for every} \quad B \in P_0\}$$
$$\vdots$$
$$P_n = \{A \in Q_n : A \subset E, A \cap B = \emptyset \quad \text{for every} \quad B \in \bigcup_{0 \leq j \leq n-1} P_j\}$$
$$\vdots$$

Clearly, the rectangles of $P = \bigcup_{n \geq 0} P_n$ are disjoint.

Let x be a point of $\prod_{1 \leq i \leq k}] - \infty, \beta_i]$ which does not belong to $\bigcup_{A \in P} A$. For each integer $n \geq 0$, if A_n is the unique element of Q_n which contains x, then A_n is not included in E. Otherwise, A_n would belong to P_n, or would be included in one element of $\bigcup_{0 \leq j \leq n-1} P_j$, which is impossible.

Now, for each $n \in \mathbf{Z}^+$, choose $x^{(n)}$ in $A_n \cap (\mathbf{R}^k - E)$. Since $|x_i^{(n)} - x_i| \leq 2^{-n}$ for all $1 \leq i \leq k$, the sequence $(x^{(n)})_{n \geq 1}$ converges to x. But the set $\prod_{1 \leq i \leq k}] - \infty, \beta_i] - E$ is closed in $\prod_{1 \leq i \leq k}] - \infty, \beta_i]$, therefore $x \notin E$, and $E = \bigcup_{A \in P} A$, as desired. $\square$

We say that a nonempty rectangle $\prod_{1 \leq i \leq k}]\alpha_i, \beta_i]$ is a square whenever $\beta_i - \alpha_i$ does not depend on i.

Let μ be a real measure on S.

For every $x \in \Omega$ and every $n \in \mathbf{N}$, let $C_n(x)$ be the class of those S-sets which are squares, with edge lengths strictly less than $1/n$, and which contain x. Put $\Delta_n(x) = \sup_{E \in C_n(x)} (\mu(E)/\lambda(E))$. If $r < \Delta_n(x)$, there exists $E \in C_n(x)$ such that $\mu(E)/\lambda(E) > r$, and we can find $E' \in C_n(x)$, $E' \supset E$, such that $\mu(E')/\lambda(E') > r$ which is a neighborhood of x. Therefore, $\Delta_n(x)$ is also the supremum of the $\mu(E)/\lambda(E)$, as E ranges over those elements of $C_n(x)$ which are neighborhoods of x. Hence, for all $n \in \mathbf{N}$, the function $x \mapsto \Delta_n(x)$ is lower semicontinuous on Ω. Finally, $\overline{D}\mu : x \mapsto \inf_{n \geq 1} \Delta_n(x)$ is a Borel function from Ω into $\overline{\mathbf{R}}$, and we set $\underline{D}\mu = -\overline{D}(-\mu)$.

Observe that $\underline{D}\mu(x) = \overline{D}\mu(x) = D\mu(x)$ at every $x \in \Omega$ for which $D\mu(x)$ exists.

From now on, we assume that μ is positive.

Proposition 12.1.1 *Let A be a subset of Ω and $\alpha > 0$ a real number. Suppose that $\underline{D}\mu(x) \geq \alpha$ for all $x \in A$. Then $\alpha\lambda^*(A) \leq \mu^*(A)$.*

PROOF: There is no restriction in assuming that $\mu^*(A)$ is finite.

Let $0 < \varepsilon < \alpha$ be given. For every $n \in \mathbf{N}$, let A_n be the set of those $x \in A$ such that $\mu(E)/\lambda(E) \geq \alpha - \varepsilon$ for all $E \in \mathcal{C}_n(x)$. Thus the sets A_n increase to A.

Choose $n \in \mathbf{N}$. Given $\delta > 0$, there exists a countable family $(E_i)_{i \in I}$ of disjoint S-sets such that $A_n \subset \bigcup_{i \in I} E_i$ and $\sum_{i \in I} \mu(E_i) \leq \mu^*(A_n) + \delta$. By Lemma 12.1.3, we may suppose that the E_i are squares and have edge lengths strictly less than $1/n$. Let $(E_i)_{i \in J}$ be the subfamily of $(E_i)_{i \in I}$ obtained by keeping only those E_i which meet A_n. Then $A_n \subset \bigcup_{i \in J} E_i$ and $\mu(E_i)/\lambda(E_i) \geq \alpha - \varepsilon$ for all $i \in J$, so

$$\mu^*(A_n) + \delta \geq \sum_{i \in J} \mu(E_i) \geq (\alpha - \varepsilon) \cdot \sum_{i \in J} \lambda(E_i) \geq (\alpha - \varepsilon) \cdot \lambda^*(A_n).$$

Since δ is arbitrary, we have $\mu^*(A_n) \geq (\alpha - \varepsilon) \cdot \lambda^*(A_n)$. Therefore, $\mu^*(A) \geq (\alpha - \varepsilon) \cdot \lambda^*(A)$. Since ε is arbitrary, we conclude that $\mu^*(A) \geq \alpha \cdot \lambda^*(A)$.   □

Proposition 12.1.2 *Let A be a subset of Ω and $\alpha > 0$ a real number. Suppose that $\overline{D}\mu(x) \leq \alpha$ for all $x \in A$. Then $\mu^*(A) \leq \alpha \cdot \lambda^*(A)$.*

PROOF: We may suppose that $\lambda^*(A)$ is finite.

Let $\varepsilon > 0$ be given. For every $n \in \mathbf{N}$, let A_n be the set of those $x \in A$ such that $\mu(E)/\lambda(E) \leq \alpha + \varepsilon$ for all $E \in \mathcal{C}_n(x)$. Thus the sets A_n increase to A.

Choose $n \in \mathbf{N}$. Given $\delta > 0$, there exists a countable family $(E_i)_{i \in I}$ of disjoint S-sets such that $A_n \subset \bigcup_{i \in I} E_i$ and $\sum_{i \in I} \lambda(E_i) \leq \lambda^*(A_n) + \delta$. Arguing as in Proposition 12.1.1, we conclude that $\mu^*(A) \leq \alpha \cdot \lambda^*(A)$.   □

Theorem 12.1.3 *If $\mu \geq 0$ is singular (i.e., if $\mu \perp \lambda$), then $\overline{D}\mu(x) = +\infty$ μ-almost everywhere.*

PROOF: Let N be a λ-negligible set which carries μ. For each integer $n \geq 1$, put $E_n = \{x \in N : \overline{D}\mu(x) \leq n\}$. Then $\mu^*(E_n) \leq n \cdot \lambda^*(E_n)$ leads to $\mu^*(E_n) = 0$. Therefore, $\{x \in N : \overline{D}\mu(x) = +\infty\}$ has μ-negligible complement.   □

Proposition 12.1.3 *If $\overline{D}\mu(x) < +\infty$ everywhere, then $\mu \ll \lambda$.*

PROOF: Let $\mu = f\lambda + \mu_s$ be the Lebesgue decomposition of μ relative to λ. Then

$$\overline{D}\mu_s \leq \overline{D}\mu + \overline{D}(-f\lambda) = \overline{D}\mu - \underline{D}(f\lambda) \leq \overline{D}\mu < +\infty,$$

and $\mu_s = 0$ by Theorem 12.1.3. $\qquad\qquad\Box$

When $k = 1$, we have a stronger result than Theorem 12.1.3.

Proposition 12.1.4 *When $k = 1$ and μ is positive singular, $\underline{D}\mu(x) = +\infty$ μ-almost everywhere.*

PROOF: μ is concentrated on a λ-negligible Borel set N. For each integer $n \geq 1$, define E_n as the set of all $x \in N$ at which $\underline{D}\mu(x) < n$. Given $\varepsilon > 0$, let V be an open subset of Ω containing N such that $\lambda(V) \leq \varepsilon$. Let K be a compact subset of E_n. Each $x \in K$ lies in an S-set $E_x \subset V$ such that E_x is a neighborhood of x and $\mu(E_x)/\lambda(E_x) \leq n$. Being compact, K is covered by finitely many of these S-sets E_{x_i}. If some point of Ω lies in three S-sets, one of these lies in the union of the other two and can be removed without changing the union. In this way, we remove the superfluous S-sets E_{x_i} and may assume that no point lies in more than two of the S-sets E_{x_i}. Then

$$\mu(K) \leq \mu\Big(\bigcup_i E_{x_i}\Big) \leq \sum_i \mu(E_{x_i}) \leq n\sum_i \lambda(E_{x_i}) \leq 2n\lambda\Big(\bigcup_i E_{x_i}\Big)$$

$$\leq 2n\lambda(V) \leq 2n\varepsilon.$$

Therefore, $\mu(E_n) \leq 2n\varepsilon$, and the result follows. $\qquad\qquad\Box$

12.2   The Modulus of an Automorphism

Let λ_k be Lebesgue measure on $\mathbf{R}^k$. If μ is a positive measure on the natural semiring S of $\mathbf{R}^k$, invariant under translations (i.e., such that $\mu(E-y) = \mu(E)$ for all $y \in \mathbf{R}^k$ and for all $E \in S$), then $\mu = c\lambda_k$ where $c = \mu(]0, 1]^k)$. Indeed, for every $q \in \mathbf{Z}^+$,

$$\mu\left(\prod_{1 \leq i \leq k} \left]\frac{p_i}{2^q}, \frac{p_i + 1}{2^q}\right]\right) = \frac{c}{2^{kq}}$$

for all $(p_1, \ldots, p_k) \in \{0, 1, \ldots, 2^q - 1\}^k$, and so for all $(p_1, \ldots, p_k) \in \mathbf{Z}^k$. Therefore, $\mu^*(U) = c\lambda_k^*(U)$ for every open subset U of $\mathbf{R}^k$ (Proposition 9.3.1), which ensures that $\mu = c\lambda_k$.

If u is a linear automorphism of $\mathbf{R}^k$, the pair (u^{-1}, S) is λ_k-suited and $u^{-1}(\lambda_k)$ is invariant under translations. Hence there exists a positive number $\mathrm{mod}(u)$, called the modulus of u, such that $u^{-1}(\lambda_k) = \mathrm{mod}(u)\lambda_k$.

Now, if u, v are two linear automorphisms of $\mathbf{R}^k$,

$$\int 1_{u(E)}\, d\left(v^{-1}(\lambda_k)\right) = \int 1_{u(E)} \circ v^{-1}\, d\lambda_k = \int 1_E \circ (v \circ u)^{-1}\, d\lambda_k$$

for all $E \in S$, whence we see that $\mathrm{mod}(v \circ u) = \mathrm{mod}(v) \cdot \mathrm{mod}(u)$.

For all $i, j \in \{1, 2, \ldots, k\}$, let $E_{i,j}$ denote the $k \times k$ matrix that has the element in the (i,j) place equal to 1 and all other elements equal to 0. If $i \neq j$ and $\alpha \in \mathbf{R}$, put $B_{i,j}(\alpha) = I_k + \alpha E_{i,j}$, where I_k is the unit matrix of order k. For any matrix X of order k, $B_{i,j}(\alpha)X$ is obtained by adding α times the jth row of X to the ith row of X. Furthermore, $B_{i,j}(\alpha)^{-1} = B_{i,j}(-\alpha)$.

Proposition 12.2.1 *Every invertible $k \times k$ matrix is a product of matrices of the form $B_{i,j}(\alpha)$ and a matrix of the form $I_k + (a - 1)E_{k,k}$.*

PROOF: Consider invertible matrices of the form

$$X = \begin{pmatrix} 1 & 0 & \ldots & 0 & \xi_{1,k-h} & \cdots & \xi_{1,k} \\ 0 & 1 & \ldots & 0 & \xi_{2,k-h} & \cdots & \xi_{2,k} \\ \ldots & \ldots & \ldots & \ldots & \ldots & & \ldots \\ 0 & 0 & \ldots & 1 & \xi_{k-h-1,k-h} & \cdots & \xi_{k-h-1,k} \\ \ldots & \ldots & \ldots & \ldots & \ldots & & \ldots \\ 0 & 0 & \ldots & 0 & \xi_{k,k-h} & \cdots & \xi_{k,k} \end{pmatrix}$$

where $0 \leq h \leq k - 1$; if $h = k - 1$, then X is an arbitrary invertible matrix. The proof is by induction on h. If $h = 0$ we must have $\xi_{k,k} \neq 0$. Hence, if we multiply on the left successively by the matrices $B_{i,k}(-\xi_{i,k} \cdot \xi_{k,k}^{-1})$ for $1 < i < k - 1$, we shall obtain the matrix $I_k + (\xi_{k,k} - 1) \cdot E_{k,k}$, and the proposition is true in the case $h = 0$. Now suppose that the result has been proved for $h = 0, 1, \ldots, n - 1$, and consider the case $h = n$ $(1 \leq n \leq k - 1)$. Since

$$\det(X) = \det(I_{k-n-1}) \cdot \det \begin{pmatrix} \xi_{k-n,k-n} & \cdots & \xi_{k-n,k} \\ \vdots & \vdots & \vdots \\ \xi_{k,k-n} & \cdots & \xi_{k,k} \end{pmatrix},$$

there certainly exists a nonzero element $\xi_{i,k-n}$, for some i such that $k - n \leq i \leq k$. Premultiplying X by $B_{j,i}\left((1 - \xi_{j,k-n})\xi_{i,k-n}^{-1}\right)$ for some index $j \neq i$ such that $k - n \leq j \leq k$, we may assume that $\xi_{j,k-n} = 1$. Multiplying successively by the matrices $B_{r,j}(-\xi_{r,k-n})$ for $r \neq j$, we end up with a matrix for which $\xi_{r,k-n} = 0$ if $r \neq j$, and $\xi_{j,k-n} = 1$. Finally, in the case $j \neq k - n$, multiplying this matrix by $B_{j,k-n}(-1) \cdot B_{k-n,j}(1)$, we obtain a matrix of the same form but with $j = k - n$. This completes the induction step, and so completes the proof of the proposition. □

Theorem 12.2.1 $\mathrm{mod}(u) = |\det(u)|$ *for every automorphism u of $\mathbf{R}^k$.*

PROOF: Clearly, if there exists $a \in \mathbf{R}^*$ such that

$$u(x_1, \ldots, x_k) = (x_1, \ldots, x_{k-1}, a x_k)$$

for all $(x_1, \ldots, x_k) \in \mathbf{R}^k$, then $\mathrm{mod}(u) = |a|$. Now suppose that u takes the form $(x_1, \ldots, x_k) \longmapsto (x_1, \ldots, x_i + \alpha x_j, \ldots, x_j, \ldots, x_k)$, where i, j are two different integers in $\{1, \ldots, k\}$, α is a real number, and $x_i + \alpha x_j$ is the term of rank i in $(x_1, \ldots, x_i + \alpha x_j, \ldots, x_k)$. We wish to prove that $\mathrm{mod}(u) = 1$. But, if $\alpha \neq 0$, then $u = w^{-1} \circ v \circ w$, where w is the automorphism $(x_1, \ldots, x_k) \longmapsto (x_1, \ldots, x_i, \ldots, \alpha x_j, \ldots, x_k)$ of $\mathbf{R}^k$ and v is the automorphism $(x_1, \ldots, x_k) \longmapsto (x_1, \ldots, x_i + x_j, \ldots, x_j, \ldots, x_k)$. So we may suppose that $\alpha = 1$. Write $(e_1, \ldots, e_k)$ for the canonical basis of $\mathbf{R}^k$. Letting

$$A = \left\{ (y_1, \ldots, y_k) \in \,]0, 1]^k : y_i > y_j \right\}$$

and

$$B = \left\{ (y_1, \ldots, y_k) \in \,]0, 1]^k : y_i \leq y_j \right\},$$

we have also

$$A = \left\{ u(x_1, \ldots, x_k) : (x_1, \ldots, x_k) \in \,]0, 1]^k, x_i + x_j \leq 1 \right\}$$

and

$$B + e_i = \left\{ u(x_1, \ldots, x_k) : (x_1, \ldots, x_k) \in \,]0, 1]^k, x_i + x_j > 1 \right\},$$

whence we deduce that $\lambda_k \left(u \left( ]0, 1]^k \right) \right) = \lambda_k ( ]0, 1]^k)$ and $\mathrm{mod}(u) = 1$. Finally, when u is arbitrary, Theorem 12.2.1 follows from Proposition 12.2.1. $\qquad \square$

Now, if u is a linear endomorphism of $\mathbf{R}^k$ which is not bijective and if p is the rank of u, there exists an orthogonal transformation of $\mathbf{R}^k$ which maps $\mathbf{R}^p \times \{0\}$ onto $u(\mathbf{R}^k)$, so that $u(\mathbf{R}^k)$ is λ_k-negligible.

12.3   Change of Variables

Given $k \in \mathbf{N}$, let λ be Lebesgue measure on $\mathbf{R}^k$. For every nonempty open subset U of $\mathbf{R}^k$, denote by λ_U Lebesgue measure on the natural semiring S_U of U.

Proposition 12.3.1 *Given an open subset V of $\mathbf{R}^k$, let T be a continuous open mapping from V into $\mathbf{R}^k$, and suppose that T is differentiable at some point $x \in V$. Put $\Delta(x) = \left| \det \left(DT(x) \right) \right|$. With notation as in Section 12.1, to every $0 < \varepsilon < 1$ there corresponds an integer $n \geq 1$ such that*

$$\left| \frac{\lambda(TE)}{\lambda(E)} - \Delta(x) \right| \leq \varepsilon$$

for all E in $\mathcal{C}_n(x)$.

PROOF: We may suppose that $x = 0$ and $T(x) = 0$.

First, assume that $DT(0) = \mathrm{id}_{\mathbf{R}^k}$, and let $0 < \delta < 1/4$ be a real number such that $1 - \varepsilon \leq (1 - 2\delta)^k < (1 + 2\delta)^k \leq 1 + \varepsilon$. We can find an integer $n \geq 1$ such that $\|T(y) - y\| \leq \delta \cdot \|y\|$ for all $y \in V$ satisfying $\|y\| \leq 1/n$ (where $\|y\| = \|(y_1, \ldots, y_k)\| = \sup_{1 \leq i \leq k}(|y_i|)$).

Let $E \in \mathcal{C}_n(0)$ with edge length l, and let E_1, E_2 be two squares concentric with E whose edges have lengths $(1 - 2\delta)l$ and $(1 + 2\delta)l$, respectively. Then

$$\|T(y) - y\| \leq \delta l \qquad \text{for every} \quad y \in \overline{E}. \tag{1}$$

If y belongs to E, (1) shows that $T(y)$ lies in E_2. Hence $T(E) \subset E_2$.

If y belongs to the boundary of E, (1) shows that $T(y)$ does not lie in the interior $\overset{\circ}{E}_1$ of E_1. Hence $\overset{\circ}{E}_1 \cap T(E) = \overset{\circ}{E}_1 \cap T(\overset{\circ}{E})$ and $\overset{\circ}{E}_1 - T(E) = \overset{\circ}{E}_1 - T(\overline{E})$ are two disjoint open sets, whose union is $\overset{\circ}{E}_1$. Since $\delta < 1/4$, (1) proves that T maps the center of E into $\overset{\circ}{E}_1$. But $\overset{\circ}{E}_1$ is connected, so $\overset{\circ}{E}_1 \subset T(E)$. Now $E_1 \subset T(\overline{E})$. Since, for all $y \in \overline{E} - E$, $T(y)$ does not lie in E_1, we conclude that $E_1 \subset T(E)$. In short, $E_1 \subset T(E) \subset E_2$, whence it follows that

$$1 - \varepsilon \leq (1 - 2\delta)^k \leq \frac{\lambda(T(E))}{\lambda(E)} \leq (1 + 2\delta)^k \leq 1 + \varepsilon,$$

as desired.

Next, assume that $u = DT(0)$ is a linear automorphism of $\mathbf{R}^k$. Then, $D(u^{-1} \circ T)(0) = \mathrm{id}_{\mathbf{R}^k}$. By what we have just shown, to every $\varepsilon > 0$ there corresponds an integer $n \geq 1$ such that

$$\left| \frac{\lambda(TE)}{\lambda(E)} - \mathrm{mod}(u) \right| = \mathrm{mod}(u) \cdot \left| \frac{\lambda(u^{-1}(TE))}{\lambda(E)} - 1 \right| \leq \varepsilon$$

for all $E \in \mathcal{C}_n(0)$.

Finally, assume that $u = DT(0)$ is singular (so that $\det(u) = 0$). Then $u(\mathbf{R}^k)$ is λ-negligible. Fix $\varepsilon > 0$. There exists $\delta > 0$ such that $\lambda(E_\delta) \leq \varepsilon$, where E_δ is the set of all points whose distance to $u([-1, 1]^k)$ is less than δ. Let $n \in \mathbf{N}$ be such that $\|T(y) - u(y)\| \leq \delta \cdot \|y\|$ for all y satisfying $\|y\| \leq 1/n$. If E is an element of $\mathcal{C}_n(0)$ with edge length l, then, for all $y \in E$, we have $\|T(y) - u(y)\| \leq l\delta$. Thus

$$d\left(\frac{T(y)}{l}, \frac{u(y)}{l} \right) \leq \delta \quad \text{and} \quad d\left(\frac{T(y)}{l}, u([-1, 1]^k) \right) \leq \delta.$$

Therefore, $T(E)/l$ is included in E_δ, and

$$\frac{\lambda(T(E))}{l^k} = \frac{\lambda(T(E))}{\lambda(E)} \leq \varepsilon.$$

$\square$

We remark that Proposition 12.3.1 is true even if the hypothesis that T is open is deleted; the proof is more difficult as might be expected.

The following change of variables formula is a central result of this book.

Theorem 12.3.1 *Let V, W be two open subsets of $\mathbf{R}^k$ and T a homeomorphism of V onto W. Assume that T is differentiable at each point $x \in V$, and put $J(x) = \det[DT(x)]$. Then $|J| : x \mapsto |J(x)|$ is locally λ_V-integrable and $\lambda_W = T(|J| \cdot \lambda_V)$.*

PROOF: Put $\mu = T^{-1}(\lambda_W)$. By Proposition 12.3.1, $\overline{D}\mu(x)$ is finite for all $x \in V$. Hence $\mu \ll \lambda_V$ (Proposition 12.1.3). Now $|J(x)| = D\mu(x)$ for λ_V-almost all x, by Proposition 12.3.1 and Theorem 12.1.1. This means that $|J|$ is locally λ_V-integrable and $\mu = |J| \cdot \lambda_V$, whence we deduce that $T(|J|\lambda_V) = \lambda_W$ (Proposition 11.1.1).   □

Observe that $\int^* f \, d\lambda_W = \int^* (f \circ T)|J| \, d\lambda_V$ for each $f : W \to [0, +\infty]$.

12.4 Polar Coordinates

Fix an integer $n \geq 1$. Write $\mathbf{S}^n$ for the unit sphere $\{z \in \mathbf{R}^{n+1} : \|z\| = 1\}$ in $\mathbf{R}^{n+1}$ (where $\|\cdot\|$ is the Euclidean norm), and $\mathcal{B}$ for the Borel σ-algebra of $\mathbf{S}^n$.

For every $A \in \mathcal{B}$, $\tilde{A} = \{tx : t \in \,]0, 1], x \in A\}$ is a Borel set in $\mathbf{R}^{n+1}$. Denote by λ_{n+1} Lebesgue measure on $\mathbf{R}^{n+1}$. The function $A \mapsto (n + 1)\lambda_{n+1}(\tilde{A})$ is σ-additive on $\mathcal{B}$, and hence a positive measure $d\mathbf{S}^n$ on $\mathcal{B}$.

Definition 12.4.1 $d\mathbf{S}^n$ (or the Radon measure on $\mathbf{S}^n$ arising from $d\mathbf{S}^n$) is called the superficial measure on $\mathbf{S}^n$, or the Riemannian volume of $\mathbf{S}^n$.

If u belongs to the group $\mathbf{O}(n + 1, \mathbf{R})$ of orthogonal transformations of $\mathbf{R}^{n+1}$, and if f is the mapping $x \mapsto u(x)$ from $\mathbf{S}^n$ onto $\mathbf{S}^n$, then

$$
\begin{aligned}
\left(f(d\mathbf{S}^n)\right)(A) &= (d\mathbf{S}^n)\left(f^{-1}(A)\right) \\
&= (n + 1)\lambda_{n+1}\left(\widetilde{f^{-1}(A)}\right) \\
&= (n + 1)\lambda_{n+1}\left(u^{-1}(\tilde{A})\right) \\
&= (n + 1)\lambda_{n+1}(\tilde{A}) = d\mathbf{S}^n(A)
\end{aligned}
$$

for all $A \in \mathcal{B}$. Therefore, $f(d\mathbf{S}^n) = d\mathbf{S}^n$, and $d\mathbf{S}^n$ is invariant under orthogonal transformations.

Let g be the homeomorphism $(t, x) \mapsto tx$ from $\,]0, +\infty[\times \mathbf{S}^n$ onto $\mathbf{R}^{n+1} - \{0\}$. We write dt for Lebesgue measure on $\,]0, +\infty[$. Now, for all real numbers r, s satisfying $0 < r < s$ and for all $A \in \mathcal{B}$,

$$(t^n \, dt \otimes d\mathbf{S}^n)(]r,s] \times A) = \frac{s^{n+1} - r^{n+1}}{n+1} \, d\mathbf{S}^n(A)$$
$$= (s^{n+1} - r^{n+1}) \lambda_{n+1}(\tilde{A})$$
$$= \lambda_{n+1}\big(g(]r,s] \times A)\big).$$

So

$$g(t^n \, dt \otimes d\mathbf{S}^n) = \lambda_{n+1/\mathbf{R}^{n+1}-\{0\}}.$$

Next, let ψ be the mapping from $\mathbf{R}^n$ into $\mathbf{S}^n$ which sends $\theta = (\theta_1, \ldots, \theta_n)$ to $(\cos\theta_n \cdots \cos\theta_2 \cos\theta_1, \cos\theta_n \cdots \cos\theta_2 \sin\theta_1, \ldots, \cos\theta_n \sin\theta_{n-1}, \sin\theta_n)$, where the $(j+1)$st coordinate is $\cos\theta_n \times \cdots \times \cos\theta_{j+1} \times \sin\theta_j$ for every $1 \le j \le n$.

Proposition 12.4.1 *Set* $P =]-\pi, \pi[\times]-\pi/2, \pi/2[^{n-1}$ *and write* μ *for the measure* $\cos^{n-1}(\theta_n) \times \cos^{n-2}(\theta_{n-1}) \times \cdots \times \cos(\theta_2) \, d\theta_1 \ldots d\theta_n$, *where* $d\theta_1$ *is Lebesgue measure on* $]-\pi, \pi[$ *and* $d\theta_i$ *is Lebesgue measure on* $]-\pi/2, \pi/2[$ *for all* $2 \le i \le n$. *Then* $\psi(\mu) = d\mathbf{S}^n$.

PROOF: Denote by $(e_1, \ldots, e_n)$ the canonical basis of $\mathbf{R}^n$. Let $J(\theta)$ be the matrix of $\big(D\psi(\theta)e_1, \ldots, D\psi(\theta)e_n, \psi(\theta)\big)$ with respect to the canonical basis of $\mathbf{R}^{n+1}$. Thus the mapping $(t, \theta) \mapsto t \cdot \psi(\theta)$ from $\mathbf{R} \times \mathbf{R}^n$ into $\mathbf{R}^{n+1}$ has Jacobian determinant $(-1)^n \cdot t^n \cdot \det\big(J(\theta)\big)$ at (t, θ).

If $n \ge 2$, write $K(\theta)$ for the $n \times n$ matrix whose first row is obtained by multiplying the first row of $J(\theta_2, \ldots, \theta_n)$ by $\sin\theta_1$, and whose ith row, for all $2 \le i \le n$, is equal to the ith row of $J(\theta_2, \ldots, \theta_n)$. Similarly, write $L(\theta)$ for the $n \times n$ matrix whose first row is obtained by multiplying the first row of $J(\theta_2, \ldots, \theta_n)$ by $\cos\theta_1$, and whose ith row, for all $2 \le i \le n$, is equal to the ith row of $J(\theta_2, \ldots, \theta_n)$. Now expanding $\det\big(J(\theta)\big)$ along the first column of $J(\theta)$, we see that

$$\det\big(J(\theta)\big) = -\cos(\theta_n) \cdots \cos(\theta_2) \sin(\theta_1) \det\big(K(\theta)\big)$$
$$- \cos(\theta_n) \cdots \cos(\theta_2) \cos(\theta_1) \det\big(L(\theta)\big),$$

and so

$$\det\big(J(\theta)\big) = -\cos(\theta_n) \cdots \cos(\theta_2)\big(\sin^2(\theta_1) + \cos^2(\theta_1)\big) \det\big(J(\theta_2, \ldots, \theta_n)\big).$$

By induction on n,

$$\det\big(J(\theta)\big) = (-1)^n \cdot \cos^{n-1}(\theta_n) \cdot \cos^{n-2}(\theta_{n-1}) \cdots \cos(\theta_2).$$

Next, put $\Omega = \mathbf{R}^{n+1} - \{0\}$. Then

$$\mathbf{S}^n - \psi(P) = \{(x_1, \ldots, x_{n+1}) \in \mathbf{S}^n : x_1 \le 0 \quad \text{and} \quad x_2 = 0\}.$$

So $\mathbf{R}^{n+1} - \{t \cdot \psi(\theta) : t \in [0, +\infty[\text{ and } \theta \in P\}$ is λ_{n+1}-negligible. By the change of variable formula, $\lambda_{n+1/\Omega}$ is the image measure of $t^n dt \otimes \mu$ under the mapping $(t,\theta) \mapsto t \cdot \psi(\theta)$. On the other hand. we know that $t^n \, dt \otimes dS^n$ is the image measure of $\lambda_{n+1/\Omega}$ under the mapping $z \mapsto (\|z\|, z/\|z\|)$. This implies that $t^n \, dt \otimes dS^n$ is the image measure of $t^n \, dt \otimes \mu$ under the mapping $(t,\theta) \mapsto (t, \psi(\theta))$. The proposition follows. $\qquad \square$

Theorem 12.4.1 λ_{n+1} *is the image measure of $t^n \, dt \otimes \mu$ under $(t, \theta) \mapsto t \cdot \psi(\theta)$ from $]0, +\infty[\times P$ into $\mathbf{R}^{n+1}$.*

PROOF: This follows from Proposition 12.4.1, because λ_{n+1} is the image measure of $t^n \, dt \otimes dS^n$ under $(t, x) \mapsto tx$ from $]0, +\infty[\times \mathbf{S}^n$ into $\mathbf{R}^{n+1}$. $\qquad \square$

If f is a λ_{n+1}-integrable function, we can, according to Theorem 12.4.1, compute $\int f \, d\lambda_{n+1}$ in polar coordinates.

As to the significance of $\theta_1, \ldots, \theta_n$, we let $(x_1, \ldots, x_{n+1})$ be a point of $\mathbf{S}^n$ such that the condition $x_1 \leq 0$ and $x_2 = 0$ is not satisfied. For each integer $2 \leq j \leq n$, the point $\left(1 - (x_{j+2}^2 + \cdots + x_{n+1}^2)\right)^{-1/2} \cdot (x_1, \ldots, x_{j+1})$ lies in $\mathbf{S}^j$, and there is a unique $\theta_j \in \,] - \pi/2, \pi/2[$ satisfying

$$\sin \theta_j = \left(1 - (x_{j+2}^2 + \cdots + x_{n+1}^2)\right)^{-1/2} \cdot x_{j+1}.$$

Finally, there is a unique $\theta_1 \in \,] - \pi, \pi[$ such that

$$(\cos \theta_1, \sin \theta_1) = \left(1 - (x_3^2 + \cdots + x_{n+1}^2)\right)^{-1/2} \cdot (x_1, x_2),$$

and we have $x = \psi(\theta)$.

When $n = 2$, θ_2 is the "latitude" of x, and θ_1 the "longitude" of x.

Proposition 12.4.2 *For all $\theta \in P$, $D\psi(\theta)$ has rank n.*

PROOF: Let $\theta \in \mathbf{R}^n$. Since $\left(D\psi(\theta)e_1, \ldots, D\psi(\theta)e_n, \psi(\theta)\right)$ is a basis of $\mathbf{R}^{n+1}$, the vectors $D\psi(\theta)e_1, \ldots, D\psi(\theta)e_n$ form a free system.

Also, let $(\varepsilon_1, \ldots, \varepsilon_{n+1})$ be the canonical basis of $\mathbf{R}^{n+1}$. Write $M(\theta)$ for the Jacobian matrix of ψ at θ, and $M_k(\theta)$ for the submatrix of $M(\theta)$ obtained by deleting the kth row of $M(\theta)$ $(1 \leq k \leq n+1)$. It is easily shown that

$$\det\left(M_k(\theta)\right) = (-1)^{k-1}\left(\psi(\theta)|\varepsilon_k\right)(\cos^{n-1}\theta_n)(\cos^{n-2}\theta_{n-1}) \cdots (\cos\theta_2),$$

where $\left(\psi(\theta)|\varepsilon_k\right)$ is the scalar product of $\psi(\theta)$ and ε_k. $\qquad \square$

By Proposition 12.4.2, ψ induces a homeomorphism from P onto $\psi(P)$, and the inverse mapping u, from $\psi(P)$ into $\mathbf{R}^n$, is a chart of the manifold $\mathbf{S}^n$.

Fix $\theta \in P$ and put $x = \psi(\theta)$. Denote by $(e_1, \ldots, e_n)$ the canonical basis of $\mathbf{R}^n$. Since $\det\left(J(\theta)\right)$ has sign $(-1)^n$, we see that the $(n+1)$-tuple $(x, D\psi(\theta)e_1, \ldots, D\psi(\theta)e_n)$ is direct in $\mathbf{R}^{n+1}$. Let j be the canonical immersion of $\mathbf{S}^n$ into $\mathbf{R}^{n+1}$ and, for all $1 \leq k \leq n$, let v_k be the unique vector in the tangent space $T_x(\mathbf{S}^n)$ such that $d_x j(v_k) = D\psi(\theta)e_k$. Then $v_1 \wedge \cdots \wedge v_n$ is in the orientation of $T_x(\mathbf{S}^n)$ (if $\mathbf{S}^n$ is oriented "toward the outside").

For each $n \in \mathbf{N}$, denote by V_n the volume $\lambda_n(B'(0,1))$ of the unit ball in $\mathbf{R}^n$, when $\mathbf{R}^n$ is given its Euclidean norm, and by Ω_n the surface area $\int 1 \, dS^{n-1}$ of $\mathbf{S}^{n-1}$. Since λ_n is the image measure of $t^{n-1} \, dt \otimes dS^{n-1}$ under $(t,x) \mapsto tx$, $\Omega_n = nV_n$. Now

$$
\begin{aligned}
V_n &= \int_{-1}^{1} dx_n \int \cdots \int 1_{B'(0,1)}(\cdot, x_n) dx_1 \ldots dx_{n-1} \\
&= V_{n-1} \int_{-1}^{1} (1 - x_n^2)^{(n-1)/2} dx_n
\end{aligned}
$$

for every $n \geq 2$. Hence $V_n = 2V_{n-1} \int_0^{\pi/2} \cos^n \theta \, d\theta$, and $\int_0^{\pi/2} \cos^n \theta \, d\theta$ is left to compute.

For all $a > 0$, put $\Gamma(a) = \int e^{-x} \cdot x^{a-1} \, dx_{/]0,+\infty[}$. The function $\Gamma : a \mapsto \Gamma(a)$ from $]0,+\infty[$ into $]0,+\infty[$ is called the Euler gamma function. Clearly, $\Gamma(a+1) = a \cdot \Gamma(a)$. Since $\Gamma(1) = 1$, we see that $\Gamma(n) = (n-1)!$ for all $n \in \mathbf{N}$. On the other hand, $\Gamma(a) = \int 2\exp(-x^2) \cdot x^{2a-1} dx_{/]0,+\infty[}$ for all $a > 0$.

Now let $a > 0$ and $b > 0$ be two real numbers, and let f be the function

$$
(x,y) \mapsto 2\exp(-x^2) \cdot x^{2a-1} \cdot 2\exp(-y^2) \cdot y^{2b-1}
$$

on $]0,+\infty[\times]0,+\infty[$. Then

$$
\iint f \, dx_{/]0,+\infty[} \, dy_{/]0,+\infty[} = \Gamma(a)\Gamma(b).
$$

But, since

$$
T : (r,\theta) \mapsto (r\cos\theta, r\sin\theta)
$$

is a diffeomorphism from $]0,+\infty[\times]0,\pi/2[$ onto $]0,+\infty[\times]0,+\infty[$,

$$
\iint f \, dx_{/]0,+\infty[} \, dy_{/]0,+\infty[} = \Gamma(a+b) \cdot 2\int (\cos^{2a-1}\theta)(\sin^{2b-1}\theta) \, d\theta_{/]0,\pi/2[}
$$

by the change of variables formula. Thus

$$
2\int (\cos^{2a-1}\theta)(\sin^{2b-1}\theta) \, d\theta_{/]0,\pi/2[} = \frac{\Gamma(a)\Gamma(b)}{\Gamma(a+b)}.
$$

Taking $a = b = 1/2$ in the last inequality, we obtain $\Gamma(1/2) = \sqrt{\pi}$. Next,

$$
\int_0^{\pi/2} \cos^n \theta \, d\theta = \frac{1}{2}\left(\Gamma\left(\frac{1}{2}\right)\Gamma\left(\frac{n+1}{2}\right) \middle/ \Gamma\left(\frac{n+2}{2}\right)\right)
$$

for every $n \in \mathbf{Z}^+$, whence we see that

$$
V_n = \pi^{n/2} \middle/ \Gamma\left(\frac{n}{2}+1\right)
$$

for every $n \in \mathbf{N}$. Thus

$$V_{2n-1} = \frac{2^n \cdot \pi^{n-1}}{1 \cdot 3 \cdots (2n-1)} \quad \text{and} \quad V_{2n} = \frac{\pi^n}{n!}$$

for every $n \in \mathbf{N}$.

Exercises for Chapter 12

1   Let G be a countable subgroup of the additive group $\mathbf{R}^n$. A subset P (respectively, C) of $\mathbf{R}^n$ is said to be a G-packing (respectively, a G-covering) if, for each $s \neq 0$ in G, we have $(s + P) \cap P = \emptyset$ (respectively, if $\mathbf{R}^n = \bigcup_{s \in G}(s + C)$). A subset P which is both a G-packing and a G-covering is called a G-tessellation. Let λ_n be Lebesgue measure on $\mathbf{R}^n$.

1. If C is a λ_n-integrable G-covering and P is a λ_n-measurable G-packing, show that $\lambda_n(C) \geq \lambda_n^*(P)$.

2. Let $\Delta(G)$ be the infimum of the numbers $\lambda_n(C)$, where C runs through all λ_n-integrable G-coverings of $\mathbf{R}^n$. If A is a λ_n-integrable set such that $\lambda_n(A) > \Delta(G)$, show that there exists $s \neq 0$ in $G \cap (A - A)$.

3. Suppose there is an integrable G-tessellation P. Let G_0 be a subgroup of finite index h in G, and let $s_1, \ldots, s_h$ be representatives of the cosets of G_0 in G. Show that $P_0 = \bigcup_{1 \leq i \leq h}(s_i + P)$ is a G_0-tessellation, and conclude that $\Delta(G_0) = h \cdot \Delta(G)$.

2   In view of the sequel, we recall some elementary properties of commutative groups.

A commutative group may be regarded as a module over the ring $\mathbf{Z}$. If G has a finite basis (over $\mathbf{Z}$), the bases of G all have the same number of elements, n, called the rank of G; then each free system in G has at most n elements, and every subgroup of G is free, of rank less than n. Finally, if G is a discrete subgroup of rank n in $\mathbf{R}^n$, then each basis of G over $\mathbf{Z}$ is free in $\mathbf{R}^n$ regarded as a vector space over $\mathbf{R}$.

Now let G be a discrete subgroup of rank n in $\mathbf{R}^n$, and let A be a compact subset of $\mathbf{R}^n$, symmetric and convex, such that $\lambda_n(A) \geq 2^n \cdot \Delta(G)$.

1. Prove that $\Delta(G) > 0$.

2. Show that there exists in $A \cap G$ a nonzero element (i.e., prove Minkowski's theorem). For this, observe that $A = \bigcap_{p \in \mathbf{N}} A_p$, where $A_p = (1 + 1/p)A$.

3   Let $1 \leq m \leq n$ be an integer and, for each $1 \leq i \leq m$, let u_i be a linear form on $\mathbf{R}^n$: $(x_j)_{1 \leq j \leq n} \mapsto \sum_{1 \leq j \leq n} c_{i,j} \cdot x_j$, where the coefficients $c_{i,j}$ are integers ($c_{i,j} \in \mathbf{Z}$). Given an integer $p > 1$, let A be a compact subset of $\mathbf{R}^n$, symmetric and convex, such that $\lambda_n(A) \geq 2^n \cdot p^m$.

1. Let θ be the canonical homomorphism of the group $\mathbf{Z}^m$ onto $\mathbf{Z}^m/p\mathbf{Z}^m$, and let $[u_i]_{1 \le i \le m}$ be the homomorphism $x \mapsto \big(u_i(x)\big)_{1 \le i \le m}$ of $\mathbf{Z}^n$ into $\mathbf{Z}^m$. If G_0 is the kernel of $\theta \circ [u_i]_{1 \le i \le m}$, show that $\Delta(G_0)$ is a divisor of p^m.

2. Deduce from Minkowski's theorem that there exists a nonzero element x in $A \cap \mathbf{Z}^n$ such that $u_i(x) \equiv 0 \pmod{p}$ for all $1 \le i \le m$.

4 In this exercise, we prove that each integer $n \ge 0$ is the sum of four squares (Lagrange's theorem).

1. Fix a prime $p \ne 2$. Let $a \in \mathbf{Z}$ and $b \in \mathbf{Z}$. Prove that there exist integers x_1, x_2, x_3, x_4, not all 0, such that

$$ax_1 + bx_2 \equiv x_3 \pmod{p}, \qquad bx_1 - ax_2 \equiv x_4 \pmod{p},$$

and

$$x_1^2 + x_2^2 + x_3^2 + x_4^2 \le \sqrt{2}\,\frac{4}{\pi}p$$

(in Exercise 3, take for A the closed ball in $\mathbf{R}^4$ with center 0 and radius $\big(\sqrt{2}\,(4/\pi)p\big)^{1/2}$).

2. Prove the existence of a and b in $\{0, 1, \dots, (p-1)/2\}$ such that $a^2 + b^2 + 1 \equiv 0 \pmod{p}$. Conclude that, if x_1, x_2, x_3, x_4 are as in part 1, then $x_1^2 + x_2^2 + x_3^2 + x_4^2 = p$.

3. For each quaternion $x = (x_1, x_2, x_3, x_4)$, put $N(x) = x_1^2 + x_2^2 + x_3^2 + x_4^2$. Show that $N(xy) = N(x)N(y)$ for all quaternions x, y. From this fact, deduce Lagrange's theorem.

5 Let F be a Banach space (over $\mathbf{R}$ or $\mathbf{C}$), E a normed space, A an open subset of E, and $f : A \times E \to F$ a continuously differentiable mapping. Suppose that $\Gamma = \{(x, y) \in A \times F : f(x, y) = 0\}$ is nonempty and that, for each $(x, y) \in \Gamma$, the partial derivative $D_2 f(x, y) = D[f(x, \cdot)](y)$ is a linear homeomorphism of F onto itself. Denote by p_1 and p_2 the projections from $A \times F$ onto A and F, respectively.

1. Show that, for all $(x_0, y_0) \in \Gamma$, there exists an open neighborhood V of (x_0, y_0) relative to Γ such that the restriction of p_1 to V is a homeomorphism from V onto an open ball centered at x_0 and contained in A (use the implicit function theorem).

2. Deduce from part 1 that every connected component G of Γ is open in Γ and that $p_1(G)$ is open in A.

6 Let f and Γ be as in Exercise 5.

A path in A is, by definition, a continuous mapping $\gamma : [a, b] \to A$ ($[a, b]$ is a compact subinterval of $\mathbf{R}$ depending on the path considered; $a < b$); a lifting of γ in Γ is a continuous mapping $u : [a, b] \to \Gamma$ such that $p_1 \circ u = \gamma$. A homotopy in A is a continuous mapping φ from $[a, b] \times [c, d]$ into A ($a < b$ and $c < d$), and a lifting of φ in Γ is a continuous mapping ψ from $[a, b] \times [c, d]$ into Γ such that $\varphi = p_1 \circ \psi$.

Henceforth, we suppose there is a connected component G of Γ such that $p_1(G) = A$, also, that for each path $\gamma : [a,b] \to A$, each t_0 in $[a,b]$, and each z_0 in $G \cap p_1^{-1}(\gamma t_0)$, there exists a lifting u of γ in Γ taking the value z_0 at t_0.

If u_1, u_2 are two liftings in Γ of a same path $\gamma : [a,b] \to A$ and if there exists $t_0 \in [a,b]$ such that $u_1(t_0) = u_2(t_0)$, then, clearly, $u_1 = u_2$.

Now let φ be a homotopy from $[a,b] \times [c,d]$ into A, and let v be a path from $[c,d]$ into G such that $p_1(v(\xi)) = \varphi(a,\xi)$ for all $\xi \in [c,d]$. For each $\xi \in [c,d]$, let u_ξ be the unique path from $[a,b]$ into G such that $p_1(u_\xi(t)) = \varphi(t,\xi)$ for all $t \in [a,b]$ and such that $u_\xi(a) = v(\xi)$. We show that the mapping $(t,\xi) \mapsto u_\xi(t)$ is continuous on $[a,b] \times [c,d]$.

Given $\zeta \in [c,d]$, let $\gamma : [a,b] \to A$ be the path $t \mapsto \varphi(t,\zeta)$. For each $t \in [a,b]$, there are real numbers $r(t) > 0$ and $s(t) > 0$ such that the intersection V_t of Γ and $B(\gamma(t), r(t)) \times B(p_2 u_\zeta(t), s(t))$ is contained in G, and such that

(a) p_1 induces a homeomorphism from V_t onto $B(\gamma(t), r(t))$;

(b) $D_2 f(x,y)^{-1} \circ D_1 f(x,y)$ is bounded on V_t.

1.   If g_t is the unique mapping from $B(\gamma(t), r(t))$ into $B(p_2 u_\zeta(t), s(t))$ such that $(x, g_t(x))$ belongs to Γ for all $x \in B(\gamma(t), r(t))$. show that g_t is continuously differentiable (use the implicit function theorem).

2.   There exists a finite family $(t_i)_{i \in I}$ in $[a,b]$ such that the

$$B\left(\gamma(t_i), \frac{1}{2}r(t_i)\right) \times B\left(p_2 u_\zeta(t_i), \frac{1}{2}s(t_i)\right)$$

cover $u_\zeta([a,b])$. To simplify the notation, put $r(t_i) = r_i$ and $s(t_i) = s_i$. Let M be an upper bound for $\|D_2 f(x,y)^{-1} \circ D_1 f(x,y)\|$ as (x,y) runs through

$$\Gamma \cap \bigcup_{i \in I} \left(B(\gamma(t_i), r_i) \times B(p_2 u_\zeta(t_i), s_i) \right).$$

Given $0 < \varepsilon \le \inf((1/2)s_i)$, let $0 < r \le \inf((1/2)r_i)$ be such that $rM \le \varepsilon$. There exists $\delta > 0$ such that, for every $\xi \in [c,d]$ satisfying $|\xi - \zeta| \le \delta$, we have on the one hand, $\|p_2 v(\xi) - p_2 v(\zeta)\| \le \varepsilon$ and, on the other hand, $\|\varphi(t,\xi) - \varphi(t,\zeta)\| \le r$ for all $t \in [a,b]$. Consider such a ξ, and denote by t_0 the supremum of those $t \in [a,b]$ such that $\|p_2 u_\xi - p_2 u_\zeta\| \le \varepsilon$ on $[a,t]$. Observe that $u_\zeta(t_0)$ belongs to $B(\gamma(t_i), (1/2)r_i) \times B(p_2 u_\zeta(t_i), (1/2)s_i)$ for some $i \in I$, and show that the hypothesis $t_0 < b$ leads to a contradiction.

3.   Conclude that $(t,\xi) \mapsto u_\xi(t)$ is continuous on $[a,b] \times [c,d]$.

4.   Now let φ be a homotopy from $[a,b] \times [c,d]$ into A. Let (t_0, ξ_0) be a point in $[a,b] \times [c,d]$, and let $z_0 \in G \cap p_1^{-1}(\varphi(t_0, \xi_0))$. Deduce from 3 that there exists a lifting ψ of φ in G which takes the value z_0 at (t_0, ξ_0).

7   In the notation of Exercise 6, a "closed path in A" is a path $\gamma : [a,b] \to A$ such that $\gamma(a) = \gamma(b)$; if $\gamma(t) = \gamma(a)$ for all $t \in [a,b]$, then γ is a "constant path". A homotopy $\varphi : [a,b] \times [c,d] \to A$ is called a homotopy of closed paths if, for all $\xi \in [c,d]$, the path $t \mapsto \varphi(t,\xi)$ is closed; in this case, we say that φ is a homotopy "from the closed path $t \mapsto \varphi(t,c)$ to the closed path $t \mapsto \varphi(t,d)$".

1. Let φ be a homotopy of closed paths in A, defined on $[a, b] \times [c, d]$, from a constant path to a closed path γ, and let ψ be a lifting of φ in G. Denote by ζ the supremum of those $\xi \in [c, d]$ such that $\psi(a, \cdot) = \psi(b, \cdot)$ on $[c, \xi]$. Show that the hypothesis $\zeta < d$ leads to a contradiction (use part 1 of Exercise 5).

2. In the notation of part 1, show that $t \mapsto \psi(t, d)$ is a closed path. Deduce that, if γ is a closed path in A, homotopic in A to a constant path, then every lifting of γ in G is a closed path.

3. Finally, suppose that A is connected and that every closed path in A is homotopic in A to a constant path (i.e., suppose that A is simply connected). From part 2, deduce that the restriction of p_1 to G is a homeomorphism, from G onto A, and that its inverse mapping, from A into $A \times F$, is continuously differentiable.

8 Let E be a normed space (over $\mathbf{C}$), A a connected open subset of E, $g : A \to \mathbf{C} - \{0\}$ a C^1-function, and f the mapping $(x, y) \mapsto e^y - g(x)$ from $A \times \mathbf{C}$ into $\mathbf{C}$.

1. In the notation of Exercise 5, let G be a connected component of Γ. Given x_0 in the closure of $p_1(G)$ relative to A, and given $y_0 \in \mathbf{C}$ for which $\exp(y_0) = g(x_0)$, there exist $r > 0$ and $0 < s \le \pi$ such that p_1 induces a homeomorphism from $V = \left(B(x_0, r) \times B(y_0, s)\right) \cap \Gamma$ onto $U = B(x_0, r)$. Show that G contains $V + (0, 2k\pi i)$ for some suitable $k \in \mathbf{Z}$. Conclude that $p_1(G) = A$.

2. Let $\gamma : [a, b] \to A$ be a path and let $u(a) \in G \cap p_1^{-1}(\gamma a)$. Write c for the supremum of those $t \in [a, b]$ such that $\gamma_{/[a, t]}$ has a lifting u_t with initial point $u(a)$. Prove that $c > a$ (use Exercise 5, part 1). Show, finally, that $c = b$.

3. Suppose A is simply connected. Given $x_0 \in A$ and $y_0 \in \mathbf{C}$ such that $\exp(y_0) = g(x_0)$, show that there is a unique continuous function $h : A \to \mathbf{C}$ such that $h(x_0) = y_0$ and $e^h = g$. Furthermore, prove that h is continuously differentiable.

4. Suppose A is simply connected. Given $x_0 \in A$ and $y_0 \in \mathbf{C}$ such that $y_0^2 = g(x_0)$, deduce from 3 that there is a unique continuous function $k : A \to \mathbf{C}$ for which $k(x_0) = y$ and $k^2 = g$.

9 Let b be an element of $]0, +\infty]$ and g, h two measurable functions from $]0, b[$ into $\mathbf{C}$ and $\mathbf{R}$, respectively. Suppose that

(a) $\int^* |g(x)| e^{h(x)} \, dx_{/]0, b[} = \int_0^b |g(x)| e^{h(x)} \, dx$ is finite;

(b) as $x \to 0$, $g(x)$ admits the asymptotic expansion

$$g(x) = \sum_{1 \le j \le r} A_j \cdot x^{\alpha_j} + o(x^{\alpha_r})$$

where $-1 < \alpha_1 < \alpha_2 < \ldots < \alpha_r$ and the A_j are complex numbers;

(c) there exist $c > 0$ and $\beta > 0$ such that $h(x) = -c \cdot x^\beta + o(x^{\beta + \alpha_r - \alpha_1})$ as $x \to 0$;

(d)   there exists $0 < \delta_0 \leq b$ such that h decreases on $]0, \delta_0[$ and $h(x) \leq h(\delta_0-)$ for all $x \in [\delta_0, b[$.

We give an asymptotic expansion of $\int_0^b g(x)e^{th(x)}\, dx$ as $t \to +\infty$ (Laplace's method).

1.   Show that h is strictly negative on $]0, b[$.

2.   Let $\delta \in]0, \delta_0]$. If $\delta < b$, show that

$$\left| t^{\frac{\alpha_r+1}{3}} \cdot \int_\delta^b g(x)e^{th(x)}\, dx \right| \leq t^{\frac{\alpha_r+1}{3}} \cdot e^{(t-1)h(\delta)} \cdot \int_0^b |g(x)|e^{h(x)}\, dx$$

as soon as $t \geq 1$. Prove that the left-hand side of the inequality tends to 0 as $t \to +\infty$.

3.   Prove that

$$\int_0^{+\infty} \exp(-cy^\beta) \cdot y^{\alpha_j}\, dy = \frac{1}{\beta} \cdot c^{-\frac{\alpha_j+1}{\beta}} \cdot \Gamma\left(\frac{\alpha_j+1}{\beta}\right)$$

for all $1 \leq j \leq r$, where Γ is the Euler gamma function.

4.   Let $t \geq 1$ be given. Show that

$$t^{\frac{\alpha_r+1}{\beta}} \cdot \int_0^\delta g(x)e^{th(x)}\, dx = t^{\frac{\alpha_r}{\beta}} \cdot \int_0^{+\infty} F_t(y)\, dy,$$

where $F_t(y)$ is equal to $\exp\left(th(yt^{-1/\beta})\right) \cdot g(yt^{-1/\beta})$ if $0 < y < \delta t^{1/\beta}$ and is 0 otherwise. Deduce that

$$t^{\frac{\alpha_r+1}{\beta}} \cdot \left[\int_0^\delta g(x)e^{th(x)}\, dx - \sum_{1 \leq j \leq r} \frac{A_j}{\beta} \cdot (ct)^{-\frac{\alpha_j+1}{\beta}} \cdot \Gamma\left(\frac{\alpha_j+1}{\beta}\right) \right]$$

is equal to $\int_0^{+\infty} t^{\alpha_r/\beta} \cdot G_t(y)\, dy$, where

$$G_t(y) = \exp\left(th(yt^{-1/\beta})\right) \cdot \left[g(yt^{-1/\beta}) - \sum_{1 \leq j \leq r} A_j(yt^{-1/\beta})^{\alpha_j} \right]$$

$$+ \exp\left(-cy^\beta\right) \cdot [\exp\left(th(yt^{-1/\beta}) + cy^\beta\right) - \vdots\, [\sum_{1 \leq j \leq r} A_j(yt^{-1/\beta})^{\alpha_j}]$$

if $0 < y < \delta \cdot t^{1/\beta}$, and where

$$G_t(y) = -\exp(-cy^\beta) \cdot \sum_{1 \leq j \leq r} A_j \cdot (yt^{-1/\beta})^{\alpha_j}$$

if $y \geq \delta \cdot t^{1/\beta}$.

5.   For all $y > 0$, show that $t^{\alpha_r/\beta} \cdot G_t(y)$ converges to 0 as $t \to +\infty$.

6. Given $d \in {]}0, c{[}$, choose $\delta \in {]}0, \delta_0{[}$ so that $|h(x)| \geq d \cdot x^\beta$, $|h(x) + cx^\beta| \leq x^{\beta + \alpha_r - \alpha_1}$, and $|g(x) - \sum_{1 \leq j \leq r} A_j x^{\alpha_j}| \leq x^{\alpha_r}$, for all x in $]0, \delta[$. For $y < \delta t^{1/\beta}$, majorize the absolute value of

$$\exp\left(th(yt^{-1/\beta}) + cy^\beta\right) - 1 = \exp(cy^\beta) \cdot \int_{-cy^\beta}^{th(yt^{-1/\beta})} e^u \, du.$$

Then apply the dominated convergence theorem to show that

$$t^{\frac{\alpha_r+1}{\beta}} \cdot \left[\int_0^b g(x) e^{th(x)} \, dx - \sum_{1 \leq j \leq r} \frac{A_j}{\beta} \cdot (ct)^{-\frac{\alpha_j+1}{\beta}} \cdot \Gamma\left(\frac{\alpha_j + 1}{\beta}\right) \right]$$

converges to 0 as $t \to +\infty$.

10 Let $b \in {]}0, +\infty]$, and let g, h be two measurable functions from $]0, b[$ into $\mathbf{C}$ and $\mathbf{R}$, respectively. Suppose

(a) $\int^* |g(x)| e^{h(x)} \, dx_{/]0, b[}$ is finite;

(b) there exist $c > 0$ and $\beta > 0$ such that $h(x) = -cx^\beta + o(x^\beta)$ as $x \to 0$;

(c) there exists $0 < \delta_0 \leq b$ such that h has a strictly negative derivative in $]0, \delta_0[$ and such that $h(x) \leq h(\delta_0-)$ for all $x \in [\delta_0, b[$.

1. The function $\varphi : x \mapsto \left(- h(x)\right)^{1/\beta}$ from $]0, \delta_0[$ onto $]0, \varphi(\delta_0-)[$ has an inverse mapping ψ. Show that, for every $t \geq 1$,

$$\int_0^{\delta_0} y(x) e^{th(x)} \, dx = \int_0^{\varphi(\delta_0-)} g(\psi y) \cdot \psi'(y) \, e^{-ty^\beta} \, dy.$$

2. Assume that, as $y \to 0$, $g(\psi y) \cdot \psi'(y)$ admits the asymptotic expansion

$$g(\psi y) \cdot \psi'(y) = \sum_{1 \leq j \leq r} A_j \cdot y^{\alpha_j} + o(y^{\alpha_r})$$

with $-1 < \alpha_1 < \ldots < \alpha_r$. Deduce from Exercise 9 that

$$\int_0^b g(x) e^{th(x)} \, dx = \sum_{1 \leq j \leq r} \frac{A_j}{\beta} (ct)^{-\frac{\alpha_j+1}{\beta}} \cdot \Gamma\left(\frac{\alpha_j + 1}{\beta}\right) + o(t^{-\frac{\alpha_r+1}{\beta}})$$

as $t \to +\infty$.

11 1. Show that $\int_0^1 e^{t[\log(1-x)+x]} \, dx + \int_0^{+\infty} e^{t[\log(1+x)-x]} \, dx$ is equal to $t^{-(t+1)} \cdot e^t \cdot \Gamma(t + 1)$ for all $t > 0$.

2. Consider the function $\theta : 0 \neq z \mapsto \dfrac{(1 - z)e^z - 1}{z^2}$, $0 \mapsto -\dfrac{1}{2}$ from $\mathbf{C}$ into $\mathbf{C}$. Show that $\theta(z)$ is the sum of a power series whose coefficients are strictly negative. Deduce that $|\theta(z)| < 1$ for all z in the disk $D = D(0, 1)$, and that $e^z \neq 1 + z$ for all $z \in D - \{0\}$. Thus the function $0 \neq z \mapsto \dfrac{-\log(1 - z) - z}{z^2}$, $0 \mapsto \dfrac{1}{2}$ from D into $\mathbf{C}$ has no zeros, which implies that

there is a unique holomorphic function f from D into $\mathbf{C}$ for which $f(0) = \sqrt{1/2}$ and $z^2 f^2(z) = -\log(1-z) - z$ for every $z \in D$ (Exercise 8). Show that $f(x) = \dfrac{1}{x}\big(-\log(1-x) - x\big)^{1/2}$ for $0 < x < 1$ and that $f(x) = -\dfrac{1}{x}\big(-\log(1-x) - x\big)^{1/2}$ for $-1 < x < 0$.

3. Let φ_1 be the function $x \mapsto \big(-\log(1-x)-x\big)^{1/2}$ from $[0,1[$ onto $[0,+\infty[$, φ_2 the function $x \mapsto \big(-\log(1+x)+x\big)^{1/2}$ from $[0,+\infty[$ onto $[0,+\infty[$, and ψ_1, ψ_2 their inverse mappings. Write u for the function $z \mapsto zf(z)$ on D, and show that u' has no zeros in D. By the inverse mapping theorem, for some $r > 0$, there is a diffeomorphism v of $D(0,r)$ into D such that $u\big(v(z)\big) = z$ for all $z \in D(0,r)$. Show that $\psi_1(y) = v(y)$ and $\psi_2(y) = -v(-y)$ for all $0 \le y < r$. Conclude that ψ_1, ψ_2 are infinitely differentiable and that $\psi_2^{(k)}(0) = (-1)^{k-1} \cdot \psi_1^{(k)}(0)$ for all $k \in \mathbf{N}$.

4. Put $\psi = \psi_1$. Deduce from Exercise 10 that, for every $n \in \mathbf{N}$,

$$t^n \cdot \left[t^{-t} \cdot t^{1/2} \cdot e^t \cdot \Gamma(t) - \sum_{0 \le k \le n} \frac{1}{(2k)!} \cdot \psi^{(2k+1)}(0) \cdot t^{-k} \cdot \Gamma\left(k + \frac{1}{2}\right) \right]$$

converges to 0 as $t \to +\infty$. Conclude that

$$\Gamma(t) = \sqrt{2\pi} \cdot t^t \cdot t^{-1/2} \cdot e^{-t} \cdot \left[\sum_{0 \le k \le n} \frac{1}{\sqrt{2}} \cdot \psi^{(2k+1)}(0) \cdot \frac{1}{2^{2k} \cdot (k!)} \cdot t^{-k} + o(t^{-n}) \right].$$

5. For all $k \in \mathbf{Z}^+$, put $\alpha_k = \dfrac{\psi^{(k)}(0)}{k!}$. By part 3, there is a neighborhood of 0 in which $v(z) = \sum_{k \ge 0} \alpha_k \cdot z^k$. From the relation $z^2 = -\log\big(1 - v(z)\big) - v(z)$, deduce that $\displaystyle\sum_{1 \le j \le n} j \cdot \alpha_j \cdot \alpha_{n-j+1} = -2\alpha_{n-1}$ for all integers $n \ge 2$. Since $\alpha_1 = \sqrt{2}$, the above formula may be used to compute the numbers α_n.

6. Deduce from part 5 that $\psi^{(2)}(0) = -\dfrac{4}{3}$, $\psi^{(3)}(0) = \dfrac{\sqrt{2}}{3}$, $\psi^{(4)}(0) = \dfrac{16}{45}$, and $\psi^{(5)}(0) = \dfrac{\sqrt{2}}{9}$. Conclude that

$$\Gamma(t) = \sqrt{2\pi} \cdot t^t \cdot t^{-1/2} \cdot e^{-t} \cdot \left[1 + \frac{t^{-1}}{12} + \frac{t^{-2}}{288} + o(t^{-2}) \right],$$

where $t^2 \cdot o(t^{-2}) \to 0$ as $t \to +\infty$. In particular,

$$n! = \sqrt{2\pi} \cdot n^{n+1/2} \cdot e^{-n} \cdot \left[1 + \frac{1}{12n} + \frac{1}{288n^2} + o\left(\frac{1}{n^2}\right) \right]$$

as $n \to +\infty$ (Stirling's formula).

12 Given an integer $n \ge 2$, let $(u,v) \mapsto (u|v)$ be the usual inner product on $\mathbf{R}^n$, and let $u \mapsto |u| = (u|u)^{1/2}$ be the usual norm on $\mathbf{R}^n$. Write $d\sigma$ for the measure $d\mathbf{S}^{n-1}/\|d\mathbf{S}^{n-1}\|$. Hence $d\sigma$ is invariant under rotations. Let $B(0,1)$ be the open ball in $\mathbf{R}^n$ with center 0 and radius 1. The function $(x,v) \mapsto \dfrac{1 - |x|^2}{|x - v|^n} = p(x,v)$ from $B(0,1) \times \mathbf{S}^{n-1}$ into $]0,+\infty[$ is called the Poisson kernel relative to $B(0,1)$.

1. Suppose $n \geq 3$. For each integer $k \geq 0$, let C_k be the Gebenbauer polynomial with degree k, relative to $\gamma = (n-2)/2$ (Chapter 10, Exercise 7). For each $k \in \mathbf{Z}^+$ and each $u \in \mathbf{S}^{n-1}$, the function $Z_k^u : v \mapsto \dfrac{2k+n-2}{n-2} \cdot C_k(u|v)$ from $\mathbf{S}^{n-1}$ into $\mathbf{R}$ is called the spherical harmonic with degree k and pole u. Show that Z_k^u is constant on each parallel with pole u, that is, on each intersection of $\mathbf{S}^{n-1}$ with a hyperplane orthogonal to u.

2. Fix $u \in \mathbf{S}^{n-1}$. Show that the spherical harmonics Z_k^u are orthogonal in $L^2_{\mathbf{C}}(d\sigma)$.

3. Let $u \in \mathbf{S}^{n-1}$ and $v \in \mathbf{S}^{n-1}$. Write f for the function

    $$r \mapsto (1 - 2r(u|v) + r^2)^{-\gamma}$$

    from $[0,1[$ into $\mathbf{R}$. Prove that $p(ru, v) = f(r) + \dfrac{2r}{n-2} f'(r)$ for every r ($0 \leq r < 1$). Deduce that $p(ru, v) = \displaystyle\sum_{k \geq 0} Z_k^u(v) r^k$. For fixed $0 \leq r < 1$ and $u \in \mathbf{S}^{n-1}$, show that the series on the right-hand side of the last equality converges to $p(ru, v)$ uniformly (note that $|Z_k^u(v)|$ is less than $Z_k^u(u)$).

4. Deduce from part 3 that $\int p(x, v) d\sigma(v) = 1$ for all $x \in B(0, 1)$.

13 $\mathbf{S}^1 = \{u \in \mathbf{R}^2 : |u| = 1\}$ is now equipped with the group structure for which $\mathbf{S}^1$ is a subgroup of the multiplicative group $\mathbf{C}^*$. Let $u \in \mathbf{S}^1$. We write Z_0^u for the constant function 1 on $\mathbf{S}^1$, and Z_k^u, for every $k \in \mathbf{N}$, for the function $v \mapsto (v/u)^k + (v/u)^{-k}$ from $\mathbf{S}^1$ into $\mathbf{R}$. Z_k^u is called the spherical harmonic with degree k and pole u.

1. Show that $p(x, v)$ is the real part of $\dfrac{v+x}{v-x} = 1 + 2\displaystyle\sum_{k \geq 1} \left(\dfrac{x}{v}\right)^k$, for all $x \in B(0, 1)$ and $v \in \mathbf{S}^1$. Deduce that $p(ru, v) = \sum_{k \geq 0} Z_k^u(v) r^k$ for all u, v of $\mathbf{S}^1$ and all $0 \leq r < 1$.

2. Show that the properties enounced in Exercise 12 remain true for $n = 2$.

14 Let $n \geq 2$ be an integer. If a function h from Ω into $\mathbf{C}$ (where Ω is an open subset of $\mathbf{R}^n$) is twice differentiable at $x \in \Omega$, we call Laplacian of h at x, and write $\Delta h(x)$, the number $\sum_{1 \leq p \leq n} (\partial^2 h / \partial x_p^2)(x)$. The function h is said to be harmonic whenever it is twice differentiable in Ω and $\Delta h(x) = 0$ for all $x \in \Omega$.

1. Given $v \in \mathbf{S}^{n-1}$, let φ be the function $x \mapsto p(x, v)$ from $B(0, 1)$ into $\mathbf{R}$. Compute $D^2\varphi(x)(x', x'')$, for every $x \in B(0, 1)$ and all x', x'' of $\mathbf{R}^n$. Deduce that φ is harmonic in $B(0, 1)$.

2. Let f be a continuous complex-valued function on $\mathbf{S}^{n-1}$. The function $P(f) : x \mapsto \int p(x, v) \, d\sigma(v)$ from $B(0, 1)$ into $\mathbf{C}$ is called the Poisson integral of f. Prove that $P(f)$ is harmonic.

3. Show that the function F, which is equal to $P(f)$ on $B(0, 1)$ and to f on $\mathbf{S}^{n-1}$, is continuous on the closed ball $B'(0, 1)$.

4. Let G be a continuous function from $B'(0.1)$ into $\mathbf{C}$ which is harmonic in $B(0,1)$ and agrees with f on $\mathbf{S}^{n-1}$. Suppose that f and G are real-valued and that $G - F$ is strictly positive at a point a of $B(0,1)$. Given $0 < \varepsilon < (G-F)(a)/4$, let h be the function $x \mapsto (G-F)(x)+\varepsilon(|x-a|^2-4)$ from $B'(0,1)$ into $\mathbf{R}$, and let b be a point at which h attains its maximum. Compute $\Delta h(b)$ and deduce that there exists an integer $1 \leq p \leq n$ for which the partial derivative $(\partial^2 h/\partial x_p^2)(b)$ is strictly positive. Prove that we so arrive at a contradiction.

5) We come back to the general case in which f is complex-valued. Deduce from part 4 that $G = F$.

15 Let $n \geq 2$ be an integer and Ω an open subset of $\mathbf{R}^n$.

1. Let f be a harmonic complex-valued function in Ω. Deduce from Exercise 14 that $f(a) = \int f(a + rv)\, d\sigma(v)$ for all $a \in \Omega$ and all $r > 0$ such that Ω contains the closed ball $B'(a, r)$.

2. Let f be a continuous complex-valued function on Ω. Suppose that, for each $b \in \Omega$, there exist arbitrarily small numbers $s > 0$ such that $f(b) = \int f(b+sv)\, d\sigma(v)$. Prove that f is harmonic in Ω. (Suppose f is real-valued; given $a \in \Omega$ and $r > 0$ such that Ω contains $B'(a.r)$, define g as the continuous function from $B'(a, r)$ into $\mathbf{R}$ which is harmonic in $B(a, r)$ and agrees with f on the sphere $a + r\mathbf{S}^{n-1}$; arguing by contradiction, show that $g = f$ on $B'(a,r)$.)

3. Let $Ha(\Omega, \mathbf{C})$ be the space of harmonic functions from Ω into $\mathbf{C}$. Show that, if a filter $\mathcal{F}$ on $Ha(\Omega, \mathbf{C})$ converges to a function f uniformly on the compact subsets of Ω, then f is harmonic.

13
Stieltjes Integral

At the end of last century Stieltjes introduced the notion of "distribution of masses" on an interval. The total mass of $]a, x]$ is an increasing function; its discontinuities correspond to masses concentrated at one point. Given such a function η, Stieltjes showed that, for any continuous function f, the "Riemann sums" $\sum f(\xi)(\eta(x_{i+1}) - \eta(x_i))$ have a limit denoted by $\int_a^b f(t)\, d\eta(t)$. The increasing function η is easily replaced by any difference of increasing functions, that is, by any function with bounded variation. This gives an explicit description of all measures on an interval.

If f is Lebesgue integrable over $[a, b]$ and $F(x) = \int_a^x f(t)\, dt$, then the integrability of $|f|$ shows that F is a function of bounded variation. However, if F is a function of bounded variation whose derivative (which exists almost everywhere) is integrable, we do not necessarily have $F(x) - F(a) = \int_a^x F'(t)\, dt$. In fact, as shown by Lebesgue, this equality holds if and only if F is *absolutely continuous*, which means that the variation of F over some open set J tends to 0 with the measure of J.

Summary

13.1 Let I be an interval in $\mathbf{R}$ and f a function from I into a metric space (E, d). If J is a subinterval of I, $f_{/J}$ is said to be of finite variation if the set $\{\sum_\Delta d(f(a_{i-1}), f(a_i))\}$, where Δ runs over the set of finite partitions of J, is bounded. When $E = \mathbf{R}$, the variation of f is locally bounded if and only if f is the difference of two increasing functions (Proposition 13.1.3).

13.2 Denote by S the natural semiring of I subinterval of $\mathbf{R}$. If μ is any measure on S, every indefinite integral of μ (see Definition 13.2.1) is a function of locally bounded variation (Theorem 13.2.1). The formula for integration by parts holds true for functions of locally bounded variation with no common point of discontinuity (Theorem 13.2.3).

13.3 In this section, which may be omitted, we define line integrals and prove some of their properties.

13.4 We prove the Lebesgue decomposition theorem: If M is a function of locally bounded variation, then $M = M_1 + M_2$, where M_1 is a singular function and M_2 is locally absolutely continuous (Theorem 13.4.4). Also, if M is increasing, then M is differentiable almost everywhere and its derivative is locally integrable (Theorem 13.4.5).

13.5 In this section, which may be omitted, we study the upper and lower derivatives of a function.

13.1 Functions of Bounded Variation

If J is a nonempty interval in $\mathbf{R}$, we call a decomposition of J any sequence $(a_0, a_1, \ldots, a_n)$ in J such that $a_0 < a_1 < \ldots < a_n$. When J is compact, a subdivision of J is a decomposition such that $a_0 = \inf(J)$ and $a_n = \sup(J)$; in this case, $\sup_{1 \le i \le n}(a_i - a_{i-1})$ is called the mesh of the subdivision.

Until further notice, I will denote an interval in $\mathbf{R}$, which is not a point.

Let E be a metric space with metric ρ, and let M be a mapping from I into E.

Definition 13.1.1 If J is a nonempty subinterval of I, the total variation of M over J, written $V(J)$ or $V(M, J)$, is the supremum of

$$\sum_\Delta = \sum_{1 \le i \le n} \rho(M(a_{i-1}), M(a_i))$$

where $\Delta = (a_0, \ldots, a_n)$ extends over the set of decompositions of J. M is said to be of bounded variation over J whenever $V(J)$ is finite. M is said to be of locally bounded variation if it is of bounded variation over all compact subintervals of I.

When J is compact, $V(J)$ is clearly the supremum of $\sum_\Delta$, where Δ extends over the set of subdivisions of J. Also, $V([a, c]) = V([a, b]) + V([b, c])$ for all points a, b, c in I satisfying $a < b < c$, because $V([a, c])$ is the supremum of the numbers $\sum_\Delta$ for all subdivisions Δ of $[a, c]$ containing b.

Lemma 13.1.1 *For any nonempty $]\alpha, \beta\rangle$ in I, $V([x, \beta\rangle)$ converges to $V(]\alpha, \beta\rangle)$ as $x \to \alpha$ in $]\alpha, \beta\rangle$. Similarly, for any nonempty $\langle\alpha, \beta[$ in I, $V(\langle\alpha, x])$ converges to $V(\langle\alpha, \beta[)$ as $x \to \beta$ in $\langle\alpha, \beta[$.*

PROOF: We will prove only the first statement. The function $x \mapsto V([x, \beta\rangle)$ from $]\alpha, \beta\rangle$ into $[0, +\infty]$ decreases, so it has a limit as $x \to \alpha$, which is less than $V(]\alpha, \beta\rangle)$. But, if r is a real number strictly less than $V(]\alpha, \beta\rangle)$, there exists a decomposition $\Delta = (a_0, \ldots, a_n)$ of $]\alpha, \beta\rangle$ such that $\sum_\Delta \ge r$. Then

$$V([x, \beta\rangle) \ge V([a_0, \beta\rangle) \ge \sum_\Delta \ge r$$

for $\alpha < x \le a_0$.   □

Lemma 13.1.2   *Assume that M is of bounded variation over a compact subinterval $[\alpha, \beta]$ of I $(\alpha < \beta)$. Then $V([\alpha, \beta]) = V(]\alpha, \beta])$ if and only if M is right-continuous at α.*

PROOF: First, suppose that M is right-continuous at α. For any $\varepsilon > 0$, there exists $0 < \eta \le \beta - \alpha$ such that $\rho(M(\alpha), M(x)) \le \varepsilon/2$ for all $x \in \ ]\alpha, \alpha + \eta]$. Now $\sum_\Delta \ge V([\alpha, \beta]) - \varepsilon/2$ for a suitable subdivision $\Delta = (a_0, \ldots, a_n)$ of $[\alpha, \beta]$, and we may suppose $a_1 \le \alpha + \eta$. Then

$$V(]\alpha, \beta]) \ \ge \ V([a_1, \beta]) \ge \sum_\Delta \ - \rho(M(\alpha), M(a_1))$$

$$\ge \ V([\alpha, \beta]) - \frac{\varepsilon}{2} - \frac{\varepsilon}{2}.$$

Since ε is arbitrary, $V(]\alpha, \beta]) = V([\alpha, \beta])$.

Conversely, suppose the equality holds. Then there exists a decomposition Δ of $]\alpha, \beta]$ such that $\sum_\Delta \ge V([\alpha, \beta]) - \varepsilon$. Therefore, if $\alpha < x \le a_0$,

$$V([\alpha, \beta]) - \varepsilon \le \sum_\Delta \le \rho(M(\alpha), M(x)) + \sum_\Delta \le V([\alpha, \beta]),$$

and so

$$\rho(M(\alpha), M(x)) \le \varepsilon.$$

□

Proposition 13.1.1 *Suppose that M is of locally bounded variation, and let x_0 be in I. Define a function V_{x_0} on I as follows:*

$$V_{x_0}(x) = \begin{cases} -V([x, x_0]) & \text{for } x \le x_0; \\ V([x_0, x]) & \text{for } x \ge x_0. \end{cases}$$

Then V_{x_0} increases on I, and is right-continuous at a point c of

$$I \cap \ ] - \infty, \sup I[$$

if and only if M is right-continuous at c.

PROOF: Let $b \in I$, $b > c$. By Lemma 13.1.1, $V([c, x]) = V([c, b]) - V([x, b])$ converges to $V([c, b]) - V(]c, b])$ as $x \to c$ in $]c, b]$. Now $V_{x_0} = V_{x_0}(c) + V_c$, so $V_{x_0}(x)$ converges to $V_{x_0}(c) + V([c, b]) - V(]c, b])$ as $x \to c$ in $]c, b]$. By Lemma 13.1.2, the last assertion of the statement follows.   □

Similarly, V_{x_0} is left-continuous at a point c of $I \cap \ ] \inf I, +\infty[$ if and only if M is left-continuous at c.

Proposition 13.1.2 *Suppose E is complete and M is of locally bounded variation. Then M has a limit from the right (respectively, from the left) at every point c of $I \cap]-\infty, \sup I[$ (respectively, of $I \cap]\inf I, +\infty[$).*

PROOF: When z tends to c in $J =]c, \sup I[$, $V([c, z])$ has a limit. Then $V([x, y]) = \left| V([c, y]) - V([c, x]) \right|$ converges to 0 when x and y tend to c in J. A fortiori, $\rho(M(x), M(y))$ converges to 0. □

Proposition 13.1.3 *Suppose that $E = \mathbf{R}$. Then M is a function of locally bounded variation if and only if it is a difference of two increasing functions.*

PROOF: The condition is clearly sufficient. Now suppose that M is of locally bounded variation, and let $x_0 \in I$. If V_{x_0} is the function considered in Proposition 13.1.1, then $M = V_{x_0} - (V_{x_0} - M)$, and $V_{x_0} - M$ increases because

$$V_{x_0}(y) - V_{x_0}(x) = V([x, y]) \geq M(y) - M(x)$$

for all $x, y \in I$ satisfying $x < y$. □

Finally, when $E = \mathbf{C}$, observe that M is of locally bounded variation if and only if its real and imaginary parts are.

13.2   Stieltjes Measures

Let $I = \langle a, b \rangle$ ($a < b$ in $\overline{\mathbf{R}}$) be a subinterval of $\mathbf{R}$, and let S be the natural semiring in I.

Definition 13.2.1 Let μ be a measure on S. $M : I \to \mathbf{C}$ is called an indefinite integral of μ whenever $M(\beta) - M(\alpha) = \mu(]\alpha, \beta])$ for all $\alpha, \beta \in I$ satisfying $\alpha \leq \beta$. For all $\alpha, \beta \in I$, we put

$$\int_\alpha^\beta d\mu = \mu(]\alpha, \beta]) \qquad \text{if } \alpha \leq \beta$$

and

$$\int_\alpha^\beta d\mu = -\mu(]\beta, \alpha]) \qquad \text{if } \beta \leq \alpha,$$

so that

$$M(\beta) - M(\alpha) = \int_\alpha^\beta d\mu.$$

Proposition 13.2.1 *Let μ be a measure on S and M one of its indefinite integrals. Then $V(M, \langle \alpha, \beta \rangle) = |\mu|^*(]\alpha, \beta))$ for any subinterval $\langle \alpha, \beta \rangle$ of I.*

PROOF: By definition of $|\mu|$, $|\mu|(]\alpha,\beta]) = V(M, [\alpha,\beta])$ for all α, $\beta \in I$ satisfying $\alpha < \beta$. Now let $J = \langle \alpha, \beta \rangle$ be a subinterval of I with endpoints α and β $(\alpha \neq \beta)$. If $J =]\alpha, \beta]$,

$$V(M, J) = \lim_{\substack{x \to \alpha \\ x \in]\alpha,\beta[}} V(M, [x, \beta]) = \lim |\mu|(]x, \beta]) = |\mu|^*(]\alpha, \beta]);$$

if $J = \langle \alpha, \beta[$,

$$V(M, J) = \lim_{\substack{x \to \beta \\ x \in]\alpha,\beta[}} V(M, \langle \alpha, x]) = \lim |\mu|^*(]\alpha, x]) = |\mu|^*(]\alpha, \beta[).$$

$\square$

By Proposition 13.2.1, if μ is a measure on S, M one of its indefinite integrals, and s an indefinite integral of $|\mu|$, then, for each S-set $\langle c, d]$, we have $s(d) - s(c) = V(M, [c, d])$. Hence, for every $x_0 \in I$, the function V_{x_0} of Proposition 13.1.1 is the indefinite integral of $|\mu|$ that vanishes at x_0.

Theorem 13.2.1 *Let μ be a measure on S. Then every indefinite integral M of μ is of locally bounded variation and is right-continuous on $\langle a, b[$. Moreover, $\mu(\{c\}) = M(c) - M(c^-)$ for all $c \subset]a, b)$, where $M(c^-) = \lim_{\substack{x \to c \\ x < c}} M(x)$.*

PROOF: We already know that M is of locally bounded variation. Let c be in $\langle a, b[$, and let $(x_i)_{i \geq 1}$ be a decreasing sequence in $]c, b)$ converging to c. Then $\mu(]c, x_n])$ converges to 0 as $n \to +\infty$, and so M is right-continuous at c.

Next, let $c \in]a, b)$, and let $(x_i)_{i \geq 1}$ be an increasing sequence in $\langle a, c]$ converging to c. Then $\mu(]x_n, c]) = M(c) - M(x_n)$ converges to $M(c) - M(c^-)$ as $n \to +\infty$, so $\mu(\{c\}) = M(c) - M(c^-)$. $\square$

Theorem 13.2.2 *Let $M : I \to \mathbf{C}$ be of locally bounded variation, and right-continuous on $\langle a, b[$. There is a unique measure μ on S admitting M as an indefinite integral and satisfying $\mu(\{a\}) = 0$ whenever $a \in I$.*

PROOF: This follows from Propositions 13.1.1 and 13.1.3, and from Proposition 2.3.1. $\square$

Let E be the vector subspace of $\mathcal{M}(S, \mathbf{C})$ consisting of those measures μ on S which have no mass at a (when $a \in I$). For $x_0 \in I$, let F be the vector space consisting of all functions of locally bounded variation from I into $\mathbf{C}$ that are right-continuous on $\langle a, b[$ and vanish at x_0. Then, by the preceding, the mapping $\mu \mapsto (x \mapsto \int_{x_0}^{x} d\mu)$ from E into F is a linear isomorphism. We now have a description of the measures on S.

Henceforth, M will denote a function of locally bounded variation over I and $M_+(x)$ will be the right-hand limit of M at x, for all $x \in \langle a, b[$. Put $M_+(b) = M(b)$ if b belongs to I. The function $M_+ : x \mapsto M_+(x)$ from I into $\mathbf{C}$

is obviously right-continuous on $\langle a, b[$. Better yet, over a compact subinterval $[c, d]$ of I $(c < d)$ its variation is less than $V(M.[c, d]) + |M_+(d) - M(d)|$. Indeed, let $(x_0, \ldots, x_n)$ be a subdivision of $[c.d]$. For each $0 \leq k \leq n - 1$, let y_k be a point of $]x_k, x_{k+1}[$. Also, put $y_n = d$. Then

$$|M_+(y_k) - M_+(y_{k+1})| \leq |M_+(y_k) - M(y_k)| + |M_+(y_{k+1}) - M(y_{k+1})|$$
$$+ |M(y_k) - M(y_{k+1})|$$

for all $0 \leq k \leq n - 1$, so

$$\sum_{0 \leq k \leq n-1} |M_+(y_k) - M_+(y_{k+1})|$$

is less than

$$\sum_{0 \leq k \leq n-1} |M_+(y_k) - M(y_k)| + \sum_{1 \leq k \leq n-1} |M_+(y_k) - M(y_k)|$$
$$+ |M_+(d) - M(d)| + V(M, [c, d]).$$

Now, letting y_k tend to x_k in $]x_k, x_{k+1}[$ (for each $0 \leq k \leq n - 1$), we obtain

$$\sum_{0 \leq k \leq n-1} |M_+(x_k) - M_+(x_{k+1})| \leq V(M, [c, d]) + |M_+(d) - M(d)|,$$

as desired.

In short, M_+ is a function of locally bounded variation over I and is right-continuous on $\langle a, b[$.

Definition 13.2.2 The unique measure μ on S admitting M_+ as an indefinite integral, and $M_+(a) - M(a)$ as its mass at a when a belongs to I, is called the Stieltjes measure defined by M, and is written dM.

Now we describe the Radon measure associated with dM.

Proposition 13.2.2 *Let f be a continuous mapping from I into a Banach space F, with compact support, hence vanishing outside a compact subinterval $[\alpha, \beta]$ of I $(\alpha < \beta)$. For every $\varepsilon > 0$, there exists $\delta > 0$ such that*

$$\left| \int f \, dM - \sum_{1 \leq i \leq n} (M(x_i) - M(x_{i-1})) \cdot f(t_i) \right| \leq \varepsilon$$

for any subdivision $(x_0, \ldots, x_n)$ of $[\alpha, \beta]$ whose mesh is smaller than δ, and for any system $(t_1, \ldots, t_n)$ of points satisfying $x_{i-1} \leq t_i \leq x_i$ for all $1 \leq i \leq n$.

PROOF: Given $\varepsilon > 0$, there exists $\delta > 0$ such that $|f(x) - f(y)| \leq \varepsilon$ for all x, y in $[\alpha, \beta]$ satisfying $|x - y| \leq 2\delta$. Let $(x_0, \ldots, x_n)$ be a subdivision of

$[\alpha, \beta]$ whose mesh is smaller than δ, and let $(t_1, \ldots, t_n)$ be a system of points satisfying $x_{i-1} \le t_i \le x_i$ for all $1 \le i \le n$. Then

$$\left| \int f\, dM - \mu(\{\alpha\}) \cdot f(\alpha) - \sum_{1 \le i \le n} \big(M_+(x_i) - M_+(x_{i-1})\big) \cdot f(t_i) \right| \le$$

$$\varepsilon \cdot |\mu|(]\alpha, \beta]).$$

Put

$$c = \mu(\{\alpha\}) \cdot f(\alpha) + \sum_{1 \le i \le n} \big(M_+(x_i) - M_+(x_{i-1})\big) \cdot f(t_i)$$

$$- \sum_{1 \le i \le n} \big(M(x_i) - M(x_{i-1})\big) \cdot f(t_i),$$

so that

$$c = \mu(\{\alpha\}) \cdot f(\alpha) - \big(M_+(x_0) - M(x_0)\big) \cdot f(t_1)$$

$$+ \sum_{1 \le i \le n-1} \big(M_+(x_i) - M(x_i)\big) \cdot \big(f(t_i) - f(t_{i+1})\big)$$

$$+ \big(M_+(x_n) - M(x_n)\big) \cdot f(t_n).$$

Now

$$\mu(\{\alpha\}) \cdot f(\alpha) - \big(M_+(x_0) - M(x_0)\big) \cdot f(t_1) =$$

$$\big(M_+(x_0) - M(x_0)\big) \cdot \big(f(\alpha) - f(t_1)\big),$$

because $f(\alpha) = 0$ if $\alpha > a$ whereas $M_+(x_0) - M(x_0) = \mu(\{\alpha\})$ if $\alpha = a$. Similarly,

$$\big(M_+(x_n) - M(x_n)\big) \cdot f(t_n) = \big(M_+(x_n) - M(x_n)\big) \cdot \big(f(t_n) - f(\beta)\big),$$

because $f(\beta) = 0$ if $\beta < b$ whereas $M_+(x_n) - M(x_n) = 0$ if $\beta = b$. This implies that

$$c = \big(M_+(x_0) - M(x_0)\big) \cdot \big(f(\alpha) - f(t_1)\big)$$

$$+ \sum_{1 \le i \le n-1} \big(M_+(x_i) - M(x_i)\big) \cdot \big(f(t_i) - f(t_{i+1})\big)$$

$$+ \big(M_+(x_n) - M(x_n)\big) \cdot \big(f(t_n) - f(\beta)\big),$$

and therefore that $|c| \le \varepsilon \cdot V(M_+ - M, [\alpha, \beta])$. Finally,

$$\left| \int f\, dM - \sum_{1 \le i \le n} \big(M(x_i) - M(x_{i-1})\big) \cdot f(t_i) \right| \le$$

$$\varepsilon \cdot |\mu|(]\alpha, \beta]) + \varepsilon \cdot V(M_+ - M, [\alpha, \beta]).$$

$$\square$$

The following result is extremely useful.

Theorem 13.2.3 (Integration by Parts) *Let u, v be two functions of locally bounded variation from I into $\mathbf{C}$. Assume that u and v have no common point of discontinuity. Then u is locally dv-integrable, v is locally du-integrable, and $d(uv) = udv + vdu$.*

PROOF: On each compact subinterval of I, u is the uniform limit of a sequence of step functions (Proposition 13.1.2). Therefore, u is locally dv-integrable.

Define the functions u_+ and v_+ in the usual way. If $x \in I$ is such that $v(x) \neq v_+(x)$, then x is a point of continuity of u, and $du_+(\{x\}) = 0$. Since v has, at most, countably many discontinuities, we conclude that $vdu_+ = v_+du_+$.

When $a \in I$, write ε_a for the measure on S defined by placing the unit mass at a.

$$vdu = vdu_+ = v_+du_+$$

if a does not belong to I, and

$$\begin{aligned} vdu &= v\big[du_+ + \big(u_+(a) - u(a)\big)\varepsilon_a\big] \\ &= v_+du_+ + \big(u_+(a) - u(a)\big)v(a)\varepsilon_a \end{aligned}$$

if a lies in I. Similarly,

$$udv = u_+dv_+$$

if a does not belong to I, and

$$udv = u_+dv_+ + u(a)\big(v_+(a) - v(a)\big)\varepsilon_a,$$

if a lies in I. But, when a belongs to I, since either u or v is continuous at a, we have $u_+(a) = u(a)$ or $v_+(a) = v(a)$, so that

$$u(a)\big(v_+(a) - v(a)\big) + \big(u_+(a) - u(a)\big)v(a) = u_+(a)v_+(a) - u(a)v(a).$$

Since

$$d(uv) = d(u_+v_+)$$

when a does not belong to I, and

$$d(uv) = d(u_+v_+) + \big(u_+v_+(a) - uv(a)\big)\varepsilon_a$$

when $a \in I$, it is enough to show that

$$d(u_+v_+) = u_+dv_+ + v_+du_+.$$

In other words, we have to prove the theorem when u and v are right-continuous on $\langle a, b[$. In this case, let α, β in I be given, such that $\alpha < \beta$. Put

$$\begin{aligned} \Delta_1 &= \{(x,y) : \alpha < x < y \leq \beta\}, \\ \Delta_2 &= \{(x,y) : \alpha < y < x \leq \beta\}, \text{ and} \\ \Delta_3 &= \{(x,y) : \alpha < x = y \leq \beta\}. \end{aligned}$$

Then

$$\int 1_{\Delta_3}(du \otimes dv) = \int du(x) \cdot 1_{]\alpha,\beta]}(x) \cdot \int 1_{\{x\}}(y)dv(y)$$

$$= \int 1_{]\alpha,\beta]}(x)dv(\{x\})du(x),$$

so

$$\int 1_{\Delta_3}(du \otimes dv) = 0$$

because du and dv have no common point mass and because $dv(\{x\}) = 0$ except perhaps at countably many points of I. On the other hand,

$$\int 1_{\Delta_1}(du \otimes dv) = \int du(x) \cdot 1_{]\alpha,\beta]}(x) \cdot \int 1_{]x,\beta]}(y)dv(y)$$

$$= \int 1_{]\alpha,\beta]}(x) \cdot (v(\beta) - v(x))du(x)$$

$$= (u(\beta) - u(\alpha))v(\beta) - (vdu)(]\alpha,\beta]).$$

Similarly,

$$\int 1_{\Delta_2}(du \otimes dv) = \int dv(y) \cdot 1_{]\alpha,\beta]}(y) \cdot \int 1_{]y,\beta]}(x)du(x)$$

$$= u(\beta)(v(\beta) - v(\alpha)) - (udv)(]\alpha,\beta]).$$

So

$$(u(\beta) - u(\alpha))(v(\beta) - v(\alpha)) = (du \otimes dv)(]\alpha,\beta]) \times (]\alpha,\beta])$$

$$= (u(\beta) - u(\alpha))v(\beta) + u(\beta)(v(\beta) - v(\alpha))$$
$$- (udv + vdu)(]\alpha,\beta]).$$

This means that

$$d(uv)(]\alpha,\beta]) = (uv)(\beta) - (uv)(\alpha) = (udv + vdu)(]\alpha,\beta]).$$

Now, since $d(uv)$ and $udv + vdu$ have no mass at a, when $a \in I$, we may conclude that $d(uv) = udv + vdu$.   □

Theorem 13.2.4 *Write λ for Lebesgue measure on I. Let $(u,v) \mapsto uv$ be a continuous bilinear mapping from $F \times G$ into H, where F, G, and H are three real Banach spaces. Also, let $f' : I \to F$ and $g' : I \to G$ be locally λ-integrable functions, and let f and g be "indefinite integrals" of f' and g', respectively. Then*

$$\int_\alpha^\beta (f'g)(t)\,dt = (fg)(\beta) - (fg)(\alpha) - \int_\alpha^\beta (fg')(t)\,dt \qquad \text{for all } \alpha,\beta \in I,$$

PROOF: Given α, β in I, let μ be $\lambda/_{[\alpha,\beta]}$. For all f' and g' in $\mathcal{L}_F^1(\mu)$ and $\mathcal{L}_G^1(\mu)$, respectively, denote by $\Phi(f', g')$ the quantity

$$\int_\alpha^\beta (f'g)(t)\, dt + \int_\alpha^\beta (fg')(t)\, dt - ((fg)(\beta) - (fg)(\alpha)),$$

which does not depend on the choice of the indefinite integrals f and g of f' and g'. The bilinear mapping $\Phi : (f', g') \mapsto \Phi(f', g')$ from $\mathcal{L}_F^1(\mu) \times \mathcal{L}_G^1(\mu)$ into H is continuous. Since it vanishes on $(\mathcal{L}_\mathbf{R}^1(\mu) \otimes F) \times (\mathcal{L}_\mathbf{R}^1(\mu) \otimes G)$ by Theorem 13.2.3, it is identically zero.   □

The following proposition is used extensively in applications.

Proposition 13.2.3 (Second Mean Value Formula) *Suppose I is compact, and let $f : I \to E$ be λ-integrable, where E is a real Banach space. If g is a positive decreasing function on I, F is the function $I \ni x \mapsto \int_a^x f(t)\, dt$, and D is a closed convex set in E containing $F(I)$, then there exists v in D such that $\int_a^b f(x)g(x)\, dx = g(a)v$.*

PROOF: Let s be a step function on I, positive and decreasing. Suppose s is right-continuous on $[a, b[$ and not identically zero. Finally. let $(a_0, \ldots, a_n)$ be a subdivision of I such that s is continuous on each $[a_{i-1}, a_i[$ $(1 \le i \le n)$. Then

$$\begin{aligned}
\int_a^b fs\, d\lambda &= s(a_0)[F(a_1) - F(a_0)] + \cdots + s(a_{n-1})[F(a_n) - F(a_{n-1})] \\
&= F(a_1)[s(a_0) - s(a_1)] + \cdots + F(a_{n-1})[s(a_{n-2}) - s(a_{n-1})] \\
&\quad + F(a_n)s(a_{n-1}).
\end{aligned}$$

Since the coefficients of the $F(a_i)$ on the right-hand side are positive with sum $s(a)$, $(1/s(a)) \int fs\, d\lambda$ lies in D.

Now replace s by the step function s_n equal to $g(b)$ at b and to $g(a_{i-1})$ on each $[a_{i-1}, a_i[$ $(1 \le i \le n)$, where $a_i = a + (i/n)(b-a)$. The sequence $(s_n)_{n \ge 1}$ converges to g at each point of continuity of g, hence λ-almost everywhere. Moreover, $|fs_n| \le g(a)|f|$. By the dominated convergence theorem, $\int fs_n\, d\lambda$ converges to $\int fg\, d\lambda$. Therefore, $(1/g(a)) \int fg\, d\lambda$ lies in D, and the proof is complete.   □

13.3 Line Integrals[†]

Let $I = \langle a, b \rangle$ be a subinterval of $\mathbf{R}$, S the natural semiring of I, and let $M : I \to \mathbf{C}$ be a function of locally bounded variation.

[†]This section may be omitted.

Definition 13.3.1 If f is a mapping from $X \supset M(I)$ into a Banach space such that $f \circ M$ is dM-integrable, then $\int (f \circ M) \, dM$ is called the integral of f over M and is written $\int_M f$.

Now, we show that the line integral $\int_M f$ is invariant under change of parameterization.

Proposition 13.3.1 *Let $J = \langle c, d \rangle$ be a subinterval of $\mathbf{R}$ and φ a continuous increasing function from J onto I. Then (φ, S) is $d(M \circ \varphi)$-suited and $dM = \varphi(d(M \circ \varphi))$. Furthermore, $|dM| = \varphi(|d(M \circ \varphi)|)$ whenever M is right-continuous on $\langle a, b[$ or φ is strictly increasing.*

PROOF: Clearly, $M \circ \varphi$ is a function of locally bounded variation. For $x \in I$, denote by x' the infimum of $\varphi^{-1}(x)$ in $\overline{\mathbf{R}}$ and by x'' the supremum of $\varphi^{-1}(x)$ in $\overline{\mathbf{R}}$. Since $\varphi = x$ on $]x', x''[$,

$$d(M \circ \varphi)(]s, t]) = (M \circ \varphi)_+(t) - (M \circ \varphi)_+(s) = 0$$

for all s, t in $]x', x''[$, and hence $\big|d(M \circ \varphi)\big|(]s, t]) = 0$. We conclude that $]x', x''[$ is $d(M \circ \varphi)$-negligible.

Now let α and β be two points in I such that $\alpha < \beta$. Clearly, $(M \circ \varphi)_+(\alpha'') = M_+(\alpha)$. When β'' lies in J, then $\varphi^{-1}(]\alpha, \beta]) =]\alpha'', \beta'']$ and $(M \circ \varphi)_+(\beta'') = M_+(\beta)$, so

$$
\begin{aligned}
d(M \circ \varphi)\big(\varphi^{-1}(]\alpha, \beta])\big) &= M_+(\beta) - M_+(\alpha) \\
&= dM(]\alpha, \beta]).
\end{aligned}
$$

On the other hand, when β'' does not belong to J, $\varphi^{-1}(]\alpha, \beta]) =]\alpha'', \beta'] \sqcup]\beta', \beta''[$ is $d(M \circ \varphi)$-integrable and

$$
\begin{aligned}
d(M \circ \varphi)\big(\varphi^{-1}(]\alpha, \beta])\big) &= d(M \circ \varphi)(]\alpha'', \beta']) \\
&= (M \circ \varphi)_+(\beta') - M_+(\alpha) \\
&= M(\beta) - M_+(\alpha)
\end{aligned}
$$

is equal to $M_+(\beta) - M_+(\alpha) = dM(]\alpha, \beta])$, because β is the right-endpoint of I. Therefore, in both cases,

$$d(M \circ \varphi)\big(\varphi^{-1}(]\alpha, \beta])\big) = dM(]\alpha, \beta]).$$

Next, suppose that a lies in I. Then a' is the left endpoint of J, and $]a', a''[$ is $d(M \circ \varphi)$-negligible. The set $\varphi^{-1}(a) - \{a''\}$ is $d(M \circ \varphi)$-negligible because, if $a' < a''$ and a' belongs to J, then

$$
\begin{aligned}
d(M \circ \varphi)(\{a'\}) &= (M \circ \varphi)_+(a') - (M \circ \varphi)(a') \\
&= M(a) - M(a) = 0.
\end{aligned}
$$

Finally, $d(M \circ \varphi)(\varphi^{-1}(a)) = d(M \circ \varphi)(\{a''\})$ is equal to $M_+(a) - M(a) = dM(\{a\})$.

In short, we have proven that $\varphi^{-1}(E)$ is $d(M \circ \varphi)$-integrable for each S-set E and that $dM(E) = d(M \circ \varphi)(\varphi^{-1}(E))$. Hence (φ, S) is $d(M \circ \varphi)$-suited and $dM = \varphi(d(M \circ \varphi))$. Clearly, $|dM| \leq \varphi(|d(M \circ \varphi)|)$.

We shall say that $s \in \langle c, d[$ is a stationary point whenever there exists $t \in J$ such that $t > s$ and $\varphi(t) = \varphi(s)$. So, if $t \in \langle c, d[$ is not a stationary point, then $(M \circ \varphi)_+(t) = M_+(\varphi(t))$.

Let $\beta \in]a, b\rangle$. For each $s \in J \cap]-\infty, \beta'[$, we have $\varphi(s) < \beta$, and the largest element of $\varphi^{-1}(\varphi(s))$ is not a stationary point. Letting t tend to β' in the set of nonstationary points of $J \cap]-\infty, \beta'[$, we see that $(M \circ \varphi)(\beta'^-) = M(\beta^-)$. In particular, if $\beta \in]a, b\rangle$ is such that β'' belongs to J and $\beta' < \beta''$, then

$$\left|d(M \circ \varphi)\right|(\varphi^{-1}(\beta)) = \left|d(M \circ \varphi)(\{\beta'\})\right| + \left|d(M \circ \varphi)(\{\beta''\})\right|$$
$$= \left|M(\beta) - M(\beta^-)\right| + \left|M_+(\beta) - M(\beta)\right|,$$

whereas

$$\left|dM(\{\beta\})\right| = \left|M_+(\beta) - M(\beta^-)\right|.$$

Now, suppose that M is right-continuous or φ is strictly increasing. It remains to be proven that $\varphi(|d(M \circ \varphi)|) \leq |dM|$.

Let α, β be two points in I satisfying $\alpha < \beta$.

Assume first that β'' lies in J, and let $(s_0, \ldots, s_n)$ be a subdivision of $[\alpha'', \beta'']$. For each $0 \leq i \leq n$, let t_i be the largest element of $\varphi^{-1}(\varphi(s_i))$. The intervals $]t_{i-1}, t_i]$ $(1 \leq i \leq n)$ are disjoint, and

$$d(M \circ \varphi)(]s_{i-1}, s_i]) = d(M \circ \varphi)(]t_{i-1}, t_i])$$

because

$$1_{]s_{i-1}, s_i]} + 1_{]s_i, t_i]} = 1_{]s_{i-1}, t_{i-1}]} + 1_{]t_{i-1}, t_i]}.$$

Therefore,

$$\sum_{1 \leq i \leq n} \left|d(M \circ \varphi)(]s_{i-1}, s_i])\right| = \sum_{1 \leq i \leq n} \left|d(M \circ \varphi)(]t_{i-1}, t_i])\right|$$
$$= \sum_{1 \leq i \leq n} \left|dM(]\varphi(s_{i-1}), \varphi(s_i)])\right|$$

is less than $|dM|(]\alpha, \beta])$. We conclude that

$$\left|d(M \circ \varphi)\right|(\varphi^{-1}(]\alpha, \beta])) \leq |dM|(]\alpha, \beta]).$$

Assume secondly that β'' does not lie in J, so that $\beta = b$. Then

$$|dM|(]\alpha, \beta[) = \lim_{\substack{x \to \beta \\ x < \beta}} |dM|(]\alpha, x])$$
$$= \lim \left|d(M \circ \varphi)\right|(\varphi^{-1}(]\alpha, x]))$$
$$= \left|d(M \circ \varphi)\right|(\varphi^{-1}(]\alpha, \beta[)).$$

On the other hand,

$$\begin{aligned}
|dM|(\{\beta\}) &= |M(\beta) - M(\beta^-)| \\
&= |d(M \circ \varphi)(\{\beta'\})| \\
&= |d(M \circ \varphi)|(\varphi^{-1}(\beta)).
\end{aligned}$$

Thus

$$|dM|(]\alpha, \beta]) = |d(M \circ \varphi)|(\varphi^{-1}(]\alpha, \beta])).$$

Finally, when a lies in I,

$$\begin{aligned}
|dM|(\{a\}) &= |M_+(a) - M(a)| \\
&= |d(M \circ \varphi)|(\{a''\}) \\
&= |d(M \circ \varphi)|(\varphi^{-1}(a)).
\end{aligned}$$

The proof is complete. □

Consequently, under the whole hypothesis of Proposition 13.3.1, and given $f : X \to F$ $(X \supset M(I), F$ Banach space), $f \circ M \circ \varphi$ is $d(M \circ \varphi)$-integrable if (and, in fact, only if) $f \circ M$ is dM-integrable. In this case, $\int_M f = \int_{M \circ \varphi} f$. Now we will juxtapose functions of locally bounded variation.

Proposition 13.3.2 *Let c be a point of $]a, b[$. Denote by M_1 (respectively, by M_2) the restriction of M to $\langle a, c]$ (respectively, to $[c, b\rangle$). Let f be a mapping from $X \supset M(I)$ into a Banach space. Then $f \circ M$ is dM-integrable if and only if $f \circ M_i$ is dM_i-integrable for all $1 \leq i \leq 2$. In this case, $\int_M f = \int_{M_1} f + \int_{M_2} f$.*

PROOF: Clearly,

$$dM_1 = (dM)_{/\langle a, c]} - (M_+(c) - M(c))\varepsilon_1$$

and

$$dM_2 = (dM)_{/[c, b\rangle} - (M(c) - M_+(c^-))\varepsilon_2,$$

where ε_1 (respectively, ε_2) is the measure on the natural semiring of $\langle a, c]$ (respectively, $[c, b\rangle$) defined by the unit mass at c. The proposition follows. □

13.4 The Lebesgue Decomposition of a Function

Let $I = \langle a, b \rangle$ be an interval in $\mathbf{R}$ $(a < b$ in $\overline{\mathbf{R}})$, and denote by λ Lebesgue measure on the natural semiring S of I.

Theorem 13.4.1 *Let μ be a complex measure on S and M an indefinite integral of μ. If $\mu = f\lambda + \mu_s$ is the Lebesgue decomposition of μ with respect to λ (so that f is locally λ-integrable and μ_s is disjoint from λ), then, for λ-almost all x, M has derivative $f(x)$ at x.*

PROOF: Define an interval J in $\mathbf{R}$ as follows: $J = I$ if I is open, $J =]a, +\infty[$ if $I =]a, b]$, $J =]-\infty, b[$ if $I = [a, b[$, and $J = \mathbf{R}$ if $I = [a.b]$. Write π for the canonical injection from I into J, T for the natural semiring in J, and λ_J for Lebesgue measure on T. For every nonempty T-set $]c, d]$, $\pi^{-1}(]c, d])$ is an S-set, because it is equal to $]\sup(a, c), \inf(b, d)]$ or to $[\sup(a, c), \inf(b, d)]$ as $c \geq a$ or $c < a$. Now the pair (π, T) is μ-suited, $\pi(|\mu|) = |\pi(\mu)|$, and μ is the measure induced by $\pi(\mu)$ on S.

Consider the function $\widetilde{M} : J \to \mathbf{C}$ which agrees with M on I, is equal to $M(a) - \mu(\{a\})$ on $]-\infty, a[$ when a lies in I, and is equal to $M(b)$ on $]b, +\infty[$ when b lies in I. Then $\widetilde{M}$ is an indefinite integral of $\pi(\mu)$.

Moreover, $\pi(\mu) = \tilde{f}\lambda_J + \pi(\mu_s)$, where $\tilde{f} : J \to \mathbf{C}$ agrees with f on I and vanishes on $J \cap I^c$. Clearly, $\tilde{f}$ is locally λ_J-integrable and $\pi(\mu_s)$ is disjoint from λ_J.

At every $x \in J$ such that $D\big(\pi(\mu)\big)(x)$ exists, $\widetilde{M}$ has derivative $D\big(\pi(\mu)\big)(x)$. Theorem 12.1.1 now shows that, at λ_J-almost all points $x \in J$, $\widetilde{M}$ has derivative $\tilde{f}(x)$. Hence, at λ-almost all $x \in I$, M has derivative $f(x)$. $\square$

Definition 13.4.1 $M : I \to \mathbf{C}$ is said to be absolutely continuous whenever, for each $\varepsilon > 0$, there exists $\delta > 0$ such that $\sum_{1 \leq i \leq p} |M(\beta_i) - M(\alpha_i)| \leq \varepsilon$ for every finite family $\big(\langle \alpha_i, \beta_i \rangle\big)_{1 \leq i \leq p}$ of mutually disjoint S-sets satisfying $\sum_{1 \leq i \leq p}(\beta_i - \alpha_i) \leq \delta$.

In this case, if I is compact, then M is of bounded variation.

Definition 13.4.2 $M : I \to \mathbf{C}$ is said to be locally absolutely continuous if its restriction to every compact subinterval of I is absolutely continuous.

In this case, M is of locally bounded variation.

Theorem 13.4.2 *Let μ be a complex measure on S and M an indefinite integral of μ. Assume that μ has no mass at a, when a lies in I. Then M is locally absolutely continuous if and only if μ is a measure with base λ.*

PROOF: Let s be a real-valued indefinite integral of $|\mu|$. We know that $s(d) - s(c) = V(M, [c, d])$ for each S-set $\langle c, d \rangle$. Hence s is locally absolutely continuous if and only if M is.

By Theorem 10.3.2, $|\mu| \ll \lambda$ if and only if s is locally absolutely continuous, which completes the proof. $\square$

Definition 13.4.3 $M : I \to \mathbf{C}$ is said to be singular whenever it is of locally bounded variation and has derivative zero λ-a.e.

Theorem 13.4.3 *Let μ be a complex measure on S and M an indefinite integral of μ. Then μ is singular (i.e., disjoint from λ) if and only if M is singular.*

PROOF: This follows immediately from Theorem 13.4.1.   □

Proposition 13.4.1 *Let M be a function of locally bounded variation from I into $\mathbf{C}$ that vanishes outside a countable subset D. Then M is singular.*

PROOF: We may suppose that I is compact. Given a real number $r > 0$, denote by E_r the set of those $x \in I - D$ such that

$$\limsup_{\substack{y \to x \\ y \neq x}} \frac{|M(y) - M(x)|}{|y - x|} > \frac{1}{r}.$$

For every $y \in D$, let I_y be the intersection of I and the compact subinterval of $\mathbf{R}$ with center y and length $2r|M(y)|$.

Each $x \in E_r$ belongs to infinitely many I_y, because there are infinitely many $y \in I$ such that $|M(y) - M(x)|/|y - x| \geq 1/r$. The function $g = \sum_{y \in D} 1_{I_y}$ is thus infinite on E_r. But $\lambda^*(g) \leq \sum_{y \in D} \lambda(I_y) \leq 2r \cdot \sum_{y \in D} |M(y)|$ is finite, because M is of bounded variation, and so we conclude that E_r is λ-negligible. The proposition follows.   □

Theorem 13.4.4 (Lebesgue Decomposition of a Function)   *Let $M : I \to \mathbf{C}$ be of locally bounded variation, and let $x_0 \in I$. Then M is the sum $M_1 + M_2$ of a locally absolutely continuous function M_1 and a singular function M_2. Furthermore, M_1 and M_2 are uniquely determined by the condition $M_1(x_0) = M(x_0)$. M is differentiable at λ-almost all points of I, its derivative M' (defined λ-almost everywhere) is locally λ-integrable, and M is an indefinite integral of $M'\lambda$ if and only if M is locally absolutely continuous.*

PROOF: For all $x \in \langle a, b[$, write $M_+(x)$ for the right-hand limit of M at x. If b lies in I, set $M_+(b) = M(b)$. Then M_+ is an indefinite integral of the Stieltjes measure dM. Now dM can be written $f\lambda + \mu_s$, where f is locally λ-integrable and μ_s is disjoint from λ. Denote by M_1 the indefinite integral of $f\lambda$ that takes the value $M(x_0)$ at x_0. Denote by M_2 the sum of $M - M_+$ and the indefinite integral of μ_s that has the value $M_+(x_0) - M(x_0)$ at x_0. Then $M_1 + [M_2 - (M - M_+)]$, as an indefinite integral of $f\lambda + \mu_s$ that has the value $M_+(x_0)$ at x_0, is equal to M_+, and $M = M_1 + M_2$. By Theorem 13.4.1, M_+ has derivative $f(x)$ at λ-almost all points $x \in I$. Proposition 13.4.1 then shows that M has derivative $f(x)$, at λ-almost all points $x \in I$.

Finally, let $\widehat{M}_1$ be a locally absolutely continuous function from I into $\mathbf{C}$ which takes the value $M(x_0)$ at x_0, and let $\widehat{M}_2 : I \to \mathbf{C}$ be singular. Suppose that $M = \widehat{M}_1 + \widehat{M}_2$. At λ-almost all $x \in I$, $\widehat{M}_1$ has derivative $f(x)$. Thus $\widehat{M}_1 = M_1$ and $\widehat{M}_2 = M - \widehat{M}_1 = M - M_1 = M_2$.

Finally, if M is locally absolutely continuous, then $dM = f\lambda$.   □

We note that the Lebesgue decomposition of measures generalizes the Lebesgue decomposition of functions.

Theorem 13.4.5 *Let M be an increasing function on I. M has a finite derivative λ-a.e., the function M' (defined λ-a.e.) is locally λ-integrable, and*

$$\int_\alpha^\beta M'(t)\, dt \le M(\beta) - M(\alpha)$$

for all α, $\beta \in I$ satisfying $\alpha < \beta$.

PROOF: By Theorem 13.4.4, the function M' is defined λ-almost everywhere and is locally λ-integrable. Since dM is positive, it can be written $M'\lambda + \mu_s$, where μ_s is singular and positive. Hence

$$\int_\alpha^\beta M'(t)\, dt = (M'\lambda)(]\alpha, \beta[) \le dM(]\alpha, \beta[) = M(\beta^-) - M_+(\alpha)$$

$$\le M(\beta) - M(\alpha)$$

for all α, $\beta \in I$ satisfying $\alpha < \beta$.   □

Observe that, if $M : [0, 1] \to \mathbf{R}$ is the Cantor singular function, the equality $\int_0^1 M'(t)\, dt = M(1) - M(0)$ fails.

Now we consider real measures on S. If μ is one such measure, then, for each $x \in I$, we define $\underline{D}\mu(x)$ and $\overline{D}\mu(x)$ as follows:

$$\underline{D}\mu(x) = \sup_{n \ge 1} \inf_{\langle c,d]} \frac{\mu(\langle c,d])}{d - c}$$

and

$$\overline{D}\mu(x) = \inf_{n \ge 1} \sup_{\langle c,d]} \frac{\mu(\langle c,d])}{d - c},$$

where $\langle c,d]$ runs through the class of those S-sets containing x such that $0 < d - c < 1/n$. Propositions 12.1.1, 12.1.2, and 12.1.4 are true for open subintervals of $\mathbf{R}$ and for arbitrary subintervals Ω of $\mathbf{R}$, by the same argument as that in Chapter 12.

Definition 13.4.4 Let $M : I \to \mathbf{R}$. At each point $x \in \langle a, b[$, M has upper and lower right-hand derivatives, defined as follows:

$$D^M(x) = \limsup_{\substack{h \to 0 \\ h > 0}} \frac{M(x + h) - M(x)}{h}$$

and

$$D_M(x) = \liminf_{\substack{h \to 0 \\ h > 0}} \frac{M(x+h) - M(x)}{h}.$$

Likewise, at each point $x \in \,]a,b\rangle$, M has upper and lower left-hand derivatives, defined as follows:

$$^M D(x) = \limsup_{\substack{h \to 0 \\ h < 0}} \frac{M(x+h) - M(x)}{h}$$

and

$$_M D(x) = \liminf_{\substack{h \to 0 \\ h < 0}} \frac{M(x+h) - M(x)}{h}.$$

Clearly, if μ is a real measure on S and M a real-valued indefinite integral of μ, then

$$_M D(x) \geq \underline{D}\mu(x) \quad \text{at each } x \in \,]a,b\rangle$$

and

$$D_M(x) \geq \underline{D}\mu(x) \quad \text{at each } x \in \langle a,b[$$

such that $\mu(\{x\}) = 0$.

Now let μ be a positive singular measure on S and M a real-valued indefinite integral of μ. By the above and Proposition 12.1.4, M has left-hand derivative $+\infty$ at μ-almost all points of $]a,b\rangle$, and it has right-hand derivative $+\infty$ at μ-almost all $x \in \langle a,b[$ such that $\mu(\{x\}) = 0$. In particular, when $\mu \neq 0$ is positive, diffuse and singular, M has derivative $+\infty$ at μ-almost all points of I, hence at uncountably many points. For example, the Cantor singular function has derivative $+\infty$ at uncountably many points of $[0,1]$.

Finally, we have the following important result.

Theorem 13.4.6 (Lebesgue) *Let f be a locally λ-integrable mapping from I into a real Banach space. For λ-almost all $x \in I$,*

$$\frac{1}{y-x} \cdot \int_x^y \left| f(t) - f(x) \right| dt$$

converges to 0 as $y \neq x$ tends to x in I.

PROOF: This follows from Theorem 12.1.2.   □

Consequently, for λ-almost all $x \in I$, $(1/(y-x)) \cdot \int_x^y f(t)\,dt$ converges to $f(x)$ as $y \neq x$ tends to x.

13.5   Upper and Lower Derivatives†

Let $I = [a, b]$ (with $a < b$) be a compact subinterval in $\mathbf{R}$, and refer to λ for Lebesgue measure on I.

Let $f : I \to \mathbf{R}$ be continuous and let $g : I \to \mathbf{R}$ be regulated. Suppose there is a countable subset D of I such that, at every $t \in I - D$, f has derivative $g(x)$. Then $f(x) - f(a) = \int_a^x g(t)\,dt$ for all $x \in I$ (Section 3.2). The following theorem, due to Lebesgue, extends this result.

Theorem 13.5.1 *Let $f : I \to \mathbf{R}$ be continuous. If there is a countable subset P of $[a, b[$ such that $D^f(x)$ (respectively, $D_f(x)$) is finite for all x in $[a, b[- P$, and if the function $x \mapsto D^f(x)$ (respectively, $x \mapsto D_f(x)$) from $[a, b[$ into $\overline{\mathbf{R}}$ is λ-integrable, then f is of bounded variation and $f(x) - f(a) \leq \int_a^x f'(t)\,dt$ (respectively, $f(x) - f(a) \geq \int_a^x f'(t)\,dt$) for all $x \in I$, where f' designates the derivative of f (defined λ-almost everywhere). Similar results are true for the upper and lower left-hand derivatives.*

PROOF: Suppose there is a countable subset P of $[a, b[$ such that $D^f(x)$ is finite for all $x \in [a, b[- P$. Also, suppose that the function $f' : x \mapsto D^f(x)$ from $[a, b[$ into $\overline{\mathbf{R}}$ is λ-integrable. Let $(a_n)_{n \geq 1}$ be an enumeration of the points in P.

Given $\varepsilon > 0$, there exists a lower semicontinuous function g from I into $\overline{\mathbf{R}}$, strictly greater than f' at every point of $[a, b[- P$, such that $\int_a^b g(t)\,dt \leq \int_a^b f'(t)\,dt + \varepsilon$. For a fixed $\delta > 0$, consider the function F_δ:

$$I \ni x \mapsto \int_a^x g(t)\,dt - f(x) + f(a) + \delta(x - a) + \sum_{a_n < x} \frac{\delta}{2^n}.$$

The set $J = \{y \in I : F_\delta(x) \geq 0 \text{ for all } a \leq x \leq y\}$ is a compact interval $[a, c]$. Suppose that $c < b$. If c does not belong to P, we can find $0 < h \leq b - c$ small enough so that $g(t)$ exceeds $f'(c)$ and $(f(t) - f(c))/(t - c)$ is smaller than $f'(c) + \delta$ for all $t \in]c, c + h]$; then $F_\delta(x) - F_\delta(c)$ is positive for all $x \in [c, c + h]$; hence $c + h$ belongs to J, which contradicts the definition of c. On the other hand, if $c = a_k$ for some index k, there exists $0 < h \leq b - c$ such that

$$\int_c^x g(t)\,dt - f(x) + f(c) + \delta(x - c) \geq -\frac{\delta}{2^k} \qquad \text{for every } x \in [c, c + h];$$

hence $c + h$ still belongs to J, which contradicts the definition of c.

We conclude that $c = b$ and that $F_\delta(b) - F_\delta(a)$ is positive. Since δ and ε are arbitrary, it follows that $f(b) - f(a) \leq \int_a^b f'(t)\,dt$.

In the above argument, we may replace a and b by arbitrary points y, z ($y < z$) of I. The function $x \mapsto \int_a^x f'(t)\,dt - (f(x) - f(a))$ is then increasing, and so f is of bounded variation.

---

†This section may be omitted.

Now suppose there is a countable subset P of $]a, b]$ such that ${}^f D(x)$ is finite for all $x \in]a, b] - P$. Also, suppose that the function $x \mapsto {}^f D(x)$ from $]a, b]$ into $\overline{\mathbf{R}}$ is λ-integrable. Let $(a_n)_{n \geq 1}$ be an enumeration of the points in P. With ε, g, and δ as before, define F_δ as the function

$$I \ni x \mapsto f(b) - f(x) - \int_x^b g(t)\, dt - \delta(b - x) - \sum_{a_n > x} \frac{\delta}{2^n}.$$

It is easy to see that $F_\delta(a)$ is negative, and the same argument as that used above proves that the function $x \mapsto f(b) - f(x) - \int_x^b f'(t)\, dt$ is increasing.

The other assertions of the statement follow from the equalities $D_f = -D^{-f}$ and ${}_f D = -{}^{-f} D$. $\square$

Proposition 13.5.1 *Let $I = \langle a, b \rangle$ be an interval in $\mathbf{R}$, and let λ be Lebesgue measure on I. Finally, let f be a continuous mapping from I into a real Banach space F. Assume that, except perhaps at countably many points, f is differentiable and that its derivative f' (defined outside a countable set) is locally λ-integrable. Then $f(\beta) - f(\alpha) = \int_\alpha^\beta f'(t)\, dt$ for all $\alpha, \beta \in I$ satisfying $\alpha < \beta$.*

PROOF: It suffices to prove that $u\big(f(\beta) - f(\alpha)\big) = \int_\alpha^\beta (u \circ f')(t)\, dt$ for every continuous linear form u on F, and the desired result now follows from Theorem 13.5.1. $\square$

Exercises for Chapter 13

1   Let $I = [a, b]$ be a compact interval in $\mathbf{R}$ ($a < b$), E a metric space, whose distance is denoted by ρ, and $M : I \to E$ a mapping of bounded variation. For every subdivision $\Delta = (a_0, \ldots, a_n)$ of I, put

$$\sum_\Delta = \sum_{1 \leq i \leq n} \rho\big(M(a_{i-1}), M(a_i)\big).$$

Suppose M is right-continuous on $]a, b[$.

Prove that, for all $0 \leq V_1 < V = V(M, [a, b])$, there exists $\delta \geq 0$ such that $\sum_\Delta \geq V_1$ for every subdivision Δ of I whose mesh is less than δ (let V_2 be in $]V_1, V[$ and let $\Delta' = (d_0, \ldots, d_p)$ be a subdivision for which $\sum_{\Delta'} \geq V_2$; choose $\delta < \inf_{0 \leq j \leq p-1}(d_{j+1} - d_j)$, such that $d_j \leq x \leq d_j + \delta$ implies

$$\rho\big(M(d_j), M(x)\big) \leq \frac{V_2 - V_1}{2(p-1)} \qquad \text{for each } 1 \leq j \leq p-1).$$

2   Let λ be Lebesgue measure on $I = [0, 1]$. Let f be a continuous increasing function from I into $\mathbf{R}$ such that $f(0) = 0$ and $f(1) = 1$

1. Equip $\mathbf{R}^2$ with its Euclidean norm. Show that the function

$$M : x \mapsto (x, f(x)) \qquad \text{from } I \text{ into } \mathbf{R}^2$$

   is of bounded variation and that $V(M, I) \leq 2$.

2. Assume now that f has derivative 0 at λ-almost all points of I. Fix $\varepsilon > 0$ and, for each integer $n \in \mathbf{N}$, denote by A_n the set of those x in $]0, 1]$ such that the relations $u \in I$, $v \in I$, $u < x \leq v$, and $v - u \leq 1/n$ imply $(f(v) - f(u))/(v - u) \leq \varepsilon$. Choose n so that $\lambda^*(A_n) \geq 1 - \varepsilon$. Write J for the set of those $k \in \{1, 2, \ldots, n\}$ such that $](k-1)/n, k/n]$ meets A_n, and write J' for the complement of J in $\{1, 2, \ldots, n\}$. Show that

$$\sum_{k \in J'} \left[f\left(\frac{k}{n}\right) - f\left(\frac{k-1}{n}\right) \right] \qquad \text{is larger than } 1 - \varepsilon$$

   and that

$$\sum_{1 \leq k \leq n} \left| M\left(\frac{k}{n}\right) - M\left(\frac{k-1}{n}\right) \right| \qquad \text{is larger than } 2(1 - \varepsilon).$$

   Conclude that $V(M, I) = 2$.

3 Let $g : I \to \mathbf{R}$ be continuous, where $I = [a, b]$ is a compact subinterval of $\mathbf{R}$. Define by Ω_+ the set of right-expansion of g in I, and by Ω_- the set of left-expansion of g in I: $x \in]a, b[$ lies in Ω_+ (respectively, Ω_-) if there exists $y \in I$ such that $y > x$ (respectively, $y < x$) and $g(y) > g(x)$.

1. Prove that Ω_+ (respectively, Ω_-) is open in $\mathbf{R}$, and that, if $]\alpha, \beta[$ is a connected component of Ω_+ (respectively, Ω_-), then $g(\alpha) \leq g(\beta)$ (respectively, $g(\alpha) \geq g(\beta)$).

2. Assume that g is increasing. Fix two real numbers r_1 and r_2 so that $0 \leq r_1 < r_2$. Denote by E' the set of left-expansion of $g(x) - r_1 x$ in I, and by E'' the union of the sets of right-expansion of $g(x) - r_2 x$ in each of the intervals $[c, d]$, the closures of the connected components of E'. If λ is Lebesgue measure on $\mathbf{R}$, prove that $\lambda(E'') \leq (r_1/r_2)\lambda(E')$.

In the following exercise, we will write $E'(g, I; r_1)$ and $E''(g, I; r_1, r_2)$ for E' and E''.

4 Let $f : I \to \mathbf{R}$ be increasing and continuous ($I = [a, b]$).

1. For all $r \in]0, +\infty[$, denote by E_r the set of right-expansion of $f(x) - rx$ in I. Prove that $\{x \in]a, b[: D^f(x) = +\infty\}$ is included in $\bigcap_{r>0} E_r$, and that it is λ-negligible.

2. Let r_1 and r_2 be two rationals such that $0 < r_1 < r_2$. Define open subsets E''_n of $]a, b[$ as follows: $E''_0 =]a, b[$ and, for every $n \in \mathbf{Z}^+$, E''_{n+1} is the union of the sets $E''(f_{/[\alpha, \beta]}, [\alpha, \beta]; r_1, r_2)$ when $]\alpha, \beta[$ runs through the class of all connected components of E''_n. Show that $\lambda(E''_n) \leq (r_1/r_2)^n \cdot (b - a)$. Deduce that the set $\{x \in]a, b[: {_f}D(x) < r_1 < r_2 < D^f(x)\}$ is λ-negligible, and conclude that $\{x \in]a, b[: {_f}D(x) < D^f(x)\}$ is λ-negligible.

3. Replace f by the function $x \mapsto -f(-x)$ from $[-b, -a]$ into $\mathbf{R}$, and deduce from part 2 that $\{x \in]a, b[: D_f(x) < {}^f D(x)\}$ is λ-negligible.

4. Using parts 1, 2, and 3, prove that f has a finite derivative at λ-almost all points of I.

5. Define $\tilde{f} : \mathbf{R} \to \mathbf{R}$ by $\tilde{f}(x) = f(x)$ for all $x \in I$, $\tilde{f}(x) = f(a)$ for all $x \in] -\infty, a]$, and $\tilde{f}(x) = f(b)$ for all $x \in [b, +\infty[$. For each $n \in \mathbf{N}$, consider the function

$$f_n : x \mapsto n \cdot \left[\tilde{f}\left(x + \frac{1}{n}\right) - \tilde{f}(x)\right] \qquad \text{from } I \text{ into } \mathbf{R}.$$

If $f' : I \to \mathbf{R}^+$ is such that $f'(x)$ is the derivative of f at x, for λ-almost all $x \in I$, show that

$$\int^* f' \, d\lambda_{/I} \leq \liminf_{n \to +\infty} \int f_n \, d\lambda_{/I} = f(b) - f(a).$$

It follows that f' is $\lambda_{/I}$-integrable.

5 Let $I = [a, b]$ be a compact subinterval of $\mathbf{R}$, μ Lebesgue measure on I, and let $f : I \to \mathbf{R}^+$ be μ-integrable. Define by F the function $I \ni x \mapsto \int f \cdot 1_{[a, x]} \, d\mu$.

1. First, assume that f is lower semicontinuous. Let $(f_n)_{n \geq 1}$ be an increasing sequence of continuous functions from I into $\mathbf{R}^+$ whose upper envelope is f. For every $n \in \mathbf{N}$, denote by F_n the function $x \mapsto \int f_n \cdot 1_{[a, x]} \, d\mu$. Finally, let $F' : I \to \mathbf{R}^+$ be such that $F'(x)$ is the derivative of F at μ-almost all points x (see Exercise 4). Show that $F' - f_n$ is positive μ-a.e., hence that $F' \geq f$ μ-a.e. Deduce from Exercise 4, part 5, that $F' = f$ μ-a.e.

2. To treat the general case, note that, by Proposition 3.7.3, there exists an increasing sequence $(g_n)_{n \geq 1}$ of upper semicontinuous functions from I into $\mathbf{R}^+$ whose upper envelope is equal to f μ-a.e. Deduce from part 1 that, for every $n \in \mathbf{N}$, the function $G_n : x \mapsto \int g_n \cdot 1_{[a, x]} \, d\mu$ has derivative $g_n(x)$ at μ-almost all $x \in I$. Then argue as in part 1, and conclude that F has derivative $f(x)$ at μ-almost all $x \in I$.

6 Let I and μ be as in Exercise 5. Also, let $(f_n)_{n \geq 1}$ be a sequence of increasing functions from I into $\mathbf{R}^+$ such that $f = \sum_{n \geq 1} f_n$ is finite. Let N be a μ-negligible subset of I such that f and the f_n are differentiable at all points of $I - N$.

1. For each $n \geq 1$, put $s_n = \sum_{1 \leq k \leq n} f_k$, and prove that $s_n'(x) \leq f'(x)$ at all $x \in I - N$.

2. Let $(s_{n_p})_{p \geq 1}$ be a subsequence of $(s_n)_{n \geq 1}$ such that $|f - s_{n_p}| \leq 2^{-p}$ for all $p \geq 1$. Deduce from part 1 that $s_{n_p}'(x)$ converges to $f'(x)$ at μ-almost all $x \in I - N$.

3. Conclude that $f'(x) = \sum_{k \geq 1} f_k'(x)$ for μ-almost all $x \in I - N$ (Fubini's theorem).

14

The Fourier Transform in $\mathbf{R}^k$

This chapter is a short introduction to Fourier analysis in $\mathbf{R}^k$ and its applications to the theory of probability. The Fourier transform of a probability P, called the characteristic function of P, is a very useful tool. It translates problems on measures in $\mathbf{R}^k$ into problems on functions defined in $\mathbf{R}^k$. This will be particularly convenient when studying convergence properties of sequences of random variables. Also, since the Fourier transform of a convolution is the product of the transforms, the characteristic function of a finite sum of independent random variables is the product of the characteristic functions (cf. Chapter 15).

Summary

14.1   In this section, we give some necessary and sufficient conditions for a quasi-measure, defined on the natural semiring of a nonempty open subset of $\mathbf{R}^k$, to be a measure (Propositions 14.1.1 to 14.1.3).

14.2   In $\mathbf{R}^k$, the distribution function of a probability defined on the Borel σ-algebra is $F : (x_1, \ldots, x_k) \mapsto P\left(\prod_{1 \le i \le k}] - \infty, x_i]\right)$. A sequence of probabilities P_n with distribution functions F_n converges weakly to a probability P with distribution function F if and only if $F_n(x)$ converges to $F(x)$ at every point of continuity of F.

14.3   In this section, we define the covariance matrix of a positive measure of order 2 and the normal distribution $N(m, \sigma^2)$.

14.4   Let μ be a bounded measure on the Borel σ-algebra of $\mathbf{R}^k$. The Fourier transform of μ is the function $\mathcal{F}\mu : x \mapsto \int \exp(-2\pi i x y)\, d\mu(y)$. The function $\mathcal{F}\mu$ is uniformly continuous and bounded (Proposition 14.4.1). Moreover, the Fourier transform is injective (Theorem 14.4.1). We next prove the Riemann–Lebesgue lemma: the Fourier transform of an integrable function vanishes at infinity. Finally, we prove the inversion theorem for bounded measures and two convergence theorems of fundamental importance in the theory of probability.

14.5   This section is devoted to a short study of the normal laws in $\mathbf{R}^k$.

14.1   Measures in $\mathbf{R}^k$

Let Ω be a nonempty open subset of $\mathbf{R}^k$ and S the natural semiring in Ω.

Now let μ be a quasi-measure on S. Assume that $\mu(A_n)$ tends to $\mu(A)$, for every (nonempty) S-rectangle $A = \prod_{1 \leq i \leq k}]a_i\,,b_i]$ and every decreasing sequence $(A_n)_{n \geq 1}$ of S-rectangles of the form $A_n = \prod_{1 \leq i \leq k}]a_i\,,b_i^n]$ such that $A = \bigcap_{n \geq 1} A_n$. We shall prove that μ is a measure. To begin, we prove the following lemma.

Lemma 14.1.1 *Let $A = \prod_{1 \leq i \leq k}]a_i\,,b_i]$ be an S-rectangle, and let $(A_n)_{n \geq 1}$ be an increasing sequence of S-rectangles of the form $A_n = \prod_{1 \leq i \leq k}]a_i^n\,,b_i]$ such that $A = \bigcup_{n \geq 1} A_n$. Then $\mu(A_n) \to \mu(A)$ as $n \to +\infty$.*

PROOF: We can find real numbers $\alpha_1, \ldots, \alpha_k$ such that $\alpha_1 < a_1, \ldots, \alpha_k < a_k$ and $\prod_{1 \leq i \leq k}[\alpha_i\,,b_i]$ is contained in Ω. By induction on $0 \leq l \leq k$, we see that, for all $c_{l+1} \in \{a_{l+1}, b_{l+1}\}, \ldots, c_k \in \{a_k, b_k\}$,

$$\mu(]a_1\,,b_1] \times \cdots \times]a_l\,,b_l] \times]a_{l+1}\,,c_{l+1}] \times \cdots \times]a_k\,,c_k]) =$$
$$\sum_{J \subset \{1,\ldots,l\}} (-1)^{|J|} \cdot \mu(]\alpha_1\,,c_{J,1}] \times \cdots \times]\alpha_k\,,c_{J,k}]),$$

where

$c_{J,i} = a_i$ for all $1 \leq i \leq l$ belonging to J,
$c_{J,i} = b_i$ for all $1 \leq i \leq l$ not belonging to J, and
$c_{J,i} = c_i$ for $l + 1 \leq i \leq k$.

In particular,

$$\mu(]a_1\,,b_1] \times \cdots \times]a_k\,,b_k]) =$$
$$\sum_{J \subset \{1,\ldots,k\}} (-1)^{|J|} \cdot \mu(]\alpha_1\,,c_{J,1}] \times \cdots \times]\alpha_k\,,c_{J,k}]),$$

where $c_{J,i} = a_i$ or $c_{J,i} = b_i$ as i belongs to J or not. Similarly, for all $n \geq 1$,

$$\mu(]a_1^n\,,b_1] \times \cdots \times]a_k^n\,,b_k]) =$$
$$\sum_{J \subset \{1,\ldots,k\}} (-1)^{|J|} \cdot \mu(]\alpha_1\,,c_{J,1}^n] \times \cdots \times]\alpha_k\,,c_{J,k}^n]),$$

where $c_{J,i}^n = a_i^n$ or $c_{J,i}^n = b_i$ as i belongs to J or not. Now, for every $J \subset \{1, \ldots, k\}$, the sequence $(c_{J,i}^n)_{n \geq 1}$ decreases to $c_{J,i}$, for all $1 \leq i \leq k$, so $\mu(]\alpha_1\,,c_{J,1}^n] \times \cdots \times]\alpha_k\,,c_{J,k}^n])$ tends to $\mu(]\alpha_1\,,c_{J,1}] \times \cdots \times]\alpha_k\,,c_{J,k}])$ as $n \to +\infty$. $\square$

Lemma 14.1.2 *Let $A = \prod_{1 \leq i \leq k}]a_i\,,b_i]$ be an S-rectangle, and let $(A_n)_{n \geq 1}$ be a decreasing sequence of S-rectangles of the form $\prod_{1 \leq i \leq k}]a_i\,,b_i^n]$ such that $A = \bigcap_{n \geq 1} A_n$. Then $V\mu(A_n)$ tends to $V\mu(A)$ as $n \to +\infty$.*

PROOF: For $\varepsilon > 0$, let $(B_p)_{1 \leq p \leq M}$ be a finite partition of A_1 into S-rectangles such that $V\mu(A_1) \leq \sum_{1 \leq p \leq M} |\mu(B_p)| + \varepsilon/3$. We may suppose that $A \cap B_p = \emptyset$ for all $1 \leq p \leq q$ and that $A \cap B_p \neq \emptyset$ for all $q+1 \leq p \leq M$, for a suitable $0 \leq q \leq M - 1$. For each $1 \leq p \leq q$, B_p has the form $\prod_{1 \leq i \leq k}]c_i, d_i]$, where $a_i \leq c_i < d_i \leq b_i^1$ for all $1 \leq i \leq k$. By Lemma 14.1.1, there is an S-rectangle B_p' of the form $\prod_{1 \leq i \leq k}]c_i', d_i]$, with $c_i' > c_i$ for all i, such that $|\mu(B_p)| \leq |\mu(B_p')| + \varepsilon/3q$. Since $A \cap B_p = \emptyset$, there exists $1 \leq i \leq k$ such that $c_i \geq b_i$, and, for n large enough, we have $b_i^n \leq c_i'$, hence $A_n \cap B_p' = \emptyset$.

Therefore, if we choose n large enough so that $A_n \cap B_p' = \emptyset$ for all $1 \leq p \leq q$, and let $(A_r')_{1 \leq r \leq N}$ (where $N \geq 0$) be a partition of $A_1 \cap A_n^c$ into S-rectangles, we see that

$$V\mu(A_n) + \sum_{1 \leq r \leq N} V\mu(A_r') = V\mu(A_1)$$

$$\leq \sum_{1 \leq p \leq q} |\mu(B_p)| + \sum_{q+1 \leq p \leq M} |\mu(B_p)| + \frac{\varepsilon}{3}$$

$$\leq \sum_{1 \leq p \leq q} |\mu(B_p')| + \sum_{q+1 \leq p \leq M} |\mu(B_p)| + \frac{2\varepsilon}{3}.$$

But $|\mu(B_p')| \leq \sum_{1 \leq r \leq N} |\mu(B_p' \cap A_r')|$ for all $1 \leq p \leq q$, because B_p' does not meet A_n, and

$$|\mu(B_p)| \leq |\mu(B_p \cap A_n)| + \sum_{1 \leq r \leq N} |\mu(B_p \cap A_r')|.$$

Hence

$$V\mu(A_n) + \sum_{1 \leq r \leq N} V\mu(A_r') \leq$$

$$\sum_{1 \leq r \leq N} \left[\sum_{1 \leq p \leq q} |\mu(B_p' \cap A_r')| + \sum_{q+1 \leq p \leq M} |\mu(B_p \cap A_r')| \right]$$

$$+ \sum_{q+1 \leq p \leq M} |\mu(B_p \cap A_n)| + \frac{2\varepsilon}{3},$$

and

$$V\mu(A_n) + \sum_{1 \leq r \leq N} V\mu(A_r') \leq \sum_{1 \leq r \leq N} V\mu(A_r') + \sum_{q+1 \leq p \leq M} |\mu(B_p \cap A_n)| + \frac{2\varepsilon}{3}.$$

It follows that

$$V\mu(A_n) \leq \sum_{q+1 \leq p \leq M} |\mu(B_p \cap A_n)| + \frac{2\varepsilon}{3}.$$

Now, for every $q+1 \leq p \leq M$, the rectangle $B_p \cap A_n = \prod_{1 \leq i \leq k}]c_i, \inf(d_i, b_i^n)]$ decreases to $B_p \cap A = \prod_{1 \leq i \leq k}]c_i, \inf(d_i, b_i)]$ as $n \to +\infty$, so $\mu(B_p \cap A_n)$ tends to $\mu(B_p \cap A)$. Then, for n large enough,

$$V\mu(A_n) \leq \sum_{q+1 \leq p \leq M} |\mu(B_p \cap A)| + \varepsilon,$$

and

$$V\mu(A_n) \leq V\mu(A) + \varepsilon.$$

$\square$

Lemma 14.1.3 *Let A be an S-rectangle and $(A_n)_{n\geq 1}$ a partition of A into S-rectangles. Then $V\mu(A) = \sum_{n\geq 1} V\mu(A_n)$.*

PROOF: Since μ is a quasi-measure, we know that $\sum_{n\geq 1} V\mu(A_n) \leq V\mu(A)$. Let $\varepsilon > 0$ be given. Applying Lemma 14.1.1 to $V\mu$, we see that there is an S-rectangle A' such that $\overline{A'} \subset A$ and $V\mu(A) \leq V\mu(A') + \varepsilon/2$. For each $n \geq 1$, there exists an S-rectangle A'_n such that $A_n \subset \overset{\circ}{A'}_n$ and $V\mu(A'_n) \leq V\mu(A_n) + \varepsilon/2^{n+1}$. Since $\overline{A'} \subset \bigcup_{n\geq 1} \overset{\circ}{A'}_n$, for n large enough, we have $\overline{A'} \subset \bigcup_{1\leq m\leq n} \overset{\circ}{A'}_m \subset \bigcup_{1\leq m\leq n} A'_m$, which now shows that

$$V\mu(A) \leq V\mu(A') + \frac{\varepsilon}{2} \leq \sum_{1\leq m\leq n} V\mu(A'_m) + \frac{\varepsilon}{2} \leq \sum_{1\leq m\leq n} V\mu(A_m) + \varepsilon,$$

because

$$V\mu(A') \leq \sum_{1\leq m\leq n} V\mu(A'_m \cap A').$$

Hence

$$V\mu(A) \leq \sum_{n\geq 1} V\mu(A_n) + \varepsilon.$$

$\square$

Lemma 14.1.4 $\mu(A) = \sum_{n\geq 1} \mu(A_n)$ *under the hypothesis of Lemma 14.1.3.*

PROOF: If $P(A)$ is a finite partition of A into S-rectangles, containing $A_1, \ldots, A_n$, then, for all $n \geq 1$,

$$\left| \mu(A) - \sum_{1\leq m\leq n} \mu(A_m) \right| \leq \sum_{\substack{B\in P(A) \\ B\neq A_1,\ldots,B\neq A_n}} |\mu(B)| \leq \sum_B V\mu(B)$$

is less than

$$V\mu(A) - \sum_{1\leq m\leq n} V\mu(A_m).$$

The lemma follows easily.

$\square$

Combining the previous four lemmas, we obtain the following proposition.

Proposition 14.1.1 *Let μ be a quasi-measure on the natural semiring S of Ω. μ is a measure if and only if $\mu(A_n)$ tends to $\mu(A)$ as $n \to +\infty$, for every S-rectangle $A = \prod_{1 \le i \le k}]a_i, b_i]$ and every decreasing sequence $(A_n)_{n \ge 1}$ of S-rectangles of the form $A_n = \prod_{1 \le i \le k}]a_i, b_i^n]$, such that $A = \bigcap_{n \ge 1} A_n$.*

If $A = \prod_{1 \le i \le k}]a_i, b_i]$ is a rectangle in $\mathbf{R}^k$, for any subset J of $\{1, \ldots, k\}$ put $c_J^A = (c_{J,i}^A)_{1 \le i \le k}$, where $c_{J,i}^A = a_i$ if i belongs to J, and $c_{J,i}^A = b_i$ if i does not belong to J. Then, for any vertex x of A, there is a unique $J \subset \{1, \ldots, k\}$ for which $x = c_J^A$. $(-1)^{|J|}$ is called the signum of x, and is written $\operatorname{sgn}_A x$.

Let Ω and S be as before, and let F be a function from Ω into $\mathbf{C}$. For every S-rectangle A, put

$$\mu(A) = \sum_{J \subset \{1, \ldots, k\}} (-1)^{|J|} \cdot F(c_J^A) = \sum_x (\operatorname{sgn}_A x) \cdot F(x),$$

where the last sum extends over the vertices of A: put $\mu(\emptyset) = 0$.

Proposition 14.1.2 *The function $\mu : A \mapsto \mu(A)$ from S into $\mathbf{C}$ is additive.*

PROOF: Suppose that each side $]a_i, b_i] = I_i$ of an S-rectangle A is partitioned into n_i subintervals $J_{i,j} =]t_{i,j-1}, t_{i,j}]$ ($1 \le j \le n_i$). where $a_i = t_{i,0} < t_{i,1} < \ldots < t_{i,n_i} = b_i$. Then the $n_1 n_2 \cdots n_k$ rectangles

$$B_{j_1, \ldots, j_k} = J_{1,j_1} \times \cdots \times J_{k,j_k} \quad (1 \le j_1 \le n_1, \ldots, 1 \le j_k \le n_k) \quad (1)$$

partition A. Such a partition is called regular. We first prove that μ is additive for regular partitions:

$$\mu(A) = \sum_{(j_1, \ldots, j_k)} \mu(B_{j_1, \ldots, j_k}) \quad (2).$$

The right-hand side of (2) is $\sum_B \sum_x \operatorname{sgn}_B x \cdot F(x)$, where the outer sum extends over the rectangles B of the form (1) and the inner sum extends over the vertices of B. Now

$$\sum_B \sum_x \operatorname{sgn}_B x \cdot F(x) = \sum_x F(x) \cdot \sum_B \operatorname{sgn}_B x \quad (3),$$

where, on the right-hand side, the outer sum extends over each x that is a vertex of one or more of the rectangles B, and. for fixed x, the inner sum extends over all B of which x is a vertex. Now suppose that x is a vertex of one or more of the rectangles B, but is not a vertex of A. Then there must be an l ($1 \le l \le k$) such that x_l is neither a_l nor b_l. Fix one l (there may be several such l). Then $x_l = t_{l,j}$ with $0 < j < n_l$. The rectangles (1) of which x is a vertex therefore occur in pairs $B' = B_{j_1, \ldots, j_{l-1}, j, j_{l+1}, \ldots, j_k}$ and $B'' = B_{j_1, \ldots, j_{l-1}, j+1, j_{l+1}, \ldots, j_k}$, and $\operatorname{sgn}_{B'} x = -\operatorname{sgn}_{B''} x$. Thus the inner sum on the right-hand side in (3) is 0 if x is not a vertex of A.

On the other hand, if x is a vertex of A, then, for each i, either $x_i = a_i = t_{i,0}$ or $x_i = b_i = t_{i,n_i}$. In this case, x is a vertex of only one B of the form (1)—the one for which $j_i = 1$ or $j_i = n_i$ as $x_i = a_i$ or $x_i = b_i$—and $\mathrm{sgn}_B x = \mathrm{sgn}_A x$. Thus the right-hand side of (3) reduces to $\mu(A)$, which proves (2).

Now suppose that $A = \bigcup_{1 \le u \le n} A_u$, where $A = I_1 \times \cdots \times I_k$ is an S-rectangle, $A_u = I_{1,u} \times \cdots \times I_{k,u}$ for $1 \le u \le n$, and the A_u are disjoint. For each i $(1 \le i \le k)$, the intervals $I_{i,1}, \ldots, I_{i,n}$ have I_i as their union, although they need not be disjoint. In any case, their endpoints split I_i into disjoint subintervals $J_{i,1}, \ldots, J_{i,n_i}$ such that each $I_{i,u}$ is the union of some $J_{i,j}$. The rectangles of the form (1) are a regular partition of A, as before; furthermore, the rectangles B contained in a single A_u form a regular partition of A_u. Since the A_u are disjoint, it follows by (2) that

$$\mu(A) = \sum_B \mu(B) = \sum_{1 \le u \le n} \sum_{B \subset A_u} \mu(B) = \sum_{1 \le u \le n} \mu(A_u).$$

$\square$

Definition 14.1.1 F is said to be right-continuous at a point x of Ω when $F(y)$ converges to $F(x)$ as y tends to $x = (x_1, \ldots, x_k)$ in $\Omega \cap \left([x_1, +\infty[\times \cdots \times [x_k, +\infty[\right)$.

Proposition 14.1.3 *Assume that the aforementioned function μ is a quasi-measure. Then μ is a measure if F is right-continuous.*

PROOF: Let $A = \prod_{1 \le i \le k}]a_i, b_i]$ be an S-rectangle and let $(A_n)_{n \ge 1}$ be a decreasing sequence of S-rectangles of the form $A_n = \prod_{1 \le i \le k}]a_i, b_i^n]$, such that $A = \bigcap_{n \ge 1} A_n$. Then $\mu(A_n) = \sum_{J \subset \{1, \ldots, k\}} (-1)^{|J|} \cdot F(c_J^{A_n})$ converges to $\mu(A)$ as $n \to +\infty$. By Proposition 14.1.1, μ is a measure. $\square$

In particular, taking for F the function $x = (x_1, \ldots, x_k) \mapsto x_1 \cdots x_k$, we obtain on S a measure λ_k satisfying

$$\lambda_k \left(\prod_{1 \le i \le k}]a_i, b_i] \right) = \prod_{1 \le i \le k} (b_i - a_i)$$

for all S-rectangles, that is, λ_k is Lebesgue measure on Ω.

Now we study vague convergence of measures on S.

For all $1 \le i \le k$, write p_i for the function $(x_1, \ldots, x_k) \mapsto x_i$ on Ω.

Proposition 14.1.4 *Let $(\mu_n)_{n \ge 1}$ be a sequence in $\mathcal{M}^+(S)$ and μ a positive measure on S. If $(\mu_n)_{n \ge 1}$ converges vaguely to μ, then $\mu_n(A)$ tends to $\mu(A)$ as $n \to +\infty$, for all μ-quadrable S-rectangles A. Conversely, if $(\mu_n(A))_{n \ge 1}$ converges to $\mu(A)$ for every S-rectangle $A = \prod_{1 \le i \le k}]a_i, b_i]$ such that $\Omega \cap p_i^{-1}(a_i)$ and $\Omega \cap p_i^{-1}(b_i)$ are μ-negligible for all $1 \le i \le k$, then $(\mu_n)_{n \ge 1}$ converges vaguely to μ.*

PROOF: The first assertion follows from Proposition 7.7.2. For the second claim, observe that, for every $1 \leq i \leq k$, the set of those $x \in \mathbf{R}$ such that $\Omega \cap p_i^{-1}(x)$ is not μ-negligible is at most countable. Denote by E_i its complement in $\mathbf{R}$.

Let C be the collection of those S-rectangles $\prod_{1 \leq i \leq k}]a_i, b_i]$ such that a_i, b_i lie in E_i for all $1 \leq i \leq k$. Now let $A_1, \ldots, A_p$ ($A_q = \prod_{1 \leq i \leq k}]a_i^q, b_i^q]$ for $1 \leq q \leq p$) be C-sets included in a same C-set $A = \prod_{1 \leq i \leq k}]a_i, b_i]$. For fixed $1 \leq i \leq k$, write $c_{i,j}$ $(0 \leq j \leq m_i)$ for the a_i, b_i, a_i^q, b_i^q arranged in increasing order. Then the S-rectangles $\prod_{1 \leq i \leq k}]c_{i,j_i-1}, c_{i,j_i}]$ $(1 \leq j_i \leq m_i)$ form a partition $P(A)$ of A, and each A_q $(1 \leq q \leq p)$ is a union of members of $P(A)$.

Consequently, for all C-sets $A_1, \ldots, A_p$ and A.

$$A \cap \left(\bigcup_{1 \leq q \leq p} A_q \right)^c = A \cap \left(\bigcup_{1 \leq q \leq p} A \cap A_q \right)^c$$

is either empty or a disjoint union of C-sets. We may conclude that $C \cup \{\emptyset\}$ is a semiring.

Next, let $\varphi : \Omega \to \mathbf{C}$ be continuous, with compact support. There exists a finite disjoint subcollection $\mathcal{D}$ of C such that $\text{supp}(\varphi) \subset \bigcup_{B \in \mathcal{D}} B$. Pick $B = \prod_{1 \leq i \leq k}]a_i, b_i]$ in $\mathcal{D}$. By hypothesis, $\mu_n(B)$ converges to $\mu(B)$ as $n \to +\infty$.

Define on $\mathbf{R}^k$ the norm $\|x\| = \sup_{1 \leq i \leq k} |x_i|$. Given $\varepsilon > 0$, there exists $\delta > 0$ such that $|\varphi(x) - \varphi(y)| \leq \varepsilon$ for all $x, y \in \overline{B}$ satisfying $\|x - y\| \leq \delta$. For all $1 \leq i \leq k$, there is a subdivision $(c_{i,j})_{0 \leq j_i \leq m_i}$ of $[a_i, b_i]$ such that the c_{i,j_i} belong to E_i and $|c_{i,j_i} - c_{i,j_i-1}| \leq \delta$ for $1 \leq j_i \leq m_i$. For each $j = (j_1, \ldots, j_k)$ in $\prod_{1 \leq i \leq k} \{1, \ldots, m_i\}$, let B_j be the rectangle $\prod_{1 \leq i \leq k}]c_{i,j_i-1}, c_{i,j_i}]$, and let s_j be one of the values taken by φ on B_j. so that $|\varphi \cdot 1_B - \sum_j s_j \cdot 1_{B_j}| \leq \varepsilon$. For every j, the sequence $(\mu_n(B_j))_{n \geq 1}$ converges to $\mu(B_j)$. Hence there is an integer $N \geq 1$ such that

$$\left| \int \sum_j s_j \cdot 1_{B_j} \, d\mu_n - \int \sum_j s_j \cdot 1_{B_j} \, d\mu \right| \leq \varepsilon \qquad \text{for all } n \geq N.$$

Since

$$\left| \int \left(\varphi \cdot 1_B - \sum_j s_j \cdot 1_{B_j} \right) d\mu_n \right| \leq \varepsilon \cdot$$

and

$$\left| \int \left(\varphi \cdot 1_B - \sum_j s_j \cdot 1_{B_j} \right) d\mu \right| \leq \varepsilon r,$$

where $r = \sup_{n \geq 1} \mu_n(B)$, we have

$$\left| \int \varphi \cdot 1_B \, d\mu_n - \int \varphi \cdot 1_B \, d\mu \right| \leq (2r + 1)\varepsilon \qquad \text{for all } n \geq N.$$

Therefore, $\int \varphi \cdot 1_B \, d\mu_n$ converges to $\int \varphi \cdot 1_B \, d\mu$ as $n \to +\infty$, whence we see that $\int \varphi \, d\mu_n = \sum_{B \in \mathcal{D}} \int \varphi \cdot 1_B \, d\mu_n$ converges to $\int \varphi \, d\mu$. $\qquad\square$

14.2   Distribution Functions

Let $\mathcal{B}$ be the Borel σ-algebra of $\mathbf{R}^k$ $(k \geq 1)$ and S the natural semiring of $\mathbf{R}^k$.

A positive measure P on $\mathcal{B}$ such that $P(\mathbf{R}^k) = 1$ is called a probability on $\mathcal{B}$. If μ is a positive measure on S such that $\mu^*(\mathbf{R}^k) = 1$, there is a unique probability P on $\mathcal{B}$ such that $P(A) = \mu(A)$ for all $A \in S$ (Section 6.4).

Definition 14.2.1 Let P be a probability on $\mathcal{B}$. The function

$$F : (x_1, \ldots, x_k) \mapsto P\Big(\prod_{1 \leq i \leq k} \,]-\infty, x_i] \Big)$$

from $\mathbf{R}^k$ into $[0, 1]$ is called the distribution function of P.

If $A = \prod_{1 \leq i \leq k}]a_i, b_i]$ is a nonempty S-rectangle and J a subset of $\{1, 2, \ldots, k\}$, we call c_J^A the vertex of A whose ith coordinate is a_i when i belongs to J and is b_i when i does not belong to J.

Proposition 14.2.1 *Let F be a function from $\mathbf{R}^k$ into $[0, 1]$. Then F is the distribution function of a probability P if and only if it has the following properties:*

(a) $\Delta_A F = \sum_{J \subset \{1, \ldots, k\}} (-1)^{|J|} \cdot F(c_J^A)$ *is positive for all S-rectangles.*

(b) *F is right-continuous.*

(c) *For every $1 \leq i \leq k$ and for every $(x_1, \ldots, \hat{x}_i, \ldots, x_k)$ in $\mathbf{R}^{k-1}$, $F(x_1, \ldots, y_i, \ldots, x_k)$ converges to 0 as $y_i \to -\infty$.*

(d) *$F(x_1, \ldots, x_k)$ tends to 1 as $x_1, \ldots, x_k$ all tend to $+\infty$.*

In this case, $P(A) = \Delta_A F$ for all S-rectangles A.

PROOF: First, suppose that F is the distribution function of a probability P, and let $A = \prod_{1 \leq i \leq k}]a_i, b_i]$ be an S-rectangle. By induction on $0 \leq l \leq k$, we have

$$P(]a_1, b_1] \times \cdots \times]a_l, b_l] \times \,]-\infty, c_{l+1}] \times \cdots \times\,]-\infty, c_k]) = \sum_{J \subset \{1, \ldots, l\}} (-1)^{|J|} \cdot F(c_J^l)$$

for all $c_{l+1} \in \{a_{l+1}, b_{l+1}\}, \ldots, c_k \in \{a_k, b_k\}$, where c_J^l is defined as follows: $c_{J,i}^l = a_i$ for $i \in J$, $c_{J,i}^l = b_i$ for $1 \leq i \leq l$, $i \notin J$, and $c_{J,i}^l = c_i$ for $l+1 \leq$

$i \leq k$. In particular, when $l = k$, we obtain $P(A) = \Delta_A F$. By the dominated convergence theorem, we see that conditions (b), (c), and (d) hold.

Conversely, assume that conditions (a), (b), (c), and (d) are satisfied, and define a measure μ on S by $\mu(A) = \Delta_A F$ for all nonempty S-rectangles A (Proposition 14.1.3). Fix $b = (b_1, \ldots, b_k) \in \mathbf{R}^k$. For every integer $0 \leq l \leq k$, every $(a_1, \ldots, a_l) \in \mathbf{R}^l$ such that $a_i < b_i$ for all $1 \leq i \leq l$, and every subset J of $\{1, \ldots, l\}$, let $c_J^{(a_1, \ldots, a_l)}$ be the point in $\mathbf{R}^k$ whose ith coordinate is a_i when i belongs to J and is b_i when i does not belong to J. By induction on $l - k$, we see that

$$\mu^*(]a_1, b_1] \times \cdots \times]a_l, b_l] \times] - \infty, b_{l+1}] \times \cdots \times] - \infty, b_k]) =$$
$$\sum_{J \subset \{1, \ldots, l\}} (-1)^{|J|} \cdot F(c_J^{(a_1, \ldots, a_l)}).$$

Therefore, $\mu^*(] - \infty, b_1] \times \cdots \times] - \infty, b_k]) = F(b_1, \ldots, b_k)$. Now, letting $b_1, \ldots, b_k$ all approach $+\infty$, we see that $\|\mu\| = 1$. □

If F is the distribution function of a probability P, a necessary and sufficient condition that F be continuous at $(x_1, \ldots, x_k)$ in $\mathbf{R}$ is that $] - \infty, x_1] \times \cdots \times] - \infty, x_k]$ have P-negligible boundary. For every $1 \leq i \leq k$, the complement in $\mathbf{R}$ of the set E_i of points x such that $p_i^{-1}(x)$ is P-negligible is at most countable. Therefore, F has a dense set of points of continuity. Nevertheless, when $k > 1$, F may be discontinuous at uncountably many points.

Proposition 14.2.2 *Let $(P_n)_{n \geq 1}$ be a sequence of probabilities on $\mathcal{B}$, and let P be a probability on $\mathcal{B}$. Call F_n the distribution function of P_n, for all $n \in \mathbf{N}$, and F the distribution function of P. Then $(P_n)_{n \geq 1}$ converges weakly to P if and only if $(F_n(x))_{n \geq 1}$ converges to $F(x)$ for every point of continuity, x, of F.*

PROOF: By Theorem 7.7.2, the condition is necessary. Conversely, suppose it is satisfied. If $A = \prod_{1 \leq i \leq k}]a_i, b_i]$ is an S-rectangle such that $p_i^{-1}(a_i)$ and $p_i^{-1}(b_i)$ are μ-negligible for all $1 \leq i \leq k$, we can express $P_n(A)$ in terms of F_n, and it follows that $(P_n(A))_{n \geq 1}$ converges to $P(A)$. Proposition 14.1.4 then shows that $(P_n)_{n \geq 1}$ converges weakly to P. □

14.3 Covariance Matrix

Equip $\mathbf{R}^k$ ($k \in \mathbf{N}$) with its usual scalar product $(x, y) \mapsto xy = \sum_{1 \leq i \leq k} x_i y_i$ and its norm $x \mapsto |x| = (xx)^{1/2}$.

Let μ be a positive measure on the natural semiring $S = S_k$ of $\mathbf{R}^k$, such that $\int^* 1 \, d\mu = 1$. Given $n \in \mathbf{N}$, μ is said to be of order n whenever $\int^* |x|^n \, d\mu(x)$

is finite. In this case, μ is of order m, for each $1 \leq m \leq n$, because $|x|^m \leq \sup(1, |x|^n)$ for all $x \in \mathbf{R}^k$.

If μ is of order 1, $E(\mu) = \int \mathrm{id}_{\mathbf{R}^k}\, d\mu$ is called the mean of μ.

In what follows, we suppose that μ is of order 2, and let $m = (m_1, \ldots, m_k)$ be its mean. The bilinear form

$$(y, z) \longmapsto \int [y(x - m)] \cdot [z(x - m)]\, d\mu(x)$$

on $\mathbf{R}^k \times \mathbf{R}^k$ is symmetric and positive, and its matrix (with respect to the canonical basis of $\mathbf{R}^k$) is called the covariance matrix of μ and is written $D(\mu)$. $D(\mu)$ has entries $c_{i,j} = \int (x_i - m_i)(x_j - m_j)\, d\mu(x)$.

If $u : \mathbf{R}^k \to \mathbf{R}^l$ is linear, write $u(\mu)$ for the measure on S_l, the image of μ under u, and A for the matrix of u (with respect to the canonical bases of $\mathbf{R}^k$ and $\mathbf{R}^l$). Then $D(u(\mu)) = A \cdot D(\mu) \cdot A'$, where A' is the transpose of A. Indeed, for all $1 \leq p, q \leq l$, let $c'_{p,q}$ be the element of $A \cdot D(\mu) \cdot A'$ situated in the pth row and the qth column; if $A = (\alpha_{p,i})_{p,i}$,

$$
\begin{aligned}
c'_{p,q} &= \sum_{1 \leq i,j \leq k} \alpha_{p,i} \cdot c_{i,j} \cdot \alpha_{q,j} \\
&= \sum_{i,j} \alpha_{p,i} \cdot \alpha_{q,j} \int (x_i - m_i)(x_j - m_j)\, d\mu(x) \\
&= \int [y_p - u(m)_p] \cdot [y_q - u(m)_q]\, d(u\mu)(y).
\end{aligned}
$$

Now, when $k = 1$, $\int (x - E(\mu))^2\, d\mu(x) = \int x^2\, d\mu(x) - E(\mu)^2$ is called the variance of μ, and is written $\mathrm{Var}(\mu)$. $\mathrm{Var}(\mu)^{1/2}$ is the standard deviation of μ.

Definition 14.3.1 For every $m \in \mathbf{R}$ and for every $\sigma > 0$,

$$\mu = \frac{1}{\sigma\sqrt{2\pi}} \exp\left(-\frac{(x - m)^2}{2\sigma^2}\right) dx$$

is a positive measure on S_1 such that $\int^* 1\, d\mu = 1$. It is called the normal measure with parameters m and σ and is written $N_1(m, \sigma^2)$. In particular, $N_1(0, 1)$ is called the standard normal measure on $\mathbf{R}$.

For every $m \in \mathbf{R}$, we denote by $N_1(m, 0)$ the measure on S_1 defined by the unit mass at m.

The measures $N_1(m, \sigma^2)$ $(m \in \mathbf{R}, \sigma \geq 0)$ are called the normal measures on $\mathbf{R}$. If $u : x \mapsto ax + b$ is affine,

$$u[N_1(m, \sigma^2)] = N_1(am + b, a^2\sigma^2).$$

In particular, $N_1(m, \sigma^2)$ is the image of $N_1(0, 1)$ under $x \longmapsto m + \sigma x$. $N_1(m, \sigma^2)$ is of order n, for all $n \geq 1$, with mean m and variance σ^2.

Let $m \in \mathbf{R}$. For every bounded continuous function $f : \mathbf{R} \to \mathbf{C}$,

$$\int f \, dN_1(m, \sigma^2) = \int f(m + \sigma y) \cdot \frac{1}{\sqrt{2\pi}} \cdot \exp\left(-\frac{y^2}{2}\right) dy$$

converges to $f(m)$ as $\sigma > 0$ approaches 0. Therefore, $N_1(m, \sigma^2)$ converges narrowly to $N_1(m, 0)$ as $\sigma > 0$ approaches 0.

Now let

$$\varphi : \mathbf{R} \ni x \longmapsto \int e^{ixy} \, d[N_1(0.1)](y) = \frac{1}{\sqrt{2\pi}} \int e^{ixy} \cdot \exp\left(-\frac{y^2}{2}\right) dy.$$

A differentiation under the integral sign shows that

$$\begin{aligned}
\varphi'(x) &= \frac{i}{\sqrt{2\pi}} \cdot \int e^{ixy} \left[y \cdot \exp\left(-\frac{y^2}{2}\right)\right] dy \\
&= -\frac{i}{\sqrt{2\pi}} \cdot \int e^{ixy} \cdot \left[\exp\left(-\frac{y^2}{2}\right)\right]' dy \\
&= -x\varphi(x),
\end{aligned}$$

so

$$\varphi(x) = \exp(-x^2/2).$$

Finally, let $m \in \mathbf{R}$ and $\sigma \in \,]0, +\infty[$, and write ψ for the function

$$x \longmapsto \int e^{ixy} \, d[N_1(m, \sigma^2)](y).$$

Since $N_1(m, \sigma^2)$ is the image of $N_1(0, 1)$ under $u : x \longmapsto m + \sigma x$, we see that

$$\psi(x) = e^{imx} \cdot \exp\left(-\frac{\sigma^2 x^2}{2}\right) \qquad \text{for all } x \in \mathbf{R}.$$

14.4   The Fourier Transform

Definition 14.4.1 Let μ be a bounded measure on the natural semiring S of $\mathbf{R}^k$. $\mathcal{F}\mu : x \longmapsto \int e^{-2i\pi xy} \, d\mu(y)$ and $\overline{\mathcal{F}}\mu : x \longmapsto \int e^{2i\pi xy} \, d\mu(y)$ are called the Fourier transform and the inverse Fourier transform of μ, respectively.

Let $C^b(\mathbf{R}^k, \mathbf{C})$ be the normed space of bounded continuous functions from $\mathbf{R}^k$ into $\mathbf{C}$.

Proposition 14.4.1 *For all $\mu \in \mathcal{M}^1(S, \mathbf{C})$, $\mathcal{F}\mu$ is uniformly continuous and bounded. The linear map $\mu \longmapsto \mathcal{F}\mu$ from $\mathcal{M}^1(S, \mathbf{C})$ into $C^b(\mathbf{R}^k, \mathbf{C})$ is continuous.*

PROOF: By the dominated convergence theorem, $\int \inf(2, \delta|z|) \, dV\mu(z)$ tends to 0 as $\delta > 0$ approaches 0. Given $\varepsilon > 0$, there thus exists $\delta > 0$ such that $\int \inf(2, \delta|z|) \, dV\mu(z) \leq \varepsilon$. Now, if x, $y \in \mathbf{R}^k$ satisfy $|x - y| \leq \delta/2\pi$, then $|\mathcal{F}\mu(x) - \mathcal{F}\mu(y)| \leq \varepsilon$, because $|e^{iu} - 1| = |\int_0^u ie^{it} \, dt| \leq |u|$ for all $u \in \mathbf{R}$. Therefore, $\mathcal{F}\mu$ is uniformly continuous.

Next, $|\mathcal{F}\mu(x)| \leq \int 1 \, dV\mu$ for all $x \in \mathbf{R}^k$, which proves the second assertion. $\qquad\square$

Let λ, or dx, be Lebesgue measure on $\mathbf{R}^k$. For all $g \in L^1_{\mathbf{C}}(dx)$, the functions $\mathcal{F}g = \mathcal{F}(g \, dx)$ and $\overline{\mathcal{F}}g = \overline{\mathcal{F}}(g \, dx)$ are called the Fourier transform and the inverse Fourier transform of g.

For example, $g_1 : y \mapsto \exp(-\pi|y|^2)$ is dy-integrable, $\int g_1(y) \, dy = 1$, and

$$
\begin{aligned}
\mathcal{F}g_1(x) &= \prod_{1 \leq j \leq k} \int e^{-2i\pi x_j y_j} \cdot \exp(-\pi y_j^2) \, dy_j \\
&= \prod_{1 \leq j \leq k} \exp(-\pi x_j^2) \\
&= g_1(x)
\end{aligned}
$$

for all x in $\mathbf{R}^k$ (by Section 14.3), so $\mathcal{F}g_1 = g_1$. For each $\sigma > 0$, define $g_\sigma : x \mapsto \sigma^{-k} \cdot g_1(x/\sigma)$. Then $\mathcal{F}g_\sigma(x) = g_1(\sigma x)$ for all $x \in \mathbf{R}^k$, and $\overline{\mathcal{F}}(\mathcal{F}g_\sigma) = g_\sigma$.

Proposition 14.4.2 Let μ be a measure on S. Then $\mu \otimes \lambda$ is invariant under the homeomorphism $(x, y) \mapsto (x, y - x)$ of $\mathbf{R}^k \times \mathbf{R}^k$ onto $\mathbf{R}^k \times \mathbf{R}^k$.

PROOF: We may suppose that μ is positive. If $F : \mathbf{R}^k \times \mathbf{R}^k \to [0, +\infty[$ is continuous with compact support, then

$$
\begin{aligned}
\iint F(x, y - x) \, d\mu(x) dy &= \int d\mu(x) \int F(x, y - x) \, dy \\
&= \int d\mu(x) \int F(x, y) \, dy \\
&= \iint F(x, y) \, d\mu(x) dy.
\end{aligned}
$$

The proposition follows. $\qquad\square$

Now, let $\mu \in \mathcal{M}^1(S, \mathbf{C})$. For all $\sigma > 0$, define

$$
\mu * g_\sigma : y \mapsto \int g_\sigma(y - x) \, d\mu(x).
$$

By Fubini's theorem,

$$
\int \overline{\mathcal{F}}(\mathcal{F}g_\sigma)(y - x) \, d\mu(x) = \int d\mu(x) \int e^{2i\pi(y-x)z} \cdot \mathcal{F}g_\sigma(z) \, dz
$$

$$= \int dz\, e^{2i\pi yz} \cdot \mathcal{F}g_\sigma(z) \cdot \int e^{-2i\pi xz}\, d\mu(x)$$

$$= \int e^{2i\pi yz} \cdot \mathcal{F}g_\sigma(z) \cdot \mathcal{F}\mu(z)\, dz$$

for $y \in \mathbf{R}^k$. Thus $\mu * g_\sigma = \overline{\mathcal{F}}(\mathcal{F}g_\sigma \cdot \mathcal{F}\mu)$, and $\mu * g_\sigma$ is continuous and bounded. Also, $\mu * g_\sigma$ is dy-integrable and

$$\begin{aligned}
\int (\mu * g_\sigma)(y)\, dy &= \int dy \int g_\sigma(y - x)\, d\mu(x) \\
&= \int d\mu(x) \int g_\sigma(y - x)\, dy \\
&= \int d\mu(x) \int g_\sigma(y)\, dy \\
&= \int 1\, d\mu.
\end{aligned}$$

Now, for each $f \in C^b(\mathbf{R}^k, \mathbf{C})$,

$$\begin{aligned}
\int f(\mu * g_\sigma)\, dy &= \int dy \int f(y)g_\sigma(y - x)\, d\mu(x) \\
&= \int d\mu(x) \int f(y)g_\sigma(y - x)\, dy \\
&= \int d\mu(x) \int f(x + y)g_\sigma(y)\, dy \\
&= \int d\mu(x) \int f(x + y) \cdot \frac{1}{\sigma^k} \cdot g_1\left(\frac{y}{\sigma}\right) dy \\
&= \int d\mu(x) \int f(x + \sigma y)g_1(y)\, dy \\
&= \int f(x + \sigma y)g_1(y)\, d\mu(x)dy.
\end{aligned}$$

Hence, by the dominated convergence theorem, $\int f(\mu * g_\sigma)\, dy$ tends to $\int f(x)g_1(y)\, d\mu(x)dy = \int f\, d\mu$ as $\sigma > 0$ approaches 0. We conclude that $(\mu * g_\sigma)\lambda$ converges narrowly to μ. Since $\mu * g_\sigma = \overline{\mathcal{F}}(\mathcal{F}\gamma_\sigma \cdot \mathcal{F}\mu)$, we have the following theorem.

Theorem 14.4.1 *The mapping $\mathcal{F}: \mu \longmapsto \mathcal{F}\mu$ from $\mathcal{M}^1(S, \mathbf{C})$ into $C^b(\mathbf{R}^k, \mathbf{C})$ is injective.*

Up to Proposition 14.4.3, let $f \in \mathcal{L}^1_{\mathbf{C}}(\lambda)$ be given. For each $x \in \mathbf{R}^k$, denote by $\gamma(x)f$ the function $y \mapsto f(y - x)$ on $\mathbf{R}^k$.

Lemma 14.4.1 *The mapping $x \mapsto \gamma(x)f$ from $\mathbf{R}^k$ into $L^1_{\mathbf{C}}(\lambda)$ is continuous.*

PROOF: First, suppose that f is continuous with compact support K. Let V_0 be a compact neighborhood of 0. Since f is uniformly continuous on $\mathbf{R}^k$, given $\varepsilon > 0$ there exists a compact neighborhood $V \subset V_0$ of 0 such that

$$|f(y - x) - f(y)| \leq \frac{\varepsilon}{\lambda(K + V_0)} \qquad \text{for all } x \in V \text{ and all } y \in \mathbf{R}^k.$$

Then

$$N_1\big(\gamma(x)f - f\big) = \int |f(y - x) - f(y)| \, dy \leq \varepsilon \qquad \text{for all } x \in V.$$

Therefore, the mapping $x \mapsto \gamma(x)f$ is continuous at 0.

Now consider the general case, in which $f \in \mathcal{L}^1_{\mathbf{C}}(\lambda)$. If $(f_n)_{n \geq 1}$ is a sequence in $\mathcal{H}(\mathbf{R}^k, \mathbf{C})$ converging to f in the mean, the relation

$$N_1\big(\gamma(x)f_n - \gamma(x)f\big) = N_1(f_n - f)$$

shows that the sequence of functions $x \mapsto \gamma(x)f_n$ converges uniformly on $\mathbf{R}^k$ to $x \mapsto \gamma(x)f$. Thus $x \mapsto \gamma(x)f$ is continuous at 0. The proposition follows easily. $\qquad \square$

Remark that, for every compact neighborhood V of 0,

$$\int_{V^c} g_\sigma \, dy = \int_{V^c} \sigma^{-k} g_1\left(\frac{y}{\sigma}\right) dy = \int_{(V/\sigma)^c} g_1(y) \, dy$$

converges to 0 as $\sigma > 0$ approaches 0.

We know that $f * g_\sigma = (f \, dx) * g_\sigma$ is λ-integrable, for each $\sigma > 0$, and we observe that

$$\begin{aligned}
(f * g_\sigma)(y) &= \int g_\sigma(y - x) f(x) \, dx \\
&= \int g_\sigma(y + x) f(-x) \, dx \\
&= \int f(y - x) g_\sigma(x) \, dx.
\end{aligned}$$

Proposition 14.4.3 $f * g_\sigma$ *converges to* f *in* $L^1_{\mathbf{C}}(\lambda)$ *as* $\sigma > 0$ *approaches* 0.

PROOF: For all $\sigma > 0$,

$$N_1(f * g_\sigma - f) \leq \int^* dy \int |f(y - x) - f(y)| g_\sigma(x) \, dx.$$

Given $\varepsilon > 0$, let V be a compact neighborhood of 0 such that $N_1\big(\gamma(x)f - f\big) \leq \varepsilon$ for all $x \in V$ (Lemma 14.4.1). Choose $\sigma_0 > 0$ so that $\int_{V^c} g_\sigma \, dx \leq \varepsilon$ for all $\sigma \in \,]0, \sigma_0]$. Then, for all $\sigma \in \,]0, \sigma_0]$,

$$\int_V g_\sigma(x) \cdot N_1\big(\gamma(x)f - f\big) \, dx \leq \varepsilon$$

and

$$\int_{V^c} g_\sigma(x) \cdot N_1(\gamma(x)f - f)\, dx \leq 2N_1(f)\varepsilon.$$

Next, the function $(x,y) \mapsto f(y-x)$ is $(g_\sigma\lambda)\otimes\lambda$-integrable, because $(x,y) \mapsto f(y)$ is $(g_\sigma\lambda)\otimes\lambda$-integrable (Proposition 14.4.2). Furthermore,

$$\int dy \int |f(y-x) - f(y)|\, d(g_\sigma\lambda)(x) = \int d(g_\sigma\lambda)(x) \int |f(y-x) - f(y)|\, dy$$

$$= \int g_\sigma(x) \cdot N_1(\gamma(x)f - f)\, dx,$$

whence we see that $N_1(f * g_\sigma - f) \leq (1 + 2N_1(f))\varepsilon$ for $\sigma \in]0, \sigma_0]$. $\square$

Theorem 14.4.2 (Riemann-Lebesgue Lemma) *For all $f \in L^1_{\mathbf{C}}(\lambda)$, $\mathcal{F}f$ vanishes at infinity.*

PROOF: For all $\sigma > 0$, the function $(x,y) \mapsto f(x)g_\sigma(y-x)$ is $\lambda \otimes \lambda$-integrable, by Proposition 14.4.2. Therefore,

$$\mathcal{F}(f * g_\sigma)(z) = \int dy\, e^{-2i\pi yz} \int g_\sigma(y-x)f(x)\, dx$$

$$= \iint e^{-2i\pi xz} f(x) e^{-2i\pi(y-x)z} g_\sigma(y-x)\, dxdy$$

$$= \iint e^{-2i\pi xz} f(x) e^{-2i\pi yz} g_\sigma(y)\, dxdy$$

$$= \mathcal{F}f(z)\mathcal{F}g_\sigma(z)$$

for all $z \in \mathbf{R}^k$.

Now $\mathcal{F}(f * g_\sigma)$ converges uniformly to $\mathcal{F}f$ as $\sigma > 0$ approaches 0 (Propositions 14.4.1 and 14.4.3), and $\mathcal{F}g_\sigma$ vanishes at infinity. We may conclude that $\mathcal{F}f$ also vanishes at infinity. $\square$

Theorem 14.4.3 (Fourier Inversion Theorem) *Let $\mu \in \mathcal{M}^1(S, \mathbf{C})$. Suppose that $\mathcal{F}\mu$ is dx-integrable. Then $\mu = \overline{\mathcal{F}}(\mathcal{F}\mu)dx$.*

PROOF: Let $C^\circ(\mathbf{R}^k, \mathbf{C})$ be the normed space of continuous functions from $\mathbf{R}^k$ into $\mathbf{C}$ which vanish at infinity.

By the dominated convergence theorem, $\mathcal{F}g_\sigma \cdot \mathcal{F}\mu$ converges to $\mathcal{F}\mu$ in $L^1_{\mathbf{C}}(dx)$, as $\sigma > 0$ approaches 0 (because $\mathcal{F}g_\sigma(x) = g_1(\sigma(x))$). Therefore, $\mu * g_\sigma = \overline{\mathcal{F}}(\mathcal{F}g_\sigma \cdot \mathcal{F}\mu)$ converges to $\overline{\mathcal{F}}(\mathcal{F}\mu)$ in $C^\circ(\mathbf{R}^k, \mathbf{C})$, and $(\mu * g_\sigma)dx$ converges vaguely to $\overline{\mathcal{F}}(\mathcal{F}\mu)dx$. On the other hand, we know that $(\mu * g_\sigma)dx$ converges narrowly to μ. Hence $\mu = \overline{\mathcal{F}}(\mathcal{F}\mu)dx$. $\square$

For all $\sigma > 0$ and for all $y \in \mathbf{R}^k$, denote by $g_\sigma(y-)$ the function $x \mapsto g_\sigma(y-x)$ from $\mathbf{R}^k$ into $\mathbf{R}$. For all $\alpha \in \mathbf{C}$ and $y \in \mathbf{R}^k$, $\alpha g_\sigma(y-)$ is of the form $x \mapsto \beta \exp(-a|x|^2 + bx)$, with $\beta \in \mathbf{C}$, $a > 0$, and $b \in \mathbf{R}^k$. Conversely, for all $\beta \in \mathbf{C}$, $a > 0$, and $b \in \mathbf{R}^k$, the function $x \mapsto \beta \exp(-a|x|^2 + bx)$ can be written $\alpha g_\sigma(y-)$, with $\sigma = \sqrt{\pi/a}$, $y = b/2a$, and α suitably chosen. Therefore, the vector subspace of $\mathcal{C}^\circ(\mathbf{R}^k, \mathbf{C})$ generated by the $g_\sigma(y-)$ is an algebra A (without unit). Moreover, the conjugate $\bar{f}$ of any $f \in A$ belongs to A, and, for all x_1, $x_2 \in \mathbf{R}^k$, there exists $f \in A$ such that $f(x_1) \neq f(x_2)$. Hence we may consider the compact space $\mathbf{R}^k \cup \{\infty\}$, apply the Stone–Weierstrass theorem, and conclude that A is dense in $\mathcal{C}^\circ(\mathbf{R}^k, \mathbf{C})$.

The following theorem is fundamental in probability theory.

Theorem 14.4.4 Let $(\mu_n)_{n \geq 1}$ be a sequence of positive bounded measures on S, and let μ be a bounded measure on S. Then $(\mu_n)_{n \geq 1}$ converges narrowly to μ if and only if $(\mathcal{F}\mu_n)_{n \geq 1}$ converges pointwise to $\mathcal{F}\mu$. In this case, $(\mathcal{F}\mu_n)_{n \geq 1}$ converges to $\mathcal{F}\mu$ uniformly on compact sets.

PROOF: First, suppose that $(\mathcal{F}\mu_n)_{n \geq 1}$ converges pointwise to $\mathcal{F}\mu$. $\int 1 \, d\mu_n = \mathcal{F}\mu_n(0)$ converges to $\mathcal{F}\mu(0) = \int 1 \, d\mu$ as $n \to +\infty$, so $\sup_n \|\mu_n\|$ is finite. Now, by the dominated convergence theorem, for every $\sigma > 0$ and for every $y \in \mathbf{R}^k$, $(\mu_n * g_\sigma)(y) = \int e^{2i\pi xy} \exp(-\pi\sigma^2|x|^2) \cdot \mathcal{F}\mu_n(x) \, dx$ converges to

$$(\mu * g_\sigma)(y) = \int e^{2i\pi xy} \exp(-\pi\sigma^2|x|^2) \cdot \mathcal{F}\mu(x) \, dx \qquad \text{as } n \to +\infty.$$

Equivalently, $\mu_n[g_\sigma(y-)]$ tends to $\mu[g_\sigma(y-)]$. It follows that $(\mu_n)_{n \geq 1}$ converges weakly to μ (Proposition 7.7.3). Thus μ is positive. Since $(\mu_n(1))_{n \geq 1}$ converges to $\mu(1)$, we may conclude that $(\mu_n)_{n \geq 1}$ converges narrowly to μ (Theorem 7.7.2).

Conversely, suppose that $(\mu_n)_{n \geq 1}$ converges narrowly to μ. Let K be a compact subset of $\mathbf{R}^k$. Put $H = \{h \in \mathcal{H}^+(\mathbf{R}^k) : h \leq 1\}$. Given $\varepsilon > 0$, there exists $h_0 \in H$ for which $\int (1 - h_0) \, d\mu < \varepsilon/3$. Now there is an integer $m \geq 1$ such that $\int (1 - h_0) \, d\mu_n \leq \varepsilon/3$ for all $n \geq m$. Finally, there exists $h \in H$, $h \geq h_0$, such that $\int (1 - h) \, d\mu_n \leq \varepsilon/3$ for all $1 \leq n \leq m$, and now $\int (1 - h) \, d\mu_n \leq \varepsilon/3$ for all $n \geq 1$. As x runs through K, the functions $y \mapsto h(y) \cdot e^{-2i\pi xy}$ form a compact subset of $\mathcal{H}(\mathbf{R}^k, \mathrm{supp}(h); \mathbf{C})$. By the Banach–Steinhaus theorem, there exists an integer $N \geq 1$ such that

$$\sup_{x \in K} \left| \int h(y) e^{-2i\pi xy} \, d\mu_n(y) - \int h(y) e^{-2i\pi xy} \, d\mu(y) \right| \leq \frac{\varepsilon}{3} \qquad \text{for all } n \geq N.$$

On the other hand,

$$\left| \int (1 - h(y)) e^{-2i\pi xy} \, d\mu_n(y) - \int (1 - h(y)) e^{-2i\pi xy} \, d\mu(y) \right| \leq 2\frac{\varepsilon}{3}.$$

Hence

$$\sup_{x \in K} |\mathcal{F}\mu_n(x) - \mathcal{F}\mu(x)| \leq \varepsilon \qquad \text{for all } n \geq N.$$

$\square$

Theorem 14.4.5 (Levy) *Let* $(\mu_n)_{n\geq 1}$ *be a sequence in* $\mathcal{M}^1_+(S)$ *such that* $\sup \|\mu_n\| < +\infty$ *and* $(\mathcal{F}\mu_n)_{n\geq 1}$ *converges to a function* φ dx-*almost everywhere. Then there exists* $\mu \in \mathcal{M}^1_+(S)$ *such that* $\varphi = \mathcal{F}\mu$ dx-*almost everywhere. Furthermore, if* φ *is continuous at* 0 *and* $(\mathcal{F}\mu_n)_{n\geq 1}$ *converges to* φ *everywhere, then* $(\mu_n)_{n\geq 1}$ *converges narrowly to* μ.

PROOF: There is a subsequence $(\mu_{n_p})_{p\geq 1} = (\mu'_p)_{p\geq 1}$ of $(\mu_n)_{n\geq 1}$ that converges weakly to $\mu \in \mathcal{M}^1_+(S)$ (Proposition 7.7.5). Since φ belongs to $\mathcal{L}^\infty_\mathbf{C}(dx)$, the function $x \mapsto \varphi(x) \cdot \exp(-\pi|x|^2)$ is dx-integrable. Now, by the dominated convergence theorem,

$$\mu'_p\big(g_1(y-)\big) \;=\; (\mu'_p * g_1)(y)$$
$$=\; \int e^{2i\pi xy} \exp\big(-\pi|x|^2\big) \cdot \mathcal{F}\mu'_p(x)\, dx$$

converges to $\int e^{2i\pi xy} \exp\big(-\pi|x|^2\big)\varphi(x)\, dx$ as $p \to +\infty$. On the other hand, $\mu'_p\big(g_1(y-)\big)$ tends to

$$\mu\big(g_1(y-)\big) \;=\; (\mu * g_1)(y)$$
$$=\; \int e^{2i\pi xy} \exp\big(-\pi|x|^2\big) \cdot \mathcal{F}\mu(x)\, dx.$$

By Theorem 14.4.1, we may conclude that $\varphi = \mathcal{F}\mu$ dx-almost everywhere.

Next, suppose that φ is continuous at 0 and $(\mathcal{F}\mu_n)_{n\geq 1}$ converges to φ everywhere. Then $E = \{x \in \mathbf{R}^k : \varphi(x) = \mathcal{F}\mu(x)\}$ is dense in $\mathbf{R}^k$. Since φ and $\mathcal{F}\mu$ are both continuous at 0, necessarily $\varphi(0) = \mathcal{F}\mu(0)$. Since $\mu'_p(1) = \mathcal{F}\mu'_p(0)$ tends to $\varphi(0) = \mathcal{F}\mu(0) = \mu(1)$ as $p \to +\infty$, the sequence $(\mu'_p)_{p\geq 1}$ converges narrowly to μ. But $(\mathcal{F}\mu'_p)_{p\geq 1}$ then converges pointwise to $\mathcal{F}\mu$, whence we see that $\varphi = \mathcal{F}\mu$ everywhere. Finally, the sequence $(\mu_n)_{n\geq 1}$ itself converges narrowly to μ, by Theorem 14.4.4. $\square$

The following proposition is useful in applications.

Proposition 14.4.4 *Let* $\mu \in \mathcal{M}^1(S, \mathbf{C})$ *and* $n \in \mathbf{N}$. *Suppose that*

$$\int^* |x|^n\, dV\mu(x) < +\infty.$$

Then

$$\varphi : x \mapsto \mathcal{F}\mu\Big(\frac{x}{2\pi}\Big) = \int e^{ixy}\, d\mu(y) \quad \text{is of class } C^n,$$

and

$$D^n\varphi(x)(u_1,\ldots,u_n) = i^n \cdot \int e^{ixy} \cdot (u_1 y)\cdots(u_n y)\, d\mu(y)$$

for all $x \in \mathbf{R}^k$ *and all* $u_1,\ldots,u_n \in \mathbf{R}^k$.

PROOF: For each point $(y_1, \ldots, y_n)$ in $(\mathbf{R}^k)^n$, denote by $\gamma_{(y_1, \ldots, y_n)}$ the function $(u_1, \ldots, u_n) \mapsto (u_1 y_1) \cdots (u_n y_n)$ from $(\mathbf{R}^k)^n$ into $\mathbf{C}$. The mapping $(y_1, \ldots, y_n) \mapsto \gamma_{(y_1, \ldots, y_n)}$ from $(\mathbf{R}^k)^n$ into $L^n(\mathbf{R}^k, \mathbf{C})$ is multilinear. Hence, for every $x \in \mathbf{R}^k$, $y \mapsto i^n \cdot e^{ixy} \cdot \gamma_{(y, \ldots, y)}$ is μ-integrable. Now, by Theorem 3.2.3, φ is of class C^n and $D^n \varphi(x) = i^n \cdot \int e^{ixy} \cdot \gamma_{(y, \ldots, y)} \, d\mu(y)$ for all $x \in \mathbf{R}^k$. $\qquad \square$

In particular, when $k = 1$ and $\int^* |x|^n \, dV\mu(x) < +\infty$,

$$\varphi^{(n)}(x) = i^n \cdot \int e^{ixy} \cdot y^n \, d\mu(y) \qquad \text{for every } x \in \mathbf{R}.$$

Definition 14.4.2 Let μ be a positive measure on S such that $\int^* 1 \, d\mu = 1$. Then

$$x \mapsto \mathcal{F}\mu\left(\frac{x}{2\pi}\right) = \int e^{ixy} \, d\mu(y)$$

is called the characteristic function of μ and is written $\text{cha}(\mu)$.

Since the characteristic function of μ determines μ (by Theorem 14.4.1), we see that the title "characteristic" is amply justified.

14.5 Normal Laws in $\mathbf{R}^n$

Let μ be a positive measure on the natural semiring S_n of $\mathbf{R}^n$ ($n \geq 1$) such that $\int^* 1 \, d\mu = 1$.

Definition 14.5.1 μ is said to be Gaussian whenever, for each linear form u on $\mathbf{R}^n$, $u(\mu)$ is a normal measure on S_1.

Proposition 14.5.1 μ is Gaussian if and only if it is of order 2 and its characteristic function is $x \mapsto e^{imx} \cdot \exp\left((-1/2)x'Dx\right)$, where $m = E(\mu)$, $D = D(\mu)$, and x is regarded as a column matrix.

PROOF: First, suppose that μ is Gaussian. For all $1 \leq i \leq n$, the image μ_i of μ under the ith projection $x \mapsto x_i$ is normal. Therefore,

$$\int |x|^2 \, d\mu(x) = \sum_{1 \leq i \leq n} \int x_i^2 \, d\mu(x) = \sum_{1 \leq i \leq n} \int x_i^2 \, d\mu_i(x_i)$$

is finite, and μ is of order 2. Put $m = E(\mu)$ and $D = D(\mu)$. Given $x \in \mathbf{R}^n$, write u for the linear form $y \mapsto xy$ on $\mathbf{R}^n$. $u(\mu)$ is normal with mean mx and variance $x'Dx$, hence

$$\int e^{itxy} \, d\mu(y) = \int e^{it\xi} \, d(u(\mu))(\xi) = \exp(imtx) \cdot \exp\left(-\frac{1}{2}t^2 \cdot x'Dx\right)$$

for all $t \in \mathbf{R}$ (Section 14.3). Taking $t = 1$, we conclude that

$$[\text{cha}(\mu)](x) = e^{imx} \cdot \exp\left(-\frac{1}{2}x'Dx\right).$$

Conversely, suppose that μ is of order 2 and

$$[\text{cha}(\mu)](x) = e^{imx} \cdot \exp\left(-\frac{1}{2}x'Dx\right)$$

for all $x \in \mathbf{R}^n$, where $m = E(\mu)$ and $D = D(\mu)$. Let $u : y \mapsto xy$ ($x \in \mathbf{R}^n$) be a linear form on $\mathbf{R}^n$. Then, for all $s \in \mathbf{R}$,

$$[\text{cha}(u(\mu))](s) = \int e^{isu(y)} \, d\mu(y) = \int e^{isxy} \, d\mu(y) = [\text{cha}(\mu)](sx).$$

Thus

$$[\text{cha}(u(\mu))](s) = e^{iE(u(\mu))s} \cdot \exp\left(-\frac{1}{2}\,\text{Var}(u(\mu)) \cdot s^2\right),$$

and $u(\mu)$ is Gaussian.   □

Clearly, if μ is Gaussian and $v : \mathbf{R}^n \to \mathbf{R}^p$ is affine, then $v(\mu)$ is Gaussian.

Let $(e_1, \ldots, e_n)$ be the canonical basis of $\mathbf{R}^n$. An $n \times n$-matrix D, with real entries $c_{i,j}$, is said to be symmetric and positive if the bilinear form ψ on $\mathbf{R}^n \times \mathbf{R}^n$, such that $\psi(e_i, e_j) = c_{i,j}$ for all $1 \leq i, j \leq n$, is symmetric and positive.

Theorem 14.5.1 *Let $m \in \mathbf{R}^n$, and let $D = (c_{i,j})_{i,j}$ be an $n \times n$-matrix with real entries, symmetric and positive. Then there is a (unique) Gaussian measure μ on S_n such that $E(\mu) = m$ and $D(\mu) = D$.*

PROOF: Let Φ be the usual scalar product on $\mathbf{R}^n$, and let ψ (respectively, γ) be the bilinear form on $\mathbf{R}^n \times \mathbf{R}^n$ (respectively, the endomorphism of $\mathbf{R}^n$) whose matrix with respect to $(e_1, \ldots, e_n)$ is D. Then

$$\Phi(\gamma(e_i), e_j) = c_{j,i} = c_{i,j} = \psi(e_i, e_j) \qquad \text{for all } 1 \leq i, j \leq n,$$

so $\Phi(\gamma(x), y) = \psi(x, y)$ for all $x, y \in \mathbf{R}^n$. Clearly, γ is Hermitian (with respect to Φ). There is an orthonormal basis $(f_1, \ldots, f_n)$ in $\mathbf{R}^n$ consisting of γ-eigenvectors. Hence $(f_1, \ldots, f_n)$ is orthonormal for Φ and orthogonal for ψ. Permuting $f_1, \ldots, f_n$ if necessary, we may suppose that there exists $0 \leq p \leq n$ for which $\psi(f_i, f_i) = \sigma_i^2 > 0$ for all $1 \leq i \leq p$ and $\psi(f_i, f_i) = 0$ for all $p + 1 \leq i \leq n$. Denote by C the diagonal $n \times n$-matrix

$$C = \begin{pmatrix} \sigma_1^2 & & & & & 0 \\ & \ddots & & & & \\ & & \sigma_p^2 & & & \\ & & & 0 & & \\ & & & & \ddots & \\ 0 & & & & & 0 \end{pmatrix}.$$

Letting u be the endomorphism of $\mathbf{R}^n$ such that $u(e_i) = f_i$ for all $1 \leq i \leq n$, its matrix A with respect to $(e_1, \ldots, e_n)$ is orthogonal. Thus $A' \cdot A = A \cdot A' = I_n$. Since C is the matrix of $\psi \circ (u \times u)$ with respect to $(e_1, \ldots, e_n)$, we see that $A' \cdot D \cdot A = C$.

Now consider the measure

$$\nu = N_1(0, \sigma_1^2) \otimes \cdots \otimes N_1(0, \sigma_p^2) \otimes \varepsilon_0 \otimes \cdots \otimes \varepsilon_0$$

on S_n, where ε_0 is the measure on S_1 defined by the unit mass at 0. Then $E(\nu) = (0, \ldots, 0)$, $D(\nu) = C$, and

$$
\begin{aligned}
[\text{cha}(\nu)](x) &= \exp\left(-\frac{1}{2}\sigma_1^2 x_1^2\right) \times \cdots \times \exp\left(-\frac{1}{2}\sigma_p^2 x_p^2\right) \\
&= \exp\left(-\frac{1}{2}x'D(\nu)x\right)
\end{aligned}
$$

for all $x \in \mathbf{R}^n$. Therefore, ν is Gaussian. The image μ of ν under the affine mapping $x \mapsto m + u(x)$ from $\mathbf{R}^n$ into $\mathbf{R}^n$ is Gaussian, and

$$E(\mu) = m + u\big(E(\nu)\big) = m,$$

$$D(\mu) = A \cdot D(\nu) \cdot A' = D.$$

$\square$

Definition 14.5.2 In the notation of Theorem 14.5.1, μ is called the normal measure on $\mathbf{R}^n$ with mean m and covariance matrix D. It is written $N_n(m, D)$.

Refer to the notation of Theorem 14.5.1. Observe that the vector subspace of $\mathbf{R}^n$ generated by $f_1, \ldots, f_p$ is the orthogonal complement of $\ker(\gamma)$, and that $\text{supp}(\mu) = m + u(\text{supp}(\nu)) = m + [\ker(\gamma)]^\perp$. Furthermore, p is the rank of D.

Proposition 14.5.2 Let m and D be as in Theorem 14.5.1. Suppose that $\det(D) \neq 0$. Then

$$N_n(m, D) = (2\pi)^{-n/2} \cdot \det(D)^{-1/2} \cdot \exp\left(-\frac{1}{2}(x-m)'D^{-1}(x-m)\right) dx,$$

where dx is Lebesgue measure on S_n.

PROOF: From the equality $C = A'DA$, in the notation of Theorem 14.5.1, it follows that

$$A'D^{-1}A = A^{-1}D^{-1}(A')^{-1} = (A'DA)^{-1}$$

is equal to

$$
C^{-1} = \begin{pmatrix}
\frac{1}{\sigma_1^2} & & & & 0 \\
& \ddots & & & \\
& & \frac{1}{\sigma_p^2} & & \\
& & & \ddots & \\
0 & & & & \frac{1}{\sigma_n^2}
\end{pmatrix}.
$$

Hence $x' \cdot A' \cdot D^{-1} \cdot A \cdot x = \sigma_1^{-2} x_1^2 \times \cdots \times \sigma_n^{-2} x_n^2$ for all $x \in \mathbf{R}^n$.
Now let $h : \mathbf{R}^n \to [0, +\infty[$ be a Borel function. Then

$$
\int^* h \, d\mu = \int^* h(m + u(x)) \, d\nu(x)
$$

is equal to

$$
(2\pi)^{-n/2} \cdot \det(D)^{-1/2} \int^* h(m + u(x)) \cdot \exp\left(-\frac{1}{2} x' A' D^{-1} A x \right) dx.
$$

The formula for change of variables shows that $\int^* h \, d\mu$ is equal to

$(2\pi)^{-n/2} \cdot \det(D)^{-1/2}$

$$
\times \int^* h(y) \cdot \exp\left[-\frac{1}{2} (u^{-1}(y - m))' \cdot A' D^{-1} A \cdot (u^{-1}(y - m)) \right] dy.
$$

Since $u^{-1}(y - m) = A^{-1}(y - m)$, the proposition follows. $\square$

Now recall the notation of Theorem 14.5.1, and let v be the linear mapping

$$
(x_1, \ldots, x_p) \mapsto (\sigma_1 x_1, \ldots, \sigma_p x_p, 0, \ldots, 0) \qquad \text{from } \mathbf{R}^p \text{ into } \mathbf{R}^n.
$$

Let M (respectively, B) be the matrix of v (respectively, $u \circ v$) relative to the canonical bases of $\mathbf{R}^p$ and $\mathbf{R}^n$. Then $B = A \cdot M$, so

$$
B \cdot B' = A \cdot M \cdot M' \cdot A' = A \cdot C \cdot A' = D.
$$

Furthermore, the rank of B is the rank p of D. Conversely, let B be an arbitrary $n \times p$-matrix of rank p, such that $B \cdot B' = D$. If w is the linear mapping from $\mathbf{R}^p$ into $\mathbf{R}^n$ with matrix B and if μ is the image of $N_p(0, I_p)$ under $x \mapsto m + w(x)$, then μ is Gaussian and $D(\mu) = B \cdot B' = D$, so $\mu = N_n(m, D)$.

At this point, we introduce some terminology that will be used in the sequel. Let P be a probability on the Borel σ-algebra of $\mathbf{R}^n$, and let μ be its restriction to the natural semiring of $\mathbf{R}^n$. The characteristic function of P is defined as the characteristic function $x \mapsto \int e^{ixy} \, dP(y)$ of μ. Likewise, we define the mean of P (whenever P is of order 1) and the covariance matrix of P (whenever P is of order 2). If $\mu = N_n(m, D)$, P is called the normal law on $\mathbf{R}^n$ with mean m and covariance matrix D, and P is still written $N_n(m, D)$.

Exercises for Chapter 14

1   Let μ be a bounded measure on the natural semiring S of $\mathbf{R}$, and let $\varphi : y \mapsto \int e^{ixy} \, d\mu(x)$. Fix $a \in \mathbf{R}$.

1.   For all $t > 0$, show that

$$\frac{1}{2t} \int_{-t}^{t} e^{-iya} \varphi(y) \, dy = \int \frac{\sin(t(x - a))}{t(x - a)} \, d\mu(x).$$

2.   Deduce from part 1 that

$$\mu(\{a\}) = \lim_{t \to +\infty} \frac{1}{2t} \int_{-t}^{t} e^{-iya} \varphi(y) \, dy.$$

2   In the notation of Exercise 1, let $\bar{\mu}$ be the measure conjugate to μ, and let ν be the measure on S, the image of $\mu \otimes \bar{\mu}$ under $(x, y) \mapsto x - y$.

1.   Show that $\int e^{isz} \, d\nu(z) = |\varphi(s)|^2$ for all $s \in \mathbf{R}$.

2.   Prove that

$$\sum_{x \in \mathbf{R}} |\mu(\{x\})|^2 = \nu(\{0\}) = \lim_{t \to +\infty} \frac{1}{2t} \int_{-t}^{t} |\varphi(s)|^2 \, ds.$$

3   Let φ_0 be the function $t \longmapsto (1 - |t|) \cdot 1_{[-1,1]}(t)$ on $\mathbf{R}$, and let $f_0 : x \mapsto \dfrac{1 - \cos x}{\pi x^2}$.

Show that $\mathcal{F}\varphi_0 \left(\dfrac{x}{2\pi} \right) = 2\pi f_0(x)$ for all $x \in \mathbf{R}$, and deduce that $\int e^{iux} f_0(x) \, dx = \varphi_0(u)$ for all $u \in \mathbf{R}$.

4   Suppose that $d_1, d_2, \ldots$ are strictly positive and $\sum_{k \geq 1} d_k = +\infty$, that $s_1 \geq s_2 \geq \ldots \geq 0$ and $\lim_{k \to +\infty} s_k = 0$, and that $\sum_{k \geq 1} s_k d_k = 1$. For each $k \geq 0$, put $t_k = d_1 + \cdots + d_k$. Define $\varphi : \mathbf{R} \to \mathbf{R}$ as follows:

(a) $\varphi(t_k) = 1 - \sum_{1 \leq j \leq k} s_j d_j$ for each $k \in \mathbf{Z}^+$;

(b) φ is affine on $[t_{k-1}, t_k]$ for each $k \in \mathbf{N}$;

(c) $\varphi(-t) = \varphi(t)$ for all $t \geq 0$.

Hence the graph of $\varphi_{/[0,+\infty[}$ is the convex polygon whose successive sides have slopes $-s_1, -s_2, \ldots$, and lengths $d_1, d_2, \ldots$ when projected on the horizontal axis.

1.   Show that $s_{n+1} t_n \to 0$ as $n \to +\infty$.

2.   For all $j \in \mathbf{N}$, let $p_j = (s_j - s_{j+1}) t_j$. Prove that $\sum_{j \geq 1} p_j = 1$ (use part 1).

3.   Let φ_0 be as in Exercise 3. Prove that

$$\varphi(t_k) = \sum_{j \geq 1} p_j \varphi_0 \left(\frac{t_k}{t_j} \right) \qquad \text{for each } k \in \mathbf{N}$$

and that

$$\varphi(t) = \sum_{j \geq 1} p_j \varphi_0 \left(\frac{t}{t_j} \right) \quad \text{for all } t \in \mathbf{R}.$$

4. Let f be the function $x \mapsto \sum_{j \geq 1} p_j t_j f_0(t_j x)$ (notation as in Exercise 3). Show that f is finite on $\mathbf{R} - \{0\}$ and continuous on $\mathbf{R}$.

5. Prove that f is dx-integrable, that $\int f \, dx = 1$, and that φ is the characteristic function of $f \, dx$.

5 Suppose that $\varphi : \mathbf{R} \to [0, +\infty[$ is continuous, $\varphi_{/[0,+\infty[}$ is convex and decreasing, $\varphi(0) = 1$, and $\varphi(-t) = \varphi(t)$ for all $t > 0$.

1. Assume that $\lim_{t \to +\infty} \varphi(t) = 0$. For each $n \in \mathbf{Z}^+$, let φ_n be the even function which agrees with φ at each $k/2^n$ ($k \in \mathbf{Z}^+$) and is affine on each $[k/2^n, (k+1)/2^n]$. Show that $\big(\varphi_n(t)\big)_{n \geq 0}$ converges to $\varphi(t)$ for all $t \in \mathbf{R}$. Deduce from Levy's theorem that there exists a positive measure μ on the natural semiring of $\mathbf{R}$ such that $\int^* 1 \, d\mu = 1$ and $\mathrm{cha}(\mu) = \varphi$.

2. Under the hypothesis of part 1, show that $\varphi_1 - \varphi$ is dx-integrable. Conclude that $\mu = f \, dx$, where $f : \mathbf{R} \to [0, +\infty]$ is continuous and is finite on $\mathbf{R} - \{0\}$. Show that $f(t) = f(-t)$ for all $t \in \mathbf{R}$.

3. In the general case, show that φ is a characteristic function (Polya's criterion).

Part III

Convergence of Random Variables; Conditional Expectation

15

The Strong Law of Large Numbers

A probability space is a triple $(\Omega, \mathcal{F}, P)$ where Ω is a nonempty set, $\mathcal{F}$ a σ-algebra in Ω, and P a probability, that is, a positive measure on $\mathcal{F}$ such that $P(\Omega) = 1$. A real-valued random variable is a measurable function from Ω into $\mathbf{R}$. Given a sequence of independent random variables $(X_n)_{n \geq 1}$ and $S_n = X_1 + \cdots + X_n$, a classical problem in probability theory is to study S_n/n and different types of convergence as $n \to +\infty$. The strong law of large numbers says that if the X_k have the same distribution then S_n/n converges almost surely to $E(X_1) = \int X_1 \, dP$ as $n \to +\infty$.

Summary

15.1 This section gives some fundamental definitions in the theory of probability, such as the definitions of a probability space and a random variable.

15.2 In this section the fundamental concept of independence is developed.

15.3 We give an example of singular function, which arises naturally from probability theory.

15.4 Given Ω, set $\Omega_n = \Omega$ for all $n \in \mathbf{N}$ and $\Omega^{\mathbf{N}} = \prod_n \Omega_n$. The one-sided shift transformation v on $\Omega^{\mathbf{N}}$ is defined by $(x_n)_{n \geq 1} \mapsto (x_{n+1})_{n \geq 1}$. The probability of a v-invariant event is either 0 or 1 (Proposition 15.4.2) and v is μ-ergodic (Proposition 15.4.3)—see Section 11.4. Then, using Birkhoff's ergodic theorem, we prove the strong law of large numbers (Theorem 15.4.1).

15.5 A number $x \neq m/b^n$ is said to be completely normal if, for every integer k and every k-tuple $(u_1, \ldots, u_k)$ of base-b digits, the k-tuple appears in the base-b expansion of x with asymptotic relative frequency $1/b^k$. Almost every number (for Lebesgue measure) in $[0, 1]$ is completely normal (Proposition 15.5.2).

15.1   Convergence in Probability

We recall that a measurable space is the pair $(\Omega, \mathcal{F})$ of a nonempty set Ω and a σ-algebra $\mathcal{F}$ in Ω. A probability on $\mathcal{F}$ is a positive measure, P, on $\mathcal{F}$ such that $P(\Omega) = 1$. Then $(\Omega, \mathcal{F}, P)$ is called a probability space.

We can represent an experiment whose results are, in some sense, random by a probability space $(\Omega, \mathcal{F}, P)$. Ω consists of all possible outcomes ω of the experiment, $\mathcal{F}$ is the set of (observable) events, and, for all $E \in \mathcal{F}$, $P(E)$ estimates the likelihood that E be realized when the experiment is performed.

For instance, consider the experiment consisting of an infinite sequence of tosses of an unbiased coin. Imagine that the sides have been labeled 1 and 0 instead of the usual heads and tails. For each $n \in \mathbf{N}$, let $\mathcal{F}_n$ be the power set of $\{0, 1\}$, and let μ_n be the probability on $\mathcal{F}_n$ such that $\mu_n(\{0\}) = \mu_n(\{1\}) = 1/2$. Finally, let $\mathcal{F} = \bigotimes_{n \geq 1} \mathcal{F}_n$ be the Borel σ-algebra of $\Omega = \{0, 1\}^{\mathbf{N}}$, and let P be the probability on $\mathcal{F}$ whose restriction to $\prod_{n \geq 1} \mathcal{F}_n$ is $\bigotimes_{n \geq 1} \mu_n$ (i.e., $P(\prod_{n \geq 1} A_n) = \prod_{n \geq 1} \mu_n(A_n)$, if $A_n \subset \{0, 1\}$ for all $n \in \mathbf{N}$ and $A_n = \{0, 1\}$ for all but finitely many indices n). Evidently, $(\Omega, \mathcal{F}, P)$ represents this experiment.

Definition 15.1.1 Let $(\Omega, \mathcal{F})$ and $(\Omega', \mathcal{F}')$ be two measurable spaces. A mapping $X : \Omega \to \Omega'$ is called a random variable (from $(\Omega, \mathcal{F})$ into $(\Omega', \mathcal{F}')$) if $X^{-1}(A')$ belongs to $\mathcal{F}$ for all $A' \in \mathcal{F}'$.

For instance, consider the measurable space $(\{0, 1\}^{\mathbf{N}}, \mathcal{F})$, which we encountered above. Let $\Omega' = \mathbf{Z}^+ = \{0, 1, \ldots\}$, and let $\mathcal{F}'$ be the power set of Ω'. Then, for fixed $k \in \mathbf{N}$, the mapping $(x_n)_{n \geq 1} \mapsto \sum_{1 \leq n \leq k} x_n$ from Ω into Ω' is a random variable. It represents the number of heads obtained after k tosses in an infinite sequence of coin tosses.

Definition 15.1.2 Let $(\Omega, \mathcal{F}, P)$ be a probability space and X a random variable from $(\Omega, \mathcal{F})$ into a measurable space $(\Omega', \mathcal{F}')$. Then the image measure of P under X (defined on $\mathcal{F}'$) is called the law (or the distribution) of X, written P_X.

If $(\Omega, \mathcal{F}, P)$ is a probability space, a property which holds P-almost everywhere is said to be true almost surely (or with probability 1).

Definition 15.1.3 Let Ω' be a separable, metrizable uniform space, and let $\mathcal{F}'$ be the Borel σ-algebra of Ω'. Let $(\Omega, \mathcal{F}, P)$ be a probability space and $(X_n)_{n \geq 1}$ a sequence of random variables from $(\Omega, \mathcal{F})$ into $(\Omega', \mathcal{F}')$. Then $(X_n)_{n \geq 1}$ is said to converge in probability to a random variable X from $(\Omega, \mathcal{F})$ into $(\Omega', \mathcal{F}')$ if it converges to X in P-measure. Equivalently, $(X_n)_{n \geq 1}$ converges to X in probability if, for each closed entourage W of Ω', $P(\{\omega : (X_n(\omega), X(\omega)) \notin W\})$ converges to 0 as $n \to +\infty$.

Proposition 15.1.1 *Let $(\Omega'', \mathcal{F}'')$ be as was $(\Omega', \mathcal{F}')$ in Definition 15.1.3. With the notation of the definition, let g be a Borel mapping from Ω' into Ω'' which is continuous at P_X-almost all points of Ω'. If $(X_n)_{n \geq 1}$ converges to X in probability, then $(g \circ X_n)_{n \geq 1}$ converges to $g \circ X$ in probability.*

PROOF: The set D of points of discontinuity of g lies in $\mathcal{F}'$, and $X^{-1}(D)$ is P-negligible. From each subsequence of $(X_n)_{n \geq 1}$, we can extract a sequence $(X_{n_k})_{k \geq 1}$ which converges to X almost surely (Proposition 5.3.1). Then $(g \circ X_{n_k})_{k \geq 1}$ converges almost surely in $\Omega - X^{-1}(D)$, which proves that $(g \circ X_n)_{n \geq 1}$ converges to $g \circ X$ in probability (Chapter 5, Exercise 5). $\square$

Let $(\Omega, \mathcal{F})$ be a measurable space. Given an integer $k \geq 1$, let $\mathcal{F}'$ be the Borel σ-algebra of $\mathbf{R}^k = \Omega'$. A random variable X from $(\Omega, \mathcal{F})$ into $(\Omega', \mathcal{F}')$ is called simply a random variable (or a random vector) from $(\Omega, \mathcal{F})$ into $\mathbf{R}^k$. Moreover, when a probability P on $\mathcal{F}$ is given, the distribution function of P_X (respectively, the characteristic function of P_X) is also called the distribution function (respectively, the characteristic function) of X. X is said to be of order $n \in \mathbf{N}$ if P_X is of order n (i.e., if $\int^* |X|^n \, dP < +\infty$). If X is of order 1, the mean $\int \mathrm{id}_{\mathbf{R}^k} \, dP_X = \int X \, dP$ of P_X is called the expected value (or mean value) of X, written $E(X)$. If X is of order 2, the covariance matrix of P_X is called the covariance matrix of X, written $D(X)$. Hence $D(X) = (c_{i,j})$, where $c_{i,j} = \int (X_i - E(X_i))(X_j - E(X_j)) \, dP$ (the X_i are the components of X). Similarly, we define the variance of X, when $k = 1$.

Let λ_k be Lebesgue measure on the natural semiring S of $\mathbf{R}^k$. In the above notation, X (or P_X) is said to have density g if $g : \mathbf{R}^k \to [0, +\infty]$ is a Borel function, $\int^* g \, d\lambda_k = 1$, and $P(X^{-1}(A)) = \int_A g \, d\lambda_k$ for all $A \in \mathcal{F}'$. Equivalently, $P_{X/S}$ is absolutely continuous with respect to λ_k and any density of $P_{X/S}$ is equal λ_k-a.c. to the Borel function $g : \mathbf{R}^k \to [0, +\infty]$. In this case, for simplicity, we still write $P_X = g\lambda_k$.

15.2   Independence of Random Variables

Let $\left((\Omega_i, \mathcal{F}_i, P_i)\right)_{i \in I}$ be a family of probability spaces. The σ-algebra $\bigotimes_{i \in I} \mathcal{F}_i$ is the σ-ring generated by the semialgebra $\prod_{i \in I} \mathcal{F}_i$. Let μ be the measure $\bigotimes_{i \in I} P_i$ on $\prod_{i \in I} \mathcal{F}_i$ defined in Chapter 9 if I is finite, and in Chapter 11 if I is infinite. Finally, let P be the extension of μ to $\bigotimes_{i \in I} \mathcal{F}_i$. Observe that μ and P have the same main prolongation.

Definition 15.2.1 In the probabilistic context, P is called the product of the P_i, written $\bigotimes_{i \in I} P_i$.

By the final remark of Section 11.2, if $(I_j)_{j \in J}$ is a partition of I into nonempty subsets, then $\bigotimes_{i \in I} P_i = \bigotimes_{j \in J} (\bigotimes_{i \in I_j} P_i)$.

Henceforth, let $(\Omega, \mathcal{F}, P)$ be a probability space. We now introduce the fundamental concept of independence.

Definition 15.2.2 Let $(\mathcal{D}_i)_{i \in I}$ be a family of σ-algebras contained in $\mathcal{F}$. The $\mathcal{D}_i$ are said to be independent (with respect to P) whenever $P(\bigcap_{i \in J} A_i) = \prod_{i \in J} P(A_i)$ for every finite subset J of I and every family $(A_i)_{i \in J}$ of events (indexed by J) such that A_i lies in $\mathcal{D}_i$ for all $i \in J$.

Until further notice, I will be an index set and, for all $i \in I$, X_i will be a random variable from $(\Omega, \mathcal{F})$ into a measurable space $(\Omega_i, \mathcal{F}_i)$. For each nonempty subset J of I, denote by X_J the random variable $[X_i]_{i \in J}$ from $(\Omega, \mathcal{F})$ into $(\Omega_J, \mathcal{F}_J)$, where $\Omega_J = \prod_{i \in J} \Omega_i$ and $\mathcal{F}_J = \bigotimes_{i \in J} \mathcal{F}_i$. The smallest σ-algebra for which the X_i are measurable for all $i \in J$ is denoted by $\sigma((X_i)_{i \in J})$, so that $\sigma((X_i)_{i \in J}) = \sigma(X_J)$.

Definition 15.2.3 The X_i are said to be independent whenever the $\sigma(X_i)$ are independent.

Theorem 15.2.1 *The X_i are independent if and only if $P_{X_J} = \bigotimes_{i \in J} P_{X_i}$ for all nonempty finite subsets J of I.*

PROOF: Let $J \subset I$ be finite and nonempty. P_{X_J} and $\bigotimes_{i \in J} P_{X_i}$ are identical if they agree on $\prod_{i \in J} \mathcal{F}_i$, that is, if

$$P\left(\bigcap_{i \in J} X_i^{-1}(B_i)\right) = P(X_J^{-1}(B_J)) = \int 1_{B_J} \, dP_{X_J}$$

$$= \prod_{i \in J} \int 1_{B_i} \, dP_{X_i}$$

$$= \prod_{i \in J} P(X_i^{-1}(B_i))$$

for each $B_J = \prod_{i \in J} B_i$ in $\prod_{i \in J} \mathcal{F}_i$. □

Consequently, in order that the X_i be independent, it is necessary and sufficient that $P_{X_I} = \bigotimes_{i \in I} P_{X_i}$.

Theorem 15.2.2 *For all $i \in I$, let C_i be a π-system in Ω_i such that $\sigma(C_i) = \mathcal{F}_i$. The X_i are independent if and only if*

$$P\left(\bigcap_{i \in J} X_i^{-1}(B_i)\right) = \prod_{i \in J} P(X_i^{-1}(B_i))$$

for every finite subset J of I and every $(A_i)_{i \in J}$ such that $A_i \in C_i$ for all $i \in J$.

PROOF: Suppose the conditions hold, and let $J \subset I$ be finite and nonempty. Denote by C_J the class consisting of the sets $B_J = \prod_{i \in J} B_i$, where $B_i \in C_i$ for all $i \in J$. Then C_J is a π-system and $\sigma(C_J) = \bigotimes_{i \in J} \mathcal{F}_i$. Since $\int 1_{B_J} \, dP_{X_J} = \int 1_{B_J} \, d(\bigotimes_{i \in J} P_{X_i})$ for all C_J-sets B_J, we conclude that $P_{X_J} = \bigotimes_{i \in J} P_{X_i}$ (Proposition 6.4.2). □

Theorem 15.2.3 *Suppose the X_i are independent. If $(I_j)_{j \in J}$ is a partition of I into nonempty sets and if, for each $j \in J$, g_j is a random variable from $\left(\prod_{i \in I_j} \Omega_i, \bigotimes_{i \in I_j} \mathcal{F}_i \right)$ into a measurable space $(\Omega'_j, \mathcal{F}'_j)$, then the $g_j \circ [X_i]_{i \in I_j}$ are independent.*

PROOF: Obvious. $\qquad\qquad\qquad\qquad\qquad\qquad\qquad\qquad\qquad\qquad$ $\square$

Theorem 15.2.4 *Suppose that, for every $i \in I$, $X_i(\Omega)$ is at most countable and $\{x_i\}$ belongs to $\mathcal{F}_i$ for all $x_i \in X_i(\Omega)$. Then the X_i are independent if and only if $P(X_i = x_i \quad \forall i \in J) = \prod_{i \in J} P(X_i = x_i)$ for every finite subset J of I and every $(x_i)_{i \in J}$ in $\prod_{i \in J} \Omega_i$.*

PROOF: Obvious. $\qquad\qquad\qquad\qquad\qquad\qquad\qquad\qquad\qquad\qquad$ $\square$

Theorem 15.2.5 *Given $k \in \mathbf{N}$, suppose that I is finite and the X_i are independent random variables from $(\Omega, \mathcal{F})$ into $\mathbf{R}^k$. For all $i \in I$, let φ_i be the characteristic function of X_i. Then $x \mapsto \prod_{i \in I} \varphi_i(x)$ is the characteristic function of $\sum_{i \in I} X_i$. If each X_i is of order 2, then $D(\sum_{i \in I} X_i) = \sum_{i \in I} D(X_i)$.*

PROOF: Obvious $\qquad\qquad\qquad\qquad\qquad\qquad\qquad\qquad\qquad\qquad$ $\square$

Definition 15.2.4 A family $(A_i)_{i \in I}$ of $\mathcal{F}$-sets is said to be independent whenever $P\left(\bigcap_{i \in J} A_i \right) = \prod_{i \in J} P(A_i)$ for every finite subset J of I.

In this case, for each $i \in I$, let $\mathcal{F}_i = \{\emptyset, \Omega, A_i, A_i^c\}$ be the σ-algebra generated by A_i. By induction on n, we find that $P\left(\bigcap_{i \in J} E_i \right) = \prod_{i \in J} P(E_i)$ for every family $(E_i)_{i \in J}$ of events, indexed by a finite subset J of I, such that $E_i \in \mathcal{F}_i$ for all $i \in J$ and $E_i = A_i^c$ for at most n indices $i \in J$. Therefore, the σ-algebras $\mathcal{F}_i$ are independent.

Theorem 15.2.6 (Second Borel–Cantelli Lemma) *If $(A_n)_{n \geq 1}$ is an independent sequence of events and $\sum_{n \geq 1} P(A_n) = +\infty$, then $P(\limsup A_n) = 1$, where $\limsup A_n = \{x \in \Omega : 1_{A_n}(x) = 1$ for infinitely many $n\}$.*

PROOF: Since $1 - e^{-x} = \int_0^x e^{-t} \, dt \leq x$ for all $x \in [0, +\infty[$,

$$P\left(\bigcap_{m \leq n \leq m+k} A_n^c \right) = \prod_{m \leq n \leq m+k} (1 - P(A_n))$$

$$\leq \prod_{m \leq n \leq m+k} \exp\left(-P(A_n) \right)$$

for all integers $m \geq 1$ and $k \geq 0$. Hence $P\left(\bigcap_{n \geq m} A_n^c \right) = 0$ for all $m \in \mathbf{N}$. But $\bigcup_{m \geq 1} \left(\bigcap_{n \geq m} A_n^c \right)$ is the complement of $\limsup A_n$, so, $P(\limsup A_n) = 1$. $\qquad\square$

15.3   An Example of Independent Random Variables

Let $b \geq 2$ be an integer, $\mathcal{F}'$ the power set of $\Omega' = \{0, 1, \ldots, b-1\}$, and let P be Lebesgue measure on the Borel σ-algebra, $\mathcal{B}$, of $\Omega = \,]0, 1]$. For each $n \in \mathbf{N}$, denote by X_n the function from Ω into Ω' such that $X_n(t) = x_n$ for every $t \in \Omega$, where $(x_i)_{i \geq 1}$ is the nonterminating base-b expansion of t. We will show that the X_n are independent random variables from $(\Omega, \mathcal{B})$ into $(\Omega', \mathcal{F}')$.

This example illustrates the notion of independence: if a real number t is chosen at random in $]0, 1]$, knowledge of the first $(n-1)$ terms of its nonterminating base-b expansion gives no indication of the term of order n.

We will, in fact, prove more.

Fix a probability μ' on $\mathcal{F}'$ such that $\mu'(\{u\}) \neq 1$ for all $0 \leq u \leq b-1$. Write p_u, or $p(u)$, instead of $\mu'(\{u\})$. Put $\Omega_n = \Omega'$, $\mathcal{F}_n = \mathcal{F}'$, and $\mu_n = \mu'$, for all integers $n \geq 1$. Thus $\bigotimes_{n \geq 1} \mathcal{F}_n$ is the Borel σ-algebra of the topological space $\Omega'^{\mathbf{N}} = \prod_{n \geq 1} \Omega_n$. Call μ the probability $\bigotimes_{n \geq 1} \mu_n$ on $\bigotimes_{n \geq 1} \mathcal{F}_n$.

$G : (x_n)_{n \geq 1} \mapsto \sum_{n \geq 1}(x_n/b^n)$ is continuous from $\Omega'^{\mathbf{N}}$ into $I = [0, 1]$, as the uniform limit of the functions $(x_n)_{n \geq 1} \mapsto \sum_{1 \leq n \leq m}(x_n/b^n)$. Let $G(\mu)$ be the image measure of μ under G (defined on the Borel σ-algebra of I), and let F be the indefinite integral of $G(\mu)$ taking the value 0 at 0. Then $F(t) = G(\mu)(\,]0, t])$ for all $t \in I$. Finally, denote by P the probability induced by $G(\mu)$ on the Borel σ-algebra $\mathcal{B}$ of $\Omega = \,]0, 1]$.

Proposition 15.3.1 *F is a continuous increasing function, strictly increasing if and only if $p_u \neq 0$ for all $0 \leq u \leq b-1$. When $p_u = 1/b$ for all $0 \leq u \leq b-1$, then $F(t) = t$ for all $t \in I$, so $G(\mu)$ is Lebesgue measure on I. When there exists $0 \leq u \leq b-1$ such that $p_u \neq 1/b$, then F is singular. Moreover,*

$$F(t) = p_0 + \cdots + p_{u-1} + p_u \cdot F(bt - u)$$

for every $0 \leq u \leq b-1$ and every $t \in [u/b, (u+1)/b]$.

PROOF: For all $t \in I$, $G^{-1}(t)$ has at most two elements, so $G(\mu)(\{t\}) = 0$. It follows that F is continuous.

Endow $\Omega'^{\mathbf{N}}$ with the lexicographic order. Now G is increasing. For $x, y \in \Omega'^{\mathbf{N}}$ satisfying $x < y$, the equality $G(x) = G(y)$ holds if and only if x, y are, respectively, the improper and the proper base-b expansion of a same $t \in \,]0, 1[\,\cap B$, where

$$B = \left\{ \frac{k}{b^n} : n \geq 1, 0 \leq k \leq b^n, \quad k \text{ integer} \right\}.$$

Let n, k be two integers satisfying $n \geq 1$ and $0 \leq k < b^n$, and write $(u_i)_{i \geq 1}$ for the proper expansion of k/b^n. If $x = (x_i)_{i \geq 1}$ is an element of $\Omega'^{\mathbf{N}}$ satisfying $k/b^n \leq G(x) \leq (k+1)/b^n$, then

(a) $(u_i)_{i \geq 1} \leq (x_i)_{i \geq 1}$ or else $k/b^n > 0$ and $(x_i)_{i \geq 1}$ is the improper expansion of k/b^n;

(b) $(x_i)_{i\geq 1} \leq (u_1, \ldots, u_n, b-1, b-1, \ldots)$, or else $(k+1)/b^n < 1$ and $(x_i)_{i\geq 1}$ is the proper expansion of $(k+1)/b^n$.

Thus we are in one of the following situations:

(c) $x_1 = u_1, \ldots, x_n = u_n$.

(d) $k/b^n > 0$ and $(x_i)_{i\geq 1}$ is the improper expansion of k/b^n.

(e) $(k+1)/b^n < 1$ and $(x_i)_{i\geq 1}$ is the proper expansion of $(k+1)/b^n$.

Consequently,

$$F\left(\frac{k+1}{b^n}\right) - F\left(\frac{k}{b^n}\right) = G(\mu)\left(\left[\frac{k}{b^n}, \frac{k+1}{b^n}\right]\right)$$

$$= \mu\left(\{(u_1, \ldots, u_n)\} \times \prod_{i \geq n+1} \Omega_i\right)$$

$$= p_{u_1} \cdots p_{u_n}.$$

We may conclude that F does not increase strictly if there exists a u such that $p_u = 0$. On the other hand, if no such u exists and if s, $t \in I$ satisfy $s < t$, let $n \geq 0$ and $0 \leq j < b^n$ be two integers such that $t - s \geq 2/b^n$ and $s \leq j/b^n < t$. Taking $k = j$ or $k = j - 1$, as $j/b^n \leq (s+t)/2$ or $j/b^n > (s+t)/2$, we find that the interval $[k/b^n, (k+1)/b^n]$ is included in $[s, t]$, and so $F(s) < F(t)$.

For every $n \geq 1$, let X_n be the function from $]0, 1]$ into Ω' such that $X_n(t) = x_n$ for all $t \in]0, 1]$, if $\sum_{i\geq 1} x_i/b^i$ is the nonterminating expansion of t (containing infinitely many nonzero elements). For each integer $0 \leq k < b^n$, X_n is constant on $]k/b^n, (k+1)/b^n]$. Hence $X_n^{-1}(A)$ is a Borel set in Ω for each $A \in \mathcal{F}_n$. Outside the μ-negligible set $G^{-1}(B)$, $X_n \circ G_{/(\Omega'^N - G^{-1}(B))}$ and the function $Z_n : (a_i)_{i\geq 1} \mapsto x_n$ agree, so

$$P(X_1 = x_1, \ldots, X_n = x_n) = \mu(Z_1 = x_1, \ldots, Z_n = x_n)$$

$$= \prod_{1\leq i \leq n} P(X_i = x_i)$$

for all $n \geq 1$ and all $x_1, \ldots, x_n$ in Ω'.

We now return to the function F.

When $p_u = 1/b$ for all $0 \leq u \leq b - 1$, for each $n \in \mathbf{N}$ we have

$$F\left(\frac{k+1}{b^n}\right) - F\left(\frac{k}{b^n}\right) = \frac{1}{b^n}$$

for all $0 \leq k < b^n$. Hence $F(k/b^n) = k/b^n$ for all $0 \leq k \leq b^n$. Since F is continuous, we conclude that $F(t) = t$ for all $t \in I$, and $G(\mu)$ is Lebesgue measure λ on the Borel σ-algebra of I.

We assume in what follows that $p_u \neq 1/b$ for at least one $0 \leq u \leq b - 1$. When β runs through Ω', the events $(X_n = \beta)$ are $\lambda_{/]0\,.1]}$-independent. So, for λ-almost all $t \in {]}0, 1]$, $X_n(t) = \beta$ infinitely often.

Let t be a point in ${]}0, 1]$ such that F has a finite derivative at t, and suppose that $p(X_n(t))$ is different from $1/b$ for infinitely many n. For every $n \geq 1$, denote by k_n the unique integer in $\{0, 1, \ldots, b^n - 1\}$ such that $k_n/b^n < t \leq (k_n + 1)/b^n$, and let $(x_i)_{i \geq 1}$ be the nonterminating expansion of t. Then

$$c_n = b^n \cdot \left[F\left(\frac{k_n + 1}{b^n}\right) - F\left(\frac{k_n}{b^n}\right) \right] = b^n \cdot p(x_1) \cdots p(x_n)$$

converges to $F'(t)$ as $n \to +\infty$. Now $F'(t)$ cannot be strictly positive, for, in this case, $c_n/(bc_{n-1}) = p(x_n)$ converges to $1/b$ as $n \to +\infty$, which contradicts the fact that $p(x_n)$ is different from $1/b$ for infinitely many n.

Hence $F'(t) = 0$ for λ-almost all $t \in I$ such that $F'(t)$ exists and is finite. that is, $F'(t) = 0$ for λ-almost all $t \in I$, and so F is singular.

Finally, given $u \in \Omega'$, let t be an element of $[u/b, (u + 1)/b]$. We have

$$F(t) - F(u/b) = G(\mu)({]}u/b, t]) = \mu\Big(\{x : G(x) \in {]}u/b, t]\}\Big).$$

But, if $x \in \Omega'^{\mathbf{N}}$ is such that $G(x)$ lies in ${]}u/b, t]$, then $G(x) = u/b + (1/b)G(vx)$, where $v(x) = (x_{i+1})_{i \geq 1}$. Therefore,

$$\begin{aligned} \mu\{x : G(x) \in {]}u/b, t]\} &= p_u \cdot \mu\{x : G(vx)/b \in {]}0, t - u/b]\} \\ &= p_u \cdot v(\mu)\Big(\{y : G(y) \in {]}0, bt - u]\}\Big) \end{aligned}$$

is equal to $p_u \cdot F(bt - u)$, and

$$\begin{aligned} F(t) &= F(u/b) + p_u \cdot F(bt - u) \\ &= p_0 + \cdots + p_{u-1} + p_u \cdot F(bt - u) \end{aligned}$$

as desired.   □

We have, in fact, proved something more: if $t \in {]}0, 1]$ is such that $F'(t)$ exists, is finite, and $p(X_n(t)) \neq 1/b$ for infinitely many n, then $F'(t) = 0$.

Note, incidentally, that the argument above contains the proof of the claim we made at the beginning of this section.

When $b = 2$, we can, in fact, regard $F(t)$ as the probability of success at "bold play" for a gambler whose initial capital is t (see Patrick Billingsley, *Probability and Measure*, Section 7, Wiley, 1986).

15.4   The One-Sided Shift Transformation

Let $(\Omega, \mathcal{F}, P)$ be a probability space and, for each $n \in \mathbf{N}$, let X_n be a random variable from $(\Omega, \mathcal{F})$ into a measurable space $(\Omega_n, \mathcal{F}_n)$.

Definition 15.4.1 $T = \bigcap_{n \geq 1} \sigma(X_n, X_{n+1}, \ldots)$ is called the tail σ-field associated with the sequence $(X_n)_{n \geq 1}$; its elements are called tail events.

The following result is Kolmogorov's zero-one law.

Proposition 15.4.1 If $(X_n)_{n \geq 1}$ is independent and A is a tail event, then either $P(A) = 0$ or $P(A) = 1$.

PROOF: Clearly, $\mathcal{G} = \bigcup_{k \geq 1} \sigma(X_1, \ldots, X_k)$ is an algebra and $\sigma(\mathcal{G}) = \sigma(X_1, \ldots, X_n, \ldots)$. Suppose that A is a tail event. Given $k \in \mathbf{N}$, if B belongs to $\sigma(X_1, \ldots, X_k)$, then A and B are independent, because $A \in \sigma(X_{k+1}, \ldots)$. Hence A is independent of $\sigma(\mathcal{G})$ by Theorem 15.2.2 (i.e., $\sigma(\mathcal{G})$ and the σ-algebra $\{\emptyset, \Omega, A, A^c\}$ are independent). But then A is independent of itself. So $P(A \cap A) = P(A) \cdot P(A)$, which implies that $P(A)$ is either 0 or 1. $\square$

In many cases it is very easy to use Proposition 15.4.1 to prove that a given set must have probability either 0 or 1. The difficulty arises hereafter in determining which of these values (0 or 1) we are dealing with. This problem, in fact, may be extremely difficult to solve.

Now let $(\Omega', \mathcal{F}', \mu')$ be a probability space. Put $\Omega_n = \Omega'$, $\mathcal{F}_n = \mathcal{F}'$, and $\mu_n = \mu'$, for all $n \in \mathbf{N}$, and consider the probability $\mu = \bigotimes_{n \geq 1} \mu_n$ on $\mathcal{C} = \bigotimes_{n \geq 1} \mathcal{F}_n$. The mapping $v : (x_n)_{n \geq 1} \mapsto (x_{n+1})_{n \geq 1}$ from $\Omega'^{\mathbf{N}}$ into $\Omega'^{\mathbf{N}}$ is called the (one-sided) shift transformation. The pair $(v, \mathcal{C})$ is μ-suited and $v(\mu) = \mu$.

Proposition 15.4.2 $\mu(A)$ is either 0 or 1 for each v invariant event A (i.e., for each $A \in \mathcal{C}$ such that $v^{-1}(A) = A$).

PROOF: For each $n \in \mathbf{N}$, define by Z_n the random variable $(x_i)_{i \geq 1} \mapsto x_n$ from $(\Omega'^{\mathbf{N}}, \mathcal{C})$ into $(\Omega_n, \mathcal{F}_n)$. Let A be a v-invariant event. Then A belongs to $\sigma(Z_1, \ldots, Z_n, \ldots)$, so $v^{-1}(A)$ lies in

$$\sigma(Z_1 \circ v, \ldots, Z_n \circ v, \ldots) = \sigma(Z_2, \ldots, Z_{n+1}, \ldots).$$

Inductively, we see that A belongs to $\sigma(Z_{k+1}, \ldots, Z_{k+n}, \ldots)$ for each integer $k \geq 0$. Therefore, A is a tail event, and $\mu(A)$ is either 0 or 1. $\square$

Proposition 15.4.3 v is μ-ergodic.

PROOF: Let A be an event such that $1_A \circ v = 1_A$ outside a μ-negligible set N. Then, for every integer $k \geq 0$, $1_A \circ v^k = 1_A$ outside $E = \bigcup_{n \geq 0} v^{-n}(N)$. Put $A' = \bigcup_{i \geq 0} v^{-i}(A)$ and $A'' = \bigcap_{j \geq 0} v^{-j}(A')$. Clearly, $1_{A'} = 1_A$ outside E, and $1_{A'} \circ v = 1_{A'}$ outside E. On the other hand, $1_{A''} = 1_{A'}$ outside E. Since $v^{-1}(A') \subset A'$, we have $A'' = v^{-1}(A'')$, so that $\mu(A) = \mu(A'')$ is either 0 or 1.

Next, let B be a μ-measurable set such that $1_B \circ v = 1_B$ μ-almost everywhere. There exists an event $A \subset B$ for which $B - A$ is μ-negligible. Then

$1_A \circ v = 1_B \circ v = 1_B = 1_A$ μ-almost everywhere, which shows that $\mu(B) = \mu(A)$ is either 0 or 1. Proposition 11.4.2 then proves that v is μ-ergodic.   □

The following result is the strong law of large numbers.

Theorem 15.4.1 *Let $(\Omega, \mathcal{F}, P)$ be a probability space, and let $(X_n)_{n\geq 1}$ be a sequence of independent random variables from $(\Omega, \mathcal{F})$ into $\mathbf{R}$. Assume that the X_n are identically distributed (i.e., $P_{X_n} = P_{X_1}$ for all $n \geq 1$) and P-integrable. Then $(1/n)S_n = (1/n)\sum_{1\leq k\leq n} X_k$ converges almost surely to $E(X_1) = \int X_1\, dP$ as $n \to +\infty$.*

PROOF: Write μ for the probability $\bigotimes_{n\geq 1} P_{X_n}$ on the Borel σ-algebra $\mathcal{C}$ of $\mathbf{R}^{\mathbf{N}}$. Let v be the shift transformation $(x_i)_{i\geq 1} \mapsto (x_{i+1})_{i\geq 1}$ from $\mathbf{R}^{\mathbf{N}}$ into itself. Finally, let Z_n be the function $(x_i)_{i\geq 1} \mapsto x_n$ from $\mathbf{R}^{\mathbf{N}}$ into $\mathbf{R}$, for all $n \geq 1$, and put $X = [X_n]_{n\geq 1}$. Then $Z_n = Z_1 \circ v^{n-1}$ for all $n \in \mathbf{N}$.

By G.D. Birkhoff's ergodic theorem, $(1/n)\sum_{1\leq k\leq n} Z_k$ converges μ-almost everywhere to $\int Z_n\, dP = E(X_1)$ as $n \to +\infty$. Therefore,

$$\frac{1}{n}S_n = \left(\frac{1}{n} \sum_{1\leq k\leq n} Z_k \right) \circ X$$

converges almost surely to $E(X_1)$.   □

Moreover, $(1/n)S_n$ converges to $E(X_1)$ in the seminormed space $\mathcal{L}^1_{\mathbf{R}}(P)$, because $(1/n)\sum_{1\leq k\leq n} X_k^+$ and $(1/n)\sum_{1\leq k\leq n} X_k^-$ converge, respectively, to $E(X_1^+)$ and $E(X_1^-)$ in $\mathcal{L}^1_{\mathbf{R}}(P)$ by Proposition 5.4.5.

The classical example is the strong law of large numbers for Bernoulli trials. Here $P(X_n = 1) = p$, $P(X_n = 0) = 1 - p$; S_n represents the number of successes in n trials, and $(1/n)S_n \to p$ almost surely. The idea of probability as frequency depends on the long-range stability of the success ratio $(1/n)S_n$.

15.5   Borel's Normal Number Theorem

Given an integer $b \geq 2$, let $\mathcal{F}'$ be the power set of $\Omega' = \{0, \dots, b-1\}$.

Proposition 15.5.1 *Let $(\Omega, \mathcal{F}, P)$ be a probability space, and let $(X_n)_{n\geq 1}$ be a sequence of independent and identically distributed random variables from $(\Omega, \mathcal{F})$ into $(\Omega', \mathcal{F}')$. Put $p_u = P(X_1 = u)$, for every $0 \leq u \leq b-1$. Finally, given an integer $k \geq 1$ and a k-tuple $(u_1, \dots, u_k)$ of base-b digits, denote by $N_n(u_1, \dots, u_k)$ the frequency of the k-tuple in the first $n+k-1$ trials, that is, the number of m such that $1 \leq m \leq n$ and $X_m = u_1, \dots, X_{m+k-1} = u_k$. Now $(1/n)N_n(u_1, \dots, u_k)$ converges almost surely to $p(u_1) \cdots p(u_k)$ as $n \to +\infty$.*

PROOF: For each integer $n \geq 1$, put $\Omega_n = \Omega'$, $\mathcal{F}_n = \mathcal{F}'$, and $\mu_n = P_{X_1}$. Write μ for the probability $\bigotimes_{n \geq 1} \mu_n$ on $\mathcal{C} = \bigotimes_{n \geq 1} \mathcal{F}_n$. Finally, let Z_n be the random variable $(x_i)_{i \geq 1} \mapsto x_n$ from $(\Omega'^{\mathbf{N}}, \mathcal{C})$ into $(\Omega', \mathcal{F}')$.

Denote by g_n the function $1_{\{(u_1, \ldots, u_k)\}} \circ [Z_n, \ldots, Z_{n+k-1}]$ from $\Omega'^{\mathbf{N}}$ into $\{0, 1\}$. So $g_n = g_1 \circ v^{n-1}$, if v is the shift transformation $(x_i)_{i \geq 1} \mapsto (x_{i+1})_{i \geq 1}$. By G.D. Birkhoff's ergodic theorem, $(1/n) \sum_{1 \leq j \leq n} g_j$ converges μ-almost everywhere to $\int g_1 \, d\mu = p(u_1) \cdots p(u_k)$ as $n \to +\infty$.

Next, for every $n \geq 1$, write Y_n for $1_{\{(u_1, \ldots, u_k)\}} \circ [X_n, \ldots, X_{n+k-1}]$. So $Y_n = g_n \circ X$, where $X = [X_i]_{i \geq 1}$. Since $P_X = \mu$, we see that $(1/n) \sum_{1 \leq j \leq n} Y_j$ converges almost surely to $p(u_1) \cdots p(u_k)$ as $n \to +\infty$. $\qquad \square$

Now, let $\Omega = \,]0, 1]$ and let P be Lebesgue measure on the Borel σ-algebra, $\mathcal{F}$, of $]0, 1]$. For each $n \in \mathbf{N}$, let X_n be the function from Ω into Ω' such that $X_n(t) = x_n$ for every $t \in \Omega$, where $(x_i)_{i \geq 1}$ is the nonterminating base-b expansion of t. We know that the X_n are independent random variables from $(\Omega, \mathcal{F})$ into $(\Omega', \mathcal{F}')$. Set $B = \{k/b^n : n \in \mathbf{N},\ 0 \leq k \leq b^n,\ k \text{ integer}\}$.

Definition 15.5.1 A number $t \in I - B$ is said to be completely normal (with respect to the base b) if, for every integer $k \geq 1$ and every k-tuple $(u_1, \ldots, u_k)$ of base-b digits, the k-tuple appears in the base-b expansion of t with asymptotic relative frequency $1/b^k$.

Proposition 15.5.2 (Borel) P-almost all $t \in I - B$ are completely normal numbers.

PROOF: Fix $k \geq 1$ and $(u_1, \ldots, u_k)$. For each integer $n \geq 1$, denote by $N_n(u_1, \ldots, u_k)$ the frequency of $(u_1, \ldots, u_k)$ in the first $n + k - 1$ trials. By Proposition 15.5.1, $(1/n) N_n(u_1, \ldots, u_k)$ converges P-almost surely to $1/b^k$. $\qquad \square$

Exercises for Chapter 15

1 Fix $n \in \mathbf{N}$ and write G_n for the group of permutations of $\Omega' = \{1, \ldots, n\}$. If $\sigma \in G_n$ and $\bar{\sigma} = \{\sigma^r : r \in \mathbf{Z}\}$ is the subgroup of G_n generated by σ, consider on Ω' the equivalence relation: $x \sim y$ whenever there exists $\sigma^r \in \bar{\sigma}$ such that $y = \sigma^r(x)$. The equivalence classes for this relation are called the orbits under $\bar{\sigma}$. Let $O_1, \ldots, O_p$ be the different orbits under $\bar{\sigma}$, enumerated so that O_q contains the smallest element of $\Omega' - \bigcup_{1 \leq i \leq q-1} O_i$, for all $1 \leq q \leq p$. For each $1 \leq q \leq p$, put $r_q = |O_q|$, and denote by $Y_{r_1 + \cdots + r_{q-1} + 1}(\sigma)$ the smallest element of O_q. Finally, put

$$Y_{r_1 + \cdots + r_{q-1} + r}(\sigma) = \sigma^{r-1} \left(Y_{r_1 + \cdots + r_{q-1} + 1}(\sigma) \right) \qquad \text{for all } 1 \leq r \leq r_q.$$

We say that $\left(Y_1(\sigma), \ldots, Y_n(\sigma) \right)$ is the standard cyclic representation of σ.

Let $\mathcal{F}'$ be the power set of Ω', $\mathcal{F}$ the power set of $\Omega = G_n$, and P the uniform probability on $\mathcal{F}$ ($P(\{\sigma\}) = 1/n!$ for all $\sigma \in \Omega$). Then, for every $1 \le k \le n$, Y_k is a random variable from $(\Omega, \mathcal{F})$ into $(\Omega', \mathcal{F}')$.

For every $1 \le k \le n$ and for all $\sigma \in \Omega$, define $X_k(\sigma) = 1$ or 0 as $Y_k(\sigma)$ completes an orbit or not (i.e.. $X_k(\sigma) = 1$ if there exists $1 \le q \le p$ such that $k = r_1 + \cdots + r_q$, and $X_k(\sigma) = 0$ if not). Then $\sum_{1 \le k \le n} X_k(\sigma)$ is the number of orbits under $\bar{\sigma}$.

For every $1 \le k \le n$, X_k is a random variable from $(\Omega, \mathcal{F})$ into $\mathbf{R}$. We show that the X_k ($1 \le k \le n$) are independent and that $P(X_k = 1) = 1/(n - k + 1)$, so that all the X_k have the same distribution.

1. For all $i, j \in \{1, \ldots, n\}$, consider the event $A_i(j)$ that σ sends i to j. If $i_1, \ldots, i_k$ are distinct and $j_1, \ldots, j_k$ are distinct, show that

    $$P\left(\bigcap_{1 \le u \le k} A_{i_u}(j_u) \right) = \frac{(n-k)!}{n!}.$$

2. Fix $k \in \mathbf{N}$. Given $x_1, \ldots, x_{k-1}$ in $\{0, 1\}$ and $y_2, \ldots, y_k$ in $\{2, \ldots, n\}$, let A be the event that

    $$X_1 = x_1, \ldots, X_{k-1} = x_{k-1} \quad \text{and} \quad Y_1 = 1, Y_2 = y_2, \ldots, Y_k = y_k.$$

    Assume that A is not empty. Set $x_0 = 1$. For all $1 \le u \le k - 1$ such that $x_u = 0$, put $j_u = y_{u+1}$. On the other hand, if $1 \le u \le k - 1$ is such that $x_u = 1$ and if $0 \le v < u$ is the largest index $< u$ for which $x_v = 1$, put $j_u = y_{v+1}$. Show that $A = \bigcap_{1 \le u \le k-1} A_{y_u}(j_u)$ and $P(A) = (n-k+1)!/n!$.

3. Using the notation of part 2, let $0 \le w < k$ be the largest integer in $\{0, 1, \ldots, k-1\}$ such that $x_w = 1$, and put $j_k = y_{w+1}$. Show that

    $$A \cap (X_k = 1) = \bigcap_{1 \le u \le k} A_{y_u}(j_u)$$

    and

    $$P\left(A \cap (X_k = 1) \right) = \frac{1}{n - k + 1} P(A).$$

4. For fixed $1 \le k \le n$, summing over the possible sets A, show that

    $$P(X_k = 1) = \frac{1}{n - k + 1}.$$

    Conclude that $P\left(A \cap (X_k = 1) \right) = P(A)P(X_k = 1)$ for each A.

5. Deduce from Theorem 15.2.4 that $[X_1, \ldots, X_{k-1}, Y_1, \ldots, Y_k]$ and X_k are independent. Then $[X_1, \ldots, X_{k-1}]$ and X_k are independent. Conclude, by induction, that $X_1, \ldots, X_k$ are independent for all $1 \le k \le n$.

2 Fix $n \in \mathbf{N}$, and let $(\Omega, \mathcal{F}, P)$ be as in Exercise 1. For each $1 \le k \le n$, if

$$\sigma = \begin{pmatrix} 1 & , & 2 & , & \cdots & , & n \\ \sigma(1) & , & \sigma(2) & , & \cdots & , & \sigma(n) \end{pmatrix} \quad \text{is a permutation of } \{1, \ldots, n\},$$

we define $X_k(\sigma)$ as the number of elements between 1 and $k-1$ lying to the right of k in the bottom row. $X_k : \sigma \mapsto X_k(\sigma)$ is a random variable from $(\Omega, \mathcal{F})$ into $\mathbf{R}$. The sum $S_n = X_1 + \cdots + X_n$ is the total number of inversions. We show that $X_1, \ldots, X_n$ are independent and that $P(X_k = i) = 1/k$ for all $0 \le i \le k - 1$.

1. Let $(i_1, \ldots, i_n)$ be a sequence of length n such that i_k belongs to $\{0, 1, \ldots, k - 1\}$ for every $1 \le k \le n$. Construct $\sigma^{-1}(n), \ldots, \sigma^{-1}(1)$ with the following properties:

    (a) $\sigma^{-1}(n) = n - i_n$.

    (b) For each $1 \le k \le n$, $\sigma^{-1}(n), \ldots, \sigma^{-1}(n-k+1)$ are distinct elements of Ω'.

    (c) For each $1 \le k \le n$ and all $1 \le l < k$, there are $n - \sigma^{-1}(n - l) - i_{n-l}$ terms $m \in \{n - l + 1, \ldots, n\}$ such that $\sigma^{-1}(m) > \sigma^{-1}(n - l)$.

    (Argue by induction on k.) Conclude that there exists $\sigma \in \Omega$ for which $X_k(\sigma) = i_k$ for all $1 \le k \le n$.

2. Deduce from part 1 that the mapping $\sigma \mapsto (X_k(\sigma))_{1 \le k \le n}$ from Ω into $\{0\} \times \{0, 1\} \times \cdots \times \{0, \ldots, n - 1\}$ is bijective.

3. For every $1 \le k \le n$ and every $(i_1, \ldots, i_k)$ in $\{0\} \times \cdots \times \{0, 1, \ldots, k - 1\}$, show that $P(X_1 = i_1, \ldots, X_k = i_k) = 1/k!$. Conclude that $X_1, \ldots, X_n$ are independent.

4. Prove that $E(S_n) = (n(n-1))/4$ and that $\operatorname{Var}(S_n) = (1/72)n(2n^2 + 3n - 5)$.

3 Let $(\Omega, \mathcal{F}, P)$ be a probability space and $(A_n)_{n \ge 1}$ a sequence of events such that $\sum_{n \ge 1} P(A_n) < +\infty$. Prove that $P(\limsup A_n) = 0$ (first Borel–Cantelli lemma).

4 Let $(\Omega, \mathcal{F}, P)$ be a probability space, and let $(A_n)_{n \ge 1}$ be a sequence of events such that $\sum_{n \ge 1} P(A_n)$ is infinite. For each integer $n \ge 1$ such that $\sum_{1 \le k \le n} P(A_k) > 0$, put

$$\theta_n = \sum_{i, j \le n} P(A_i \cap A_j) \Big/ \Big(\sum_{k \le n} P(A_k) \Big)^2.$$

1. For every $n \in \mathbf{N}$, put $N_n = 1_{A_1} + \cdots + 1_{A_n}$ and $m_n = E(N_n) = p_1 + \cdots + p_n$ (where $p_k = P(A_k)$ for all $1 \le k \le n$). When $m_n > 0$, show that

$$\operatorname{Var}(N_n) = \int (N_n - m_n)^2 \, dP = (\theta_n - 1)m_n^2$$

and hence $\theta_n \ge 1$. Now set $c = \liminf_{n \to +\infty} \theta_n$.

2. Let $x > 0$ be given. For each integer n such that $m_n > x$, prove that

$$P(N_n \le x) \le P(|N_n - m_n| \ge m_n - x) \le (\theta_n - 1)\frac{m_n^2}{(m_n - x)^2}.$$

Conclude that $\liminf_{n \to +\infty} P(N_n \le x) \le c - 1$ and $P(\sup_k N_k \le x) \le c - 1$.

3. Deduce from part 2 that $P(A_n \text{ infinitely often}) \geq 1 - (c - 1) = 2 - c$ (Rényi–Lamperti lemma).

4. When the events A_n are independent, we have, in fact, the second Borel–Cantelli lemma. Prove this.

5 Let $(\Omega, \mathcal{F}, P)$ be a probability space. Write $\mathcal{F}'$ for the class of all subsets of $\Omega' = \{0, 1\}$. Let $(X_n)_{n \geq 1}$ be a sequence of independent and identically distributed random variables from $(\Omega, \mathcal{F})$ into $(\Omega', \mathcal{F}')$, such that $0 < p = P(X_1 = 0) < 1$. For simplicity, assume that $(X_n(\omega))_{n \geq 1}$ contains infinitely many nonzero elements, for every $\omega \in \Omega$.

For each $n \in \mathbf{N}$, define a function $l_n : \Omega \to \mathbf{Z}^+$ as follows: $l_n(\omega) = k$ if $X_n(\omega) = \ldots = X_{n+k-1}(\omega) = 0$ and $X_{n+k}(\omega) \neq 0$. Hence $l_n(\omega)$ is the length of the run of zeros starting at n. Consider the σ-algebra in $\mathbf{Z}^+$ consisting of all subsets of $\mathbf{Z}^+$. Clearly, l_n is a random variable and $P(l_n = k) = p^k(1 - p)$ for all $k \geq 0$. So $P(l_n \geq k) = p^k$.

Finally, let $(r_n)_{n \geq 1}$ be a sequence of positive integers.

1. If $\sum_{n \geq 1} p^{r_n} < +\infty$, deduce from Exercise 3 that

$$P\big(\{\omega : l_n(\omega) \geq r_n \quad \text{i.o.}\}\big) = 0$$

(where i.o. is an abbreviation for "infinitely often").

2. Assume, in what follows, that $\sum_{n \geq 1} p^{r_n} = +\infty$. Let

$$A_n = (l_n \geq r_n) = (X_n = \ldots = X_{n+r_n-1} = 0).$$

For every $n \in \mathbf{N}$, prove that

$$\sum_{i,j \leq n} P(A_i \cap A_j) \leq \sum_{k \leq n} P(A_k) + \left(\sum_{k \leq n} P(A_k)\right)^2 + 2\frac{p}{1-p}\sum_{k \leq n} P(A_k).$$

3. If $\sum_{n \geq 1} p^{r_n} = +\infty$, deduce from Exercise 4 that $P\big(l_n \geq r_n \quad \text{i.o.}\big) = 1$.

4. Prove that $(l_n \geq r_n \quad \text{i.o.})$ is a tail event for $(X_n)_{n \geq 1}$.

6 Write Ω for $]0, 1]$, $\mathcal{F}$ for the Borel σ-algebra in Ω, and P for Lebesgue measure on $\mathcal{F}$.

Fix an integer $b \geq 2$. For each $n \in \mathbf{N}$, define $X_n :]0, 1] \to \{0, 1\}$ as follows: $X_n(t) = 0$ or 1 as $x_n = 0$ or not, if $(x_i)_{i \geq 1}$ is the nonterminating base-b expansion of t. Use Exercise 5 to obtain results on base-b expansions.

7 We refer to the notation of Section 15.3, and assume that $b = 2$. Show that $F(t) \leq t$ for all $t \in I$ or $F(t) \geq t$ for all $t \in I$, as $p_0 \leq 1/2$ or $p_0 \geq 1/2$.

8 In the notation of Section 15.3, assume that $b = 3$, $p_0 = p_2 = 1/2$, and $p_1 = 0$. Show that F is the Cantor singular function.

9 In the notation of Section 15.3, let $Q : I \to \mathbf{R}$ be bounded, such that $Q(t) = p_0 + \cdots + p_{u-1} + p_u \cdot Q(bt - u)$ for every integer $0 \le u \le b - 1$ and every $t \in [u/b, (u + 1)/b]$.

1. Prove that $Q(t) = Q(k/b^n) + p(u_1) \cdots p(u_n) \cdot Q(b^n t - k)$ for all integers $n \ge 1$, $0 \le k < b^n$, and all t in $[k/b^n, (k+1)/b^n]$, where $u_1/b + \cdots + u_n/b^n$ is the proper expansion of k/b^n.

2. Show that $Q = F$.

16

The Central Limit Theorem

Let $(X_n)_{n \geq 1}$ be a sequence of independent and identically distributed random variables, taking their values in $\mathbf{R}^d$ (with $d \geq 1$). Assume that each X_n is square-integrable. Denote by D the covariance matrix of X_n. Then $(1/\sqrt{n})(S_n - nE(X_1))$, where $S_n = X_1 + \cdots + X_n$, converges in law to the normal law $N_d(0, D)$. This (central limit) theorem and its generalization, the Lindeberg theorem, are of utmost importance in probability theory.

Summary

16.1 We study some properties of the convergence in law of random variables.

16.2 Let $(X_{n,k})_{\substack{n \geq 1 \\ 1 \leq k \leq r_n}}$ be a triangular array of independent, centered, and square-integrable random variables. For every $n \geq 1$, write $s_n = \left[\sum_{1 \leq k \leq r_n} \mathrm{Var}(X_{n,k}) \right]^{1/2}$ and $S_n = \sum_{1 \leq k \leq r_n} X_{n,k}$. The Lindeberg condition is sufficient for S_n/s_n to converge in law to the normal law (Theorem 16.2.1). Note, incidentally, that this condition is also necessary in the most usual cases (as a consequence of a Feller's theorem, which is not proved here).

16.3 We prove the central limit theorem (Theorem 16.3.1), as well as some refinements of this theorem.

16.1 Convergence in Law

Definition 16.1.1 Let Ω' be a locally compact Hausdorff space whose topology has a countable basis, and let μ be a probability on the Borel σ-algebra $\mathcal{F}'$ of Ω'. Finally, let $((\Omega_n, \mathcal{F}_n, P_n))_{n \geq 1}$ be a sequence of probability spaces and, for all $n \in \mathbf{N}$, let X_n be a random variable from $(\Omega_n, \mathcal{F}_n)$ into $(\Omega', \mathcal{F}')$. Then $(X_n)_{n \geq 1}$ is said to converge in law (or in distribution) to μ if $P_n(X_n)$ converges narrowly to μ. When μ is the law of a given random variable X, the sequence $(X_n)_{n \geq 1}$ is also said to converge in law to X, and we write $X_n \Rightarrow \mu$ or $X_n \Rightarrow X$.

Hence $(X_n)_{n \geq 1}$ converges to μ in law if $(\int f \circ X_n \, dP_n)_{n \geq 1}$ converges to $\int f \, d\mu$ for each $f \in C^b(\Omega', \mathbf{C})$ (or, by Theorem 7.7.2, for each $f \in \mathcal{H}(\Omega', \mathbf{C})$).

Theorem 16.1.1 (Mapping Theorem) *Let Ω' (respectively, Ω'') be a locally compact Hausdorff space whose topology has a countable basis, and let $\mathcal{F}'$ (respectively, $\mathcal{F}''$) be the Borel σ-algebra of Ω' (respectively, Ω''). Let g be a random variable from $(\Omega', \mathcal{F}')$ into $(\Omega'', \mathcal{F}'')$. Finally, let μ be a probability on $\mathcal{F}'$, and assume that g is continuous at μ-almost all points of Ω'. If $(X_n)_{n \geq 1}$ converges in law to μ, then $(g \circ X_n)_{n \geq 1}$ converges in law to $g(\mu)$.*

PROOF: For all $n \in \mathbf{N}$, put $\mu_n = P_n(X_n)$. Denote by D the set of discontinuities of g. If F is a closed subset of Ω'', $\overline{g^{-1}(F)}$ is included in $g^{-1}(F) \cup D$, so

$$\limsup g(\mu_n)(F) \leq \limsup \mu_n \left(\overline{g^{-1}(F)} \right) \leq \mu \left(\overline{g^{-1}(F)} \right)$$
$$= \mu(g^{-1}(F)) = g(\mu)(F),$$

and the conclusion follows from Proposition 7.7.6. $\qquad\qquad\square$

Theorem 16.1.1, though elementary, is useful.

Proposition 16.1.1 *Let $(\Omega', \mathcal{F}')$ and the X_n be as in Definition 16.1.1. Given $c \in \Omega'$, let ε_c be the measure on $\mathcal{F}'$ defined by the unit mass at c. Then, in order that $(X_n)_{n \geq 1}$ converge in law to ε_c, it is necessary and sufficient that $P_n(X_n \notin V)$ converge to 0 as $n \to +\infty$, for each compact neighborhood V of c.*

PROOF: Suppose the condition holds, and let $f \in \mathcal{H}(\Omega', \mathbf{C})$. Given $\varepsilon > 0$, there exists a compact neighborhood V of c such that $|f(x) - f(c)| \leq \varepsilon/2$ for all $x \in V$. Then

$$\int f \circ X_n \, dP_n - f(c) = \int_{(X_n \in V)} (f \circ X_n - f(c)) \, dP_n$$
$$+ \int_{(X_n \notin V)} (f \circ X_n - f(c)) \, dP_n,$$

so

$$\left| \int f \circ X_n \, dP_n - f(c) \right| \leq \frac{\varepsilon}{2} + 2\|f\| P_n(X_n^{-1}(V^c)) \quad \text{for all } n \in \mathbf{N},$$

and $\left| \int f \circ X_n \, dP_n - f(c) \right|$ is less than ε if n is large enough.

Conversely, assume that $(X_n)_{n\geq 1}$ converges in law to ε_c, and let V be a compact neighborhood of c. If $f : \Omega' \to [0,1]$ is continuous, vanishes at c, and is equal to 1 on V^c, then

$$P(X_n \notin V) \leq \int f \circ X_n \, dP_n = \int f \circ X_n \, dP_n - f(c) \quad \text{for each } n \in \mathbf{N},$$

so $P(X_n \notin V)$ converges to 0 as $n \to +\infty$. $\square$

Theorem 16.1.2 *Let Ω' be a locally compact, metrizable, separable uniform space, and let $\mathcal{F}'$ be its Borel σ-algebra. Given a probability space $(\Omega, \mathcal{F}, P)$, let $(X_n)_{n\geq 1}$ be a sequence of random variables from $(\Omega, \mathcal{F})$ into $(\Omega', \mathcal{F}')$, and let X be a random variable from $(\Omega, \mathcal{F})$ into $(\Omega', \mathcal{F}')$. If $(X_n)_{n\geq 1}$ converges in probability to X, then it converges to X in law. Conversely, if X is almost surely equal to a constant and if $(X_n)_{n\geq 1}$ converges to X in law, then it converges to X in probability.*

PROOF: Assume that $(X_n)_{n\geq 1}$ converges in probability to X, and let $f \in \mathcal{H}(\Omega', \mathbf{C})$. f is uniformly continuous on Ω', so, given $\varepsilon > 0$, there is a closed entourage W of Ω' such that $|f(x) - f(y)| \leq \varepsilon/2$ for all $(x,y) \in W$. Choose $n_0 \in \mathbf{N}$ so that

$$P\left(\left\{\omega : (X_n(\omega), X(\omega)) \notin W\right\}\right) \leq \frac{\varepsilon}{4\|f\|} \quad \text{for all } n \geq n_0.$$

Then, for all $n \geq n_0$, we have

$$\left|\int (f \circ X_n - f \circ X)\, dP\right| \leq \int_{[X_n, X]^{-1}(W)} |f \circ X_n - f \circ X|\, dP$$

$$+ \int_{[X_n, X]^{-1}(W^c)} |f \circ X_n - f \circ X|\, dP \leq \varepsilon,$$

which proves that $(X_n)_{n\geq 1}$ converges in law to X.

The second assertion of the statement follows from Proposition 16.1.1. $\square$

Now take for Ω' the real line and for X a random variable from $(\Omega, \mathcal{F})$ into $\mathbf{R}$ uniformly distributed over $[-1, 1]$ ($P_{X/S} = \frac{1}{2} \cdot 1_{[-1,1]}\, dx$). For all $n \in \mathbf{N}$, put $X_n = -X$. Observe that $(X_n)_{n\geq 1}$ converges to X in law, but not in probability, and that $X_n - X$ does not converge to 0 in law.

In what follows, we take for Ω' a Euclidean space $\mathbf{R}^k$ ($k \in \mathbf{N}$) and for $\mathcal{F}'$ the Borel σ-algebra of $\mathbf{R}^k$.

In the notation of Definition 16.1.1, let F_n be the distribution function of X_n, for all $n \in \mathbf{N}$, and let F be the distribution function of μ. Then $(X_n)_{n\geq 1}$ converges to μ in law if and only if $\big(F_n(x)\big)_{n\geq 1}$ converges to $F(x)$ for every point x at which F is continuous (Proposition 14.2.2). On the other hand,

$(X_n)_{n\geq 1}$ converges to μ in law if and only if $(\int e^{iX_n y}\, dP_n)_{n\geq 1}$ converges to $\int e^{ixy}\, d\mu(x)$ for every $y \in \mathbf{R}^k$ (Theorem 14.4.4).

Now we consider simple propositions, which are, nonetheless, very useful. For simplicity, we give $\mathbf{R}^k$ the norm $|x| = \sup |x_i|$.

Proposition 16.1.2 *Let $\left(\mu^{(v)}\right)_{v\geq 1}$ be a sequence of probabilities on $\mathcal{F}'$ converging weakly to a probability μ. For each $n \in \mathbf{N}$, let $(\Omega_n, \mathcal{F}_n, P_n)$ be a probability space and let Y_n be a random variable from $(\Omega_n, \mathcal{F}_n)$ into $\mathbf{R}^k$. Finally, for each $v \geq 1$, let $\left(X_n^{(v)}\right)_{n\geq 1}$ be a sequence of random variables from $(\Omega_n, \mathcal{F}_n)$ into $\mathbf{R}^k$ which converges in law to $\mu^{(v)}$. Assume that*

$$\lim_{v\to +\infty}\left[\limsup_{n\to +\infty} P_n\left(|X_n^{(v)} - Y_n| > \varepsilon\right)\right] = 0 \quad \text{for every } \varepsilon > 0.$$

Then $(Y_n)_{n\geq 1}$ converges in law to μ.

PROOF: Define on $\mathbf{R}^k$ the order $x \leq y$ if $x_i \leq y_i$ for all $1 \leq i \leq k$. Let $F^{(v)}$ be the distribution function of $\mu^{(v)}$, for all $v \in \mathbf{N}$, and let F be the distribution function of μ.

Suppose F is continuous at $x \in \mathbf{R}^k$. For each real number h, put $B_h = \{y \in \mathbf{R}^k : y_i \leq x_i + h \;\; \forall\, 1 \leq i \leq k\}$. Except for countably many $h \in \mathbf{R}$, the boundary ∂B_h of B_h is negligible for μ and all the $\mu^{(v)}$. Now choose $h' > 0$ and $h'' > 0$ such that F and all the $F^{(v)}$ are continuous at $y' = (x_1 - h', \ldots, x_k - h')$ and $y'' = (x_1 + h'', \ldots, x_k + h'')$. Then, for all $n \in \mathbf{N}$ and for all $v \in \mathbf{N}$,

$$P_n\left(X_n^{(v)} \leq y'\right) - P_n\left(|X_n^{(v)} - Y_n| > h'\right) \leq P_n(Y_n \leq x)$$
$$\leq P_n\left(X_n^{(v)} \leq y''\right)$$
$$+ P_n\left(|X_n^{(v)} - Y_n| > h''\right).$$

Letting n, and then v, tend to $+\infty$, we obtain

$$\mu(\,]-\infty, y'\,]) \leq \liminf P_n(Y_n \leq x)$$
$$\leq \limsup P_n(Y_n \leq x)$$
$$\leq \mu(\,]-\infty, y''\,]).$$

Finally, letting h', h'' tend to 0, we see that

$$\mu(\,]-\infty, x\,]) = \lim_{n\to +\infty} P_n(Y_n \leq x).$$

$\square$

Theorem 16.1.3 *For each $n \in \mathbf{N}$, let $(\Omega_n, \mathcal{F}_n, P_n)$ be a probability space, and let X_n, Y_n be two random variables from $(\Omega_n, \mathcal{F}_n)$ into $\mathbf{R}^k$. If $(X_n)_{n\geq 1}$ converges in law to a probability μ and if $(X_n - Y_n)_{n\geq 1}$ converges in law to ε_0, then $(Y_n)_{n\geq 1}$ converges in law to μ.*

PROOF: Take $X_n^{(v)} = X_n$ and $\mu^{(v)} = \mu$ in Proposition 16.1.2. $\qquad\qquad\square$

Proposition 16.1.3 Let $\left((\Omega_n, \mathcal{F}_n, P_n)\right)_{n \geq 1}$ be a sequence of probability spaces and, for all $n \in \mathbf{N}$, let X_n be a random variable from $(\Omega_n, \mathcal{F}_n)$ into $\mathbf{R}^k$. Assume that $(X_n)_{n \geq 1}$ converges in law to a probability μ. Then $a_n X_n \Rightarrow \varepsilon_0$ for each sequence $(a_n)_{n \geq 1}$ of real numbers converging to 0.

PROOF: Fix $\delta > 0$. Given $\varepsilon > 0$, choose $r > 0$ so that $B = [-r, r]^k$ is μ-quadrable and $\mu(\mathbf{R}^k - B) \leq \varepsilon/2$. For a suitably chosen integer $n_0 \geq 1$, we have

$$|a_n| \leq \delta/r \quad \text{and} \quad \left| P_n(|X_n| > r) - \mu(\mathbf{R}^k - B) \right| \leq \varepsilon/2 \quad \text{for all } n \geq n_0,$$

so $P_n(|a_n X_n| > \delta) \leq P_n(|X_n| > r) \leq \varepsilon$. In short, $P_n(|a_n X_n| > \delta)$ converges to 0 as $n \to +\infty$. Therefore, $a_n X_n \Rightarrow \varepsilon_0$ (Proposition 16.1.1). $\qquad\square$

Theorem 16.1.4 Let $\left((\Omega_n, \mathcal{F}_n, P_n)\right)_{n \geq 1}$, $(X_n)_{n \geq 1}$, and μ be as in Proposition 16.1.3. Let $(a_n)_{n \geq 1}$ be a sequence going to a in $]0, +\infty[$. and let $(b_n)_{n \geq 1}$ be a sequence converging to b in $\mathbf{R}^k$. Then $(a_n X_n + b_n)_{n \geq 1}$ converges in law to the probability ν on $\mathcal{F}'$, the image of μ under $x \mapsto ax + b$.

PROOF: $aX_n + b \Rightarrow \nu$ by the mapping theorem. Next, $(a_n - a)X_n \Rightarrow \varepsilon_0$ by Proposition 16.1.3. Hence $(a_n X_n + b_n) - (aX_n + b) = (a_n - a)X_n + (b_n - b)$ converges in law to ε_0 as $n \to +\infty$ (Proposition 16.1.1). Then, by Theorem 16.1.3, $a_n X_n + b_n \Rightarrow \nu$. $\qquad\square$

16.2   The Lindeberg Theorem

Fix $d \in \mathbf{N}$. Write $\mathcal{F}'$ for the Borel σ-algebra of $\Omega' = \mathbf{R}^d$. Give $\mathbf{R}^d$ its Euclidean inner product and, for each $t \in \mathbf{R}^d$, define $h_t : \mathbf{R}^d \ni x \mapsto tx$.

Proposition 16.2.1 Let $\left((\Omega_n, \mathcal{F}_n, P_n)\right)_{n \geq 1}$ be a sequence of probability spaces and, for each $n \geq 1$, let X_n be a random variable from $(\Omega_n, \mathcal{F}_n)$ into $\mathbf{R}^d$. Finally, let μ be a probability on $\mathcal{F}'$. Then $X_n \Rightarrow \mu$ if and only if $tX_n \Rightarrow h_t(\mu)$ for each $t \in \mathbf{R}^d$ such that $|t| = 1$.

PROOF: Suppose the condition holds. Then $tX_n \Rightarrow h_t(\mu)$ for each $t \in \mathbf{R}^d$, by the mapping theorem, and hence

$$E(e^{istX_n}) \to \int e^{istx} \, d\mu(x) \quad \text{for all real } s.$$

Taking $s = 1$ shows that the characteristic function of X_n converges pointwise to that of μ. Therefore, $X_n \Rightarrow \mu$.

Conversely, if $X_n \Rightarrow \mu$, then $tX_n \Rightarrow h_t(\mu)$ for each $t \in \mathbf{R}^d$, by the mapping theorem. $\qquad\qquad\qquad\qquad\qquad\qquad\qquad\qquad\qquad\qquad\qquad\qquad\quad$ □

By means of Proposition 16.2.1, certain limit theorems can be reduced routinely to the one-dimensional case.

In view of the sequel, we need the following lemma.

Lemma 16.2.1 *For every $n \in \mathbf{Z}^+$ and all $x \in \mathbf{R}$,*

$$\left| e^{ix} - \sum_{0 \leq k \leq n} \frac{(ix)^k}{k!} \right| \leq \inf \left(\frac{|x|^{n+1}}{(n+1)!}, \frac{2|x|^n}{n!} \right).$$

PROOF: Let $x \in \mathbf{R}$. Integration by parts shows that

$$\int_0^x (x - s)^n e^{is} \, ds = \frac{x^{n+1}}{n+1} + \frac{i}{n+1} \int_0^x (x - s)^{n+1} e^{is} \, ds \qquad (1),$$

and it follows by induction that

$$e^{ix} = \sum_{0 \leq k \leq n} \frac{(ix)^k}{k!} + \frac{i^{n+1}}{n!} \int_0^x (x - s)^n e^{is} \, ds \qquad (2)$$

for $n \geq 0$. Replace n by $n - 1$ in (1), solve for the integral on the right, and substitute this for the integral in (2); this gives

$$e^{ix} = \sum_{0 \leq k \leq n} \frac{(ix)^k}{k!} + \frac{i^n}{(n-1)!} \int_0^x (x - s)^{n-1} (e^{is} - 1) \, ds \qquad (3)$$

for $n \geq 1$. Estimating the integrals in (2) and (3) (consider separately the cases $x \geq 0$ and $x < 0$) now leads to

$$\left| e^{ix} - \sum_{0 \leq k \leq n} \frac{(ix)^k}{k!} \right| \leq \min \left(\frac{|x|^{n+1}}{(n+1)!}, 2\frac{|x|^n}{n!} \right).$$

$\qquad\qquad\qquad\qquad\qquad\qquad\qquad\qquad\qquad\qquad\qquad\qquad\qquad\qquad\qquad\quad$ □

In the last inequality, the first term on the right gives a sharp estimate for $|x|$ small, the second an estimate for $|x|$ large.

Now suppose that, for each $n \in \mathbf{N}$, $X_{n,1}, \ldots, X_{n,r_n}$ are random variables from a probability space $(\Omega_n, \mathcal{F}_n, P_n)$ into $\mathbf{R}^d$; $(\Omega_n, \mathcal{F}_n, P_n)$ may change with n. Such a collection is called a triangular array of random variables. Exercises 1 and 2 of Chapter 15 show the usefulness of triangular arrays.

For all $n \in \mathbf{N}$, put $S_n = X_{n,1} + \cdots + X_{n,r_n}$. Suppose that $X_{n,1}, \ldots, X_{n,r_n}$ are independent, each $X_{n,k}$ $(1 \leq k \leq r_n)$ is of order 2, $E(X_{n,k}) = 0$, and $D_n = D(S_n) = I_d$.

To establish the asymptotic normality of S_n, it is necessary to expand the characteristic function of each $X_{n,k}$ to the second-order terms, and to estimate the remainder. The following Lindeberg condition assures a successful result:

$$\lim_{n \to +\infty} \sum_{1 \le k \le r_n} \int_{|X_{n,k}| > \varepsilon} |X_{n,k}|^2 \, dP_n = 0 \qquad \text{for } \varepsilon > 0.$$

Theorem 16.2.1 (Lindeberg) *Suppose that, for each n, the sequence $X_{n,1}, \ldots, X_{n,r_n}$ is independent and satisfies $E(X_{n,k}) = 0$ and $D(S_n) = I_d$. If Lindeberg condition holds for all $\varepsilon > 0$, then $S_n \Rightarrow N_d(0, I_d)$.*

PROOF:

First part of the proof:

Suppose that $d = 1$. For each $n \ge 1$ and each $1 \le k \le r_n$, put $\sigma_{n,k}^2 = E(X_{n,k}^2)$. By hypothesis, $\sum_{1 \le k \le r_n} \sigma_{n,k}^2 = 1$.

To begin, notice that

$$|z_1 \cdots z_m - w_1 \cdots w_m| \le \sum_{1 \le k \le m} |z_k - w_k| \qquad (1)$$

for all complex numbers $z_1, \ldots, z_m, w_1, \ldots, w_m$ of modulus at most 1. Indeed, this follows by induction from

$$z_1 \cdots z_m - w_1 \cdots w_m = (z_1 - w_1) z_2 \cdots z_m + w_1 (z_2 \cdots z_m - w_2 \cdots w_m).$$

Now, by Lemma 16.2.1,

$$\left| e^{itx} - \left(1 + itx - \frac{1}{2} t^2 x^2\right) \right| \le \min\left(|tx|^2, |tx|^3 \right).$$

Therefore, the characteristic function $\varphi_{n,k}$ of $X_{n,k}$ satisfies

$$\left| \varphi_{n,k}(t) - \left(1 - \frac{1}{2} t^2 \sigma_{n,k}^2\right) \right| \le E\left[\min(|tX_{n,k}|^2, |tX_{n,k}|^3) \right]. \qquad (2)$$

For $0 < \varepsilon \le 1$, the right-hand side of (2) is at most

$$\int_{|X_{n,k}| \le \varepsilon} |tX_{n,k}|^3 \, dP_n + \int_{|X_{n,k}| > \varepsilon} |tX_{n,k}|^2 \, dP_n \le$$

$$\varepsilon |t|^3 \sigma_{n,k}^2 + t^2 \int_{|X_{n,k}| > \varepsilon} X_{n,k}^2 \, dP_n.$$

Since the $\sigma_{n,k}^2$ total to 1 and ε is arbitrary, it follows by the Lindeberg condition that

$$\sum_{1 \le k \le r_n} \left| \varphi_{n,k}(t) - \left(1 - \frac{1}{2} t^2 \sigma_{n,k}^2\right) \right| \to 0 \quad \text{for each fixed } t \qquad (3)$$

The objective now is to show that

$$\prod_{1 \leq k \leq r_n} \varphi_{n,k}(t) = \prod_{1 \leq k \leq r_n} \left(1 - \frac{1}{2} t^2 \sigma_{n,k}^2\right) + o(n^0)$$

$$= \prod_{1 \leq k \leq r_n} \exp\left(-\frac{1}{2} t^2 \sigma_{n,k}^2\right) + o(n^0)$$

$$= \exp\left(-\frac{t^2}{2}\right) + o(n^0) \qquad (4).$$

For $\varepsilon > 0$, $\sigma_{n,k}^2 \leq \varepsilon^2 + \int_{|X_{n,k}| > \varepsilon} X_{n,k}^2 \, dP_n$, and so it follows by the Lindeberg condition that

$$\max_{1 \leq k \leq r_n} \sigma_{n,k}^2 \to 0 \qquad (5).$$

For large enough n, $1 - (1/2) t^2 \sigma_{n,k}^2$ are all between 0 and 1, and by (1) $\prod_{1 \leq k \leq r_n} \varphi_{n,k}(t)$ and $\prod_{1 \leq k \leq r_n}(1 - (1/2) t^2 \sigma_{n,k}^2)$ differ by at most the sum in (3). This establishes the first of the asymptotic relations in (4).

Next, (1) also implies that

$$\left| \prod_{1 \leq k \leq r_n} \exp\left(-\frac{1}{2} t^2 \sigma_{n,k}^2\right) - \prod_{1 \leq k \leq r_n} \left(1 - \frac{1}{2} t^2 \sigma_{n,k}^2\right) \right| \leq$$

$$\sum_{1 \leq k \leq r_n} \left| \exp\left(-\frac{1}{2} t^2 \sigma_{n,k}^2\right) - 1 + \frac{1}{2} t^2 \sigma_{n,k}^2 \right|.$$

For $x \geq 0$, $|e^{-x} - 1 + x| = \left| \int_0^x e^{-u}(x - u) \, du \right| \leq \int_0^x (x - u) \, du = x^2/2$. Using this in the right-hand member of the preceding inequality, we obtain the bound $(1/8) t^4 \sum_{1 \leq k \leq r_n} \sigma_{n,k}^4$; by (5) and the fact that $\sum_{1 \leq k \leq r_n} \sigma_{n,k}^2 = 1$, this sum goes to 0, from which the second equality in (4) follows.

Now, the characteristic function $\prod_{1 \leq k \leq r_n} \varphi_{n,k}$ of S_n converges pointwise to that of $N_1(0, 1)$. This completes the proof when $d = 1$.

Second part of the proof:

We now pass to the case where d is not necessarily equal to 1.

Let $t \in \mathbf{R}^d$ be given, such that $|t| = 1$. For every $n \in \mathbf{N}$, put $\sigma_{n,k}^2 = \mathrm{Var}(tX_{n,k})$ for all $1 \leq k \leq r_n$, so that

$$\sum_{1 \leq k \leq n} \sigma_{n,k}^2 = D(tS_n) = t' D(S_n) t = |t|^2 = 1.$$

The Lindeberg condition shows that

$$\lim_{n \to +\infty} \sum_{1 \leq k \leq r_n} \int_{|tX_{n,k}| > \varepsilon} |tX_{n,k}|^2 \, dP_n \to 0 \qquad \text{for } \varepsilon > 0.$$

Therefore, $tS_n \Rightarrow N_1(0, 1) = h_t(N_d(0, I_d))$. By Proposition 16.2.1, we may conclude that $S_n \Rightarrow N_d(0, I_d)$. $\qquad \square$

Now we illustrate the Lindeberg theorem by two examples. First, we retain the notation of Chapter 15, Exercise 1, but we write $X_{n,k}$ (for $1 \leq k \leq n$) instead of X_k, and P_n instead of P. For each $n \in \mathbf{N}$, we put

$$S_n = \sum_{1 \leq k \leq n} X_{n,k},$$

$$L_n = E(S_n) = \sum_{1 \leq k \leq n} (1/k),$$

$$s_n^2 = \mathrm{Var}(S_n) = \sum_{1 \leq k \leq n} \frac{1}{k} - \sum_{1 \leq k \leq n} (1/k^2),$$

and

$$s_n = \left[s_n^2\right]^{1/2}.$$

Then we apply the Lindeberg theorem to $\left(X_{n.k} - E(X_{n,k})\right)/s_n$, and see that $(S_n - L_n)/s_n$ converges in law to $N_1(0,1)$. Since $a_n = s_n/\sqrt{\log n}$ converges to 1 and $b_n = (L_n - \log n)/\sqrt{\log n}$ converges to 0 as $n \to +\infty$, we conclude that $(S_n - \log n)/\sqrt{\log n} = a_n(S_n - L_n)/s_n + b_n \Rightarrow N_1(0,1)$. This result is known as Goncharov's theorem. In particular,

$$\text{for every } \varepsilon > 0 \qquad P_n\left(\left|\frac{S_n}{\log n} - 1\right| > \varepsilon\right) \text{ approaches 0 as } n \to +\infty,$$

that is, most permutations of $\{1, \ldots, n\}$ have about $\log n$ cycles.

Next, we use the notation of Chapter 15, Exercise 2, but we write P_n instead of P, and $X_{n,k}$ (for $1 \leq k \leq n$) instead of X_k. We put $s_n = \sqrt{\mathrm{Var}(S_n)}$. Then $(n-1)/s_n$ converges to 0 as $n \to +\infty$, whereas $|X_{n.k} - E(X_{n,k})| \leq n - 1$ for all $1 \leq k \leq n$. By the Lindeberg theorem, $(1/s_n)[S_n - n(n-1)/4] \Rightarrow N_1(0,1)$. Therefore,

$$\sigma n^{-3/2}\left(S_n - \frac{n^2}{4}\right) \Rightarrow N_1(0,1).$$

16.3   The Central Limit Theorem

We begin with the following result.

Proposition 16.3.1 *Let $(\Omega, \mathcal{F}, P)$ be a probability space, and let Y be a random variable of order 2 from $(\Omega, \mathcal{F})$ into $\mathbf{R}^d$ $(d \in \mathbf{N})$ such that $E(Y) = 0$. Let B be a $d \times p$ matrix, with rank p, such that $B \cdot B' = D(Y)$, and let $(i_1, \ldots, i_p)$ be a subsequence of $(1, \ldots, d)$ such that the submatrix, M, of B formed with the rows of indices $i_1, \ldots, i_p$ is nonsingular. Write $X_1, \ldots, X_p$ for the random variables from $(\Omega, \mathcal{F})$ into $\mathbf{R}$ such that*

$$\begin{pmatrix} Y_{i_1}(\omega) \\ \vdots \\ Y_{i_p}(\omega) \end{pmatrix} = M \cdot \begin{pmatrix} X_1(\omega) \\ \vdots \\ X_p(\omega) \end{pmatrix} \qquad \forall \omega \in \Omega.$$

and put $X = [X_1, \ldots, X_p]$. Then $D(X) = I_p$ and $Y = BX$ almost surely.

PROOF: For all $1 \leq j \leq d$, denote by l_j the jth row of $B = (\alpha_{h,k})$.

Given $j \in \{1, \ldots, d\} - \{i_1, \ldots, i_p\}$, there exist real numbers $t_1, \ldots, t_p$ such that $t_1 l_{i_1} + \cdots + t_p l_{i_p} = l_j$. Let B_j be the $(p+1) \times p$ matrix whose rows are $l_{i_1}, \ldots, l_{i_p}, l_j$. Then

$$\text{Var}(t_1 Y_{i_1} + \cdots + t_p Y_{i_p} - Y_j)$$

$$= (t_1, \ldots, t_p, -1) \cdot D([Y_{i_1}, \ldots, Y_{i_p}, Y_j]) \cdot \begin{pmatrix} t_1 \\ \vdots \\ t_p \\ -1 \end{pmatrix}$$

$$= (t_1, \ldots, t_p, -1) \cdot B_j \cdot B_j' \cdot \begin{pmatrix} t_1 \\ \vdots \\ t_p \\ -1 \end{pmatrix}$$

$$= 0,$$

because $(t_1, \ldots, t_p, -1) \cdot B_j = 0$. Therefore, $Y_j = t_1 Y_{i_1} + \cdots + t_p Y_{i_p}$ almost surely.

From the equality $l_j = t_1 l_{i_1} + \cdots + t_p l_{i_p}$ follows $\alpha_{j,k} = t_1 \alpha_{i_1,k} + \cdots + t_p \alpha_{i_p,k}$ for all $1 \leq k \leq p$, and so

$$\alpha_{j,1} X_1 + \cdots + \alpha_{j,p} X_p = t_1(\alpha_{i_1,1} X_1 + \cdots + \alpha_{i_1,p} X_p) + \cdots$$
$$+ t_p(\alpha_{i_p,1} X_1 + \cdots + \alpha_{i_p,p} X_p).$$

Finally,

$$(\alpha_{j,1} X_1 + \cdots + \alpha_{j,p} X_p)(\omega) = (t_1, \ldots, t_p) \cdot M \cdot \begin{pmatrix} X_1(\omega) \\ \vdots \\ X_p(\omega) \end{pmatrix}$$

$$= (t_1, \ldots, t_p) \cdot \begin{pmatrix} Y_{i_1}(\omega) \\ \vdots \\ Y_{i_p}(\omega) \end{pmatrix}$$

$$= Y_j(\omega)$$

almost surely, whence we see that $Y = BX$ almost surely.

Clearly, X is a random variable of order 2, and

$$D([Y_{i_1}, \ldots, Y_{i_p}]) = M \cdot D(X) \cdot M'.$$

Thus

$$M \cdot M' = M \cdot D(X) \cdot M'$$

and

$$D(X) = I_p.$$

□

Now let $(\Omega, \mathcal{F}, P)$ be a probability space and $(Y_n)_{n \geq 1}$ an independent sequence of identically distributed random variables from $(\Omega, \mathcal{F})$ into $\mathbf{R}^d$. Suppose that $\int^* |Y_1|^2 \, dP < +\infty$ and put $D = D(Y_1)$. Write S_n for $Y_1 + \cdots + Y_n$.

Theorem 16.3.1 (Central Limit Theorem)

$$\frac{1}{\sqrt{n}} \Big(S_n - n E(Y_1) \Big) \Rightarrow N_d(0, D)$$

PROOF: We may suppose that $E(Y_1) = 0$. To each Y_n, associate X_n as in Proposition 16.3.1. The X_n are independent. identically distributed, and $D(X_n) = I_p$. If $\frac{1}{\sqrt{n}}(X_1 + \cdots + X_n) \Rightarrow N_p(0, I_p)$, then $\frac{1}{\sqrt{n}}(Y_1 + \cdots + Y_n)$ converges in law to $w\big(N_p(0, I_p)\big) = N_d(0, D)$ (where w is the linear mapping from $\mathbf{R}^p$ into $\mathbf{R}^d$ with matrix B).

Therefore, it is enough to prove the theorem when $E(Y_1) = 0$ and $D(Y_1) = I_d$. But, in this case, put $X_{n,k} = Y_k/\sqrt{n}$ for each $n \in \mathbf{N}$ and each $1 \leq k \leq n$, so that $D(X_{n,1} + \cdots + X_{n,n}) = I_d$. By the Lindeberg theorem $S_n/\sqrt{n} \Rightarrow N_d(0, I_d)$, as desired.

□

In fact, there is a simple proof of the central limit theorem, which we give now.

By Proposition 14.4.4, the characteristic function φ of $Y_1 - E(Y_1)$ is C^2. Furthermore, $D\varphi(0) = 0$ and $-D$ is the matrix of $D^2\varphi(0)$. Thus

$$\varepsilon(x) = |x|^{-2} \big(\varphi(x) - 1 + (1/2) x' D x \big)$$

converges to 0 as x approaches 0 in $\mathbf{R}^d - \{0\}$. We put $\varepsilon(0) = 0$. Now the characteristic function ψ_n of $(1/\sqrt{n})(S_n - n E(Y_1))$ is $x \mapsto \big(\varphi(x/\sqrt{n}) \big)^n$. For fixed $x \in \mathbf{R}^d$ and for large enough $n \in \mathbf{N}$, $\varphi(x/\sqrt{n})$ is near $\varphi(0) = 1$, so $|1 - \varphi(x/\sqrt{n})| < 1$. Finally,

$$n \log \varphi \left(x/\sqrt{n} \right) = n \log \left(1 - \left(\frac{1}{2n} \right) x' D x + \varepsilon \left(\frac{x}{\sqrt{n}} \right) \frac{|x|^2}{n} \right)$$

converges to $(-1/2) x' D x$ as $n \to +\infty$; hence $v_n(x) = \exp \left(n \log \varphi(x/\sqrt{n}) \right)$ approaches $\exp \left(-(1/2) x' D x \right)$, and $(\psi_n)_{n \geq 1}$ converges pointwise to the characteristic function of $N_d(0, D)$.

If we reinforce the hypothesis of the central limit theorem, the conclusions are strengthened. To see this, we use the following result.

Proposition 16.3.2 *Let $(\Omega, \mathcal{F}, P)$ be a probability space, and let X be a random variable of order 2 from $(\Omega, \mathcal{F})$ into $\mathbf{R}^d$ such that $E(X) = 0$ and the characteristic function φ of X is dx-integrable. Put $D = D(X)$. Then*

(a) *D is invertible and, for each $\ 0 < a < \inf_{|t|=1}(t'Dt)$, there exists $\rho > 0$ such that $|\varphi(t)| \leq \exp((-1/2)a|t|^2)$ for all $t \in \mathbf{R}^d$ satisfying $|t| \leq \rho$;*

(b) *for each $r > 0$, $\ \sup_{|t| \geq r} |\varphi(t)| < 1$.*

PROOF: By the Fourier inversion formula, X has a continuous density.

Let γ be the bilinear form on $\mathbf{R}^d \times \mathbf{R}^d$ whose matrix is D. Assume that $\det(D) = 0$. Then γ is degenerate, and there exists t in $\mathbf{R}^d - \{0\}$ such that $\gamma(t,t) = t'Dt = D(tX) = 0$. Hence $tX = 0$ almost surely, and X takes its values in the hyperplane orthogonal to t, almost surely, in contradiction with the fact that X has a continuous density. Therefore, D is invertible.

Now define $\psi : \mathbf{R}^d \ni t \mapsto \exp\left((1/2)a|t|^2\right)\varphi(t)$, so that $D\psi(0) = 0$ and $aI_d - D$ is the matrix of $D^2\psi(0)$. Then

$$\psi(t) = 1 + \frac{1}{2}t'(aI_d - D)t + |t|^2\varepsilon(t),$$

where $\varepsilon(t) \to 0$ as t approaches 0 in $\mathbf{R}^d$. Put $m = \inf_{|t|=1}(t'Dt)$. Clearly,

$$t'(aI_d - D)t \leq -(m-a)|t|^2$$

for all $t \in \mathbf{R}^d$. If we take t so that $|(1/2)t'(aI_d - D)t| \leq 1$, we obtain

$$|\psi(t)| \leq \left|1 + \frac{1}{2}t'(aI_d - D)t\right| + |t|^2|\varepsilon(t)| \leq 1 - |t|^2\left[\frac{m-a}{2} - |\varepsilon(t)|\right].$$

Thus there exists $\rho > 0$ such that $|\psi(t)| \leq 1$ for $|t| \leq \rho$, which proves 1.

Next, suppose that $|\varphi(t)| = 1$ for some $t \neq 0$. Then there exists $s \in \mathbf{R}$ such that $E(e^{itX}) = e^{is}$, or equivalently $E(e^{i(tX-s)} - 1) = 0$. A fortiori,

$$E\big(1 - \cos(tX - s)\big) = 0$$

and

$$\cos(tX - s) = 1$$

almost surely. Hence X takes its values in $\bigcup_{n \in \mathbf{Z}}\{x \in \mathbf{R}^d : tx = s + 2n\pi\}$, almost surely. Since this last set is dx-negligible and X has a continuous density, we arrive at a contradiction.

In short, $|\varphi(t)| < 1$ for all $t \in \mathbf{R}^d - \{0\}$. Since φ vanishes at infinity, we conclude that $\sup_{|t| \geq r} |\varphi(t)| < 1$ for each $r > 0$. $\qquad\square$

Theorem 16.3.2 *Let $(\Omega, \mathcal{F}, P)$ be a probability space and $(X_n)_{n \geq 1}$ an independent sequence of identically distributed random variables from $(\Omega, \mathcal{F})$ into $\mathbf{R}^d$. Suppose that $\int^* |X_1|^2 \, dP < +\infty$ and the characteristic function φ of X_1 is dx-integrable. Then $(1/\sqrt{n})\big(S_n - nE(X_1)\big)$ has a continuous density g_n, for all $n \in \mathbf{N}$, and $(g_n)_{n \geq 1}$ converges uniformly to the continuous density g of $N_d(0, D)$, where $D = D(X_1)$.*

PROOF: We may suppose that $E(X_1) = 0$.

Since the characteristic function $\psi_n : t \longmapsto [\varphi(t/\sqrt{n})]^n$ of $S_n/\sqrt{n}$ is dx-integrable, $S_n/\sqrt{n}$ has a continuous density.

We show that $(\psi_n)_{n \geq 1}$ converges to $t \longmapsto \exp\left((-1/2)t'Dt\right)$ in $L^1_{\mathbf{C}}(\mathbf{R}^d, dx)$. For this, write $\psi_n = \psi_n \cdot 1_{A_n} + \psi_n \cdot 1_{B_n}$, where

$$
\begin{aligned}
A_n &= \{t \in \mathbf{R}^d : |t| < \rho\sqrt{n}\}, \\
B_n &= \{t \in \mathbf{R}^d : |t| \geq \rho\sqrt{n}\}.
\end{aligned}
$$

and ρ is given by Proposition 16.3.2.

For every $t \in \mathbf{R}^d$, $\psi_n \cdot 1_{A_n}(t)$ approaches $\exp\left(-(1/2)t'Dt\right)$ as n tends to $+\infty$, by the central limit theorem. Furthermore,

$$
\left|\psi_n \cdot 1_{A_n}(t)\right| = \left|\varphi\left(\frac{t}{\sqrt{n}}\right)\right|^n \cdot 1_{A_n}(t) \leq \exp\left(-\frac{a}{2}\left|\frac{t}{\sqrt{n}}\right|^2\right)^n = \exp\left(-\frac{a}{2}|t|^2\right),
$$

so $(\psi_n \cdot 1_{A_n})_{n \geq 1}$ converges to $t \mapsto \exp(-(1/2)t'Dt)$ in the mean, by the dominated convergence theorem.

Now let $c = \sup\{|\varphi(t)| : |t| \geq \rho\}$. Then

$$
\int_{B_n} |\psi_n(t)|\, dt \leq c^{n-1} \int \left|\varphi\left(\frac{t}{\sqrt{n}}\right)\right| dt = c^{n-1} n^{d/2} \int |\varphi(t)|\, dt,
$$

so $\psi_n \cdot 1_{B_n}$ converges to 0 in $L^1_{\mathbf{C}}(\mathbf{R}^d, dx)$.

Hence $(\psi_n)_{n \geq 1}$ converges to $t \mapsto \exp\left(-(1/2)t'Dt\right)$ in $L^1_{\mathbf{C}}(\mathbf{R}^d, dx)$, as desired. The Fourier inversion formula shows that

$$
g_n(x) - g(x) = (2\pi)^{-d} \int e^{-itx}\left[\psi_n(t) - \exp\left(-\frac{1}{2}t'Dt\right)\right] dt \qquad \text{for } x \in \mathbf{R}^d,
$$

and now

$$
|g_n(x) - g(x)| \leq (2\pi)^{-d} \int \left|\psi_n(t) - \exp\left(-\frac{1}{2}t'Dt\right)\right| dt.
$$

Therefore, $\sup_{x \in \mathbf{R}^d} |g_n(x) - g(x)|$ converges to 0 as $n \to +\infty$. $\qquad \square$

From the central limit theorem, we deduce the following result, due to De Moivre and Laplace.

Theorem 16.3.3 *Fix $p \in \left]0, 1\right[$. For each $n \in \mathbf{N}$, let $B(n, p)$ be the binomial law with parameters n and p (Chapter 6, Exercise 2), and let μ_n be the probability on $\mathbf{R}$, the image of $B(n, p)$ under $x \mapsto (x - np)/\sqrt{np(1 - p)}$. Then $(\mu_n)_{n \geq 1}$ converges weakly to $N_1(0, 1)$.*

PROOF: Let $B(1, p)$ be the probability on the Borel σ-algebra of $\mathbf{R}$ such that $B(1, p)(\{1\}) = p$ and $B(1, p)(\{0\}) = 1 - p$. Let $(X_n)_{n \geq 1}$ be an independent sequence of random variables (defined on $(\Omega, \mathcal{F}, P)$) whose common law is

$B(1,p)$. For every $n \in \mathbf{N}$, the law of $X_1 + \cdots + X_n$ is $B(n,p)$. Now, by the central limit theorem,

$$\frac{1}{\sqrt{np(1-p)}}(X_1 + \cdots + X_n - np) \Rightarrow N_1(0,1),$$

and so $(\mu_n)_{n \geq 1}$ converges weakly to $N_1(0,1)$. □

For example, consider an exchange which serves n telephones. At a given time, a caller has the probability p of using his telephone. The calls are assumed to be independent. What is the number of calls the exchange must be able to handle simultaneously so that, at a given time, there is a probability $\geq 1 - \alpha$ that all callers are satisfied?

Let $I = \{1, \ldots, n\}$ and $\Omega = \{0,1\}^I$. The outcomes of the present experiment are the recordings of calls at the exchange, that is, the $(x_k)_{1 \leq k \leq n} \in \Omega$ ($x_k = 1$ if the kth caller phones). For all $1 \leq k \leq n$, let μ_k be the probability on $\{0,1\}$ such that $\mu_k(\{1\}) = p$, and define $P = \bigotimes_{1 \leq k \leq n} \mu_k$ (on the power set of Ω). Finally, let X_k be the random variable $(x_j)_{1 \leq j \leq n} \mapsto x_k$ from $(\Omega, \mathcal{F})$ into $\mathbf{R}$. Then $S_n = X_1 + \cdots + X_n$ is the number of calls recorded at a given instant; the problem is to find an integer c such that $P(S_n > c) < \alpha$, that is, such that $B(n,p)(]c, +\infty[) < \alpha$. But, for each real number u, $B(n,p)(]np + u\sqrt{np(1-p)}, +\infty[)$ converges to

$$\frac{1}{\sqrt{2\pi}} \int_u^{+\infty} \exp\left(-\frac{1}{2}x^2\right) dx$$

by the De Moivre–Laplace theorem. Therefore, if n is large enough and u is such that

$$\int_u^{+\infty} \frac{1}{\sqrt{2\pi}} \exp\left(-\frac{1}{2}x^2\right) dx = \alpha,$$

we need only take $c > np + u\sqrt{np(1-p)}$.

Exercises for Chapter 16

1   Let $\mathbf{N}$ be the set of strictly positive integers, $\mathcal{F}$ the power set of $\mathbf{N}$, and, for each $n \in \mathbf{N}$, let P_n be the probability on $\mathcal{F}$ such that $P_n(\{i\}) = 1/n$ for all $1 \leq i \leq n$.

1.   Put $M_a = \{ka : k \in \mathbf{N}\}$ for each $a \in \mathbf{N}$, so that $P_n(M_a) = (1/n)[n/a]$ for all $n \geq 1$. If m, n are two integers ≥ 1, we write $m|n$ whenever m divides n, and reserve the letter p for primes. Now fix $n \in \mathbf{N}$, and prove that $1 - P_n\left(\bigcup_{p|n} M_p\right) = \prod_{p|n}(1 - 1/p)$. If $\varphi(n)$ is the number of strictly positive integers less than and relatively prime to n, conclude that $\varphi(n)/n = \prod_{p|n}(1 - 1/p)$.

2. For each prime p, define δ_p on $\mathbf{N}$ as follows: $\delta_p(m) = 1$ or 0 as p divides m or not. Let $(p_i)_{i \in I}$ be a finite family of distinct primes. For each $(d_i)_{i \in I}$ in $\{0, 1\}^I$, show that

$$P_n(\delta_{p_i} = d_i \quad \forall i \in I) \quad \text{converges to} \quad \prod_{i \in I}\left[\frac{d_i}{p_i} + (1 - d_i)\left(1 - \frac{1}{p_i} \right) \right]$$

as $n \to +\infty$.

3. Let f_u be the function

$$m \mapsto \sum_{p \leq u} \delta_p(m) \log\left(1 - \frac{1}{p} \right) \quad \text{on } \mathbf{N}, \text{ for each } u \in \mathbf{N},$$

and let f be the function

$$m \mapsto \sum_p \delta_p(m) \log\left(1 - \frac{1}{p} \right) = \log\left(\frac{\varphi(m)}{m} \right).$$

Prove that $\sup_{n \geq 1} \int (f_u - f)\, dP_n$ approaches 0 as $u \to +\infty$.

4. Let $(X_p)_p$ be a family of independent random variables from a common probability space $(\Omega, \mathcal{A}, P)$ into $\{0, 1\}$, such that $P(X_p = 1) = 1/p$. For each v in $\mathbf{N}$, let $F^{(v)}$ be the distribution function of $X^{(v)} = \sum_{p \leq v} X_p \log(1 - 1/p)$. Prove that, for every $x \in \mathbf{R}$, $P_n(\{m : f_v(m) \leq x\})$ converges to $F^{(v)}(x)$ as $n \to +\infty$. In other words, for fixed $v \in \mathbf{N}$, the measure $\mu_n^{(v)}$ (on the Borel σ-algebra of $\mathbf{R}$), the image of P_n under f_v, converges weakly as $n \to +\infty$ to the image $\mu^{(v)}$ of P under $X^{(v)}$.

5. Show that

$$\int \left(X_p - \frac{1}{p} \right)^2 dP = \frac{1}{p}\left(1 - \frac{1}{p} \right) \qquad \text{for each prime } p.$$

Deduce that $X^{(v)}$ converges P-almost surely as $v \to +\infty$, and so, in fact, converges in law (see P. Billingsley, Theorem 22.6).

6. For each $n \in \mathbf{N}$, let μ_n be the image of P_n under f. Using Proposition 16.1.2, deduce that $(\mu_n)_{n \geq 1}$ converges weakly.

7. Conclude that $\big(\varphi(m)/m \big)_{m \geq 1}$ has a limit distribution in the sense of Chapter 8, Exercise 6.

2 Let $\big(X_{n,k} \big)_{\substack{n \geq 1 \\ 1 \leq k \leq r_n}}$ be a triangular array of real random variables such that $X_{n,1}, \ldots, X_{n,r_n}$ are independent for every $n \in \mathbf{N}$. Suppose that the $|X_{n,k}|^{2+\delta}$ are integrable for some $\delta > 0$, $E(X_{n,k}) = 0$, and Lyapounov's condition

$$\lim_{n \to +\infty} \sum_{1 \leq k \leq r_n} \frac{1}{s_n^{2+\delta}} E(|X_{n,k}|^{2+\delta}) = 0$$

holds. Show that Lindeberg's condition follows.

3 Let μ be a probability on $\mathbf{R}$, concentrated on $\mathbf{Z}^+$. For each $n \in \mathbf{Z}^+$, define $p_n = \mu(\{n\})$. Let g be the function

$$u \longmapsto \int e^{x \log u} \, d\mu(x) = \sum_{n \geq 0} p_n u^n \quad \text{on }]0,1].$$

Given $r \in \mathbf{N}$, suppose that $\int^* x^{r-1} \, d\mu(x) < +\infty$. Prove that g is of class C^{r-1}, that $g^{(r-1)}$ is continuously differentiable on $]0,1]$, and that

$$g^{(r)}(1) = \int^* x(x-1)\cdots(x-r+1) \, d\mu(x)$$

($g^{(r)}(1)$ may be $+\infty$).

4 Elements are drawn from a population of size n, randomly and with replacement, until the number of distinct elements that have been sampled is r_n, where $1 \leq r_n \leq n$. Let S_n be the drawing for which this first happens. Suppose that r_n varies with n so that $r_n/n \to \rho$ $(0 < \rho < 1)$. What is the approximate distribution of S_n?

To answer this question, we let Ω be the set of those $\omega = (\omega(k))_{k \geq 1}$ in $\{1,\ldots,n\}^{\mathbf{N}}$ for which $\big|\{\omega(k) : k \geq 1\}\big| = n$. For each $k \in \mathbf{N}$, let μ_k be the uniform probability on $\{1,\ldots,n\}$. Then Ω is an open subset of the topological space $\{1,\ldots,n\}^{\mathbf{N}}$ and Ω^c is $\bigotimes_{k \geq 1} \mu_k$-negligible. Let P be the probability induced by $\bigotimes_{k \geq 1} \mu_k$ on the Borel σ-algebra $\mathcal{F}$ of Ω. Now we define random variables $X_{n,k}$ $(1 \leq k \leq r_n)$ from $(\Omega, \mathcal{F})$ into $\mathbf{R}$ as follows: $X_{n,1}(\omega) = 1$; $(X_{n,1} + \cdots + X_{n,k})(\omega)$ is the smallest integer j such that $\big|\{\omega(1),\ldots,\omega(j)\}\big| = k$.

1. Let Y_p be the trial on which success first occurs in a Bernoulli sequence with probability p $(p \in]0,1])$ for success: $P(Y_p = k) = q^{k-1}p$, where $q = 1 - p$. Show that the $X_{n,k}$ $(1 \leq k \leq r_n)$ are independent and that $X_{n,k}$ is distributed as $Y_{(n-k+1)/n}$.

2. Put $S_n = X_{n,1} + \cdots + X_{n,r_n}$, $m_n = E(S_n)$, and $s_n^2 = \text{Var}(S_n)$. Show that

$$s_n^2 = \sum_{1 \leq k \leq r_n} \frac{k-1}{n} \left(1 - \frac{k-1}{n}\right)^{-2}.$$

Let g be the function $x \longmapsto \left(1 - (x/n)\right)^{-2} (x/n)$ from $[0, n[$ into $\mathbf{R}$. By the Euler–MacLaurin summation formula,

$$s_n^2 = n \int_0^\rho \frac{y}{(1-y)^2} \, dy + n \int_\rho^{(r_n-1)/n} \frac{y}{(1-y)^2} \, dy$$

$$+ \frac{1}{2} \frac{r_n-1}{n} \left(1 - \frac{r_n-1}{n}\right)^{-2} + R_0,$$

where

$$0 \leq R_0 \leq \frac{1}{12n}\left[\left(1 + \frac{r_n-1}{n}\right)\left(1 - \frac{r_n-1}{n}\right)^{-3} - 1\right].$$

Deduce that $s_n^2 \sim n \int_0^\rho \left(y/(1-y)^2 \right) dy$ as $n \to +\infty$. Similarly, show that $m_n \sim n \int_0^\rho \left(1/(1-x) \right) dx$.

3. For all $p \in]0,1]$, show that $E\left((Y_p - 1/p)^4 \right) = qp^{-4}(1 + 7q + q^2)$ (use Exercise 3).

4. Deduce from part 3 that Lyapounov's condition holds for the random variables $X_{n,k} - n/(n-k+1)$, and for $\delta = 2$. Conclude that $(S_n - m_n)/s_n \Rightarrow N_1(0,1)$.

5 We refer the reader to the notation of Chapter 15, Exercise 5.

1. Let $k \in \mathbf{N}$, $n \in \mathbf{N}$, and let $i_1, \dots, i_k$ be positive integers such that $u + i_u < k + n$ for all $1 \le u \le k$. Denote by $\mathcal{L}$ the class of those $B \in \sigma(l_{k+n}, l_{k+n+1}, \dots)$ such that

$$P[(l_u = i_u \quad \forall u \le k) \cap B] = P(l_u = i_u \quad \forall u \le k) \cdot P(B).$$

For all integers $m \ge k + n$ and $i_{k+n}, \dots, i_m \ge 0$, show that the event $(l_u = i_u \quad \forall k + n \le u \le m)$ belongs to $\mathcal{L}$. Deduce that $\mathcal{L}$ contains the ring $\bigcup_{m \ge k+n} \sigma(l_{k+n}, \dots, l_m)$, and finally that $\mathcal{L} = \sigma(l_{k+n}, \dots)$ (use the $\pi - \lambda$-theorem).

2. Let $k \ge 1, n \ge 1$ be two integers, A an element of $\sigma(l_1, \dots, l_k)$, and B an element of $\sigma(l_{k+n}, \dots)$. There exists a subset H of $\mathbf{N}^k$ such that $A = \{\omega : [l_1, \dots, l_k](\omega) \in H\}$. Then

$$|P(A \cap B) - P(A)P(B)| \le$$
$$\sum \left| P((l_u = i_u \quad \forall u \le k) \cap B) - P(l_u = i_u \quad \forall u \le k)P(B) \right|,$$

where the sum extends over the elements $(i_1, \dots, i_k)$ of H. From part 1, deduce that

$$\left| P(A \cap B) - P(A)P(B) \right| \le \sum_{1 \le u \le k} P(l_u \ge k + n - u) \le \frac{p^n}{1-p}.$$

3. Prove that the law of $[l_n, l_{n+1}, \dots]$ does not depend on n.

4. Show that $E(l_1) = \int_0^{+\infty} P(l_1 \ge t)\, dt = p/(1-p)$ and that $\text{Var}(l_1) = p/(1-p)^2$.

5. Let $n \in \mathbf{N}$. Then $\int l_1 l_{n+1}\, dP = \sum_{i \ge 0} \int_{l_1 = i} l_1 l_{n+1}\, dP$. Deduce that

$$\int l_1 l_{n+1}\, dP = \sum_{1 \le i \le n-1} ip^{i+1} + \sum_{i \ge n} i(i-n)p^i(1-p).$$

6. Denote by F the function $x \mapsto \sum_{1 \le i \le x} p^{i-1}(1-p)$ from $[0, +\infty[$ into $\mathbf{R}$ and by dF the Stieltjes measure associated with F. Thus $F(k) = 1 - p^k$ for all $k \in \mathbf{Z}^+$. Given $n \in \mathbf{N}$, show that

$$\int_0^{n-1} x\, dF(x) = \frac{1 - p^n}{1-p} - np^{n-1}.$$

and that

$$\int_n^{+\infty} x(x-n)\,dF(x) = \frac{2p}{1-p}\int_{n-1}^{+\infty} x\,dF(x) - (n-1)\frac{p^n}{1-p}.$$

Compute

$$\int l_1 l_{n+1}\,dP = \frac{p^2}{1-p}\int_0^{n-1} x\,dF(x) + p\int_n^{+\infty} x(x-n)\,dF(x),$$

and deduce that

$$\int \big(l_1 - E(l_1)\big)\big(l_{n+1} - E(l_{n+1})\big)\,dP = \frac{p^{n+1}}{(1-p)^2}(2-p).$$

Conclude that

$$\begin{aligned}
\sigma^2 &= \operatorname{Var}(l_1) + 2\sum_{n\geq 1}\int \big(l_1 - E(l_1)\big)\big(l_{n+1} - E(l_{n+1})\big)\,dP \\
&= p(1-p)^{-3}\left[\frac{17}{8} - 2\left(p-\frac{3}{4}\right)^2\right].
\end{aligned}$$

7. Show that $\dfrac{1}{\sigma\sqrt{n}}\left(\displaystyle\sum_{1\leq k\leq n} l_k - n\frac{p}{1-p}\right)$ converges in law to $N_1(0,1)$ as $n \to +\infty$ (see P. Billingsley, Theorem 27.5).

6 1. Define $\alpha = \inf_{|t|\geq 1}\big(1 - (\sin t)/t\big)$. For any probability μ on $\mathbf{R}$, with characteristic function φ, and any $A > 0$, show that

$$\mu\left(|x| > \frac{1}{A}\right) \leq \frac{1}{\alpha}\left(1 - \frac{1}{2A}\int_{-A}^{A}\varphi(t)\,dt\right).$$

2. Let $(\varphi_n)_{n\geq 1}$ be a sequence of characteristic functions such that $\big(\varphi_n(t)\big)_{n\geq 1}$ converges to 1 for all t in some neighborhood V of 0. Prove that $\big(\varphi_n(t)\big)_{n\geq 1}$ converges to 1 for every t. (If μ_n is the probability on $\mathbf{R}$ whose characteristic function is φ_n, show that $(\mu_n)_{n\geq 1}$ is tight.)

7 Let $(\Omega, \mathcal{F}, P)$ be a probability space and $(X_n)_{n\geq 1}$ an independent sequence of random variables from $(\Omega, \mathcal{F})$ into $\mathbf{R}$. For all $n \in \mathbf{N}$, set $S_n = \sum_{1\leq k\leq n} X_k$. Suppose that $(S_n)_{n\geq 1}$ converges in law. Show that $(S_n)_{n\geq 1}$ converges in probability.

17

Order Statistics

One of the first goals of statistics is to obtain, given identically distributed random variables $X_1, \ldots, X_n$, as much information as possible on their common distribution μ. The n-tuple $(X_1, \ldots, X_n)$ is called an n-sample of μ. In this chapter we give some relatively elementary results in this direction.

Summary

17.1 Given an n-sample, for almost every $\omega \in \Omega$, there is a permutation $\sigma = \sigma_\omega$ of $\{1, \ldots, n\}$ such that $X_{\sigma(1)} < \ldots < X_{\sigma(n)}$. The coordinates of the random variable defined almost everywhere by $X(\omega) = \left(X_{\sigma_\omega(1)}(\omega), \ldots, X_{\sigma_\omega(n)}(\omega) \right)$ are called the order statistics of the sample. We compute the distributions of $X_{\sigma(k)}$ and X.

17.2 We introduce the notion of median of a sample and give some more convergence theorems.

17.1 Definition of the Order Statistics

Let μ be a probability on the Borel σ-algebra of $\mathbf{R}^d$ $(d \geq 1)$.

Definition 17.1.1 If $(\Omega, \mathcal{F}, P)$ is a probability space and $X_1, \ldots, X_n$ are independent random variables from $(\Omega, \mathcal{F})$ into $\mathbf{R}^d$ whose common law is μ, then $(X_1, \ldots, X_n)$ is called an n-sample of μ.

Henceforth, suppose that $d = 1$ and μ is diffuse. Write F for the distribution function of μ. Let an n-sample $(X_1, \ldots, X_n)$ of μ be given. Since $P(X_i = X_j) = P_{[X_i, X_j]}\left(\{(t, t) : t \in \mathbf{R}\} \right) = 0$ for all distinct i, j in $\{1, \ldots, n\}$, the set of those $\omega \in \Omega$ such that $X_1(\omega), \ldots, X_n(\omega)$ are all distinct has P-negligible complement. For every permutation σ of $\{1, \ldots, n\}$, we put $\Omega_\sigma = \{\omega : X_{\sigma(1)}(\omega) < \ldots < X_{\sigma(n)}(\omega)\}$. The sets Ω_σ are mutually disjoint, they belong to $\mathcal{F}$, and their union has P-negligible complement. Now let X be a random variable from $(\Omega, \mathcal{F})$ into $\mathbf{R}^n$ equal to $[X_{\sigma(1)}, \ldots, X_{\sigma(n)}]$ on each Ω_σ. Then X can be written $[X_{(1)}, \ldots, X_{(n)}]$, and the $X_{(k)}$, for $1 \le k \le n$, which are almost surely determined, are called the order statistics of the sample. $X_{(k)}$ is the kth order statistic of the sample.

Proposition 17.1.1 *For every $1 \le k \le n$, the distribution function of $X_{(k)}$ is*

$$\Phi_k : t \mapsto \frac{n!}{(k-1)!(n-k)!} \int_0^{F(t)} u^{k-1}(1-u)^{n-k}\, du.$$

If μ has a (Borelian) density f with respect to dx, then $P_{X_{(k)}}$ has the density

$$\varphi_k : t \mapsto \frac{n!}{(k-1)!(n-k)!} F(t)^{k-1}\big(1-F(t)\big)^{n-k} f(t)$$

with respect to dx.

PROOF: Fix $t \in \mathbf{R}$. For all $1 \le j \le n$, $X'_j = 1_{]-\infty, t]} \circ X_j$ is a random variable from $(\Omega, \mathcal{F})$ into $\mathbf{R}$, and $P_{X'_j}(\{1\}) = F(t)$, $P_{X'_j}(\{0\}) = 1 - F(t)$. Since the X'_j are independent, $Y = \sum_{1 \le j \le n} 1_{]-\infty, t]} \circ X_j$ has $B(n, F(t))$ as law (Chapter 6, Exercise 2). For every $\omega \in \Omega$, $Y(\omega)$ is the cardinal number of $\{1 \le j \le n : X_j(\omega) \le t\}$, so that $\{\omega : Y(\omega) \ge k\}$ and $\{\omega : X_{(k)}(\omega) \le t\}$ differ only by a P-negligible set. Therefore, if Φ_k is the distribution function of $X_{(k)}$,

$$\begin{aligned}
\Phi_k(t) &= B\big(n, F(t)\big)\big(\{k, \ldots, n\}\big) \\
&= \sum_{k \le r \le n} \frac{n!}{r!(n-r)!} F(t)^r \big(1 - F(t)\big)^{n-r}.
\end{aligned}$$

Now let s be the function

$$\theta \mapsto \sum_{k \le r \le n} \frac{n!}{r!(n-r)!} \theta^r (1-\theta)^{n-r}$$

from $[0,1]$ into $\mathbf{R}$. Since

$$s(0) = 0$$

and

$$s'(\theta) = \frac{n!}{(k-1)!(n-k)!} \theta^{k-1}(1-\theta)^{n-k}$$

for every $\theta \in [0,1]$, we see that

$$s(\theta) = \frac{n!}{(k-1)!(n-k)!} \int_0^\theta u^{k-1}(1-u)^{n-k}\, du.$$

Hence

$$\Phi_k(t) = \frac{n!}{(k-1)!(n-k)!} \int_0^{F(t)} u^{k-1}(1-u)^{n-k}\, du.$$

Next, assume that μ has a density f with respect to dx. Then F is increasing and absolutely continuous. On the other hand, s is absolutely continuous. It follows that $\Phi_k = s \circ F$ is absolutely continuous. At dx-almost all $t \in \mathbf{R}$, F has derivative $f(t)$ (Theorems 13.4.1 and 13.4.2). So, at dx-almost all $t \in \mathbf{R}$, Φ_k has derivative

$$\varphi_k(t) = \frac{n!}{(k-1)!(n-k)!} F(t)^{k-1}(1-F(t))^{n-k} f(t),$$

and the proof is complete.   □

Proposition 17.1.2 *Let* $B = \{(x_1,\dots,x_n) \in \mathbf{R}^n : x_1 < x_2 < \dots < x_n\}$. *Then* $P_X = n! \cdot 1_B \cdot (\mu \otimes \cdots \otimes \mu)$.

PROOF: For each Borel function h from $\mathbf{R}^n$ into $[0,+\infty[$,

$$\int^* h \circ X\, dP = \sum_{\sigma \in G_n} \int_{\Omega_\sigma}^* h \circ [X_{\sigma(1)},\dots,X_{\sigma(n)}]\, dP$$

$$= \sum_{\sigma \in G_n} \int^* (h \cdot 1_B)(x_{\sigma(1)},\dots,x_{\sigma(n)})\, dP_{[X_1,\dots,X_n]}(x_1,\dots,x_n);$$

but, for all $\sigma \in G_n$, $P_{[X_1,\dots,X_n]} = P_{X_1} \otimes \cdots \otimes P_{X_n}$ is invariant under the automorphism $(x_1,\dots,x_n) \mapsto (x_{\sigma(1)},\dots,x_{\sigma(n)})$ of $\mathbf{R}^n$, so

$$\int^* h\, dP_X = n! \int^* (h \cdot 1_B)(x_1,\dots,x_n)\, d\mu(x_1)\cdots d\mu(x_n),$$

and the proposition follows.   □

Now suppose $n \geq 2$.

Proposition 17.1.3 *Put* $D = \{(s,t) \in \mathbf{R}^2 : s < t\}$. *Then*

$$P_{[X_{(1)},X_{(n)}]} = n(n-1)(F(t)-F(s))^{n-2} \cdot 1_D(s,t)\, d\mu(s)\, d\mu(t).$$

PROOF: For all $k \in \mathbf{N}$ and for all $\alpha \in [0,1]$, the function $t \mapsto (F(t)-\alpha)^k$ is an indefinite integral of $k(F(t)-\alpha)^{k-1}\, dF(t)$. For every $(x,y) \in D$,

$$\iint_{]-\infty,x]\times]-\infty,y]} n(n-1)(F(t)-F(s))^{n-2} \cdot 1_D(s,t)\, d\mu(s)\, d\mu(t)$$

$$= \int_{-\infty}^x n\, d\mu(s) \int_s^y (n-1)[F(t)-F(s)]^{n-2}\, dF(t)$$

$$= \int_{-\infty}^{x} n[F(y) - F(s)]^{n-1} \, dF(s)$$

$$= -\left(F(y) - F(s)\right)^{n} \Big|_{-\infty}^{x}$$

$$= F(y)^{n} - \left(F(y) - F(x)\right)^{n}.$$

On the other hand, for every $(x, y) \in D^c$,

$$\iint_{]-\infty,x] \times]-\infty,y]} n(n-1)\left(F(t) - F(s)\right)^{n-2} \cdot 1_D(s,t) \, d\mu(s) \, d\mu(t)$$

$$= \int_{-\infty}^{y} n \, d\mu(t) \int_{-\infty}^{t} (n-1)[F(t) - F(s)]^{n-2} \, dF(s)$$

$$= \int_{-\infty}^{y} nF(t)^{n-1} \, dF(t)$$

$$= F(y)^{n}.$$

Now let $F_{(1,n)}$ be the distribution function of $[X_{(1)}, X_{(n)}]$. Then

$$F_{(1,n)}(x, y) = P\left((X_{(1)} \le x) \cap (X_{(n)} \le y)\right)$$

$$= P(X_{(n)} \le y) - P\left((X_{(1)} > x) \cap (X_{(n)} \le y)\right)$$

$$= F(y)^{n} - P\left((X_{(1)} > x) \cap (X_{(n)} \le y)\right)$$

for all $(x, y) \in \mathbf{R}^2$. Hence

$$F_{(1,n)}(x, y) = F(y)^{n} - P\left(\bigcap_{1 \le i \le n} (x < X_i \le y)\right)$$

$$= F(y)^{n} - \left(F(y) - F(x)\right)^{n}$$

if $x < y$, and $F_{(1,n)}(x, y) = F(y)^{n}$ if $x \ge y$. We conclude that $F_{(1,n)}$ is the distribution function of

$$n(n-1)\left(F(t) - F(s)\right)^{n-2} \cdot 1_D(s,t) \, d\mu(s) \, d\mu(t),$$

as desired.   □

17.2   Convergence of the Empirical Median

Let $(\Omega, \mathcal{F}, P)$ be a probability space.

Definition 17.2.1 If $X_1, \ldots, X_n$ are random variables from $(\Omega, \mathcal{F})$ into $\mathbf{R}$, the function $F_n : (t, \omega) \mapsto (1/n) \sum_{1 \le k \le n} 1_{]-\infty, t]}(X_k(\omega))$ from $\mathbf{R} \times \Omega$ into $\mathbf{R}$ is called the empirical distribution function for $X_1, \ldots, X_n$.

For all $x \in \mathbf{R}$, denote by ε_x the measure (on the Borel σ-algebra of $\mathbf{R}$) defined by placing the mass 1 at x. Next, observe that. for every $\omega \in \Omega$, $F_n(\cdot, \omega)$ is the distribution function of $(1/n)\sum_{1 \leq k \leq n} \varepsilon_{X_k(\omega)}$.

Now let $(X_k)_{k \geq 1}$ be an independent sequence of random variables from $(\Omega, \mathcal{F})$ into $\mathbf{R}$ having the same diffuse law μ. Write F for the distribution function of μ.

Definition 17.2.2 For all $n \geq 1$, let $X_{(1)}, \ldots, X_{(2n+1)}$ be the order statistics associated with the sample $(X_1, \ldots, X_{2n+1})$ of μ. Then $M_n = X_{(n+1)}$ is called the nth empirical median of $(X_k)_{k \geq 1}$.

Thus M_n is almost surely determined.

Definition 17.2.3 Every real number t such that $F(t) = 1/2$ is called a median of μ.

Proposition 17.2.1 *Suppose there exists a unique $M \in \mathbf{R}$ such that $F(M) = 1/2$. Then M_n converges almost surely to M as $n \to +\infty$.*

PROOF: Denote by D the set of those $\omega \in \Omega$ such that $X_i(\omega) \neq X_j(\omega)$ for all distinct $i, j \in \mathbf{N}$. Recall that D has a P-negligible complement.

Now let s and t be two rational numbers such that $s < M < t$. For every $\omega \in D$ and for every $n \in \mathbf{N}$, the relation $s < M_n(\omega) \leq t$ is equivalent to

$$F_{2n+1}(s, \omega) \leq \frac{n}{2n+1} \quad \text{and} \quad F_{2n+1}(t, \omega) \geq \frac{n+1}{2n+1}.$$

For almost all $\omega \in D$, $F_{2n+1}(s, \omega)$ converges to $F(s) < 1/2$ and $F_{2n+1}(t, \omega)$ converges to $F(t) > 1/2$ as $n \to +\infty$, by the strong law of large numbers. Hence, for almost all $\omega \in D$, $1_{]s,t]}(M_n(\omega))$ converges to 1 as $n \to +\infty$.

For almost all $\omega \in D$, therefore,

$$\limsup M_n(\omega) \leq M \leq \liminf M_n(\omega).$$

and the proof is complete. $\qquad \square$

Proposition 17.2.2 *Suppose that μ has a unique median M and a density f with respect to dx which is continuous at M and strictly positive at M. Then $Z_n = \sqrt{2n+1}(M_n - M)$ converges in law to $N_1\left(0, 1/(4f^2(M))\right)$.*

PROOF: The function

$$t \mapsto \frac{(2n+1)!}{(n!)^2} F(t)^n [1 - F(t)]^n f(t)$$

is a density of P_{M_n} with respect to dt, by Proposition 17.1.1. Hence P_{Z_n} has a density g_n given by

$$
g_n(u) = \frac{(2n+1)!}{(n!)^2\sqrt{2n+1}} \cdot \frac{1}{4^n}
$$

$$
\times \left(4F\left(M + \frac{u}{\sqrt{2n+1}}\right)\left[1 - F\left(M + \frac{u}{\sqrt{2n+1}}\right)\right]\right)^n f\left(M + \frac{u}{\sqrt{2n+1}}\right)
$$

$$
= \alpha_n \cdot [\psi_n(u)]^n f\left(M + \frac{u}{\sqrt{2n+1}}\right).
$$

We show that, for all $A > 0$, $(g_n)_{n\geq 1}$ converges uniformly on $[-A, A]$ to the continuous density of $N_1\left(0, 1/(4f^2(M))\right)$. This will prove that $(P_{Z_n})_{n\geq 1}$ converges vaguely to $N_1\left(0, 1/(4f^2(M))\right)$.

Since $n! \sim (n/e)^n \sqrt{2\pi n}$ as $n \to +\infty$ (Chapter 13, Exercise 11), we see first that

$$
\alpha_n \sim \frac{(2n+1)^{2n+1}\sqrt{2\pi(2n+1)}\,e^{2n}}{4^n\sqrt{2n+1}\,n^{2n}2\pi n e^{2n+1}} = \frac{1}{4^n}2^{2n}\left(1 + \frac{1}{2n}\right)^{2n}\frac{1}{e\sqrt{2\pi}}\frac{2n+1}{n}
$$

$$
\sim \sqrt{\frac{2}{\pi}}.
$$

Denote by r the function

$$
0 \mapsto 0 \qquad 0 \neq x \mapsto \frac{1}{f(M)}\left[\frac{F(M+x) - F(M)}{x} - f(M)\right]
$$

from $\mathbf{R}$ into $\mathbf{R}$. Then, for $n \in \mathbf{N}$ and for $u \in \mathbf{R}$,

$$
2F\left(M + \frac{u}{\sqrt{2n+1}}\right) = 1 + \frac{2u}{\sqrt{2n+1}}f(M)\left[1 + r\left(\frac{u}{\sqrt{2n+1}}\right)\right]
$$

and

$$
4F\left(M + \frac{u}{\sqrt{2n+1}}\right)\left[1 - F\left(M + \frac{u}{\sqrt{2n+1}}\right)\right]
$$

$$
= 1 - \frac{4u^2}{2n+1}f^2(M)\left[1 + r\left(\frac{u}{\sqrt{2n+1}}\right)\right]^2.
$$

Recall that, for all $x \in [0, 1[$,

$$
\log(1 - x) = -x - x^2 \int_0^1 \frac{t}{1 - tx}\,dt,
$$

where

$$
\int_0^1 \frac{t}{1 - tx}\,dt \leq \frac{1}{2(1 - x)}.
$$

If we take n large enough that

$$\frac{4u^2}{2n+1} f^2(M) \left[1 + r\left(\frac{u}{\sqrt{2n+1}}\right)\right]^2 = x_n(u) < 1$$

for all $u \in [-A, A]$, we obtain

$$\left| n \log\left(\psi_n(u)\right) + \frac{4nu^2}{2n+1} f^2(M) \left[1 + r\left(\frac{u}{\sqrt{2n+1}}\right)\right]^2 \right|$$

$$\leq \frac{nx_n(u)}{2(1 - x_n(u))} x_n(u)$$

for all $u \in [-A, A]$. So

$$n \log\left(\psi_n(u)\right) + \frac{4nu^2}{2n+1} f^2(M) \left[1 + r\left(\frac{u}{\sqrt{2n+1}}\right)\right]^2$$

converges uniformly to 0 on $[-A, A]$ as $n \to +\infty$. Then, evidently, $n \log\left(\psi_n(u)\right) + 2u^2 f^2(M)$ converges uniformly to 0, and $[\psi_n(u)]^n$ converges uniformly to $\exp\left(-2u^2 f^2(M)\right)$.

Therefore, $g_n(u)$ converges uniformly to $\sqrt{2/\pi} \exp\left(-2u^2 f^2(M)\right) f(M)$ on $[-A, A]$ as $n \to +\infty$. The proof is complete. $\square$

If μ has a unique median M and if M is unknown, Propositions 17.2.1 and 17.2.2 provide an estimate of M by means of the empirical median. In the particular case that $\mu = N_1(m, \sigma^2)$ is normal, we obtain an estimate of m.

Exercises for Chapter 17

1   Let $(\Omega, \mathcal{F}, P)$ be a probability space and $(X_n)_{n \geq 1}$ an independent sequence of random variables from $(\Omega, \mathcal{F})$ into $\mathbf{R}$ with common distribution function F. For all $n \in \mathbf{N}$, let F_n be the empirical distribution function for $X_1, \ldots, X_n$.

   1.   For every $n \in \mathbf{N}$, show that $D_n : \omega \mapsto \sup_{x \in \mathbf{R}} |F_n(x, \omega) - F(x)|$ is a random variable from $(\Omega, \mathcal{F})$ into $\mathbf{R}$.

   2.   For every $x \in \mathbf{R}$, show that $\left(F_n(x. \omega)\right)_{n \geq 1}$ converges to $F(x)$ for all $\omega \in A_x^c$, where A_x is P-negligible. Similarly, $\left(F_n(x_-, \omega)\right)_{n \geq 1}$ converges to $F(x_-)$ for all $\omega \in B_x^c$, where B_x is P-negligible.

   3.   For every $u \in]0, 1[$, put $\varphi(u) = \inf\{x : u \leq F(x)\}$. Show that

$$F\left(\varphi(u)_-\right) \leq u \leq F\left(\varphi(u)\right).$$

For all integers $m \geq 2$ and $1 \leq k \leq m-1$, put $x_{m,k} = \varphi(k/m)$. Next, for all integers $n \geq 1$ and $m \geq 2$, denote by $D_{m,n}(\omega)$ the maximum of the quantities

$$\left|F_n(x_{m,k}, \omega) - F(x_{m,k})\right| \quad \text{and} \quad \left|F_n(x_{m,k_-}, \omega) - F(x_{m,k_-})\right|$$

for $1 \leq k \leq m-1$. Show that $D_n(\omega) \leq D_{m,n}(\omega) + 1/m$ for all $\omega \in \Omega$.

4. Show that, for P-almost all $\omega \in \Omega$, the functions $F_n(\cdot, \omega)$ converge uniformly to F (Glivenko–Cantelli theorem).

5. Deduce from part 4 that, for P-almost all ω, $(1/n)\sum_{1 \leq k \leq n} \varepsilon_{X_k(\omega)}$ converges weakly to $\mu = P_{X_1}$ as $n \to +\infty$.

2 Let $(\Omega, \mathcal{F}, P)$ be a probability space and $(X_1, \ldots, X_n)$ a sample of $\alpha e^{-\alpha x} \cdot 1_{]0,+\infty[}(x)\,dx$, the exponential law with parameter α ($\alpha > 0$). Write $X_{(1)}, \ldots, X_{(n)}$ for the order statistics of $(X_1, \ldots, X_n)$, and set $Z_1 = X_{(1)}$, $Z_i = X_{(i)} - X_{(i-1)}$ for all $2 \leq i \leq n$.

1. Prove that $P_{[Z_1,\ldots,Z_n]}$ has a density, and compute it.

2. Show that $Z_1, \ldots, Z_n$ are independent and that

$$P_{Z_k} = \alpha(n-k+1)\exp[-\alpha(n-k+1)x] \cdot 1_{]0,+\infty[}(x)\,dx$$

for every $1 \leq k \leq n$. In other words, P_{Z_k} is the exponential law with parameter $\alpha(n-k+1)$.

3 Given $t > 0$, let $(\Omega, \mathcal{F}, P)$ be a probability space, and let $X_1, \ldots, X_n$ (with $n \geq 2$) be independent random variables from $(\Omega, \mathcal{F})$ into $\mathbf{R}$ whose common law is $\mu = (1/t) \cdot 1_{]0,t[}\,dx$. Write $X_{(1)}, \ldots, X_{(n)}$ for the order statistics of the sample $(X_1, \ldots, X_n)$. Put $L_1 = X_{(1)}$, $L_i = X_{(i)} - X_{(i-1)}$ for all $2 \leq i \leq n$, and $L_{n+1} = t - X_{(n)}$. So, for every ω such that the points $X_k(\omega)$ (with $1 \leq k \leq n$) are distinct and lie in $]0,t[$, these $X_k(\omega)$ partition $]0,t[$ into successive intervals of lengths $L_1(\omega), \ldots, L_{n+1}(\omega)$.

For all $x \in \mathbf{R}$, put $x_+ = \sup(x,0)$. Let $x_1, \ldots, x_{n+1}$ be positive. Show that

$$P(L_1 > x_1, \ldots, L_{n+1} > x_{n+1}) = \frac{1}{t^n}[(t - x_1 - \cdots - x_{n+1})_+]^n.$$

18

Conditional Probability

This chapter is devoted to one of the most fundamental objects of the theory of probability: conditional expectations. Given two random variables X and Y, and two events A and B, we wish to determine the probability that $Y \in A$ knowing that $X \in B$. This problem is easily solved when the range of X is finite; the probability that $Y \in C$ knowing that $X = x$, denoted by $P(Y \in C | X = x)$, is simply $P\big((Y \in C) \cap (X = x)\big)/P(X = x)$.

Summary

18.1   Let $(\Omega, \mathcal{F}, P)$ be a probability space and let $\mathcal{D}$ be a sub-σ-algebra of $\mathcal{F}$. For any random variable Y from $(\Omega, \mathcal{F})$ into $\mathbf{R}^k$, there exists a unique (up to a.e. equality) random variable Z from $(\Omega, \mathcal{D})$ into $\mathbf{R}^k$ such that $\int_A Z \, dP = \int_A Y \, dP$ for all A in $\mathcal{D}$. Z is called the conditional expected value of Y given $\mathcal{D}$. If Y is of order 2, Z is the orthogonal projection of Y on a closed subspace of $L^2(\Omega, P; \mathbf{R}^k)$ (Theorem 18.1.2).

18.2   In this section we prove a measure theoretic converse of the mean value theorem.

18.3   Here, we generalize Jensen's inequality to conditional expected values: if φ is convex and Y is P-integrable, $\varphi \circ E(Y|\mathcal{D}) \leq E\big(\varphi(Y)|\mathcal{D}\big)$ (Proposition 18.3.1).

18.4   We define the conditional expected value of Y given the random variable X. When the law of X is absolutely continuous with respect to Lebesgue measure, we can compute $E(Y|X)$ (as a limit) outside some P_X-negligible set (Proposition 18.4.1).

18.5   This section is devoted to the study of the conditional law of Y given X.

18.6   We compute some conditional laws when P_X is defined by masses and when $P_{[X,Y]}$ is absolutely continuous with respect to a product $\mu \otimes \nu$.

18.7   We prove the existence of conditional laws when Y is an $\mathbf{R}^k$ valued random variable (Theorem 18.7.1).

18.1   Conditional Expectation

Let $(\Omega, \mathcal{F}, P)$ be a probability space, $\mathcal{D}$ a sub-σ-algebra of $\mathcal{F}$, and $P_{/\mathcal{D}}$ the probability $E \mapsto \int 1_E\, dP$ on $\mathcal{D}$.

Let $Z : \Omega \to \mathbf{R}^k$ $(k \geq 1)$ be measurable $\mathcal{D}$. P_Z is both the image measure of P under Z and the image measure of $P_{/\mathcal{D}}$ under Z. Since the identity of $\mathbf{R}^n$ is P_Z-measurable, to say it is P_Z-integrable means that Z is P-integrable, or that Z is $P_{/\mathcal{D}}$-integrable. In short, Z is $P_{/\mathcal{D}}$-integrable if and only if it is P-integrable, and then $\int Z\, d(P_{/\mathcal{D}}) = \int Z\, dP$.

Theorem 18.1.1 *Let Y be a P-integrable random variable from $(\Omega, \mathcal{F})$ into $\mathbf{R}^k$ $(k \geq 1)$. There exists a P-integrable random variable Z from $(\Omega, \mathcal{D})$ into $\mathbf{R}^k$ such that $\int_A Z\, dP = \int_A Y\, dP$ for all $\mathcal{D}$-sets A. Any random variable Z' from $(\Omega, \mathcal{D})$ into $\mathbf{R}^k$ having the same properties is equal to Z P-almost surely.*

PROOF: We may suppose that $k = 1$. The measure $(YP)_{/\mathcal{D}}$ is absolutely continuous with respect to $P_{/\mathcal{D}}$. By the Radon–Nikodym theorem, there exists, therefore, a measurable $\mathcal{D}$ and $P_{/\mathcal{D}}$-integrable function Z from Ω into $\mathbf{R}$ such that $Z(P_{/\mathcal{D}}) = (YP)_{/\mathcal{D}}$. Then $\int Z \cdot 1_A\, d(P_{/\mathcal{D}}) = \int Y \cdot 1_A\, dP$ for all $A \in \mathcal{D}$; equivalently, $\int Z \cdot 1_A\, dP = \int Y \cdot 1_A\, dP$ for all $A \in \mathcal{D}$.

Now, if Z' has the same properties than Z, then

$$Z'(P_{/\mathcal{D}}) = (YP)_{/\mathcal{D}} = Z(P_{/\mathcal{D}}),$$

hence $Z' = Z$ $P_{/\mathcal{D}}$-almost surely, and $Z' = Z$ P-almost surely.   □

Definition 18.1.1 In the notation of Theorem 18.1.1, the class of Z (for the relation of almost sure equality between measurable $\mathcal{D}$ functions) is called the conditional expected value of Y given $\mathcal{D}$, and is written $E(Y|\mathcal{D})$. Z is a version of $E(Y|\mathcal{D})$.

Improperly speaking, we shall often say that Z is the conditional expected value of Y given $\mathcal{D}$ (so identifying Z with its class), and put $Z = E(Y|\mathcal{D})$.

Proposition 18.1.1 *Let $\mathcal{C}$ be a π-system in Ω such that $\sigma(\mathcal{C}) = \mathcal{D}$. If Z is a P-integrable random variable from $(\Omega, \mathcal{D})$ into $\mathbf{R}^k$ such that $\int_A Z\, dP = \int_A Y\, dP$ for all $A \in \mathcal{C}$, then $Z = E(Y|\mathcal{D})$ almost surely.*

PROOF: We may suppose that $k = 1$. Since $Z(P_{/\mathcal{D}})$ and $(YP)_{/\mathcal{D}}$ agree on $\mathcal{C}$, they are identical by Proposition 6.4.2.   □

Proposition 18.1.2 *Give $\mathbf{R}^k$ the Euclidean norm, and let Z be a version of $E(Y|\mathcal{D})$. Then $|Z| \leq E(|Y|\,|\mathcal{D})$ almost surely.*

PROOF: For all $A \in \mathcal{D}$,

$$\left| \int_A Z \, d(P_{/\mathcal{D}}) \right| = \left| \int_A Z \, dP \right| = \left| \int_A Y \, dP \right| \leq \int_A |Y| \, dP$$

$$= [E(|Y| \,|\mathcal{D}) \cdot (P_{/\mathcal{D}})](A).$$

By Theorem 6.2.1, we conclude that

$$\int_A |Z| \, d(P_{/\mathcal{D}}) \leq [E(|Y| \,|\mathcal{D}) \cdot (P_{/\mathcal{D}})](A)$$

for all $A \in \mathcal{D}$, and finally that $|Z| \cdot (P_{/\mathcal{D}}) \leq E(|Y| \,|\mathcal{D}) \cdot (P_{/\mathcal{D}})$. Hence $|Z| \leq E(|Y| \,|\mathcal{D})$ $P_{/\mathcal{D}}$-almost surely, and P-almost surely.   □

Consider the relation of P-almost sure equality between P-integrable random variables from $(\Omega, \mathcal{F})$ into $\mathbf{R}^k$ (respectively, from $(\Omega, \mathcal{D})$ into $\mathbf{R}^k$). It is an equivalence relation, and the corresponding set of equivalence classes can be identified with $L^1(\Omega, P; \mathbf{R}^k)$ (respectively, with a subspace of $L^1(\Omega, P; \mathbf{R}^k)$). Since $\int |Z| \, dP \leq \int |Y| \, dP$, for any version Z of $E(Y|\mathcal{D})$, the linear mapping $\dot{Y} \mapsto E(Y|\mathcal{D})$ from $L^1(\Omega, P; \mathbf{R}^k)$ into itself is continuous.

When Y is of order 2, there is a simple characterization of $E(Y|\mathcal{D})$, which we give now.

Theorem 18.1.2 *The classes $\dot{Z}$, for the random variables Z from $(\Omega, \mathcal{D})$ into $\mathbf{R}^k$ which are of order 2, constitute a closed vector subspace H of $L^2(\Omega, P; \mathbf{R}^k)$ and, for every $\dot{Y} \in L^2(\Omega, P; \mathbf{R}^k)$, the orthogonal projection of $\dot{Y}$ on H is $E(Y|\mathcal{D})$.*

PROOF: As H is isometric to $L^2(\Omega, P_{/\mathcal{D}}; \mathbf{R}^k)$, it is complete, and hence closed in $L^2(\Omega, P; \mathbf{R}^k)$. The projection $\dot{Z}$ of $\dot{Y}$ on H is characterized by the property that $\int \langle Y - Z, \varphi \rangle \, dP = 0$ for all $\varphi \in H$. Fix $A \in \mathcal{D}$. For every $1 \leq i \leq k$, taking for φ the function whose ith component is 1_A and whose other components are identically zero, we see that $\int_A (Y_i - Z_i) \, dP = 0$, if Y_i and Z_i are the ith components of Y, Z. Thus $\int_A Z \, dP = \int_A Y \, dP$ for all $A \in \mathcal{D}$, and $Z = E(Y|\mathcal{D})$.   □

If Y is a positive integrable random variable from $(\Omega, \mathcal{F})$ into $\mathbf{R}$, then $E(Y|\mathcal{D})$ is clearly almost surely positive.

If Y is a P-integrable random variable from $(\Omega, \mathcal{F})$ into $\mathbf{R}^q$ $(q \geq 1)$ and $u : \mathbf{R}^q \to \mathbf{R}^k$ is a linear mapping, then $E(u \circ Y|\mathcal{D}) = u \circ E(Y|\mathcal{D})$ almost surely.

Proposition 18.1.3 *Let Φ be a bilinear mapping from $\mathbf{R}^p \times \mathbf{R}^q$ into $\mathbf{R}^k$, where p, q, k are strictly positive integers. Let Y be a random variable from $(\Omega, \mathcal{D})$ into $\mathbf{R}^p$ and Z a random variable from $(\Omega, \mathcal{F})$ into $\mathbf{R}^q$. Assume that Z and $|Y||Z|$ are P-integrable. Then $E(\Phi \circ [Y, Z]|\mathcal{D}) = \Phi \circ [Y, E(Z|\mathcal{D})]$ almost surely.*

PROOF: First, suppose that Y is $\mathcal{D}$-simple. Then $Y = \sum_{i \in I} y_i \cdot 1_{A_i}$, where I is finite, y_i belongs to $\mathbf{R}^p$ for every $i \in I$, and the A_i are disjoint $\mathcal{D}$-sets. Now

$$\int_A \Phi \circ [Y, E(Z|\mathcal{D})] \, dP \;=\; \sum_{i \in I} \int_{A \cap A_i} \Phi(y_i, E(Z|\mathcal{D})) \, dP$$

$$=\; \sum_{i \in I} \int_{A \cap A_i} \Phi(y_i, Z) \, dP$$

is equal to $\int_A \Phi \circ [Y, Z] \, dP$, for all $\mathcal{D}$-sets A, which proves that

$$\Phi \circ [Y, E(Z|\mathcal{D})] = E(\Phi \circ [Y, Z]|\mathcal{D})$$

almost surely.

Passing now to the general case, we observe that, since $|Y|$ is the upper envelope of an increasing sequence $(X_n)_{n \geq 1}$ of positive $\mathcal{D}$-simple functions, $|Y| \cdot E(|Z| \,|\mathcal{D})$ is the upper envelope of the $X_n \cdot E(|Z| \,|\mathcal{D})$ (if we choose $E(|Z| \,|\mathcal{D})$ positive). But

$$\int X_n \cdot E(|Z| \,|\mathcal{D}) \, dP = \int X_n |Z| \, dP,$$

so

$$\int X_n \cdot E(|Z| \,|\mathcal{D}) \, dP \leq \int |Y| \,|Z| \, dP,$$

and the monotone convergence theorem shows that $|Y| \cdot E(|Z| \,|\mathcal{D})$ is P-integrable.

Next, there exists a sequence $(Y_n)_{n \geq 1}$ of $\mathcal{D}$-simple functions from Ω into $\mathbf{R}^p$ which converges pointwise to Y. Replacing Y_n, if necessary, by $Y_n \cdot 1_{(|Y_n| \leq 2|Y|)}$, we may suppose that $|Y_n| \leq 2|Y|$. Then

$$|\Phi \circ [Y_n, E(Z|\mathcal{D})]| \;\leq\; \|\Phi\| \cdot |Y_n| \cdot |E(Z|\mathcal{D})|$$

$$\leq\; 2\|\Phi\| \cdot |Y| \cdot E(|Z| \,|\mathcal{D})$$

almost surely. Because $|Y| \cdot E(|Z| \,|\mathcal{D})$ is P-integrable, the dominated convergence theorem proves that $\Phi \circ [Y, E(Z|\mathcal{D})]$ is P-integrable. Finally,

$$\int_A \Phi \circ [Y, E(Z|\mathcal{D})] \, dP \;=\; \lim_{n \to +\infty} \int_A \Phi \circ [Y_n, E(Z|\mathcal{D})] \, dP$$

$$=\; \lim_{n \to +\infty} \int_A \Phi \circ [Y_n, Z] \, dP$$

is equal to $\int_A \Phi \circ [Y, Z] \, dP$ for all $\mathcal{D}$-sets A, and the proof is complete. $\quad\square$

Proposition 18.1.4 *Let $(Y_n)_{n \geq 1}$ be a sequence of random variables from $(\Omega, \mathcal{F})$ into $\mathbf{R}^k$, converging almost surely to a random variable Y. Suppose that the sequence $(|Y_n|)_{n \geq 1}$ is almost surely dominated by a P-integrable random variable from $(\Omega, \mathcal{F})$ into $\mathbf{R}$. Then $E(Y_n|\mathcal{D})$ converges P-almost surely to $E(Y|\mathcal{D})$ as $n \to +\infty$.*

PROOF: Suppose that $\sup_{n\geq 1}|Y_n|$ is finite everywhere and, for every $n \in \mathbf{N}$, put $X_n = \sup_{p\geq n}|Y_p - Y|$. Then $(X_n)_{n\geq 1}$ converges to 0 almost surely, and

$$\left|E(Y_n|\mathcal{D}) - E(Y|\mathcal{D})\right| \leq E(X_n|\mathcal{D})$$

almost surely. It suffices, therefore, to prove that $E(X_n|\mathcal{D})$ converges to 0 almost surely.

We may suppose that $E(X_n|\mathcal{D}) \geq 0$ for all $n \in \mathbf{N}$ and that the sequence $\left(E(X_n|\mathcal{D})\right)_{n\geq 1}$ is decreasing. Let then f be its lower envelope. We have

$$\int f\, dP \leq \int E(X_n|\mathcal{D})\, dP = \int X_n\, dP \qquad \text{for all } n \geq 1,$$

and the monotone convergence theorem shows that $\int X_n\, dP \to 0$ as $n \to +\infty$. Hence $\int f\, dP = 0$, and $f = 0$ almost surely. □

Now our objective is to define the conditional expected value of a quasi-integrable random variable.

Proposition 18.1.5   Let Y be a P-quasi-integrable random variable from $(\Omega, \mathcal{F})$ into $\overline{\mathbf{R}}$. There exists a random variable Z from $(\Omega, \mathcal{D})$ into $\overline{\mathbf{R}}$ such that $\int_A^* Z\, dP = \int_A^* Y\, dP$ for all $\mathcal{D}$-sets A. If Z' is another random variable from $(\Omega, \mathcal{D})$ into $\overline{\mathbf{R}}$ with the same properties, then $Z' = Z$ almost surely. Moreover, Z is P-quasi-integrable.

PROOF: First, suppose that Y is positive. For every integer $n \geq 0$, put $Y_n = \inf(Y, n)$ and $Z_n = E(Y_n|\mathcal{D})$. Then $Z_m \leq Z_n$ almost surely, for all $0 \leq m \leq n$. The function $Z = \sup_{n\geq 1} Z_n$ is measurable $\mathcal{D}$, and

$$\int_A^* Z\, dP = \sup_n \int_A Z_n\, dP = \sup_n \int_A Y_n\, dP = \int_A^* Y\, dP$$

for all $A \in \mathcal{D}$. Now let Z' be another measurable $\mathcal{D}$ function from Ω into $\overline{\mathbf{R}}$ such that $\int_A^* Z'\, dP = \int_A^* Y\, dP$ for all $\mathcal{D}$-sets A. Taking for A the set $\{\omega : Z'(\omega) < 0\}$, we see that Z' is almost surely positive. For each integer $n \geq 0$, put $A_n = Z^{-1}([-\infty, n])$. The function $Y \cdot 1_{A_n}$ is P-integrable, because $\int_{A_n}^* Y\, dP = \int_{A_n}^* Z\, dP$ is finite. From the equality

$$\int_A^* Z' \cdot 1_{A_n}\, dP = \int_{A\cap A_n}^* Z'\, dP$$

$$= \int_{A\cap A_n}^* Y\, dP$$

$$= \int_A^* Y \cdot 1_{A_n}\, dP$$

$$= \int_A^* Z \cdot 1_{A_n}\, dP$$

for all $A \in \mathcal{D}$, it follows that $Z' \cdot 1_{A_n}$ is P-integrable and $Z' \cdot 1_{A_n} = Z \cdot 1_{A_n}$ almost surely. Hence $Z' \le Z$ almost surely. Likewise, $Z \le Z'$ almost surely, so that $Z' = Z$ almost surely.

We now pass to the general case, in which Y is a P-quasi-integrable random variable from $(\Omega, \mathcal{F})$ into $\overline{\mathbf{R}}$. To fix ideas, suppose that Y^- is P-integrable. There exist a measurable $\mathcal{D}$ function $Z_1 : \Omega \to [0, +\infty]$ such that $\int_A^* Z_1 \, dP = \int_A^* Y^+ \, dP$ for all $\mathcal{D}$-sets A, and a measurable $\mathcal{D}$ and P-integrable function $Z_2 : \Omega \to [0, +\infty[$ such that $\int_A Z_2 \, dP = \int_A Y^- \, dP$ for all $\mathcal{D}$-sets A. Then $Z = Z_1 - Z_2$ is measurable $\mathcal{D}$ and

$$
\begin{aligned}
\int_A^* Z \, dP &= \int_A^* Z_1 \, dP - \int_A Z_2 \, dP \\
&= \int_A^* Y^+ \, dP - \int_A Y^- \, dP \\
&= \int_A^* Y \, dP
\end{aligned}
$$

for all $\mathcal{D}$-sets A.

Next, let Z' be another measurable $\mathcal{D}$ function from Ω into $\overline{\mathbf{R}}$ such that $\int_A^* Z' \, dP = \int_A^* Y \, dP$ for all $\mathcal{D}$-sets A. Then

$$
\begin{aligned}
\int_A^* (Z' + Z_2) \, dP &= \int_A^* Z' \, dP + \int_A Z_2 \, dP \\
&= \int_A^* Y \, dP + \int_A Y^- \, dP \\
&= \int_A^* Y^+ \, dP
\end{aligned}
$$

for all $\mathcal{D}$-sets A. Hence $Z' + Z_2 = Z_1$ almost surely, and $Z' = Z$ almost surely. $\square$

Definition 18.1.2 In the notation of Proposition 18.1.5, the class of Z (for the relation of almost sure equality) is still called the conditional expected value of Y given $\mathcal{D}$, and is written $E(Y|\mathcal{D})$. Z is a version of $E(Y|\mathcal{D})$.

Proposition 18.1.6 *If Y, Y' are two quasi-integrable random variables from $(\Omega, \mathcal{F})$ into $\overline{\mathbf{R}}$ and if $Y \le Y'$ almost surely, then, $E(Y|\mathcal{D}) \le E(Y'|\mathcal{D})$ almost surely.*

PROOF: First, suppose that Y, Y' are random variables from $(\Omega, \mathcal{F})$ into $[0, +\infty]$ such that $Y \le Y'$. Since $Y_n = \inf(Y, n) \le Y'_n = \inf(Y', n)$ for every integer $n \ge 0$, we have $E(Y_n|\mathcal{D}) \le E(Y'_n|\mathcal{D})$ almost surely, whence the result $E(Y|\mathcal{D}) \le E(Y'|\mathcal{D})$ almost surely.

We now consider the case in which Y, Y' are quasi-integrable random variables from $(\Omega, \mathcal{F})$ into $\overline{\mathbf{R}}$ such that $Y \le Y'$. Then $Y^+ \le Y'^+$ and $Y^- \ge Y'^-$, so that $E(Y^+|\mathcal{D}) \le E(Y'^+|\mathcal{D})$ almost surely and $E(Y^-|\mathcal{D}) \ge E(Y'^-|\mathcal{D})$ almost surely. Since each of $E(Y^+|\mathcal{D}) - E(Y^-|\mathcal{D})$ and $E(Y'^+|\mathcal{D}) - E(Y'^-|\mathcal{D})$ is defined almost surely, we conclude that $E(Y|\mathcal{D}) \le E(Y'|\mathcal{D})$ almost surely. $\square$

Proposition 18.1.7 *Let* $(Y_n)_{n \ge 1}$ *be a sequence of quasi-integrable random variables from* $(\Omega, \mathcal{F})$ *into* $\bar{\mathbf{R}}$, *which increases almost surely. If* $\int^* Y_n \, dP > -\infty$ *for* n *large enough, then the sequence* $(E(Y_n|\mathcal{D}))_{n \ge 1}$ *increases almost surely to* $E((\sup_{n \ge 1} Y_n)|\mathcal{D})$.

PROOF: By hypothesis, Y_n^- is integrable for n large enough. If $Z = \sup_{n \ge 1} E(Y_n|\mathcal{D})$ and $Y = \sup Y_n$, then

$$\int_A^* Z \, dP = \sup_{n \ge 1} \int_A^* E(Y_n|\mathcal{D}) \, dP = \sup_{n \ge 1} \int_A^* Y_n \, dP = \int_A^* Y \, dP$$

for all $\mathcal{D}$-sets A. Hence $Z = E(Y|\mathcal{D})$ almost surely. $\square$

18.2   The Converse of the Mean-Value Theorem

The following additional results on measures will be used to prove Jensen's inequality for conditional expected values (Section 18.3).

Suppose $\Omega \ne \emptyset$.

Definition 18.2.1 A nonempty class $\mathcal{C}$ of subsets of Ω is said to be monotone if it is closed under the formation of monotone unions and intersections:

(a) For any increasing sequence $(A_n)_{n \ge 1}$ in $\mathcal{C}$, $\bigcup_{n \ge 1} A_n$ lies in $\mathcal{C}$.

(b) For any decreasing sequence $(A_n)_{n \ge 1}$ in $\mathcal{C}$, $\bigcap_{n \ge 1} A_n$ lies in $\mathcal{C}$.

In particular, a σ-ring is a monotone class.

Proposition 18.2.1 *If* $\mathcal{R}$ *is a ring in* Ω *and* $\mathcal{C}$ *a monotone class containing* $\mathcal{R}$, *then* $\sigma(\mathcal{R}) \subset \mathcal{C}$.

PROOF: Let m be the minimal monotone class over $\mathcal{R}$, the intersection of all monotone classes containing $\mathcal{R}$. It suffices to prove $\sigma(\mathcal{R}) \subset m$: this will follow as soon as m is shown to be a ring, because a monotone ring is a σ-ring.

For any $X \subset \Omega$, put

$$Z_X = \{Y \subset \Omega : X \cap Y^c \in m, \ X^c \cap Y \in m, \ X \cup Y \in m\}.$$

Clearly, $Y \in Z_X$ is equivalent to $X \in Z_Y$. Let $(Y_i)_{i \geq 1}$ be a monotone (increasing, for example) sequence in $\mathcal{P}(\Omega)$. For all $X \subset \Omega$,

$$X \cap \left(\bigcup_{i \geq 1} Y_i \right)^c = \bigcap_{i \geq 1} (X \cap Y_i^c),$$

$$X^c \cap \left(\bigcup_{i \geq 1} Y_i \right) = \bigcup_{i \geq 1} (X^c \cap Y_i),$$

and

$$X \cup \left(\bigcup_{i \geq 1} Y_i \right) = \bigcup_{i \geq 1} (X \cup Y_i),$$

which shows that Z_X is a monotone class, whenever it is nonempty. If X lies in $\mathcal{R}$, Z_X clearly contains $\mathcal{R}$, and therefore m as well. Now, for every $Y \in m$, Y lies in Z_X, so X lies in Z_Y. Since Z_Y is a monotone class over $\mathcal{R}$, it contains m. Hence m is a ring. □

Proposition 18.2.2 *Let S be a semiring in Ω, $\mathcal{R}$ the ring generated by S, and μ a positive measure on S. Also, let f be an essentially μ-integrable mapping from Ω into a real Banach space F, and let D be a closed subset of F. If $(1/\mu(A)) \int_A f \, d\mu$ lies in D for all $A \in \mathcal{R}$ such that $\mu(A) > 0$, then $f^{-1}(F - D)$ is locally μ-negligible.*

PROOF: Fix $E \in S$. The class of those $A \in \widetilde{S}$ such that $A \subset E$ and $\int_A f \, d\mu$ belongs to $\mu(A)D$ is a monotone class over $\mathcal{R}_{/E}$. So it is equal to $\widetilde{S}_{/E}$, that is, $(1/\mu(A)) \int_A f \, d\mu$ lies in D for each $A \in \widetilde{S}$ such that $A \subset E$ and $\mu(A) > 0$.

There exists a μ-negligible subset $N \in \widetilde{S}$ of E such that $f_{/E-N}$ is measurable $\widetilde{S}_{/E-N}$ and $f(E - N)$ is separable (Theorem 6.1.2). Hence $f^{-1}(B) \cap (E - N)$ lies in $\widetilde{S}$ for every Borel set $B \subset F$. Suppose that $f(E - N) \cap D^c$ is nonempty, and let $\{y_m : m \geq 1\}$ be dense in $f(E - N) \cap D^c$. Define r_m, for all $m \geq 1$, as the largest of those real numbers $r > 0$ such that $B(y_m, r)$ does not meet D; put

$$Z_m = (E - N) \cap f^{-1}\big(B(y_m, r_m)\big).$$

Then $f(E - N) \cap D^c$ is included in $\bigcup_{m \geq 1} B(y_m, r_m)$, so $\bigcup_{m \geq 1} Z_m$ contains $(E - N) \cap f^{-1}(D^c)$. First, we show that $\mu(Z_m) = 0$ for all $m \geq 1$; it will then be obvious that $E \cap f^{-1}(D^c)$ is negligible.

Assume that $\mu(Z_m) \neq 0$ for one $m \geq 1$. Then $y = (1/\mu(Z_m)) \int_{Z_m} f \, d\mu$ lies in D. But

$$|y - y_m| = \left| \frac{1}{\mu(Z_m)} \int_{Z_m} (f - y_m) \, d\mu \right| < r_m,$$

because $|f - y_m| < r_m$ on Z_m. Thus we have a contradiction. □

18.3   Jensen's Inequality

Let $(\Omega, \mathcal{F}, P)$ be a probability space and $\mathcal{D}$ a sub-σ-algebra of $\mathcal{F}$.

Proposition 18.3.1 *Let H be a closed convex set in $\mathbf{R}^k$, φ a lower semicontinuous, convex function on H, and Y a P-integrable random variable from $(\Omega, \mathcal{F})$ into $\mathbf{R}^k$, with values in H. Then $E(Y|\mathcal{D})$ takes, almost surely, its values in H, $\varphi \circ Y$ is quasi-integrable, and $\varphi \circ E(Y|\mathcal{D}) \leq E(\varphi \circ Y|\mathcal{D})$ almost surely.*

PROOF: By Lemma 3.7.1, there exists an affine function $\psi : \mathbf{R}^k \to \mathbf{R}$ such that $\psi_{/H} \leq \varphi$. Hence $\varphi \circ Y$ is bounded below by the integrable function $\psi \circ Y$. This proves that $\varphi \circ Y$ is quasi-integrable.

Let Z be a version of $E(Y|\mathcal{D})$. By Proposition 3.6.4,

$$\frac{1}{P(A)} \int_A Z \, dP = \frac{1}{P(A)} \int_A Y \, dP \quad \text{lies in } H$$

for every $\mathcal{D}$-set A satisfying $P(A) > 0$. Now we may choose Z with values in H (Proposition 18.2.2). We prove that $\varphi \circ Z \leq E(\varphi \circ Y|\mathcal{D})$ almost surely.

First, suppose that $\varphi \circ Y$ is integrable. Since $[Y, \varphi \circ Y]$ takes its values in the closed convex subset $G = \{(x,t) \in H \times \mathbf{R} : t \geq \varphi(x)\}$ of $\mathbf{R}^{k+1}$, the mapping $[Z, E(\varphi \circ Y|\mathcal{D})] = E([Y, \varphi \circ Y]|\mathcal{D})$ takes, almost surely, its values in G. Hence $\varphi \circ Z \leq E(\varphi \circ Y|\mathcal{D})$ almost surely, as desired.

Next, we treat the general case, in which $\varphi \circ Y$ is not necessarily integrable. Given an integer $n \geq 0$, put $A_n = \{\omega \in \Omega : E(\varphi \circ Y|\mathcal{D})(\omega) \leq n\}$ and assume that $P(A_n)$ is strictly positive. Denote by P' the probability $A \mapsto P(A)/P(A_n)$ on $\{A \in \mathcal{F} : A \subset A_n\}$, and by $\mathcal{D}'$ the σ-ring $\{A \in \mathcal{D} : A \subset A_n\}$. Finally, write Y' for the restriction of Y to A_n, and Z' for the restriction of $E(Y|\mathcal{D})$ to A_n. Then

$$\int_A^* Y' \, dP' = \frac{1}{P(A_n)} \int_A^* Y \, dP = \frac{1}{P(A_n)} \int_A^* Z \, dP = \int_A^* Z' \, dP'$$

for all $A \in \mathcal{D}'$. Thus $Z' = E(Y'|\mathcal{D}')$ almost surely. Likewise, the restriction of $E(\varphi \circ Y|\mathcal{D})$ to A_n is equal to $E(\varphi \circ Y'|\mathcal{D}')$, almost surely. Since

$$\int^* \varphi \circ Y' \, dP' = \frac{1}{P(A_n)} \int_{A_n}^* \varphi \circ Y \, dP = \frac{1}{P(A_n)} \int_{A_n} E(\varphi \circ Y|\mathcal{D}) \, dP$$

is finite, $\varphi \circ Y'$ is P'-integrable. Now

$$\varphi \circ E(Y'|\mathcal{D}') \leq E(\varphi \circ Y'|\mathcal{D}')$$

P'-almost surely, so $\varphi \circ E(Y|\mathcal{D}) \leq E(\varphi \circ Y|\mathcal{D})$ almost surely in A_n. We conclude that $\varphi \circ E(Y|\mathcal{D}) \leq E(\varphi \circ Y|\mathcal{D})$ almost surely.   □

Proposition 18.3.2   *Assume that φ is strictly convex on H. Then $\varphi \circ E(Y|\mathcal{D}) = E(\varphi \circ Y|\mathcal{D})$ almost surely if and only if Y is, almost surely, equal to a random variable from $(\Omega, \mathcal{D})$ into $\mathbf{R}$.*

PROOF: Let Z be a version of $E(Y|\mathcal{D})$ with values in H, and assume that $\varphi \circ Z = E(\varphi \circ Y|\mathcal{D})$ almost surely.

First, suppose that $\varphi \circ Y$ is integrable, and fix a real number $\alpha \in]0,1[$. Then $\varphi \circ (\alpha Y + (1-\alpha)Z) \leq \alpha \cdot \varphi \circ Y + (1-\alpha) \cdot \varphi \circ Z$, which proves that $\varphi \circ (\alpha Y + (1-\alpha)Z)$ is P-integrable. Since

$$\varphi \circ Z = \varphi \circ E\Big(\alpha Y + (1-\alpha)Z|\mathcal{D}\Big)$$
$$\leq E\Big(\varphi \circ [\alpha Y + (1-\alpha)Z]|\mathcal{D}\Big) \leq E\Big([\alpha \cdot \varphi \circ Y + (1-\alpha)\varphi \circ Z]|\mathcal{D}\Big) = \varphi \circ Z$$

almost surely, we have

$$E\Big(\varphi \circ [\alpha Y + (1-\alpha)Z]|\mathcal{D}\Big) = E\Big([\alpha \cdot \varphi \circ Y + (1-\alpha)\varphi \circ Z]|\mathcal{D}\Big)$$

almost surely. Hence

$$\int \varphi \circ [\alpha Y + (1-\alpha)Z]\, dP = \int [\alpha \cdot \varphi \circ Y + (1-\alpha) \cdot \varphi \circ Z]\, dP,$$

and

$$\varphi \circ [\alpha Y + (1-\alpha)Z] = \alpha \cdot \varphi \circ Y + (1-\alpha) \cdot \varphi \circ Z$$

almost surely. As φ is strictly convex, we conclude that $Y = Z$ almost surely.

Now to prove the general case, in which $\varphi \circ Y$ is not necessarily integrable, set $A_n = \{\omega \in \Omega : E(\varphi \circ Y|\mathcal{D})(\omega) \leq n\}$ for every integer $n \geq 0$. Arguing as in Proposition 18.3.1, we see that, if $P(A_n) > 0$, then $Y_{/A_n}$ is almost surely equal in A_n to a measurable $\mathcal{D}_{/A_n}$-function, and, since $\Omega - \bigcup_{n \geq 0} A_n$ is P-negligible, we conclude that Y is almost surely equal to a measurable $\mathcal{D}$ function. $\square$

When $k = 1$, we have a slightly more general result.

Proposition 18.3.3   *Let $I = \langle a, b \rangle$ (with $a < b$ in $\bar{\mathbf{R}}$) be an interval in $\mathbf{R}$, φ a convex function on I, and Y a P-integrable random variable from $(\Omega, \mathcal{F})$ into $\mathbf{R}$, with values in I. Then $E(Y|\mathcal{D})$ takes, almost surely, its values in I, $\varphi \circ Y$ is quasi-integrable, and $\varphi \circ E(Y|\mathcal{D}) \leq E(\varphi \circ Y|\mathcal{D})$ almost surely. When φ is strictly convex, the equality $\varphi \circ E(Y|\mathcal{D}) = E(\varphi \circ Y|\mathcal{D})$ holds almost surely if and only if Y is almost surely equal to a random variable from $(\Omega, \mathcal{D})$ into $\mathbf{R}$.*

PROOF: Since φ is convex, the limits $\varphi(a_+)$ and $\varphi(b_-)$ exist in $\bar{\mathbf{R}}$. When a belongs to I, choose $s \in I$, $s > a$; then

$$\frac{\varphi(s) - \varphi(a)}{s - a} \leq \frac{\varphi(s) - \varphi(t)}{s - t}$$

for all $t \in]a, s[$; letting t tend to a, we see that

$$\frac{\varphi(s) - \varphi(a)}{s - a} \leq \frac{\varphi(s) - \varphi(a_+)}{s - a},$$

and finally that $\varphi(a_+) \leq \varphi(a)$. Similarly, $\varphi(b_-) \leq \varphi(b)$ when b lies in I. So φ is upper semicontinuous, and hence is a Borel function.

Clearly, $u = \int Y \, dP$ lies in I. If u belongs to $]a, b[$, then

$$\frac{\varphi(t) - \varphi(u)}{t - u} \leq \varphi'_l(u) = \beta$$

(where $\varphi'_l(u)$ is the left-hand derivative of φ at u) for all t in $I \cap]-\infty, u[$, and $(\varphi(t) - \varphi(u))/(t - u) \geq \varphi'_r(u)$ for all t in $I \cap]u, +\infty[$; hence $\varphi(t) \geq \varphi(u) + \beta(t-u)$ for all $t \in I$. Therefore, $\varphi \circ Y$ is bounded below by the integrable function $\varphi(u) + \beta(Y - u)$. We conclude that $\varphi \circ Y$ is quasi-integrable and that $\int^* \varphi \circ Y \, dP \geq \varphi(\int Y \, dP)$. These conclusions persist when $u = a$ or $u = b$.

$E(Y|\mathcal{D})$ takes, almost surely, its values in the closure $\bar{I}$ of I relative to $\mathbf{R}$. If a belongs to $\bar{I} - I$ and if we put $A = E(Y|\mathcal{D})^{-1}(a)$, then $aP(A) = \int_A Y \, dP$, so $\int_A (Y - a) \, dP = 0$ and $P(A) = 0$. Similarly, if b belongs to $\bar{I} - I$ and if we put $B = E(Y|\mathcal{D})^{-1}(b)$, then $P(B) = 0$. Hence $E(Y|\mathcal{D})$ takes, almost surely, its values in I.

Denote by $\bar{\varphi}$ the continuous function from $\bar{I}$ into $\bar{\mathbf{R}}$ equal to φ on $]a, b[$, and define G by:

$$\begin{aligned} G \;&=\; \{(x, t) \in \bar{I} \times \mathbf{R} : t \geq \bar{\varphi}(x)\} \\ &=\; \{(x, t) \in \bar{I} \times \mathbf{R} : \bar{\varphi}(x) < +\infty \text{ and } t \geq \bar{\varphi}(x)\}. \end{aligned}$$

Clearly, G is a closed convex subset of $\mathbf{R}^2$.

First, suppose that $\varphi \circ Y$ is integrable. $[Y, \varphi \circ Y]$ taking its values in G, $[E(Y|\mathcal{D}), E(\varphi \circ Y|\mathcal{D})]$ takes, almost surely, its values in G. Thus

$$\bar{\varphi} \circ E(Y|\mathcal{D}) \leq E(\varphi \circ Y|\mathcal{D}) \quad \text{almost surely.}$$

When a lies in I and $P(A) > 0$, where $A = E(Y|\mathcal{D})^{-1}(a)$, then $Y = a$ almost surely in A. If P' is the probability $E \mapsto P(E)/P(A)$ on $\{E \in \mathcal{F} : E \subset A\}$, if $\mathcal{D}' = \{E \in \mathcal{D} : E \subset A\}$, and if Y' is the restriction of Y to A, then $E(Y|\mathcal{D})_{/A} = E(Y'|\mathcal{D}') = a$ P'-almost surely and $E(\varphi \circ Y|\mathcal{D})_{/A} = E(\varphi \circ Y'|\mathcal{D}') = \varphi(a)$ P'-almost surely. Thus $\varphi \circ E(Y|\mathcal{D}) \leq E(\varphi \circ Y|\mathcal{D})$ almost surely in A. Likewise, $\varphi \circ E(Y|\mathcal{D}) \leq E(\varphi \circ Y|\mathcal{D})$ almost surely in $B = E(Y|\mathcal{D})^{-1}(b)$, when b lies in I. In short, $\varphi \circ E(Y|\mathcal{D}) \leq E(\varphi \circ Y|\mathcal{D})$ almost surely.

Now, arguing as in Proposition 18.3.1, we see that the last inequality persists even if $\varphi \circ Y$ is not integrable. Finally, we note that the last assertion follows from the same argument as that in Proposition 18.3.2. $\qquad \square$

18.4   Conditional Expected Value Given a Random Variable

Definition 18.4.1 Let $(\Omega, \mathcal{F})$ be a measurable space and let X, Z be two random variables from $(\Omega, \mathcal{F})$ into measurable spaces $(F, \mathcal{B})$ and $(G, \mathcal{C})$, respectively. Z is said to be a function of X if there exists a random variable h from $(F, \mathcal{B})$ into $(G, \mathcal{C})$ such that $Z = h \circ X$.

Theorem 18.4.1 *When* $G = \mathbf{R}^k$ *(respectively, when* $G = \bar{\mathbf{R}}$*) and* $\mathcal{C}$ *is the Borel* σ*-algebra of* $\mathbf{R}^k$ *(respectively, of* $\bar{\mathbf{R}}$*), then* Z *is a function of* X *if and only if the* σ*-algebra* $\sigma(Z)$ *(consisting of the* $Z^{-1}(C)$ *for* $C \in \mathcal{C}$*) is included in* $\sigma(X)$.

PROOF: We will consider only the case $G = \mathbf{R}$. Assume that $\sigma(Z) \subset \sigma(X)$.

First, suppose that Z is simple: $Z = \sum_{i \in I} \alpha_i \cdot 1_{A_i}$, where I is finite and the A_i are disjoint $\sigma(X)$-sets. For every $i \in I$, there exists $B_i \in \mathcal{B}$ such that $A_i = X^{-1}(B_i)$. Then $h = \sum_{i \in I} \alpha_i \cdot 1_{B_i}$ is a random variable from $(F, \mathcal{B})$ into $(G, \mathcal{C})$, and $h \circ X = Z$.

Now, for the general case, observe that there exists a sequence $(Z_n)_{n \geq 1}$ of simple random variables from $(\Omega, \sigma(X))$ into $\mathbf{R}$ which converges pointwise to Z. Since Z_n is simple, we can find, for every $n \geq 1$, a random variable h_n from $(F, \mathcal{B})$ into $(G, \mathcal{C})$ such that $h_n \circ X = Z_n$. Let L be the set of convergence of the sequence $(h_n)_{n \geq 1}$. Since

$$L = (\limsup h_n = \liminf h_n) \cap (|\liminf h_n| < +\infty),$$

it lies in $\mathcal{B}$. Moreover, $X(\Omega)$ is included in L. If we define h by $h = \lim h_n$ on L and $h = 0$ on L^c, then $h \circ X = Z$, as desired. $\square$

Now let $(\Omega, \mathcal{F}, P)$ be a probability space, X a random variable from $(\Omega, \mathcal{F})$ into a measurable space $(F, \mathcal{B})$, and $\sigma(X)$ the σ-algebra generated by X. Given $k \in \mathbf{N}$, the relation of P_X-almost sure equality between random variables from $(F, \mathcal{B})$ into $\mathbf{R}^k$ is an equivalence relation. If Y is a P-integrable random variable from $(\Omega, \mathcal{F})$ into $\mathbf{R}^k$, any P_X-integrable random variable h from $(F, \mathcal{B})$ into $\mathbf{R}^k$ such that $\int_B h \, dP_X = \int_{X^{-1}(B)} Y \, dP$ for all $B \in \mathcal{B}$ is called a version of the conditional expected value of Y given X. Thus a random variable h from $(F, \mathcal{B})$ into $\mathbf{R}^k$ is a version of the conditional expected value of Y given X if $h \circ X$ is a version of $E(Y|\sigma(X))$; the class of h is also called the conditional expected value of Y given X, and is written $E(Y|X)$. We often identify h with its class. Given a π-system T in F such that $\sigma(T) = \mathcal{B}$, a P_X-integrable random variable h from $(F, \mathcal{B})$ into $\mathbf{R}^k$ is a version of $E(Y|X)$ whenever $\int_B h \, dP_X = \int_{X^{-1}(B)} Y \, dP$ for all $B \in T$. If X' is a random variable from $(\Omega, \mathcal{F})$ into $(F, \mathcal{B})$ which is almost surely equal to X, then $E(Y|X) = E(Y|X')$, although $\sigma(X) \neq \sigma(X')$.

Similarly, we can define $E(Y|X)$ for any P-quasi-integrable random variable Y from $(\Omega, \mathcal{F})$ into $\bar{\mathbf{R}}$. $h : F \to \bar{\mathbf{R}}$ is a version of $E(Y|X)$ if it is measurable

$\mathcal{B}$ and

$$\int_B^* h \, dP_X = \int_{X^{-1}(B)}^* Y \, dP \qquad \text{for all } B \in \mathcal{B}.$$

In this case, let Z be a version of $E(Y|\sigma(X))$. To fix ideas, we suppose that the negative part of Z is P-integrable. Putting $A = \{x \in F : h(x) \leq 0\}$, we observe that $\int_A^* h \, dP_X = \int_{X^{-1}(A)}^* Z \, dP$ is not $-\infty$ and that the negative part of h is P_X-integrable. Now $\int_B^* h \, dP_X = \int_{X^{-1}(B)}^* (h \circ X) \, dP$ for all $B \in \mathcal{B}$; thus $h \circ X$ is a version of $E(Y|\sigma(x))$.

There exist results analogous to those of Sections 18.1 and 18.3 for expected values given X.

We can compute $E(Y|X)$ as follows.

Proposition 18.4.1 *Let $(\Omega, \mathcal{F}, P)$ be a probability space and X a random variable from $(\Omega, \mathcal{F})$ into $\mathbf{R}^j$ $(j \geq 1)$ whose law P_X is absolutely continuous with respect to Lebesgue measure λ_j on $\mathbf{R}^j$. Finally, let Y be a P-integrable random variable from $(\Omega, \mathcal{F})$ into $\mathbf{R}^k$ $(k \geq 1)$ and h a version of $E(Y|X)$. Then we can find a P_X-negligible Borel set N with the following property: for all $x \notin N$, and for every sequence $(B_n)_{n \geq 1}$ of Borel subsets of $\mathbf{R}^j$ which shrinks to x nicely, $(1/\lambda_j(B_n))P_X(B_n)$ tends to a strictly positive number and $(1/P_X(B_n)) \int_{X^{-1}(B_n)} Y \, dP$ tends to $h(x)$ as $n \to +\infty$.*

PROOF: Let f be a λ_j-integrable Borel function from $\mathbf{R}^j$ into $[0, +\infty[$ such that $P_X = f\lambda_j$. Since f and hf are λ_j-integrable, there is a λ_j-negligible Borel set N' such that, for all $x \notin N'$, $P_X(B_n)/\lambda_j(B_n)$ and $(1/\lambda_j(B_n)) \int_{B_n} hf \, d\lambda_j$ converge to $f(x)$ and $h(x)f(x)$ as $n \to +\infty$, for every sequence $(B_n)_{n \geq 1}$ of Borel sets which shrinks to x nicely. Now take $N = N' \cup f^{-1}(0)$. $\square$

18.5   Conditional Law of Y Given X

Let $(\Omega, \mathcal{F}, P)$ be a probability space and X, Y two random variables from $(\Omega, \mathcal{F})$ into the measurable spaces $(F, \mathcal{B})$ and $(G, \mathcal{C})$, respectively.

Proposition 18.5.1 *Let $(\mu_x)_{x \in F'}$, where $F' \in \mathcal{B}$ and F' carries P_X, be a family of probabilities on $\mathcal{C}$. Fix a π-system Ψ such that $\sigma(\Psi) = \mathcal{C}$. If the function $F' \ni x \mapsto \mu_x(C)$, $F' \not\ni x \mapsto 0$ is a version of $E(1_C \circ Y|X)$ for every $C \in \Psi$, then it is a version of $E(1_C \circ Y|X)$ for every $C \in \mathcal{C}$.*

PROOF: Designate by $\mathcal{L}$ the collection of those $C \in \mathcal{C}$ for which the function $F' \ni x \mapsto \mu_x(C)$, $F' \not\ni x \mapsto 0$ is a version of $E(1_C \circ Y|X)$. Then $\mathcal{L}$ is a λ-system and it contains Ψ, so $\mathcal{L} = \mathcal{C}$. $\square$

Definition 18.5.1 In the notation of Proposition 18.5.1, if, for all $C \in \mathcal{C}$, the function $F' \ni x \mapsto \mu_x(C)$, $F' \not\ni x \mapsto 0$ is a version of $E(1_C \circ Y | X)$, then $(\mu_x)_{x \in F'}$ is called a conditional law of Y given X. We shall often write $\mu_x = P(Y | X = x)$.

Theorem 18.5.1 Let Φ (respectively, Ψ) be a π-system in F (respectively, in G) such that $\sigma(\Phi) = \mathcal{B}$ (respectively, $\sigma(\Psi) = \mathcal{C}$), and let $(\mu_x)_{x \in F'}$, where $F' \in \mathcal{B}$ and F' carries P_X, be a family of probabilities on $\mathcal{C}$. Then $(\mu_x)_{x \in F'}$ is a conditional law of Y given X if and only if

(a) for all $C \in \Psi$, the function $F' \ni x \mapsto \mu_x(C)$, $F' \not\ni x \mapsto 0$ is measurable $\mathcal{B}$;

(b) $P_{[X,Y]}(B \times C) = \int_B \mu_x(C) \, dP_X(x)$ for all $B \in \Phi$ and $C \in \Psi$.

PROOF: Clearly, the conditions are necessary. Conversely, suppose they are satisfied. For every $C \in \Psi$, the function $F' \ni x \longmapsto \mu_x(C)$, $F' \not\ni x \longmapsto 0$ is measurable $\mathcal{B}$ and

$$\int_B \mu_x(C) \, dP_X(x) = P\Big(X^{-1}(B) \cap Y^{-1}(C)\Big) = \int_{X^{-1}(B)} 1_C \circ Y \, dP$$

for all $B \in \Phi$. Hence, for every $C \in \Psi$, the function $F' \ni x \longmapsto \mu_x(C)$, $F' \not\ni x \mapsto 0$ is a version of $E(1_C \circ Y | X)$ (Proposition 18.1.1). From Proposition 18.5.1, we deduce that it is, in fact, a version of $E(1_C \circ Y | X)$ for all $C \in \mathcal{C}$. $\square$

Until further notice, we shall suppose that, for given X and Y, there exists a conditional law of Y given X. We let $(\mu_x)_{x \in F'}$ be such a conditional law.

Proposition 18.5.2 Let $f : F \times G \to [0, +\infty]$ be measurable $\mathcal{B} \otimes \mathcal{C}$ (respectively, $P_{[X,Y]}$-measurable). Then the function $F' \ni x \mapsto \int^* f(x, y) \, d\mu_x(y)$, $F' \not\ni x \mapsto 0$ is measurable $\mathcal{B}$ (respectively, P_X-measurable), and

$$\int^* f \, dP_{[X,Y]} = \int^* dP_X(x) \int^* f(x, y) \, d\mu_x(y).$$

PROOF: For all $A \in \mathcal{B} \otimes \mathcal{C}$, we know that $1_A(x, \cdot)$ is measurable $\mathcal{C}$ for every $x \in F$. Denote by $\mathcal{L}$ the collection of those $A \in \mathcal{B} \otimes \mathcal{C}$ such that the function $F' \ni x \mapsto \int 1_A(x, y) \, d\mu_x(y)$, $F' \not\ni x \mapsto 0$ is measurable $\mathcal{B}$, and such that $\int 1_A \, dP_{[X,Y]} = \int dP_X(x) \int 1_A(x, y) \, d\mu_x(y)$. Then $\mathcal{L}$ is a λ-system and it contains $\mathcal{B} \times \mathcal{C}$. Hence $\mathcal{L} = \mathcal{B} \otimes \mathcal{C}$.

Now, if $f : F \times G \to [0, +\infty]$ is measurable $\mathcal{B} \otimes \mathcal{C}$, it is the upper envelope of an increasing sequence in $St^+(\mathcal{B} \otimes \mathcal{C})$. The first assertion of the statement follows.

Finally, suppose that f is $P_{[X,Y]}$-measurable. There exists a measurable $\mathcal{B} \otimes \mathcal{C}$ function, g, from $F \times G$ into $[0, +\infty]$ which agrees with f outside a

$P_{[X,Y]}$-negligible set $A \in \mathcal{B} \otimes \mathcal{C}$. Since

$$\int^* dP_X(x) \int^* 1_A(x,y) \, d\mu_x(y) = 0,$$

the section $A(x, \cdot)$ is μ_x-negligible for all $x \in F'$ which do not belong to some P_X-negligible subset N of F'. For every $x \in F' - N$, we have

$$\int^* f(x,y) \, d\mu_x(y) = \int^* g(x,y) \, d\mu_x(y).$$

Hence

$$\varphi : \begin{cases} F' \ni x \longmapsto \int^* f(x,y) \, d\mu_x(y), \\ F' \not\ni x \longmapsto 0 \end{cases}$$

is P_X-measurable, and

$$\begin{aligned}
\int^* \varphi \, dP_X &= \int^* dP_X(x) \int^* g(x,y) \, d\mu_x(y) \\
&= \int^* g \, dP_{[X \, Y]} \\
&= \int^* f \, dP_{[X,Y]}.
\end{aligned}$$

$\square$

Proposition 18.5.3 *Let f be a $P_{[X,Y]}$-integrable mapping from $F \times G$ into a real Banach space H. Then the mapping $x \longmapsto \int f(x,y) \, d\mu_x(y)$ is P_X-almost surely defined on F', is P_X-integrable, and*

$$\int f \, dP_{[X,Y]} = \int dP_X(x) \int f(x,y) \, d\mu_x(y).$$

PROOF: If f is $\mathcal{B} \otimes \mathcal{C}$-simple, $x \longmapsto \int f(x,y) \, d\mu_x(y)$ is defined on F' and is P_X-integrable. Furthermore,

$$\int f \, dP_{[X,Y]} = \int dP_X(x) \int f(x,y) \, d\mu_x(y).$$

Now, for the general case, we find a measurable $\mathcal{B} \otimes \mathcal{C}$ mapping g from $F \times G$ into H which agrees with f $P_{[X,Y]}$-almost surely. Let $(f_n)_{n \geq 1}$ be a sequence of $\mathcal{B} \otimes \mathcal{C}$ simple mappings from $F \times G$ into H, converging $P_{[X,Y]}$-almost surely to g and such that $|f_n| \leq 2|g|$. There is a $P_{[X,Y]}$-negligible set $A \in \mathcal{B} \otimes \mathcal{C}$ such that

(a) $\big(f_n(x,y)\big)_{n \geq 1}$ converges to $f(x,y)$ for each $(x,y) \in A^c$;

(b) $|f_n| \leq 2|f|$ on A^c.

Since $\int^* dP_X(x) \int^* 1_A(x,y) \, d\mu_x(y) = 0$, the section $A(x, \cdot)$ is μ_x-negligible for all $x \in F'$ which do not belong to some P_X-negligible subset N' of F'. On the other hand, $\int^* |f|(x,y) \, d\mu_x(y)$ is finite for all $x \in F'$ which do not belong to some P_X-negligible subset N'' of F'. Put $N = N' \cup N''$. For every $x \in F' - N$, the sequence $\big(f_n(x, \cdot)\big)_{n \geq 1}$ converges to $f(x, \cdot)$ μ_x-almost surely, and $|f_n|(x, \cdot)$ is dominated by $2|f|(x, \cdot)$ μ_x-almost surely. Hence $f(x, \cdot)$ is μ_x-integrable and $\int f(x,y) \, d\mu_x(y) = \lim_{n \to +\infty} \int f_n(x,y) \, d\mu_x(y)$.

Next, observe that the functions $x \mapsto \big| \int f_n(x,y) \, d\mu_x(y) \big|$ from $F' - N$ into $\mathbf{R}$ are dominated by $x \mapsto 2 \int^* |f|(x,y) \, d\mu_x(y)$. So the mapping $x \mapsto \int f(x,y) \, d\mu_x(y)$ from $F' - N$ into H is P_X-integrable and

$$
\int dP_X(x) \int f(x,y) \, d\mu_x(y) = \lim_{n \to +\infty} \int dP_X(x) \int f_n(x,y) \, d\mu_x(y)
$$

$$
= \int f \, dP_{[X,Y]}.
$$

$\square$

Now we use Proposition 18.5.3 to compute conditional expected values.

Theorem 18.5.2 *Let* $f : F \times G \to \mathbf{R}^k$ *be measurable* $\mathcal{B} \otimes \mathcal{C}$ *and* $P_{[X,Y]}$ *integrable. Then* $x \mapsto \int f(x,y) \, d\mu_x(y)$ *is* P_X-*almost surely defined on* F', *and it is* P_X-*almost surely equal to* $E(f \circ [X,Y]|X)$.

PROOF: By Proposition 18.5.3, $x \mapsto \int f(x,y) \, d\mu_x(y)$ is P_X-almost surely defined on F', is P_X-integrable, and

$$
\int_B dP_X(x) \int f(x,y) \, d\mu_x(y) = \int dP_X(x) \int f(x,y) \cdot 1_{B \times G}(x,y) \, d\mu_x(y)
$$

$$
= \int f \cdot 1_{B \times G} \, dP_{[X,Y]}
$$

$$
= \int_{X^{-1}(B)} f \circ [X,Y] \, dP
$$

for all $B \in \mathcal{B}$. Thus this mapping is P_X-almost surely equal to $E\big(f \circ [X,Y]|X\big)$.
$\square$

Similarly, we have the following result.

Proposition 18.5.4 *Let* $f : F \times G \to \overline{\mathbf{R}}$ *be measurable* $\mathcal{B} \otimes \mathcal{C}$ *and* $P_{[X,Y]}$-*quasi-integrable. Then the function* $x \mapsto \int^* f(x,y) \, d\mu_x(y)$ *from* F' *into* $\overline{\mathbf{R}}$ *is equal to* $E(f \circ [X,Y]|X)$, P_X-*almost surely.*

PROOF: First, suppose that f is positive. Then, by Proposition 18.5.2, the function

$$
F' \ni x \mapsto \int^* f(x,y) \, d\mu_x(y) \qquad F' \not\ni x \mapsto 0
$$

is measurable $\mathcal{B}$. Moreover,

$$
\begin{aligned}
\int_B^* dP_X(x) \int^* f(x,y)\, d\mu_x(y) &= \int^* dP_X(x) \int^* f(x,y) \cdot 1_B(x)\, d\mu_x(y) \\
&= \int^* f \cdot 1_{B \times G}\, dP_{[X,Y]} \\
&= \int_{X^{-1}(B)}^* f \circ [X,Y]\, dP
\end{aligned}
$$

for all $B \in \mathcal{B}$, whence the desired result.

Now, if f is $P_{[X,Y]}$-quasi-integrable, we may suppose, without loss of generality, that f^- is $P_{[X,Y]}$-integrable. Then $f^-(x,\cdot)$ is μ_x-integrable for all $x \in F'$ which do not belong to some suitable P_X-negligible subset N of F'. For all $x \in F' - N$, the function $f(x,\cdot) = f^+(x,\cdot) - f^-(x,\cdot)$ is μ_x-quasi-integrable and

$$
\int^* f(x,y)\, d\mu_x(y) = \int^* f^+(x,y)\, d\mu_x(y) - \int f^-(x,y)\, d\mu_x(y).
$$

So

$$
\begin{aligned}
\int^* f(x,y)\, d\mu_x(y) &= E(f^+ \circ [X,Y]|X)(x) - E(f^- \circ [X,Y]|X)(x) \\
&= E(f \circ [X,Y]|X)(x)
\end{aligned}
$$

P_X-almost surely in F'. $\qquad \square$

The following corollaries to Theorem 18.5.2 and Proposition 18.5.4 justify the terminology regarding the conditional expected value, establishing its relationship with the expected value.

Corollary 1 *Let $h : G \to \mathbf{R}^k$ be measurable $\mathcal{C}$ and P_Y-integrable. Then $\int h\, d\mu_x$ is P_X-almost surely defined on F', and*

$$
E(h \circ Y|X)(x) = \int h\, d\mu_x
$$

for P_X-almost all points x.

PROOF: This is an instant consequence of Theorem 18.5.2, because $(x,y) \mapsto h(y)$ is measurable $\mathcal{B} \otimes \mathcal{C}$ and $P_{[X,Y]}$-integrable. $\qquad \square$

Corollary 2 *Let $h : G \to \overline{\mathbf{R}}$ be measurable $\mathcal{C}$ and P_Y-quasi-integrable. Then $E(h \circ Y|X)(x) = \int^* h\, d\mu_x$ for P_X-almost all points $x \in F'$.*

Definition 18.5.2 Let $(\Omega, \mathcal{F}, P)$ be a probability space, $\mathcal{D}$ a sub-σ-algebra of $\mathcal{F}$, and Y a random variable from $(\Omega, \mathcal{F})$ into a measurable space $(G, \mathcal{C})$. Write X for the random variable $\omega \mapsto \omega$ from $(\Omega, \mathcal{F})$ into $(\Omega, \mathcal{D})$. Every conditional law of Y given X is called a condition law of Y given $\mathcal{D}$.

18.6 Computation of Conditional Laws

Let $(\Omega, \mathcal{F}, P)$ be a probability space.

Theorem 18.6.1 *Let X and Y be two random variables from $(\Omega, \mathcal{F})$ into the measurable spaces $(F, \mathcal{B})$ and $(G, \mathcal{C})$, respectively. Assume that P_X is defined by masses α_x at points x of a subset F' of F. For each $x \in F'$, denote by μ_x the probability $C \mapsto P\big((Y \in C) \cap (X = x)\big)/P(X = x)$ on C. Then $(\mu_x)_{x \in F'}$ is a conditional law of Y given X.*

PROOF: Obvious.   □

Theorem 18.6.2 *Let X and Y be two random variables from $(\Omega, \mathcal{F})$ into the measurable spaces $(F, \mathcal{B})$ and $(G, \mathcal{C})$, respectively. Also, let S and T be two semirings in F and G, respectively, such that $\sigma(S) = \mathcal{B}$ and $\sigma(T) = \mathcal{C}$. Finally, let μ and ν be two positive measures on S and T, respectively. Assume that $P_{[X,Y]}$ is absolutely continuous with respect to $\mu \otimes \nu$. Let $f : F \times G \to [0, +\infty[$ be measurable $\mathcal{B} \otimes \mathcal{C}$ and $\mu \otimes \nu$-integrable, such that $P_{[X,Y]} = f \cdot (\mu \otimes \nu)$. Now define φ as the function $x \mapsto \int^{*} f(x, y)\, d\nu(y)$ from F into $[0, +\infty]$. For all x in $F' = F - \varphi^{-1}(0) - \varphi^{-1}(+\infty)$, set $\mu_x = \big(f(x, \cdot)/\varphi(x)\big)\nu$. Then $(\mu_x)_{x \in F'}$ is a conditional law of Y given X.*

PROOF: By Proposition 9.2.7, φ is measurable $\mathcal{B}$ and finite μ-almost everywhere. Let ψ be a measurable $\mathcal{B}$ function from F into $[0, +\infty[$ which agrees with φ on $F - \varphi^{-1}(+\infty)$. Then $P_X = \psi\mu$. Clearly, $\varphi^{-1}(0) = \psi^{-1}(0)$ is P_X-negligible.

Since $\int^{*}_{\varphi^{-1}(0) \times G} f\, d(\mu \otimes \nu) = \int^{*}_{\varphi^{-1}(0)} \varphi\, d\mu = 0$, f vanishes $\mu \otimes \nu$-almost everywhere on $\varphi^{-1}(0) \times G$. For all $B \in S$ and all $C \in T$,

$$
\begin{aligned}
P_{[X,Y]}(B \times C) &= \iint_{B \times C} f(x, y)\, d\mu(x)\, d\nu(y) \\
&= \iint_{(B \cap F') \times C} f(x, y)\, d\mu(x)\, d\nu(y) \\
&= \int_{B \cap F'} \left[\int_C \frac{f(x, y)}{\varphi(x)}\, d\nu(y) \right] \psi(x)\, d\mu(x) \\
&= \int_{B \cap F'} \left[\int_C \frac{f(x, y)}{\varphi(x)}\, d\nu(y) \right] dP_X(x).
\end{aligned}
$$

Furthermore, for every $C \in T$,

$$
F' \ni x \mapsto \mu_x(C) = (1/\varphi(x)) \int^{*} f \cdot 1_{F \times C}(x, y)\, d\nu(y)
$$

is measurable $\mathcal{B}_{/F'}$. Theorem 18.5.1 now shows that $(\mu_x)_{x \in F'}$ is a conditional law of Y given X.   □

Proposition 18.6.1 *Let X and Y be two random variables from $(\Omega, \mathcal{F})$ into $\mathbf{R}^j$ and $\mathbf{R}^k$ $(j, k \in \mathbf{N})$, respectively. Suppose there exists a family $(\mu_x)_{x \in B}$ of probabilities on $\mathbf{R}^k$ with the following properties:*

(a) B is a Borel subset of $\mathbf{R}^j$, on which P_X is concentrated.

(b) For all $t \in \mathbf{R}^k$, $B \ni x \mapsto \mu_x(]-\infty, t])$, $B \not\ni x \mapsto 0$ is a Borel function.

(c) $P_{[X,Y]}(]-\infty, s] \times]-\infty, t]) = \int_{]-\infty, s]} \mu_x(]-\infty, t]) \, dP_X(x)$ for all $s \in \mathbf{R}^j$ and $t \in \mathbf{R}^k$.

Then $(\mu_x)_{x \in B}$ is a conditional law of Y, given X.

PROOF: This follows from Theorem 18.5.1.   □

As an example, let $(\Omega, \mathcal{F}, P)$ be a probability space, and let $X_1, \ldots, X_n$ be independent random variables from $(\Omega, \mathcal{F})$ into $\mathbf{R}$ with common continuous distribution function F. Put $X = \sup(X_1, \ldots, X_n)$ and let $1 \leq k \leq n$ be an integer. We wish to compute the conditional law of $Y = X_k$ given X.

For all real numbers s, t, put $G(s,t) = P(X \leq s, Y \leq t)$. If $s \leq t$, since $X_k \leq X$, we have $G(s,t) = P(X \leq s) = F(s)^n$. If $s > t$, we have

$$G(s,t) = P(X_1 \leq s, \ldots, X_k \leq t, \ldots, X_n \leq s) = F(t) F^{n-1}(s).$$

Now $dF^n = nF^{n-1} \, dF$ and $dF^{n-1} = (n-1)F^{n-2} \, dF$, where dF is the Stieltjes measure associated with F. Hence, for every $t \in \mathbf{R}$, the function $s \mapsto G(s,t)$ is an indefinite integral of

$$nF^{n-1}(x) \cdot 1_{]-\infty, t]}(x) \, dF(x) + (n-1)F(t)F^{n-2}(x) \cdot 1_{]t, +\infty[}(x) \, dF(x).$$

For all x such that $F(x) > 0$, put

$$\mu_x = \frac{n-1}{n} \cdot \frac{1}{F(x)} \cdot 1_{]-\infty, x[} \, dF + \frac{1}{n} \varepsilon_x,$$

with ε_x the measure defined by the unit mass at x. Write $\mathbf{R}'$ for $\{x \in \mathbf{R} : F(x) > 0\}$. Then, for each $x \in \mathbf{R}'$,

$$\mu_x(]-\infty, t]) = \frac{n-1}{n} \cdot \frac{F(t)}{F(x)} \cdot 1_{]t, +\infty[}(x) + 1_{]-\infty, t]}(x)$$

for all $t \in \mathbf{R}$; so

$$\mu_x(]-\infty, t]) \, dP_X(x) = nF^{n-1}(x) \cdot 1_{]-\infty, t]}(x) \, dF(x)$$
$$+ (n-1)F(t)F^{n-2}(x) \cdot 1_{]t, +\infty[}(x) \, dF(x).$$

We conclude that $(\mu_x)_{x \in \mathbf{R}'}$ is a conditional law of X_k given $\sup(X_1, \ldots, X_n)$. We remark that, for each $t \in \mathbf{R}$,

$$\mu_x(]-\infty, t]) = E(1_{]-\infty, t]} \circ X_k | X)(x)$$

for P_X-almost all $x \in \mathbf{R}'$. Hence, for each $t \in \mathbf{R}$, the function

$$\mathbf{R}' \ni x \mapsto \frac{n-1}{n} \cdot \frac{F(t)}{F(x)} \cdot 1_{]t,+\infty[}(x) + 1_{]-\infty,t]}(x), \qquad \mathbf{R}' \not\ni x \mapsto 0$$

is a version of $E(1_{]-\infty,t]} \circ X_k | X)$.

18.7 Existence of Conditional Laws when $G = \mathbf{R}^k$

Theorem 18.7.1 *Let* $(\Omega, \mathcal{F}, P)$ *be a probability space,* X *a random variable from* $(\Omega, \mathcal{F})$ *into a measurable space* $(F, \mathcal{B})$, *and* Y *a random variable from* $(\Omega, \mathcal{F})$ *into* $\mathbf{R}^k$. *Then there exists a conditional law of* Y *given* X.

PROOF: For each $t = (t_1, \ldots, t_k) \in \mathbf{R}^k$, put $] - \infty, t] = \prod_{1 \leq i \leq k}] - \infty, t_i]$. Let $A = \prod_{1 \leq i \leq k}]a_i, b_i]$ be a nonempty rectangle. By induction on $0 \leq l \leq k$, we have

$$E(1_{]a_1,b_1] \times \cdots \times]a_l,b_l] \times]-\infty,c_{l+1}] \times \cdots \times]-\infty,c_k]} \circ Y | X)$$
$$= \sum_{J \subset \{1,\ldots,l\}} (-1)^{|J|} \cdot E(1_{]-\infty,c_J^l]} \circ Y | X)$$

for all $c_{l+1} \in \{a_{l+1}, b_{l+1}\}, \ldots, c_k \in \{a_k, b_k\}$, where c_J^l is defined as follows: $c_{J,i}^l = a_i$ for $i \in J$, $c_{J,i}^l = b_i$ for $1 \leq i \leq l$, $i \notin J$, and $c_{J,i}^l = c_i$ for $l+1 \leq i \leq k$. In particular,

$$E(1_A \circ Y | X) = \sum_{J \subset \{1,\ldots,k\}} (-1)^{|J|} \cdot E(1_{]-\infty,c_J^A]} \circ Y | X),$$

where $c_{J,i}^A = a_i$ for $i \in J$ and $c_{J,i}^A = b_i$ for $i \notin J$.

For every $r \in \mathbf{Q}^k$ (with rational coordinates), choose a version $x \mapsto F_x(r)$ of $E(1_{]-\infty,r]} \circ Y | X)$. By the preceding paragraph and Proposition 18.1.4, for all x in some suitable $B' \in \mathcal{B}$ carrying P_X, the following are true:

(a) $\Delta_A F_x = \sum_{J \subset \{1,\ldots,k\}} (-1)^{|J|} \cdot F_x(c_J^A) \geq 0$ for every nonempty rectangle $A = \prod_{1 \leq i \leq k}]a_i, b_i]$ whose vertices have rational coordinates.

(b) For every $r \in \mathbf{Q}^k$, $F_x(r + (1/n, \ldots, 1/n))$ converges to $F_x(r)$ as $n \to +\infty$.

(c) For every $r \in \mathbf{Q}^k$ and for every $1 \leq i \leq k$, $F_x(r_1, \ldots, r_i - n, \ldots, r_k)$ converges to 0 as $n \to +\infty$.

(d) $F_x(n, \ldots, n)$ converges to 1 as $n \to +\infty$.

Now fix $x \in B'$. Let $s \in \mathbf{Q}^k$, $1 \leq l \leq k$ an integer, and r_l a rational number in $] - \infty, s_l[$. For all $n_1, \ldots, n_{l-1}, n_{l+1}, \ldots, n_k$ in $\mathbf{N}$, consider the rectangle

$$A_{(n_1,\ldots,\hat{n}_l,\ldots,n_k)} =]s_1 - n_1, s_1] \times \cdots \times]r_l, s_l] \times \ldots \times]s_k - n_k, s_k].$$

Noting that $\Delta_{A(n_1,...,\hat{n}_l,...,n_k)} F_x$ is positive, and letting successively $n_1, ...,$ $n_{l-1}, n_{l+1}, ..., n_k$ tend to $+\infty$, we see that $F_x(s) - F_x(s_1, ..., r_l, ..., s_k)$ is positive. Hence

$$F_x(s) - F_x(r) =$$

$$\sum_{1 \leq l \leq k} [F_x(r_1, ..., r_{l-1}, s_l, s_{l+1}, ..., s_k) - F_x(r_1, ..., r_{l-1}, r_l, s_{l+1}, ..., s_k)]$$

is positive for all $r, s \in \mathbf{Q}^k$ such that $r_i \leq s_i$ for all $1 \leq i \leq k$.

For each $t \in \mathbf{R}^k$, set $F_x(t) = \inf_{s \in \mathbf{Q}^k, s \geq t} F_x(s)$. By the preceding paragraph, we extend to $\mathbf{R}^k$ the function F_x initially defined on $\mathbf{Q}^k$. For each $t \in \mathbf{R}^k$ and for every $\varepsilon > 0$, there exists $r \in \mathbf{Q}^k$, $r \geq t$, such that $F_x(r) \leq F_x(t) + \varepsilon/2$. Next, there exists $n \in \mathbf{N}$ such that

$$F_x(r + (1/n, ..., 1/n)) \leq F_x(r) + \varepsilon/2 \leq F_x(t) + \varepsilon.$$

Then, for all $s \in \mathbf{R}^k$ satisfying $t \leq s \leq r + (1/n, ..., 1/n)$, we have $F_x(t) \leq F_x(s) \leq F_x(t) + \varepsilon$, which proves that F_x is right-continuous at t. Now it is easy to see that, for every nonempty rectangle $A = \prod_{1 \leq i \leq k}]a_i, b_i]$, $\Delta_A F_x$ is positive.

By Proposition 14.2.1, F_x is the distribution function of a probability μ_x. For all $r \in \mathbf{Q}^k$, the function

$$B' \ni x \mapsto \mu_x(]-\infty, r]) = F_x(r) \qquad B' \not\ni x \mapsto 0$$

is, by construction, a version of $E(1_{]-\infty,r]} \circ Y|X)$. Since the sets $]-\infty, r]$, for $r \in \mathbf{Q}^k$, constitute a π-system and since the σ-ring generated by this π-system is the Borel σ-algebra of $\mathbf{R}^k$, Proposition 18.5.1 shows that $(\mu_x)_{x \in B'}$ is a conditional law of Y given X. $\qquad \square$

Theorem 18.7.2 *Let X and Y be as in Theorem 18.7.1. Let $(\mu_x)_{x \in B'}$ and $(\nu_x)_{x \in B'}$, where $B' \in \mathcal{B}$ carries P_X, be two conditional laws of Y given X. Then $\mu_x = \nu_x$ for P_X-almost all $x \in B'$.*

PROOF: For every $r \in \mathbf{Q}^k$, the function $F' \ni x \mapsto \mu_x(]-\infty, r])$, $F' \not\ni x \mapsto 0$ and the function $F' \ni x \mapsto \nu_x(]-\infty, r])$, $F' \not\ni x \mapsto 0$ are two versions of $E(1_{]-\infty,r]} \circ Y|X)$. Hence there exists a P_X-negligible subset $N_r \in \mathcal{B}$ of B' such that $\mu_x(]-\infty, r]) = \nu_x(]-\infty, r])$ for all $x \in B' - N_r$. Thus, if x lies in $B' - \bigcup_{r \in \mathbf{Q}^k} N_r$, we have $\mu_x(]-\infty, r]) = \nu_x(]-\infty, r])$ for all $r \in \mathbf{Q}^k$, and so for all $r \in \mathbf{R}^k$; it follows that $\mu_x = \nu_x$. $\qquad \square$

Even though the results of Theorems 18.7.1 and 18.7.2 persist when $\mathbf{R}^k$ is replaced by a complete separable metric space, we shall neither prove nor use these more general statements.

Finally, we give a classical example in which there is no conditional law of Y given X.

Let λ be Lebesgue measure on the Borel σ-algebra $\mathcal{B}$ of $\Omega = [0, 1]$. Given a subset E of Ω such that

$$\lambda_*(B \cap E) = 0 \qquad \text{and} \qquad \lambda^*(B \cap E) = \lambda(B)$$

for all $B \in \mathcal{B}$ (Section 4.3), denote by $\mathcal{F}$ the collection of the sets $(B_1 \cap E) \cup (B_2 \cap E^c)$, where B_1 and B_2 run through $\mathcal{B}$. Hence $\mathcal{F}$ is the smallest σ-algebra containing both $\mathcal{B}$ and E. Clearly, $P : A \mapsto \lambda^*(A \cap E)$ is a probability on $\mathcal{F}$ extending λ. Write X for the random variable $\omega \mapsto \omega$ from $(\Omega, \mathcal{F})$ into $(\Omega, \mathcal{B})$ and Y for the random variable $\omega \mapsto \omega$ from $(\Omega, \mathcal{F})$ into $(\Omega, \mathcal{F})$. Suppose there is a conditional law $(\mu_\omega)_{\omega \in \Omega - D_1}$ of Y given X, where $D_1 \in \mathcal{B}$ is λ-negligible. Then $\int \mu_\omega(E) \, d\lambda(\omega) = P(E) = 1$, so $D_2 = \{\omega \in \Omega - D_1 : \mu_\omega(E) < 1\}$ is λ-negligible.

Let $\mathcal{C}$ be the class consisting of those sets $]p, q]$ and $[0, r]$ such that p, q, r are rationals in $[0, 1]$ and $p \le q$. Then $\mathcal{C}$ is a countable π-system and it generates the Borel σ-algebra $\mathcal{B}$ of $[0, 1]$. For every $B \in \mathcal{C}$, the function $D_1 \ni \omega \mapsto \mu_\omega(B)$, $D_1 \not\ni \omega \mapsto 0$ is a version of $E(1_B | X)$. Since $\mu_\omega(B) = 1_B(\omega)$ for λ-almost all $\omega \in \Omega - D_1$, there exists a λ-negligible subset $D_3 \in \mathcal{B}$ of $\Omega - (D_1 \cup D_2)$ such that $\mu_\omega(B) = 1_B(\omega)$ for all ω in $\Omega - (D_1 \cup D_2 \cup D_3)$ and all $B \in \mathcal{C}$. Now, for every $\omega \in \Omega - (D_1 \cup D_2 \cup D_3)$, let ε_ω be the measure on $\mathcal{F}$ defined by the unit mass at ω; μ_ω and ε_ω agree on $\mathcal{C}$, and so on $\mathcal{B}$. In particular, $\mu_\omega(\{\omega\}) = 1$. But, since $\mu_\omega(E) = 1$, we conclude that ω lies in E. In short, $\Omega - (D_1 \cup D_2 \cup D_3) \subset E$, which is a contradiction because $\lambda(\Omega - D_1 \cup D_2 \cup D_3) = 1$ whereas $\lambda_*(E) = 0$.

Exercises for Chapter 18

1   Let $(\Omega, \mathcal{F}, P)$ be a probability space and X a random variable from $(\Omega, \mathcal{F})$ into a measurable space $(F, \mathcal{B})$.

1.   Show that the constant function $x \mapsto \int Y \, dP$ is a version of $E(Y|X)$ if $Y : (\Omega, \mathcal{F}) \to \mathbf{R}^k$ is P-integrable and X and Y are independent (note that Y is the pointwise limit of a sequence $(Z_n)_{n \ge 1}$ of $\sigma(Y)$-simple mappings such that $|Z_n| \le 2|Y|$).

2.   Show that the constant function $x \mapsto \int^* Y \, dP$ is a version of $E(Y|X)$ if $Y : (\Omega, \mathcal{F}) \to \overline{\mathbf{R}}$ is P-quasi-integrable and X and Y are independent (assume that Y^- is P-integrable).

2   Let $(\Omega, \mathcal{F}, P)$ be a probability space, and let $X_1, \ldots, X_n$ $(n \ge 2)$ be independent random variables from $(\Omega, \mathcal{F})$ into $\mathbf{R}$ with a common continuous distribution function F. Consider the order statistics $X_{(1)}, \ldots, X_{(n)}$ arising from $(X_1, \ldots, X_n)$. By Proposition 17.1.2,

$$P_{[X_{(1)}, \ldots, X_{(n)}]} = n! \cdot 1_B(x_1, \ldots, x_n) \, dF(x_1) \cdots dF(x_n),$$

where $B = \big\{(x_1,\ldots,x_n) \in \mathbf{R}^n : x_1 < x_2 < \ldots < x_n\big\}$. Henceforth, assume that $n \geq 2$.

1. Put $\mathbf{R}' = \big\{x \in \mathbf{R} : F(x) > 0\big\}$. For all $x \in \mathbf{R}'$, write

$$B_x = \big\{(x_1,\ldots,x_{n-1}) \in \mathbf{R}^{n-1} : x_1 < x_2 < \ldots < x_{n-1} < x\big\}$$

and

$$\mu_x = \frac{(n-1)!}{F(x)^{n-1}} \cdot 1_{B_x}(x_1,\ldots.x_{n-1})\, dF(x_1)\cdots dF(x_{n-1}).$$

Deduce from Theorem 18.6.2 that $(\mu_x)_{x\in\mathbf{R}'}$ is a conditional law of $[X_{(1)},\ldots,X_{(n-1)}]$ given $X_{(n)}$. In other words, given the event $X_{(n)} = x$, the random variable $[X_{(1)},\ldots,X_{(n-1)}]$ is distributed as $[X'_{(1)},\ldots,X'_{(n-1)}]$, if $(X'_1,\ldots,X'_{n-1})$ is an $(n-1)$-sample of the law $\big(1/F(x)\big) \cdot 1_{]-\infty,x[}\, dF$.

2. Now, put $X = \sup(X_1,\ldots,X_n)$ and $Y = \inf(X_1,\ldots,X_n)$. Write D for $\big\{(x,y) \in \mathbf{R}^2 : y < x\big\}$. By Proposition 17.1.3, $P_{[X,Y]}$ has the density $(x,y) \mapsto n(n-1)\big(F(x) - F(y)\big)^{n-2} \cdot 1_D(x,y)$ with respect to $dF \otimes dF$. For each $x \in \mathbf{R}'$, put

$$\mu_x = \frac{n-1}{F(x)^{n-1}} \big(F(x) - F(y)\big)^{n-2} \cdot 1_{]-\infty,x[}(y)\, dF(y).$$

Deduce from Theorem 18.6.2 that $(\mu_x)_{x\in\mathbf{R}'}$ is a conditional law of Y given X. Show that $\mu_x\big(]-\infty,y]\big) = 1 - \big(1 - F(y)/F(x)\big)_{+}^{n-1}$ for every y in $\mathbf{R}$. Conclude that, if $(X'_1,\ldots,X'_{n-1})$ is an $(n-1)$-sample of $\big(1/F(x)\big){\cdot}1_{]0,x[}\, dF$, then $\inf(X'_1,\ldots,X'_{n-1})$ has μ_x as law.

3. In what follows, assume that $X_1,\ldots,X_n$ are integrable. Show that $E\big(X_k|X_{(n)}\big)$ does not depend on k $(1 \leq k \leq n)$. Using the notation of part 1, and defining by h the function $(x_1,\ldots,x_{n-1}) \mapsto x_1 + \cdots + x_{n-1}$ from $\mathbf{R}^{n-1}$ into $\mathbf{R}$, prove that $E\big(X_{(1)} + \cdots + X_{(n-1)}|X_{(n)}\big)(x) = \int h\, d\mu_x$ is equal to $E(X'_1 + \cdots + X'_{n-1})$, hence to $(n-1)\int_{-\infty}^{x} \big(t/F(x)\big)\, dF(t)$. Conclude that, for each $1 \leq k \leq n$,

$$E\big(X_k|X_{(n)}\big)(x) = \frac{n-1}{n} \cdot \frac{1}{F(x)} \int_{-\infty}^{x} t\, dF(t) + \frac{1}{n}x$$

for dF-almost all $x \in \mathbf{R}'$. Compare with the final example of Section 18.6.

4. Now take $X = \sup(X_1,\ldots,X_n)$ and $Y = \inf(X_1,\ldots,X_n)$, as before. Show that

$$E(Y|X)(x) = \frac{n-1}{F(x)} \int_{-\infty}^{x} y \left(1 - \frac{F(y)}{F(x)}\right)^{n-2} dF(y)$$

for dF-almost all $x \in \mathbf{R}'$.

3 Let $(\Omega, \mathcal{F}, P)$ be a probability space, and let $X_1,\ldots,X_n$ $(n \geq 2)$ be independent random variables from $(\Omega, \mathcal{F})$ into $\mathbf{R}$ whose common law is $(1/t) \cdot 1_{]0,t[}\lambda$ (for a given $t > 0$). For every $0 < x < t$, put

$$B_x = \big\{(x_1,\ldots,x_{n-1}) \in \mathbf{R}^{n-1} : 0 < x_1 < \ldots < x_{n-1} < x\big\}$$

and $\mu_x = \big((n-1)!/x^{n-1}\big){\cdot}1_{B_x}\lambda_{n-1}$ (where λ_{n-1} is Lebesgue measure on $\mathbf{R}^{n-1}$).

1. Deduce from Exercise 2 that $(\mu_x)_{x \in]0,t[}$ is a conditional law of $[X_{(1)}, \ldots, X_{(n-1)}]$ given $X_{(n)}$.

2. Choose positive numbers $x_1, \ldots, x_{n+1}$ so that $x_1 + \cdots + x_{n+1} < t$, and put

$$A_n(t) = \left\{ (y_1, \ldots, y_n) \in \mathbf{R}^n : \right.$$
$$\left. y_1 > x_1, y_2 - y_1 > x_2, \ldots, y_n - y_{n-1} > x_n, t - y_n > x_{n+1} \right\}.$$

For every y in $]x_1 + \cdots + x_n, t - x_{n+1}[$, denote by $A_{n-1}(y)$ the set

$$\left\{ (y_1, \ldots, y_{n-1}) \in \mathbf{R}^{n-1} : y_1 > x_1, y_2 - y_1 > x_2, \ldots, y - y_{n-1} > x_n \right\}.$$

Finally, put $r_n(t) = P_{[X_{(1)}, \ldots, X_{(n)}]}\big(A_n(t)\big) = (n!/t^n)\lambda_n\big(A_n(t)\big)$ and $r_{n-1}(y) = \big((n-1)!/y^{n-1}\big)\lambda_{n-1}\big(A_{n-1}(y)\big)$. From Proposition 18.5.2 and part 1 deduce that

$$r_n(t) = \int^* \frac{n y^{n-1}}{t^n} \, r_{n-1}(y) \cdot 1_{]x_1 + \cdots + x_n, \, t - x_{n+1}[}(y) \, d\lambda(y).$$

By induction, conclude that $r_n(t) = (1/t^n)(t - x_1 - \cdots - x_{n+1})^n$.

3. From part 2, deduce that

$$P\big(X_{(1)} > x_1, X_{(2)} - X_{(1)} > x_2, \ldots, X_{(n)} - X_{(n-1)} > x_n, t - X_{(n)} > x_{n+1}\big)$$
$$= \frac{1}{t^n}(t - x_1 - \cdots - x_{n+1})_+^n$$

for all $x_1 \geq 0, \ldots, x_{n+1} \geq 0$.

4. Fix a real number h such that $0 < 2h < t$, and consider the event $A = (h < X_i < t - h \ \ \forall 1 \leq i \leq n)$. Show that

$$\mu_1 = \frac{1}{(t - 2h)^n} \cdot 1_{]h, \, t - h[^n} \lambda_n$$

is the measure

$$C \longmapsto \frac{P\big(A \cap ([X_1, \ldots, X_n] \in C)\big)}{P(A)}$$

on the Borel σ-algebra of $\mathbf{R}^n$. Interpret this result.

4 Let $(\Omega, \mathcal{F}, P)$ be a probability space, and let X and Y be two random variables from $(\Omega, \mathcal{F})$ into the measurable spaces $(F, \mathcal{B})$ and $(G, \mathcal{C})$, respectively. Suppose that there is a conditional law $(\mu_x)_{x \in F'}$ of Y given X.

Let $h : G \to [0, +\infty]$ be measurable $\mathcal{C}$. Show that

$$E(h \circ Y | X)(x) = \int^* \mu_x\big(h^{-1}(]t, +\infty])\big) \, d\lambda_{]0, +\infty[}(t)$$

for P_X-almost all $x \in F'$ (use Proposition 3.6.5).

5 Let $p \geq 1$ and $q \geq 1$ be two real numbers such that $1/p + 1/q = 1$. Given a probability space $(\Omega, \mathcal{F}, P)$, let X be a random variable from $(\Omega, \mathcal{F})$ into a measurable space $(F, \mathcal{B})$, and let Y and Z be two random variables from $(\Omega, \mathcal{F})$ into $\mathbf{R}$ such that $\int^{\bullet} |Y|^p \, dP < +\infty$ and $\int^{\bullet} |Z|^q \, dP < +\infty$. Prove that

$$E(|YZ| \,|X) \leq \left[E(|Y|^p|X) \right]^{1/p} \cdot \left[E(|Z|^q|X) \right]^{1/q},$$

P_X-almost surely (consider a conditional law of $[Y, Z]$ given X).

6 Let $(\Omega, \mathcal{F}, P)$ be a probability space. A sequence $(A_n)_{n \geq 1}$ of $\mathcal{F}$-sets is said to be mixing with constant $\alpha \geq 0$ if $\lim_{n \to +\infty} P(A_n \cap E) = \alpha P(E)$ for all $E \in \mathcal{F}$.

1. If $(A_n)_{n \geq 1}$ is mixing with constant α, show that $\lim_{n \to +\infty} \int_{A_n} X \, dP = \alpha \int X \, dP$ for every P-integrable random variable X from $(\Omega, \mathcal{F})$ into $\mathbf{R}$ (first, assume that X is positive).

2. Let $\mathcal{C}$ be a π-system in $\mathcal{F}$ such that Ω and the sets A_n belong to $\sigma(\mathcal{C})$. Suppose that $\lim_{n \to +\infty} P(A_n \cap E) = \alpha P(E)$ for all $E \in \mathcal{C}$. Show that $\lim_{n \to +\infty} P(A_n \cap E) = \alpha P(E)$ for all $E \in \sigma(\mathcal{C})$ (use the π-λ theorem) and that $\int_{A_n} X \, dP = \int_{A_n} E(X|\sigma(\mathcal{C})) \, dP$ converges to $\alpha \int X \, dP$ as $n \to +\infty$, for every P-integrable random variable from $(\Omega, \mathcal{F})$ into $\mathbf{R}$. Conclude that $(A_n)_{n \geq 1}$ is α-mixing.

3. Let P_0 be a probability on $\mathcal{F}$, absolutely continuous with respect to P. Show that mixing is preserved if P is replaced by P_0.

7 Let $(\Omega, \mathcal{F}, P)$ be a probability space, and let $(X_n)_{n \geq 1}$ be an independent sequence of identically distributed random variables from $(\Omega, \mathcal{F})$ into $\mathbf{R}$. Suppose that X_1 is of order 2 and $E(X_1) = 0$. Set $\sigma^2 = \text{Var}(X_1)$. For each $n \in \mathbf{N}$, put $S_n = X_1 + \cdots + X_n$, $Y_n = S_n/(\sigma\sqrt{n})$, and $Z_n = \left(1/(\sigma\sqrt{n})\right)(S_n - S_{[\log n]})$ (where $[\log n]$ is the integral part of $\log n$). Finally, define $\mathcal{C}$ as $\bigcup_{k \geq 1} \sigma(X_1, \ldots, X_k)$.

1. Show that $(Y_n - Z_n)_{n \geq 1}$ converges in P-probability to 0.

2. Deduce from part 1 that $P_{Z_n} \Rightarrow N_1(0, 1)$ (use Theorem 16.1.3).

3. For all $x \in \mathbf{R}$, write α_x for $N_1(0, 1)(]-\infty, x])$. Show that, for each $E \in \mathcal{C}$, $P\left((Z_n \leq x) \cap E\right)$ converges to $\alpha_x P(E)$ as $n \to +\infty$ (use the independence of the X_n).

4. Prove that, for each $E \in \mathcal{F}$, $P\left((Z_n \leq x) \cap E\right)$ converges to $\alpha_x P(E)$ as $n \to +\infty$ (use Exercise 6).

5. Let P_0 be a probability on $\mathcal{F}$, absolutely continuous with respect to P. Show that, for all $x \in \mathbf{R}$, $P_0(Z_n \leq x) \to \alpha_x$ (use Exercise 6).

6. Prove that $(Y_n - Z_n)_{n \geq 1}$ converges in P_0-probability to 0 (use part 1 and Theorem 10.3.2).

7. Conclude that $(P_0)_{Y_n} \Rightarrow N_1(0, 1)$.

Part IV

Operations on Radon Measures

19

μ-Adequate Family of Measures

From now on, our treatment of measure theory follows closely that of N. Bourbaki (*Intégration*, chapters 5, 7, and 8). In this chapter, we introduce the concept of μ-adequate family of Radon measures, which, for example, simplifies the study of integration on homogeneous spaces.

Summary

19.1 For Radon measures, the notion of induced measure is quite natural. If X is a locally compact Hausdorff space and Y a locally compact subspace, any Radon measure on X induces a Radon measure on Y as follows. If f is continuous on Y with compact support, define g by $g(x) = f(x)$ when $x \in Y$ and $g(x) = 0$ otherwise. Then $f \mapsto \int g \, d\mu$ is the measure induced by μ on Y.

19.2 μ-dense families of compact sets, introduced in this section, will be needed later in the text.

19.3 Given a positive Radon measure μ and a μ-dense class $\mathcal{D}$ of compact sets, there is a summable family $(\mu_\alpha)_{\alpha \in A}$ of measures such that $\mu = \sum_{\alpha \in A} \mu_\alpha$, that supp($\mu_\alpha$) belongs to $\mathcal{D}$, and that the sets supp(μ_α) form a locally countable class (Theorem 19.3.1).

19.4 In this section, we study integration with respect to $\int \lambda_t \, d\mu(t)$, where $t \mapsto \lambda_t$ is a μ-adequate family of measures. We prove a result which is similar to Fubini's theorem (Theorem 19.4.2).

19.5 We specialize the results of Section 19.4 to μ-adapted pairs.

19.1   Induced Radon Measure

Let X be a locally compact Hausdorff space, μ a Radon measure on X, and Y a locally compact subspace of X. For every $x \in Y$, there is a neighborhood W_x of x in X such that $W_x \cap Y$ is compact, and hence closed in W_x. Choose an open neighborhood V_x of x contained in W_x; then $V_x \cap Y = V_x \cap (W_x \cap Y)$ is closed in V_x. Putting $U = \bigcup_{x \in Y} V_x$, we see that U is open and Y is closed in U. In short, Y is the intersection of a closed subset of X and an open subset of X.

For all $g \in \mathcal{H}(Y, \mathbf{C})$, denote by g' the function from X into $\mathbf{C}$ which agrees with g on Y and vanishes outside Y. For all $g \in \mathcal{H}^+(Y)$, g' is upper semi-continuous, with compact support. It follows, for $g \in \mathcal{H}(Y, \mathbf{C})$, that g' is μ-integrable. Furthermore, $|\int g' \, d\mu| \le \int |g'| \, dV\mu \le \|g\| \, V\mu(\operatorname{supp} g)$.

Definition 19.1.1 The Radon measure $g \mapsto \int g' \, d\mu$ on Y is called the measure induced by μ on Y and is written μ_Y or $\mu_{/Y}$.

We shall study in detail, in Section 20.1, the integration with respect to an induced measure. For now, we shall require only the following result.

Theorem 19.1.1 $|\mu_Y| = |\mu|_Y$.

PROOF: Let $f \in \mathcal{H}^+(Y)$. There exists, for every $\varepsilon > 0$, a function $g \in \mathcal{H}(Y, \mathbf{C})$ such that $|g| \le f$ and $|\mu_Y|(f) \le |\mu_Y(g)| + \varepsilon$. Since $|g'| \le f'$, we have

$$\left| \int g' \, d\mu \right| \le \int |g'| \, d|\mu| \le \int f' \, d|\mu| = |\mu|_Y(f),$$

whence results $|\mu_Y|(f) \le |\mu|_Y(f) + \varepsilon$ and, as ε is arbitrary, $|\mu_Y|(f) \le |\mu|_Y(f)$.

On the other hand, let K be the support of f, and let U be an open neighborhood of K in X such that $|\mu|(U - K) \le \varepsilon$. For each $x \in K$, there exists a compact neighborhood V_x of x in U such that $V_x \cap Y$ is compact. We let $(x_i)_{i \in I}$ be a finite family in K such that the $\overset{\circ}{V}_{x_i}$ cover K. Hence $V = \bigcup_{i \in I} V_{x_i}$ is a compact neighborhood of K, included in U, and $V \cap Y$ is compact. By the Tietze extension theorem, there is a continuous function f_2 from V into $[0, \|f\|]$ which extends $f_{/V \cap Y}$. If φ is a continuous function from X into $[0, 1]$ which vanishes outside V and is equal to 1 on K, then define f_1 by $f_1(x) = f_2(x)\varphi(x)$ for all $x \in V$ and by $f_1(x) = 0$ for all $x \in X - V$. Clearly, f_1 belongs to $\mathcal{H}^+(X)$, extends f, and is such that $\|f_1\| = \|f\|$. Now, given $\varepsilon > 0$, we can find $h_1 \in \mathcal{H}(X, \mathbf{C})$ such that $|h_1| \le f_1$ and $|\mu|(f_1) \le |\mu(h_1)| + \varepsilon$. Writing h for the restriction of h_1 to Y, we have $h \in \mathcal{H}(Y, \mathbf{C})$, $|h| \le f$, and $\mu(h_1) - \mu_Y(h) = \mu(h_1 \cdot 1_{U-K})$. Thus $|\mu(h_1) - \mu_Y(h)| \le \|f\| \cdot |\mu|(U - K) \le \varepsilon \|f\|$. On the other hand, $|\mu|(f_1) - |\mu|_Y(f) = |\mu|(f_1 \cdot 1_{U-K})$ is positive. We conclude that

$$|\mu|_Y(f) \le |\mu_Y(h)| + \varepsilon(1 + \|f\|)$$

$$\leq |\mu_Y|(f) + \varepsilon(1 + \|f\|)$$

and, as ε is arbitrary, that $|\mu|_Y(f) \leq |\mu_Y|(f)$. $\square$

Lemma 19.1.1 *Let X be a locally compact Hausdorff space, μ a positive (Radon) measure on X, and K a nonempty compact subset of X.*

(a) For every compact (respectively, open) set H in K, $\mu_K(H) = \mu(H)$.

(b) A subset N of K is μ_K-negligible if and only if it is μ-negligible.

(c) If the support S of μ_K is nonempty, $\mathrm{supp}(\mu_S) = S$.

PROOF:

(a) We may consider only the case in which H is compact. Denote by f the indicator function of H on K. f is upper semicontinuous, and so the lower envelope of a downward-directed subset Φ of $\mathcal{H}^+(K)$. Since the indicator function 1_H of H on X is the lower envelope of the downward-directed set $\Phi' = \{g' : g \in \Phi\}$ of upper semicontinuous functions (g' the extension of g by zero),

$$\mu_K(H) = \inf_{g \in \Phi} \int g\, d\mu_K = \inf_{g \in \Phi} \int g'\, d\mu = \mu(H).$$

(b) If N is μ-negligible, there exists, for each $\varepsilon > 0$, an open neighborhood U of N in X such that $\mu(U) \leq \varepsilon$; since $K - (U \cap K)$ is compact, we deduce from part 1 that $\mu_K(U \cap K) \leq \mu(U) \leq \varepsilon$; thus N is μ_K-negligible. Conversely, if N is μ_K-negligible, there exists an open neighborhood V of N in X such that $\mu_K(V \cap K) \leq \varepsilon$; by part 1, $\mu(V \cap K) = \mu_K(V \cap K)$; since ε is arbitrary, $\mu(N) = 0$.

(c) For each open set U in K intersecting S, $\mu_K(U \cap S) \neq 0$, hence

$$
\begin{aligned}
\mu(U \cap S) &= \mu(S) - \mu(S - U \cap S) \\
&= \mu_K(S) - \mu_K(S - U \cap S) \\
&= \mu_K(U \cap S) \\
&\neq 0.
\end{aligned}
$$

Since every nonempty open set in S can be written $U \cap S$ for a suitable open set U in K, we conclude that $\mathrm{supp}(\mu_S) = S$.

$\square$

19.2 μ-Dense Families of Compact Sets

Let X be a locally compact Hausdorff space and μ a positive Radon measure on X.

Now let A be a μ-measurable subset of X, and let $\mathcal{D}$ be a class of compact sets in A with the following properties:

1. Each closed subset of a $\mathcal{D}$-set is a $\mathcal{D}$-set.

2. Each finite union of $\mathcal{D}$-sets is a $\mathcal{D}$-set.

Proposition 19.2.1 *The following conditions are equivalent:*

(a) *For every nonnegligible compact set K in A. there exists $L \in \mathcal{D}$ such that $L \subset K$ and $\mu(L) > 0$.*

(b) *For every compact set K in A and every $\varepsilon > 0$, there exists $L \in \mathcal{D}$ such that $L \subset K$ and $\mu(K - L) \le \varepsilon$.*

(c) *For every compact set K in A, there exists a sequence $(H_n)_{n \ge 1}$ of disjoint $\mathcal{D}$-sets contained in K such that $K - \bigcup_{n \ge 1} H_n$ is μ-negligible.*

(d) *A subset B of A is locally μ-negligible whenever $\mu^*(B \cap L) = 0$ for all $L \in \mathcal{D}$.*

PROOF: First, assume that (a) holds, and let K be a compact subset of A. Put $\alpha = \sup_{L \in \mathcal{D}, L \subset K} \mu(L)$, and let $(L_n)_{n \ge 1}$ be an increasing sequence of $\mathcal{D}$-sets included in K such that $\alpha = \sup_{n \ge 1} \mu(L_n)$. Then $\bigcup_{n \ge 1} L_n$ is μ-integrable and $\mu(\bigcup_{n \ge 1} L_n) = \alpha$. Putting $B = K - \bigcup_{n \ge 1} L_n$, there cannot exist $L \in \mathcal{D}$ such that $L \subset B$ and $\mu(L) > 0$: otherwise, $\mu(L \cup L_n) > \alpha$ for n large enough. Therefore, B is μ-negligible, $\alpha = \mu(K)$, and (b) is satisfied.

Now, if (b) holds and K is a compact subset of A, we can inductively define a sequence $(H_n)_{n \ge 1}$ of $\mathcal{D}$-sets such that $H_{n+1} \subset K - \bigcup_{p \le n} H_p$ and $\mu(K - \bigcup_{p \le n} H_p) \le 1/n$; hence condition (c) is also satisfied.

For every compact set H in X, $A \cap H$ is a union of countably many disjoint compact sets and a μ-negligible set, so (c) implies (d).

Finally, (d) obviously implies (a).   □

Definition 19.2.1 When the conditions of Proposition 19.2.1 hold, $\mathcal{D}$ is said to be μ-dense in A.

If $A = X$ and the conditions of Proposition 19.2.1 are satisfied, $\mathcal{D}$ is said, simply, to be μ-dense.

Proposition 19.2.2 *Suppose $\mathcal{D}$ is μ-dense in A. Let $\mathcal{D}'$ be a class of compact sets in A such that each closed subset of a $\mathcal{D}'$-set is a $\mathcal{D}'$-set and such that each finite union of $\mathcal{D}'$-sets is a $\mathcal{D}'$-set. If, for each $K \in \mathcal{D}$, the $\mathcal{D}'$-sets which are included in K form a μ-dense class in K, then $\mathcal{D}'$ is μ-dense in A.*

PROOF: Let L be a compact subset of A. For every $\varepsilon > 0$, there exists $K \in \mathcal{D}$ such that $K \subset L$ and $\mu(L - K) \leq \varepsilon/2$. Next, there exists $H \in \mathcal{D}'$ such that $H \subset K$ and $\mu(K - H) \leq \varepsilon/2$. Now $H \subset L$ and $\mu(L - H) \leq \varepsilon$, which proves the proposition. $\square$

Definition 19.2.2 A class $\mathcal{C}$ of sets in a topological space Ω is said to be locally countable if, for each $x \in \Omega$, there exists a neighborhood V of x which meets only a countable number of the $\mathcal{C}$-sets. If $\mathcal{C}$ is locally countable, every compact subset of Ω intersects only a countable number of the $\mathcal{C}$-sets.

Clearly, the union of a locally countable class of μ-measurable sets is μ-measurable.

The following theorem is very useful.

Theorem 19.2.1 *Suppose that $\mathcal{D}$ is μ-dense in A. There exists a locally countable class $\mathcal{C} \subset \mathcal{D}$, consisting of nonempty and mutually disjoint sets, such that $A - \bigcup_{K \in \mathcal{C}} K$ is locally μ-negligible and that $\operatorname{supp}(\mu_K) = K$ for all $K \in \mathcal{C}$.*

PROOF: Consider the classes $\mathcal{L} \subset \mathcal{D}$, consisting of mutually disjoint nonempty sets, such that $\operatorname{supp}(\mu_K) = K$ for all $K \in \mathcal{L}$. These classes form a subset Φ of the set of all subclasses of $\mathcal{D}$, and Φ is nonempty since it contains the empty class. Order Φ by inclusion. Then, by Zorn's lemma, Φ has a maximal element $\mathcal{C}$.

For every $x \in X$, let V be an open neighborhood of x, with compact closure. If $(K_i)_{i \in I}$ is a finite family of distinct $\mathcal{C}$-sets and if each K_i meets V, then $\sum_{i \in I} \mu(K_i \cap V) \leq \mu(V)$; moreover, $\mu(K_i \cap V) > 0$ for all $i \in I$ because $\operatorname{supp}(\mu_{K_i}) = K_i$. Thus V can intersect only a countable number of the $\mathcal{C}$-sets, which proves that $\mathcal{C}$ is locally countable.

We know that $N = A - \bigcup_{K \in \mathcal{C}} K$ is μ-measurable. Suppose that it is not locally μ-negligible. N contains a nonnegligible compact set, and hence contains a $\mathcal{D}$-set K that is not μ-negligible. If L is the support of μ_K, then $\operatorname{supp}(\mu_L) = L$, and $\mathcal{C} \cup \{L\}$ belongs to Φ, which contradicts the maximality of $\mathcal{C}$. $\square$

Proposition 19.2.3 *Let f be a mapping from A into a topological space F, and let $\mathcal{D}$ consist of those compact subsets K of A such that $f_{/K}$ is continuous. The following conditions are equivalent:*

(a) *$\mathcal{D}$ is μ-dense in A.*

(b) *Every extension of f to a mapping g from X into F that is constant outside A is μ-measurable.*

(c) There is a μ-measurable mapping $g : X \to F$ that extends f and is constant outside A.

PROOF: Assume that condition (a) holds, and let g be an extension of f to X, constant on A^c. For each compact subset K of X, $K \cap A$ and $K \cap (X - A)$ are μ-integrable. Given $\varepsilon > 0$, there exist compact sets $P \subset K \cap A$ and $Q \subset K \cap (X-A)$ such that $\mu(K \cap A - P) < \varepsilon/2$ and $\mu(K \cap A^c - Q) \leq \varepsilon/2$. On the other hand, there exists a compact subset H of P such that $\mu(K \cap A - H) \leq \varepsilon/2$ and $f_{/H}$ is continuous. Then $\mu(K - H \cup Q) \leq \varepsilon$ and $g_{/H \cup Q}$ is continuous, which shows that g is μ-measurable.

Conversely, (c) obviously implies (a).   □

Given the conditions of Proposition 19.2.3, f is said to be μ-measurable on A. When F is metrizable, this definition agrees with Definition 5.3.1.

Proposition 19.2.4 *Let C be a locally countable class of μ-measurable subsets of A, such that $A - \bigcup_{B \in C} B$ is locally μ-negligible. Then a mapping f from A into a topological space is μ-measurable on A if (and only if) $f_{/B}$ is μ-measurable on B, for all $B \in C$.*

PROOF: For all $B \in C$, those compact subsets H of B for which $f_{/H}$ is continuous form a μ-dense class $\mathcal{D}_B$ in B (Proposition 19.2.3). Now let K be a compact subset of A. By hypothesis, there exists a sequence $(B_n)_{n \geq 1}$ of C-sets such that $N = K \cap (A - \bigcup_{n \geq 1} B_n)$ is μ-negligible. For all $n \geq 1$, put $C_n = K \cap B_n \cap (\bigcup_{i < n} B_i)^c$. So N and the C_n form a partition of K into μ-integrable sets. Since $\mathcal{D}_{B_n}$ is μ-dense in B_n, there is a sequence $(C_{m,n})_{m \geq 1}$ of disjoint compact subsets of C_n such that $N_n = C_n - \bigcup_{m \geq 1} C_{m,n}$ is μ-negligible and each $f_{/C_{m,n}}$ is continuous. Then $N \cup (\bigcup_{n \geq 1} N_n)$ is μ-negligible. This proves that f is μ-measurable on A.   □

A mapping f from X into a topological space is said to be universally measurable whenever it is ν-measurable for every Radon measure ν on X.

Proposition 19.2.5 *Let f be a μ-measurable mapping from X into a topological space F. Then there exists $g : X \to F$. universally measurable, such that $f = g$ locally μ-almost everywhere.*

PROOF: Let $\mathcal{D}$ be the class of those compact sets K in X for which $f_{/K}$ is continuous. Since $\mathcal{D}$ is μ-dense, there exists a locally countable subclass C of $\mathcal{D}$, consisting of mutually disjoint sets, such that $N = X - \bigcup_{K \in C} K$ is locally μ-negligible (Theorem 19.2.1). Let y be an element of F. Put $g(x) = f(x)$ if x belongs to $\bigcup_{K \in C} K$ and $g(x) = y$ otherwise. Then g has the desired property.   □

19.3   Sums of Radon Measures

Let X be a locally compact Hausdorff space. For all positive Radon measures μ on X and all $f : X \to [0, +\infty]$, recall that $\int^\bullet f \, d\mu = \sup_K \int^* f \cdot 1_K \, d\mu$, where K runs through the class of compact subsets of X.

Let μ_1, μ_2 be two positive Radon measures on X, and put $\mu = \mu_1 + \mu_2$. Then $\int^* f \, d\mu = \int^* f \, d\mu_1 + \int^* f \, d\mu_2$ and $\int^\bullet f \, d\mu = \int^\bullet f \, d\mu_1 + \int^\bullet f \, d\mu_2$, for $f : X \to [0, +\infty]$. A mapping f from X into a topological space is μ-measurable if and only if it is both μ_1-measurable and μ_2-measurable.

Now let μ_1, μ_2 be two complex Radon measures on X, and put $\mu = \mu_1 + \mu_2$. A mapping f from X into a Banach space is $(|\mu_1| + |\mu_2|)$-integrable (respectively, essentially integrable) if and only if it is integrable (respectively, essentially integrable) with respect to each of μ_1 and μ_2, in which case $\int f \, d\mu = \int f \, d\mu_1 + \int f \, d\mu_2$.

Proposition 19.3.1 *Let H be an upward-directed set of positive Radon measures on X, bounded above in $\mathcal{M}(X, \mathbf{R})$, and let μ be its supremum. Then $\mu^\bullet(f) = \sup_{\nu \in H} \nu^\bullet(f)$ for $f : X \to [0, +\infty]$.*

PROOF: When f belongs to $\mathcal{H}^+(X)$, the relation $\mu(f) = \sup_{\nu \in H} \nu(f)$ follows from Lemma 1.3.1.

Now suppose that f is bounded and vanishes outside a compact set, and choose $g \in \mathcal{H}^+$ so that $f \le g$. Given $\varepsilon > 0$, there exists $\nu \in H$ such that $\nu(g) \ge \mu(g) - \varepsilon$, and now $(\mu - \nu)^*(f) \le (\mu - \nu)(g) \le \varepsilon$. So $\nu^*(f) \ge \mu^*(f) - \varepsilon$. This yields $\sup_{\nu \in H} \nu^*(f) \ge \mu^*(f)$, and the reverse inequality is clear.

Next, suppose that f vanishes outside a compact set, but that it is not necessarily bounded. For all $n \ge 0$, put $f_n = \inf(f, n)$. Then,

$$\mu^*(f) = \sup_n \mu^*(f_n) = \sup_n \sup_{\nu \in H} \nu^*(f_n) = \sup_{\nu \in H} \sup_n \nu^*(f_n) = \sup_{\nu \in H} \nu^*(f).$$

Finally, removing all restrictions on f,

$$
\begin{aligned}
\mu^\bullet(f) &= \sup_K \mu^*(f \cdot 1_K) \\
&= \sup_K \sup_{\nu \in H} \nu^*(f \cdot 1_K) \\
&= \sup_\nu \sup_K \nu^*(f \cdot 1_K) \\
&= \sup_{\nu \in H} \nu^\bullet(f).
\end{aligned}
$$

$\square$

We observe that $\mu^*(f)$ is not necessarily equal to $\sup_{\nu \in H} \nu^*(f)$ (cf. Exercise 1).

Proposition 19.3.2 *Let H and μ be as in Proposition 19.3.1, and let f be a mapping from X into a topological space. Then f is μ-measurable if (and only if) it is ν-measurable for all $\nu \in H$.*

PROOF: Let K be a compact set in X. Given $\varepsilon > 0$, there exists $\nu \in H$ such that $\nu(K) \geq \mu(K) - \varepsilon/2$, and we can find a compact subset L of K such that $\nu(K - L) \leq \varepsilon/2$ and $f_{/L}$ is continuous. Then

$$\mu(K - L) = (\mu - \nu)(K - L) + \nu(K - L) \leq \varepsilon.$$

which proves that f is μ-measurable.   □

Definition 19.3.1 A family $(\mu_\alpha)_{\alpha \in A}$ of positive Radon measures on X is said to be summable whenever $(\mu_\alpha(f))_{\alpha \in A}$ is summable for all $f \in \mathcal{H}^+(X)$. The resulting Radon measure $f \mapsto \sum_{\alpha \in A} \mu_\alpha(f)$ $(f \in \mathcal{H}(X, \mathbf{C}))$ is called the sum of the μ_α, and is written $\sum_{\alpha \in A} \mu_\alpha$.

Suppose that $(\mu_\alpha)_{\alpha \in A}$ is a summable family of positive Radon measures on X, and define $\mu = \sum_{\alpha \in A} \mu_\alpha$. Then $\mu^\bullet(f) = \sum_{\alpha \in A} \mu_\alpha^\bullet(f)$ for all $f : X \rightarrow [0, +\infty]$; a mapping g from X into a topological space is μ-measurable if and only if it is μ_α-measurable for all $\alpha \in A$; a mapping f from X into a real Banach space is essentially μ-integrable if and only if it is essentially μ_α-integrable for all $\alpha \in A$ and the sum $\sum_{\alpha \in A} \int^\bullet f | \, d\mu_\alpha$ is finite, in which case $\int f \, d\mu = \sum_{\alpha \in A} \int f \, d\mu_\alpha$.

Theorem 19.3.1 *Let μ be a positive Radon measure on X, and let $\mathcal{D}$ be a μ-dense class of compact sets in X. Then there exists a summable family $(\mu_\alpha)_{\alpha \in A}$ of positive Radon measures on X, such that $\mu = \sum_{\alpha \in A} \mu_\alpha$ and such that the supports of the measures μ_α belong to $\mathcal{D}$ and form a locally countable family of mutually disjoint compact sets.*

PROOF: Consider a locally countable family, $(K_\alpha)_{\alpha \in A}$, of nonempty and mutually disjoint $\mathcal{D}$-sets such that $N = X - \bigcup_{\alpha \in A} K_\alpha$ is locally μ-negligible (Theorem 19.2.1). Define a measure μ_α on X by $\mu_\alpha(f) = \int f \cdot 1_{K_\alpha} \, d\mu$ for $f \in \mathcal{H}(X, \mathbf{C})$. The support of μ_α is included in K_α, and therefore is a $\mathcal{D}$-set. It remains to be shown that $\sum_{\alpha \in A} \mu_\alpha(f) = \mu(f)$ for each $f \in \mathcal{H}^+(X)$. Write S for the support of f and A' for the (countable) subset of A consisting of those $\alpha \in A$ such that K_α intersects S. Since $N \cap S$ is μ-negligible,

$$\mu(f) = \int f \cdot 1_S \, d\mu$$

$$= \sum_{\alpha \in A'} \int f \cdot 1_{S \cap K_\alpha} \, d\mu$$

$$= \sum_{\alpha \in A'} \int f \cdot 1_{K_\alpha} \, d\mu$$

$$= \sum_{\alpha \in A} \int f \cdot 1_{K_\alpha} \, d\mu,$$

and so $\mu(f) = \sum_{\alpha \in A} \mu_\alpha(f)$, as desired.   □

19.4 μ-Adequate Families

Let X be a locally compact Hausdorff space and $\mathcal{M}^+(X)$ the set of positive Radon measures on X. In this section, $\mathcal{M}^+(X)$ will always be equipped with the topology induced by the vague topology of $\mathcal{M}(X, \mathbf{C})$. To say that a mapping $\wedge : t \mapsto \lambda_t$ from the locally compact Hausdorff space T into $\mathcal{M}^+(X)$ is continuous means, therefore, that, for every $f \in \mathcal{H}^+(X)$, the real-valued function $t \mapsto \lambda_t(f)$ is continuous. We shall say, in this case, that $\wedge$ is vaguely continuous. To say that $\wedge$ is μ-measurable (where μ is a positive Radon measure on T) means that the collection of all compact subsets K of T such that the restriction of $\wedge$ to K is vaguely continuous is μ-dense (Proposition 19.2.3); in this case, we shall say that $\wedge$ is vaguely μ-measurable.

Let $\wedge : t \mapsto \lambda_t$ be a mapping from T into $\mathcal{M}^+(X)$. We shall say that $\wedge$ is scalarly essentially integrable for μ if, for every $f \in \mathcal{H}(X, \mathbf{C})$, the function $t \mapsto \lambda_t(f)$ is essentially μ-integrable. Then, clearly, $\nu : f \mapsto \int \lambda_t(f)\,d\mu(t)$ is a positive linear form on $\mathcal{H}(X, \mathbf{C})$. Thus it is a Radon measure. We shall write $\nu = \int \lambda_t \, d\mu(t)$. For $f \in \mathcal{H}(X, \mathbf{C})$, the essential integral $\int \lambda_t(f)\,d\mu(t)$ shall also be written $\int d\mu(t) \int f(x)\,d\lambda_t(x)$.

Proposition 19.4.1 *Assume that μ is the supremum of an upward-directed subset H of $\mathcal{M}^+(T)$. Then $\wedge$ is scalarly essentially μ-integrable if and only if it is scalarly essentially μ'-integrable, for every $\mu' \in H$, and if the set $\{\int \lambda_t \, d\mu'(t) : \mu' \in H\}$ is bounded above in $\mathcal{M}^+(X)$. In this case,*

$$\int \lambda_t \, d\mu(t) = \sup_{\mu' \in H} \int \lambda_t \, d\mu'(t).$$

PROOF: Verifying that $\wedge$ is scalarly essentially integrable for a positive Radon measure ρ on T amounts to verifying that $t \mapsto \lambda_t(g)$ is ρ-measurable and that $\int^{\bullet} \lambda_t(g)\,d\rho(t)$ is finite, for every $g \in \mathcal{H}^+(X)$. Hence the proposition follows from Propositions 19.3.1 and 19.3.2.   □

Proposition 19.4.2 *Assume that μ is the sum of a summable family $(\mu_\alpha)_{\alpha \in A}$ of positive measures on T. Then $\wedge$ is scalarly essentially μ-integrable if and only if it is scalarly essentially μ_α-integrable for all $\alpha \in A$ and if the family $\left(\int \lambda_t \, d\mu_\alpha(t) \right)_{\alpha \in A}$ of measures is summable. In this case, $\int \lambda_t \, d\mu(t) = \sum_{\alpha \in A} \int \lambda_t \, d\mu_\alpha(t)$.*

PROOF: This follows from Proposition 19.4.1.   □

Definition 19.4.1 Let X be a locally compact Hausdorff space, $\wedge : t \mapsto \lambda_t$ a scalarly essentially μ-integrable mapping from T into $\mathcal{M}^+(X)$, and $\nu = \int \lambda_t \, d\mu(t)$ the integral of $\wedge$. $\wedge$ is said to be μ-preadequate whenever, for every lower semicontinuous function $f \geq 0$ defined on X, the function $t \mapsto \int^* f \, d\lambda_t$ is μ-measurable on T and

$$\int^{\bullet} f(x) \, d\nu(x) = \int^{\bullet} d\mu(t) \int^* f(x) \, d\lambda_t(x). \qquad (1)$$

$\wedge$ is said to be μ-adequate whenever it is μ'-preadequate for every positive measure $\mu' \leq \mu$.

The following proposition often makes it possible to check that a given mapping is μ-adequate.

Proposition 19.4.3 *Let* $\wedge : t \mapsto \lambda_t$ *be a scalarly essentially μ-integrable mapping from T into $\mathcal{M}^+(X)$, and put* $\nu = \int \lambda_t \, d\mu(t)$.

(a) *If $\wedge$ is vaguely continuous, then $t \mapsto \lambda_t^*(f)$ is lower semicontinuous for every lower semicontinuous function $f \geq 0$ on X, and*

$$\int^* f(x) \, d\nu(x) = \int^* d\mu(t) \int^* f(x) \, d\lambda_t(x) \qquad (2).$$

(b) *If $\wedge$ is vaguely μ-measurable, it is μ-adequate.*

PROOF: Let $f : X \to [0, +\infty]$ be lower semicontinuous. and let F be the upward-directed set of functions $g \in \mathcal{H}^+(X)$ such that $g \leq f$. For $g \in F$, designate by h_g the function $t \mapsto \lambda_t(g)$ on T. Similarly, put $h_f(t) = \lambda_t^*(f) = \sup_{g \in F} h_g(t)$. We introduce the following hypothesis, weaker than that in (a): suppose only that the restriction of $\wedge$ to S is vaguely continuous, S being a closed subset of T containing the support of μ. For $g \in F$, define by $\overline{h}_g$ the function which agrees with h_g on S and takes the value $+\infty$ on S^c. Put $\overline{h}_f = \sup_{g \in F} \overline{h}_g$, so that $\overline{h}_f = h_f$ on S. $\overline{h}_g$ is lower semicontinuous, for every $g \in F$. Thus $\overline{h}_f$ is lower semicontinuous, and the family $(\overline{h}_g)_{g \in F}$, being upwardly directed, satisfies

$$\mu^*(\overline{h}_f) = \sup_{g \in F} \mu^*(\overline{h}_g) = \sup_{g \in F} \mu^*(h_g) = \sup_{g \in F} \nu(g) = \nu^*(f)$$

(Theorem 7.2.2). Since f and $\overline{h}_f$ are lower semicontinuous. the preceding relation gives the equality $\mu^{\bullet}(\overline{h}_f) = \nu^{\bullet}(f)$ (Proposition 3.6.2). Since $\overline{h}_f = h_f$ μ-almost everywhere, we conclude that $\nu^{\bullet}(f) = \mu^{\bullet}(h_f)$. as desired. Thus $\wedge$ is μ-preadequate. In this argument, we could have replaced μ by $\mu' \leq \mu$, and ν by $\nu' = \int \lambda_t \, d\mu'(t)$, because $\wedge$ is also scalarly essentially μ'-integrable and S contains the support of μ'. It follows that $\wedge$ is μ-adequate.

Assume that $\wedge$ is vaguely continuous. We can take $S = T$; then $h_f = \overline{h}_f$ is lower semicontinuous, and the proof of (a) is complete.

Now suppose that $\wedge$ is vaguely μ-measurable. As the collection $\mathcal{D}$, of those compact subsets K of T such that the restriction of $\wedge$ to K is continuous, is μ-dense, there exists a summable family $(\mu_\alpha)_{\alpha \in A}$ of positive measures on T such that $\mu = \sum_{\alpha \in A} \mu_\alpha$ and the support S_α of μ_α belongs to $\mathcal{D}$ (Theorem 19.3.1). For every $\alpha \in A$, $\wedge$ is scalarly essentially μ_α-integrable, and we put $\nu_\alpha = \int \lambda_t \, d\mu_\alpha(t)$; the family $(\nu_\alpha)_{\alpha \in A}$ is summable, with sum ν (Proposition 19.4.2). If $f : X \to [0, +\infty]$ is lower semicontinuous, the first part of the proof applied to the measures μ_α and the closed sets S_α shows that

(c) h_f is μ_α-measurable for all $\alpha \in A$, and therefore μ-measurable; and

(d) $\int^{\bullet} f(x) \, d\nu_\alpha(x) = \int^{\bullet} d\mu_\alpha(t) \int^{*} f(x) \, d\lambda_t(x)$.

Formula (1) follows by summation over α. Applying the preceding argument to any measure $\mu' \leq \mu$, we find that $\wedge$ is μ-adequate. This proves (b). □

In the remainder of this section, we define by $\wedge : t \longmapsto \lambda_t$ a μ-preadequate mapping from T into $\mathcal{M}^+(X)$, and by ν the integral $\int \lambda_t \, d\mu(t)$ of $\wedge$.

Proposition 19.4.4 *Let f be a function from X into $[0, +\infty]$. Then*

(a) $\int^{*} f(x) \, d\nu(x) \geq \int^{\bullet} d\mu(t) \int^{*} f(x) \, d\lambda_t(x)$;

(b) if $\wedge$ is vaguely continuous, $\int^{*} f(x) \, d\nu(x) \geq \int^{*} d\mu(t) \int^{*} f(x) \, d\lambda_t(x)$.

PROOF: Let g be a lower semicontinuous function on X such that $f \leq g$. Since $\int^{*} f(x) \, d\lambda_t(x) \leq \int^{*} g(x) \, d\lambda_t(x)$ for every $t \in T$, we have

$$\int^{\bullet} d\mu(t) \int^{*} f(x) \, d\lambda_t(x) \leq \int^{\bullet} d\mu(t) \int^{*} g(x) \, d\lambda_t(x) = \int^{*} g(x) \, d\nu(x)$$

by formula (1), which proves (a).

The inequality (b) is just as easy to establish, with formula (2) (instead of formula (1)). □

Proposition 19.4.5 *Let f be a ν-measurable mapping from X into a topological space, which is constant outside a ν-moderate set. Denote by M the set of those $t \in T$ for which f is not λ_t-measurable. Then M is locally μ-negligible, and it is μ-negligible if $\wedge$ is vaguely continuous.*

PROOF: Since each ν-integrable set is included in a ν-integrable open set, f is constant on the complement B of a countable union of ν-integrable open sets. There exists a partition of $X - B$ consisting of a ν-negligible set N and countably many compact sets K_i, such that the restriction of f to each K_i is continuous.

Denote by S the set of those $t \in T$ for which N is not λ_t-negligible. By Proposition 19.4.4, S is locally μ-negligible, and it is μ-negligible if $\wedge$ is vaguely continuous. The sets K_i, N, B are universally measurable, and the restriction of f to each of them is λ_t-measurable for every $t \notin S$. Thus f is λ_t-measurable for every $t \notin S$. □

Theorem 19.4.1 *Let* $f : X \rightarrow [0, +\infty]$ *be* ν-measurable and ν-moderate. *Denote by* M *the set of those* $t \in T$ *for which* f *is not* λ_t-measurable and λ_t-moderate.

(a) *M is locally* μ-negligible, $t \mapsto \int^* f(x) \, d\lambda_t(x)$ *is* μ-measurable, and

$$\int^* f(x) \, d\nu(x) = \int^\bullet d\mu(t) \int^* f(x) \, d\lambda_t(x) \qquad (3).$$

(b) *If* $\wedge$ *is vaguely continuous, then* M *is* μ-negligible. *the function* $t \mapsto \int^* f(x) \, d\lambda_t(x)$ *is* μ-measurable and μ-moderate, and

$$\int^* f(x) \, d\nu(x) = \int^* d\mu(t) \int^* f(x) \, d\lambda_t(x) \qquad (4).$$

PROOF: Replacing f by $\inf(f, n)$ (n integer ≥ 1), we may suppose that f is bounded.

Let N and $(K_i)_{i \in I}$ be as in the argument of Proposition 19.4.5. $\int^* f(x) \, d\lambda_t(x) = \sum_{i \in I} \int^* (f \cdot 1_{K_i})(x) \, d\lambda_t(x)$ for every $t \notin S$, because $f = \sum_{i \in I} f \cdot 1_{K_i}$ λ_t-almost everywhere. Hence we need only prove the theorem when there exists a compact set K such that f vanishes on K^c and such that the restriction of f to K is finite and continuous.

Let G be a ν-integrable open set containing K, c a constant which dominates f, h the lower semicontinuous function $c \cdot 1_G$, and g the function $h - f$. f is upper semicontinuous, so g is lower semicontinuous. Moreover, f, g, h are ν-integrable. Apply formula (1) (respectively, formula (2)) to the lower semicontinuous functions h and g. Then, evidently, $t \mapsto \int^* f(x) \, d\lambda_t(x)$ is μ-measurable and we have formula (3) (respectively, formula (4)).

Finally, if $\wedge$ is vaguely continuous, $t \mapsto \int^* f(x) \, d\lambda_t(x)$ has a finite upper integral; thus it is μ-moderate. $\qquad \square$

Proposition 19.4.6 *Let* f *be a mapping from* X *into a real Banach space*, ν-measurable and ν-moderate. *Then* f *is* ν-integrable if and only if $\int^\bullet d\mu(t) \int^* |f(x)| \, d\lambda_t(x)$ *is finite*.

PROOF: This follows immediately from Theorem 19.4.1 $\qquad \square$

Theorem 19.4.2 *Let* f *be a* ν-integrable mapping from X into a real Banach space F, *and let* H *be the set of those* $t \in T$ *for which* f *is not* λ_t-integrable.

(a) *H is locally* μ-negligible, $t \mapsto \int f(x) \, d\lambda_t(x)$ *(defined for* $t \notin H$*) is essentially* μ-integrable, and

$$\int f(x) \, d\nu(x) = \int d\mu(t) \int f(x) \, d\lambda_t(x) \qquad (5).$$

(b) If $\wedge$ is vaguely continuous, then H is μ-negligible and $t \mapsto \int f(x)\, d\lambda_t(x)$ (defined for $t \notin H$) is μ-integrable.

PROOF: The result is true when f is a positive function (Theorem 19.4.1), and therefore when f is real-valued.

Let $\mathcal{H}(X, \mathbf{R}) \otimes F$ be the subspace of $\mathcal{L}_F^1(\nu)$ consisting of the linear combinations, with coefficients in F, of functions in $\mathcal{H}(X, \mathbf{R})$. By linearity, the statement is valid when f belongs to $\mathcal{H}(X, \mathbf{R}) \otimes F$.

Now, for every $f \in \mathcal{L}_F^1(\nu)$, there exists a sequence $(f_n)_{n \geq 1}$ in $\mathcal{H}(X, \mathbf{R}) \otimes F$ with the following properties:

(c) $(f_n)_{n \geq 1}$ converges to f in $\mathcal{L}_F^1(\nu)$ and ν-almost everywhere.

(d) $g = |f_1| + \sum_{n \geq 1} |f_{n+1} - f_n|$ is such that $\nu^*(g) < +\infty$ (Theorem 3.1.2).

Denote by N_1 the set of those $t \in T$ such that $\lambda_t^*(g) = +\infty$. By formula (3) (respectively, formula (4)), N_1 is locally μ-negligible (respectively, μ-negligible). For $t \notin N_1$, the mappings f_n belong to $\mathcal{L}_F^1(\lambda_t)$ and the sequence $(f_n)_{n \geq 1}$ converges in $\mathcal{L}_F^1(\lambda_t)$. Let M be the set of those $x \in X$ such that $(f_n(x))_{n \geq 1}$ does not converge to $f(x)$. As M is ν-negligible, the set N_2, of those $t \in T$ for which M is not λ_t-negligible, is locally μ-negligible (respectively, μ-negligible), by Proposition 19.4.4.

Suppose that t does not belong to $N_1 \cup N_2$. Then the sequence $(f_n)_{n \geq 1}$ converges in $\mathcal{L}_F^1(\lambda_t)$ and it converges λ_t-almost everywhere to f. Hence f belongs to $\mathcal{L}_F^1(\lambda_t)$ and $\int f\, d\lambda_t = \lim_{n \to +\infty} \int f_n\, d\lambda_t$ (Proposition 3.1.8). So the set H is contained in $N_1 \cup N_2$. Finally, for every $t \notin N_1 \cup N_2$ and every n, $\left| \int f_n(x)\, d\lambda_t(x) \right| \leq \int^* g(x)\, d\lambda_t(x)$. The function $t \mapsto \int^* g(x)\, d\lambda_t(x)$ being essentially μ-integrable (respectively, μ-integrable) by Theorem 19.4.1, we can apply the dominated convergence theorem. Then $t \mapsto \int f(x)\, d\lambda_t(x)$ (defined for $t \notin N_1 \cup N_2$) is essentially μ-integrable (respectively, μ-integrable) and

$$\int d\mu(t) \int f(x)\, d\lambda_t(x) = \lim_{n \to +\infty} \int d\mu(t) \int f_n(x)\, d\lambda_t(x)$$

$$= \lim_{n \to +\infty} \int f_n(x)\, d\nu(x).$$

Since $\int f_n(x)\, d\nu(x)$ converges to $\int f(x)\, d\nu(x)$ as n tends to $+\infty$, formula (5) follows. $\square$

19.5 μ-Adapted Pairs

Let X and T be locally compact Hausdorff spaces, π a mapping from T into X, and g a function from T into $[0, +\infty[$. For each $t \in T$, write $\varepsilon_{\pi(t)}$ for the Radon measure on X defined by the mass 1 at $\pi(t)$, and λ_t for $g(t)\varepsilon_{\pi(t)}$. Finally, let μ be a positive Radon measure on T.

Definition 19.5.1 The pair (π, g) is said to be μ-adapted whenever the following conditions hold:

(a) π and g are μ-measurable.

(b) For all $f \in \mathcal{H}^+(X)$, the function $t \mapsto f(\pi(t))g(t)$ is essentially μ-integrable.

Remark. This definition does not relate to that of Section 11.1.

Proposition 19.5.1 *If (π, g) is μ-adapted, then $\wedge : t \mapsto \lambda_t$ is scalarly essentially μ-integrable and vaguely μ-measurable. Conversely, if $\wedge$ is scalarly essentially μ-integrable and vaguely μ-measurable, then g is μ-measurable and the restriction of π to $S = \{t \in T : g(t) > 0\}$ is μ-measurable.*

PROOF: Suppose that (π, g) is μ-adapted. Clearly, $\wedge$ is scalarly essentially μ-integrable. The compact subsets K of T such that the restrictions of π and g to K are continuous form a μ-dense class in T. If K is such a compact set, the restriction of $\wedge$ to K is vaguely continuous. so $\wedge$ is vaguely μ-measurable.

Conversely, suppose that $\wedge$ is scalarly essentially μ-integrable and vaguely μ-measurable; then it is μ-adequate (Proposition 19.4.3). As the function 1 is lower semicontinuous on X, $t \mapsto \lambda_t(1) = g(t)$ is μ-measurable, and so S is μ-measurable. Now the compact sets $K \subset S$ such that $g_{/K}$ is continuous and $\wedge_{/K}$ is vaguely continuous form a μ-dense class $\mathcal{D}$ in S; if $K \in \mathcal{D}$, the restriction of $t \mapsto \varepsilon_{\pi(t)} = (1/g(t))\lambda_t$ to K is vaguely continuous, which implies that $\pi_{/K}$ is continuous. Therefore, $\pi_{/S}$ is μ-measurable. □

We will see that, when (π, g) is μ-adapted, we can improve the results of Section 19.4 for $\wedge : t \mapsto \lambda_t = g(t)\varepsilon_{\pi(t)}$.

First, we need the following result.

Lemma 19.5.1 *Assume that π is continuous and proper (i.e., π is closed and $\pi^{-1}(x)$ is compact for all $x \in X$). Finally, let $h : T \to \bar{\mathbf{R}}$ be lower semicontinuous. Then $v : x \mapsto \inf_{t \in \pi^{-1}(x)} h(t)$ is lower semicontinuous on X.*

PROOF: For each real number a, put

$$A_a = \{t \in T : h(t) \le a\} \quad \text{and} \quad B_a = \{x \in X : v(x) \le a\}.$$

Since $v(\pi(t)) \le h(t)$ for all $t \in T$, we have $\pi(A_a) \subset B_t$. On the other hand, for each $x \in B_a$, $\pi^{-1}(x)$ is nonempty and compact, and hence there exists $t \in \pi^{-1}(x)$ for which $v(x) = h(t)$; t lies in A_a and $\pi(t) = x$, so $\pi(A_a) = B_a$. B_a is therefore closed, which proves that v is lower semicontinuous. □

Theorem 19.5.1   *Suppose that (π, g) is μ-adapted, and set $\nu = \int g(t)\varepsilon_{\pi(t)} \, d\mu(t)$. For every $f : X \to [0, +\infty]$,*

$$\int^{\bullet} f(x) \, d\nu(x) = \int^{\bullet} f(\pi t)g(t) \, d\mu(t). \quad (1)$$

PROOF:

A. First, suppose that μ has compact support K and the restrictions of g and π to K are continuous. By formula (1) of Section 19.4, $\nu^{\bullet}(1) = \int_K g(t)\, d\mu(t) < +\infty$ and so all the measures which occur in formula (1) are bounded. Thus we may replace $\int^{\bullet}$ by $\int^{*}$ in the left-hand side of the equation (1). By Proposition 19.4.4, we need only show that

$$\int^{*} f(x)\, d\nu(x) \leq \int^{\bullet} f(\pi t)g(t)\, d\mu(t),$$

where the symbol $\int^{\bullet}$ on the right-hand side of the equation may be, in turn, replaced by $\int^{*}$. By definition of the upper integral, it suffices to verify the inequality

$$\int^{*} f(x)\, d\nu(x) \leq \int^{*} h(t)\, d\mu(t) \qquad (2)$$

for every function h, lower semicontinuous on T, which dominates $t \mapsto f(\pi t)g(t)$.

Now, given $\varepsilon > 0$, let u be the function $(h + \varepsilon)/g$, which is lower semicontinuous on K. For every $t \in T$, $u(t) \geq f(\pi t)$; indeed, $u(t) = +\infty$ if $g(t) = 0$ whereas $u(t)g(t) = h(t) + \varepsilon \geq f(\pi t)g(t)$ if $g(t) > 0$. Write $v(x)$, for all $x \in X$, for the infimum of $\big\{ u(t) : t \in K \cap \pi^{-1}(\{x\})\big\}$. The function v dominates f by the preceding, it is lower semicontinuous on X by Lemma 19.5.1 (applied to $\pi_{/K}$), and we have $v(\pi t)g(t) \leq h(t) + \varepsilon$ for every $t \in K$. Applying to v formula (1) of Section 19.4, we obtain

$$\int^{*} f(x)\, d\nu(x) \leq \int^{*} v(x)\, d\nu(x) = \int^{*} v(\pi t)g(t)\, d\mu(t)$$

$$\leq \int_{K}^{*} (h(t) + \varepsilon)\, d\mu(t) = \int^{*} h\, d\mu + \varepsilon\mu(1).$$

Since μ is bounded and ε is arbitrary, inequality (2) results.

B. Now we pass to the general case. Those compact sets K in T for which the restrictions of π and g to K are continuous form a μ-dense class, $\mathcal{D}$, in T. By Theorem 19.3.1, μ is the sum of a summable family $(\mu_{\alpha})_{\alpha \in A}$ of measures carried by $\mathcal{D}$-sets. As (π, g) is μ_{α}-adapted, we can consider $\nu_{\alpha} = \int g(t)\varepsilon_{\pi(t)}\, d\mu_{\alpha}(t)$. By part A, $\int^{\bullet} f(x)\, d\nu_{\alpha}(x) = \int^{\bullet} f(\pi t)g(t)\, d\mu_{\alpha}(t)$. Finally, since the measures ν_{α} are summable, with sum ν (Proposition 19.4.2),

$$\int^{\bullet} f(x)\, d\nu(x) = \sum_{\alpha \in A} \int^{\bullet} f(x)\, d\nu_{\alpha}(x)$$

$$= \sum_{\alpha} \int^{\bullet} f(\pi t)g(t)\, d\mu_{\alpha}(t)$$

$$= \int^{\bullet} f(\pi t)g(t)\, d\mu(t),$$

as was to be shown. $\square$

Proposition 19.5.2 *Suppose that π is continuous and proper, g is continuous, and $g^{-1}(0)$ is μ-integrable. Then (π, g) is μ-adapted and $\int^* f(x)\, d\nu(x) = \int^* f(\pi t)g(t)\, d\mu(t)$ for every $f : X \to [0, +\infty]$ (where $\nu = \int g(t)\varepsilon_{\pi(t)}\, d\mu(t)$).*

PROOF: For every $\psi \in \mathcal{H}^+(X)$, $\psi \circ \pi$ has compact support (remark after Definition 23.5.1). Thus (π, g) is μ-adapted. Moreover, $\wedge : t \mapsto g(t)\varepsilon_{\pi(t)}$ is vaguely continuous. Now let h be a lower semicontinuous function on T, such that $f(\pi t)g(t) \leq h(t)$ for all $t \in T$. We show that

$$\int^* f(x)\, d\nu(x) \leq \int^* h(t)\, d\mu(t) \qquad (3).$$

Fix a μ-integrable open set U containing $g^{-1}(0)$. Given $\varepsilon > 0$, put $h' = h + \varepsilon \cdot 1_U$. As h'/g is lower semicontinuous, the function $\overline{f}$ defined on X by $\overline{f}(x) = \inf\left\{h'(t)/g(t) : t \in \pi^{-1}(x)\right\}$ is lower semicontinuous (Lemma 19.5.1). Moreover, the following properties hold:

(a) $\overline{f}(x) \geq f(x)$ for all $x \in X$.

(b) $\overline{f}(\pi t)g(t) \leq h'(t)$ for all $t \in T$.

Proposition 19.4.3 shows, therefore, that

$$\int^* f(x)\, d\nu(x) \leq \int^* \overline{f}(x)\, d\nu(x) = \int^* \overline{f}(\pi t)g(t)\, d\mu(t)$$

$$\leq \int^* h'(t)\, d\mu(t) = \int^* h(t)\, d\mu(t) + \varepsilon\mu^*(U).$$

Since ε is arbitrary, we obtain $\int^* f(x)\, d\nu(x) \leq \int^* h(t)\, d\mu(t)$. as desired.

Next, from (3) results $\int^* f(x)\, d\nu(x) \leq \int^* f(\pi t)g(t)\, d\mu(t)$, and the reverse inequality follows from Proposition 19.4.4. $\square$

Henceforth, suppose that (π, g) is μ-adapted and write ν for $\int \varepsilon_{\pi(t)}g(t)\, d\mu(t)$.

Theorem 19.5.2 *Let f be a mapping from X into a topological space. Then f is ν-measurable if and only if the restriction of $f \circ \pi$ to $S = g^{-1}(]0, +\infty[)$ is μ-measurable.*

PROOF: First, assume that f is ν-measurable. Since π is μ-measurable, the compact subsets K of S such that $\pi_{/K}$ is continuous form a μ-dense class $\mathcal{D}$ in S. Let $K \in \mathcal{D}$ be given. By hypothesis, for each compact subset H of K, there exists a partition of $\pi(H)$ consisting of a ν-negligible set N and a sequence $(C_n)_{n \geq 1}$ of compact sets such that the restriction of f to each C_n is continuous. Under these conditions, $H \cap \pi^{-1}(N)$ and the sets $H \cap \pi^{-1}(C_n)$ partition H.

Now $H \cap \pi^{-1}(N)$ is μ-negligible (Theorem 19.5.1), the sets $H \cap \pi^{-1}(C_n)$ are compact, and the restriction of $f \circ \pi$ to each of them is continuous. Thus the compact sets H in K for which $f \circ \pi_{/H}$ is continuous form a μ-dense class in K (Proposition 19.2.1), which proves that $f \circ \pi_{/S}$ is μ-measurable (Proposition 19.2.2).

Conversely, suppose that $f \circ \pi_{/S}$ is μ-measurable. Define $\mathcal{L}$ as the collection of those compact sets L in X for which $f_{/L}$ is continuous. It is enough to show that $\mathcal{L}$ is ν-dense in X. Let N be a subset of X such that $N \cap L$ is ν-negligible for all $L \in \mathcal{L}$. We have to prove that N is locally ν-negligible or, equivalently, that $\pi^{-1}(N) \cap S$ is locally μ-negligible (Theorem 19.5.1). Now the compact subsets H of S such that the restrictions of π and $f \circ \pi$ to H are continuous form a μ-dense class $\mathcal{D}$ in S. It suffices, therefore, to prove that $\pi^{-1}(N) \cap H$ is μ-negligible for every $H \in \mathcal{D}$ (Proposition 19.2.1). If $\mathcal{F}$ is an ultrafilter on $\pi(H)$ converging to $x \in \pi(H)$ and if $\mathcal{U}$ is an ultrafilter on H containing $\{H \cap \pi^{-1}(A) : A \in \mathcal{F}\}$, then $\pi(\mathcal{U}) = \mathcal{F}$ and $\mathcal{U}$ converges to a point $t \in H$; thus $\pi(t) = x$, and, since $(f \circ \pi)(\mathcal{U})$ converges to $f(\pi t) = f(x)$, we conclude that $f(\mathcal{F})$ converges to $f(x)$. Hence the restriction of f to $\pi(H)$ is continuous. In other words, $\pi(H)$ belongs to $\mathcal{L}$, and so $N \cap \pi(H)$ is ν-negligible. By Theorem 19.5.1, $\pi^{-1}(N \cap \pi(H)) \cap S$ is locally μ-negligible. Consequently, $H \cap \pi^{-1}(N)$, which is included in $\pi^{-1}(N \cap \pi(H)) \cap S$, is μ-negligible, and the proof is complete. $\square$

If f takes its values in a Banach space (or in $\bar{\mathbf{R}}$), the statements "$f \circ \pi_{/S}$ is μ-measurable" and "$(f \circ \pi)g$ is μ-measurable" are equivalent, because g is μ-measurable, does not vanish in S, and vanishes on $T - S$ (Proposition 19.2.3).

Theorem 19.5.3 *Let f be a mapping from X into a real Banach space. Then f is essentially ν-integrable if and only if $t \mapsto f(\pi t)g(t)$ is essentially μ-integrable; in this case,*

$$\int f(x)\, d\nu(x) = \int f(\pi t)g(t)\, d\mu(t) \quad (4).$$

Assume, moreover, that π is continuous and proper, g is continuous, and $g^{-1}(0)$ is μ-integrable. Then f is ν-integrable if and only if $t \mapsto f(\pi t)g(t)$ is μ-integrable.

PROOF:
A. First, assume that μ is carried by a compact set K on which g is bounded. In this case, ν is bounded (Theorem 19.5.1) and we can replace, in the statement, "essentially integrable" by "integrable". If f is ν-integrable, then, by Theorem 19.4.2, $t \mapsto f(\pi t)g(t)$ is μ-integrable and relation (4) holds. Conversely, if $t \mapsto f(\pi t)g(t)$ is μ-integrable, then f is ν-measurable (Theorem 19.5.2) and $\int^{\bullet} |f(x)|\, d\nu(x) = \int^{\bullet} |f(\pi t)|g(t)\, d\mu(t) < +\infty$ (Theorem 19.5.1); thus f is essentially ν-integrable and, in fact, ν-integrable.

B.   Next, we consider the general case. The class $\mathcal{D}$, consisting of those compact subsets K of T such that $g_{/K}$ is continuous is μ-dense, and μ is the sum of a family $(\mu_\alpha)_{\alpha \in A}$ of measures carried by $\mathcal{D}$-sets (Theorem 19.3.1). (π, g) is, clearly, μ_α-adapted for every $\alpha \in A$, and ν is the sum of the family $(\nu_\alpha)_{\alpha \in A} = \left(\int \varepsilon_{\pi(t)} g(t) \, d\mu_\alpha(t) \right)_{\alpha \in A}$ of measures (Proposition 19.4.2). The argument of A applies to the measures μ_α and ν_α. and the first part of the statement now results from Section 19.3.

C.   The second part of the statement follows from the first part and Proposition 19.5.2.   □

Exercises for Chapter 19

1   Let Ω be the locally compact space considered in Exercise 3 of Chapter 7. For each integer $n \geq 1$, let μ_n be the Radon measure on Ω defined by the masses $1/m^3$ at the points $(1/m, k/m^2)$ for $1 \leq m \leq n$ and $k \in \mathbf{Z}$.

   1.   Prove that the supremum of $\{\mu_n : n \geq 1\}$ in $\mathcal{M}^+(\Omega)$ is the Radon measure considered in Exercise 4 of Chapter 7.

   2.   Show that $\mu_n^\bullet(D) = 0$ for all $n \geq 1$, but that $\mu^\bullet(D) = +\infty$. Compare with the statement of Proposition 19.3.1.

2   Let Ω and μ be as in Exercise 1.

   1.   Let A be the class of finite sets in $\mathbf{R}$. For every $\alpha \in A$, denote by f_α the indicator function of $\{0\} \times \alpha$ on Ω. Show that the mapping $\omega \mapsto \left(f_\alpha(\omega) \right)_{\alpha \in A}$ from Ω into $\mathbf{R}^A$ is μ-measurable.

   2.   Show that $\int^* f_\alpha \, d\mu = 0$ for all $\alpha \in A$, but that $\int^* (\sup_{\alpha \in A} f_\alpha) \, d\mu = +\infty$.

3   Let X be a locally compact Hausdorff space, μ a positive Radon measure on X, and $(f_\alpha)_{\alpha \in A}$ an upward-directed family of positive functions on X. Suppose that the mapping $x \mapsto \left(f_\alpha(x) \right)_{\alpha \in A}$ from X into $\bar{\mathbf{R}}^A$ is μ-measurable.

   1.   Show that $f = \sup_{\alpha \in A} f_\alpha$ is μ-measurable and that

$$\int^\bullet f \, d\mu = \sup_{\alpha \in A} \int^\bullet f_\alpha \, d\mu.$$

   2.   Compare the result of part 1 with that of Exercise 2.

20

Radon Measures Defined by Densities

We define the Radon measures with a given base μ and we study integration with respect to these measures.

Summary

20.1 Let μ be a Radon measure on a locally compact space X. Let Y be a locally compact subspace of X. Integration with respect to the induced measure μ_Y is done in the most natural way (Theorem 20.1.1). Piecing together Radon measures, as done in Theorem 20.1.2, is extremely useful in analysis and differential geometry.

20.2 Let μ be a Radon measure on a locally compact space T. A function g from T into $\mathbf{C}$ is said to be locally integrable when gh is μ-integrable for every continuous function h with compact support. In this case $g\mu : h \mapsto \int gh\,d\mu$ is a Radon measure. We study how to integrate with respect to $g\mu$ (Theorems 20.2.1 and 20.2.2).

20.3 The Radon–Nikodym theorem and Lebesgue's decomposition theorem still hold for Radon measures.

20.4 From Chapter 10 follow some results on the duality of $L_{\mathbf{C}}^p(\mu)$ spaces, when μ is a Radon measure.

20.1 Integration with Respect to Induced Measures

Let T be a locally compact Hausdorff space, X a locally compact subspace of T, μ a complex Radon measure on T, and μ_X the measure induced by μ on X. Recall that $|\mu|_X = |\mu_X|$ (Theorem 19.1.1). Define a mapping π from T into X by $\pi(t) = t$ for $t \in X$ and $\pi(t) = t_0$ for $t \in X^c$, where $t_0 \in X$ is fixed. Clearly, the pair $(\pi, 1_X)$ is $|\mu|$-adapted and $|\mu|_X = \int \varepsilon_{\pi(t)} \cdot 1_X(t)\,d|\mu|(t)$. For every $f : X \rightarrow [0, +\infty]$, therefore, $\int^* f\, d|\mu_X| = \int^* f'\, d|\mu|$ (where f' is the function which agrees with f on X and vanishes outside X).

A mapping g from X into a topological space is μ_X-measurable if and only if it is μ-measurable on X (Theorem 19.5.2).

Theorem 20.1.1 *Let f be a mapping from X into a Banach space. f is essentially μ_X-integrable if and only if its (canonical) extension f' is essentially μ-integrable; in this case, $\int f\, d\mu_X = \int f'\, d\mu$.*

PROOF: This results from Theorem 19.5.3, when μ is positive. The first assertion is then obvious for complex μ, and the second one follows from the fact that μ is a linear combination of four positive Radon measures smaller than $|\mu|$. □

Observe that $\mu_Y = (\mu_X)_Y$, whenever Y is a locally compact subspace of X.

Notice that $\int^* f\, d|\mu_X| \leq \int^* f'\, d|\mu|$ for all functions f from X into $[0, +\infty]$, and that the equality holds when X is open. It does not necessarily hold when X is closed (Chapter 7, Exercise 4).

The following theorem is used in many branches of mathematics.

Theorem 20.1.2 *Let T be a locally compact Hausdorff space and $(U_\alpha)_{\alpha \in A}$ an open covering of T. For each α, let μ_α be a Radon measure on U_α such that, for each pair of indices α, β for which $U_\alpha \cap U_\beta \neq \emptyset$, the measures induced by μ_α and μ_β on $U_\alpha \cap U_\beta$ are equal. Then there exists a unique Radon measure μ on T whose restriction $\mu_{/U_\alpha}$ to U_α is μ_α for each $\alpha \in A$.*

PROOF: First, we shall show that each function $f \in \mathcal{H}(T, \mathbf{C})$ can be written in the form $f = \sum_{1 \leq i \leq n} f_i$ where, for each index i, there exists $\alpha_i \in A$ such that $f_i \in \mathcal{H}(T, \mathbf{C})$ and $\mathrm{supp}(f_i) \subset U_{\alpha_i}$. For this purpose, we observe that, if K is a compact subset of T containing the support of f, then there exist finitely many indices $\alpha_i \in A$ $(1 \leq i \leq n)$ such that the U_{α_i} cover K. Hence (Theorem 7.1.1) there exist n continuous mappings $h_i : T \rightarrow [0, 1]$ such that $\mathrm{supp}(h_i)$ is compact and is contained in U_{α_i} for $1 \leq i \leq n$, and such that $\sum_{1 \leq i \leq n} h_i(t) = 1$ for all $t \in K$. Then the functions $f_i = f h_i$ satisfy the required conditions.

This already proves the uniqueness of μ: for, by definition, we must have $\mu(f) = \sum_{1 \leq i \leq n} \mu(f h_i) = \sum_{1 \leq i \leq n} \mu_{\alpha_i}(f h_i)$.

To prove the existence of a linear form μ on $\mathcal{H}(T, \mathbf{C})$ whose restriction to $\mathcal{H}(U_\alpha, \mathbf{C})$ is μ_α for each $\alpha \in A$ ($\mathcal{H}(U_\alpha, \mathbf{C})$ is identified with $\mathcal{H}(T, U_\alpha; \mathbf{C})$), it suffices to establish the following assertion: given two finite sequences $(g_i)_{1 \leq i \leq m}$ and $(h_j)_{1 \leq j \leq n}$ of functions in $\mathcal{H}(T, \mathbf{C})$ such that $\mathrm{supp}(g_i) \subset U_{\alpha_i}$ for $1 \leq i \leq m$ and $\mathrm{supp}(h_j) \subset U_{\beta_j}$ for $1 \leq j \leq n$, and such that $\sum_{1 \leq i \leq m} g_i(t) = \sum_{1 \leq j \leq n} h_j(t) = 1$ for all $x \in \mathrm{supp}(f)$, then

$$\sum_{1 \leq i \leq m} \mu_{\alpha_i}(f g_i) = \sum_{1 \leq j \leq n} \mu_{\beta_j}(f h_j).$$

Now $fg_i = \sum_{1 \leq j \leq n} fg_i h_j$, and therefore

$$\sum_{1 \leq i \leq m} \mu_{\alpha_i}(fg_i) = \sum_{1 \leq i \leq m} \left(\sum_{1 \leq j \leq n} \mu_{\alpha_i}(fg_i h_j) \right).$$

Similarly

$$\sum_{1 \leq j \leq n} \mu_{\beta_j}(fh_j) = \sum_{1 \leq j \leq n} \left(\sum_{1 \leq i \leq m} \mu_{\beta_j}(fg_i h_j) \right).$$

Since $\text{supp}(fg_i h_j)$ is contained in $U_{\alpha_i} \cap U_{\beta_j}$, it follows from the hypotheses that $\mu_{\alpha_i}(fg_i h_j) = \mu_{\beta_j}(fg_i h_j)$, and our assertion results.

It remains to be shown that the linear form μ so defined is a Radon measure. Let K be a compact subset of T, and define the U_{α_i} and the h_i as at the beginning of the proof. If $H_i = \text{supp}(h_i)$, then, by hypothesis, there exists $a_i \geq 0$ such that $|\mu_{\alpha_i}(g)| \leq a_i \|g\|$ for all $g \in \mathcal{H}(T, H_i; \mathbf{C})$. Hence, for each $f \in \mathcal{H}(T, K; \mathbf{C})$, we have

$$\left| \mu_{\alpha_i}(fh_i) \right| \leq a_i \|fh_i\| \leq a_i \|f\|,$$

so that $|\mu(f)| \leq \left(\sum_{1 \leq i \leq n} a_i \right) \|f\|.$ $\square$

20.2   Radon Measures with Base μ

Let T be a locally compact Hausdorff space and μ a complex Radon measure on T.

Proposition 20.2.1 *Let g be a mapping from T into a Banach space. Then the following conditions are equivalent:*

(a) gh is μ-integrable for all $h \in \mathcal{H}^+(T)$.

(b) $g \cdot 1_K$ is μ-integrable for all compact subsets K of T.

PROOF: Assume that condition (a) is satisfied, and let K be a compact subset of T. There exists $h \in \mathcal{H}^+(T)$ such that $h = 1$ on K, and so $g \cdot 1_K = (gh) \cdot 1_K$ is μ-integrable.

Conversely, assume that condition (b) holds. Then g is μ-measurable, so gh is μ-measurable for all $h \in \mathcal{H}^+(T)$. Since

$$\int^* |gh| \, d|\mu| \leq \|h\| \int^* |g| \cdot 1_{\text{supp}(h)} \, d|\mu| < +\infty,$$

gh is μ-integrable. $\square$

Definition 20.2.1 Under the conditions of Proposition 20.2.1, g is said to be locally μ-integrable.

In this case, the linear form $f \mapsto \int fg\,d\mu$ on $\mathcal{H}(T, \mathbf{C})$ is clearly a Radon measure on T, called the Radon measure with density g relative to μ, and is written $g\mu$ (or $g \cdot \mu$).

Proposition 20.2.2 *Assume that* $g : T \to \mathbf{C}$ *is locally μ-integrable. Then* $|g\mu| = |g| \cdot |\mu|$.

PROOF: Write B for $\{h \in \mathcal{H}(T. \mathbf{C}) : |h| \leq 1\}$. Then, for every $f \in \mathcal{H}^+(T)$,

$$\int |fg|\,d|\mu| = \sup_{h \in B} \left| \int fgh\,d\mu \right| = \sup_{h \in B} \left| \int fh\,d(g\mu) \right| = \int f\,d|g\mu|$$

(Theorem 7.4.3). Now $\int |fg|\,d|\mu| = \int f\,d(|g| \cdot |\mu|)$, and so $|g\mu| = |g| \cdot |\mu|$. $\quad\Box$

The following theorems are of constant use.

Theorem 20.2.1 *Assume that* $g : T \to \mathbf{C}$ *is locally μ-integrable. Then* $\int^\bullet f\,d|g\mu| = \int^\bullet f|g|\,d|\mu|$ *for all* $f : T \to [0, +\infty]$. *Moreover. if g is continuous and $g^{-1}(0)$ is μ-integrable,* $\int^* f\,d|g\mu| = \int^* f|g|\,d|\mu|$ *for all* $f : T \to [0, +\infty]$.

PROOF: Write I for the identity of T. Then the pair $(I. |g.)$ is $|\mu|$-adapted and $|g\mu| = \int \varepsilon_t |g|(t)\,d|\mu|(t)$. Theorem 20.2.1 follows. therefore, from Theorem 19.5.1 and Proposition 19.5.2. $\quad\Box$

Theorem 20.2.2 *Assume that g is locally μ-integrable. Then*

(a) *a mapping f from T into a topological space is $g\mu$-measurable if and only if f is μ-measurable on $S = \{t \in T : g(t) \neq 0\}$:*

(b) *a mapping f from T into a Banach space is essentially $g\mu$-integrable if and only if fg is essentially μ-integrable; in this case, $\int f\,d(g\mu) = \int fg\,d\mu$.*

PROOF: Assertion (a) and the first part of assertion (b) result from Theorems 19.5.2 and 19.5.3 . Now $\int f\,d(g\mu) = \int fg\,d\mu$ is true for positive g and positive μ (same reference), whence it follows that it remains valid for complex g and complex μ because $\mu = (\operatorname{Re}\mu)^+ - (\operatorname{Re}\mu)^- + i(\operatorname{Im}\mu)^+ - i(\operatorname{Im}\mu)^-$. $\quad\Box$

As a consequence of Theorem 20.2.2, in order that a mapping f from T into a Banach space be $g\mu$-measurable, it is necessary and sufficient that fg be μ-measurable.

Theorem 20.2.3 *Let $g_1 : T \to \mathbf{C}$ be locally μ-integrable. Then $g_2 : T \to \mathbf{C}$ is locally $g_1\mu$-integrable if and only if $g_1 g_2$ is locally μ-integrable. In this case, $g_2(g_1\mu) = (g_1 g_2)\mu$.*

PROOF: Obvious.   □

The set $\mathcal{L}^1_{\text{loc}}(\mu)$ of locally μ-integrable functions from T into $\mathbf{C}$ is a complex vector space, and the mapping $g \mapsto g\mu$ from $\mathcal{L}^1_{\text{loc}}(\mu)$ into $\mathcal{M}(T, \mathbf{C})$ is linear. $g_1\mu = g_2\mu$ if and only if $|g_1 - g_2|V\mu = 0$, that is, if $\int^{\bullet} |g_1 - g_2|\, dV\mu = \int^{\bullet} 1\, d(|g_1 - g_2|V\mu) = 0$. Equivalently, $g_1\mu = g_2\mu$ if and only if $g_1 = g_2$ locally μ-almost everywhere.

Definition 20.2.2 A Radon measure ν on T is said to be a measure with base μ, or absolutely continuous with respect to μ, whenever there is a locally μ-integrable function g from T into $\mathbf{C}$ such that $\nu = g\mu$. Then g, which is defined up to a locally μ-negligible set, is called a density of ν relative to μ.

Henceforth, suppose that μ is positive. If $g : T \to \mathbf{C}$ is locally μ-integrable and $g\mu$ is positive, then $g\mu = |g|\mu$, and so $g = |g|$ locally μ-almost everywhere. In other words, a necessary and sufficient condition that $g\mu$ be positive is that g be positive locally μ-almost everywhere. Finally, notice that

$$\sup(g_1\mu, g_2\mu) = \sup(g_1, g_2)\mu \quad \text{and} \quad \inf(g_1\mu, g_2\mu) = \inf(g_1, g_2)\mu$$

for all locally μ-integrable functions g_1, g_2 from T into $\mathbf{R}$.

A family $(g_\alpha)_{\alpha \in A}$ of μ-measurable functions from T into $[0, +\infty[$ is said to be locally countable whenever the family $(g_\alpha^{-1}(]0, +\infty[))_{\alpha \in A}$ is locally countable (Definition 19.2.2).

Proposition 20.2.3 *Let $(g_\alpha)_{\alpha \in A}$ be a locally countable family of locally μ-integrable functions from T into $[0, +\infty[$. The following conditions are equivalent:*

(a) There exists $g : T \to [0, +\infty[$, locally μ-integrable, which agrees with $\sum_{\alpha \in A} g_\alpha$ locally μ-almost everywhere.

(b) The family $(g_\alpha\mu)_{\alpha \in A}$ of measures is summable.

In this case, $g\mu = \sum_{\alpha \in A} g_\alpha\mu$.

PROOF: By Egorov's theorem, $h = \sum_{\alpha \in A} g_\alpha$ is μ-measurable. Thus condition (a) is satisfied if and only if $\mu^{\bullet}(fh)$ is finite for every $f \in \mathcal{H}^+(T)$. Since $\{\alpha \in A : fg_\alpha \neq 0\}$ is countable, $\mu^{\bullet}(fh) = \sum_{\alpha \in A} \mu^{\bullet}(fg_\alpha)$ (Proposition 3.6.1). Now put $\nu_\alpha = g_\alpha\mu$; the condition $\mu^{\bullet}(fh) < +\infty$ is equivalent to the condition $\sum_{\alpha \in A} \nu_\alpha(f) < +\infty$; in other words, (a) holds if and only if the family $(\nu_\alpha)_{\alpha \in A}$ is summable. Then, writing ν for $\sum_{\alpha \in A} \nu_\alpha$, the preceding computation gives $\nu(f) = \mu^{\bullet}(fh)$, and the proof is complete.   □

20.3   The Radon–Nikodym Theorem

Let T be a locally compact Hausdorff space. We need the following two lemmas.

Lemma 20.3.1 *Let α be a positive bounded measure on T. and let β be a complex measure on T such that $|\beta| \leq M\alpha$. where M is a positive constant. Then there exists an α-integrable complex-valued function u such that $\beta = u\alpha$.*

PROOF: Let $g \in \mathcal{L}^2_{\mathbf{C}}(T, \alpha)$; since g is β-measurable and

$$\int^* |g|^2 \, d|\beta| \leq M \int^* |g|^2 \, d\alpha$$

is finite, g lies in $\mathcal{L}^2_{\mathbf{C}}(T, \beta)$, and so in $\mathcal{L}^1_{\mathbf{C}}(T, \beta)$; moreover.

$$
\begin{aligned}
\left| \int g \, d\beta \right|^2 &\leq \left(\int |g| \, d|\beta| \right)^2 \\
&\leq \left(\int 1 \, d|\beta| \right) \left(\int |g|^2 \, d|\beta| \right) \\
&\leq M^2 \left(\int 1 \, d\alpha \right) \left(\int |g|^2 \, d\alpha \right).
\end{aligned}
$$

Now the mapping $g \mapsto \int g \, d\beta$ is a continuous linear form on $\mathcal{L}^2_{\mathbf{C}}(T, \alpha)$. Since $L^2_{\mathbf{C}}(T, \alpha)$ is a Hilbert space, there exists u in $\mathcal{L}^2_{\mathbf{C}}(T, \alpha)$. and therefore in $\mathcal{L}^1_{\mathbf{C}}(T, \alpha)$, such that $\int g \, d\beta = \int gu \, d\alpha$ for every $g \in \mathcal{L}^2_{\mathbf{C}}(T, \alpha)$. Applying this relation for $g \in \mathcal{H}(T, \mathbf{C})$, we see that $\beta = u\alpha$. $\square$

Lemma 20.3.2 *Let μ and ν be two positive Radon measures on T. Assume that each μ-negligible compact set is ν-negligible. Write $\mathcal{D}$ for the class of all the compact subsets K of T that have the following property:*

(a)   There exists a constant $M \geq 0$ such that $1_K \cdot \nu \leq M \cdot 1_K \cdot \mu$.

Then $\mathcal{D}$ is μ-dense in T.

PROOF: If K is a $\mathcal{D}$-set and A is a Borel subset of K. then

$$1_A \cdot \nu = 1_A \cdot (1_K \cdot \nu) \leq M \cdot 1_A \cdot \mu;$$

now the union of two arbitrary $\mathcal{D}$-sets K and K' is a $\mathcal{D}$-set. because $1_{K \cup K'} = 1_K + 1_{K' \cap K^c}$. To establish the lemma, we need only show that each compact set L verifying $\mu(L) > 0$ contains a compact set $K \in \mathcal{D}$ such that $\mu(K) > 0$ (Proposition 19.2.1). Choose $M > \nu(L)/\mu(L)$ and apply Lemma 20.3.1 to the positive bounded measure $\alpha = 1_L \cdot (\nu + M\mu)$ and to the measure $\beta = 1_L \cdot (\nu - M\mu)$. Modifying, if necessary, the function u such that $\beta = u\alpha$, we may suppose that u is real-valued, universally measurable (Proposition 19.2.5), and that it vanishes outside L. Since $H = \{t \in T : u(t) < 0\}$ is ν-measurable and contained in L, $\nu(H)$ is the supremum of the numbers $\nu(E)$, when E ranges over the class of all compact subsets of H. Now H cannot be μ-negligible: otherwise, it would be ν-negligible (and so α-negligible), $\beta = u\alpha$ would be positive, and $\beta(L)$ would be positive, in contradiction with the choice of M. Therefore, there exists a compact subset K of H such that $\mu(K) > 0$. Since

$$1_K \cdot \nu - M \cdot 1_K \cdot \mu = 1_K \cdot (\nu - M\mu) = 1_K \cdot \beta = 1_K \cdot (u\alpha) = (1_K \cdot u)\alpha$$

is negative, K is a $\mathcal{D}$-set, and the proof is complete. $\square$

Theorem 20.3.1 (Radon–Nikodym) *Let μ and ν be two positive measures on T. The following conditions are equivalent:*

(a) ν is a measure with base μ.

(b) Every locally μ-negligible set is locally ν-negligible.

(c) Every μ-negligible compact set is ν-negligible.

PROOF: Clearly, (a) implies (b) (Theorem 20.2.1) and (b) implies (c).

Next, assume that condition (c) holds, and define $\mathcal{D}$ as in Lemma 20.3.2. Let $(K_\alpha)_{\alpha \in A}$ be a locally countable family of mutually disjoint $\mathcal{D}$-sets such that $N = T - \bigcup_{\alpha \in A} K_\alpha$ is locally μ-negligible. Since the family $(K_\alpha)_\alpha$ is locally countable, N is universally measurable, and so locally ν-negligible. As the functions 1_{K_α} form a locally countable family, whose sum is equal to 1 locally almost everywhere for μ and for ν, Proposition 20.2.3 implies that $\mu = \sum_{\alpha \in A} \mu_\alpha$ and $\nu = \sum_{\alpha \in A} \nu_\alpha$ (where $\mu_\alpha = 1_{K_\alpha} \cdot \mu$ and $\nu_\alpha = 1_{K_\alpha} \cdot \nu$). On the other hand, by definition of $\mathcal{D}$, there exists, for each α, a constant M_α such that $\nu_\alpha \leq M_\alpha \mu_\alpha$; thus, by Lemma 20.3.1, there exists a function g_α, which we may suppose to be zero outside K_α and positive, such that $\nu_\alpha = g_\alpha \mu_\alpha$. Therefore,

$$\nu_\alpha = g_\alpha \mu_\alpha = g_\alpha (1_{K_\alpha} \cdot \mu) = (g_\alpha \cdot 1_{K_\alpha})\mu = g_\alpha \mu.$$

By Proposition 20.2.3, there exists a locally μ-integrable function g from T into $[0, +\infty[$ such that $\nu = g\mu$, as was to be shown. $\square$

Proposition 20.3.1 *Let θ be a complex Radon measure on T. Then there exists a universally measurable function v such that $|v| = 1$, $\theta = v|\theta|$, and $|\theta| = \bar{v}\theta$.*

PROOF: Define θ_i, for $1 \leq i \leq 4$, by $\theta_1 = (\operatorname{Re}\theta)^+$, $\theta_2 = (\operatorname{Re}\theta)^-$, $\theta_3 = (\operatorname{Im}\theta)^+$, and $\theta_4 = (\operatorname{Im}\theta)^-$. By Theorem 20.3.1, the θ_i are all measures with base $|\theta|$. Hence there exists a locally $|\theta|$-integrable function v such that $\theta = v|\theta|$. Now $|\theta| = |v||\theta|$, and so $|v| = 1$ locally $|\theta|$-almost everywhere. Moreover,

$\bar{v}\theta = (v\bar{v})|\theta| = |\theta|$. Finally, we may suppose that v is universally measurable and that $|v| = 1$ everywhere (same argument as that of Proposition 19.2.5).   □

Now, it is easily seen that the Radon–Nikodym theorem remains valid for complex Radon measures. In Exercise 2, we give an other proof of this result.

Theorem 20.3.2 *Let μ be a positive Radon measure on T and let ν be a real Radon measure on T. The following conditions are equivalent:*

(a) *For every essentially ν-integrable function f from T into $[0, +\infty[$ and for every $\varepsilon > 0$, there exists $\delta > 0$ such that the relations $0 \le h \le f$ and $\int^\bullet h \, d\mu \le \delta$ imply $\int^\bullet h \, d|\nu| \le \varepsilon$.*

(b) *ν belongs to the band generated by μ in the Riesz space $\mathcal{M}(T, \mathbf{R})$.*

(c) *Each μ-negligible compact set is ν-negligible.*

(d) *ν is a measure with base μ.*

PROOF: We can restrict ourselves to the case where ν is positive. By Proposition 1.3.5, ν belongs to the band generated by μ if and only if, for all $f \in \mathcal{H}^+(T)$ and for all $\varepsilon > 0$, there exists $\delta > 0$ such that the relations $h \in \mathcal{H}^+(T)$, $0 \le h \le f$, and $\int h \, d\mu \le \delta$ imply $\int h \, d\nu \le \varepsilon$. Thus condition (a) implies condition (b).

Next, assume that condition (b) holds, and let K be a μ-negligible compact set. Choose $f \in \mathcal{H}^+(T)$ such that $f = 1$ on K. Given $\varepsilon > 0$, let $\delta > 0$ be as above. We can find $h \in \mathcal{H}^+(T)$ such that $h \le f$, $h = 1$ on K, and $\int h \, d\mu \le \delta$. Then $\nu(K) \le \int h \, d\nu \le \varepsilon$. This shows that K is ν-negligible.

Now (c) implies (d), by the Radon–Nikodym theorem.

Finally, (d) implies (a), by the same argument as that of Theorem 10.3.2.   □

Two Radon measures μ and ν on T are said to be equivalent whenever the locally μ-negligible sets are exactly the locally ν-negligible sets. μ and ν are equivalent if and only if there exists a locally μ-integrable function g from T into $\mathbf{C}$ such that $\nu = g\mu$ and $g \ne 0$ locally μ-almost everywhere.

Definition 20.3.1 Two Radon measures μ, ν on T are said to be disjoint (or mutually singular) if $\inf\left(|\mu|, |\nu|\right) = 0$, and, in this case, we write $\mu \perp \nu$.

Clearly, Propositions 10.3.1, 10.3.2, and Theorem 10.3.3 remain valid for Radon measures.

Definition 20.3.2 A Radon measure μ on T is said to be concentrated on $E \subset T$ (or carried by E) whenever $T - E$ is locally μ-negligible.

If two Radon measures μ, ν on T are concentrated on disjoint sets, then they are disjoint, and the reverse property is true by the following Proposition 20.3.2.

A set $E \subset T$ which intersects every compact set $K \subset T$ in a Borel set is said to be locally Borelian. Let $\mathcal{A}$ be the σ-algebra of locally Borelian sets, which is included in the σ-algebra of all universally measurable sets.

Proposition 20.3.2 *Let $(\mu_n)_{n \geq 1}$ be a sequence of Radon measures on T such that μ_p and μ_q are mutually singular for all distinct $p \in \mathbf{N}$ and $q \in \mathbf{N}$. Then there exists a sequence $(E_n)_{n \geq 1}$ of disjoint locally Borelian sets such that each μ_n is concentrated on E_n.*

PROOF: Assume that μ_n is positive.

Putting $\mu = \mu_1 + \mu_2$, there exist locally μ-integrable functions f_1, f_2 from T into $[0, +\infty[$ such that $\mu_i = f_i \mu$. We may suppose that f_1 and f_2 are measurable $\mathcal{A}$ (Proposition 19.2.5). Then μ_1 and μ_2 are concentrated on $E_1^2 = \{t \in T : f_1(t) > 0, f_2(t) = 0\}$ and $F_1^2 = \{t \in T : f_1(t) = 0, f_2(t) > 0\}$, respectively.

For each integer $j > 1$, there exist disjoint $\mathcal{A}$-sets E_1^j and F_1^j which carry μ_1 and μ_j, respectively. Then μ_1 is concentrated on $E_1 = \bigcap_{j \geq 1} E_1^j$ and the μ_j are concentrated on $F_1 = \bigcup_{j \geq 1} F_1^j$.

In the same way, we can find $E_2 \subset F_1$ and $F_2 \subset F_1$ such that μ_2 is carried by E_2 and μ_j carried by F_2 for all $j > 2$.

Proceeding with the construction step-by-step, we obtain the sequence $(E_n)_{n \geq 1}$, that has the desired properties.   □

Now, for each $t \in T$, let ε_t be the Radon measure on T defined by the unit mass at t. If μ_a is an atomic Radon measure on T, then $|\mu_a|$ is the sum of the summable family $\left(|\mu_a(\{t\})| \varepsilon_t \right)_{t \in T}$. Each positive diffuse measure on T is disjoint from all the ε_t. Therefore, it is disjoint from $|\mu_a|$ (Theorem 1.2.1). Hence an atomic Radon measure and a diffuse Radon measure are disjoint.

Next, let μ be a Radon measure on T, and let μ_a be the Radon measure defined by the mass $\mu(\{t\})$ at each $t \in T$. Then $(\mu - \mu_a)(\{t\}) = 0$ for every $t \in T$, so $\mu - \mu_a$ is diffuse. We conclude that μ can be written $\mu_a + \mu_d$, where μ_a is atomic and μ_d diffuse; moreover, this decomposition is unique.

Finally, notice that results similar to Propositions 10.4.1, 10.4.2, 10.4.3, and Theorem 10.4.1 exist for Radon measure, and are, in fact, easier to enounce.

20.4   Duality of L^p Spaces

Let Ω be a locally compact Hausdorff space and μ a Radon measure on Ω. Given $p \in [1, +\infty]$, let q be its conjugate exponent. For all $g \in L^q_{\mathbf{C}}(\mu)$, write θ_g for the continuous linear form $f \mapsto \int fg \, d\mu$ on $L^p_{\mathbf{C}}(\mu)$. Then, $\theta : g \mapsto \theta_g$ is an isometry from $L^q_{\mathbf{C}}(\mu)$ onto a subspace of $\left(L^p_{\mathbf{C}}(\mu) \right)'$.

Let $\hat{\mu}$ be the main prolongation of μ.

When $1 < p < +\infty$, applying Theorem 10.5.1, we see that θ maps $L_C^q(\mu) = L_C^q(\hat{\mu})$ onto $\left(L_C^p(\mu)\right)' = \left(L_C^p(\hat{\mu})\right)'$. Similarly, when $p = 1$, θ maps $L_C^\infty(\mu)$ onto $\left(L_C^1(\mu)\right)'$.

Next, the results of Section 10.6 remain valid for μ. Moreover, $\left(L_C^\infty(\mu)\right)' = L_C^1(\mu)$ if and only if the support of μ is finite.

All these results can be proved directly, without resorting to Chapter 10.

Exercises for Chapter 20

1 Let μ be a complex Radon measure on a locally compact Hausdorff space, and let $g : \Omega \to C$ be locally μ-integrable. We give a new proof of the following assertions:

   (a) $|g\mu| = |g| \cdot |\mu|$.

   (b) $\int^\bullet f \, d|g\mu| = \int^\bullet f|g| \, d|\mu|$ for all $f : \Omega \to [0, +\infty]$.

   (c) A mapping f from Ω into a complex Banach space is essentially $g\mu$-integrable if and only if fg is essentially μ-integrable; in this case, $\int f \, d(g\mu) = \int fg \, d\mu$.

   (d) A mapping f from Ω into a topological space is $g\mu$-measurable if and only if f is μ-measurable on $A = \{x \in \Omega : g(x) \neq 0\}$.

   1. Let S be the semiring in Ω consisting of the $K \cap L^c$, where K and L range over the class of compact sets in Ω. Define a measure α on S by $E \mapsto \int 1_E \, d\mu$. Hence μ and α have the same essential prolongation (Theorem 8.1.1), and g is locally α-integrable. For a fixed compact set K in Ω, there exists a sequence $(U_n)_{n \geq 1}$ of open neighborhoods of K such that $|\mu|^*(U_n) \leq |\mu|(K) + 1/n$ and $|g\mu|^*(U_n) \leq |g\mu|(K) + 1/n$ for all $n \geq 1$. Construct a decreasing sequence $(f_n)_{n \geq 1}$ of continuous functions from Ω into $[0, 1]$, such that each f_n is equal to 1 on K and has compact support contained in U_n. Deduce that $\int 1_K \, d(g\mu) = \int g \cdot 1_K \, d\mu$. Conclude that $g\mu$ and $g\alpha$ have the same essential prolongation, whence follow the first three assertions.

   2. Prove, now, that assertion (d) is true.

2 Let μ and ν be two complex Radon measures on a locally compact Hausdorff space. Assume that each μ-negligible compact set is ν-negligible. We give a new proof that ν is a measure with base μ.

   1. Let E be a μ-negligible relatively compact set, and let $(U_n)_{n \geq 1}$ be a sequence of open neighborhoods of E, with compact closures, such that $V\mu(U_n)$ converges to 0 as $n \to +\infty$. Show that $\bigcap_{n \geq 1} U_n$, and so E itself, is ν-negligible. Conclude that each locally μ-negligible set is locally ν-negligible.

   2. Let S be the semiring in Ω consisting of the $K \cap L^c$ (K, L compact subsets of Ω). Write α for the measure $E \mapsto \int 1_E \, d\mu$ on S, and β for the measure

$E \mapsto \int 1_E \, d\nu$ on S. We know that $\overline{\alpha} = \overline{\mu}$ and $\overline{\beta} = \overline{\nu}$ (Theorem 8.1.1). Deduce from Theorem 19.2.1 and Theorem 10.3.1 that there exists a locally α-integrable function g for which $\beta = g\alpha$. Now show that $\overline{g\mu} = \overline{g\alpha} = \overline{\nu}$ and that $g\mu = \nu$.

21

Images of Radon Measures and Product Measures

Summary

21.1 Let T, X be two locally compact spaces, μ a positive Radon measure on T, and π a μ-measurable mapping from T into X such that $g \circ \pi$ is essentially μ-integrable for every $g \in \mathcal{H}(X, \mathbf{C})$. Denote by ν the image measure, $f \mapsto \int (f \circ \pi) \, d\mu$, of μ under π. Then $\int^{\bullet} f \, d\nu = \int^{\bullet} (f \circ \pi) \, d\mu$ for every function $f : X \to [0, +\infty]$. A mapping f from X into a topological space is ν-measurable when $f \circ \pi$ is μ-measurable (Theorem 21.1.1).

21.2 We study the decomposition of a measure in slices. This decomposition will be used in Chapter 23.

21.3 We define the product of two Radon measures.

21.1 Images of Radon Measures

Henceforth, we let T and X be two locally compact Hausdorff spaces, μ a complex Radon measure on T, and π a μ-measurable mapping from T into X.

Definition 21.1.1 π is said to be μ-proper whenever the following equivalent conditions hold:

(a) For all $f \in \mathcal{H}^{+}(X)$, $f \circ \pi$ is essentially μ-integrable.

(b) For all compact sets $K \subset X$, $\pi^{-1}(K)$ is essentially μ-integrable.

In this case, the Radon measure $\mathcal{H}(X, \mathbf{C}) \ni f \mapsto \int f \circ \pi \, d\mu$ is called the image of μ under π and is written $\pi(\mu)$.

Until further notice, assume that μ is positive and π is μ-proper. Then the pair $(\pi, 1)$ is μ-adapted (in the sense of Chapter 19), and

$$\pi(\mu) = \int \varepsilon_{\pi(t)} \, d\mu(t).$$

Consequently, we obtain the following result.

Theorem 21.1.1 *Write ν for $\pi(\mu)$. Then*

(a) $\int^{\bullet} f(x) \, d\nu(x) = \int^{\bullet} f(\pi t) \, d\mu(t)$ *for all $f : X \to [0, +\infty]$;*

(b) *a mapping f from X into a topological space is ν-measurable if and only if $f \circ \pi$ is μ-measurable;*

(c) *a mapping f from X into a real Banach space is essentially ν-integrable if and only if $f \circ \pi$ is essentially μ-integrable; in this case, $\int f \, d\nu = \int (f \circ \pi) \, d\mu$.*

Moreover, if π is continuous and proper, then $\int^{*} f \, d\nu = \int^{*} f \circ \pi \, d\mu$ for all $f : X \to [0, +\infty]$.

We now return to the case in which μ is a complex Radon measure on X and observe that $|\pi(\mu)| \leq \pi(|\mu|)$.

Theorem 21.1.2 *Let f be a mapping from X into a Banach space F. If $f \circ \pi$ is essentially μ-integrable, then f is essentially $\pi(\mu)$-integrable and $\int f \, d(\pi(\mu)) = \int f \circ \pi \, d\mu$.*

PROOF: The linear mappings $g \longmapsto \int g \, d(\pi(\mu))$ and $g \longmapsto \int g \circ \pi \, d\mu$ from $\overline{\mathcal{L}}_F^1(\pi(|\mu|))$ into F are continuous and they agree on $\mathcal{H}(X, \mathbf{R}) \otimes F$, so they are identical.   □

Notice that, if f is essentially $\pi(\mu)$-integrable, $f \circ \pi$ is not essentially μ-integrable, in general.

Results similar to Propositions 11.1.1 to 11.1.5 exist for Radon measures. We leave the details to the reader.

For example, we obtain the following result.

Proposition 21.1.1 *Let T, T', T'' be three locally compact Hausdorff spaces, μ a positive measure on T, π a μ-measurable mapping from T into T', π' a mapping from T' into T'', and $\pi'' = \pi' \circ \pi$.*

(a) *Suppose that π is μ-proper, and define $\mu' = \pi(\mu)$. For π' to be μ'-proper, it is necessary and sufficient that π'' be μ-proper, and, in this case, $\pi''(\mu) = \pi'(\pi(\mu))$.*

(b) *Assume that π' is continuous and π'' is μ-proper: then π is μ-proper, π' is $\pi(\mu)$-proper, and $\pi''(\mu) = \pi'(\pi(\mu))$.*

PROOF: Assertion (a) is obvious.

For (b), let K' be a compact subset of T'. Then $K'' = \pi'(K')$ is compact. Hence $\pi''^{-1}(K'')$ is essentially μ-integrable as well as its subset $\pi^{-1}(K')$. Thus π is μ-proper, and we can use assertion (a) to conclude.   □

21.2   Decomposition of a Measure in Slices[†]

Let X be a locally compact Hausdorff space, π a mapping from X into a locally compact Hausdorff space T, μ a positive Radon measure on T, $\wedge : t \mapsto \lambda_t$ a mapping from T into $\mathcal{M}^+(X)$, scalarly essentially μ-integrable and vaguely μ-measurable. Put $\nu = \int \lambda_t \, d\mu(t)$. If λ_t is concentrated on $\pi^{-1}(t)$, we say that the formula $\nu = \int \lambda_t \, d\mu(t)$ gives a decomposition of ν in slices, with respect to π.

The following result is sometimes useful.

Proposition 21.2.1 *Suppose that λ_t is concentrated on $\pi^{-1}(t)$, for every $t \in T$, and that π is ν-measurable.*

(a) *If $N \subset T$ is locally μ-negligible, $\pi^{-1}(N)$ is locally ν-negligible.*

(b) *If f is a μ-measurable mapping from T into a topological space, $f \circ \pi$ is ν-measurable.*

PROOF: Let H consist of those functions $h \in \mathcal{H}^+(X)$ such that $h \leq 1$; for every $h \in H$, we put $g_h(t) = \lambda_t(h)$ for all $t \in T$, designate by $\wedge_h$ the μ-adequate mapping $t \mapsto h\lambda_t$, and denote by ν_h the measure $\int h\lambda_t \, d\mu(t)$ on X. Now

$$\int^* 1 \, d\nu_h = \int^\bullet d\mu(t) \int^* 1 \, d(h\lambda_t) = \int \lambda_t(h) \, d\mu(t) < +\infty,$$

hence ν_h is bounded and π is ν_h-proper. On the other hand, g_h is essentially μ-integrable. For every $f \in \mathcal{H}^+(T)$,

$$\int (f \circ \pi) \, d\nu_h = \int^* (f \circ \pi) \, d\nu_h$$

$$= \int^\bullet d\mu(t) \int^* (f \circ \pi) \, d(h\lambda_t)$$

$$= \int^\bullet f(t)(h\lambda_t)^*(1) \, d\mu(t)$$

$$= \int f(t) g_h(t) \, d\mu(t)$$

[†]This section may be omitted.

by Theorem 19.4.1 and the fact that $h\lambda_t$ is concentrated on $\pi^{-1}(t)$. Therefore, $\pi(\nu_h) = g_h\mu$.

The measures ν_h, for $h \in H$, constitute an upward-directed subset of $\mathcal{M}^+(X)$. Let $f \in \mathcal{H}^+(X)$. Since the collection $\mathcal{D}$, of those compact sets K in T such that $\wedge_{/K}$ is vaguely continuous, is μ-dense,

$$\nu(f) = \int^{\bullet} \lambda_t(f)\,d\mu(t) = \sup_{K \in \mathcal{D}} \int_K \lambda_t(f)\,d\mu(t).$$

Along the filter of sections of H, the mapping $t \mapsto (h\lambda_t)(f) = \lambda_t(fh)$ converges to $t \mapsto \lambda_t(f)$ uniformly on every $K \in \mathcal{D}$ (by Dini's theorem). Hence, for all $K \in \mathcal{D}$,

$$\int_K \lambda_t(f)\,d\mu(t) = \sup_{h \in H} \int_K (h\lambda_t)(f)\,d\mu(t).$$

We conclude that

$$\nu(f) = \sup_{h \in H} \sup_{K \in \mathcal{D}} \int_K (h\lambda_t)(f)\,d\mu(t) = \sup_{h \in H} \int (h\lambda_t)(f)\,d\mu(t) = \sup_h \nu_h(f).$$

Therefore, $\nu = \sup_{h \in H} \nu_h$ (Lemma 1.3.1).

If $N \subset T$ is locally μ-negligible, then, for every $h \in H$, it is locally $g_h\mu$-negligible, and $\pi^{-1}(N)$ is locally ν_h-negligible, whence we deduce that $\pi^{-1}(N)$ is locally ν-negligible (Proposition 19.3.1).

Similarly, if f is μ-measurable, then $f \circ \pi$ is ν_h-measurable, so $f \circ \pi$ is ν-measurable.   ⊔

21.3   Product of Radon Measures

In this section, X and Y are two locally compact Hausdorff spaces.

For any two compact subsets K, L of X, Y, define a linear isometry ω from $\mathcal{H}(X \times Y, K \times L; \mathbf{C})$ onto $\mathcal{H}(X, K; \mathcal{H}(Y, L; \mathbf{C}))$ by $(\omega f)(x)(y) = f(x, y)$.

Lemma 21.3.1 *The closed vector subspace of* $\mathcal{H}(X \times Y, K \times L; \mathbf{C})$ *generated by* $\{g \otimes h : g \in \mathcal{H}(X, K; \mathbf{C}), h \in \mathcal{H}(Y, L; \mathbf{C})\}$ *is* $\mathcal{H}(X \times Y, K \times L; \mathbf{C})$ *itself.*

PROOF: Let $u \in \mathcal{H}(X, K; F)$, where $F = \mathcal{H}(Y, L; \mathbf{C})$. Proposition 7.1.3 ensures that, given $\varepsilon > 0$, there exist $g_1, \dots, g_n$ in $\mathcal{H}(X, K; \mathbf{C})$ and $h_1, \dots, h_n$ in F such that $\|u(x) - \sum_{1 \le i \le n} g_i(x)h_i\| \le \varepsilon$ for all $x \in X$. The result follows.   □

Lemma 21.3.2 *Let* μ *be a complex Radon measure on* X. *For every* f *in* $\mathcal{H}(X \times Y, K \times L; \mathbf{C})$, *the function* $y \mapsto \int f(x, y)\,d\mu(x)$ *belongs to* $\mathcal{H}(Y, L; \mathbf{C})$.

PROOF: Put $u = \omega(f)$. Then $\int u(x)\,d\mu(x)$ belongs to $\mathcal{H}(Y, L; \mathbb{C})$, and, since the evalutation map $h \mapsto h(y)$ from $\mathcal{H}(Y, L; \mathbb{C})$ into $\mathbb{C}$ is linear and continuous for all $y \in Y$,

$$\left(\int u(x)\,d\mu(x)\right)(y) = \int (u(x)(y))\,d\mu(x) = \int f(x, y)\,d\mu(x).$$

The proof is complete. $\qquad\qquad\qquad\qquad\qquad\qquad\qquad\qquad\qquad\qquad\qquad$ $\square$

Theorem 21.3.1 *Let μ and ν be two complex Radon measures on X and Y, respectively. There exists a unique Radon measure λ on $X \times Y$ such that $\lambda(g \otimes h) = \mu(g)\nu(h)$ for all $g \in \mathcal{H}(X, \mathbb{C})$ and all $h \in \mathcal{H}(Y, \mathbb{C})$. Moreover, $\lambda(f) = \int d\nu(y) \int f(x, y)\,d\mu(x)$ for all $f \in \mathcal{H}(X \times Y, \mathbb{C})$.*

PROOF: Define by λ the function $f \longmapsto \int d\nu(y) \int f(x, y)\,d\mu(x)$ from $\mathcal{H}(X \times Y, \mathbb{C})$ into $\mathbb{C}$. If K (respectively, L) is compact in X (respectively, Y), there is a real number a_K (respectively, b_L) such that $|\mu(g)| \leq a_k\|g\|$ (respectively, $|\nu(h)| \leq b_L\|h\|$) for all g (respectively, h) in $\mathcal{H}(X, K; \mathbb{C})$ (respectively, in $\mathcal{H}(Y, L; \mathbb{C})$). Let $f \in \mathcal{H}(X \times Y, K \times L; \mathbb{C})$. Then $|\int f(x, y)\,d\mu(x)| \leq a_K\|f\|$ for all $y \in Y$, so $|\lambda(f)| \leq a_K b_L\|f\|$, which proves that λ is a Radon measure. As a consequence of Lemma 21.3.1, we see that λ is the unique Radon measure such that $\lambda(g \otimes h) = \mu(g)\nu(h)$ for all g and all h. $\qquad\qquad$ $\square$

Definition 21.3.1 The measure λ, written $\mu \otimes \nu$, is said to be the product of μ and ν.

Theorem 21.3.2 *In this notation, $|\mu \otimes \nu| = |\mu| \otimes |\nu|$.*

PROOF: Let $f \in \mathcal{H}^+(X \times Y)$, and let $g \in \mathcal{H}(X \times Y, \mathbb{C})$ be such that $|g| \leq f$. We have

$$|\lambda(g)| = \left|\int d\mu(x) \int g(x, y)\,d\nu(y)\right| \leq \int d|\mu|(x) \int |g(x, y)|\,d|\nu|(y)$$

$$= (|\mu| \otimes |\nu|)(|g|) \leq (|\mu| \otimes |\nu|)(f).$$

Therefore, $|\lambda|(f) \leq (|\mu| \otimes |\nu|)(f)$ and, finally, $|\lambda| \leq |\mu| \otimes |\nu|$.

Now let $u \in \mathcal{H}^+(X)$ and $v \in \mathcal{H}^+(Y)$. Given $\varepsilon > 0$, there exist $u_1 \in \mathcal{H}(X, \mathbb{C})$ and $v_1 \in \mathcal{H}(Y, \mathbb{C})$ such that $|u_1| \leq u$, $|v_1| \leq v$, and

$$|\mu(u_1)| \geq |\mu|(u) - \varepsilon,$$
$$|\nu(v_1)| \geq |\nu|(v) - \varepsilon.$$

Then

$$|u_1 \otimes v_1| \leq u \otimes v$$

and

$$|\lambda|(u \otimes v) \geq |\lambda(u_1 \otimes v_1)| = |\mu(u_1)\nu(v_1)| \geq \left(|\mu|(u) - \varepsilon\right)\left(|\nu|(v) - \varepsilon\right).$$

Since this is true for all $\varepsilon > 0$,

$$|\lambda|(u \otimes v) \geq |\mu|(u)|\nu|(v) = \left(|\mu| \otimes |\nu|\right)(u \otimes v)$$

and also

$$|\lambda|(u \otimes v) = \left(|\mu| \otimes |\nu|\right)(u \otimes v).$$

This equality, in fact, holds for u in $\mathcal{H}(X, \mathbf{C})$ and v in $\mathcal{H}(Y, \mathbf{C})$, and Lemma 21.3.1 now proves that $|\lambda| = |\mu| \otimes |\nu|$. $\quad\square$

We now change our notation. Let Ω', Ω'' be two locally compact Hausdorff spaces, let μ' and μ'' be two complex Radon measures on Ω' and Ω'', respectively, and denote by μ the Radon measure $\mu' \otimes \mu''$ on $\Omega = \Omega' \times \Omega''$. If $f : \Omega \to [0, +\infty]$ is lower semicontinuous, then $x' \mapsto \int^* f(x', \cdot)\, dV\mu''$ is lower semicontinuous and

$$\int^* f\, dV\mu = \int^* dV\mu'(x') \int^* f(x', \cdot)\, dV\mu''$$

(Lemma 7.2.1 and Theorem 7.2.2).

The results of Section 9.2 (with the exception of Proposition 9.2.7), as well as Theorem 10.4.2 and Proposition 11.1.5, are true even for Radon measures. We need only return to the proofs of the following propositions.

Proposition 21.3.1 *Let A' (respectively, A'') be a μ'-moderate subset of Ω' (respectively, a μ''-moderate subset of Ω''). Then $A' \times A''$ is μ-moderate.*

PROOF: We may restrict our attention to integrable subsets A' and A''. Let V' and V'' be integrable open sets containing A' and A'', respectively. Since $\int^* 1_{V' \times V''}\, dV\mu$ is finite, $V' \times V''$ is μ-integrable. $\quad\square$

Proposition 21.3.2 *Let f' be a μ'-measurable mapping from Ω' into a topological space F. Then $f : (x', x'') \mapsto f'(x')$ is μ-measurable.*

PROOF: Let K be compact in Ω and denote by K' the projection of K on Ω'. There exists a sequence $(K'_n)_{n \geq 1}$ of disjoint compact subsets of K' such that $f'_{/K'_n}$ is continuous for all $n \geq 1$ and such that $N' = K' - \bigcup_{n \geq 1} K'_n$ is μ'-negligible. Now $K \cap (N' \times \Omega'')$ and the $K_n = K \cap (K'_n \times \Omega'')$ form a partition of K, and $f_{/K_n}$ is continuous for all $n \geq 1$. $\quad\square$

Proposition 21.3.3 *Let f be a μ-measurable mapping from Ω into a topological space, and assume that f is constant on the complement of a μ-moderate set. Write N' for the set of those $x' \in \Omega'$ such that $f(x', \cdot)$ is not μ''-measurable. Then N' is μ'-negligible.*

PROOF: Let $(K_n)_{n \geq 1}$ be a sequence of disjoint compact sets such that f takes the value z a.e. on the complement of $\bigcup_{n \geq 1} K_n$ and such that $f_{/K_n}$ is continuous for all $n \geq 1$. Define

$$N = \left\{ x \in \Omega - \bigcup K_n : f(x) \neq z \right\}.$$

There is a μ'-negligible subset A' of Ω' such that $N(x', \cdot)$ is μ''-negligible for all $x' \notin A'$. Then, for all $x' \notin A'$, $f(x', \cdot)$ is μ''-measurable. □

What we have done for the product of two measures extends easily to finite product of Radon measures.

Exercises for Chapter 21

1   Let Ω be a compact Hausdorff space, and let μ be a diffuse positive Radon measure on Ω such that $\int^* 1 \, d\mu = 1$.

   1.   If K is a compact set in Ω and W an open set containing K, prove the existence, for every $\varepsilon > 0$, of a μ-quadrable open set U such that $K \subset U \subset \overline{U} \subset W$ and $\mu(U) \leq \mu(K) + \varepsilon$.

   2.   Let K be a compact set in Ω, W an open set containing K such that $\mu(K) < \mu(W)$, and δ a real number satisfying $\mu(K) < \delta < \mu(W)$. For every $\varepsilon > 0$, prove the existence of μ-quadrable open sets U, V such that $K \subset U \subset \overline{U} \subset V \subset \overline{V} \subset W$ and $\delta - \varepsilon \leq \mu(U) < \delta < \mu(V) \leq \delta + \varepsilon$ (construct V first, and next U).

   3.   Let K, W, δ be as in part 2. Show that there exists a μ-quadrable open set U for which $K \subset U \subset \overline{U} \subset W$ and $\mu(U) = \delta$.

   4.   Let D be the set of dyadic numbers $k/2^n$ ($n \geq 0$, $0 \leq k \leq 2^n$). Define inductively a family $\big(U(r)\big)_{r \in D}$ of μ-quadrable open sets such that

      (a) $U(0) = \emptyset$ and $U(1) = \Omega$,

      (b) $\mu(U(r)) = r$ for all $r \in D$,

      (c) $\overline{U}(r) \subset U(r')$ for all $r, r' \in D$ satisfying $r < r'$.

      Now, for every $t \in I = [0, 1]$, put $U(t) = \bigcup_{r \in D, r \leq t} U(r)$, and show that the boundary of $U(t)$ is μ-negligible and that $\mu\big(U(t)\big) = t$. Prove that $\overline{U}(t) \subset U(t')$ for all $t, t' \in I$ satisfying $t < t'$.

   5.   Retaining the notation of part 4, show that the function

$$\pi : x \mapsto \inf \big\{ t \in I : x \in U(t) \big\}$$

      maps Ω onto $[0, 1]$ and that it is continuous.

   6.   Prove that $\mu\big(\pi^{-1}(\langle a, b]) \big) = b - a$ for all $\langle a, b]$ in the natural semiring of I. Conclude that Lebesgue measure on I is the image of μ under π.

2   Let Ω and μ be as in Exercise 1. Show that there exist subsets of Ω that are not μ-measurable.

3 Let Ω and Ω' be two locally compact Hausdorff spaces, and let μ, μ' be two positive Radon measures on Ω and Ω', respectively. Assume that each compact subset of Ω is metrizable.

Let f be a mapping from $\Omega \times \Omega'$ into a metrizable space F such that

    (a) $f(x, \cdot)$ is μ'-measurable, for every $x \in \Omega$,

    (b) $f(\cdot, x')$ is continuous, for every $x' \in \Omega'$.

We prove that f is $\mu \otimes \mu'$-measurable.

1. Fix a compact subset K of Ω and a distance d on K compatible with the topology of K. For each integer $n \geq 1$, let $\left(x_i^{(n)}\right)_{1 \leq i \leq k_n}$ be a sequence in K such that K is the union of the open balls $B\left(x_i^{(n)}, 1/n\right)$ $(1 \leq i \leq k_n)$. For each $x' \in \Omega'$, construct a mapping $f_n(\cdot, x')$ from Ω into F so that f_n is $(1_K \cdot \mu) \otimes \mu'$-measurable and $(f_n)_{n \geq 1}$ converges to f. Conclude that f is $(1_K \cdot \mu) \otimes \mu'$-measurable.

2. Let $(K_\alpha)_{\alpha \in A}$ be a locally countable family of disjoint compact sets in Ω such that $\Omega - \bigcup_{\alpha \in A} K_\alpha$ is locally μ-negligible. Show that $\mu = \sum_{\alpha \in A} 1_{K_\alpha} \cdot \mu$, that $\mu \otimes \mu' = \sum_{\alpha \in A} (1_{K_\alpha} \cdot \mu) \otimes \mu'$, and that f is $\mu \otimes \mu'$-measurable.

4 Let Ω be the interval $[0, 1]$, with the usual topology, and let Ω' be the interval $[0, 1]$ equipped with the discrete topology. Let μ be Lebesgue measure on Ω, and let μ' be the measure on Ω' such that $\mu'(\{x'\}) = 1$ for all $x' \in \Omega'$.

1. For any $x \in \Omega$, show that $\{x\} \times \Omega'$ is not $\mu \otimes \mu'$-negligible.

2. Let $\Delta = \{(x, x) : x \in \Omega\}$ be the diagonal of Ω, and write f for its indicator function on $\Omega \times \Omega'$. Show that $\int^\bullet d\mu(x) \int^\bullet f(x, x') \, d\mu'(x') \neq 0$, even though Δ is locally $\mu \otimes \mu'$-negligible. Compare this result with Theorem 9.2.5.

22

Operations on Regular Measures

Using measures on semirings, we gave some applications of measure theory to orthogonal polynomials, the differentiation of functions, and the formula for change of variables ... that we wish, now, to recover in the context of Radon measures.

For this, the operations on abstract measures and the operations on Radon measures have to be connected.

Summary

22.1   We have seen in Chapter 8 that we can associate a regular measure μ with a Radon measure λ, that is, a continuous linear functional on a space of continuous functions with compact support. We describe herein the functionals associated with the induced measures μ_Y, the measures $g\mu$ that have a density with respect to μ, and the image measures $\pi(\mu)$.

22.2   We define the σ-ring of Baire sets in a locally compact space.

22.3   We use the results of Section 22.2 to show that the product, μ, of two regular measures, μ' and μ'', is also a regular measure, provided that the compact sets of the product space are μ-measurable (Theorem 22.3.1).

22.4   Section 22.1 allows us to complete the results obtained in Chapter 11 on the change of variable formula.

22.1   Some Operations on Regular Measures

To begin, let X be a locally compact Hausdorff space. S a semiring in X, μ a regular measure on S, Y a locally compact subspace of X, and let T be a semiring in Y such that each compact subset of Y is contained in a countable union of T-sets. If $\mu_{/T}$ exists, it is regular. Indeed, write λ for the

Radon measure on X arising from μ; by the rules of essential integration with respect to $\mu_{/T}$ and λ_Y, $f : Y \to \mathbf{C}$ is essentially $\mu_{/T}$-integrable if and only if it is essentially λ_Y-integrable, and then $\int f\, d(\mu_{/T}) = \int f\, d\lambda_Y$.

On the other hand, the following proposition holds.

Proposition 22.1.1 *Let Ω be a locally compact Hausdorff space, S a semiring in Ω, μ a regular measure on S, and λ the Radon measure arising from μ. Define ν a measure on S, absolutely continuous with respect to μ. Then ν is regular if (and only if) $|\nu|^*(K) < +\infty$ for all compact sets K. In this case, there exists a locally μ-integrable function g from Ω into $\mathbf{C}$ such that $\nu = g\mu$; moreover, g is locally λ-integrable and $g\lambda$ is the Radon measure arising from $g\mu$.*

PROOF: Let $E \subset \Omega$ be μ-measurable. For all $A \in S$, there exists $B \in \widetilde{S}$ such that $B \subset E \cap A$ and $E \cap A - B$ is μ-negligible. Since $E \cap A - B$ is also ν-negligible, we conclude that $E \cap A$ is ν-integrable and E is ν-measurable.

By Theorem 21.2.1, there exists a locally countable class $\mathcal{C}$ of disjoint compact sets such that $\Omega - \bigcup_{K \in \mathcal{C}} K$ is locally μ-negligible. By the above, each $K \in \mathcal{C}$ is ν-integrable. On the other hand, each $A \in S$ is a union of disjoint compact sets K_n $(n \geq 1)$ and a μ-negligible set (because μ is regular), so A is contained in a union of countably many $\mathcal{C}$-sets and a μ-negligible set. The Radon–Nikodym theorem now proves the existence of a locally μ-integrable function g such that $\nu = g\mu$.

Since 1_K is ν-integrable, $g \cdot 1_K$ is μ-integrable for each compact set K and hence is λ-integrable. Thus g is locally λ-integrable. The rules of essential integration with respect to $g\mu$ and $g\lambda$ now show that $\overline{g\mu} = \overline{g\lambda}$, and finally that $\nu = g\mu$ is regular. □

Corollary *Let Ω, S, and μ be as in Proposition 22.1.1. Define ν a measure on S such that $|\nu| \leq |\mu|$. Then ν is regular.*

In particular, if μ_1, μ_2 are regular measures on S, then $\mu_1 + \mu_2$ is regular, because $|\mu_1 + \mu_2| \leq |\mu_1| + |\mu_2|$. Hence the regular measures on S form a vector subspace of $\mathcal{M}(S, \mathbf{C})$.

Proposition 22.1.2 *Let Ω be a locally compact Hausdorff space, μ a regular measure on a semiring S in Ω, λ the Radon measure arising from μ, and π a λ-proper mapping from Ω into a locally compact Hausdorff space Ω'. Define S' a semiring in Ω' such that each compact subset L of Ω' is included in a countable union of S'-sets, and suppose that $\pi^{-1}(B)$ is essentially μ-integrable for all $B \in S'$. Then (π, S') is μ-suited. Moreover, if the compact subsets of Ω' are $\pi(|\mu|)$-measurable, then $\pi(\mu)$ is regular and gives rise to the Radon measure $\pi(\lambda)$.*

PROOF: First, assume that μ is positive.

Write $\mathcal{D}$ for the class of those compact sets $K \subset \Omega$ such that $\pi_{/K}$ is continuous. For all $K \in \mathcal{D}$, there exists a sequence $(B_n)_{n \geq 1}$ of S'-sets such that K is included in $\bigcup_{n \geq 1} \pi^{-1}(B_n)$. Since $\mathcal{D}$ is λ-dense in Ω. each compact subset of Ω is a union of countably many disjoint $\mathcal{D}$-sets and a μ-negligible set. Finally, each $E \in S$ is a disjoint union of countably many compact sets and a μ-negligible set. This implies that (π, S') is μ-suited.

Next, assume that the compact subsets of Ω' are $\pi(\mu)$-measurable. Let $(K_\alpha)_{\alpha \in A}$ be a locally countable class of disjoint compact sets such that $N = \Omega - \bigcup_{\alpha \in A} K_\alpha$ is locally μ-negligible, and, for all $\alpha \in A$. put $\mu_\alpha = 1_{K_\alpha} \cdot \mu$ and $\lambda_\alpha = 1_{K_\alpha} \cdot \lambda$. Since $\int^\bullet 1 \, d\pi(\lambda_\alpha) = \int^\bullet 1 \, d\lambda_\alpha = \mu(K_\alpha)$. the Radon measure $\pi(\lambda_\alpha)$ is bounded. Each $B \in S'$ is essentially $\pi(\lambda_\alpha)$-integrable. and so $\pi(\lambda_\alpha)$-integrable, because $\pi^{-1}(B)$ is essentially λ_α-integrable: moreover,

$$\int 1_B \, d\pi(\lambda_\alpha) = \mu\big(\pi^{-1}(B) \cap K_\alpha\big) = \mu_\alpha\big(\pi^{-1}(B)\big) = \pi(\mu_\alpha)(B).$$

Now each compact subset L of Ω' is $\pi(\mu_\alpha)$-integrable. because it is $\pi(\mu_\alpha)$-measurable and $\pi(\mu_\alpha)$ is bounded. Therefore. there exist B_1. B_2 in $\tilde{S}'$ such that $B_1 \subset L \subset B_2$ and $B_2 - B_1$ is $\pi(\mu_\alpha)$-negligible. Under these notations, $B_2 - B_1$ is $\pi(\lambda_\alpha)$-negligible because $\pi^{-1}(B_2 - B_1)$ is λ_α-negligible. By Theorem 8.1.1, $\pi(\mu_\alpha)$ is regular and gives rise to $\pi(\lambda_\alpha)$.

Next,

$$\pi(\mu)(B) = \mu\big(\pi^{-1}(B)\big) = \sum_{\alpha \in A} \mu_\alpha\big(\pi^{-1}(B)\big) = \sum_\alpha \pi(\mu_\alpha)(B)$$

for all $B \in S'$, so the family $\big(\pi(\mu_\alpha)\big)_{\alpha \in A}$ is summable with sum $\pi(\mu)$. On the other hand, the family $\big(\pi(\lambda_\alpha)\big)_{\alpha \in A}$ of Radon measures is summable with sum $\pi(\lambda)$. The rules of essential integration with respect to these two sums $\pi(\mu)$ and $\pi(\lambda)$ prove, now, that $\pi(\mu)$ is regular and gives rise to $\pi(\lambda)$.

Finally, pass to the general case, in which μ is complex. We have $|\pi(\mu)| \leq \pi(|\mu|)$, so $\pi(\mu)$ is regular. Each $B \in S'$ is essentially $\pi(\lambda)$-integrable, and $\int 1_B \, d\big(\pi(\lambda)\big) = \lambda\big(\pi^{-1}(B)\big) = \pi(\mu)(B)$. The linear forms $f \mapsto \int f \, d\pi(\mu)$ and $f \mapsto \int f \, d\pi(\lambda)$ on $\overline{\mathcal{L}_{\mathbf{C}}^1}\big(\pi(|\mu|)\big) = \overline{\mathcal{L}_{\mathbf{C}}^1}\big(\pi(|\lambda|)\big)$ are continuous and they agree on $St(S', \mathbf{C})$. So they are identical. In particular, $\int f \, d\pi(\mu) = \int f \, d\pi(\lambda)$ for all $f \in \mathcal{H}(\Omega', \mathbf{C})$, which proves that $\pi(\lambda)$ arises from $\pi(\mu)$. □

In the notation of Proposition 22.1.2, suppose that the compact sets in Ω' are $\pi(|\mu|)$-measurable. A mapping f from Ω' into a metrizable space is $\pi(\mu)$-measurable whenever $f \circ \pi$ is μ-measurable. A mapping f from Ω' into a Banach space is essentially $\pi(\mu)$-integrable whenever $f \circ \pi$ is essentially μ-integrable; then $\int f \, d\pi(\mu) = \int f \circ \pi \, d\mu$.

22.2 Baire Sets†

Let Ω be a locally compact Hausdorff space. A subset E of Ω is said to be a G_δ if it is the intersection of a countable family of open sets. An intersection of countably many G_δ sets is itself a G_δ; the union of a finite family of G_δ sets is a G_δ.

Proposition 22.2.1 *If f is a continuous real-valued function on Ω and $c \in$ $\mathbf{R}$, the set $\{x \in \Omega : f(x) \geq c\}$ is a closed G_δ. If K is a compact G_δ, there exists $f \in \mathcal{H}^+(\Omega)$ such that $0 \leq f \leq 1$ and $K = \{x \in \Omega : f(x) = 1\}$.*

PROOF: Let $(V_n)_{n\geq 1}$ be a decreasing sequence of open sets such that $K = \bigcap_{n\geq 1} V_n$, and assume that V_1 has compact closure. For every $n \geq 1$, there exists $f_n \in \mathcal{H}^+$ such that $0 \leq f_n \leq 1$, $f_{n/K} = 1$, and $f_n = 0$ on $\Omega - V_n$. Then $f = \sum_{n\geq 1} 2^{-n} f_n$ has the desired properties. $\quad\square$

Definition 22.2.1 The σ-ring in Ω generated by the class of all compact G_δ is called the class of Baire sets in Ω.

Proposition 22.2.2 *Every compact Baire set is a G_δ.*

PROOF: Let $\mathcal{C}$ be the class of all compact G_δ and denote by Φ the class of those Baire sets E with the following property: there exists a countable subclass $\mathcal{C}_E$ of $\mathcal{C}$ such that E belongs to the σ-ring generated by $\mathcal{C}_E$. Then Φ is a σ-ring and is therefore equal to the class of all Baire sets. Now let K be a compact Baire set. There exists a sequence $(K_n)_{n\geq 1}$ of compact G_δ such that K belongs to the σ-ring generated by $\{K_n : n \geq 1\}$ (Proposition 2.1.5). Moreover, for every $n \geq 1$, there is a function f_n in $\mathcal{H}^+(\Omega)$ such that $0 \leq f_n \leq 1$ and $K_n = \{x : f_n(x) = 1\}$. Letting d be the pseudometric

$$\Omega^2 \ni (x,y) \mapsto \sum_{n\geq 1} 2^{-n} |f_n(x) - f_n(y)|,$$

the relation $x \sim y$ if $d(x,y) = 0$ is an equivalence relation on Ω. The set of classes will be denoted by $\Omega/\sim$ and the canonical projection from Ω into $\Omega/\sim$ by T.

If $T(x_1) = T(y_1)$ and $T(x_2) = T(y_2)$, then

$$d(x_1, x_2) \leq d(x_1, y_1) + d(y_1, y_2) + d(y_2, x_2) = d(y_1, y_2)$$

and, by symmetry,

$$d(y_1, y_2) \leq d(x_1, x_2).$$

---

†This section may be omitted.

Therefore,

$$d(x_1, x_2) = d(y_1 . y_2).$$

It follows that we may define a distance δ on the quotient space $\Omega/\sim$ by $\delta(\varepsilon_1, \varepsilon_2) = d(x_1, x_2)$ if $\varepsilon_1 = T(x_1)$ and $\varepsilon_2 = T(x_2)$. Then T is continuous from Ω onto $\Omega/\sim$.

A subset of Ω is the inverse image (under T) of a subset of $\Omega/\sim$ if and only if, whenever it contains a point x. it also contains the class of x. Each K_n has this property and the class of all inverse images is a σ-ring. Now K belongs to the σ-ring generated by the K_n; hence there exists a subset L of $\Omega/\sim$ such that $K = T^{-1}(L)$. Moreover. $L = T(K)$ is compact and, since every closed subset of a metric space is a G_δ, there exists a sequence of open sets $(V_n)_{n \geq 1}$ in $\Omega/\sim$ with $L = \bigcap_{n \geq 1} V_n$. Then $K = T^{-1}(L) = \bigcap_{n \geq 1} T^{-1}(V_n)$ is a G_δ. $\square$

Proposition 22.2.3 *If C is any collection of open sets that generates the topology of X, then the σ-ring generated by C contains every Baire set.*

PROOF: Let $\mathcal{A}$ be the class of all finite intersections of C-sets. By hypothesis, each open set in Ω is the union of some subclass of $\mathcal{A}$. We need show that $\sigma(C)$ contains every compact G_δ.

Suppose $K \subset U$, where K is compact and U is open. Since $\mathcal{A}$ is a basis for the topology, there exists a finite sequence $V_1, \ldots, V_n$ in $\mathcal{A}$ such that $K \subset V_1 \cup \cdots \cup V_n \subset U$, and then $V_1 \cup \cdots \cup V_n$ lies in $\sigma(C)$. Now, if K is any compact G_δ, let U_n be a sequence of open sets such that $K = \bigcap_n U_n$. Since $K \subset U_n$, there exists $S_n \in \sigma(C)$ such that $K \subset S_n \subset U_n$. Therefore, $K = \bigcap_n S_n$, and K belongs to $\sigma(C)$. $\square$

Proposition 22.2.4 *Let U be open in Ω. For K compact, $K \subset U$, there exist Baire sets V and L such that*

(a) $K \subset V \subset L \subset U$;

(b) V is open and is the union of a sequence of compact G_δ;

(c) L is a compact G_δ.

PROOF: There exists $f \in \mathcal{H}^+(\Omega)$ such that $0 \leq f \leq 1$, $f_{/K} = 1$, and $f = 0$ on $\Omega - U$. Define $V = \{x \in \Omega : f(x) > 1/2\}$ and $L = \{x \in \Omega : f(x) \geq 1/2\}$. Then, since $V = \bigcup_n \{x : f(x) \geq 1/2 + 1/n\}$, it is the union of a sequence of compact G_δ. $\square$

If U is open and $x \in U$, taking $K = \{x\}$ in Proposition 22.2.4, we find an open Baire set V such that $x \in V$ and $V \subset U$. It follows that the class of open Baire sets is a basis for the topology of Ω.

Proposition 22.2.5 *Let Ω' and Ω'' be two locally compact Hausdorff spaces, and let $\mathcal{A}'$, $\mathcal{A}''$ and $\mathcal{A}$ be the σ-rings of Baire sets in Ω', Ω'', and $\Omega' \times \Omega''$. Then $\mathcal{A}$ is the σ-ring generated by $\mathcal{A}' \times \mathcal{A}''$.*

PROOF: If E' and E'' are compact Baire sets in Ω' and Ω'', respectively, then $E' \times E''$ is a compact G_δ, so that $\mathcal{A}' \times \mathcal{A}'' \subset \mathcal{A}$. Now the class of all sets of the form $V' \times V''$, where V' and V'' are open Baire sets, is a basis for the topology. By Proposition 22.2.3, the σ-ring generated by $\mathcal{A}' \times \mathcal{A}''$ contains $\mathcal{A}$. $\qquad\square$

22.3 Product of Regular Measures[†]

Let Ω' and Ω'' be two locally compact Hausdorff spaces, S' and S'' two semi-rings in Ω' and Ω'', respectively, and let μ', μ'' be two regular measures on S', S'' associated with the Radon measures λ', λ''. Write μ for $\mu' \otimes \mu''$ and λ for $\lambda' \otimes \lambda''$.

Proposition 22.3.1 *Suppose that μ' and μ'' are positive. If f is a μ-measurable function from $\Omega = \Omega' \times \Omega''$ into $[0, +\infty]$, then $\int^\bullet f \, d\mu = \int^\bullet f \, d\lambda$.*

PROOF: For all $E' \in S'$ and all $E'' \in S''$, the set $E = E' \times E''$ is essentially λ-integrable and $\lambda(E) = \mu(E)$; hence $\lambda^*(E) = \mu^*(E)$ for all $E \in \tilde{S}$ (Proposition 6.1.1). Each μ-negligible set is locally λ-negligible because it is contained in a μ-negligible $\tilde{S}$-set. Now, if E is μ-measurable and μ-moderate, there exists $F \in \tilde{S}$ with $F \subset E$ such that $E - F$ is μ-negligible; by the preceding, E is λ-measurable and $\lambda^*(E) = \lambda^*(F) = \mu^*(F) = \mu^*(E)$. Next, let f be a μ-measurable and μ-moderate function from Ω into $[0, +\infty]$. For each integer $n \geq 1$, define $f_n = \sum_{1 \leq k \leq n2^n + 1}(k-1)2^{-n}1_{A_{k,n}}$, where

$$A_{k,n} = \begin{cases} f^{-1}\big([(k-1)/2^n, k/2^n[\big) & \text{if } 1 \leq k \leq n \cdot 2^n \\ f^{-1}\big([n, +\infty]\big) & \text{if } k = n \cdot 2^n + 1. \end{cases}$$

Then

$$\int^* f \, d\mu = \sup_n \int^* f_n \, d\mu = \sup_n \int^\bullet f_n \, d\lambda = \int^\bullet f \, d\lambda.$$

Finally, let f be μ-measurable from Ω into $[0, +\infty]$. We have

$$\int^* f 1_E \, d\mu = \int^\bullet f 1_E \, d\lambda \leq \int^\bullet f \, d\lambda$$

[†]This section may be omitted.

for all μ-measurable and μ-moderate sets E, which proves that $\int^\bullet f\,d\mu \le \int^\bullet f\,d\lambda$. On the other hand, if K' and K'' are compact in Ω' and Ω'', respectively, then $K = K' \times K''$ is μ-moderate and $\int^* f1_K\,d\lambda = \int^\bullet f1_K\,d\lambda = \int^* f1_K\,d\mu$ is less than $\int^\bullet f\,d\mu$, whence $\int^\bullet f\,d\lambda \le \int^\bullet f\,d\mu$.   □

Proposition 22.3.2 *Let F be a Banach space. Then $\overline{\mathcal{L}_F^1}(\mu) \subset \overline{\mathcal{L}_F^1}(\lambda)$, and $\int f\,d\mu = \int f\,d\lambda$ for all $f \in \overline{\mathcal{L}_F^1}(\mu)$.*

PROOF: Let $f \in \overline{\mathcal{L}_F^1}(\mu)$. For every $\varepsilon > 0$, there exists $g \in St(S, F)$ such that $\int^\bullet |f - g|\,dV\mu \le \varepsilon$; therefore, $\int^\bullet |f - g|\,dV\lambda \le \varepsilon$ by Proposition 22.3.1; further, $g \in \overline{\mathcal{L}_F^1}(\lambda)$, which proves that $f \in \overline{\mathcal{L}_F^1}(\lambda)$. The linear mappings $f \mapsto \int f\,d\mu$ and $f \mapsto \int f\,d\lambda$ from $\overline{\mathcal{L}_F^1}(\mu)$ into F are continuous and agree on $St(S, F)$. Therefore, they are equal. The result follows.   □

Proposition 22.3.3 *Each $f \in \mathcal{H}^+(\Omega)$ is μ-measurable and, for all compact sets $K \subset \Omega$, $V\mu^*(K) = V\lambda^*(K)$.*

PROOF: The Baire sets in Ω' are μ'-measurable and the Baire sets in Ω'' are μ''-measurable; hence the Baire sets in Ω are μ-measurable (Proposition 22.2.5). If $f \in \mathcal{H}^+(\Omega)$, then $\{x \in \Omega : f(x) \ge c\}$ is a compact G_δ for $c > 0$ and is equal to Ω for $c \le 0$. It follows that f is μ-measurable, and now, by Proposition 22.3.2, $\int f\,dV\lambda = \int f\,dV\mu$. This implies that

$$V\lambda^*(K) = \inf_{\substack{f \in \mathcal{H}^+ \\ f \ge 1_K}} \int f\,dV\lambda = \inf_{\substack{f \in \mathcal{H}^+ \\ f \ge 1_K}} \int f\,dV\mu$$

and $V\lambda^*(K) \ge V\mu^*(K)$ for all compact K. However, by definition of $\int^* f\,dV\mu$,

$$\int^\bullet f\,dV\lambda \le \int^* f\,dV\mu$$

for all functions f from Ω into $[0, +\infty]$. Hence $V\lambda^*(K) \le V\mu^*(K)$.   □

Theorem 22.3.1 *The measure μ is regular if and only if all compact subsets of Ω are μ-measurable. In this case, μ gives rise to λ.*

PROOF: First, assume that all compact subsets of Ω are μ-measurable. Since $U = \overline{U} - (\overline{U} - U)$, each open set U with compact closure is μ-integrable and $V\lambda(U) = V\mu(U)$. For any λ-negligible set B with compact closure and for $\varepsilon > 0$, there exists an open set U with compact closure such that $B \subset U$ and $V\lambda(U) \le \varepsilon$. It follows that B is μ-negligible.

Next, let A be a locally λ-negligible set. If $E \in S$, there is an increasing sequence $(K_n)_{n \geq 1}$ of compact subsets of E such that $E - \bigcup_n K_n$ is locally λ-negligible. But, since

$$
\begin{aligned}
V\mu(E - K_n) &= V\mu(E) - V\mu(K_n) \\
&= V\lambda(E) - V\lambda(K_n) \\
&= V\lambda(E - K_n)
\end{aligned}
$$

converges to 0 as n tends to $+\infty$, $E - \bigcup_n K_n$ is also μ-negligible. On the other hand, since $A \cap K_n$ is λ-negligible for all $n \geq 1$, it is also μ-negligible. This proves that $A \cap E$ is μ-negligible, and finally that A is locally μ-negligible.

Now, let $f \in \overline{\mathcal{L}^1_{\mathbf{C}}}(\lambda)$ and let $(f_n)_{n \geq 1}$ be a sequence in $\mathcal{H}(\Omega, \mathbf{C})$ that converges to f locally λ-a.e. Then $(f_n)_{n \geq 1}$ converges to f locally μ-a.e. and each f_n is μ-measurable; thus f is μ-measurable. We see therefore that $f \in \overline{\mathcal{L}^1_{\mathbf{C}}}(\mu)$, which completes the proof. $\square$

When Ω' and Ω'' are second countable, Ω is second countable, each compact subset of Ω is a G_δ, and μ is regular.

22.4   Change of Variable Formula

To clarify ideas, we give one application of Proposition 22.1.2.

We retain the notation of Section 11.3, and prove the following result.

Proposition 22.4.1 *Suppose that $G(a+)$ and $G(b-)$ exist in $\bar{\mathbf{R}}$. Let $J = \langle G(a+), G(b-) \rangle$ be an interval included in $G(I)$, with endpoints $G(a+)$ and $G(b-)$. If $f : G(I) \to F$ is such that $(f \circ G)g$ is λ-integrable, then $f 1_J$ is ν-integrable, and $\int (f \circ G)g \, d\lambda$ is equal to $\int f 1_J \, d\nu$ or to $- \int f 1_J \, d\nu$ as $G(a+) \leq G(b-)$ or $G(a+) \geq G(b-)$.*

PROOF: Fix $x_0 \in I$. First, assume that the supremum M for $G_{/[x_0,b)}$ is not attained, so that $G(b-) = M$. Let c be a point in $[x_0, b)$ such that $G(\beta) \geq G(x_0)$ for all $\beta \in [c, b)$. For every $\beta \in [c, b)$, the function $(|f| \circ G)g 1_{[x_0, \beta]}$ is λ-integrable; hence $|f| 1_{[G(x_0), G(\beta)]}$ is ν-integrable and

$$
\int |f| 1_{[G(x_0), G(\beta)]} \, d\nu = \int (|f| \circ G)g 1_{[x_0, \beta]} \, d\lambda
$$

(to see this, replace I by $[x_0, \beta]$ in Proposition 11.3.1, and use the remark just following Proposition 22.1.2). Letting β converge to b, we see that $|f| 1_{[G(x_0), G(b-))}$ is ν-integrable. For all $\beta \in [c, b)$, the function $f 1_{[G(x_0), G(\beta)]}$ is dominated (in absolute value) by $|f| 1_{[G(x_0), G(b-))}$, so $\int f 1_{[G(x_0), G(\beta)]} \, d\nu = \int (f \circ G)g 1_{[x_0, \beta]} \, d\lambda$ converges to $\int f 1_{[G(x_0), G(b-))} \, d\nu$, whence

$$
\int f 1_{[G(x_0), G(b-))} \, d\nu = \int (f \circ G)g 1_{[x_0, b)} \, d\lambda.
$$

Similarly, if the infimum m for $G_{/[x_0,b\rangle}$ is not attained, then $f1_{\langle G(b-),G(x_0)]}$ is ν-integrable and

$$\int (f \circ G)g1_{[x_0,b\rangle}\,d\lambda = -\int f1_{\langle G(b-),G(x_0)]}\,d\nu.$$

Next, assume that the bounds m and M for $G_{/[x_0,b\rangle}$ are attained at points x_1, x_2, and let $(b_n)_{n\geq 1}$ be a sequence in $[x_0,b\rangle$ converging to b. For each $n \in \mathbf{N}$, $f1_{[G(x_0),G(b_n)]}$ is ν-integrable and dominated (in absolute value) by $|f|1_{[G(x_1),G(x_2)]}$. Thus $\int f1_{[G(x_0),G(b_n)]}\,d\nu$ converges to $\int f1_{[G(x_0),G(b-)\rangle}\,d\nu$ as $n \to \infty$. If $G(x_0) \leq G(b-)$, the sequence of the

$$\theta_n = \int (f \circ G)g1_{[x_0,b_n]}\,d\lambda = \pm\int f1_{[G(x_0),G(b_n)]}\,d\nu$$

has the same limit as the sequence of the $\int f1_{[G(x_0),G(b_n)\rangle}\,d\nu$, hence

$$\int (f \circ G)g1_{[x_0,b\rangle}\,d\lambda = +\int f1_{[G(x_0),G(b-)\rangle}\,d\nu.$$

On the other hand, if $G(b-) \leq G(x_0)$, then

$$\int (f \circ G)g1_{[x_0,b\rangle}\,d\lambda = -\int f1_{\langle G(b-),G(x_0)\rangle}\,d\nu.$$

Replacing $[x_0,b\rangle$ by $\langle a,x_0]$ and arguing as above, we obtain similar results.

Now the proposition is obvious if two or more of $G(x_0)$. $G(a+)$, and $G(b-)$ are equal. Therefore, we may suppose that they are all distinct. Define real numbers ε, ε', and ε'' as follows: $\varepsilon = +1$ if $G(a+) < G(b-)$ and $\varepsilon = -1$ if $G(a+) > G(b-)$; $\varepsilon' = +1$ if $G(a+) < G(x_0)$ and $\varepsilon' = -1$ if $G(a+) > G(x_0)$; $\varepsilon'' = +1$ if $G(x_0) < G(b-)$ and $\varepsilon'' = -1$ if $G(x_0) > G(b-)$. Then

$$\varepsilon 1_{\langle G(a+),G(b-)\rangle} = \varepsilon' 1_{\langle G(a+),G(x_0)]} + \varepsilon'' 1_{[G(x_0),G(b-)\rangle}$$

ν-almost everywhere, so $f1_{\langle G(a+),G(b-)\rangle}$ is ν-integrable, and

$$\int (f \circ G)g\,d\mu = \varepsilon\int f1_{\langle G(a+),G(b-)\rangle}\,d\nu.$$

$\square$

Proposition 22.4.1 is a general "change of variable" formula.

23

Haar Measures

The traditional theory of Fourier series and Fourier integrals is a particular case of harmonic analysis with respect to a Gelfand pair. The latter theory, as well as the neighboring one of group representations, starts out with the notions of Haar measure and convolution of measures on a locally compact group, that we introduce now.

Summary

23.1   In this section we define invariant, relatively invariant, and quasi-invariant measures on a locally compact group.

23.2   On every locally compact group there exists a left-invariant measure. This measure is unique, up to a multiplicative constant (Theorem 23.2.1).

23.3   We define the modular function on a locally compact group.

23.4   Quasi-invariant measures on a locally compact group are all equivalent (Proposition 23.4.3).

23.5   Let G be a locally compact group and H a closed subgroup of G. With every measure λ on the homogeneous space G/H we can associate a measure $\lambda^\sharp$ on G (Theorem 23.5.2) that possesses interesting properties.

23.6   We study integration with respect to $\lambda^\sharp$ (Theorem 23.6.1 and Proposition 23.6.1).

23.7   We reconstitute λ from $\lambda^\sharp$ (Proposition 23.7.4).

23.8   There is only one class of quasi-invariant measures on G/H (Theorems 23.8.1 and 23.8.2).

23.9   In contrast, invariant measures exist on G/H only if the modular functions Δ_G and Δ_H of G and H coincide on H (Theorem 23.9.2).

23.10   By means of Euler angles we describe the invariant measure on the group $SO(n+1, \mathbf{R})$ of rotations of $\mathbf{R}^{n+1}$ (Proposition 23.10.3).

23.11 We describe Haar measure on the group $SH(n + 1, \mathbf{R})$ of hyperbolic rotations of $\mathbf{R}^{n+1}$.

In this chapter and the following one, all locally compact spaces will be taken to be Hausdorff, unless otherwise stated.

23.1 Invariant Measures

Let G be a topological group with unit e and X a locally compact space. A left action of G on X is a continuous mapping $(g, x) \mapsto gx$ from $G \times X$ into X such that

(a) $ex = x$ for all $x \in X$;

(b) $(st)x = s(tx)$ for all $s, t \in G$ and all $x \in X$.

Then G is said to act on X from the left. The action is said to be transitive whenever, for all $x, y \in X$, there exists $s \in G$ such that $y = sx$.

Denote by $\gamma(s)$ the mapping $x \mapsto sx$ from X into X. We have $\gamma(st) = \gamma(s) \circ \gamma(t)$; in particular, for every $s \in G$. $\gamma(s)$ is a homeomorphism from X onto X.

For every mapping f from X into a set Y, we put $\gamma(s)f = f \circ \gamma(s^{-1})$, so that $(\gamma(s)f)(\gamma(s)x) = f(x)$ for all $x \in X$. Also. if μ is a Radon measure on X, we shall denote by $\gamma(s)\mu$ the image measure of μ under $\gamma(s)$, so that

$$\langle f, \gamma(s)\mu \rangle = \langle \gamma(s^{-1})f, \mu \rangle \qquad \text{for all } f \in \mathcal{H}(X, \mathbf{C})$$

or

$$\int f(x)\, d(\gamma(s)\mu) = \int f(sx)\, d\mu(x) \qquad (1).$$

Instead of $d(\gamma(s)\mu)$ we shall sometimes write $d\mu(s^{-1}x)$; then (1) reads

$$\int f(x)\, d\mu(s^{-1}x) = \int f(sx)\, d\mu(x).$$

If A is a $\gamma(s)\mu$-integrable set, then $s^{-1}A = \{s^{-1}a : a \in A\}$ is μ-integrable and $(\gamma(s)\mu)(A) = \mu(s^{-1}A)$.

Definition 23.1.1 Let μ be a Radon measure on X. μ is said to be

(a) invariant under G (i.e., under the action of G) if $\gamma(s)\mu = \mu$ for every $s \in G$;

(b) relatively invariant under G if $\gamma(s)\mu$ is proportional to μ, for every $s \in G$;

(c) quasi-invariant under G if $\gamma(s)\mu$ is equivalent to μ, for every $s \in G$ (Chapter 20).

If μ is invariant, then $|\mu|$, $\mathrm{Re}(\mu)$, and $\mathrm{Im}(\mu)$ are invariant. If μ is a real invariant measure, then $\mu^+ = (1/2)(|\mu| + \mu)$ and $\mu^- = (1/2)(|\mu| - \mu)$ are invariant.

Assume that $\mu \neq 0$ is relatively invariant. For every $s \in G$ there exists a unique $\chi(s) \in \mathbf{C}^*$ such that $\gamma(s)\mu = \chi(s^{-1})\mu$, and the function χ is a group homomorphism from G into $\mathbf{C}^*$, called the multiplier of μ.

Formula (1) then becomes

$$\int f(sx)\, d\mu(x) = \chi(s)^{-1} \int f(x)\, d\mu(x) \quad (2),$$

and $\mu(sA) = \chi(s)\mu(A)$ for every μ-integrable set A. The relation $\gamma(s)\mu = \chi(s)^{-1}\mu$ can also be written $d\mu(sx) = \chi(s)\, d\mu(x)$.

Since $|\gamma(s)\mu| = \gamma(s)(|\mu|)$, a measure μ on X is quasi-invariant if and only if $|\mu|$ is quasi-invariant. This amounts to saying that the class of μ is invariant under G. A measure μ on X is quasi-invariant if and only if the collection of all locally μ-negligible subsets of X is invariant under G, or, equivalently, if sK is μ-negligible for every μ-negligible compact subset of X and every $s \in G$ (Theorem 20.3.1). If μ is quasi-invariant under G, then the support of μ is invariant under G. In particular, whenever G acts transitively on X, this support is either empty (for $\mu = 0$) or else equal to X.

Let X, Y, and Z be three topological spaces. If $u : (x, y) \mapsto u(x, y)$ is a continuous mapping from $X \times Y$ into Z, then, for every $x \in X$, we shall denote by u_x the mapping from Y into Z defined by $u_x(y) = u(x, y)$.

Proposition 23.1.1 *Let X, Y, and Z be three topological spaces. Assume that Y is locally compact and let $u : X \times Y \to Z$ be continuous. Furthermore, let f be a continuous mapping from Z into a Banach space, with support S, and let μ be a measure on Y.*

If, for every $x_0 \in X$, there exists a neighborhood V of x_0 in X such that $\bigcup_{x \in V} u_x^{-1}(S)$ has compact closure in Y, then

(a) for every $x \in X$, $f \circ u_x$ is continuous with compact support in Y,

(b) $x \mapsto \int_Y f(u(x, y))\, d\mu(y)$ is continuous on X.

PROOF: The assertion (a) is obvious. We prove (b): since continuity is a local property, we may suppose that $\bigcup_{x \in X} u_x^{-1}(S)$ is contained in a compact subset Y' of Y. Since the function $(x, y) \mapsto f(u(x, y))$ is continuous on $X \times Y$, $f \circ u_x$ converges to $f \circ u_{x_0}$, uniformly on Y', as x tends to x_0; thus $\mu(f \circ u_x)$ tends to $\mu(f \circ u_{x_0})$. $\qquad \square$

We retain the above notation.

Proposition 23.1.2 *Assume that G is a locally compact group. Let $\mu \neq 0$ be a relatively invariant measure on X. Then its multiplier χ is a continuous function on G.*

PROOF: Let $f \in \mathcal{H}(X, \mathbf{C})$, S be the support of f, s_0 a point of G, and V a compact neighborhood of s_0 in G; then $\bigcup_{s \in V} \gamma(s)^{-1}(S) = V^{-1}S$ is compact in X. By Proposition 23.1.1 and formula (2), $\chi(s)^{-1}\langle f, \mu \rangle$ depends continuously on s. If we choose f so that $\langle f, \mu \rangle \neq 0$, we see that χ is continuous. $\square$

Now let G be a topological group acting on a locally compact space X via the continuous operation $(x, s) \mapsto xs$ from $X \times G$ into X, with $x(st) = (xs)t$ for all s, $t \in G$, $x \in X$, and $xe = x$ for all $x \in X$. Denote by $\delta(s)$ the homeomorphism $x \mapsto xs^{-1}$ from X onto X. For every mapping f from X into a set Y, we put $\delta(s)f = f \circ \delta(s^{-1})$, so that $(\delta(s)f)(\delta(s)x) = f(x)$ for all $x \in X$. Also, if μ is a Radon measure on X, we shall denote by $\delta(s)\mu$ the image measure of μ under $\delta(s)$, so that

$$\langle f, \delta(s)\mu \rangle = \langle \delta(s^{-1})f, \mu \rangle \qquad \text{for all } f \in \mathcal{H}(X, \mathbf{C})$$

or

$$\int f(x) \, d(\delta(s)\mu)(x) = \int f(xs^{-1}) \, d\mu(x) \qquad (3).$$

Instead of $d(\delta(s)\mu)$, we shall sometimes write $d\mu(xs)$; then (3) reads

$$\int f(x) \, d\mu(xs) = \int f(xs^{-1}) \, d\mu(x).$$

If A is a $\delta(s)\mu$-integrable set, then $As = \{as : a \in A\}$ is μ-integrable and $(\delta(s)\mu)(A) = \mu(As)$.

Invariant, relatively invariant, and quasi-invariant measures (under G) on X are defined in a similar way. If $\mu \neq 0$ is relatively invariant on X, we define its multiplier χ by the formula $\delta(s)\mu = \chi(s)\mu$ for all $s \in G$; then (3) becomes

$$\int f(xs) \, d\mu(x) = \chi(s^{-1}) \int f(x) \, d\mu(x),$$

and $\mu(As) = \chi(s)\mu(A)$ for every μ-integrable set A. The relation $\delta(s)\mu = \chi(s)\mu$ can also be written $d\mu(xs) = \chi(s) \, d\mu(x)$.

Let G° be the topological group that has the same underlying set as G, the same topology, and whose multiplication is given by $(s, t) \mapsto ts$ ("the opposite group of G"). Now consider G° as operating on X by $(s, x) \mapsto xs$. If $\mu \neq 0$ is relatively invariant under G (operating on X from the right), then it is relatively invariant under G° (operating on X from the left), with the same multiplier χ.

Finally, let G be a locally compact group. It acts on itself by left translations and right translations, according to the formulas $\gamma(s)x = sx$ and $\delta(s)x = xs^{-1}$. We have

$$\gamma(s)\delta(t) = \delta(t)\gamma(s) \qquad (4).$$

If we apply the preceding remarks, now, we have on G the notions of left invariant measures, right invariant measures, measures relatively invariant with respect to left (or right) translations... (however, see Section 23.4).

For every mapping f from G into a set Y, we define $\check{f}$ by $\check{f}(x) = f(x^{-1})$. Also, for every Radon measure μ on G, $\check{\mu}$ is the image measure of μ under the homeomorphism $x \mapsto x^{-1}$. Hence $\check{\mu}(f) = \mu(\check{f})$ for $f \in \mathcal{H}(G, \mathbf{C})$. In other words,

$$\int f(x)\, d\check{\mu}(x) = \int f(x^{-1})\, d\mu(x) \qquad (5).$$

If we agree to write $d\mu(x^{-1})$ instead of $d\check{\mu}(x)$, then (5) becomes

$$\int f(x)\, d\mu(x^{-1}) = \int f(x^{-1})\, d\mu(x).$$

23.2   Existence and Uniqueness of Left Haar Measure

Definition 23.2.1 Let G be a locally compact group. We call left Haar measure on G (respectively, right Haar measure on G) any positive nonzero Radon measure on G that is left invariant (respectively, right invariant).

Theorem 23.2.1 *On every locally compact group G, there exists a left Haar measure (respectively, a right Haar measure), which is unique up to a multiplicative constant.*

PROOF:

A. Existence

Designate by $\mathcal{H}_+$ the set of continuous functions from G into $[0, +\infty[$, with compact support, and put $\mathcal{H}_+^* = \mathcal{H}_+ - \{0\}$. If K is a compact subset of G, we denote by $\mathcal{H}_+^*(K)$ the set of those $f \in \mathcal{H}_+^*$ whose support is contained in K.

For $f \in \mathcal{H}_+$ and $g \in \mathcal{H}_+^*$, there exist numbers $c_1 \geq 0, \ldots, c_n \geq 0$ and elements $s_1, \ldots, s_n$ of G such that $f \leq \sum_{1 \leq i \leq n} c_i \gamma(s_i) g$; indeed, there exists a nonempty open subset U of G such that $\inf_{s \in U} g(s) > 0$, and the support of f can be covered by finitely many sets $\gamma(s)U$. Then, writing $(f : g)$ for the infimum of the numbers $\sum_{1 \leq i \leq n} c_i$, for all systems $(c_1, \ldots, c_n, s_1, \ldots, s_n)$ of positive numbers and elements of G such that $f \leq \sum_{1 \leq i \leq n} c_i \gamma(s_i) g$, we have

(a) $(\gamma(s)f : g) = (f : g)$ for $f \in \mathcal{H}_+, g \in \mathcal{H}_+^*, s \in G$;

(b) $(\lambda f : g) = \lambda(f : g)$ for $f \in \mathcal{H}_+, g \in \mathcal{H}_+^*, \lambda \geq 0$;

(c) $((f + f') : g) \leq (f : g) + (f' : g)$ for $f \in \mathcal{H}_+, f' \in \mathcal{H}_+, g \in \mathcal{H}_+^*$;

(d) $(f : g) \geq \sup f / \sup g$ for $f \in \mathcal{H}_+, g \in \mathcal{H}_+^*$;

(e) $(f : h) \leq (f : g)(g : h)$ for $f \in \mathcal{H}_+, g \in \mathcal{H}_+^*, h \in \mathcal{H}_+^*$;

(f) $0 < 1/(f_0 : f) \leq (f : g)/(f_0 : g) \leq (f : f_0)$ for f, f_0, g in $\mathcal{H}_+^*$;

(g) given f, f', h in $\mathcal{H}_+$, with $h(s) \geq 1$ on the support of $f + f'$, and given $\varepsilon > 0$, there exists a compact neigborhood V of e such that, for every $g \in \mathcal{H}_+^*(V)$,

$$(f : g) + (f' : g) \leq ((f + f') : g) + \varepsilon(h : g).$$

Properties (a), (b), and (c) are clear. Let $f \in \mathcal{H}_+$ and $g \in \mathcal{H}_+^*$; if

$$f \leq \sum_{1 \leq i \leq n} c_i \gamma(s_i) g$$

with $c_i \geq 0$, we have

$$\sup(f) \leq \sum_{1 \leq i \leq n} c_i g(s_i^{-1} s)$$

for an $s \in G$; thus

$$\sup(f) \leq \left(\sum_{1 \leq i \leq n} c_i \right) \sup(g),$$

whence (d). For (e), let $f \in \mathcal{H}_+$, $g \in \mathcal{H}_+^*$, $h \in \mathcal{H}_+^*$; if

$$f \leq \sum_{1 \leq i \leq n} c_i \gamma(s_i) g \quad \text{and} \quad g \leq \sum_{1 \leq j \leq p} d_j \gamma(t_j) h$$

(with $c_i \geq 0$, $d_j \geq 0$, s_i, t_j in G), then

$$f \leq \sum_{i,j} c_i d_j \gamma(s_i t_j) h,$$

and so

$$(f : h) \leq \sum_{i,j} c_i d_j = \left(\sum_i c_i \right) \left(\sum_j d_j \right):$$

therefore,

$$(f : h) \leq (f : g)(g : h).$$

If we apply (e) to f_0, f, g and to f, f_0, g, we obtain (f). Finally, let f, f', h be in $\mathcal{H}_+$, with $h(s) \geq 1$ on the support of $f + f'$, and let $\varepsilon > 0$ be given. Put $F = f + f' + (1/2)\varepsilon h$; the functions φ and φ' that agree with f/F and f'/F, respectively, on the support of $f + f'$, and that vanish outside this support, belong to $\mathcal{H}_+$. For every $\eta > 0$, there exists a compact neighborhood V of e such that $|\varphi(s) - \varphi(t)| \leq \eta$ and $|\varphi'(s) - \varphi'(t)| \leq \eta$ for $s^{-1}t \in V$. Then let g be an element of $\mathcal{H}_+^*(V)$; for all $s \in G$, we have $\varphi \cdot \gamma(s)g \leq (\varphi(s) + \eta)\gamma(s)g$. Indeed, at points where $\gamma(s)g$ vanishes, this is clear; thus this inequality holds outside sV; in sV, $\varphi \leq \varphi(s) + \eta$. Similarly, $\varphi' \cdot \gamma(s)g \leq (\varphi'(s) + \eta)\gamma(s)g$. Now, if we let $c_1, \ldots, c_n$ be positive numbers and $s_1, \ldots, s_n$ be elements of G such that $F \leq \sum_{1 \leq i \leq n} c_i \gamma(s_i)g$, we see that

$$f = \varphi F \leq \sum_{1 \leq i \leq n} c_i \varphi \gamma(s_i) g \leq \sum_{1 \leq i \leq n} c_i (\varphi(s_i) + \eta) \gamma(s_i) g$$

and similarly for f'; hence

$$(f : g) + (f' : g) \le \sum_{1 \le i \le n} c_i (\varphi(s_i) + \varphi'(s_i) + 2\eta) \le (1 + 2\eta) \sum_{1 \le i \le n} c_i,$$

since $\varphi + \varphi' \le 1$. Applying the definition of F and, next, (b), (c), and (e), we conclude that

$$
\begin{aligned}
(f : g) + (f' : g) &\le (1 + 2\eta)(F : g) \\
&\le (1 + 2\eta)\left(((f + f') : g) + \frac{1}{2}\varepsilon(h : g)\right) \\
&\le ((f + f') : g) + \frac{1}{2}\varepsilon(h : g) \\
&\quad + 2\eta((f + f') : h)(h : g) + \varepsilon\eta(h : g).
\end{aligned}
$$

Furthermore, if we have chosen η so that $\eta[2((f + f') : h) + \varepsilon] \le (1/2)\varepsilon$, we obtain (g).

When V runs through the set of compact neighborhoods of e, the sets $\mathcal{H}_+^*(V)$ form a filter basis $\mathcal{B}$ on $\mathcal{H}_+^*$. Let $\mathcal{F}$ be an ultrafilter on $\mathcal{H}_+^*$ finer than $\mathcal{B}$. Also, fix $f_0 \in \mathcal{H}_+^*$ and, for all $f \in \mathcal{H}_1^*$, $g \in \mathcal{H}_+^*$, put

$$I_g(f) = \frac{(f : g)}{(f_0 : g)}.$$

By (f), $\lim_{\mathcal{F}}[g \mapsto I_g(f)] = I(f)$ exists in the compact space

$$\left[\frac{1}{(f_0 : f)}, (f : f_0)\right].$$

By (c), we have $I(f + f') \le I(f) + I(f')$. By (g), we have $I(f) + I(f') \le I(f + f') + \varepsilon I(h)$ for every $\varepsilon > 0$, if h is greater than 1 on the support of $f + f'$. It follows that $I(f + f') = I(f) + I(f')$. By Proposition 1.3.3, I can be extended to a linear form on $\mathcal{H}(G, \mathbf{R})$; this linear form is a positive nonzero Radon measure on G, and, by (a), is left invariant. This is the desired left Haar measure. Applying the foregoing analysis to the opposite group, we deduce the existence of a right Haar measure.

B. Uniqueness

Let μ be a left Haar measure and let ν be a right Haar measure. Then $\check{\nu}$ is left invariant. We show that μ and $\check{\nu}$ are proportional. This will prove, in fact, that two left Haar measures are proportional.

Let $f \in \mathcal{H}_+$ be such that $\mu(f) \ne 0$. By Proposition 23.1.1, the function D_f defined on G by

$$D_f(s) = \mu(f)^{-1} \int f(t^{-1}s)d\nu(t) \quad (1)$$

is continuous. Choose $g \in \mathcal{H}_+$. The function $(s,t) \mapsto f(s)g(ts)$ is continuous on $G \times G$, with compact support. Hence

$$
\begin{aligned}
\mu(f)\nu(g) &= \left(\int f(s)\, d\mu(s) \right) \left(\int g(t)\, d\nu(t) \right) \\
&= \int d\mu(s) \int f(s)g(ts)\, d\nu(t) \\
&= \int d\nu(t) \int f(s)g(ts)\, d\mu(s) \\
&= \int d\nu(t) \int f(t^{-1}s)g(s)\, d\mu(s) \\
&= \int g(s) \left(\int f(t^{-1}s)\, d\nu(t) \right) d\mu(s) \\
&= \mu(g\mu(f)D_f),
\end{aligned}
$$

and $\nu(g) = (D_f\mu)g$. This proves that D_f does not depend on f. For, if $f' \in \mathcal{H}_+$ is such that $\mu(f') \neq 0$, then $D_f\mu = D_{f'}\mu$, and $D_f = D_{f'}$ locally μ-almost everywhere; finally, $D_f = D_{f'}$ everywhere, because D_f and $D_{f'}$ are continuous and the support of μ is G. Therefore, put $D_f = D$. Formula (1) gives $\mu(f)D(e) = \check{\nu}(f)$. Next, if $f \in \mathcal{H}^+$ is such that $\mu(f) = 0$, then $f = 0$ everywhere, and the formula $\mu(f)D(e) = \check{\nu}(f)$ remains valid. Since $\check{\nu} \neq 0$, we have $D(e) \neq 0$. This proves that μ and $\check{\nu}$ are proportional.   □

Corollary *Every left invariant (respectively, right invariant) complex measure on G is proportional to any left Haar measure (respectively, to any right Haar measure).*

In the above proof, the axiom of choice was used. It is possible, however, to construct left Haar measure on G without resorting to this axiom (see H. Cartan's proof in *Compte-rendu de l'Académie des Sciences de Paris*, 1940).

On the additive group $\mathbf{R}^k$, Lebesgue measure λ_k is a Haar measure. On the multiplicative group $\mathbf{R}^*_+$, $d\lambda(x)/x$ is a Haar measure; moreover, the homeomorphism $x \mapsto \log x$ from $\mathbf{R}^*_+$ onto $\mathbf{R}$ transforms $d\lambda(x)/x$ into λ. On the torus $\mathbf{T}$, the measure ν, the image of $\lambda_{/[0,1]}$ under $x \mapsto e^{2i\pi x}$, is a Haar measure.

Theorem 23.2.2 *Let G be a locally compact group, and let μ be a left Haar measure or a right Haar measure on G. G is discrete if and only if $\mu(\{e\}) > 0$. G is compact if and only if $\mu^*(G) < +\infty$.*

PROOF: The conditions are clearly necessary. For sufficiency, let V be a compact neighborhood of e. If $\mu(\{e\}) > 0$, V is a finite set, because $\mu(V) < +\infty$; as G is Hausdorff, it is discrete. Now suppose that $\mu^*(G) < +\infty$. We may assume that μ is left invariant. Consider the collection $\mathcal{F}$ of finite subsets $\{s_1, \ldots, s_n\}$ of G such that $s_i V \cap s_j V = \emptyset$ for $i \neq j$; we have

$$
n\mu(V) = \mu(s_1 V \cup \cdots \cup s_n V) \leq \mu^*(G).
$$

hence

$$n \leq \frac{\mu^*(G)}{\mu(V)}.$$

We can, therefore, choose a maximal element $\{s_1, \ldots, s_n\}$ in $\mathcal{F}$. For every $s \in G$, there is an index i such that $sV \cap s_iV \neq \emptyset$, and therefore such that s lies in s_iVV^{-1}. Then G is union of the compact sets s_iVV^{-1}, and is compact.
$\square$

23.3   Modular Function on G

Let μ be a left Haar measure on G. For every $s \in G$, $\delta(s)\mu$ is still left invariant, because $\gamma(t)\delta(s) = \delta(s)\gamma(t)$ for all $t \in G$. So there exists a unique number $\Delta_G(s) > 0$ such that $\delta(s)\mu = \Delta_G(s)\mu$. By Theorem 23.2.1, the number $\Delta_G(s)$ does not depend on the choice of μ.

Definition 23.3.1 The function $\Delta_G : s \mapsto \Delta_G(s)$ is called the modular function on G. If $\Delta_G = 1$, the group G is said to be unimodular.

Notice that μ is relatively invariant under right translations, with multiplier Δ_G. Thus Δ_G is a continuous homomorphism from G into $\mathbf{R}_+^*$. In particular, $\Delta_G(sts^{-1}t^{-1}) = \Delta_G(e) = 1$ for all $s, t \in G$; therefore, if the subgroup G' of G generated by the commutators is dense in G, then G is unimodular; this shows that any connected semisimple Lie group is unimodular.

If φ is an isomorphism from G onto a locally compact group G', $\Delta_{G'} \circ \varphi = \Delta_G$. In particular, $\Delta_G \circ \varphi = \Delta_G$ if φ is an automorphism of G. Henceforth, we set $\Delta = \Delta_G$. Since

$$\delta(s)(\Delta^{-1}\mu) = \left(\delta(s)\Delta^{-1}\right)\left(\delta(s)\mu\right) = \left(\Delta(s)^{-1}\Delta^{-1}\right)\left(\Delta(s)\mu\right) = \Delta^{-1}\mu$$

for all $s \in G$, $\Delta^{-1}\mu = \mu'$ is a right Haar measure. But

$$\gamma(s)\mu' = \left(\gamma(s)\Delta^{-1}\right)\mu = \Delta(s)(\Delta^{-1}\mu) = \Delta(s)\mu'.$$

Hence, for every right Haar measure ν, $\gamma(s)\nu = \Delta(s)\nu$. Since $\check{\mu}$ is a right Haar measure, $\check{\mu} = a\Delta^{-1}\mu$ for some $a > 0$. We deduce that $\mu = a(\Delta^{-1}\mu)\check{} = a\Delta\check{\mu} = a^2\mu$; hence $a = 1$, and finally $\check{\mu} = \Delta^{-1}\mu$. Similarly, $\check{\nu} = \Delta\nu$. In short, we have the following result.

Theorem 23.3.1 Let G be a locally compact group, Δ its modular function, μ a left Haar measure, and ν a right Haar measure. Then $\delta(s)\mu = \Delta(s)\mu$ and $\gamma(s)\nu = \Delta(s)\nu$ for every $s \in G$. Moreover, $\check{\mu} = \Delta^{-1}\mu$ and $\check{\nu} = \Delta\nu$.

For any μ-integrable mapping f from G into a real Banach space, its left translates and right translates are μ-integrable, $\int f(sx) \, d\mu(x) = \int f(x) \, d\mu(x)$, and $\int f(xs) \, d\mu(x) = \Delta(s)^{-1} \int f(x) \, d\mu(x)$. Moreover. $\check{f}$ is $\Delta^{-1}\mu$-integrable and $\int f(x^{-1})\Delta^{-1}(x) \, d\mu(x) = \int f(x) \, d\mu(x)$.

Likewise, for any ν-integrable mapping f from G into a real Banach space, its left translates and right translates are ν-integrable. $\int f(xs) \, d\nu(x) = \int f(x) \, d\nu(x)$, and $\int f(sx) \, d\nu(x) = \Delta(s) \int f(x) \, d\nu(x)$. Moreover, $\check{f}$ is $\Delta\nu$-integrable and $\int f(x^{-1})\Delta(x) \, d\nu(x) = \int f(x) \, d\nu(x)$.

Proposition 23.3.1 *If there exists in G a compact neighborhood V of e that is invariant under inner automorphisms, then G is unimodular.*

PROOF: Let μ be a left Haar measure on G. For every $s \in G$,

$$\mu(V) = \mu(s^{-1}Vs) = \Delta(s)\mu(V).$$

whence it follows that $\Delta(s) = 1$. $\qquad\square$

As a consequence, if G is commutative, or discrete. or compact, it is unimodular. If G is discrete, the measure on G for which each point has mass 1 is plainly a Haar measure. called the normalized Haar measure on G. If G is compact, there exists one and only one Haar measure on G such that $\mu(G) = 1$; this measure is the normalized Haar measure on G. These two conventions do not agree when G is finite: in this case, we shall always specify the measure that we call "normalized Haar measure".

23.4   Relatively Invariant Measures on a Group

Let G be a locally compact group.

Proposition 23.4.1 *Let $\mu \neq 0$ be a measure on G. relatively invariant under left translations, with multiplier χ. If χ_1 is a continuous homomorphism from G into $\mathbf{C}^*$, then the measure $\chi_1\mu$ is relatively invariant under left translations, with multiplier $\chi_1\chi$.*

PROOF:

$$
\begin{aligned}
\gamma(s)(\chi_1\mu) &= (\gamma(s)\chi_1)(\gamma(s)\mu) \\
&= (\chi_1(s^{-1})\chi_1)\chi(s^{-1})\mu \\
&= (\chi_1\chi)(s^{-1})(\chi_1\mu).
\end{aligned}
$$

$\qquad\square$

Theorem 23.4.1 *Let μ be a left Haar measure on G. A measure $\nu \neq 0$ on G is relatively invariant under left translations if and only if it is of the form $a\chi\mu$, where $a \in \mathbf{C}^*$ and χ is a continuous homomorphism from G into $\mathbf{C}^*$. In this case, its multiplier is χ.*

PROOF: By Proposition 23.4.1, the condition is sufficient. On the other hand, if $\nu \neq 0$ is relatively invariant under left translations, with multiplier χ, then $\chi^{-1}\nu$ is left invariant (Proposition 23.4.1) and hence is of the form $a\mu$ for some $a \in \mathbf{C}^*$. □

Proposition 23.4.2 *Every measure on G that is relatively invariant under left translations is also relatively invariant under right translations.*

PROOF: In the notation of Theorem 23.4.1,

$$\delta(s)(\chi\mu) = (\delta(s)\chi)(\delta(s)\mu) = (\chi(s)\chi)(\Delta(s)\mu) = (\chi\Delta)(s)(\chi\mu)$$

for all $s \in G$. □

On account of Proposition 23.4.2, we may refer unambiguously to relatively invariant measures on G. Given a relatively invariant measure $\nu \neq 0$ on G, we must distinguish its left multiplier χ and its right multiplier χ', defined by $\gamma(s)\nu = \chi(s^{-1})\nu$ and $\delta(s)\nu = \chi'(s)\nu$. We have $\chi' = \chi\Delta$, by Proposition 23.4.2. In the notation of Theorem 23.4.1, $\check{\nu} = a(\chi\mu)\check{} = a\check{\chi}\check{\mu} = a(\chi^{-1}\Delta^{-1})\mu$; therefore, $\check{\nu}$ is relatively invariant, with left multiplier $\chi^{-1}\Delta^{-1}$ and with right multiplier χ^{-1}.

By Theorem 23.4.1, negligible, locally negligible, measurable, and locally integrable functions are the same for all nonzero relatively invariant measures.

Proposition 23.4.3 *Let μ be a left Haar measure on G. In order that a measure $\nu \neq 0$ on G be quasi-invariant under left translations, it is necessary and sufficient that it be equivalent to μ.*

PROOF: Sufficiency is obvious. For necessity, we let $\nu \neq 0$ be a measure on G, quasi-invariant under left translations, and show that ν is equivalent to μ. We need only consider the case where ν is positive.

Let B be a compact subset of G. By Theorem 20.3.1, it suffices to show that the conditions $\mu(B) = 0$ and $\nu(B) = 0$ are equivalent.

If A is a compact subset of G, the function $(x, y) \mapsto 1_A(x)1_B(xy)$ on $G \times G$ is $\mu \otimes \nu$-integrable, because it is upper semicontinuous with support contained in $A \times A^{-1}B$. Hence,

$$\int d\nu(y) \int 1_A(x)1_B(xy)\, d\mu(x) = \int d\mu(x)1_A(x) \int 1_B(xy)\, d\nu(y) \qquad (1).$$

First, assume that $\nu(B) = 0$. By hypothesis, $\nu(x^{-1}B) = 0$ for every $x \in G$, and so the right-hand side of (1) is 0, that is, there exists a ν-negligible set N_A such that, for $y \notin N_A$,

$$0 = \int 1_A(x)1_B(xy)\, d\mu(x) = \Delta(y)^{-1} \int 1_A(xy^{-1})1_B(x)\, d\mu(x) \qquad (2).$$

If K is a compact subset of G such that $\nu(K) \neq 0$, we may take for A the set BK^{-1}. There exists a point $y \in K$ for which (2) holds. Since $1_A(xy^{-1})1_B(x) = 1_B(x)$ for all $x \in G$, this proves that $\mu(B) = 0$.

Assume now that $\mu(B) = 0$. For every compact subset A of G, the left-hand side of (1) is 0, and hence the right-hand side is also 0. As a consequence, there exists a locally μ-negligible set M such that $\int 1_B(xy) \, d\nu(y) = 0$ for $x \notin M$. Since $\mu \neq 0$, we conclude that $\nu(x^{-1}B) = 0$ for some $x \in G$, and finally that $\nu(B) = 0$. $\qquad\square$

We can apply Proposition 23.4.3 to G°. Since $\Delta^{-1}\mu$ is a right Haar measure for every left Haar measure μ, right Haar measures and left Haar measures are equivalent. Hence a measure on G is quasi-invariant under right translations if and only if it is quasi-invariant under left translations. In this case, we shall simply say that it is quasi-invariant.

23.5   Homogeneous Spaces

Let X be a topological space, R an equivalence relation on X, and π the canonical projection from X onto the quotient space X/R. We give X/R the quotient topology. Then a necessary and sufficient condition that a subset V of X/R be open is that $\pi^{-1}(V)$ be open in X. By definition, the relation R is open whenever the saturate $\mathrm{Sat}(A) = \pi^{-1}(\pi A)$ of every open set A is open.

Proposition 23.5.1   *If X/R is Hausdorff, the graph C of R is closed in $X \times X$. Conversely, if the relation R is open and C is closed in $X \times X$, then X/R is Hausdorff.*

PROOF: C is the inverse image of the diagonal Δ of $X/R \times X/R$ under $(x, y) \mapsto (\pi(x), \pi(y))$. Hence C is closed if X/R is Hausdorff. Conversely, assume that the relation R is open and that C is closed in $X \times X$. Let $\pi(x)$ and $\pi(y)$ be two distinct points in X/R. Since (x, y) does not lie in C, there exist open neighborhoods V of x and W of y such that $(V \times W) \cap C = \emptyset$. Then $\pi(V) \cap \pi(W) = \emptyset$, which proves that X/R is Hausdorff. $\qquad\square$

Proposition 23.5.2 *Assume that X is locally compact. Then C is closed if and only if $\mathrm{Sat}(K)$ is closed in X for every compact subset K of X.*

PROOF: Suppose that the condition holds, and let (a, b) be a point in the closure of C. Let V (respectively, W) be a compact neighborhood of a (respectively, a neighborhood of b) in X. Since $C \cap (V \times W)$ is nonempty, $W \cap \mathrm{Sat}(V)$ is nonempty. Hence b lies in $\mathrm{Sat}(V)$. Now $B = \mathrm{Sat}(\{b\})$ is closed in X and it intersects every compact neighborhood V of a. So a lies in B, and (a, b) belongs to C.

Conversely, assume that C is closed in $X \times X$. Let K be a compact subset of X and a a point in the closure of $\mathrm{Sat}(K)$. For every neighborhood V of a, the set $V \cap \mathrm{Sat}(K)$ and also $\mathrm{Sat}(V) \cap K$ are nonempty. The filter $\mathcal{F}$ on K generated by the sets $\mathrm{Sat}(V) \cap K$ has a cluster point $b \in K$ (lying in the closure of every $M \in \mathcal{F}$). If V, W are neighborhoods of a, b in X, then $W \cap \mathrm{Sat}(V) \neq \emptyset$; thus $C \cap (V \times W)$ is nonempty. Since C is closed, (a, b) lies in C, whence it follows that a belongs to $\mathrm{Sat}(K)$. $\qquad \square$

Proposition 23.5.3 *Assume that X is locally compact, R is open, and X/R is Hausdorff. Then X/R is locally compact and, for every compact subset K' of X/R, there exists a compact subset K of X such that $K' = \pi(K)$.*

PROOF: The first assertion is obvious. Now, for every $y \in K'$, let $V(y)$ be a compact neighborhood of a point of $f^{-1}(y)$, so that $\pi(V(y))$ is a compact neighborhood of y. There exist finitely many points $y_i \in K'$ such that the sets $\pi(V(y_i))$ cover K'. If L is the compact subset $\bigcup_i V(y_i)$ of X, then $K' \subset \pi(L)$. Hence $K = L \cap \pi^{-1}(K')$ is compact, and $\pi(K) = K'$. $\qquad \square$

Proposition 23.5.4 *Let f be a continuous mapping from a topological space Y into a topological space Z. The following conditions are equivalent:*

(a) *If $\mathcal{F}$ is a filter on Y and $z \in Z$ is a cluster point of $f(\mathcal{F})$, there exists $y \in Y$, a cluster point of $\mathcal{F}$, such that $f(y) = z$.*

(b) *If $\mathcal{U}$ is an ultrafilter on Y and $z \in Z$ is a limit point of $f(\mathcal{U})$, then, $\mathcal{U}$ has a limit point y such that $f(y) = z$.*

(c) *f is closed and, for every $z \in Z$, $f^{-1}(z)$ is compact (not necessarily Hausdorff).*

PROOF: To see that (b) implies (c), we consider a closed subset A of Y and a point z in the closure of $f(A)$. When V runs through the collection of all neighborhoods of z, the sets $A \cap f^{-1}(V)$ constitute a filter basis $\mathcal{B}$. If $\mathcal{U}$ is an ultrafilter finer than $\mathcal{B}$, there exists a limit point y of $\mathcal{U}$ such that $f(y) = z$. Since y lies in A, z lies in $f(A)$.

To see that (c) implies (a), we consider a filter $\mathcal{F}$ on Y such that $f(\mathcal{F})$ has a cluster point z. For each $M \in \mathcal{F}$, $f(\overline{M})$ is closed and contains $f(M)$. Therefore, z lies in $f(\overline{M})$. Now, when M runs through $\mathcal{F}$, the sets $\overline{M} \cap f^{-1}(z)$ constitute a filter basis. Since $f^{-1}(z)$ is compact, this filter basis has a cluster point $y \in f^{-1}(z)$. Thus $\mathcal{F}$ admits y as cluster point. $\qquad \square$

Definition 23.5.1 Under the conditions of Proposition 23.5.4, f is said to be proper.

If f is proper and K is a compact subset of Z, then $f^{-1}(K)$ is compact. Indeed, when $f^{-1}(K) \neq \emptyset$, every ultrafilter on $f^{-1}(K)$ has a limit point.

Henceforth, we let H be a locally compact group operating continuously from the right on a locally compact space X, and we assume that H acts properly on X, that is, that the mapping $\theta : (x, \xi) \mapsto (x, x\xi)$ from $X \times H$ into $X \times X$ is proper. Then the image of $X \times H$ under θ is closed in $X \times X$. In other words, if we consider on X the equivalence relation $R: x \sim y$ whenever there exists $\xi \in H$ such that $y = x\xi$, then the graph of R is closed. The equivalence classes for this relation R are called the orbits. For every open subset V of X, $\mathrm{Sat}(V) = VH = \bigcup_{\xi \in H} V\xi$ is open in X. This means that R is an open relation. By Propositions 23.5.1 and 23.5.3, $X/H = X/R$ is locally compact. We shall denote by π the canonical projection from X onto X/H. Every compact subset of X/H is the image under π of a compact subset of X. If K, L are two compact subsets of X, then $P(K, L) = \{\xi \in H : K\xi \cap L \neq \emptyset\}$ is compact in G; indeed, since θ is proper, $\theta^{-1}(K \times L)$ is compact in $X \times H$, and its image $P(K, L)$ under the second projection from $X \times H$ onto H is also compact.

Let χ be a continuous homomorphism of H into $\mathbf{R}_+^*$. If a mapping g from X into a real vector space satisfies $g(x\xi) = \chi(\xi)g(x)$ for all $x \in X$ and all $\xi \in H$, its support S is invariant under H, and therefore can be written $\pi^{-1}(\pi(S))$. We denote by $\mathcal{H}^\chi(X, \mathbf{C})$ the vector space of those continuous functions g from X into $\mathbf{C}$ that satisfy $g(x\xi) = \chi(\xi)g(x)$ $(x \in X, \xi \in H)$ and whose support is included in $\pi^{-1}(\pi(K))$ for some compact subset K of X (depending on g). Then $S = \pi^{-1}(\pi(S \cap K))$, and so S is the saturate of a compact set.

In what follows, we fix a left Haar measure β on H.

Theorem 23.5.1 *Let $f : X \to \mathbf{C}$ be a continuous function, whose support S meets the saturate of every compact subset of X in a compact set.*

(a) *For every $x \in X$, the function $\xi \mapsto f(x\xi)$ belongs to $\mathcal{H}(H, \mathbf{C})$.*

(b) *The function $f^\chi : x \mapsto \int_H f(x\xi)\chi(\xi)^{-1} \, d\beta(\xi)$ is continuous, vanishes outside SH, and satisfies $f^\chi(x\xi) = \chi(\xi)f^\chi(x)$.*

(c) *If $g : X \to \mathbf{C}$ is any continuous function satisfying $g(x\xi) = \chi(\xi)g(x)$, then $(fg)^\chi = f^1 g$ (where $f^1(x) = \int f(x\xi) \, d\beta(\xi)$).*

(d) *For each $\eta \in H$, $(\delta(\eta)f)^\chi = \chi(\eta)\Delta_H(\eta)^{-1}f^\chi$.*

PROOF: Let $x_0 \in X$ and let V be a compact neighborhood of x_0 in X. The set of those $\xi \in H$ such that $V\xi$ meets S is also the set of those $\xi \in H$ such that $V\xi$ meets $S \cap VH$. Hence it is compact, since $S \cap VH$ is compact. This proves assertion (a) and the continuity of f^χ (take $\mu = \chi^{-1}\beta$ in Proposition 23.1.1). The rest of (b) is obvious. Finally, (c) and (d) follow from the computations:

$$(fg)^\chi(x) = \int_H f(x\xi)g(x\xi)\chi(\xi)^{-1} \, d\beta(\xi)$$

$$= \int_H f(x\xi)g(x)\chi(\xi)\chi(\xi)^{-1}\,d\beta(\xi)$$

$$= f^1(x)g(x)$$

and

$$(\delta(\eta)f)^\chi(x) = \int_H f(x\xi\eta)\chi(\xi)^{-1}\,d\beta(\xi)$$

$$= \Delta_H(\eta)^{-1}\int_H f(x\xi)\chi(\xi\eta^{-1})^{-1}\,d\beta(\xi)$$

$$= \chi(\eta)\Delta_H(\eta)^{-1}f^\chi(x).$$

$\square$

Proposition 23.5.5 *Let K be a compact subset of X and u an element of $\mathcal{H}^+(X)$ which is strictly positive on K. Then $\inf_{x\in KH} u^1(x)$ is strictly positive. For every $g \in \mathcal{H}^\chi(X, \mathbf{C})$ whose support is contained in KH, denote by $\psi(g)$ the function equal to $u \cdot (g/u^1)$ on KH and to 0 on $X - KH$. Then $\psi(g)$ lies in $\mathcal{H}(X, \mathbf{C})$ and $(\psi(g))^\chi = g$.*

PROOF: For all $x \in K$, we have $u^1(x) > 0$. Hence $\inf_{x\in KH} u^1(x) = \inf_{x\in K} u^1(x)$ is strictly positive. The function h, equal to g/u^1 on KH and which vanishes on $X - KH$, lies in $\mathcal{H}^\chi(X, \mathbf{C})$. Moreover, $(uh)^\chi = u^1 h = g$ by part (c) of Theorem 23.5.1. $\square$

Thus $f \mapsto f^\chi$ maps $\mathcal{H}(X, \mathbf{C})$ onto $\mathcal{H}^\chi(X, \mathbf{C})$, and $\mathcal{H}^+(X)$ onto $\mathcal{H}^\chi_+(X, \mathbf{C})$. If f_1 and f_2 belong to $\mathcal{H}^+(X)$ and if $g \in \mathcal{H}^\chi(X, \mathbf{C})$ satisfies $|g| \le f_1^\chi + f_2^\chi$, then, for each $1 \le i \le 2$, we define g_i as the function equal to

$$g(x) \cdot \frac{f_i^\chi(x)}{f_1^\chi(x) + f_2^\chi(x)}$$

at points $x \in X$ such that $(f_1^\chi + f_2^\chi)(x) > 0$, and 0 elsewhere. Then g_i belongs to $\mathcal{H}^\chi(X, \mathbf{C})$, $|g_i| \le f_i^\chi$, and $g = g_1 + g_2$.

If I is a linear form of finite variation over $\mathcal{H}^\chi(X, \mathbf{C})$ (Section 1.4), then $\mu_I : f \mapsto I(f^\chi)$ is a linear form of finite variation over $\mathcal{H}(X, \mathbf{C})$ (because $|f^\chi| \le |f|^\chi$ for every $f \in \mathcal{H}(X, \mathbf{C})$). Hence μ_I is a Radon measure on X. The mapping $I \mapsto \mu_I$ is an injection. The measures μ_I so obtained may be characterized as follows.

Proposition 23.5.6 *Let μ be a measure on X. The following conditions are equivalent:*

(a) *There exists a linear form I of finite variation over $\mathcal{H}^\chi(X, \mathbf{C})$ such that $I(f^\chi) = \mu(f)$ for all $f \in \mathcal{H}(X, \mathbf{C})$.*

(b) $\delta(\xi)\mu = \chi(\xi)^{-1}\Delta_H(\xi)\mu$ *for all* $\xi \in H$.

(c) *For all* f, g *in* $\mathcal{H}(X, \mathbf{C})$, $\mu(fg^1) = \mu(f^\times g)$.

(d) *If* $f \in \mathcal{H}(X, \mathbf{C})$ *is such that* $f^\times = 0$, *then* $\mu(f) = 0$.

PROOF: If condition (a) is satisfied, then, for every $\xi \in H$ and every f in $\mathcal{H}(X, \mathbf{C})$,

$$
\begin{aligned}
\langle f, \delta(\xi)\mu \rangle &= \langle \delta(\xi^{-1})f, \mu \rangle \\
&= I\big((\delta(\xi^{-1})f)^\times\big) \\
&= I\big(\chi(\xi)^{-1}\Delta_H(\xi)f^\times\big) \\
&= \chi(\xi)^{-1}\Delta_H(\xi)\langle f \cdot \mu \rangle;
\end{aligned}
$$

hence

$$
\delta(\xi)\mu = \chi(\xi)^{-1}\Delta_H(\xi)\mu.
$$

Assume that condition (b) is satisfied. If f, g lie in $\mathcal{H}(X, \mathbf{C})$, the functions $(x, \xi) \mapsto f(x)g(x\xi)$ and $(x, \xi) \mapsto f(x\xi^{-1})g(x)\chi(\xi)\Delta_H(\xi)^{-1}$ are continuous on $X \times H$ and compactly supported (because H acts properly on X). Therefore, we see that

$$
\begin{aligned}
\int_X d\mu(x)f(x)\int_H g(x\xi)\,d\beta(\xi) &= \int_H d\beta(\xi)\int_X f(x)g(x\xi)\,d\mu(x) \\
&= \int_H d\beta(\xi)\int_X f(x\xi^{-1})g(x)\chi(\xi)\Delta_H(\xi)^{-1}\,d\mu(x) \\
&= \int_X d\mu(x)g(x)\int_H f(x\xi^{-1})\chi(\xi)\Delta_H(\xi)^{-1}\,d\beta(\xi) \\
&= \int_X d\mu(x)g(x)\int_H f(x\xi)\chi(\xi)^{-1}\,d\beta(\xi),
\end{aligned}
$$

which proves that $\mu(fg^1) = \mu(f^\times g)$.

If condition (c) holds and $f \in \mathcal{H}(X, \mathbf{C})$ is such that $f^\times = 0$, then $\mu(fg^1) = 0$ for every $g \in \mathcal{H}(X, \mathbf{C})$. Choosing g so that $g^1 = 1$ on supp(f) (which is possible by Proposition 23.5.5), we conclude that $\mu(f) = 0$.

Finally, suppose that condition (d) holds. Then there exists a linear form I on $\mathcal{H}^\times(X, \mathbf{C})$ such that $\mu(h) = I(h^\times)$ for all $h \in \mathcal{H}(X, \mathbf{C})$. Let $f \in \mathcal{H}_+^\times(X, \mathbf{C})$. There exists a compact subset K of X such that f vanishes on $X - KH$. In the notation of Proposition 23.5.5, for every $g \in \mathcal{H}^\times(X, \mathbf{C})$ satisfying $|g| \leq f$, we have $|\psi(g)| \leq \psi(f)$. Therefore, $\{|I(g)| = |\mu(\psi(g))| : g \in \mathcal{H}^\times(X, \mathbf{C}), |g| \leq f\}$ is bounded. $\square$

Now assume that $\chi = 1$. Then $f \mapsto f \circ \pi$ is a one-to-one mapping from $\mathcal{H}(X/H, \mathbf{C})$ onto $\mathcal{H}^1(X, \mathbf{C})$. The preceding results can be reformulated as follows:

Let $f : X \to \mathbf{C}$ be a continuous function whose support meets the saturate of every compact subset of X in a compact set. The formula $f^b(\dot{x}) = \int_H f(x\xi)\, d\beta(\xi)$ (where $\dot{x} = \pi(x)$) defines a continuous function from X/H into $\mathbf{C}$. If g is any continuous function from X/H into $\mathbf{C}$, then $\big(f \cdot (g \circ \pi)\big)^b = f^b g$. For every $\eta \in H$, we have $\big(\delta(\eta)f\big)^b = \Delta_H(\eta)^{-1} f^b$. If f belongs to $\mathcal{H}(X, \mathbf{C})$, then f^b lies in $\mathcal{H}(X/H, \mathbf{C})$. The mapping $f \mapsto f^b$ sends $\mathcal{H}(X, \mathbf{C})$ onto $\mathcal{H}(X/H, \mathbf{C})$, and it sends $\mathcal{H}^+(X)$ onto $\mathcal{H}^+(X/H)$.

Theorem 23.5.2

(a) *For every measure λ on X/H, the functional $f \mapsto \lambda(f^b)$ from $\mathcal{H}(X, \mathbf{C})$ into $\mathbf{C}$ is a measure $\lambda^\sharp$ on X, and $\delta(\xi)\lambda^\sharp = \Delta_H(\xi)\lambda^\sharp$ for all $\xi \in H$.*

(b) *Conversely, let μ be a measure on X such that $\delta(\xi)\mu = \Delta_H(\xi)\mu$ for all $\xi \in H$. Then there exists a unique measure λ on X/H such that $\mu = \lambda^\sharp$.*

PROOF: Apply Proposition 23.5.6.   □

$\lambda^\sharp$ is real (respectively, positive) if and only if λ is. The formula $\lambda^\sharp(f) = \lambda(f^b)$ can be written

$$\int_X f(x)\, d\lambda^\sharp(x) = \int_{X/H} d\lambda(\dot{x}) \int_H f(x\xi)\, d\beta(\xi).$$

This notation is improper, because $\int_H f(x\xi)\, d\beta(\xi)$ is considered a function of $\dot{x}$, and not a function of x.

Proposition 23.5.7

(a) *Let $(\lambda_i)_{i \in I}$ be a family of real measures on X/H. The family $(\lambda_i)_{i \in I}$ is bounded above in $\mathcal{M}(X/H, \mathbf{R})$ if and only if the family $(\lambda_i^\sharp)_{i \in I}$ is bounded above in $\mathcal{M}(X, \mathbf{R})$. Then $\sup_i(\lambda_i^\sharp) = (\sup_i \lambda_i)^\sharp$.*

(b) *For every real measure λ on X/H, $(\lambda^+)^\sharp = (\lambda^\sharp)^+$ and $(\lambda^-)^\sharp = (\lambda^\sharp)^-$.*

(c) *For every complex measure λ on X/H, $|\lambda|^\sharp = |\lambda^\sharp|$.*

PROOF: Suppose that the family $(\lambda_i)_{i \in I}$ is bounded above, and put $\mu = \sup \lambda_i$. Since $\mu^\sharp \geq \lambda_i^\sharp$ for all i, the family $(\lambda_i^\sharp)_i$ is bounded above and $(\sup \lambda_i)^\sharp \geq \sup(\lambda_i^\sharp)$.

Conversely, suppose that $(\lambda_i^\sharp)_i$ is bounded above, and put $\nu = \sup(\lambda_i^\sharp)$. Since $\delta(\xi)\lambda_i^\sharp = \Delta_H(\xi)\lambda_i^\sharp$ for every $\xi \in H$, clearly $\delta(\xi)\nu = \Delta_H(\xi)\nu$. Hence there exists a real measure μ' on X/H such that $\nu = \mu'^\sharp$. From $\mu'^\sharp \geq \lambda_i^\sharp$ for all i, we deduce that $\mu' \geq \lambda_i$. The family $(\lambda_i)_i$ is thus bounded above, and we have $\nu = \mu'^\sharp \geq (\sup \lambda_i)^\sharp$, or $\sup(\lambda_i^\sharp) \geq (\sup \lambda_i)^\sharp$. This completes the proof of (a).

Assertion (b) follows immediately. Finally, if λ is a complex measure on X/H, then $|\lambda| = \sup_\alpha \text{Re}(\alpha\lambda)$, where α extends over the set of complex numbers of modulus 1 (Section 1.4). Therefore.

$$|\lambda|^\sharp = \sup_\alpha [\text{Re}(\alpha\lambda)]^\sharp = \sup_\alpha \text{Re}[(\alpha\lambda)^\sharp] = \sup_\alpha \text{Re}(\alpha\lambda^\sharp) = |\lambda^\sharp|$$

by assertion (a). $\square$

Definition 23.5.2 If μ is a measure on X such that $\delta(\xi)\mu = \Delta_H(\xi)\mu$ for all $\xi \in H$, the unique measure λ on X/H such that $\mu = \lambda^\sharp$ is called the quotient of μ by β and is written $\mu_{/\beta}$.

23.6   Integration with Respect to $\lambda^\sharp$

Let H be a locally compact group which acts from the right, continuously and properly, on a locally compact space X, and let β be a left Haar measure on H. Write θ for the mapping $(x, \xi) \mapsto (x, x\xi)$ from $X \times H$ into $X \times X$.

Fix $x \in X$. If A is a closed subset of H, then $\{x\} \times A$ is closed in $X \times H$. Therefore, $\{x\} \times xA = \theta(\{x\} \times A)$ is closed in $X \times X$, and xA is closed in X. On the other hand, for all $y \in X$, the set $\{\xi \in H : x\xi = y\}$ is compact in H. Hence the mapping $\xi \mapsto x\xi$ from H into X is proper.

The image measure of β under this mapping is concentrated on the orbit xH. Since β is left invariant, this image measure depends only on the class $u = \pi(x)$ of x. We denote it by β_u. By definition, for $f \in \mathcal{H}(X, \mathbf{C})$,

$$\int_X f(y)\, d\beta_u(y) = \int_H f(x\xi)\, d\beta(\xi) = f^\flat(u) \qquad (1).$$

Since $f^\flat$ lies in $\mathcal{H}(X/H, \mathbf{C})$, formula (1) proves that the mapping $\wedge : u \mapsto \beta_u$ from X/H into $\mathcal{M}^+(X)$ is vaguely continuous. For every positive measure λ on X/H, $\wedge$ is λ-adequate (by Proposition 19.4.3) and $\lambda^\sharp = \int_{X/H} \beta_u\, d\lambda(u)$.

Theorem 23.6.1 *Let λ be a complex measure on X/H.*

(a) *If f is a $\lambda^\sharp$-measurable mapping from X into a topological space, such that f is constant outside a $\lambda^\sharp$-moderate set, then the set of those $\dot{x}$ in X/H such that $\xi \mapsto f(x\xi)$ is not β-measurable is λ-negligible.*

(b) *If $f : X \mapsto [0, +\infty]$ is $\lambda^\sharp$-measurable and $\lambda^\sharp$-moderate, then the function $\dot{x} \mapsto \int^* f(x\xi)\, d\beta(\xi)$ on X/H is λ-measurable and*

$$\int_X^* f(x)\, d|\lambda^\sharp|(x) = \int_{X/H}^* d|\lambda|(\dot{x}) \int_H^* f(x\xi)\, d\beta(\xi).$$

(c) If f is a $\lambda^\sharp$-integrable mapping from X into a Banach space, then the set of those $\dot{x} \in X/H$ such that $\xi \mapsto f(x\xi)$ is not β-integrable is λ-negligible. Moreover, the mapping $f^\flat$, defined λ-almost everywhere by $f^\flat(\dot{x}) = \int_H f(x\xi)\,d\beta(\xi)$, is λ-integrable and

$$\int_{X/H} f^\flat\,d\lambda = \int_X f\,d\lambda^\sharp \qquad (2).$$

PROOF: Assertions (a) and (b) follow from Proposition 23.5.7, and from Proposition 19.4.5 and Theorem 19.4.1. Assertion (c) follows from Theorem 19.4.2 and the fact that λ can be written

$$(\operatorname{Re}\lambda)^+ - (\operatorname{Re}\lambda)^- + i(\operatorname{Im}\lambda)^+ - i(\operatorname{Im}\lambda)^-.$$

$\square$

Proposition 23.6.1 *Fix λ, a complex measure on X/H.*

(a) *Let g be a complex valued function on X/H. g is locally λ-integrable if and only if $g \circ \pi$ is locally $\lambda^\sharp$-integrable, in which case $(g\lambda)^\sharp = (g \circ \pi)\lambda^\sharp$.*

(b) *Let N be a subset of X/H. N is locally λ-negligible if and only if $\pi^{-1}(N)$ is locally $\lambda^\sharp$-negligible.*

(c) *Let h be a mapping from X/H into a topological space. Then h is λ-measurable if and only if $h \circ \pi$ is $\lambda^\sharp$-measurable.*

PROOF: Assume that $g \circ \pi$ is locally $\lambda^\sharp$-integrable. For every f in $\mathcal{H}(X, \mathbf{C})$, $f \cdot (g \circ \pi)$ is $\lambda^\sharp$-integrable; thus $(f \cdot (g \circ \pi))^\flat = f^\flat g$ is λ-integrable (Theorem 23.6.1) and $\int_{X/H} f^\flat g\,d\lambda = \int_X f \cdot (g \circ \pi)\,d\lambda^\sharp$. Since $f \mapsto f^\flat$ maps $\mathcal{H}(X, \mathbf{C})$ onto $\mathcal{H}(X/H, \mathbf{C})$, we see that g is locally λ-integrable and that $(g\lambda)^\sharp = (g \circ \pi)\lambda^\sharp$.

Conversely, if g is locally λ-integrable, then, by Proposition 21.2.1, $g \circ \pi$ is $\lambda^\sharp$-measurable. For each $f \in \mathcal{H}^+(X)$,

$$\int_X^* f(x)|g \circ \pi|(x)\,d|\lambda^\sharp|(x) = \int_{X/H}^* f^\flat(u)|g|(u)\,d|\lambda|(u) < +\infty$$

(Theorem 23.6.1). Hence $g \circ \pi$ is locally $\lambda^\sharp$-integrable. This proves (a).

If N is locally λ-negligible, then $\pi^{-1}(N)$ is locally $\lambda^\sharp$-negligible (Proposition 21.2.1). Conversely, if $\pi^{-1}(N)$ is locally $\lambda^\sharp$-negligible, then $(1_N\lambda)^\sharp = (1_N \circ \pi)\lambda^\sharp = 0$ and $1_N\lambda = 0$, so N is locally λ-negligible.

If h is λ-measurable, then $h \circ \pi$ is $\lambda^\sharp$-measurable by Proposition 21.2.1.

Conversely, assume that $h \circ \pi$ is $\lambda^\sharp$-measurable, and let K' be a compact subset of X/H. Choose $f \in \mathcal{H}^+(X)$ so that $f^\flat = 1$ on K', and let $K = \operatorname{supp}(f)$. Since $f^\flat$ vanishes outside $\pi(K)$, K' is included in $\pi(K)$. There exists a partition of K consisting of a $\lambda^\sharp$-negligible set M and a sequence $(K_n)_{n\geq 1}$ of compact sets such that $h \circ \pi_{/K_n}$ is continuous for all n. As the restriction of $h \circ \pi$ to $K_1 \cup \cdots \cup K_n$ is continuous, the restriction of h to $\pi(K_1 \cup \cdots \cup K_n)$ is continuous. Denote by P the set of those points of K' that do not belong to $\bigcup_{n\geq 1} \pi(K_n)$. Then $\pi^{-1}(P) \cap K$ is included in M, which says that $f1_{\pi^{-1}(P)}$ is $\lambda^\sharp$-negligible. From Theorem 23.6.1, we deduce that

$$0 = \int_X f1_{\pi^{-1}(P)} \, d|\lambda^\sharp| = \int_{X/H} f^\flat 1_P \, d|\lambda| \geq \int_{X/H}^* 1_P \, d|\lambda|;$$

hence P is λ-negligible and h is λ-measurable. $\square$

Proposition 23.6.2 *Let λ and λ' be two complex measures on X/H. Then, in order that λ' be a measure with base λ, it is necessary and sufficient that $\lambda'^\sharp$ be a measure with base $\lambda^\sharp$.*

PROOF: This follows from (a) and (b) of Proposition 23.6.1. $\square$

Proposition 23.6.3 *Let $\lambda \neq 0$ be a complex measure on X/H. π is $\lambda^\sharp$-proper if and only if H is compact. In this case. $\pi(\lambda^\sharp) = \beta(H)\lambda$.*

PROOF: Suppose that λ is positive. Since $\wedge : u \mapsto \beta_u$ is λ-adequate, for every $g \in \mathcal{H}^+(X/H)$ we have

$$\int^* (g \circ \pi) \, d\lambda^\sharp = \int_{X/H}^* d\lambda(\dot{x}) \int_H^* (g \circ \pi)(x\xi) \, d\beta(\xi) = \beta^*(H) \int g(\dot{x}) \, d\lambda(\dot{x}).$$

Hence π is $\lambda^\sharp$-proper if and only if $\beta^*(H)$ is finite or, equivalently, if H is compact. In this case, by the above equality, $\pi(\lambda^\sharp) = \beta(H)\lambda$. $\square$

23.7   Reconstitution of $\lambda_{/\beta}^\sharp$ [†]

From topology, we know that a locally compact space is paracompact if and only if it can be partitioned into σ-compact open sets.

Proposition 23.7.1 *Let X be a locally compact space, and let R be an open equivalence relation on X such that the quotient space is paracompact. Denote by π the canonical projection from X onto X/R. There exists a positive continuous function F on X with the following properties:*

[†]This section may be omitted.

(a) F is not identically zero on any equivalence class.

(b) For every compact subset K of X/R, the intersection of supp(F) and $\pi^{-1}(K)$ is compact.

PROOF: With each $u \in X/R$, we associate a function $f_u \in \mathcal{H}^+(X)$ that does not vanish identically on $\pi^{-1}(u)$. Then u lies in $\pi(\Omega_u)$, where

$$\Omega_u = \{x \in X : f_u(x) > 0\}.$$

Since π is open, the sets $\pi(\Omega_u)$ form an open covering of X/R. There exist an open covering $(V_i)_{i \in I}$, locally finite, which refines the covering $\big(\pi(\Omega_u)\big)_u$, and a continuous partition of unity $(g_i)_{i \in I}$ on X/H subordinate to the covering $(V_i)_{i \in I}$. For each $i \in I$, we choose a point $u_i \in X/R$ so that $V_i \subset \pi(\Omega_{u_i})$. The function $F_i = (g_i \circ \pi) f_{u_i}$ belongs to $\mathcal{H}^+(X)$ and its support is included in $\pi^{-1}(V_i)$. The supports of the F_i thus form a locally finite family, and $F = \sum_{i \in i} F_i$ is continuous. For every $u \in X/R$, there exists an index i such that $g_i(u) > 0$, and u lies in V_i. Then there is a point $x \in \Omega_{u_i}$ for which $\pi(x) = u$. As $f_{u_i}(x) > 0$ and $g_i(\pi x) > 0$, we have $F_i(x) > 0$, and, evidently, $F(x) > 0$. This proves that F satisfies (a).

Finally, let K be a compact subset of X/R. There exists a finite subset J of I such that, for $i \in I - J$, $V_i \cap K$ is empty. Thus $\pi^{-1}(K) \cap \text{supp}(F_i) = \emptyset$ for all $i \in I - J$. Since $\bigcup_{i \in I} \text{supp}(F_i)$ is closed, it is equal to supp(F). Then

$$\pi^{-1}(K) \cap \text{supp}(F) - \pi^{-1}(K) \cap \left(\bigcup_{i \in J} \text{supp}(F_i) \right)$$

is compact.   □

Again, let H be a locally compact group that acts from the right, continuously and properly, on a locally compact space X, and let β be a left Haar measure on H.

Proposition 23.7.2 *Assume that X/H is paracompact. Let χ be a continuous homomorphism from H into $\mathbf{R}^*_+$.*

(a) *There exists a continuous function ρ from X into $]0, +\infty[$ such that $\rho(x\xi) = \chi(\xi)\rho(x)$ for all $x \in X$ and all $\xi \in H$.*

(b) *The mapping $g \mapsto g/\rho$ is a linear isomorphism from $\mathcal{H}^\chi(X, \mathbf{C})$ onto $\mathcal{H}^1(X, \mathbf{C})$.*

PROOF: We apply Theorem 23.5.1, taking for f a positive function that does not vanish identically on any orbit (this is possible by Proposition 23.7.1). Then $\rho = f^\chi$ satisfies $\rho(x\xi) = \chi(\xi)\rho(x)$. Assertion (b) is obvious.   □

Proposition 23.7.3 *Assume that X/H is paracompact. There exists a positive continuous function h on X, whose support meets the saturate of every compact set in a compact set, such that $h^\flat = 1$.*

PROOF: We apply Theorem 23.5.1, with $\chi = 1$, taking for f a positive function that does not vanish identically on any orbit. For every $x \in X$, $f^1(x)$ is strictly positive. Now put $h = f/f^1$. Then, $h^1 = f^1/f^1 = 1$, and so $h^\flat = 1$. $\qquad\square$

Proposition 23.7.4 *Suppose there exists h as in Proposition 23.7.3. Let λ be a positive measure on X/H.*

(a) *The pair (π, h) is $\lambda^\sharp$-adapted and $\int_X h(x)\varepsilon_{\pi(x)}\, d\lambda^\sharp(x) = \lambda$.*

(b) *π is proper for the measure $h\lambda^\sharp$, and $\pi(h\lambda^\sharp) = \lambda$.*

(c) *A function k from X/H into a real Banach space is λ-integrable if and only if $h(k \circ \pi)$ is $\lambda^\sharp$-integrable.*

PROOF: Let $f \in \mathcal{H}(X/H, \mathbf{C})$. Then $h \cdot (f \circ \pi)$ belongs to $\mathcal{H}(X, \mathbf{C})$ and

$$\int_X h(x)f(\pi x)\, d\lambda^\sharp(x) = \int_{X/H} d\lambda(\dot{x})f(\dot{x})\int_H h(x\xi)\, d\beta(\xi)$$

$$= \int_{X/H} f(\dot{x})\, d\lambda(\dot{x})$$

(Theorem 23.6.1). This proves (a). (b) can be proved similarly. If k is λ-integrable, then $h \cdot (k \circ \pi)$ is $\lambda^\sharp$-integrable (Theorem 19.4.2). If $h \cdot (k \circ \pi)$ is $\lambda^\sharp$-integrable, Theorem 23.6.1 proves that $\left(h \cdot (k \circ \pi)\right)^\flat = h^\flat k = k$ is λ-integrable. $\qquad\square$

23.8   Quasi-Invariant Measures on Homogeneous Spaces

Let G be a locally compact group and H a closed subgroup of G. Consider the homogeneous space G/H of left cosets of H in G. Likewise, $H\backslash G$ will denote the space of right cosets of H in G. G operates continuously, from the left, on G/H. On the other hand, H acts continuously and properly, from the right, on G, and the quotient space is none other than G/H.

Proposition 23.8.1 *G/H is paracompact.*

PROOF: Let V be a symmetric compact neighborhood of e in G and $G_0 = \bigcup_{n \geq 1} V^n$ the subgroup of G generated by V. Since G_0 is open in G, $\pi^{-1}(G_0 z) = G_0 \pi^{-1}(z) H$ is open in G (for $z \in G/H$). Hence each orbit $G_0 z$ is open in G/H. On the other hand, $G_0 z$ is the union of the compact sets $V^n z$.   □

We can apply the results of the preceding sections, with $X = G$, to obtain the mappings $f \mapsto f^{\flat}$ from $\mathcal{H}(G, \mathbf{C})$ onto $\mathcal{H}(G/H, \mathbf{C})$, and $\lambda \mapsto \lambda^{\sharp}$ from $\mathcal{M}(G/H, \mathbf{C})$ into $\mathcal{M}(G, \mathbf{C})$ (once a left Haar measure β on H has been fixed). The fact that G acts continuously on G/H leads to some additional results. If f belongs to $\mathcal{H}(G, \mathbf{C})$ and s lies in G, then $\gamma_{G/H}(s)f^{\flat} = \left(\gamma_G(s)f\right)^{\flat}$; indeed, for each $x \in G$,

$$
\begin{aligned}
\left(\gamma_{G/H}(s)f^{\flat}\right)(\pi(x)) &= f^{\flat}\left(s^{-1}\pi(x)\right) \\
&= f^{\flat}\left(\pi(s^{-1}x)\right) \\
&= \int_H f(s^{-1}x\xi)\, d\beta(\xi) \\
&= \int_H \left(\gamma_G(s)f\right)(x\xi)\, d\beta(\xi).
\end{aligned}
$$

Therefore, for every $\lambda \in \mathcal{M}(G/H, \mathbf{C})$ and every $s \in G$,

$$
\left(\gamma_{G/H}(s)\lambda\right)^{\sharp} = \gamma_G(s)\lambda^{\sharp} \qquad (1).
$$

Proposition 23.8.2 *Let $\lambda \neq 0$ be a measure on G/H and μ a left Haar measure on G. The following conditions are equivalent:*

(a) λ is quasi-invariant under G.

(b) The locally λ-negligible subsets of G/H are those subsets A of G/H for which $\pi^{-1}(A)$ is locally μ-negligible.

(c) The measure $\lambda^{\sharp}$ is equivalent to μ.

Assume that these conditions hold, and let ρ be a locally μ-integrable function from G into $\mathbf{C}^{*}$ such that $\lambda^{\sharp} = \rho\mu$. Then, for all $s \in G$, if θ_s is a density of $\gamma_{G/H}(s)\lambda$ with respect to λ, we have $\theta_s(\pi x) = \rho(s^{-1}x)/\rho(x)$ locally μ-almost everywhere.

PROOF: (c) implies (b), by Proposition 23.6.1. If condition (b) holds, the class of all locally λ-negligible subsets of G/H is invariant under G, and so λ is quasi-invariant.

Now assume that λ is quasi-invariant. For every $s \in G$, $\gamma_G(s)\lambda^{\sharp}$ and $\lambda^{\sharp}$ are equivalent (Proposition 23.6.2 and formula (1)). Therefore $\lambda^{\sharp}$ is equivalent to μ (Proposition 23.4.3). By Proposition 23.6.1, $(\theta_s \circ \pi)\lambda^{\sharp} = (\theta_s\lambda)^{\sharp}$; hence

$$
(\theta_s \circ \pi)\lambda^{\sharp} = \left(\gamma_{G/H}(s)\lambda\right)^{\sharp} = \gamma_G(s)\lambda^{\sharp} = \left(\gamma_G(s)\rho\right)\mu = \frac{\gamma_G(s)\rho}{\rho}\lambda^{\sharp},
$$

and $\theta_s(\pi x) = \rho(s^{-1}x)/\rho(x)$ locally μ-almost everywhere. $\square$

Theorem 23.8.1 *Two nonzero quasi-invariant measures on G/H are necessarily equivalent. A subset A of G/H is locally negligible with respect to these measures if and only if $\pi^{-1}(A)$ is locally negligible with respect to Haar measure on G.*

PROOF: Apply Proposition 23.8.2. $\square$

Proposition 23.8.3 *Let μ be a left Haar measure on G and ρ a locally μ-integrable complex-valued function on G. $\rho\mu$ has the form $\lambda^\cdot$ if and only if, for every $\xi \in H$, $\rho(x\xi) = \bigl(\Delta_H(\xi)/\Delta_G(\xi)\bigr)\rho(x)$ locally μ-almost everywhere in G.*

PROOF: By Theorem 23.5.2, $\rho\mu$ has the form $\lambda^\cdot$ if and only if. for each $\xi \in H$, $\delta(\xi)(\rho\mu) = \Delta_H(\xi)\rho\mu$. But

$$\delta(\xi)(\rho\mu) = \bigl(\delta(\xi)\rho\bigr)\bigl(\delta(\xi)\mu\bigr) = \Delta_G(\xi)\bigl(\delta(\xi)\rho\bigr)\mu.$$

whence the result. $\square$

Theorem 23.8.2 *Let μ be a left Haar measure on G.*

(a) *There exist continuous functions $\rho > 0$ on G such that $\rho(x\xi) = \bigl(\Delta_H(\xi)/\Delta_G(\xi)\bigr)\rho(x)$ for all $x \in G$ and all $\xi \in H$.*

(b) *For such a function ρ, $\delta(\xi)\rho\mu = \Delta_H(\xi)\mu$, and $\lambda = (\rho\mu)_{/\beta}$ is positive, nonzero, and quasi-invariant.*

(c) *There exists a continuous function χ from $G \times (G/H)$ into $]0, +\infty[$ such that $\chi\bigl(s, \pi(x)\bigr) = \rho(sx)/\rho(x)$ for all s, x in G; moreover, $\gamma_{G/H}(s)\lambda = \chi(s^{-1}, \cdot)\lambda$ for every $s \in G$.*

PROOF: Assertion (a) follows from Proposition 23.7.2. Assertion (b) is then a consequence of Propositions 23.8.3 and 23.8.2. Finally, assertion (c) follows from Proposition 23.8.2. $\square$

In short, there is one and only one class of nonzero quasi-invariant measures on G/H.

23.9 Relatively Invariant Measures on G/H

Henceforth, G will always denote a locally compact group, H a closed subgroup of G, β a left Haar measure on H, and μ a left Haar measure on G.

Proposition 23.9.1 *Let $\lambda \neq 0$ be a measure on G/H and χ a continuous homomorphism from G into $\mathbf{C}^*$. The following properties are equivalent:*

(a) λ is relatively invariant on G/H, with multiplier χ.

(b) $\lambda^\sharp$ is relatively invariant on G, with left multiplier χ.

(c) $\lambda^\sharp$ has the form $a\chi\mu$ (where $a \in \mathbf{C}^*$).

PROOF: Condition (a) means that $\gamma_{G/H}(s)\lambda = \chi(s)^{-1}\lambda$ for every $s \in G$, that $\left(\gamma_{G/H}(s)\lambda\right)^\sharp = \chi(s)^{-1}\lambda^\sharp$, or that $\gamma_G(s)\lambda^\sharp = \chi(s)^{-1}\lambda^\sharp$. Hence conditions (a) and (b) are equivalent. By Theorem 23.4.1, conditions (b) and (c) are also equivalent. $\square$

Theorem 23.9.1 *Let χ be a continuous homomorphism from G into $\mathbf{C}^*$.*

(a) There exists, on G/H, a nonzero relatively invariant measure with multiplier χ if and only if $\chi(\xi) = \Delta_H(\xi)/\Delta_G(\xi)$ for all $\xi \in H$.

(b) In this case, the measure is unique up to a multiplicative constant; more precisely, it is proportional to $(\chi\mu)_{/\beta}$.

PROOF: In order that there exist on G/H a nonzero relatively invariant measure with multiplier χ, a necessary and sufficient condition is that $\chi\mu$ have the form $\lambda^\sharp$ (Proposition 23.9.1), or that $\delta(\xi)(\chi\mu) = \Delta_H(\xi)\chi\mu$ for all $\xi \in H$. This condition can also be written $\chi(\xi)\chi\Delta_G(\xi)\mu = \Delta_H(\xi)\chi\mu$, that is, $\chi(\xi) = \Delta_H(\xi)/\Delta_G(\xi)$ for all $\xi \in H$. This proves (a). Assertion (b) follows immediately from Proposition 23.9.1 and the fact that the mapping $\lambda \mapsto \lambda^\sharp$ is injective. $\square$

There are simple examples in which the mapping $\xi \mapsto \Delta_H(\xi)/\Delta_G(\xi)$ from H into $\mathbf{C}^*$ cannot be extended to a continuous homomorphism from G into $\mathbf{C}^*$. In this case, there exists no nonzero relatively invariant measure on G/H.

Theorem 23.9.2 *In order that there exist on G/H a positive invariant measure, it is necessary and sufficient that Δ_G agree with Δ_H on H. In this case, the measure is unique up to a multiplicative constant.*

PROOF: Apply Theorem 23.9.1. $\square$

For instance, if we take for G the group $SO(n+1, \mathbf{R})$ of rotations of $\mathbf{R}^{n+1}$, equipped with its canonical topology, and for H the subgroup of G consisting of those $u \in SO(n+1, \mathbf{R})$ which fix $e_{n+1} = (0, \ldots, 0, 1)$, then H can be identified with $SO(n, \mathbf{R})$. The mapping $\dot{u} \mapsto u(e_{n+1})$ is a homeomorphism of $SO(n+1, \mathbf{R})/SO(n, \mathbf{R})$ onto the sphere $\mathbf{S}^n$. The group $SO(n+1, \mathbf{R})$ acts on $\mathbf{S}^n$, from the left, as follows: $u \cdot x = u(x)$. Up to a multiplicative constant, the measure $d\mathbf{S}^n$ defined in Section 12.6 (or the associated Radon measure) is the unique nonzero complex measure on $\mathbf{S}^n$ which remains invariant under $SO(n+1, \mathbf{R})$.

Theorem 23.9.3 *Let G be a locally compact group, G' a closed normal subgroup of G, G'' the group G/G', $\pi : x \mapsto \dot{x}$ the canonical projection from G onto G'', and let α, α' and α'' be left Haar measures on G, G', and G'', respectively.*

(a) Up to some constant multiple of α, $\alpha'' = \alpha/\alpha'$. Then

$$\int_G f(x) \, d\alpha(x) = \int_{G''} d\alpha''(\dot{x}) \int_{G'} f(x\xi) \, d\alpha'(\xi)$$

for every $f \in \mathcal{H}(G, \mathbf{C})$.

(b) $\Delta_G(\xi) = \Delta_{G'}(\xi)$ for all $\xi \in G'$.

PROOF: There exists on G'' a measure invariant under G, namely α''. So, if we apply Theorem 23.9.1, we obtain (a) and (b).   □

23.10   Haar Measure on $SO(n+1, \mathbf{R})$

Given an integer $n \geq 1$, denote by $(e_1, \ldots, e_{n+1})$ the canonical basis of $\mathbf{R}^{n+1}$. Let ψ be the function from $\mathbf{R}^n$ into $\mathbf{S}^n$ considered in Section 12.4. Write p_n for the mapping

$$(\theta_1, \ldots, \theta_n) \mapsto \psi\left(\frac{\pi}{2} - \theta_1, \frac{\pi}{2} - \theta_2, \ldots, \frac{\pi}{2} - \theta_n\right)$$

from $\mathbf{R}^n$ onto $\mathbf{S}^n$. Thus

$$p_n(\theta_1, \ldots, \theta_n) = \big(\sin(\theta_n) \cdots \sin(\theta_1), \ldots, \cos(\theta_n)\big),$$

where, for each $1 \leq j \leq n$, the $(j+1)$st coordinate of $p_n(\theta_1, \ldots, \theta_n)$ is

$$\sin(\theta_n) \cdots \sin(\theta_{j+1}) \cos(\theta_j).$$

If $J'(\theta)$ is the matrix of $\big(Dp_n(\theta)e_1, \ldots, Dp_n(\theta)e_n, p_n(\theta)\big)$ in the canonical basis of $\mathbf{R}^{n+1}$, then

$$\det J'(\theta) = \sin^{n-1}(\theta_n) \sin^{n-2}(\theta_{n-1}) \cdots \sin(\theta_2).$$

Put $A_n = [0, 2\pi[\times [0, \pi]^{n-1}$. Clearly, p_n is one-to-one from $[0, 2\pi[\times]0, \pi[^{n-1}$ onto $\mathbf{S}^n - \{x \in \mathbf{S}^n : x_1 = 0 = x_2\}$. For $x = p_n(\theta)$ (with θ in $[0, 2\pi[\times]0, \pi[^{n-1})$ and for $2 \leq j \leq n$, θ_j is the geodesic distance in $\mathbf{S}^j$ between $(0, \ldots, 0, 1)$ (the north pole of $\mathbf{S}^j$) and $(1 - x_{j+2}^2 - \cdots - x_{n+1}^2)^{-1/2}(x_1, \cdots, x_{j+1})$. Likewise, the rotation of $\mathbf{R}^2$ with angle $-\theta_1$ maps the north pole $(0, 1)$ of $\mathbf{S}^1$ to $(1 - x_3^2 - \cdots - x_{n+1}^2)^{-1/2}(x_1, x_2)$. Furthermore, dS^n is the image measure of $\sin^{n-1}(\theta_n) \sin^{n-2}(\theta_{n-1}) \cdots \sin(\theta_2)(d\theta_1 \cdots d\theta_n)_{/A_n}$ under p_{n/A_n}.

For all $1 \leq k \leq n$ and for all $\alpha \in \mathbf{R}$, define a rotation $g_k(\alpha)$ of $\mathbf{R}^{n+1}$ as follows:

(a) $g_k(\alpha)$ maps the plane $\mathbf{R}e_k + \mathbf{R}e_{k+1}$ onto itself.

(b) The restriction of g to this plane (oriented so that (e_k, e_{k+1}) is direct) is the rotation with angle $-\alpha$.

(c) The restriction of $g_k(\alpha)$ to the orthogonal space $(\mathbf{R}e_k + \mathbf{R}e_{k+1})^\perp$ is the identity.

We identify $A = \prod_{1 \leq k \leq n} A_k$ with $\prod_{\{(j,k):1 \leq j \leq k \leq n\}} A_j^k$, where $A_j^k = [0, \pi]$ for $2 \leq j \leq k$ and $A_1^k = [0, 2\pi[$. For each $\omega = (\theta_j^k)_{1 \leq j \leq k \leq n}$ in A, define $q(\omega)$ to be $g^{(n)} \circ \cdots \circ g^{(1)}$, where

$$g^{(k)} = g_1(\theta_1^k)g_2(\theta_2^k) \cdots g_k(\theta_k^k) \qquad (1).$$

Proposition 23.10.1 $q : \omega \mapsto q(\omega)$ maps A onto $SO(n+1, \mathbf{R})$.

PROOF: Assume that $n \geq 2$ and that the proposition is true for $SO(n, \mathbf{R})$. Given $g \in SO(n+1, \mathbf{R})$, let $(\theta_1^n, \ldots, \theta_n^n)$ be an element of A_n such that $g(e_{n+1}) = p_n(\theta_1^n, \ldots, \theta_n^n)$. By induction on $0 \leq j \leq n-1$, we see that $[g_{n-j}(\theta_{n-j}^n) \circ \cdots \circ g_n(\theta_n^n)](e_{n+1})$ is equal to

$$\sin(\theta_n^n) \cdots \sin(\theta_{n-j}^n)e_{n-j} + \sum_{n-j \leq k \leq n} \sin(\theta_n^n) \cdots \sin(\theta_{k+1}^n)\cos(\theta_k^n)e_{k+1}.$$

Therefore, $[g^{(n)}]^{-1} \circ g$ fixes e_{n+1}. By hypothesis, $[g^{(n)}]^{-1} \circ g = g^{(n-1)} \circ \cdots \circ g^{(1)}$, where each $g^{(k)}$ is as in (1), and so $g = g^{(n)} \circ g^{(n-1)} \circ \cdots \circ g^{(1)}$. □

Proposition 23.10.2 Let $\omega = (\theta_j^k)_{1 \leq j \leq k \leq n}$ be an element of A such that θ_j^k belongs to $]0, \pi[$ whenever $j \geq 2$. Then $g = q(\omega)$ is the image under q of a unique element of A. Moreover,

$$\left[g^{(n)} \circ \cdots \circ g^{(k+1)}\right]^{-1}\left(g(e_{k+1})\right) = \left(p_k(\theta_1^k, \ldots, \theta_k^k), 0, \ldots, 0\right)$$

for all $1 \leq k \leq n$.

PROOF: Let $\omega' = (\theta'^k_j)_{1 \le j \le k \le n}$ be an element of A such that $q(\omega') = g$. Given $1 \le k \le n$, suppose it has been established that $\theta^l_j = \theta'^l_j$ for all $k < l \le n$ and all $1 \le j \le l$. Then

$$\left[g^{(n)} \circ \cdots \circ g^{(k+1)}\right]^{-1}\left(g(e_{k+1})\right) = \left[g^{(k)} \circ \cdots \circ g^{(1)}\right](e_{k+1})$$
$$= g^{(k)}(e_{k+1})$$
$$= \left(p_k(\theta^k_1, \ldots, \theta^k_k), 0, \ldots, 0\right).$$

Similarly,

$$\left[g'^{(n)} \circ \cdots \circ g'^{(k+1)}\right]^{-1}\left(g(e_{k+1})\right) = \left(p_k(\theta'^k_1, \ldots, \theta'^k_k), 0, \ldots, 0\right).$$

Since $g'^{(l)} = g^{(l)}$ for all $k < l \le n$, we see that $p_k(\theta'^k_1, \ldots, \theta'^k_k) = p_k(\theta^k_1, \ldots, \theta^k_k)$. But, as θ^k_j belongs to $]0, \pi[$ for $2 \le j \le k$, the first two coordinates of $p_k(\theta^k_1, \ldots, \theta^k_k)$ do not vanish simultaneously; thus θ'^k_j belongs to $]0, \pi[$ for $2 \le j \le k$. Since p_k is one-to-one on $[0, 2\pi[\times]0, \pi[^{k-1}$, we conclude that $(\theta'^k_j)_{1 \le j \le k} = (\theta^k_j)_{1 \le j \le k}$. So the proposition can be proved inductively. $\square$

Definition 23.10.1 Under the hypothesis of Proposition 23.10.2, the numbers θ^k_j are called the Euler angles of the rotation g.

Proposition 23.10.3 *Define μ_{n+1} as the measure*

$$c_{n+1} \cdot \bigotimes_{\{(j,k):1 \le j \le k \le n\}} \sin^{j-1}(\theta^k_j)\left(d\theta^k_{j / A^k_j}\right)$$

on A, where $c_{n+1} = \prod_{2 \le k \le n+1}\left(\Gamma(k/2)/2\pi^{k/2}\right)$. Then the normalized Haar measure on $SO(n+1, \mathbf{R})$ is the image measure of μ_{n+1} under the mapping $\omega \mapsto q(\omega)$.

PROOF: Assume that $n \ge 2$ and the proposition is true for $SO(n, \mathbf{R})$. Denote by K the subgroup of $SO(n+1, \mathbf{R})$ consisting of those rotations that fix e_{n+1}. Write β for the normalized Haar measure on K and λ for the uniform measure on $\mathbf{S}^n$ with mass 1, that is,

$$\lambda = \frac{\Gamma((n+1)/2)}{2\pi^{(n+1)/2}} \, d\mathbf{S}^n.$$

For each continuous complex-valued function f on $SO(n+1, \mathbf{R})$, define the mapping $f^\flat$ on $\mathbf{S}^n$ by

$$f^\flat\left(g(e_{n+1})\right) = \int f(g\xi) \, d\beta(\xi) \qquad \text{for all } g \in SO(n+1, \mathbf{R}).$$

Then

$$f^\flat \left(p_n(\theta_1^n, \ldots, \theta_n^n) \right) = \int f(g^{(n)}\xi) \, d\beta(\xi)$$

$$= \int f(g^{(n)} g^{(n-1)} \cdots g^{(1)}) \, d\mu_n \left((\theta_j^k)_{1 \le j \le k \le n-1} \right)$$

for all $(\theta_1^n, \ldots, \theta_n^n)$ in A_n. Now dS^n is the image measure of

$$\sin^{n-1}(\theta_n^n) \cdots \sin(\theta_2^n)(d\theta_1^n \cdots d\theta_n^n)/A_n$$

under p_n. Therefore,

$$\int f^\flat \, d\lambda = \frac{\Gamma((n+1)/2)}{2\pi^{(n+1)/2}}$$

$$\times \int \cdots \int (f^\flat \circ p_n)(\theta_1^n, \ldots, \theta_n^n) \sin^{n-1}(\theta_n^n) \cdots \sin(\theta_2^n) \, (d\theta_1^n \cdots d\theta_n^n)/A_n$$

is equal to

$$\int f(g^{(n)} g^{(n-1)} \cdots g^{(1)}) \, d\mu_{n+1} \left((\theta_j^k)_{1 \le j \le k \le n} \right).$$

Since $\lambda^\sharp : f \mapsto \int f^\flat \, d\lambda$ is the normalized Haar measure on $SO(n+1, \mathbf{R})$ (Proposition 23.9.1), the proof is complete. $\square$

By Proposition 23.10.3, the Euler angles of g are defined for almost all g in $SO(n+1, \mathbf{R})$.

Now, for $SO(3, \mathbf{R})$, we give the geometric meaning of Euler angles.

Let k be a unitary vector in $\mathbf{R}^3$, and let the plane H orthogonal to k be endowed with the orientation for which $(\varepsilon_1, \varepsilon_2, k)$ is direct if $(\varepsilon_1, \varepsilon_2)$ is a direct orthonormal basis of H. For every real number θ, there is a unique rotation $g \in SO(3, \mathbf{R})$ fixing k, whose restriction to H is the rotation of θ with angle θ. It is called the rotation around k with angle θ and is written $r(k, \theta)$.

Let $g \in SO(3, \mathbf{R})$ be such that $g(e_3) \ne e_3$ and $g(e_3) \ne -e_3$. Denote by a the unique vector orthogonal to e_3 and $g(e_3)$ such that $\|a\| = 1$ and $(e_3, g(e_3), a)$ is direct. Now there exist $0 \le \varphi < 2\pi$, $0 \le \psi < 2\pi$, and $0 < \theta < \pi$, such that $r(e_3, \varphi)$ sends e_1 to a, $r(a, \theta)$ sends e_3 to $g(e_3)$, and $r(g(e_3), \psi)$ sends a to $g(e_1)$. Since $r(g(e_3), \psi) \circ r(a, \theta) \circ r(e_3, \varphi)$ maps e_1 to $g(e_1)$ and e_3 to $g(e_3)$, it is equal to g. From the relations

$$r(g(e_3), \psi) = \left[r(a, \theta) \circ r(e_3, \varphi) \right] \circ r(e_3, \psi) \circ \left[r(a, \theta) \circ r(e_3, \varphi) \right]^{-1}$$

and

$$r(a, \theta) = r(e_3, \varphi) \circ r(e_1, \theta) \circ r(e_3, \varphi)^{-1},$$

we deduce that

$$g = r(e_3, \varphi) \circ r(e_1, \theta) \circ r(e_3, \psi).$$

Consequently, $\theta_1^2 = -\varphi$, $\theta_2^2 = -\theta$, and $\theta_1^1 = -\psi$ are the Euler angles of g.

23.11   Haar Measure on $SH(n, \mathbf{R})^{\dagger}$

Let E be a real vector space of finite dimension $n + 1$ (where $n \geq 1$) and Φ be a nondegenerate symmetric bilinear form on $E \times E$, whose signature is $(n, 1)$. Write q for the quadratic form $E \ni x \mapsto \Phi(x, x)$.

Choose an orthogonal basis $(a_1, \ldots, a_{n+1})$ of E such that $q(a_j) = 1$ for $1 \leq j \leq n$ and $q(a_{n+1}) = -1$. Write M for the matrix of Φ in this basis, so that

$$M = \begin{pmatrix} I_n & 0 \\ 0 & -1 \end{pmatrix} = M^{-1}.$$

Finally, let g be an endomorphism of E and g^* its adjoint with respect to Φ. The matrix $U = (\alpha_{i,j})_{1 \leq i,j \leq n+1}$ of g in the basis $(a_1, \ldots, a_{n+1})$ can be written as a matrix of matrices:

$$U = \begin{pmatrix} A & D \\ C & B \end{pmatrix}$$

where $A = (\alpha_{i,j})_{1 \leq i.j \leq n}$ and $B = (\alpha_{n+1.n+1})$. Then the matrix of g^* in the basis $(a_1, \ldots, a_{n+1})$ is

$$U^* = M^{-1} \cdot {}^t U \cdot M = M \cdot {}^t U \cdot M = \begin{pmatrix} {}^t A & -{}^t C \\ -{}^t D & {}^t B \end{pmatrix}.$$

Now ${}^t U \cdot M \cdot U$ is the matrix of $\Phi \circ (g \times g)$ in $(a_1, \ldots, a_{n+1})$ and $U^* \cdot U$ is the matrix $\left(q(a_i) \Phi(g(a_i), g(a_j)) \right)_{1 \leq i,j \leq n+1}$. Therefore, the following conditions are equivalent:

(a) g is a unitary operator with respect to Φ.

(b) $(g(a_1), \ldots, g(a_{n+1}))$ is an orthogonal basis of E such that $q(g(a_j)) = 1$ for $1 \leq j \leq n$ and $q(g(a_{n+1})) = -1$.

(c) $U^* \cdot U = I_{n+1}$.

(d) $U \cdot U^* = I_{n+1}$.

Designate by $O(\Phi)$ the unitary group relative to Φ. For every g in $O(\Phi)$, $\det({}^t U \cdot M \cdot U) = \det(M)$, and so $\det(U)$ is 1 or -1. Since

$$\alpha_{1,n+1}^2 + \cdots + \alpha_{n,n+1}^2 - \alpha_{n+1,n+1}^2 = -1,$$

we have $|\alpha_{n+1,n+1}| \geq 1$.

From the relation $U^* \cdot U = I_{n+1}$, we deduce that

$$\det(U) \cdot {}^t U^* = \det(U) \cdot \begin{pmatrix} A & -D \\ -C & B \end{pmatrix}$$

$$= \det(U) \cdot \left(q(a_i) q(a_j) \alpha_{i,j} \right)_{1 \leq i.j \leq n+1}$$

---

†This section may be omitted.

is the matrix of cofactors of U. Hence $\det(U) \cdot \alpha_{n+1,n+1}$ is the cofactor of $\alpha_{n+1,n+1}$, and $\det(A) \cdot \alpha_{n+1,n+1}$ has the same sign as $\det(U)$.

Let h be the mapping $g \mapsto \big(\mathrm{sgn}(\det A), \mathrm{sgn}(\alpha_{n+1,n+1})\big)$ from $O(\Phi)$ onto $\{-1, 1\} \times \{-1, 1\}$. If g and g' are two unitary operators on E, with matrices U and U', respectively, then

$$\sum_{1 \le k \le n} \alpha_{n+1,k}\alpha'_{k,n+1} + \alpha_{n+1,n+1}\alpha'_{n+1,n+1}$$

is the coefficient of $U \cdot U'$ situated in the $(n+1)$st row and the $(n+1)$st column. Now, from the inequalities

$$\left| \sum_{1 \le k \le n} \alpha_{n+1,k}\alpha'_{k,n+1} \right| \le \left(\sum_{1 \le k \le n} \alpha_{n+1,k}^2 \right)^{1/2} \left(\sum_{1 \le k \le n} \alpha'^2_{k,n+1} \right)^{1/2}$$
$$\le (\alpha_{n+1,n+1}^2 - 1)^{1/2}(\alpha'^2_{n+1,n+1} - 1)^{1/2}$$
$$\le |\alpha_{n+1,n+1}\alpha'_{n+1,n+1}| - 1,$$

follows that

$$\mathrm{sgn}\left(\sum_{1 \le k \le n+1} \alpha_{n+1,k}\alpha'_{k,n+1} \right) = \mathrm{sgn}(\alpha_{n+1,n+1})\,\mathrm{sgn}(\alpha'_{n+1,n+1}).$$

Therefore, $h(gg') = h(g)h(g')$, and h is a group homomorphism.

Next, let $(a'_1, \dots, a'_{n+1})$ be an other orthogonal basis of E such that $q(a'_j) = 1$ for $1 \le j \le n$ and $q(a'_{n+1}) = -1$, and denote by v the endomorphism of E which sends a_j to a'_j for all $1 \le j \le n+1$. Then, for every $g \in O(\Phi)$, the matrix of g in the basis $(a'_1, \dots, a'_{n+1})$ is the matrix of $v^{-1} \circ g \circ v$ in $(a_1, \dots, a_{n+1})$. As $h(g) = h(v^{-1} \circ g \circ v)$, we see that h does not depend on the choice of $(a_1, \dots, a_{n+1})$.

From now on, we take for Φ the bilinear form

$$(x, y) \mapsto \sum_{1 \le j \le n} x_j y_j - x_{n+1}y_{n+1}$$

on $\mathbf{R}^{n+1} \times \mathbf{R}^{n+1}$. The kernel of h is called the group of hyperbolic rotations of $\mathbf{R}^{n+1}$ and is written $SH(n+1, \mathbf{R})$. For simplicity, we also write G instead of $SH(n+1, \mathbf{R})$. If $g \in G$ lies in the stabilizer K of e_{n+1} and $U = (\alpha_{i,j})_{i,j}$ is the matrix of g in the canonical basis of $\mathbf{R}^{n+1}$, then $g(e_{n+1}) = e_{n+1}$ and $g^*(e_{n+1}) = e_{n+1}$ (because $g^* = g^{-1}$), hence $\alpha_{n+1,n+1} = 1$ and $\alpha_{k,n+1} = \alpha_{n+1,k} = 0$ for all $1 \le k \le n$. Therefore, the topological group K is canonically isomorphic to $SO(n, \mathbf{R})$.

Since the mapping $(x_1, \dots, x_n) \mapsto (1 + x_1^2 + \cdots + x_n^2)^{1/2}$ is infinitely differentiable in $\mathbf{R}^n$, its graph

$$H_n = \big\{ x \in \mathbf{R}^{n+1} : q(x) = -1 \text{ and } x_{n+1} \ge 1 \big\}$$

is a connected submanifold of $\mathbf{R}^{n+1}$, and $\varphi : x \mapsto (x_1, \ldots, x_n)$ is a chart of H_n.

Every $g \in O(\Phi)$ maps H_n onto H_n or

$$\{x \in \mathbf{R}^{n+1} : q(x) = -1 \text{ and } x_{n+1} \leq -1\}.$$

Define $O^+(\Phi)$ as the group of those $g \in O(\Phi)$ that map H_n onto H_n. Clearly, $g \in O^+(\Phi)$ lies in G if and only if $\det(g) = 1$.

For every $x \in H_n$, equip the space

$$(Rx)^\circ = \{y \in \mathbf{R}^{n+1} : \Phi(x, y) = 0\}$$

tangent to H_n at x with the orientation for which

$$((d_x\varphi)^{-1}(e_1), \ldots, (d_x\varphi)^{-1}(e_n))$$

is direct ($d_x\varphi$ is the mapping from $(Rx)^\circ$ onto $\mathbf{R}^n$ whose inverse is $D\varphi^{-1}(\varphi x)$). The matrix of $((d_x\varphi)^{-1}(e_1), \ldots, (d_x\varphi)^{-1}(e_n).x)$ with respect to $(e_1, \ldots, e_{n+1})$ is

$$T = \begin{pmatrix} I_n & & & x_1 \\ & & & \vdots \\ & & & x_n \\ \dfrac{x_1}{x_{n+1}} & \dfrac{x_2}{x_{n+1}} & \cdots & \dfrac{x_n}{x_{n+1}} & x_{n+1} \end{pmatrix},$$

and

$$\begin{aligned}
\det(T) &= \frac{1}{x_{n+1}} \det \begin{pmatrix} I_n & & & x_1 \\ & & & \vdots \\ & & & x_n \\ x_1 & x_2 & \cdots & x_n & x_{n+1}^2 \end{pmatrix} \\
&= \frac{1}{x_{n+1}}\left(x_{n+1}^2 - x_1^2 - \cdots - x_n^2\right) \\
&= \frac{1}{x_{n+1}}
\end{aligned}$$

is positive. Therefore, if $(a_1, \ldots, a_n)$ is any direct 'asis of $(Rx)^\circ$, then $(a_1, \ldots, a_n, x)$ is direct for the usual orientation of $\mathbf{R}^{n+1}$.

Now, given $g \in O^+(\Phi)$ and $x \in H_n$, let $(a_1, \ldots, a_n)$ be a direct orthogonal basis of $(Rx)^\circ$ such that $q(a_j) = 1$ for all $1 \leq j \leq n$. Denote by A the matrix of $(a_1, \ldots, a_n, x)$ with respect to $(e_1, \ldots, e_{n+1})$, and by B the matrix of g in the basis $(a_1, \ldots, a_n, x)$. Then AB is the matrix of $(g(a_1), \ldots, g(a_n), g(x))$ with respect to $(e_1, \ldots, e_{n+1})$. Therefore, $(g(a_1), \ldots, g(a_n), g(x))$ is direct in $\mathbf{R}^{n+1}$ if and only if $\det(AB) = \det(A)\det(g)$ is positive, that is, if $\det(g) = 1$. In short, g is compatible with the orientations of $(Rx)^\circ$ and $(Rg(x))^\circ$ if and only if it lies in G.

For all x, y in H_n, if $(a_1, \ldots, a_n)$ and $(b_1, \ldots, b_n)$ are direct orthonormal bases of $(\mathbf{R}x)^\circ$ and $(\mathbf{R}y)^\circ$, respectively, then the endomorphism of $\mathbf{R}^{n+1}$ that sends x to y and each a_j to b_j belongs to G. Hence G acts transitively on H_n.

The mapping $\pi : \dot{g} \mapsto g(e_{n+1})$ from G/K onto H_n is continuous. On the other hand, if, for each $x \in H_n$, we define $s(x)$ as the endomorphism of $\mathbf{R}^{n+1}$ that sends e_{n+1} to x and e_j to $\left(1 - x_j^2/x_{n+1}^2\right)^{-1/2}(d_x\varphi)^{-1}(e_j)$ for $1 \leq j \leq n$, then $s : x \mapsto s(x)$ is continuous from H_n into G and $\pi\big(\overline{s(x)}\big) = x$. Therefore, π is a homeomorphism. As K and G/K are connected, the topological group G is connected (J. Dieudonné, *Treatise on Analysis*, 12.10.12), and so G is the connected component of $O(\Phi)$ containing the unit e of $O(\Phi)$.

Now we construct a positive measure on H_n, invariant under G. For this, we need the following result.

Lemma 23.11.1 *Let A be a commutative ring, and let $M(X_1, \ldots, X_n)$ be the matrix $I_n + (X_i X_j)_{1 \leq i, j \leq n}$, with entries in the ring $A[X_1, \ldots, X_n]$ of polynomials. Then*
$$\det M(X_1, \ldots, X_n) = 1 + X_1^2 + \cdots + X_n^2.$$

PROOF: First, let $M_n'(X_1, \ldots, X_{n-1})$ be the matrix

$$\begin{pmatrix} I_{n-1} + (X_i X_j)_{1 \leq i, j \leq n-1} & \begin{matrix} X_1 \\ X_2 \\ \vdots \\ X_{n-1} \end{matrix} \\ \begin{matrix} X_1 & & X_2 & \cdots & X_{n-1} \end{matrix} & 1 \end{pmatrix}.$$

Subtracting from the first row of $M_n'(X_1, \ldots, X_{n-1})$ the product of X_1 and the last row, we see that

$$\det M_n'(X_1, \ldots, X_{n-1}) = \det M_{n-1}'(X_2, \ldots, X_{n-1})$$

for all $n \geq 2$. Hence $\det M_n'(X_1, \ldots, X_{n-1}) = 1$ for all $n \geq 1$.

Now assume that $n \geq 2$. Let $2 \leq j < n$ be given. Permuting in $M(X_1, \ldots, X_n)$ the columns of indices j and n, and next withdrawing from the matrix so obtained the jth row and the jth column, we obtain the matrix

$$\begin{pmatrix} 1 + X_1^2 & X_1 X_2 & \cdots & X_1 X_{j-1} & X_1 X_{j+1} & \cdots & X_1 X_{n-1} & X_1 X_j \\ \vdots & 1 + X_2^2 & \vdots & \vdots & \vdots & \vdots & \vdots & \vdots \\ \vdots & \vdots & \vdots & 1 + X_{j-1}^2 & \vdots & \vdots & \vdots & \vdots \\ \vdots & \vdots & \vdots & & 1 + X_{j+1}^2 & \vdots & \vdots & \vdots \\ \vdots & \vdots & \vdots & \vdots & \vdots & \vdots & 1 + X_{n-1}^2 & \vdots \\ X_n X_1 & X_n X_2 & \cdots & X_n X_{j-1} & X_n X_{j+1} & \cdots & X_n X_{n-1} & X_n X_j \end{pmatrix}$$

whose determinant is

$$X_j X_n \det M'_{n-1}(X_1, \ldots, X_{j-1}, X_{j+1}, \ldots, X_{n-1}) = X_j X_n.$$

Therefore, the cofactor of $X_j X_n$ in the matrix $M(X_1, \ldots, X_n)$ is $-X_j X_n$. Expanding $\det M(X_1, \ldots, X_n)$ along the last column of $M(X_1, \ldots, X_n)$, we see that

$$\det M(X_1, \ldots, X_n) = (1 + X_n^2) \det M(X_1, \ldots, X_{n-1}) - \sum_{1 \le j \le n-1} X_j^2 X_n^2.$$

Now the lemma follows, by induction on n. □

Taking for A the field $\mathbf{C}$ of complex numbers.

$$\det \left(I_n - \lambda (X_j X_k)_{1 \le j.k \le n} \right) = \det \left(I_n + (i\sqrt{\lambda} X_j i \sqrt{\lambda} X_k)_{1 \le j,k \le n} \right)$$
$$= 1 - \lambda (X_1^2 + \cdots + X_n^2)$$

for all $\lambda > 0$. In particular,

$$\det \left(I_n - \frac{1}{1 + x_1^2 + \cdots + x_n^2} (x_i x_j)_{1 \le i,j \le n} \right) = (1 + x_1^2 + \cdots + x_n^2)^{-1}$$

for all real numbers $x_1, \ldots, x_n$.

For all $x \in H_n$, write $m(x)$ for the restriction of Φ to $(\mathbf{R}x)^\circ \times (\mathbf{R}x)^\circ$, so that $x \mapsto m(x)$ is a Riemannian metric on H_n. Then

$$\tilde{m}(\varphi x) = m(x) \circ \left((d_x \varphi)^{-1} \times (d_x \varphi)^{-1} \right)$$

is the bilinear form

$$(v, w) \mapsto \langle v, w \rangle - \frac{1}{x_{n+1}^2} \langle v, \varphi x \rangle \langle w, \varphi x \rangle$$

on $\mathbf{R}^n \times \mathbf{R}^n$ (where $\langle \, , \rangle$ is the usual scalar product of $\mathbf{R}^n$). The matrix of $\tilde{m}(\varphi x)$ in the canonical basis of $\mathbf{R}^n$ is

$$I_n - \frac{1}{x_{n+1}^2} (x_i x_j)_{1 \le i,j \le n}.$$

Hence the discriminant of $\tilde{m}(\varphi x)$ is $(1 + x_1^2 + \cdots + x_n^2)^{-1}$.

Fix $g \in G$ and denote by g' the mapping $x \mapsto g(x)$ from H_n onto H_n. For all $x \in H_n$, we have $m(y) \circ \left(g_{/(\mathbf{R}x)^\circ} \times g_{/(\mathbf{R}x)^\circ} \right) = m(x)$ (where $y = g(x)$), and so

$$\tilde{m}(\varphi y) \circ \left[D(\varphi g' \varphi^{-1})(\varphi x) \times D(\varphi g' \varphi^{-1})(\varphi x) \right] = \tilde{m}(\varphi x),$$

whence it follows that the discriminant $\text{dis}(\tilde{m}(\varphi x))$ of $\tilde{m}(\varphi x)$ is equal to

$$\text{dis}(\tilde{m}(\varphi y)) \cdot \left| \det D(\varphi g' \varphi^{-1})(\varphi x) \right|^2.$$

In short,

$$\frac{1}{y_{n+1}} \left| \det D(\varphi g' \varphi^{-1})(\varphi x) \right| = \frac{1}{x_{n+1}},$$

and the formula for change of variables proves that the measure

$$(1 + x_1^2 + \cdots + x_n^2)^{-1/2} dx_1 \cdots dx_n$$

on $\mathbf{R}^n$ is invariant under $\varphi g' \varphi^{-1}$. Therefore, dH_n, the image measure of $(1 + x_1^2 + \cdots + x_n^2)^{-1/2} dx_1 \cdots dx_n$ under φ^{-1}, is invariant under g'.

Definition 23.11.1 dH_n is called the Riemannian volume of H_n.

We retain the notation of Section 23.10.

For each integer $n \geq 2$, let f_n be the mapping from $\mathbf{R}^n$ onto H_n that sends every $(\theta_1, \ldots, \theta_n)$ to $\big(\sinh(\theta_n) p_{n-1}(\theta_1, \ldots, \theta_{n-1}), \cosh(\theta_n) \big)$. Also, let f_1 be the function $\theta \mapsto (\sinh \theta, \cosh \theta)$ from $\mathbf{R}$ onto H_1. For every $n \geq 1$ and for every $\theta \in \mathbf{R}^n$, write $J''(\theta)$ for the matrix of $\big(Df_n(\theta)e_1, \ldots, Df_n(\theta)e_n, f_n(\theta) \big)$ in the canonical basis of $\mathbf{R}^{n+1}$.

If $n \geq 2$, write $K''(\theta)$ for the $n \times n$ matrix whose first row is obtained by multiplying the first row of $J''(\theta_2, \ldots, \theta_n)$ by $\cos \theta_1$, and whose ith row, for all $2 \leq i \leq n$, is equal to the ith row of $J''(\theta_2, \ldots, \theta_n)$. Similarly, write $L''(\theta)$ for the $n \times n$ matrix whose first row is obtained by multiplying the first row of $J''(\theta_2, \ldots, \theta_n)$ by $\sin \theta_1$, and whose ith row, for all $2 \leq i \leq n$, is equal to the ith row of $J''(\theta_2, \ldots, \theta_n)$. Now, developing $\det \big(J''(\theta) \big)$ along the first column of $J''(\theta)$, we see that $\det \big(J''(\theta) \big)$ is equal to

$$\sinh(\theta_n) \sin(\theta_{n-1}) \cdots \sin(\theta_2) \cos(\theta_1) \det K''(\theta)$$
$$+ \sinh(\theta_n) \sin(\theta_{n-1}) \cdots \sin(\theta_2) \sin(\theta_1) \det L''(\theta),$$

and so

$$\det J''(\theta) = \sinh(\theta_n) \sin(\theta_{n-1}) \cdots \sin(\theta_2) \det J''(\theta_2, \ldots, \theta_n).$$

By induction on n,

$$\det J''(\theta) = \sinh^{n-1}(\theta_n) \sin^{n-2}(\theta_{n-1}) \cdots \sin(\theta_2).$$

For every $n \geq 1$ and for every $\theta \in \mathbf{R}^n$, let $J(\varphi \circ f_n)(\theta)$ be the Jacobian matrix of $\varphi \circ f_n$ at θ. For all $1 \leq i \leq n+1$ and $1 \leq j \leq n+1$, we let $\alpha_{i,j}$ be the entry of $J''(\theta)$ situated in the ith row and the jth column. Then, for $n \geq 2$, the cofactor of $\alpha_{n+1,n+1} = \cosh(\theta_n)$ in $J''(\theta)$ is $\det \big(J(\varphi \circ f_n)(\theta) \big)$, whereas the cofactor of $\alpha_{n+1,n} = \sinh(\theta_n)$ is $- \det \big(\sinh(\theta_n) J'(\theta_1, \ldots, \theta_{n-1}) \big)$. Developing $\det J''(\theta)$ along the last row of $J''(\theta)$, we see, therefore, that

$$\det J''(\theta) = -\sinh^{n+1}(\theta_n) \sin^{n-2}(\theta_{n-1}) \cdots \sin(\theta_2)$$
$$+ \cosh(\theta_n) \det \big(J(\varphi \circ f_n)(\theta) \big)$$

and

$$\det\big(J(\varphi \circ f_n)(\theta)\big) =$$
$$\frac{1}{\cosh(\theta_n)}\big(1 + \sinh^2(\theta_n)\big)\sinh^{n-1}(\theta_n)\sin^{n-2}(\theta_{n-1})\cdots\sin(\theta_2).$$

Finally,

$$\det\big(J(\varphi \circ f_n)(\theta)\big) = \cosh(\theta_n)\sinh^{n-1}(\theta_n)\sin^{n-2}(\theta_{n-1})\cdots\sin(\theta_2).$$

From this last equality follows, for instance, that f_n is a diffeomorphism from $P =]0,2\pi[\times]0,\pi[^{n-2}\times]0,+\infty[$ onto

$$H_n - \{x \in H_n : x_1 = 0 \text{ and } x_2 \geq 0\}.$$

The inverse diffeomorphism is a chart of H_n that preserves the orientation of H_n.

More important, the previous calculus of $\det\big(J(\varphi \circ f_n)(\theta)\big)$ shows that dH_n is the image measure of

$$\sinh^{n-1}(\theta_n)\sin^{n-2}(\theta_{n-1})\cdots\sin(\theta_2)(d\theta_1\cdots d\theta_n)_{/P}$$

under f_n.

For all $n \geq 2$ and for all $\alpha \in \mathbf{R}$, define $h_n(\alpha)$ as the hyperbolic rotation of $\mathbf{R}^{n+1}$ whose matrix with respect to $(e_1,\ldots,e_{n+1})$ is

$$\begin{pmatrix} I_{n-1} & 0 & 0 \\ 0 & \cosh(\alpha) & \sinh(\alpha) \\ 0 & \sinh(\alpha) & \cosh(\alpha) \end{pmatrix}.$$

Set $B = \prod_{\{(j,k):1\leq j\leq k\leq n\}} B_j^k$, where $B_1^k = [0,2\pi[$, $B_j^k = [0,\pi]$ for $2 \leq j \leq \inf(k,n-1)$, and $B_n^n = [0,+\infty[$. For each $\omega = (\theta_j^k)_{1\leq j\leq k\leq n}$ in B, put $r(\omega) = h^{(n)} \circ \cdots \circ h^{(1)}$, where

$$h^{(k)} = g_1(\theta_1^k)\cdots g_k(\theta_k^k) \qquad \text{for } 1 \leq k \leq n-1$$

and

$$h^{(n)} = g_1(\theta_1^n)\cdots g_{n-1}(\theta_{n-1}^n)h_n(\theta_n^n).$$

Arguing exactly as in Section 23.10, we see that $r : \omega \mapsto r(\omega)$ maps B onto $SH(n+1,\mathbf{R})$. Moreover, if $\omega = (\theta_j^k)_{1\leq j\leq k\leq n}$ is an element of B such that $\theta_n^n \in]0,+\infty[$ and $\theta_j^k \in]0,\pi[$ for $2 \leq j \leq \inf(k,n-1)$, then $g = r(\omega)$ is the image under r of a unique element of B.

Finally, for $n \geq 2$, define ν_{n+1}, as the measure

$$\left[\sinh^{n-1}(\theta_n^n)\sin^{n-2}(\theta_{n-1}^n)\cdots\sin(\theta_2^n)(d\theta_1^n\cdots d\theta_n^n)_{/B_n}\right] \otimes \mu_n$$

on B, where $B_n = [0,2\pi[\times[0,\pi]^{n-2}\times[0,+\infty[$ and μ_n has been defined in Section 23.10. By Proposition 23.9.1, the image measure of ν_{n+1} under $\omega \mapsto r(\omega)$ is a left Haar measure on $SH(n+1,\mathbf{R})$.

We shall show in the next chapter that (G,K) is a Gelfand pair, whence it follows that $SH(n+1,\mathbf{R})$ is unimodular.

Exercises for Chapter 23

1 Let G be a locally compact group, μ a left Haar measure on G, and A a μ-integrable subset of G. Show that the function $s \mapsto \mu(A \triangle sA)$ is continuous on G and uniformly continuous with respect to the left uniform structure of G (first, suppose that A is compact).

2 Let G be a locally compact group, μ a left Haar measure on G, A and B two subsets of G, and f the function $s \mapsto \mu^*(sA \cap B)$ from G into $[0, +\infty]$.

1. Suppose that A is μ-integrable. Then, for all $s, t \in G$,

$$\left| \mu^*(sA \cap B) - \mu^*(tA \cap B) \right| \le \mu^*\big((sA \cap B) \triangle (tA \cap B)\big) \le \mu^*(sA \triangle tA).$$

Deduce from this inequality and Exercise 1 that f is uniformly continuous with respect to the left uniform structure of G.

2. Suppose that $\mu^*(A)$ is finite and B is μ-measurable. Let $(A_n)_{n \ge 1}$ be a decreasing sequence of μ-integrable sets containing A, such that $\mu^*(A) = \inf \mu(A_n)$. Given $\varepsilon > 0$, let $n \in \mathbf{N}$ be such that $\mu(A_n) \le \mu^*(A) + \varepsilon/3$. Show that $0 \le \mu(sA_n \cap B) - \mu^*(sA \cap B) \le \varepsilon/3$. Then deduce from part 1 that f is uniformly continuous with respect to the left uniform structure of G.

3. Suppose that A is μ-measurable and $\mu^*(B)$ is finite. Deduce from part 2 that f is uniformly continuous with respect to the right uniform structure of G.

4. Suppose that A^{-1} and B are μ-integrable. Deduce from Lemma 24.2.1 that $\mu \otimes \mu$ is invariant under $(s, x) \mapsto (sx, x^{-1})$ and that

$$\int^* f(s)\, d\mu(s) = \int^* d\mu(s) \int^* 1_B(st) 1_{A^{-1}}(t^{-1})\, d\mu(t) = \mu(A^{-1}) \mu(B).$$

5. Suppose that A^{-1} is μ-integrable and $\mu^*(B)$ is finite. Let $(B_n)_{n \ge 1}$ be a decreasing sequence of μ-integrable sets containing B such that $\mu^*(B) = \inf_{n \ge 1} \mu(B_n)$. Show that $\mu^*(sA \cap B) = \inf \mu(sA \cap B_n)$. Then, from parts 3 and 4, deduce that $\int f(s)\, d\mu(s) = \mu(A^{-1}) \mu^*(B)$.

6. Suppose that A is μ-measurable and $\mu^*(B)$ is finite. Prove that $\int^* f(s)\, d\mu(s) = \mu^\bullet(A^{-1}) \mu^*(B)$ (distinguish the cases $\mu^\bullet(A^{-1}) < +\infty$ and $\mu^\bullet(A^{-1}) = +\infty$).

7. Suppose that A is μ-measurable, and that neither A nor B is locally μ-negligible. Choose a subset C of B such that $0 < \mu^*(C) < +\infty$. By parts 3 and 6, the function $g : s \mapsto \mu^*(sA \cap C)$ is continuous on G and $\int^* g\, d\mu = \mu^\bullet(A^{-1}) \mu^*(C)$. Moreover, CA^{-1} contains $\{s \in G : g(s) > 0\}$. Deduce that BA^{-1} has an interior point. Conclude that BA and AB each have an interior point.

3 Write $\mathbf{T}$ for the multiplicative group of complex numbers with modulus 1 and ν for the normalized Haar measure on $\mathbf{T}$. Let θ_1 and θ_2 be two real numbers such that the equation $n_1 \theta_1 + n_2 \theta_2 = p$ has no other solution (n_1, n_2, p) in $\mathbf{Z}^3$ than $(0, 0, 0)$. Denote by B the subgroup $\left\{ \exp\big(2i\pi(n_1 \theta_1 + n_2 \theta_2)\big) : n_1 \in 2\mathbf{Z}, n_2 \in \mathbf{Z} \right\}$ of $\mathbf{T}$, by C the subset $e^{2i\pi\theta_1} B$ of $\mathbf{T}$, and put $A = B \cup C$.

1. We define on $\mathbf{T}$ an equivalence relation $\sim$ as follows: $\xi \sim \zeta$ if and only if $\xi^{-1}\zeta$ lies in A. Let E be a subset of T having one and only one element in each equivalence class. Show that EB and EC partition $\mathbf{T}$ and that B and C are dense in $\mathbf{T}$ (use Section 4.3).

2. Show that the function $s \mapsto \nu^*(sE \cap BE)$ from $\mathbf{T}$ into $\mathbf{R}^+$ is everywhere discontinuous, though $\nu^*(E)$ and $\nu^*(BE)$ are finite. Compare this result with part 3 of Exercise 2.

4 Let G be a locally compact group, μ a left Haar measure on G, and A a μ-integrable subset of G such that $\mu(A) > 0$. Denote by $H(A)$ the set of those $s \in G$ for which $\mu(A) = \mu(A \cap sA)$.

1. Show that $H(A)$ is a closed subgroup of G (use part 1 of Exercise 2).

2. Let B be a compact subset of A such that $\mu(B) > (1/2)\mu(A)$. Show that, for $s \in H(A)$, B and sB cannot be disjoint. Conclude that $H(A)$ is compact.

5 Let G be a commutative locally compact group, written additively. Let μ be a Haar measure on G, and let A, B be two integrable subsets of G.

1. For each $s \in G$, let $A' = A \cup (B + s)$ and $B' = (A - s) \cap B$. Show that $\mu(A') + \mu(B') = \mu(A) + \mu(B)$ and that $A' + B' \subset A + B$.

2. Henceforth, suppose that A is nonempty and 0 lies in B. Construct a sequence $((A_n, B_n))_{n \geq 0}$ with the following properties:

    (a) for all $k \geq 0$, A_k and B_k are nonempty μ-integrable subsets of G, and 0 belongs to B_k;

    (b) $(A_0, B_0) = (A, B)$;

    (c) for all $k \geq 0$, there exists s_{k+1} in A_k such that $A_{k+1} = A_k \cup (B_k + s_{k+1})$ and $B_{k+1} = (A_k - s_{k+1}) \cap B_k$;

    (d) $\mu(B_{k+1}) \leq 2^{-k} + \inf_{s \in A_k} \mu((A_k - s) \cap B_k)$.

    Next, defining A_∞ and B_∞ as $\bigcup_{n \geq 0} A_n$ and $\bigcap_{n \geq 0} B_n$, respectively, show that $\mu(A_\infty) + \mu(B_\infty) = \mu(A) + \mu(B)$. Finally, prove that $\mu((A_\infty - s) \cap B_\infty) = \mu(B_\infty)$ for every $s \in A_\infty$.

3. Suppose that $\mu(B_\infty) > 0$. By parts 1 and 4 of Exercise 2, the function $f : s \mapsto \mu((A_\infty - s) \cap B_\infty) = \mu(A_\infty \cap (s + B_\infty))$ is continuous on G and $\int f \, d\mu = \mu(-B_\infty)\mu(A_\infty) = \mu(A_\infty)\mu(B_\infty)$. Moreover, f takes the value $\mu(B_\infty)$ on A_∞. Prove that f takes only the values 0 and $\mu(B_\infty)$, and that $C = f^{-1}(\mu(B_\infty))$ is a clopen subset of G. Show that $\mu(C) = \mu(A_\infty)$ and that C is the closure of A_∞.

4. Suppose that $\mu(B_\infty) > 0$. We define D as the set of all $s \in B_\infty$ such that the intersection of B_∞ with every neighborhood of s is nonnegligible. So $D = \mathrm{supp}(1_{B_\infty}\mu) \cap B_\infty$. Prove that $\mu(D) = \mu(B_\infty)$. If $s \in A_\infty$ and $t \in D$, show that $(s + V) \cap A_\infty \neq \emptyset$ for every compact neighborhood V of t. Conclude that $A_\infty + D \subset C$ and that $C + D \subset C$. Consider $H(C) = \{s \in G : \mu((C - s) \cap C) = \mu(C)\}$. For each $s \in H(C)$, prove that

the complement of $(C - s) \cap C$ in C is empty and that an element $s \in G$ lies in $H(C)$ if and only if $C + s$ is included in C. Hence D is included in $H(C)$ and $C + H(C) = C$. Deduce from part 7 of Exercise 2 that $H(C)$ has nonempty interior. Conclude that $H(C)$ is both open and compact.

5. Suppose that $\mu(B_\infty) > 0$. If s lies in C, then $[C + H(C)] \cap (s - D) = s - D$ (because $-D \subset H(C)$), therefore,

$$\mu\big(A_\infty \cap (s - B_\infty)\big) = \mu\big(C \cap (s - D)\big) = \mu(s - D) = \mu(B_\infty).$$

Deduce that C is included in $A + B$. From the fact that $\mu(C) + \mu(D) = \mu(A_\infty) + \mu(B_\infty) = \mu(A) + \mu(B)$, and since $D \subset H(C)$, deduce that $\mu(C) \geq \mu(A) + \mu(B) - \mu\big(H(C)\big)$.

6. If $\mu(B_\infty) = 0$, show that $\mu_*(A + B) \geq \mu(A) + \mu(B)$.

7. Suppose now that A, B are nonempty. We do not require that B contains 0. Choose $t \in B$. Then $\mu_*(A + B) = \mu_*\big(A + (B - t)\big)$ and $\mu(A) + \mu(B) = \mu(A) + \mu(B - t)$. Deduce from parts 4, 5, and 6 that

   (a)   either $\mu_*(A + B) \geq \mu(A) + \mu(B)$ or

   (b)   there exists an open compact subgroup H of G such that $A + B$ contains a class of H in G, and $\mu_*(A + B) \geq \mu(A) + \mu(B) - \mu(H)$.

8. Assume that G is connected and not compact. Prove that $\mu_*(A + B) \geq \mu(A) + \mu(B)$ for all nonempty μ-integrable subsets A and B of G.

6 Let A (respectively, B) be the set of real numbers x whose proper dyadic expansion $x = x_0 + \sum_{n \geq 1}(x_n/2^n)$ (with $x_0 \in \mathbf{Z}$) is such that $x_n = 0$ for $n > 0$ even (respectively, for n odd).

1. Let π be the mapping $(x_n)_{n \geq 1} \mapsto \sum_{n \geq 1}(x_n/2^n)$ from $\Omega = \{0, 1\}^{\mathbf{N}}$ onto $[0, 1]$, and let μ be the usual product measure on the locally compact space Ω. If λ designates Lebesgue measure on $\mathbf{R}$, we recall that $\pi(\mu) = \lambda_{/[0,1]}$. Prove that $A \cap [0, 1[$ is λ-negligible, hence that A is λ-negligible. Similarly, B is λ-negligible, and yet $A + B = \mathbf{R}$.

2. Regarding $\mathbf{R}$ as a vector space over the field $\mathbf{Q}$ of rational numbers, show that $A \cup B$ contains a basis H of $\mathbf{R}$ and that $P_1 = \bigcup_{r \in \mathbf{Q}} rH$ is λ-negligible.

3. For each $n \in \mathbf{N}$, we denote by P_n the set of those real numbers x which have at most n nonzero coordinates relative to the basis H. Given $n \in \mathbf{N}$, assume that P_n is λ-negligible and that P_{n+1} is λ-measurable but nonnegligible. Choose $h \in H$, and write S for the set of those $x \in P_{n+1}$ whose coordinate with respect to h is nonzero. Prove that S is λ-negligible. Thus, for some suitable $k \in \mathbf{Z}$, $E = P_{n+1} \cap S^c \cap [k, k+1[$ is nonnegligible. Since the function $x \mapsto \lambda\big((x + E) \cap E\big)$ is continuous on $\mathbf{R}$ by Exercise 2, it cannot vanish identically on $(\mathbf{Q} - \{0\})h$. Prove that this constitutes a contradiction with our original assumption.

4. Deduce from part 3 that, for some suitable integer $k \geq 2$, the sets $P_1, \ldots, P_{k-1}$ are λ-negligible, but P_k is not λ-measurable. Put $C = P_1$, $D = P_{k-1}$, and prove that $C + D$ is not λ-measurable, even though C and D are λ-negligible.

7 Let G be the multiplicative group of matrices $\begin{pmatrix} a & b \\ 0 & 1 \end{pmatrix}$ such that $a > 0$ and $b \in \mathbf{R}$. We may identify G with the group consisting of the mappings $\mathbf{R} \ni x \mapsto ax + b$.

Now we identify G with the half-plane $\{(a, b) : a > 0, \ b \in \mathbf{R}\} = P$. Show that $(1/a^2)da\,db_{/P}$ is a left Haar measure on G and that $(1/a)da\,db_{/P}$ is a right Haar measure on G. Prove that $\Delta_G\left(\begin{pmatrix} a & b \\ 0 & 1 \end{pmatrix}\right) = 1/a$ for all $a > 0$ and $b \in \mathbf{R}$.

8 We denote by $GL(2, \mathbf{R})$ (respectively, $SL(2, \mathbf{R})$) the multiplicative group of matrices $M = \begin{pmatrix} a & b \\ c & d \end{pmatrix}$ (a, b, c, d real numbers) such that $\det(M) \neq 0$ (respectively, $\det(M) = 1$), by $SO(2, \mathbf{R})$ the subgroup of $SL(2, \mathbf{R})$ consisting of all matrices $\begin{pmatrix} a & b \\ c & d \end{pmatrix}$ such that $a = d$, $b = -c$ and $ad - bc = 1$. Since $SL(2, \mathbf{R})$ is the subgroup of $GL(2, \mathbf{R})$ generated by the commutators of $GL(2, \mathbf{R})$, it is unimodular. We write P for the half-plane $\{z \in \mathbf{C} : \mathrm{Im}(z) > 0\}$.

1. Show that $SL(2, \mathbf{R})$ operates on P from the left via the mapping
$$\left(\begin{pmatrix} a & b \\ c & d \end{pmatrix}, z\right) \longmapsto \frac{az + b}{cz + d}.$$

   that $SO(2, \mathbf{R})$ is the stabilizer of i ($i^2 = -1$), and that $SL(2, \mathbf{R})$ acts transitively on P.

2. Write λ for the measure $(1/y^2)(dx \otimes dy)_{/P}$ on P. Fix $\begin{pmatrix} a & b \\ c & d \end{pmatrix}$ in $SL(2, \mathbf{R})$, and let u be the mapping $z \mapsto (az + b)/(cz + d)$ from P onto P. Show that $\det\big(Du(z)\big) = 1/|cz + d|^4$ for all $z \in P$ and that λ is invariant under u.

3. Denote by β the normalized Haar measure on $SO(2, \mathbf{R})$. For every f in $\mathcal{H}\big(SL(2, \mathbf{R}), \mathbf{C}\big)$, define a function f' on P by
$$f'\left(\frac{ai + b}{ci + d}\right) = \int f\left(\begin{pmatrix} a & b \\ c & d \end{pmatrix}\xi\right) d\beta(\xi).$$

   Deduce from Proposition 23.9.1 that $\lambda^\sharp : f \mapsto \lambda(f')$ is a Haar measure on $SL(2, \mathbf{R})$.

In fact, if K is any commutative nondiscrete, locally compact field, it is possible to describe Haar measure on $SL(n, K)$ (N. Bourbaki, *Intégration*, Chapter 7).

24

Convolution of Measures

Convolution of measures on a locally compact group G is of first importance in analysis. $L^1(G)$ being a Banach algebra, one can apply general theorems pertaining to these algebras. Besides, it is very interesting to decompose the left regular representation of G on $L^1(G)$.

Summary

24.1   We define convolution of measures. Two bounded measures are always convolvable (Proposition 24.1.2).

24.2   We supply some necessary and sufficient conditions for a measure and a function to be convolvable (Propositions 24.2.2 and 24.2.3).

24.3   If μ is a measure on G (respectively, a measure with compact support) and if f is a continuous function on G with compact support (respectively, a continuous function), then μ and f are convolvable.

24.4   For every bounded measure μ on G and for every function f in $\overline{L^p}(G)$ (with $1 \leq p \leq +\infty$), μ and f are convolvable (Theorems 24.4.1 and 24.4.2).

24.5   For $1 \leq p < +\infty$, the endomorphism $g \mapsto \check{\mu} * g$ of $L_C^p(G)$ is the transpose of $f \mapsto \mu * f$ (Proposition 24.5.2).

24.6   We study convolution of functions on G.

24.7   We enunciate a few results on regularization of functions (Theorems 24.7.1 and 24.7.2).

24.8   We define Gelfand pairs (important in harmonic analysis). We show that $\big(SO(n+1, \mathbf{R}), SO(n, \mathbf{R})\big)$ and $\big(SH(n+1, \mathbf{R}), SO(n, \mathbf{R})\big)$ are Gelfand pairs.

In this chapter, all locally compact spaces are supposed to be Hausdorff.

24.1   Convolvable Measures

Definition 24.1.1 Let $(X_i)_{i \in I}$ be a finite family of locally compact spaces $(I \neq \emptyset)$ and φ a mapping from $\prod_{i \in I} X_i$ into a locally compact space Y. For each $i \in I$ let μ_i be a Radon measure on X_i, and write μ for $\bigotimes_{i \in I} \mu_i$. We say that the family $(\mu_i)_{i \in I}$ is φ-convolvable if φ is μ-proper (Definition 21.1.1); in this case, $\nu = \varphi(\mu)$ is called the convolution (or convolution product) of the family $(\mu_i)_{i \in I}$ (relative to φ) and is written $*^\varphi (\mu_i)_{i \in I}$ or $*_{i \in I} \mu_i$.

Of course, the notation $*_{i \in I} \mu_i$ can be used only when there is no ambiguity on φ.

$(\mu_i)_{i \in I}$ is φ-convolvable if and only if $(|\mu_i|)_{i \in I}$ is. In this case, $| *_{i \in I} \mu_i | \leq *_{i \in I} |\mu_i|$; moreover, $(\nu_i)_{i \in I}$ is φ-convolvable for all measures ν_i such that $|\nu_i| \leq |\mu_i|$.

Now let $(X_i)_{i \in I}$ be a finite family of locally compact spaces $(I \neq \emptyset)$, and let $(I_j)_{j \in J}$ be a partition of I into nonempty sets. For each $j \in J$, let φ_j be a mapping from $\prod_{i \in I_j} X_i$ into a locally compact space Y_j; also, let ψ be a mapping from $\prod_{j \in J} Y_j$ into a locally compact space Z. Write φ for the mapping

$$\prod_{i \in I} X_i \ni (x_i)_{i \in I} \longmapsto \psi\left(\left(\varphi_j((x_i)_{i \in I_j})\right)_{j \in J}\right).$$

For each $i \in I$, consider μ_i, a measure on X_i.

Proposition 24.1.1

(a) *Suppose that* $(\mu_i)_{i \in I_j}$ *is* φ_j-*convolvable, for every* j *in* J, *and that* $\left(*_{i \in I_j} |\mu_i|\right)_{j \in J}$ *is* ψ-*convolvable. Then* $(\mu_i)_{i \in I}$ *is* φ-*convolvable and*

$$*^\varphi (\mu_i)_{i \in I} = *^\psi \left(*^{\varphi_j} (\mu_i)_{i \in I_j} \right)_{j \in J} \qquad (1).$$

(b) *Suppose that* ψ *and the* φ_j *are continuous and that* $(\mu_i)_{i \in I}$ *is* φ-*convolvable. If* $(\mu_i)_{i \in I_j}$ *is* φ_j-*convolvable for every* j *in* J, *then* $\left(*_{i \in I_j} |\mu_i|\right)_{j \in J}$ *is* ψ-*convolvable and formula (1) holds.*

PROOF: This results from Proposition 21.1.1.   □

Under the hypothesis of Proposition 24.1.1, part (b), it can be shown, for fixed $j \in J$, that $(\mu_i)_{i \in I_j}$ is φ_j-convolvable whenever $\bigotimes_{i \in I \cap I_j^c} \mu_i$ is nonzero.

Henceforth, let φ be as in Definition 24.1.1.

Proposition 24.1.2 *Let* μ_i *be a bounded measure on* X_i *(where* $i \in I$). *Write* μ *for* $\bigotimes_{i \in I} \mu_i$, *and suppose that* φ *is* μ-*measurable. Then* $(\mu_i)_{i \in I}$ *is* φ-*convolvable and* $\| *_{i \in I} \mu_i \| \leq \prod_{i \in I} \|\mu_i\|$. *In fact,* $\| *_{i \in I} \mu_i \| = \prod_{i \in I} \|\mu_i\|$ *whenever the* μ_i *are positive.*

PROOF: Clearly, φ is μ-proper and $\left\| *_{i \in I} \mu_i \right\| \leq \int^* 1 \, d\left(\bigotimes_i |\mu_i| \right) = \prod_{i \in I} \|\mu_i\|$. $\qquad \square$

Proposition 24.1.3 *Suppose φ is continuous. Let $\mu_i \in \mathcal{M}^1(X_i, \mathbf{C})$ (where $i \in I$) and, given $j \in I$, define $\mathcal{F}$ as a filter on $\mathcal{M}^1(X_j, \mathbf{C})$ converging narrowly to μ_j. Let p be the mapping $\mathcal{M}^1(X_j, \mathbf{C}) \ni \nu_j \mapsto *_{i \in I} \nu_i$, where $\nu_i = \mu_i$ for $i \neq j$. Then $p(\mathcal{F})$ converges narrowly to $*_{i \in I} \mu_i$.*

PROOF: Let $f : Y \to \mathbf{C}$ be bounded and continuous. For each $x_j \in X_j$, denote by f_{x_j} the function

$$(x_i)_{i \in I - \{j\}} \longmapsto (f \circ \varphi)((x_i)_{i \in I}) \qquad \text{from } X_j' = \prod_{i \in I - \{j\}} X_i \text{ into } \mathbf{C}.$$

Fix $a_j \in X_j$. Now, for every $\varepsilon > 0$, there exists a compact subset K of X_j' such that $|\nu|(X_j' - K) \leq \varepsilon/(4\|f\|)$ (where $\nu = \bigotimes_{i \in I - \{j\}} \mu_i$). For x_j close enough to a_j, we have $\left| \int_K (f_{x_j} - f_{a_j}) \, d\nu \right| \leq \varepsilon/2$, hence $\left| \int (f_{x_j} - f_{a_j}) \, d\nu \right| \leq \varepsilon$.

In short, $g : x_j \mapsto \int f_{x_j} \, d\nu$ is continuous (and bounded).

We conclude that $\int f \, d(*_{i \in I} \nu_i) = \int g \, d\nu_j$ converges to $\int f \, d(*_{i \in I} \mu_i)$ as $\nu_j \to \mu_j$ along $\mathcal{F}$. $\qquad \square$

Proposition 24.1.4 *For all $i \in I$, let μ_i be a measure on X_i with support S_i. Suppose that φ is continuous and the restriction of φ to $S = \prod_{i \in I} S_i$ is proper (Definition 23.5.1). Then $(\mu_i)_{i \in I}$ is φ-convolvable and $\operatorname{supp}(*_{i \in I} \mu_i)$ is included in the closure of $\varphi(S)$.*

PROOF: If K is a compact subset of Y, then $\varphi^{-1}(K) \cap S$ is compact. Since S^c is μ-negligible (where $\mu = \bigotimes_{i \in I} \mu_i$), we see that $\varphi^{-1}(K)$ is μ-integrable. Therefore, φ is μ-proper. $\qquad \square$

24.2 Convolution of a Measure and a Function

Let G be a locally compact group acting from the left on a locally compact space X.

Let $I = \{i_1, \ldots, i_n\}$ be a finite totally ordered set ($i_1 < i_2 < \cdots < i_n$). Write $X_i = G$ for $i \in I - \{i_n\}$, and $X_{i_n} = X$. For all $i \in I$, let μ_i be a measure on X_i. The measures $\mu_{i_1}, \ldots, \mu_{i_n}$ are said to be convolvable whenever $(\mu_i)_{i \in I}$ is convolvable with respect to $\varphi : (x_i)_{i \in I} \mapsto \left(\prod_{1 \leq m \leq n-1} x_{i_m} \right) x_{i_n}$. In this case, $*^\varphi(\mu_i)_{i \in I}$ is written $\mu_{i_1} * \mu_{i_2} * \cdots * \mu_{i_n}$.

This holds, in particular, when $X = G$ and the action of G on itself is $(s, t) \mapsto st$.

Under this notation, we leave to the reader the translation of Proposition 24.1.1.

Now suppose there is a relatively invariant, positive measure β on X (with $\beta \neq 0$). We write χ for the multiplier of β.

Proposition 24.2.1 *Let $\mu \in \mathcal{M}(G, \mathbf{C})$, and let ν be a measure on X with base β. Suppose that μ and ν are convolvable. Then $\mu * \nu$ is a measure with base β.*

PROOF: We may suppose that μ and ν are positive.

Let $h : X \to [0, +\infty]$ be lower semicontinuous. Since the function $(s, x) \mapsto h(sx)$ is lower semicontinuous,

$$\int^* h \, d(\mu * \nu) = \int^\bullet h \, d(\mu * \nu) = \int^\bullet h(sx) \, d(\mu \otimes \nu)(s, x)$$

$$= \int^* h(sx) \, d(\mu \otimes \nu)(s, x)$$

$$= \int^* d\mu(s) \int^* h(sx) \, d\nu(x).$$

Furthermore, $s \mapsto \int^* h(sx) \, d\nu(x)$ is lower semicontinuous.

Now let K be a compact subset of X, and let $g \in \mathcal{H}^+(X)$ be such that $g \geq 1_K$. Then

$$\int^* g \, d(\mu * \nu) = \int^* d\mu(s) \int^* g(sx) \, d\nu(x),$$

and

$$\int^* (g - 1_K) \, d(\mu * \nu) = \int^* d\mu(s) \int^* (g - 1_K)(sx) \, d\nu(x).$$

Since $s \mapsto \int g(sx) \, d\nu(x)$ and $s \mapsto \int (g - 1_K)(sx) \, d\nu(x)$ are μ-integrable, we obtain

$$\int 1_K \, d(\mu * \nu) = \int g \, d(\mu * \nu) - \int (g - 1_K) \, d(\mu * \nu)$$

$$= \int d\mu(s) \int 1_K(sx) \, d\nu(x)$$

$$= \int d\mu(s) \int 1_K \, d\big(\gamma(s)\nu\big).$$

But, for each $s \in G$, $\gamma(s)\nu$ is a measure with base β. Hence, if K is β-negligible, it is also $\mu * \nu$-negligible. This proves that $\mu * \nu$ is a measure with base β. □

Definition 24.2.1 Let μ be a measure on G and let $f : X \to \mathbf{C}$ be locally β-integrable. μ and f are said to be convolvable (with respect to β) whenever

μ and $f\beta$ are convolvable. In this case, any density of $\mu * (f\beta)$ with respect to β is called a convolution product of μ and f (relative to β), written $\mu \overset{\beta}{*} f$ (or $\mu * f$, when no ambiguity on β is possible).

$\mu \overset{\beta}{*} f$ is defined up to a locally β-negligible set. If $\operatorname{supp}(\beta) = X$ and if there exists a convolution product of μ and f that is continuous, this last one is uniquely determined, and it is called the convolution product of μ and f (relative to β).

Lemma 24.2.1 *Let G be a locally compact group acting continuously from the left on a locally compact space X, and let $\beta \neq 0$ be a positive measure on X, quasi-invariant under G. Also, let χ be a continuous function from $G \times X$ into $]0, +\infty[$ such that $\gamma_X(s)\beta = \chi(s^{-1}, \cdot)\beta$ for each $s \in G$. Then, for any measure μ on G, $\chi(\mu \otimes \beta)$ is the image measure of $\mu \otimes \beta$ under the homeomorphism $(s, x) \mapsto (s, s^{-1}x)$ from $G \times X$ onto $G \times X$.*

PROOF: Let $F \in \mathcal{H}^+(G \times X)$. Then

$$
\begin{aligned}
\int F(s, s^{-1}x)\, d(\mu \otimes \beta)(s, x) &= \int d\mu(s) \int F(s, s^{-1}x)\, d\beta(x) \\
&= \int d\mu(s) \int F(s, x)\, d(\gamma(s^{-1})\beta)(x) \\
&= \int d\mu(s) \int F(s, x)\chi(s, x)\, d\beta(x) \\
&= \int F(s, x)\chi(s, x)\, d(\mu \otimes \beta)(s, x).
\end{aligned}
$$

$\square$

Proposition 24.2.2 *Let μ be a measure on G and let $f : X \to \mathbf{C}$ be locally β-integrable. Suppose that $s \mapsto f(s^{-1}x)\chi^{-1}(s)$ is essentially μ-integrable except for a locally β-negligible set of values of x, and that the function $x \mapsto \int |f(s^{-1}x)| \chi^{-1}(s)\, d|\mu|(s)$, defined locally almost everywhere for β, is locally β-integrable. Then μ and f are convolvable.*

PROOF: We may suppose that μ and f are positive.

Let $h \in \mathcal{H}^+(X)$. We prove that $(s, x) \mapsto h(sx)$ is essentially integrable for $\mu \otimes (f\beta)$, or that $\iint^\bullet h(sx)f(x)\, d\mu(s)\, d\beta(x) < +\infty$. We need only show there exists $a \geq 0$ such that $\iint^\bullet h(sx)f(x)1_K(s)\, d\mu(s)\, d\beta(x) \leq a$ for any compact subset of G.

Since $\mu \otimes \beta$ is the image measure of $(\chi^{-1}\mu) \otimes \beta$ under the homeomorphism $(s, x) \mapsto (s, s^{-1}x)$,

$$
\iint^\bullet h(sx)f(x)1_K(s)\, d\mu(s)\, d\beta(x) =
$$

$$
\iint^* h(x)f(s^{-1}x)1_K(s)\chi^{-1}(s)\, d\mu(s)\, d\beta(x) \qquad (1).
$$

Now $(s, x) \mapsto h(x)f(s^{-1}x)1_K(s)\chi^{-1}(s)$ is $\mu \otimes \beta$-measurable. with compact support. Therefore, the right-hand side of (1) is equal to

$$\int^* d\beta(x)\, h(x) \int^* f(s^{-1}x)1_K(s)\chi^{-1}(s)\, d\mu(s),$$

and so less than

$$\|h\| \int^* d\beta(x)\, 1_S(x) \int^\bullet f(s^{-1}x)\chi^{-1}(s)\, d\mu(s)$$

(where $S = \operatorname{supp}(h)$). The proof is complete.   □

Proposition 24.2.3 *Let μ be a measure on G and let $f : X \to \mathbf{C}$ be locally β-integrable. Suppose that one of the following conditions holds:*

(a) *f is continuous.*

(b) *G acts properly on X and f vanishes outside a union of countably many compact sets.*

(c) *μ is carried by a union of countably many compact sets.*

Furthermore, assume that μ and f are convolvable. Then

$$s \longmapsto f(s^{-1}x)\chi^{-1}(s)$$

is essentially μ-integrable except for a locally β-negligible set of values of x, and

$$\left(\mu \overset{\beta}{*} f\right)(x) = \int f(s^{-1}x)\chi^{-1}(s)\, d\mu(s) \qquad (*)$$

locally β-almost everywhere.

PROOF: Let $h \in \mathcal{H}(X, \mathbf{C})$. Since $(s, x) \longmapsto h(sx)f(x)$ is essentially $\mu \otimes \beta$-integrable, the function $(s, x) \longmapsto h(x)f(s^{-1}x)\chi^{-1}(s)$ is essentially $\mu \otimes \beta$-integrable. In fact, under condition (a) or (b) of the statement, it is $\mu \otimes \beta$-integrable, because in the first case it is continuous, and in the second case it vanishes outside a union of countably many compact sets. Then, by Fubini's theorem,

$$\iint h(sx)\, d\mu(s)\, d(f\beta)(x) = \iint h(x)f(s^{-1}x)\chi^{-1}(s)\, d\mu(s)\, d\beta(x)$$

$$= \int d\beta(x) \int h(x)f(s^{-1}x)\chi^{-1}(s)\, d\mu(s).$$

We conclude that $g : x \mapsto \int f(s^{-1}x)\chi^{-1}(s)\, d\mu(s)$ is defined locally β-almost everywhere, locally β-integrable, and that $\langle h, \mu * (f\beta)\rangle = (g\beta)(h)$. Hence $g = \mu \overset{\beta}{*} f$.

Next, we suppose that μ is carried by a union S of countably many compact sets. The function $(s, x) \mapsto h(x)f(s^{-1}x)\chi^{-1}(s)1_S(s)$ is essentially $\mu \otimes \beta$-integrable and vanishes outside a union of countably many compact sets, hence it is $\mu \otimes \beta$-integrable. Since $\mu = 1_S\mu$, we may complete the argument as before.
$\square$

24.3 Convolution of a Measure and a Continuous Function

Let G, X, β, and χ be as in Section 24.2.

Proposition 24.3.1 *Let μ be a measure on G with compact support, and let $f : X \to \mathbf{C}$ be continuous. Then*

(a) μ and f are convolvable;

(b) $x \mapsto \int f(s^{-1}x)\chi^{-1}(s)\,d\mu(s)$ is a continuous density of $\mu * (f\beta)$ with respect to β.

PROOF: Given x_0 in X, the function $s \mapsto f(s^{-1}x)\chi^{-1}(s)$ converges uniformly on supp(μ) to $s \mapsto f(s^{-1}x_0)\chi^{-1}(s)$ as $x \to x_0$, and $\int f(s^{-1}x)\chi^{-1}(s)\,d\mu(s)$ tends to $\int f(s^{-1}x_0)\chi^{-1}(s)\,d\mu(s)$. Therefore, $g : x \mapsto \int f(s^{-1}x)\chi^{-1}(s)\,d\mu(s)$ is continuous.

Now, by Proposition 24.1.4, μ and $f\beta$ are convolvable. Moreover, by Proposition 24.2.3, g is a density of $\mu * (f\beta)$ with respect to β.
$\square$

Observe that, if μ is a measure on G with compact support and f belongs to $\mathcal{H}(X, \mathbf{C})$, then $\mu \overset{\beta}{*} f$ vanishes outside $(\text{supp}\,\mu)(\text{supp}\,f)$, so that $\mu * f$ has compact support.

Proposition 24.3.2 *Suppose that G acts properly on X. Let μ be in $\mathcal{M}(G, \mathbf{C})$ and f in $\mathcal{H}(X, \mathbf{C})$.*

(a) μ and f are convolvable.

(b) $x \mapsto \int f(s^{-1}x)\chi^{-1}(s)\,d\mu(s)$ is a continuous density of $\mu * (f\beta)$ with respect to β.

PROOF: Given x_0 in X, let V be a compact neighborhood of x_0. The set K, of those s in G such that $s^{-1}x$ belongs to supp(f) for at least one x in V, is compact. Furthermore, $f(s^{-1}x)\chi^{-1}(s)$ converges uniformly on K to $f(s^{-1}x_0)\chi^{-1}(s)$ as $x \to x_0$ in V. Thus $\int f(s^{-1}x)\chi^{-1}(s)\,d\mu(s)$ converges to $\int f(s^{-1}x_0)\chi^{-1}(s)\,d\mu(s)$ as $x \to x_0$, which proves that

$$g : x \mapsto \int f(s^{-1}x)\chi^{-1}(s)\,d\mu(s)$$

is continuous.

Next, μ and $f\beta$ are convolvable (Proposition 24.1.4) and g is a density of $\mu * (f\beta)$ with respect to β (Proposition 24.2.3).   □

24.4   Convolution of $\mu \in \mathcal{M}(G, \mathbf{C})$ and $f \in \overline{\mathcal{L}^p}(\beta)$

Let G, X, β, and χ be as in Section 24.2. For all $p \in [1, +\infty]$, we write N_p for the usual seminorm on $\overline{\mathcal{L}^p}(\beta) = \overline{\mathcal{L}^p_{\mathbf{C}}}(\beta)$.

Now let p and q (where $p \geq 1$) be two conjugate exponents. Until further notice, we suppose that p is finite.

For each $s \in G$ and each $f \in \overline{\mathcal{L}^p}(\beta)$, we let $\gamma_{\chi,p}(s)f$ be the function $x \mapsto f(s^{-1}x)\chi^{-1/p}(s)$ on X. Since $\gamma(s)f : x \mapsto f(s^{-1}x)$ is $\chi^{-1}(s)\beta$-measurable, $\gamma_{\chi,p}(s)f$ is β-measurable; moreover,

$$\int^{\bullet} \left| f(s^{-1}x)\chi^{-1/p}(s) \right|^p d\beta(x) = \int^{\bullet} \left| f(s^{-1}x) \right|^p d(\gamma(s)\beta)(x) = \int^{\bullet} |f|^p \, d\beta.$$

Hence $\gamma_{\chi,p}(s)f$ lies in $\overline{\mathcal{L}^p}(\beta)$ and $N_p(\gamma_{\chi,p}(s)f) = N_p(f)$.

For each $s \in G$, we denote by $\gamma_{\chi,p}(s)$ the continuous endomorphism $\dot{f} \mapsto \widehat{\gamma_{\chi,p}(s)}f$ of $L^p_{\mathbf{C}}(\beta)$. The mapping $s \mapsto \gamma_{\chi,p}(s)$ is a representation of G on $L^p_{\mathbf{C}}(\beta)$.

Proposition 24.4.1 *For all $f \in \overline{\mathcal{L}^p}(\beta)$, the mapping $s \mapsto \gamma_{\chi,p}(s)\dot{f}$ from G into $L^p_{\mathbf{C}}(\beta)$ is continuous.*

PROOF: First, suppose that $f \in \mathcal{H}(X, \mathbf{C})$, and put $K = \mathrm{supp}(f)$. Given $s_0 \in G$, let V be a compact neighborhood of s_0 in G; $\gamma_{\chi,p}(s)f$ converges to $\gamma_{\chi,p}(s_0)f$ in $\mathcal{H}(X, VK; \mathbf{C})$, and hence in $\overline{\mathcal{L}^p}(\beta)$, as $s \to s_0$ in V. Therefore, $s \mapsto \gamma_{\chi,p}(s)\dot{f}$ is continuous.

Now we pass to the general case. Since

$$N_p\big(\gamma_{\chi,p}(s)f - \gamma_{\chi,p}(s)g\big) = N_p(f - g)$$

for all $g \in \mathcal{H}(X, \mathbf{C})$ and all $s \in G$, the mapping $s \mapsto \gamma_{\chi,p}(s)g$ converges to $s \mapsto \gamma_{\chi,p}(s)f$ uniformly on compact sets, as $g \in \mathcal{H}(X, \mathbf{C})$ converges to f in $\overline{\mathcal{L}^p}(\beta)$. The proposition follows.   □

Similarly, for each $s \in G$ and each $f \in \overline{\mathcal{L}^p}(\beta)$, we let $\gamma_\chi(s)f$ be the function $x \mapsto f(s^{-1}x)\chi^{-1}(s)$ on X, so that $\gamma_\chi(s)f$ lies in $\overline{\mathcal{L}^p}(\beta)$ and $N_p(\gamma_\chi(s)f) = \chi^{-1/q}(s)N_p(f)$. We denote by $\gamma_\chi(s)$ the continuous endomorphism $\dot{f} \mapsto \widehat{\gamma_\chi(s)}f$ of $L^p_{\mathbf{C}}(\beta)$. Then $s \mapsto \gamma_\chi(s)$ is a representation of G on $L^p_{\mathbf{C}}(\beta)$. Moreover, for all $f \in L^p_{\mathbf{C}}(\beta)$, $s \mapsto \gamma_\chi(s)f$ is continuous.

Theorem 24.4.1 *Let μ be a measure on G such that $\chi^{-1/q}$ is μ-integrable, and let $f \in \overline{\mathcal{L}^p}(\beta)$. Then*

(a) *μ and f are convolvable;*

(b) *$\mu * (f\beta)$ has a density $\mu * f \in \overline{\mathcal{L}^p}(\beta)$ with respect to β, given locally β-almost everywhere by formula $(*)$ of Proposition 24.2.3;*

(c) *$N_p(\mu * f) \le N_p(f)\left(\int \chi^{-1/q} \, d|\mu| \right).$*

PROOF: $\chi^{-1/q}\mu$, and so μ, is carried by a countable union S of compact sets.

Let $h \in \mathcal{H}(X, \mathbf{C})$. Since $\mu \otimes \beta$ is the image measure of $(\chi^{-1}\mu) \otimes \beta$ under $(s, x) \mapsto (s, s^{-1}x)$, the function $(s, x) \mapsto f(s^{-1}x)\chi^{-1}(s)$ is $\mu \otimes \beta$-measurable. Then $(s, x) \mapsto h(x)f(s^{-1}x)\chi^{-1}(s)$ is $\mu \otimes \beta$-measurable, and

$$\int^* |h(x)f(s^{-1}x)|\chi^{-1}(s)1_S(s) \, d(|\mu| \otimes \beta)(s, x)$$
$$= \int^* d|\mu|(s)1_S(s) \int^* |h(x)f(s^{-1}x)|\chi^{-1}(s) \, d\beta(x).$$

But

$$\int^* |h(x)f(s^{-1}x)|\chi^{-1}(s) \, d\beta(x) = \int^* |h(x)f(s^{-1}x)| \, d(\gamma(s)\beta)(x)$$
$$= \int^* |h(sy)f(y)| \, d\beta(y)$$

is less than $N_p(f)N_q\big(\gamma_{s^{-1}}(h)\big) = N_p(f)\chi^{-1/q}(s)N_q(h)$. Therefore,

$$\int^\bullet |h(x)f(s^{-1}x)|\chi^{-1}(s) \, d(|\mu| \otimes \beta)(s, x) \le N_p(f)\left(\int \chi^{-1/q} \, d|\mu| \right) N_q(h),$$

and $(s, x) \mapsto h(x)f(s^{-1}x)\chi^{-1}(s)$ is essentially $\mu \otimes \beta$-integrable. Consequently, $(s, y) \mapsto h(sy)f(y)$ is essentially $\mu \otimes \beta$-integrable. In short, μ and $f\beta$ are convolvable.

Next, the mapping $s \mapsto \gamma_\chi(s)\dot{f}$ from G into $L^p_{\mathbf{C}}(\beta)$ is continuous and $\int^* N_p(\gamma_\chi(s)\dot{f}) \, d|\mu|(s) \le N_p(f)\left(\int \chi^{-1/q} \, d|\mu| \right) = c$. Hence $s \mapsto \gamma_\chi(s)\dot{f}$ is μ-integrable.

Denote by $\langle , \rangle$ the continuous bilinear mapping $(\dot{g}, \dot{h}) \mapsto \int gh \, d\beta$ from $L^p_{\mathbf{C}}(\beta) \times L^q_{\mathbf{C}}(\beta)$ into $\mathbf{C}$, and put $\dot{g} = \int (\gamma_\chi(s)\dot{f}) \, d\mu(s)$. Let $h \in \mathcal{H}(X, \mathbf{C})$. Then

$$\int h \, d(g\beta) = \int gh \, d\beta = \langle \dot{g}, \dot{h} \rangle = \int \langle \gamma_\chi(s)\dot{f}, \dot{h} \rangle \, d\mu(s),$$

so

$$\int h \, d(g\beta) = \int d\mu(s) \int h(x)f(s^{-1}x)\chi^{-1}(s) \, d\beta(x)$$
$$= \int d\mu(s)1_S(s) \int h(x)f(s^{-1}x)\chi^{-1}(s) \, d\beta(x).$$

In other words,

$$
\begin{aligned}
\int h\,d(g\beta) &= \int h(x)f(s^{-1}x)\chi^{-1}(s)1_S(s)\,d(\mu\otimes\beta)(s,x) \\
&= \int h(sy)f(y)1_S(s)\,d(\mu\otimes\beta)(s,y) \\
&= \int h(sy)\,d(\mu\otimes f\beta)(s,y) \\
&= (\mu * f\beta)(h).
\end{aligned}
$$

We conclude that g is a density of $\mu * f\beta$ with respect to β.   □

Write $\mathcal{L}^\infty(\beta)$ for $\mathcal{L}^\infty_{\mathbf{C}}(\beta)$, and $C^b(X)$ (respectively, $C^0(X)$) for the set of those continuous functions from X into $\mathbf{C}$ that are bounded (respectively, that vanish at infinity).

Theorem 24.4.2 *Let μ be a measure on G such that χ^{-1} is μ-integrable, and let $f \in \mathcal{L}^\infty(\beta)$.*

(a) *μ and f are convolvable, and $\mu * f\beta$ has a density $\mu * f \in \mathcal{L}^\infty(\beta)$ with respect to β, given locally β-almost everywhere by formula (*) of Proposition 24.2.3; moreover, $N_\infty(\mu * f) \le N_\infty(f)(\int \chi^{-1}\,d|\mu|)$.*

(b) *If $f \in C^b(X)$, the function $g : x \mapsto \int f(s^{-1}x)\chi^{-1}(s)\,d\mu(s)$ lies in $C^b(X)$.*

(c) *If $f \in C^0(X)$, $\mu * f$ lies in $C^0(X)$.*

PROOF: By the argument of Theorem 24.4.1, μ and f are convolvable. Proposition 24.2.3 shows that $g : x \mapsto \int f(s^{-1}x)\chi^{-1}(s)\,d\mu(s)$ is defined locally β-almost everywhere and is a density of $\mu*(f\beta)$ with respect to β. If $f' : X \to \mathbf{C}$ is such that $f' = f$ locally β-almost everywhere and $|f'| \le N_\infty(f)$, then $(\mu*f)(x) = (\mu*f')(x) = \int f'(s^{-1}x)\chi^{-1}(s)\,d\mu(s)$ locally β-almost everywhere. Hence $N_\infty(\mu * f) \le N_\infty(f)\int \chi^{-1}\,d|\mu|$.

Next, suppose that f belongs to $C^b(X)$. Clearly, $s \mapsto f(s^{-1}x)\chi^{-1}(s)$ is μ-integrable, for every $x \in X$.

Let $x_0 \in X$. Given $\varepsilon > 0$, define K as a compact subset of G such that $\int_{G-K}\chi^{-1}\,d|\mu| \le \varepsilon/N_\infty(f)$. There exists a neighborhood V of x_0 in X such that $x \in V$ implies $|f(s^{-1}x)\chi^{-1}(s) - f(s^{-1}x_0)\chi^{-1}(s)| \le \varepsilon/(|\mu|(K))$ for all $s \in K$. Then, for $x \in V$,

$$
\begin{aligned}
\left| \int f(s^{-1}x)\chi^{-1}(s)\,d\mu(s) - \int f(s^{-1}x_0)\chi^{-1}(s)\,d\mu(s) \right| \\
\le \int_K \frac{\varepsilon}{|\mu|(K)}\,d|\mu| + 2\int_{G-K} N_\infty(f)\chi^{-1}\,d|\mu| \\
\le 3\varepsilon.
\end{aligned}
$$

Consequently, $g : x \mapsto \int f(s^{-1}x)\chi^{-1}(s)\,d\mu(s)$ is continuous.

Finally, assume that $f \in \mathcal{C}^0(X)$. Let ε and K be as above. Also, let H be a compact subset of X such that $|f(y)| \leq \varepsilon$ for $y \notin H$. Fix $x \notin KH$. Since $s^{-1}x \notin H$ for $s \in K$, we have

$$|g(x)| \leq \int_K \varepsilon \chi^{-1} \, d|\mu| + \int_{G-K} N_\infty(f) \chi^{-1} \, d|\mu| \leq \varepsilon \left(1 + \int \chi^{-1} d|\mu|\right),$$

so g vanishes at infinity. $\square$

Proposition 24.4.2 *Let $\rho : G \to \mathbf{C}^*$ be a continuous homomorphism. Also, let $\lambda \in \mathcal{M}(G, \mathbf{C})$ and $\mu \in \mathcal{M}(G, \mathbf{C})$. If $\rho\lambda$ and $\rho\mu$ are convolvable (with respect to $(s, t) \mapsto st$), then λ and μ are convolvable and $\rho(\lambda * \mu) = (\rho\lambda) * (\rho\mu)$.*

PROOF: Let $f \in \mathcal{H}(G, \mathbf{C})$. Since $(s, t) \mapsto (f\rho^{-1})(st)$ is $(\rho\lambda) \otimes (\rho\mu)$-integrable, the function $(s, t) \mapsto f(st)$ is $\lambda \otimes \mu$-integrable. Hence λ and μ are convolvable. Furthermore,

$$
\begin{aligned}
(\rho\lambda * \rho\mu)(f) &= \iint f(st)\rho(s)\rho(t) \, d\lambda(s) \, d\mu(t) \\
&= \iint (f\rho)(st) \, d\lambda(s) \, d\mu(t) \\
&= (\lambda * \mu)(f\rho)
\end{aligned}
$$

for all $f \in \mathcal{H}(G, \mathbf{C})$, so

$$\rho\lambda * \rho\mu = \rho(\lambda * \mu).$$

$\square$

Proposition 24.4.3 *Let $\rho : G \to \mathbf{R}_+^*$ be a continuous homomorphism. Write $\mathcal{M}^\rho(G)$ for the set of those measures μ on G such that ρ is μ-integrable. For all $\mu \in \mathcal{M}^\rho(G)$, we put $\|\mu\|_\rho = \int \rho \, d|\mu|$. Then*

(a) two arbitrary elements of $\mathcal{M}^\rho(G)$ are convolvable;

(b) for the convolution and the norm $\mu \mapsto \|\mu\|_\rho$, $\mathcal{M}^\rho(G)$ is a Banach algebra.

PROOF: If $\lambda \in \mathcal{M}^\rho(G)$ and $\mu \in \mathcal{M}^\rho(G)$, then $\rho\lambda$ and $\rho\mu$ are convolvable (Proposition 24.1.1), so λ and μ are convolvable; moreover,

$$\|\lambda * \mu\|_\rho = \left\|\rho(\lambda * \mu)\right\|_1 = \|\rho\lambda * \rho\mu\|_1 \leq \|\rho\lambda\|_1 \|\rho\mu\|_1 = \|\lambda\|_\rho \|\mu\|_\rho.$$

Since the mapping $\mu \mapsto \rho\mu$ is a linear isometry of $\mathcal{M}^\rho(G)$ onto $\mathcal{M}^1(G)$, $\mathcal{M}^\rho(G)$ is a Banach space. Finally, by Section 24.1, the convolution in $\mathcal{M}^\rho(G)$ is associative. $\square$

Clearly, the measure ε_e, defined by placing the unit mass at the unit element e of G, is a unit in $\mathcal{M}^\rho(G)$.

Now let G, X, β, χ be as in Section 24.2, and let p, q (with $1 \leq p \leq +\infty$) be two conjugate exponents. For each μ in $A = \mathcal{M}^{\chi^{-1/q}}(G)$, we denote by $\gamma_\chi(\mu)$ the continuous endomorphism $\dot{f} \mapsto \widehat{\mu * f}$ of $L^p_{\mathbf{C}}(\beta)$ (Theorem 24.4.1).

Theorem 24.4.3 $\gamma_\chi : \mu \mapsto \gamma_\chi(\mu)$ *is a representation of A on $L^p_{\mathbf{C}}(\beta)$.*

PROOF: The result follows from the fact that $(\lambda * \mu) * (f\beta) = \lambda * (\mu * f\beta)$ for all $\lambda \in A$, $\mu \in A$, and $f \in \overline{\mathcal{L}^p}(\beta)$. $\square$

Now consider the case in which $X = G$, the action of G on X is $(s, t) \mapsto st$, and β is a left Haar measure on G. Then $\chi = 1$. The mapping γ_χ (which does not depend on the choice of β) is called the left regular representation of $\mathcal{M}^1(G)$ on $L^p_{\mathbf{C}}(\beta)$.

24.5 Convolution and Transposition

Let G, X, β and χ be as in Section 24.2.

Proposition 24.5.1 *Let $\mu \in \mathcal{M}(G, \mathbf{C})$, and write $\check{\mu}$ for the image measure of μ under $s \mapsto s^{-1}$. Let $f : X \to \mathbf{C}$ and $g : X \to \mathbf{C}$ be locally β-integrable. Suppose that*

(a) *μ and f are convolvable and formula (*) of Proposition 24.2.3 defines, locally β-almost everywhere, a convolution product $\mu \overset{\beta}{*} f$;*

(b) *$\chi\check{\mu}$ and g are convolvable and formula (*) of Proposition 24.2.3 (where μ is replaced by $\chi\check{\mu}$ and f by g) defines, locally β-almost everywhere, a convolution product $(\chi\check{\mu}) \overset{\beta}{*} g$;*

(c) *there exists a set S carrying μ such that the function $h : (s, x) \mapsto g(x)f(s^{-1}x)\chi^{-1}(s)1_S(s)$ is $\mu \otimes \beta$-integrable.*

Then $g(\mu * f)$ and $f(\chi\check{\mu} * g)$ are essentially β-integrable and

$$\int f(\chi\check{\mu} * g) \, d\beta = \int g(\mu * f) \, d\beta \quad (1).$$

PROOF: Put $\psi = 1_S$ and $I = \int h \, d(\mu \otimes \beta)$. The function

$$x \longmapsto g(x) \int f(s^{-1}x)\chi^{-1}(s) \, d\mu(s) = \int g(x)f(s^{-1}x)\chi^{-1}(s)\psi(s) \, d\mu(s)$$

is defined locally β-almost everywhere. By Fubini's theorem, it is essentially β-integrable and $\int g(\mu * f) \, d\beta = I$.

Next, $(s, x) \longmapsto g(x)f(sx)\chi(s)\psi(s^{-1}) = h(s^{-1}x)$ is $\check{\mu} \otimes \beta$-integrable. Therefore, $(s, x) \longmapsto g(x)f(sx)\psi(s^{-1})$ is $(\chi\mu) \otimes \beta$-integrable, and $(s, x) \longmapsto g(s^{-1}x)f(x)\psi(s^{-1})$ is $\check{\mu} \otimes \beta$-integrable. Moreover,

$$I = \iint g(s^{-1}x)f(x)\psi(s^{-1})\, d\check{\mu}(s)\, d\beta(x).$$

From Fubini's theorem, we deduce that $f(\chi\check{\mu} * g)$ is essentially β-integrable and that $I = \int f(\chi\check{\mu} * g)\, d\beta$. $\square$

Proposition 24.5.2 *Let p and q be two conjugate exponents $(1 \le p \le +\infty)$. Also, let μ be a measure on G such that $\chi^{-1/q}$ is μ-integrable. Then*

$$\int f(\chi\check{\mu} * g)\, d\beta = \int g(\mu * f)\, d\beta$$

for all $f \in \overline{\mathcal{L}^p}(\beta)$ and $g \in \overline{\mathcal{L}^q}(\beta)$.

PROOF: We may suppose that either f or g vanishes outside a union of countably many compact sets.

Since μ is moderate, it is carried by a countable union S of compact sets. We put $\psi = 1_S$. The function

$$h : (s, x) \longmapsto g(x)f(s^{-1}x)\chi^{-1}(s)\psi(s)$$

is $\mu \otimes \beta$-measurable, since $(s, x) \mapsto f(s^{-1}x)\chi^{-1}(s)$ is. Therefore,

$$
\begin{aligned}
\int^* h\, d|\mu \otimes \beta| &= \int^\bullet h\, d|\mu \otimes \beta| \\
&= \int^\bullet d\beta(x) \int^\bullet |g(x)f(s^{-1}x)|\chi^{-1}(s)\psi(s)\, d|\mu|(s) \\
&= \int^\bullet |g|(|\mu| * |f|)\, d\beta
\end{aligned}
$$

is finite, and we may apply Proposition 24.5.1. $\square$

If $1 \le p < +\infty$, Proposition 24.5.2 means that the endomorphism $\dot{g} \mapsto (\widehat{\chi\check{\mu}}) * g$ of $L^q_{\mathbf{C}}(\beta)$ is the transpose of the endomorphism $\dot{f} \mapsto \widehat{\mu * f}$ of $L^p_{\mathbf{C}}(\beta)$. Now suppose that $X = G$.

Proposition 24.5.3 *Let f and g be two locally β-integrable functions from G into $\mathbf{C}$. Also, let μ be a measure on G. Suppose there exists $S \subset G$, carrying μ, such that $(s, x) \mapsto f(x)g(s^{-1}x)1_S(s)$ is $\mu \otimes \beta$-moderate. Then*

$$
\begin{aligned}
\int^\bullet d|\mu|(s) \int^\bullet |f(x)g(s^{-1}x)|\, d\beta(x) \\
= \int^\bullet d\beta(x)|g(x)| \int^\bullet |f(sx)|\chi(s)\, d|\mu|(s) \\
= \int^* |f(x)g(s^{-1}x)|1_S(s)\, d|\mu \otimes \beta|(s, x).
\end{aligned}
$$

If this number is finite, then

(a) *the function $\chi f * \check{g} : s \mapsto \int f(x)g(s^{-1}x)\,d\beta(x)$ is defined locally μ-almost everywhere and is essentially μ-integrable;*

(b) *$\check{\mu} * f : x \mapsto \int f(sx)\chi(s)\,d\mu(s)$ is defined locally $g\beta$-almost everywhere and is essentially $g\beta$-integrable;*

(c) *$\int (\chi f * \check{g})\,d\mu = \int (\check{\mu} * f)g\,d\beta.$*

PROOF: As $(s,x) \mapsto g(x)$ is $(\chi\mu) \otimes \beta$-measurable, the function $(s,x) \mapsto g(s^{-1}x)$ is $\mu \otimes \beta$-measurable. Therefore, $(s,x) \mapsto f(x)g(s^{-1}x)1_S(s)$ is measurable and moderate for $\mu \otimes \beta$. Now, since $x \mapsto f(x)g(s^{-1}x)1_S(s)$ is β-moderate for μ-almost all $s \in G$, and $s \mapsto \int^* |f(x)g(s^{-1}x)|1_S(s)\,d\beta(x)$ is μ-moderate,

$$\int^{\bullet} d|\mu|(s) \int^{\bullet} |f(x)g(s^{-1}x)|\,d\beta(x)$$

$$= \int^{\bullet} d|\mu|(s) \int^{\bullet} |f(x)g(s^{-1}x)|1_S(s)\,d\beta(x)$$

$$= \int^{*} d|\mu|(s) \int^{*} |f(x)g(s^{-1}x)|1_S(s)\,d\beta(x)$$

$$= \int^{*} |f(x)g(s^{-1}x)|1_S(s)\,d(|\mu| \otimes \beta)(s,x).$$

Next, $(s,x) \mapsto f(sx)g(x)1_S(s)$ is measurable and moderate for the image measure $(\chi\mu) \otimes \beta$ of $\mu \otimes \beta$ under $(s,x) \mapsto (s,s^{-1}x)$, and

$$\int^{*} |f(x)g(s^{-1}x)|1_S(s)\,d|\mu \otimes \beta|(s,x)$$

$$= \int^{*} |f(sx)g(x)|1_S(s)\,d|(\chi\mu) \otimes \beta|(s,x).$$

The right-hand side in the last equality is

$$\int^{\bullet} d\beta(x)|g(x)| \int^{\bullet} |f(sx)|\chi(s)\,d|\mu|(s).$$

This proves the first assertion of the statement.

Finally, suppose that

$$\int^{*} |f(x)g(s^{-1}x)|1_S(s)\,d|\mu \otimes \beta|(s,x)$$

is finite, that is, $(s,x) \mapsto f(x)g(s^{-1}x)1_S(s)$ is $\mu \otimes \beta$-integrable. For μ-almost all s in S, $x \mapsto f(x)g(s^{-1}x)$ is β-integrable. Thus, for locally μ-almost all s in G, $x \mapsto f(x)g(s^{-1}x)$ is β-integrable. Moreover, the function

$$\chi f * \check{g} : s \mapsto \int f(x)g(s^{-1}x)\,d\beta(x)$$

is essentially μ-integrable and

$$\int (\chi f * \check{g}) \, d\mu = \int f(x)g(s^{-1}x)1_S(s) \, d(\mu \otimes \beta)(s, x).$$

On the other hand, $(s, x) \mapsto f(sx)g(x)1_S(s)\chi(s)$ is $\mu \otimes \beta$-integrable. Hence, for locally β-almost all $x \in G$, $s \mapsto f(sx)g(x)1_S(s)\chi(s)$ is μ-integrable, and $s \mapsto f(sx)g(x)\chi(s)$ is essentially μ-integrable. Now $s \mapsto f(sx)\chi(s)$ is essentially μ-integrable for locally $g\beta$-almost all $x \in G$, and $x \mapsto \int f(sx)\chi(s) \, d\mu(s)$ is essentially $g\beta$-integrable. Furthermore,

$$\begin{aligned}
\int (\check{\mu} * f) \, d(g\beta) &= \int f(sx)g(x)1_S(s)\chi(s) \, d(\mu \otimes \beta)(s, x) \\
&= \int f(x)g(s^{-1}x)1_S(s) \, d(\mu \otimes \beta)(s, x).
\end{aligned}$$

The proof is complete. $\qquad\qquad\qquad\qquad\qquad\qquad\qquad\qquad\qquad\qquad$ □

24.6   Convolution of Functions on a Group

Let G be a locally compact group acting from the right on a locally compact space X. Suppose there is a relatively invariant, positive measure β ($\beta \neq 0$) on X, with multiplier χ'.

Let $f : X \to \mathbf{C}$ be locally β-integrable and let μ be a measure on G. If $f\beta$ and μ are convolvable (for $(x, s) \mapsto xs$), $f\beta * \mu$ is a measure with base β. Then f and μ are said to be convolvable (with respect to β); any density of $(f\beta) * \mu$ relative to β is called a convolution product of f and μ with respect to β and is written $f \overset{\beta}{*} \mu$, or simply $f * \mu$.

Define G^0 as the topological group whose topology is that of G and whose composition law is $(s, t) \mapsto ts$. The mapping $(s, x) \mapsto xs$ from $G^0 \times X$ into X is a left action of G^0 on X. Moreover, β is relatively invariant for G^0, with left multiplier χ'. The convolution products $f \overset{\beta}{*} \mu$ are the convolution products $\mu \overset{\beta}{*} f$ for G^0 acting from the left on X. We may therefore apply the results of the preceding sections. In particular, if f and μ are convolvable and if one of the conditions of Proposition 24.2.3 holds, then $f \overset{\beta}{*} \mu$ is given locally β-almost everywhere by the formula

$$(f \overset{\beta}{*} \mu)(x) = \int f(xs^{-1})\chi'^{-1}(s) \, d\mu(s).$$

Henceforth, let β (where $\beta > 0$) be a relatively invariant measure on a locally compact group G. We write χ (respectively, χ') for the left multiplier (respectively, the right multiplier) of β. Recall that the property, for a complex-valued function on G, to be locally β-integrable does not depend on the choice of β. We let $\mathcal{L}(G)$ be the set of functions with this property.

When we regard the composition law of G as a left action (respectively, a right action) of G onto itself, we can define, for μ in $\mathcal{M}(G, \mathbf{C})$ and f in $\mathcal{L}(G)$, the convolvability of μ and f and the products $\mu \overset{\beta}{*} f$ (respectively, the convolvability of f and μ and the products $f \overset{\beta}{*} \mu$).

For $f \in \mathcal{L}(G)$ and $g \in \mathcal{L}(G)$, the relationship "$f\beta$ and $g\beta$ are convolvable" does not depend on any particular choice of β (Proposition 24.4.2). f and g are then said to be convolvable. $(f\beta) * (g\beta)$ can be written $h\beta$. with $h \in \mathcal{L}(G)$, and h is determined up to the locally β-negligible sets. h is written $f \overset{\beta}{*} g$ and is called a convolution product of f and g with respect to β (β is omitted when no ambiguity is possible). If ψ is a continuous homomorphism from G into $\mathbf{R}_+^*$ and if $a > 0$, then

$$f \overset{a\psi\beta}{*} g = a f \overset{\beta}{*} g.$$

Clearly,

$$f \overset{\beta}{*} g = (f\beta) \overset{\beta}{*} g = f \overset{\beta}{*} (g\beta).$$

If one of the convolution products of f and g is continuous, it is uniquely determined, and it is called the convolution product of f and g with respect to β.

Proposition 24.6.1   Let $f \in \mathcal{L}(G)$, $g \in \mathcal{L}(G)$, and $x \in G$. The function $s \longmapsto g(s^{-1}x)f(s)\chi^{-1}(s)$ is essentially β-integrable if and only if $s \longmapsto f(xs^{-1})g(s)\chi'^{-1}(s)$ is essentially β-integrable. In this case,

$$\int g(s^{-1}x)f(s)\chi^{-1}(s)\,d\beta(s) = \int f(xs^{-1})g(s)\chi'^{-1}(s)\,d\beta(s).$$

PROOF: Recall that $\chi' = \chi\Delta$, where Δ is the modular function on G. Furthermore, $\chi^{-1}\beta$ is a left Haar measure on G and $\chi'^{-1}\beta$ is a right Haar measure on G.

If $s \mapsto g(s^{-1}x)f(s)\chi^{-1}(s)$ is essentially β-integrable, then

$$\begin{aligned}
\int g(s^{-1}x)f(s)\chi^{-1}(s)\,d\beta(s) &= \int g(sx)f(s^{-1})\,d(\chi^{-1}\beta)(s) \\
&= \int g(sx)f(s^{-1})\Delta^{-1}(s)\chi^{-1}(s)\,d\beta(s) \\
&= \int g(sx)f(s^{-1})\chi'^{-1}(s)\,d\beta(s) \\
&= \int g(sx)f(s^{-1})\,d(\chi'^{-1}\beta)(s) \\
&= \int g(s)f(xs^{-1})\chi'^{-1}(s)\,d\beta(s).
\end{aligned}$$

Similarly, if $s \mapsto g(s)f(xs^{-1})\chi'^{-1}(s)$ is essentially β-integrable, then

$$\int g(s)f(xs^{-1})\chi'^{-1}(s)\,d\beta(s) = \int g(s^{-1}x)f(s)\chi^{-1}(s)\,d\beta(s).$$

$\square$

Proposition 24.6.2 Let $f \in \mathcal{L}(G)$ and $g \in \mathcal{L}(G)$. Suppose that one of these functions is continuous or vanishes outside a union of a locally β-negligible set and countably many compact sets. When f and g are convolvable, the function $f * g$ is given locally β-almost everywhere by

$$(f * g)(x) = \int g(s^{-1}x)f(s)\chi^{-1}(s)\,d\beta(s)$$

$$= \int f(xs^{-1})g(s)\chi'^{-1}(s)\,d\beta(s) \qquad (1).$$

PROOF: This follows from Proposition 24.2.3. $\square$

Theorem 24.6.1 Let p and q be two conjugate exponents $(1 \leq p \leq +\infty)$. If $f\chi^{-1/q} \in \overline{\mathcal{L}^1}(G)$ and $g \in \overline{\mathcal{L}^p}(G)$, then f and g are convolvable, $f * g$ is given locally β-almost everywhere by formula (1), $f * g \in \overline{\mathcal{L}^p}(\beta)$, and $N_p(f * g) \leq N_1(f\chi^{-1/q})N_p(g)$. If $f \in \overline{\mathcal{L}^p}(\beta)$ and $g\chi'^{-1/q} \in \overline{\mathcal{L}^1}(G)$, then f and g are convolvable, $f*g$ is given locally β-almost everywhere by formula (1), $f * g \in \overline{\mathcal{L}^p}(\beta)$, and $N_p(f * g) \leq N_p(f)N_1(g\chi'^{-1/q})$.

PROOF: See Theorems 24.4.1 and 24.4.2. $\square$

Proposition 24.6.3 If $f\chi^{-1} \in \overline{\mathcal{L}^1}(\beta)$ and $g \in C^0(G)$, or if $f \in C^0(G)$ and $g\chi'^{-1} \in \overline{\mathcal{L}^1}(\beta)$, then f and g are convolvable and formula (1) defines, everywhere on G, a convolution product $f * g$ that lies in $C^0(G)$.

PROOF: See Theorem 24.4.2. $\square$

When V runs through the filter of neighborhoods of e, the sets

$$\{(x, x') \in G \times G : x^{-1}x' \in V\}$$

(respectively, the sets $\{(x, x') \in G \times G : x'x^{-1} \in V\}$) form a filter basis on $G \times G$. The filter generated by this basis is a uniform structure on G, called the left uniform structure of G (respectively, the right uniform structure of G).

Theorem 24.6.2 If $f\chi^{-1} \in \overline{\mathcal{L}^1}(\beta)$ and $g \in \mathcal{L}^\infty(\beta)$, formula (1) defines, everywhere on G, a convolution product $f * g$ that is bounded and uniformly continuous for the right uniform structure of G.

PROOF: Since $f\chi^{-1} \in \overline{\mathcal{L}^1}(\beta)$ and $(\chi^{-1}\beta)\check{} = \chi'^{-1}\beta$, the function $\check{f}\chi'^{-1}$ lies in $\overline{\mathcal{L}^1}(\beta)$. Given $x \in G$, $\delta(x^{-1})(\check{f}\chi'^{-1})$ belongs to $\overline{\mathcal{L}^1}(\chi'^{-1}(x)\beta)$; thus $s \mapsto f(xs^{-1})\chi'(s^{-1})$ belongs to $\overline{\mathcal{L}^1}(\beta)$, and $s \mapsto f(xs^{-1})g(s)\chi'(s^{-1})$ is essentially β-integrable. Denote by $f * g$ the function

$$ x \longmapsto \int f(xs^{-1})g(s)\chi'(s^{-1})\,d\beta(s) $$

and by ν the measure $\chi'^{-1}\beta$.

For all x, $x' \in G$, we have

$$ |(f * g)(x) - (f * g)(x')| \le N_\infty(g) \int |f(xs^{-1}) - f(x's^{-1})|\,d\nu(s). $$

But

$$ \int |f(xs^{-1}) - f(x's^{-1})|\,d\nu(s) = \int |f(xs) - f(x's)|\,d\check{\nu}(s) $$

$$ = \int |f(xx'^{-1}s) - f(s)|\,d\check{\nu}(s) $$

$$ = \int |(\gamma_{x'x^{-1}}f) - f|\,d\check{\nu}. $$

Hence $|(f * g)(x) - (f * g)(x')| \le \epsilon$ as soon as $x'x^{-1}$ is close enough to e (Proposition 24.4.1). Consequently, $f * g$ is uniformly continuous for the right uniform structure of G.

Next, $f * g$ is a convolution product of f and g, and $f * g$ belongs to $\mathcal{L}^\infty(\beta)$, by Theorem 24.4.2. □

Theorem 24.6.3 *Let p and q be two conjugate exponents $(1 < p < +\infty)$. Suppose that β is left invariant. Let $f \in \overline{\mathcal{L}^p}(\beta)$ and $g \in \overline{\mathcal{L}^q}(\check{\beta})$. Then f and g are convolvable. Formula (1) defines, everywhere on G, a convolution product that belongs to $\mathcal{C}^0(G)$ and satisfies $N_\infty(f * g) \le N_p(f)N_q(\check{g})$.*

PROOF: For every x in G, $\gamma_x(\check{g})$ lies in $\overline{\mathcal{L}^q}(\gamma_x\beta) = \overline{\mathcal{L}^q}(\beta)$, so $s \mapsto g(s^{-1}x)$ belongs to $\overline{\mathcal{L}^q}(\beta)$, and $s \mapsto g(s^{-1}x)f(s)$ is essentially β-integrable; moreover,

$$ \int |g(s^{-1}x)f(s)|\,d\beta(s) \le N_p(f)N_q(\gamma_x(\check{g})) = N_p(f)N_q(\check{g}). $$

Therefore, f and g are convolvable (Proposition 24.2.2), and formula (1) defines a convolution product $f * g$ such that $|f * g|(x) \le N_p(f)N_q(\check{g})$ for every x (Proposition 24.6.2). For f, g in $\mathcal{H}(G, \mathbf{C})$, $f * g$ lies in $\mathcal{H}(G, \mathbf{C})$ (Proposition 24.3.1); thus, for $f \in \overline{\mathcal{L}^p}(\beta)$ and $g \in \overline{\mathcal{L}^q}(\check{\beta})$, the convolution product given by formula (1) is a uniform limit of functions with compact support, and $f * g \in \mathcal{C}^0(G)$. □

Observe that, in general, $f * g$ does not belong to $\mathcal{C}^0(G)$ for $f \in \overline{\mathcal{L}^1}(\beta)$ and $g \in \mathcal{L}^\infty(\check{\beta})$.

24.7 Regularization

Let G, X, β, and χ be as in Section 24.2, and let p, q be two conjugate exponents $(1 \leq p \leq +\infty)$.

Define $\mathcal{F}$ as a filter on an index set I. Finally, let $(\mu_i)_{i \in I}$ be a family of measures on G with the following properties:

(a) $\chi^{-1/q}$ is μ_i-integrable, for all $i \in I$.

(b) $a = \sup_{i \in I} \int \chi^{-1/q} d|\mu_i|$ is finite.

(c) $\int \chi^{-1/q} d\mu_i$ converges to 1 along $\mathcal{F}$.

(d) For every compact neighborhood V of e, $\int_{V^c} \chi^{-1/q} d|\mu_i|$ converges to 0 along $\mathcal{F}$.

Proposition 24.7.1 $\chi^{-1/q}\mu_i$ converges narrowly to ε_e along $\mathcal{F}$.

PROOF: Fix $g \in C^b(G)$. Given $\varepsilon > 0$, let V be a compact neighborhood of e such that $|g(x) - g(e)| \leq \varepsilon$ for all $x \in V$, and choose $A \in \mathcal{F}$ so that $|1 - \int \chi^{-1/q} d\mu_i| \leq c$ and $\int_{V^c} \chi^{-1/q} d|\mu_i| \leq \varepsilon$ for all $i \in A$. Then, for every $i \in A$, we have

$$g(e) - \int g \, d(\chi^{-1/q}\mu_i) = g(e)\left(1 - \int_V \chi^{-1/q} d\mu_i\right)$$
$$+ \int_V (g(e) - g(s)) \, d(\chi^{-1/q}\mu_i)(s) - \int_{V^c} g \, d(\chi^{-1/q}\mu_i).$$

Thus

$$\left| g(e) - \int g \, d(\chi^{-1/q}\mu_i) \right| \leq 2\varepsilon|g(e)| + \varepsilon a + \varepsilon \|g\|.$$

$\square$

Theorem 24.7.1 *Suppose that p is finite, and let $g \in \overline{L^p}(\beta)$. Then $\mu_i * g$ converges to g in $\overline{L^p}(\beta)$ along $\mathcal{F}$.*

PROOF: We may suppose that g is β-moderate. For each $i \in I$, put $c_i = \int \chi^{-1/q} d\mu_i$.

For every $i \in I$,

$$N_p(\mu_i * g - c_i g)^p \leq$$
$$\int^{\bullet} d\beta(x) \left(\int^{\bullet} |g(s^{-1}x)\chi^{-1/p}(s) - g(x)| \, d(\chi^{-1/q}|\mu_i|)(s) \right)^p.$$

Since $t \mapsto t^p$ is convex on $[0, +\infty[$, Jensen's inequality (Theorem 3.7.3) shows that

$$N_p(\mu_i * g - c_i g)^p \leq$$
$$a^{p-1} \int^\bullet d\beta(x) \int^\bullet |g(s^{-1}x)\chi^{-1/p}(s) - g(x)|^p \, d(\chi^{-1/q}|\mu_i|)(s).$$

Given $\varepsilon > 0$, there exists a compact neighborhood V of e such that

$$N_p(\gamma_{\chi,p}(s)g - \gamma_{\chi,p}(e)g)^p \leq \varepsilon \qquad \text{for all } s \in V$$

(Proposition 24.4.1). Now choose $A \in \mathcal{F}$ so that $\int_{V^c} d(\chi^{-1/q}|\mu_i|) \leq \varepsilon$ for all $i \in A$. Then, for every $i \in A$,

$$\int_V d(\chi^{-1/q}|\mu_i|)(s)N_p(\gamma_{\chi,p}(s)g - \gamma_{\chi,p}(e)g)^p \leq a\varepsilon$$

and

$$\int_{V^c} d(\chi^{-1/q}|\mu_i|)(s)N_p(\gamma_{\chi,p}(s)g - \gamma_{\chi,p}(e)g)^p \leq 2^p N_p(g)^p \varepsilon.$$

Since $(s, y) \mapsto |g|^p(y)\chi^{-1}(s)$ is $\chi(\chi^{-1/q}\mu_i) \otimes \beta$-integrable and the measure $\chi(\chi^{-1/q}\mu_i) \otimes \beta$ is the image of $(\chi^{-1/q}\mu_i) \otimes \beta$ under $(s, x) \mapsto (s, s^{-1}x)$, the function $(s, x) \mapsto g(s^{-1}x)\chi^{-1/p}(s)$ lies in $\mathcal{L}_{\mathbb{C}}^p((\chi^{-1/q}\mu_i) \otimes \beta)$, as well as $(s, x) \mapsto g(s^{-1}x)\chi^{-1/p}(s) - g(x)$. Thus, by Fubini's theorem,

$$\int^\bullet d\beta(x) \int^\bullet |g(s^{-1}x)\chi^{-1/p}(s) - g(x)|^p \, d(\chi^{-1/q}|\mu_i|)(s)$$
$$= \int d(\chi^{-1/q}|\mu_i|)(s)N_p(\gamma_{\chi,p}(s)g - \gamma_{\chi,p}(e)g)^p.$$

We conclude that

$$N_p(\mu_i * g - c_i g)^p \leq a^p\varepsilon + a^{p-1}2^p N_p(g)^p\varepsilon \qquad \text{for all } i \in A.$$

The proof is complete. $\square$

Now suppose that $p = +\infty$.

Proposition 24.7.2 *Let $g \in \mathcal{L}^\infty(\beta)$. Along $\mathcal{F}$, $\mu_i * g$ converges weakly to g in $\mathcal{L}^\infty(\beta)$ ($\mathcal{L}^\infty(\beta)$ regarded as the dual space of $\mathcal{L}^1(\beta)$).*

PROOF: Fix $h \in \mathcal{L}^1(\beta)$. For each $s \in G$,

$$\langle \gamma_\chi(s^{-1})h, g \rangle = \chi(s) \int h(sx)g(x) \, d\beta(x) = \int h(x)g(s^{-1}x) \, d\beta(x),$$

and $x \mapsto h(x)g(s^{-1}x)$ is β-integrable.

For every $i \in I$, the function $(s, y) \mapsto g(y)$ is $\chi(\chi^{-1}\mu_i) \otimes \beta$-measurable. Thus $(s, x) \mapsto g(s^{-1}x)$ is $(\chi^{-1}\mu_i) \otimes \beta$-measurable. Since

$$\int^* |h(x)g(s^{-1}x)| \, d(\chi^{-1}|\mu_i| \otimes \beta)(s, x)$$

$$= \int^* d\beta(x)|h(x)| \int^* |g(s^{-1}x)| \, d(\chi^{-1}|\mu_i|)(s)$$

$$= \langle |h|, |\mu_i| * |g| \rangle,$$

we see that $(s, x) \mapsto h(x)g(s^{-1}x)$ is $(\chi^{-1}\mu_i) \otimes \beta$-integrable. Now, by Fubini's theorem,

$$\langle h, \mu_i * g \rangle = \int d(\chi^{-1}\mu_i)(s) \int h(x)g(s^{-1}x) \, d\beta(x).$$

Therefore, if V is a compact neighborhood of e, $\langle h, g - (\mu_i * g) \rangle$ is equal to

$$\left(1 - \int_V d(\chi^{-1}\mu_i)\right) \langle h, g \rangle + \int_V d(\chi^{-1}\mu_i)(s) \int h(x)(g(x) - g(s^{-1}x)) \, d\beta(x)$$

$$- \int_{V^c} d(\chi^{-1}\mu_i)(s) \int h(x)g(s^{-1}x) \, d\beta(x).$$

Given $\varepsilon > 0$, we can choose V so that $N_1(\gamma_\chi(e)h - \gamma_\chi(s^{-1})h) \leq \varepsilon$ for all $s \in V$. Hence

$$\left| \int h(x)(g(x) - g(s^{-1}x)) \, d\beta(x) \right| = \left| \langle h, g \rangle - \chi(s) \int h(sx)g(x) \, d\beta(x) \right|$$

$$= \left| \langle \gamma_\chi(e)h - \gamma_\chi(s^{-1})h, g \rangle \right|$$

is less than $\varepsilon N_\infty(g)$ for $s \in V$. Then, taking $A \in \mathcal{F}$ so that

$$\left| 1 - \int d(\chi^{-1}\mu_i) \right| \leq \varepsilon \quad \text{and} \quad \int_{V^c} d(\chi^{-1}|\mu_i|) \leq \varepsilon$$

for all $i \in A$, we have

$$\left| \langle h, g - (\mu_i * g) \rangle \right| \leq 2\varepsilon |\langle h, g \rangle| + a\varepsilon N_\infty(g) + \varepsilon N_1(h)N_\infty(g)$$

for all $i \in A$. The proposition follows. $\square$

Proposition 24.7.3 Let $g \in C^b(X)$. Then $\mu_i * g$ converges to g along $\mathcal{F}$, uniformly on each compact set.

PROOF: $\mu_i * g$ lies in $C^b(X)$, by Theorem 24.4.2. Let K be compact. Given $\varepsilon > 0$, there exists a compact neighborhood V of e such that $|g(s^{-1}x) - g(x)| \leq \varepsilon$

for all $s \in V$ and $x \in K$. Choose $A \in \mathcal{F}$ so that $\left|1 - \int 1\, d(\chi^{-1}\mu_i)\right| \leq \varepsilon$ and $\int_{V^c} \chi^{-1}\, d|\mu_i| \leq \varepsilon$ for all $i \in A$. Then, for all $i \in A$ and for all $x \in K$,

$$g(x) - (\mu_i * g)(x) = g(x)\left(1 - \int_V \chi^{-1}\, d\mu_i\right)$$
$$+ \int_V (g(x) - g(s^{-1}x))\chi^{-1}(s)\, d\mu_i(s) - \int_{V^c} g(s^{-1}x)\chi^{-1}(s)\, d\mu_i(s).$$

Thus

$$\left|g(x) - (\mu_i * g)(x)\right| \leq 2\varepsilon\|g\| + a\varepsilon + \varepsilon\|g\|,$$

which completes the proof.   □

Similarly, if $g \in C^b(X)$ and if, for every $\varepsilon > 0$, there exists a compact neighborhood V of e such that $\left|g(s^{-1}x) - g(x)\right| \leq \varepsilon$ for $s \in V$ and $x \in X$, then $\mu_i * g$ converges uniformly to g along $\mathcal{F}$.

Proposition 24.7.4 *Suppose there exists a compact set $S \subset G$ containing $\bigcup_{i \in I} \mathrm{supp}(\mu_i)$. If $g \in C(X, \mathbf{C})$, then $\mu_i * g$ converges to g along $\mathcal{F}$, uniformly on each compact set.*

PROOF: By Proposition 24.3.1, $\mu_i * g$ is continuous for all $i \in I$. Let K be compact. Given $\varepsilon > 0$, there exists a compact neighborhood V of e such that $\left|g(s^{-1}x) - g(x)\right| \leq \varepsilon$ for all $s \in V$ and all $x \in K$. Choose $A \in \mathcal{F}$ so that $\left|1 - \int d(\chi^{-1}\mu_i)\right| \leq \varepsilon$ and $\int_{V^c} d(\chi^{-1}|\mu_i|) \leq \varepsilon$ for $i \in A$. Then

$$g(x) - (\mu_i * g)(x) = g(x)\left(1 - \int_V d(\chi^{-1}\mu_i)\right)$$
$$+ \int_V (g(x) - g(s^{-1}x))\, d(\chi^{-1}\mu_i)(s) - \int_{V^c} g(s^{-1}x)\, d(\chi^{-1}\mu_i)(s).$$

Thus, for all $i \in A$ and for all $x \in K$,

$$\left|g(x) - (\mu_i * g)(x)\right| \leq 2\varepsilon \sup_{z \in K} |g(z)| + \varepsilon a + \varepsilon \sup_{z \in S^{-1}K} |g(z)|.$$

The proof is complete.   □

Henceforth, we suppose that $X = G$ and $(s, x) \mapsto sx$ is the multiplication in G. Also, we assume that $p = +\infty$ and μ_i is a measure $f_i\beta$ with base β for all $i \in I$.

Proposition 24.7.5 *Let $K \subset G$ be compact. If $g \in \mathcal{L}^\infty(\beta)$ is continuous at each point of K, then $f_i * g$ converges to g along $\mathcal{F}$, uniformly on K.*

PROOF: For all $i \in I$, $f_i * g$ is bounded and continuous, by Theorem 24.6.2. Given $\varepsilon > 0$, let V and A be as in Proposition 24.7.3. Then, for all $i \in A$

and for all $x \in K$, $g(x) - (\mu_i * g)(x)$ can be written as above. Write j for the mapping $s \mapsto s^{-1}$ from G onto itself. Since $|g \circ \delta_{x^{-1}} \circ j| \leq N_\infty(g)$ locally almost everywhere for $(j \circ \delta_x)(\beta)$, we have $|g(s^{-1}x)| \leq N_\infty(g)$ locally $\chi'(x)\check\beta$-almost everywhere, hence locally β-almost everywhere. Therefore,

$$\left| \int_{V^c} g(s^{-1}x)\chi^{-1}(s)\, d\mu_i(s) \right| \leq \varepsilon N_\infty(g),$$

whence it follows that

$$|g(x) - (f_i * g)(x)| \leq 2\varepsilon \sup_{z \in K} |g(z)| + \varepsilon a + \varepsilon N_\infty(g).$$

$\square$

Proposition 24.7.6 *Suppose that $f_i \in \mathcal{H}(G, \mathbf{C})$ for all $i \in I$. Moreover, assume there exists, for each compact neighborhood V of e, a set $A \in \mathcal{F}$ such that V contains $\bigcup_{i \in A} \operatorname{supp}(f_i)$. Let $g \in \mathcal{L}(\beta)$ be continuous at each point of a given compact set K. Then $f_i * g$ converges to g along $\mathcal{F}$, uniformly on K.*

PROOF: Given $\varepsilon > 0$, let V be as in Proposition 24.7.3. Choose $A \in \mathcal{F}$ so that V contains $\bigcup_{i \in A} \operatorname{supp}(f_i)$ and $|1 - \int \chi^{-1}\, d\mu_i| \leq \varepsilon$ for all $i \in A$. Then, for all $i \in A$ and for all $x \in K$,

$$g(x) - (f_i * g)(x) = g(x)\left(1 - \int \chi^{-1}\, d\mu_i\right) + \int \left(g(x) - g(s^{-1}x)\right) d(\chi^{-1}\mu_i)(s),$$

hence

$$|g(x) - (f_i * g)(x)| \leq \varepsilon \sup_{z \in K} |g(z)| + \varepsilon a.$$

$\square$

Now, we have proven the usual results relative to the regularization of measures on a locally compact group. Next, we give an application of Theorem 24.7.1.

Theorem 24.7.2 *Let G be a locally compact group, and let β (where $\beta > 0$) be a relatively invariant measure on G. Equip $L^1(\beta)$ with the multiplication $(\dot{f}, \dot{g}) \mapsto \dot{\widehat{f * g}}$. Then a closed vector subspace W of $L^1(\beta)$ is a left ideal of the Banach algebra $L^1(\beta)$ if and only if $\widehat{\gamma(s)}g$ belongs to W for all $s \in G$ and all $g \in W$.*

PROOF: First, suppose that $\widehat{\gamma(s)}g$ belongs to W for all $s \in G$ and all $g \in W$. Let $\mu \in \mathcal{M}^1(G, \mathbf{C})$. Since $\widehat{\mu * g} = \int \left(\gamma_\chi(s)\dot{g}\right) d\mu(s)$ belongs to W for all $\dot{g} \in W$, we see that W is a left ideal of $L^1(\beta)$.

Conversely, suppose W is a left ideal of $L^1(\beta)$. Let $\mathcal{U}$ be a basis for the filter of neighborhoods of e. For each $V \in \mathcal{U}$, choose $f_V \cdot \in \mathcal{H}^+(G)$ so that $f_V = 0$ on V^c and $\int f_V \, d\beta = 1$. Finally. write $\mathcal{F}$ for the filter of sections of $\mathcal{U}$. Fix $\dot{g} \in W$ and $s \in G$. By Theorem 24.7.1, $f_V * (\varepsilon_s * g)$ converges to $\varepsilon_s * g$ in $\overline{L^1}(\beta)$, along $\mathcal{F}$. But $f_V * (\varepsilon_s * g) = (f_V * \varepsilon_s) * g$ belongs to W. Thus $\varepsilon_s * g = \chi(s^{-1})\gamma_s(g)$ also lies in W. We conclude that $\overset{\cdot}{\widehat{\gamma_s(g)}}$ belongs to W. $\square$

24.8   Definition of Gelfand Pair

Let G be a locally compact group and K a compact subgroup of G. The continuous complex-valued functions on G/K are in one-to-one correspondence, via $f \mapsto f \circ \pi$ (where $\pi : G \to G/K$ is the canonical mapping) with the continuous complex-valued functions g on G such that $g(st) = g(s)$ for all $s \in G$ and all $t \in K$, or equivalently such that

$$\delta(t^{-1})g = g \qquad (t \in K) \qquad (1).$$

We shall identify the vector space $\mathcal{C}(G/K)$ of all continuous functions on G/K with the subspace of functions $g \in \mathcal{C}(G, \mathbf{C})$ that satisfy (1). Likewise, we shall identify $\mathcal{C}(K\backslash G)$ with the subspace of $\mathcal{C}(G) = \mathcal{C}(G, \mathbf{C})$ consisting of all functions that satisfy

$$\gamma(t)g = g \qquad (t \in K) \qquad (2)$$

(or equivalently $g(ts) = g(s)$ for all $s \in G$ and $t \in K$). We denote by $\mathcal{C}(K\backslash G/K)$ the intersection $\mathcal{C}(G/K) \cap \mathcal{C}(K\backslash G)$. which consists of the continuous functions that are constant on each double coset KsK.

Since K is compact, the inverse image $\pi^{-1}(A)$ of every compact subset A of G/K is compact (Section 23.5); the mapping $f \mapsto f \circ \pi$ is therefore a bijection of the subspace $\mathcal{H}(G/K)$ of $\mathcal{C}(G/K)$, consisting of those continuous functions on G/K with compact support, onto a vector subspace of $\mathcal{H}(G) = \mathcal{H}(G, \mathbf{C})$. We shall therefore identify $\mathcal{H}(G/K)$ with this subspace, which can be written as $\mathcal{H}(G) \cap \mathcal{C}(G/K)$. Likewise, we denote by $\mathcal{H}(K\backslash G)$ the intersection $\mathcal{H}(G) \cap \mathcal{C}(K\backslash G)$, and by $\mathcal{H}(K\backslash G/K)$ the intersection

$$\mathcal{H}(G/K) \cap \mathcal{H}(K\backslash G) = \mathcal{H}(G) \cap \mathcal{C}(K\backslash G/K).$$

Equip G with a left Haar measure m_G. $\mathcal{H}(G)$ is a subalgebra (with respect to convolution) of the algebra $L^1_{\mathbf{C}}(G)$ (because the support of m_G is the whole of G). It follows from (1) and (2) that $\mathcal{H}(G/K)$ is a left ideal and $\mathcal{H}(K\backslash G)$ is a right ideal in $\mathcal{H}(G)$. The intersection $\mathcal{H}(K\backslash G/K)$ is a subalgebra of $\mathcal{H}(G)$ (and also of $L^1_{\mathbf{C}}(G)$).

Let m_K be the normalized Haar measure on K. If we put

$$f^\sharp(s) = \iint_{K \times K} f(tst') \, dm_K(t) \, dm_K(t')$$

for all functions $f \in \mathcal{C}(G)$, the mapping $f \mapsto f^{\sharp}$ is a projection of the vector space $\mathcal{C}(G)$ onto the vector space $\mathcal{C}(K\backslash G/K)$: this follows directly from the left and right invariance of m_K. Furtermore, the projection $f \mapsto f^{\sharp}$ maps $\mathcal{H}(G)$ onto $\mathcal{H}(K\backslash G/K)$.

Definition 24.8.1   If the algebra $\mathcal{H}(K\backslash G/K)$ is commutative, (G, K) is called a Gelfand pair.

Proposition 24.8.1 *Suppose that (G, K) is a Gelfand pair. Then G is unimodular.*

PROOF: Let Δ be the modular function on G. As $\Delta(K)$ is a compact subgroup of $\mathbf{R}_+^*$, $\Delta(K) = \{1\}$, and $\Delta = 1$ on K.

For each $f \in \mathcal{H}(G)$,

$$\int f^{\sharp} \, dm_G \;=\; \iiint f(tst') \, dm_G(s) \, dm_K(t) \, dm_K(t')$$

$$=\; \iint dm_K(t) \, dm_K(t') \int f(tst') \, dm_G(s).$$

But

$$\int f(tst') \, dm_G(s) \;=\; \int f(st') \, dm_G(s)$$

$$=\; \Delta(t'^{-1}) \int f(s) \, dm_G(s)$$

$$=\; \int f \, dm_G.$$

Hence

$$\int f^{\sharp} \, dm_G = \int f \, dm_G.$$

Similarly,

$$\int f^{\sharp} \Delta^{-1} \, dm_G \;=\; \iiint f(tst') \Delta^{-1}(s) \, dm_G(s) \, dm_K(t) \, dm_K(t')$$

$$=\; \iiint (f\Delta^{-1})(tst') \, dm_G(s) \, dm_K(t) \, dm_K(t')$$

$$=\; \iint dm_K(t) \, dm_K(t') \int (f\Delta^{-1})(tst') \, dm_G(s)$$

$$=\; \int (f\Delta^{-1}) \, dm_G.$$

Now, given f in $\mathcal{H}(K\backslash G/K)$, let h in $\mathcal{H}(G)$ be such that $h = 1$ on $K \operatorname{supp}(f) K$. Then $g = h^{\sharp}$ belongs to $\mathcal{H}(K\backslash G/K)$ and $g = 1$ on $\operatorname{supp}(f)$. Moreover,

$$\int f \, dm_G = \int \check{f}(s^{-1}) \, dm_G(s) = (g * \check{f})(e) = (\check{f} * g)(e) = \int f\Delta^{-1} \, dm_G.$$

This proves that $\int f \, dm_G = \int f\Delta^{-1} \, dm_G$ for all $f \in \mathcal{H}(K\backslash G/K)$ and, in fact, by the first part of the proof, for all $f \in \mathcal{H}(G)$. Therefore. $m_G = \Delta^{-1}m_G$, and $\Delta = 1$. □

Proposition 24.8.2 (Gelfand) *Suppose there exists an involutory automorphism θ of G such that, for every $s \in G$, s^{-1} belongs to $K\theta(s)K$. Then (G, K) is a Gelfand pair.*

PROOF: For every $f \in \mathcal{H}(G)$, put $f^{\theta} = f \circ \theta$.

There exists $c > 0$ such that $\theta(m_G) = cm_G$ (because $\theta(m_G)$ is left invariant). Since θ^2 is the identity, $c^2 = 1$, and $c = 1$. This implies that $f^{\theta} * g^{\theta} = (f * g)^{\theta}$ for all $f \in \mathcal{H}(G)$ and all $g \in \mathcal{H}(G)$. On the other hand,

$$
\begin{aligned}
(f * g)\check{}(x) &= (f * g)(x^{-1}) \\
&= \int g(s^{-1}x^{-1})f(s) \, dm_G(s) \\
&= \int g(s^{-1})f(x^{-1}s) \, dm_G(s) \\
&= (\check{g} * \check{f})(x)
\end{aligned}
$$

for all $x \in G$, so $(f * g)\check{} = \check{g} * \check{f}$.

By hypothesis, $\check{f} = f^{\theta}$ for all $f \in \mathcal{H}(K\backslash G/K)$. Therefore, for all f in $\mathcal{H}(K\backslash G/K)$ and all $g \in \mathcal{H}(K\backslash G/K)$,

$$(f * g)\check{} = \check{g} * \check{f} = g^{\theta} * f^{\theta} = (g * f)^{\theta} = (g * f)\check{}:$$

hence $f * g = g * f$, as was to be shown. □

Now suppose that $X = G/K$ is equipped with an invariant distance d:

$$\text{For all } x, y \in X \text{ and for all } s \in G, \qquad d(sx, sy) = d(x, y)$$

(d compatible with the topology of G/K). The action of G on X is said to be doubly transitive if, for all (x, y) and (x', y') in $X \times X$ satisfying $d(x, y) = d(x', y')$, there exists $s \in G$ such that $x' = sx$ and $y' = sy$.

Proposition 24.8.3 *If the action of G on G/K is doubly transitive, then (G, K) is a Gelfand pair.*

PROOF: Let $x_0 = eK$. For each $s \in G$, $d(x_0, sx_0) = d(x_0, s^{-1}x_0)$. By the double transitivity, there exists $t \in K$ such that s^{-1} belongs to tsK. Thus s^{-1} belongs to KsK, and we may apply Proposition 24.8.2. □

Now we give two examples of Gelfand pairs.

Given an integer $n \geq 2$, write G for $SO(n + 1, \mathbf{R})$ and K for the stabilizer of $e_{n+1} = (0, \ldots, 0, 1)$, so that G/K is homeomorphic to $\mathbf{S}^n$. Let $(x, y) \mapsto$

$\langle x, y \rangle$ be the usual scalar product in $\mathbf{R}^{n+1}$. For all x, y in $\mathbf{S}^n$, put $d(x, y) = \arccos\langle x, y \rangle$. We prove that d is a distance on $\mathbf{S}^n$.

Let x, y, and z be points in $\mathbf{S}^n$ such that $z \neq x$ and $z \neq y$. Denote $d(x, z)$, $d(y, z)$, and $d(x, y)$ by α, β, and γ, respectively. $x - \langle x, z \rangle z$ and $y - \langle y, z \rangle z$ are the orthogonal projections of $x - z$ and $y - z$, respectively, on $(\mathbf{R}z)^\perp$. Define u and v as $(1/\sin(\alpha))(x - \langle x, z \rangle z)$ and $(1/\sin(\beta))(y - \langle y, z \rangle z)$, respectively. Then $\langle u, u \rangle = \langle v, v \rangle = 1$. Now

$$
\begin{aligned}
\cos(\gamma) &= \langle x, y \rangle \\
&= \big\langle \sin(\alpha)u + \langle x, z \rangle z, \sin(\beta)v + \langle y, z \rangle z \big\rangle \\
&= \cos(\alpha)\cos(\beta) + \sin(\alpha)\sin(\beta)\langle u, v \rangle,
\end{aligned}
$$

hence $\cos(\gamma) \geq \cos(\alpha + \beta)$. This proves that $\gamma \leq \alpha + \beta$, and that d is actually a distance.

Next, let x, y and x_0, y_0 be points in $\mathbf{S}^n$ such that $d(x, y) = d(x_0, y_0)$. Suppose that x, y are distinct, and define γ as $d(x, y)$. Then

$$
\frac{y_0 - \cos(\gamma)x_0}{\sin(\gamma)} \text{ belongs to } (\mathbf{R}x_0)^\perp \text{ and } \left\| \frac{y_0 - \cos(\gamma)x_0}{\sin(\gamma)} \right\| = 1.
$$

Likewise,

$$
\frac{y - \cos(\gamma)x}{\sin(\gamma)} \text{ belongs to } (\mathbf{R}x)^\perp \text{ and } \left\| \frac{y - \cos(\gamma)x}{\sin(\gamma)} \right\| = 1.
$$

There exists $g \in G$ that sends x_0 to x and $(y_0 - \cos(\gamma)x_0)/\sin(\gamma)$ to $(y - \cos(\gamma)x)/\sin(\gamma)$. Now

$$
y - \cos(\gamma)x = g(y_0 - \cos(\gamma)x_0) = g(y_0) - \cos(\gamma)x,
$$

hence $y = g(y_0)$. We conclude that the action of G on $\mathbf{S}^n$ is doubly transitive, and (G, K) is a Gelfand pair (Proposition 24.8.3).

d is the geodesic distance on the Riemannian submanifold $\mathbf{S}^n$ of $\mathbf{R}^{n+1}$.

Now take for G the group $SH(n + 1, \mathbf{R})$ and for K the stabilizer of e_{n+1}, so that G/K is homeomorphic to H_n (Section 23.11). Let Φ be the bilinear form

$$
(x, y) \longmapsto \sum_{1 \leq j \leq n} x_j y_j - x_{n+1} y_{n+1}
$$

on $\mathbf{R}^{n+1} \times \mathbf{R}^{n+1}$. For all x, y in H_n, put $d(x, y) = \operatorname{arg\,cosh}\big(-\Phi(x, y)\big)$. We prove that d is a distance on H_n, called the hyperbolic distance.

Let x, y, and z be points in H_n such that $z \neq x$ and $z \neq y$. Denote $d(x, z)$, $d(y, z)$, and $d(x, y)$ by α, β, and γ, respectively. $x + \Phi(x, z)z$ is the orthogonal projection of $x - z$ on $(\mathbf{R}z)^\circ$, and

$$
\begin{aligned}
q\big(x + \Phi(x, z)z\big) &= q(x) + \Phi(x, z)^2 q(z) + 2\Phi(x, z)^2 \\
&= \Phi(x, z)^2 - 1 \\
&= \sinh^2(\alpha).
\end{aligned}
$$

Likewise, $y + \Phi(y, z)z$ is the orthogonal projection of $y - z$ on $(\mathbf{R}z)^\circ$, and $q(y + \Phi(y, z)z) = \sinh^2(\beta)$. Define u and v as

$$\frac{1}{\sinh\alpha}(x + \Phi(x, z)z) \quad \text{and} \quad \frac{1}{\sinh\beta}(y + \Phi(y, z)z),$$

respectively. Then $q(u) = q(v) = 1$. Now

$$\begin{aligned}
\Phi(x, y) &= \Phi\big(\sinh(\alpha)u - \Phi(x, z)z, \sinh(\beta)v - \Phi(y, z)z\big) \\
&= \sinh(\alpha)\sinh(\beta)\Phi(u, v) - \cosh(\alpha)\cosh(\beta),
\end{aligned}$$

hence $\cosh(\gamma) = \cosh(\alpha)\cosh(\beta) - \sinh(\alpha)\sinh(\beta)\Phi(u, v)$. Since the restriction of Φ to $(\mathbf{R}z)^\circ \times (\mathbf{R}z)^\circ$ is a positive bilinear form, $|\Phi(u, v)|$ is less than $q(u)q(v) = 1$. Therefore, we obtain

$$\cosh(\gamma) \leq \cosh(\alpha)\cosh(\beta) + \sinh(\alpha)\sinh(\beta) = \cosh(\alpha + \beta),$$

and finally $\gamma \leq \alpha + \beta$. This proves that d is actually a distance on H_n.

Next, let x, y and x_0, y_0 be points in H_n such that

$$d(x, y) = d(x_0, y_0).$$

Suppose that x, y are distinct, and define γ as $d(x, y)$. Then

$$\frac{y_0 - \cosh(\gamma)x_0}{\sinh(\gamma)} \text{ belongs to } (\mathbf{R}x_0)^\circ \quad \text{and} \quad q\left(\frac{y_0 - \cosh(\gamma)x_0}{\sinh(\gamma)}\right) = 1.$$

Likewise,

$$\frac{y - \cosh(\gamma)x}{\sinh(\gamma)} \text{ belongs to } (\mathbf{R}x)^\circ \quad \text{and} \quad q\left(\frac{y - \cosh(\gamma)x}{\sinh(\gamma)}\right) = 1.$$

There exists $g \in G$ that sends x_0 to x and $(y_0 - \cosh(\gamma)x_0)/\sinh(\gamma)$ to $(y - \cosh(\gamma)x)/\sinh(\gamma)$. Since $y = g(y_0)$, the action of G on H_n is doubly transitive, and (G, K) is a Gelfand pair.

When H_n is equipped with the Riemannian metric

$$x \longmapsto \Phi_{/(\mathbf{R}x)^\circ \times (\mathbf{R}x)^\circ},$$

d is the geodesic distance on H_n. In fact, if x and y are two distinct points of H_n and if we put $\gamma = \arg\cosh\big(-\Phi(x, y)\big)$, there exists a unique geodesic $\rho : [0, 1] \to H_n$ such that $\rho(0) = x$ and $\rho(1) = y$. ρ is given by

$$\begin{aligned}
\rho(t) &= \cosh(t\gamma)x + \sinh(t\gamma)\frac{y - \cosh(\gamma)x}{\sinh(\gamma)} \\
&= \frac{\sinh\big((1-t)\gamma\big)}{\sinh(\gamma)}x + \frac{\sinh(t\gamma)}{\sinh(\gamma)}y.
\end{aligned}$$

Exercises for Chapter 24

1 Let G be a locally compact group and let β be a left Haar measure on G. By abuse of notation, we identify essentially β-integrable complex-valued functions on G and their classes in $L_{\mathbf{C}}^1(\beta)$; likewise for the $L_{\mathbf{C}}^p(\beta)$.

1. Let A be a continuous endomorphism of $L_{\mathbf{C}}^1(\beta)$ such that $A(f * \varepsilon_s) = A(f) * \varepsilon_s$ for all $f \in L_{\mathbf{C}}^1(\beta)$ and all $s \in G$. Show that, for every $g \in \mathcal{H}(G, \mathbf{C})$, we have $A(f * g) = A(f) * g$ (remark that $s \mapsto g(s)A(f * \varepsilon_s)$ is a β-integrable mapping from G into $L_{\mathbf{C}}^1(\beta)$). Deduce that there exists a unique bounded measure μ on G such that $A(f) = \mu * f$ for all $f \in L_{\mathbf{C}}^1(\beta)$, and show that $\|A\| = \|\mu\|$ (use the compactness of the unit ball of $\mathcal{M}^1(G, \mathbf{C})$ in the weak topology and, in the notation of Theorem 24.7.2, consider the limit of $A(f_V)$ along an ultrafilter finer than the filter of sections of $\mathcal{U}$).

2. Let $\mu \in \mathcal{M}^1(G, \mathbf{C})$. Show directly that the norm of the continuous endomorphism $\gamma(\mu) : f \mapsto \mu * f$ of $L_{\mathbf{C}}^\infty(\beta)$ is equal to $\|\mu\|$ (bring yourself to the case in which there exists $g \in \mathcal{H}(G, \mathbf{C})$ such that $\mu = g|\mu|$). Deduce once more that the continuous endomorphism $f \mapsto \mu * f$ of $L_{\mathbf{C}}^1(\beta)$ has norm $\|\mu\|$.

3. Suppose G is compact and μ is positive. Show that, for $1 < p < +\infty$, the norm of the continuous endomorphism $\gamma(\mu) : f \mapsto \mu * f$ of $L_{\mathbf{C}}^p(\beta)$ is equal to $\|\mu\|$.

4. Take for G the cyclic group $\mathbf{Z}/3\mathbf{Z}$. Let μ be a measure on G, with support G, such that $i \mapsto \mu(i)/|\mu|(i)$ does not take a constant value. Show that the norm of the endomorphism $\gamma(\mu) : f \mapsto \mu * f$ of $L_{\mathbf{C}}^p(\beta)$ is strictly less than $\|\mu\|$, for $1 < p < +\infty$.

2 Notation is that of Exercise 1.

Show that a bounded measure μ on G is such that $N_1(\mu * f) = N_1(f)$ for all $f \in L_{\mathbf{C}}^1(\beta)$ if and only if $\text{supp}(\mu)$ is a point and $\|\mu\| = 1$. (Show that we must have $\left| \int f \, d\mu \right| = \int |f| \, d|\mu|$ for every $f \in \mathcal{H}(G, \mathbf{C})$; first, deduce that $\mu = c|\mu|$, where c is a complex number with modulus 1, and next, that $\text{supp}(\mu)$ is a point.)

3 For each $n \in \mathbf{N}$, let μ_n be the measure on $\{0, 1\}$ such that $\mu_n(\{0\}) = \mu_n(\{1\}) = 1/2$. Write μ for $\bigotimes_{n \geq 1} \mu_n$, μ' for the image measure of μ under

$$(x_n)_{n \geq 1} \longmapsto \sum_{n \geq 1} \frac{x_{2n-1}}{2^{2n-1}},$$

and μ'' for the image measure of μ under

$$(x_n)_{n \geq 1} \mapsto \sum_{n \geq 1} \frac{x_{2n}}{2^{2n}}.$$

Show that μ' and μ'' are singular, even though $\mu' * \mu'' = 1_{[0,1]}\lambda$ (λ Lebesgue measure on $\mathbf{R}$).

4   Given $k \in \mathbb{N}$, write λ_k or dx for Lebesgue measure on $\mathbb{R}^k$. Let g_1 be a complex-valued continuous function on $\mathbb{R}^k$, such that $|y|^{k+1}|g_1(y)|$ is bounded (by $M \geq 0$) in $\mathbb{R}^k$ and $\int g_1(y)\,dy = 1$. For all $\sigma > 0$, write g_σ for the function $y \mapsto (1/\sigma^k)g_1(y/\sigma)$. Finally, let f be a λ_k-integrable complex-valued function on $\mathbb{R}^k$, and let $x \in \mathbb{R}^k$ be such that

$$\frac{1}{\lambda_k(B'_a)} \int_{|z| \leq a} |f(x+z) - f(x)|\,dz$$

approaches 0 as $a \to 0$ in $]0, +\infty[$ (B'_a the closed ball in $\mathbb{R}^k$ with center 0 and radius a). We prove that $(f * g_\sigma)(x)$ converges to $f(x)$ as $\sigma \to 0$. Write dt for Lebesgue measure on $]0, +\infty[$ and $d\mathbf{S}^{k-1}$ for the superficial measure on $\mathbf{S}^{k-1}$ (if $k = 1$, $d\mathbf{S}^0$ is the measure on $\{-1, 1\}$ defined by $d\mathbf{S}^0(\{1\}) = d\mathbf{S}^0(\{-1\}) = 1$). Since λ_k is the image measure of $(t^{k-1}dt) \otimes d\mathbf{S}^{k-1}$ under $(t, x) \mapsto tx$, the function $\varphi : t \mapsto t^{k-1}\int^* |f(x - tu) - f(x)|\,d\mathbf{S}^{k-1}(u)$ is locally dt-integrable, and $\psi : a \mapsto \int_{B'_a} |f(x-y) - f(x)|\,dy$ is an indefinite integral of φ.

Given $\varepsilon > 0$, choose $a > 0$ so that

$$\frac{1}{\sigma^k V} \int_{B'_\sigma} |f(x-y) - f(x)|\,dy \leq \varepsilon \qquad \text{for all } 0 < \sigma \leq a$$

$(V = \lambda_k(B'_1))$. Next, fix $0 < \sigma \leq a$.

1.   Prove that

$$|(f * g_\sigma)(x) - f(x)| \leq \int_{]0,\sigma]} \frac{1}{\sigma^k} \theta\left(\frac{t}{\sigma}\right) \varphi(t)\,dt$$

$$+ \int^*_{[\sigma, +\infty[} \frac{1}{\sigma^k} M\left(\frac{\sigma}{t}\right)^{k+1} \varphi(t)\,dt,$$

where $\theta(t) = \sup\left\{|g_1(tu)| : u \in \mathbf{S}^{k-1}\right\}$. Deduce that

$$|(f * g_\sigma)(x) - f(x)| \leq \varepsilon V N_\infty(g_1) + \sigma M \int^*_{[\sigma, +\infty[} \frac{\varphi(t)}{t^{k+1}}\,dt.$$

2.   Show that

$$\int^{+\infty}_\sigma \frac{\varphi(t)}{t^{k+1}}\,dt \leq (k+1) \int^a_\sigma \frac{\psi(t)}{t^{k+2}}\,dt + (k+1) \int^{+\infty}_a \frac{\psi(t)}{t^{k+2}}\,dt.$$

Deduce that

$$|(f * g_\sigma)(x) - f(x)| \leq \varepsilon V N_\infty(g_1) + \varepsilon V M(k+1)$$

$$+ \sigma M(k+1) \int^{+\infty}_a \frac{\psi(t)}{t^{k+2}}\,dt.$$

3.   From part 2, deduce that $(f * g_\sigma)(x)$ converges to $f(x)$ as $\sigma \to 0$ in $]0, +\infty[$.

5 Let μ be a positive bounded measure on $\mathbf{R}^k$, and let $x \in \mathbf{R}^k$ be such that $\left(1/\lambda_k(B_a')\right)\mu(x + B_a')$ approaches 0 as $a \to 0$ in $]0, +\infty[$. In the notation of Exercise 4, we prove that $(\mu * g_\sigma)(x)$ converges to 0 as $\sigma \to 0$ in $]0, +\infty[$.

Write m for the image measure of μ under $y \mapsto y - x$. For all $\sigma > 0$, let $\check{g}_\sigma$ be $y \mapsto g_\sigma(-y)$; hence $(\mu * g_\sigma)(x) = \int \check{g}_\sigma(y)\, dm(y)$.

Given $\varepsilon > 0$, choose $a > 0$ so that $(1/\sigma^k V)\mu(x + B_\sigma') \le \varepsilon$ for all $0 < \sigma \le a$. Next, fix $0 < \sigma \le a$.

1. From the relation

$$(\mu * g_\sigma)(x) = \int_{B_\sigma'} \frac{1}{\sigma^k}\check{g}_1\left(\frac{y}{\sigma}\right) dm(y) + \int_{(B_\sigma')^c} \frac{1}{\sigma^k}\check{g}_1\left(\frac{y}{\sigma}\right) dm(y),$$

deduce that

$$|\mu * g_\sigma|(x) \le \varepsilon V N_\infty(g_1) + \sigma M \int_{(B_\sigma')^c} \frac{1}{|y|^{k+1}}\, dm(y).$$

2. Define a function h on $\mathbf{R}^k$ by $h(y) = 0$ for $y \in B_\sigma'$ and $h(y) = 1/|y|^{k+1}$ for $y \in (B_\sigma')^c$. Show that

$$\int h(y)\, dm(y) = \int m(h^{-1}[t, +\infty[)\, dt$$
$$= \int (k+1)u^{-(k+2)}c(u)\, du$$
$$= \int_\sigma^{+\infty} (k+1)u^{-(k+2)}m(B_u')\, du - m(B_\sigma')\sigma^{-(k+1)},$$

where $c(u) = m\left(h^{-1}[u^{-(k+1)}, +\infty[\right)$.

3. From part 2, deduce that $\int_{B_a' \cap (B_\sigma')^c} \left(1/|y|^{k+1}\right) dm(y)$ is equal to

$$\int_\sigma^a (k+1)u^{-(k+2)}m(B_u')\, du + m(B_a')a^{-(k+1)} - m(B_\sigma')\sigma^{-(k+1)}$$

and less than $\int_\sigma^a (k+1)u^{-(k+2)}\varepsilon u^k V\, du + m(B_a')a^{-(k+1)}$.

4. From parts 1 and 3, deduce that $|\mu * g_\sigma|(x)$ is less than

$$\varepsilon V N_\infty(g_1) + \varepsilon V M(k+1) + \sigma M m(B_a')a^{-(k+1)} + \sigma M \int_{(B_a')^c} \frac{1}{|y|^{k+1}}\, dm(y).$$

Conclude that $(\mu * g_\sigma)(x)$ converges to 0 as $\sigma \to 0$.

6 Let μ be a bounded measure on $\mathbf{R}^k$, and let $\mu = f\lambda_k + (\mu - f\lambda_k)$ be the Lebesgue decomposition of μ with respect to λ_k. In the notation of Exercise 4, show that, for λ_k-almost all $x \in \mathbf{R}^k$, $(\mu * g_\sigma)(x)$ converges to $f(x)$ as $\sigma \to 0$ in $]0, +\infty[$.

7 Let $\mathbf{T}$ be the multiplicative group of complex numbers with modulus 1 and let ν be the normalized Haar measure on $\mathbf{T}$. For every $n \in \mathbf{Z}^+$, denote by D_n the function $\mathbf{T} \ni \zeta \mapsto \sum_{-n \leq k \leq n} \zeta^k$, and by Δ_n the function

$$\mathbf{T} \ni \zeta \longmapsto \frac{1}{n+1} \sum_{0 \leq j \leq n} D_j(\zeta) = \sum_{-n \leq k \leq n} \left(1 - \frac{|k|}{n+1}\right)\zeta^k.$$

D_n (respectively, Δ_n) is called the Dirichlet kernel (respectively, the Féjer kernel) of order n.

1. Prove that $D_k(e^{2i\pi x}) = \sin\big((2k+1)\pi x\big)/\sin(\pi x)$ for all $k \in \mathbf{Z}^+$ and for all $x \in \mathbf{R} - \mathbf{Z}$.

2. For every $n \in \mathbf{Z}^+$ and every $x \in \mathbf{R} - \mathbf{Z}$, show that

$$\sum_{0 \leq k \leq n} \sin\big((2k+1)\pi x\big) = z + \bar{z},$$

where $z = \big(e^{i\pi x}/2i\big) \sum_{0 \leq k \leq n} e^{2i\pi k x}$. Deduce that

$$\Delta_n(e^{2i\pi x}) = \frac{\sin^2\big((n+1)\pi x\big)}{(n+1)\sin^2(\pi x)}.$$

3. Prove that $\int \Delta_n \, d\nu = 1$.

4. Given $0 < \delta \leq 1/2$, write $V(\delta)$ for the set $\exp\big(2i\pi[-\delta, \delta]\big)$. Show that $\int_{T-V(\delta)} \Delta_n \, d\nu$ converges to 0 as $n \to +\infty$ (use part 2).

5. Let μ be a measure on $\mathbf{T}$ and let $\mathcal{F}\mu$ be the Fourier transform

$$\mathbf{Z} \ni k \longmapsto \int \zeta^{-k} \, d\mu(\zeta)$$

of μ. For every $n \in \mathbf{Z}^+$, prove that

$$(\mu * D_n)(\zeta) = \sum_{-n \leq k \leq n} \mathcal{F}\mu(k)\zeta^k$$

and

$$(\mu * \Delta_n)(\zeta) = \big(1/(n+1)\big) \sum_{0 \leq j \leq n} (\mu * D_j)(\zeta)$$

for all $\zeta \in \mathbf{T}$. $\mu * D_n$ and $\mu * \Delta_n$ are called the nth Fourier sum and the nth Féjer sum of μ, respectively.

6. Let $f \in \mathcal{L}_{\mathbf{C}}^p(\nu)$ $(1 \leq p < +\infty)$. Deduce from parts 3 and 4 that $f * \Delta_n$ converges to f in $\mathcal{L}_{\mathbf{C}}^p(\nu)$ as $n \to +\infty$. In fact, for $1 < p < +\infty$, it can be shown that $f * D_n$ converges to f in $\mathcal{L}_{\mathbf{C}}^p(\nu)$.

8 Given $f \in \mathcal{L}_{\mathbf{C}}^1(\nu)$, the function $g : \mathbf{R} \ni x \mapsto f(e^{2i\pi x})$ is locally dx-integrable. Let $x \in \mathbf{R}$ be such that $(1/a) \int_0^a \big|g(x+y) + g(x-y) - 2g(x)\big| \, dy$ approaches 0 as $a \to 0$ in $]0, +\infty[$. Set $\zeta = e^{2i\pi x}$. We prove that $(f * \Delta_n)(\zeta)$ converges to $f(\zeta)$ as $n \to +\infty$.

Define h and H as the functions $y \mapsto \big|g(x+y) + g(x-y) - 2g(x)\big|$ and $a \mapsto \int_0^a h(y) \, dy$ on $\mathbf{R}$. Let $n \geq 2$ be an integer.

1. Show that

$$\left|(f * \Delta_{n-1})(\zeta) - f(\zeta)\right| = \left|\int_{-1/2}^{1/2} \Delta_{n-1}(e^{2i\pi y})\big(g(x-y) - g(x)\big) \, dy\right|$$

is less than

$$\frac{1}{n}\int_0^{1/n} h(y)\frac{\sin^2(n\pi y)}{\sin^2(\pi y)} \, dy + \frac{1}{n}\int_{1/n}^{1/2} h(y)\frac{1}{\sin^2(\pi y)} \, dy.$$

2. Prove that

$$\frac{\sin(n\pi y)}{n\sin(\pi y)} \le 1$$

for $y \in [0, 1/n]$ and

$$\frac{\sin(\pi y)}{\pi y} \ge \frac{\sin(\pi/2)}{\pi/2}$$

for $y \in [0, 1/2]$. Deduce that $\left|(f * \Delta_{n-1})(\zeta) - f(\zeta)\right|$ is less than

$$nH\left(\frac{1}{n}\right) + \frac{1}{n}\frac{H(y)}{4y^2}\bigg|_{1/n}^{1/2} + \frac{1}{2n}\int_{1/n}^{1/2} H(y)\frac{dy}{y^3}.$$

3. Given $\varepsilon > 0$, there exists $0 < \delta \le 1/2$ such that $H(y) \le \varepsilon y$ for all $y \in [0, \delta]$. Choose n so that $1/n \le \delta$. Show that $\left|(f * \Delta_{n-1})(\zeta) - f(\zeta)\right|$ is less than

$$nH\left(\frac{1}{n}\right) + \frac{1}{n}H\left(\frac{1}{2}\right) + \frac{1}{2n}\int_{1/n}^{\delta} H(y)\frac{1}{y^3} \, dy + \frac{1}{2n}\int_{\delta}^{1/2} H(y)\frac{1}{y^3} \, dy.$$

Conclude that $\left|(f * \Delta_{n-1})(\zeta) - f(\zeta)\right|$ is less than ε for large enough n.

9 Let μ be a positive measure on $\mathbf{T}$, and let ζ be a point of $\mathbf{T}$ such that $(1/a)\mu\big(\zeta \exp(2i\pi[-a, a])\big)$ approaches 0 as a tends to 0 in $]0, 1/2]$. We prove that $(\mu * \Delta_n)(\zeta)$ converges to 0 as $n \to +\infty$.

For each $a \in]0, +\infty[$, put $V_a = \zeta \exp\big(2i\pi[-a, a]\big)$. Let n be an integer $n \ge 2$.

1. Show that $\Delta_{n-1}(\zeta/\xi) \le n$ for all points ξ in $V_{1/n}$. Deduce that $\int_{V_{1/n}} \Delta_{n-1}(\zeta/\xi) \, d\mu(\xi)$ converges to 0 as $n \to +\infty$.

2. Prove that $\Delta_{n-1}(\zeta/\zeta e^{2i\pi y}) \le (1/4ny^2)$ for all y in $E_n = [-1/2, 1/2] - [-1/n, 1/n]$.

3. For all $t \in]0, +\infty[$, put $F_t = \big\{\zeta e^{2i\pi y} : y \in E_n, \, 1/(4ny^2) \ge t\big\}$. Show that

$$\int_{\mathbf{T} - V_{1/n}} \Delta_{n-1}(\zeta/\xi) \, d\mu(\xi) \le \int_0^{+\infty} \mu(F_t) \, dt$$

$$= \int_{1/n}^{+\infty} \mu\big(F_{1/4nu^2}\big)\big(1/(2nu^3)\big) \, du$$

is less than $\int_{1/n}^{+\infty} \mu(V_u)\big(1/(2nu^3)\big) \, du - (1/2n)\mu(V_{1/n})\int_{1/n}^{+\infty}(1/u^3) \, du$.

4. Given $\varepsilon > 0$, let $0 < \delta \leq 1/2$ be such that $\mu(V_u) \leq \varepsilon u$ for all $u \in]0, \delta]$. Choose $n \geq 2$ so that $1/n \leq \delta$. Prove that

$$\int_{\mathbf{T} - V_{1/n}} \Delta_{n-1}(\zeta/\xi) \, d\mu(\xi) \leq \int_{1/n}^{\delta} \mu(V_u)\left(1/(2nu^3)\right) du$$

$$+ \int_{\delta}^{+\infty} \|\mu\|\left(1/(2nu^3)\right) du - (1/4)n\mu(V_{1/n}).$$

Deduce that $(\mu * \Delta_{n-1})(\zeta)$ is less than ε for large enough n.

10 Let μ be a measure on T, and let $\mu = f\nu + \mu_s$ be the Lebesgue decomposition of μ with respect to ν. From Exercises 8 and 9, deduce that, for ν-almost all ζ, $(\mu * \Delta_n)(\zeta)$ converges to $f(\zeta)$ as $n \to +\infty$.

Index

Symbol Index

A^c: complement of A (in a given set)

$A - B$: difference $A \cap B^c$

1_A: indicator function of A (on a given set Ω)

$f_{/A}$: restriction of function f to A

$\langle a, b \rangle$: interval in $\mathbf{R}$ with endpoints a and b in $\overline{\mathbf{R}}$

$]a, b[$: open interval in $\mathbf{R}$

$\mathcal{H}(\Omega, \mathbf{C})$: vector space consisting of complex-valued functions; 16

$\mathcal{H}(\Omega, \mathbf{R})$: space of real-valued functions in $\mathcal{H}(\Omega, \mathbf{C})$; 16

$\mathcal{H}^+$: set of positive functions in $\mathcal{H}(\Omega, \mathbf{R})$; 16

$Q\mathcal{M}(\Omega, \mathbf{C})$: space of quasi-measures; 17

$\mathcal{M}(\Omega, \mathbf{C})$: space of Daniell measures; 17

$Q\mathcal{M}(\Omega, \mathbf{R})$: space of real quasi-measures; 17

$\mathcal{M}(\Omega, \mathbf{R})$: space of real Daniell measures; 17

$\mathcal{M}^+$: set of positive Daniell measures; 18

$\bar{\mu}$: conjugate of the Daniell measure μ; 17

$|\mu|$ or $V\mu$: absolute value, or variation, of μ; 16

μ^+ and μ^-: positive and negative parts of the real measure μ; 17

$\sigma(\mathcal{C})$ or $\tilde{\mathcal{C}}$: σ-ring generated by $\mathcal{C}$; 29

or $\langle f, \mu \rangle$: essential integral of $f \in \overline{\mathcal{L}}^1_F(\mu)$; 68

$\int_A f \, d\mu$: $\int f 1_A \, d\mu$ for $f \in \mathcal{L}^1_F(\mu)$ and $A \in \mathcal{M}$, integral of f over the set A 68

$\|\mu\|$: total mass $\int^\bullet 1 \, dV\mu$ of μ; 69

$\mathcal{J}$: set of functions h with the following property: there exists $g \in \mathcal{H}(\Omega, \mathbf{R})$ such that $h - g$ belongs to $\mathcal{J}^+$; 70

$\int^* f \, d\mu$ and $\int_* f \, d\mu$: upper integral and lower integral of the extended real-valued function f; 72

$\mu^*(A)$: outer measure $\int^* 1_A \, d\mu$ of the set A; 74

$\mu_*(A)$: inner measure of A; 74

$\hat{\mu}$: main prolongation of μ; 81

$\overline{\mathcal{R}}$: ring of essentially integrable sets; 81

$\bar{\mu}$: essential prolongation of μ; 82

$N_p(f)$: $[\int^* f^p \, dV\mu]^{1/p}$ for $p \in]0, +\infty[$ and positive f; 94

$N_p(f)$: $[\int^* |f|^p \, dV\mu]^{1/p}$ for the vector-valued function f; 98

$\mathcal{F}^p_F(\mu)$: space of vector-valued functions f such that $N_p(f) < +\infty$ $(1 \le p < +\infty)$; 98

$\mathcal{L}^p_F(\mu)$: space of μ-measurable functions in $\mathcal{F}^p_F(\mu)$; 98

$L^p_F(\mu)$: quotient space of $\mathcal{L}^p_F(\mu)$; 99

$\overline{N_p}(f)$: $[\int^\bullet |f|^p \, dV\mu]^{1/p}$ for the vector-valued function f; 99

$\overline{\mathcal{F}}^p_F(\mu)$: space of vector-valued functions such that $\overline{N_p}(f) < +\infty$; 99

$\overline{\mathcal{L}}^p_F(\mu)$: space of μ-measurable functions in $\overline{\mathcal{F}}^p_F(\mu)$; 99

$N_\infty(f)$: $M_\infty(|f|)$, for the vector-valued function f; 101

$\mathcal{L}^\infty_F(\mu)$: space of essentially bounded functions; 101

$L^\infty_F(\mu)$: quotient space of $\mathcal{L}^\infty_F(\mu)$; 101

$\mathcal{L}(A, \mu; F)$: space of functions that are μ-measurable on A; 105

$\mathrm{supp}(f)$: support of the function f; 135

$\mathcal{H}(\Omega, F)$: space of compactly supported continuous functions from Ω into F; 137

$\mathcal{H}(\Omega, K; F)$: space of those continuous functions from Ω into F that vanish outside the compact set K; 137

$\mathcal{H}^+(\Omega)$: set of compactly supported continuous functions from Ω into $\mathbf{R}^+$; 145

$V(M, J)$ or $V(J)$: variation of M over J; 266

dM: Stieltjes measure associated with M; 270

$\int_M f$: line integral; 275

$E(\mu)$: mean of the measure μ; 295

$D(\mu)$: covariance matrix of μ; 295

$\text{Var}(\mu)$: variance of μ; 295

$N_1(m, \sigma^2)$: normal law on $\mathbf{R}$ with mean m and variance σ^2; 295

$\mathcal{F}\mu$: Fourier transform of the bounded measure μ; 296

$\overline{\mathcal{F}}\mu$: inverse Fourier transform of μ; 296

$\mathcal{F}f$: Fourier transform of the Lebesgue integrable function f; 297

$\overline{\mathcal{F}}f$: inverse Fourier transform of f; 297

$\text{cha}(\mu)$: characteristic function of μ; 303

A': transpose of the matrix A; 305

$N_n(m, D)$: normal law on $\mathbf{R}^n$ with mean m and covariance matrix D; 305

P_X or $P(X)$: law of the random variable X; 312

$E(X)$: expected value of X; 313

$D(X)$: covariance matrix of X; 313

$\text{Var}(X)$: variance of X; 313

$\bigotimes_{i \in I} P_i$: product of the probabilities P_i; 313

$\sigma((X_i)_{i \in I})$: σ-algebra generated by the random variables X_i; 314

$X_n \Rightarrow \mu$ and $X_n \Rightarrow X$: convergence in law; 327

$E(Y|\mathcal{D})$: conditional expected value of Y given the σ-algebra $\mathcal{D}$; 353

$E(Y|X)$: conditional expected value of Y given the random variable X; 363

μ_Y or $\mu_{/Y}$: Radon measure induced by μ on Y; 380

$\sum_{\alpha \in A} \mu_\alpha$: sum of the summable family $(\mu_\alpha)_{\alpha \in A}$ of Radon measures; 386

$\int \lambda_t \, d\mu(t)$: integral of scalarly essentially μ-integrable family of Radon measures; 387

$g \cdot \mu$ or $g\mu$: Radon measure with density g relative to μ; 400

$\mu \perp \nu$: relation of mutual singularity for Radon measures; 404

$\pi(\mu)$: image measure under π of the Radon measure μ; 409

$\mu \otimes \nu$: product of Radon measures; 412

Universitext

(continued)